5th International Conference on Fisheries, Aquatic, and Environmental Sciences (ICFAES 2023)

BIO Web of Conferences Volume 87 (2024)

Banda Aceh, indonesia
25 October 2023

Editors:

C.D. Dewi
M. Nanda
S. Ali Akbar
V. Prajaputra
Y. Haditiar

ISBN: 978-1-7138-8811-6

Printed from e-media with permission by:

Curran Associates, Inc.
57 Morehouse Lane
Red Hook, NY 12571

Some format issues inherent in the e-media version may also appear in this print version.

For additional information, please contact EDP Sciences – Web of Conferences
at the address below.

EDP Sciences – Web of Conferences
17, Avenue du Hoggar
Parc d'Activité de Courtabœuf
BP 112
F-91944 Les Ulis Cedex A
France

Phone: +33 (0) 1 69 18 75 75
Fax: +33 (0) 1 69 28 84 91

contact-edps@webofconferences.org

Additional copies of this publication are available from:

Curran Associates, Inc.
57 Morehouse Lane
Red Hook, NY 12571 USA
Phone: 845-758-0400
Fax: 845-758-2633
Email: curran@proceedings.com
Web: www.proceedings.com

TABLE OF CONTENTS

AQUATIC (OCEANOGRAPHY, MARINE SCIENCE, PROTECTING MARINE LIFE, ENDANGERED SPECIES CONSERVATION, WATER PROTECTION, FLOOD RISK MANAGEMENT, URBAN ADAPTATION ON GLOBAL CLIMATE CHANGE)

ENVIRONMENT (ECOSYSTEM, HABITAT CONSERVATION, CLIMATE, HABITAT CONSULTATION, ENVIRONMENT MODELING, WATER RESOURCES AND MANAGEMENT)

FISHERIES (BIODIVERSITY, AQUACULTURE, FOOD AND NUTRITION, SUSTAINABLE SEAFOOD, NATURAL RESOURCES, CONSERVATION)

Preface

It is with great pleasure that we welcome esteemed participants, scholars, and researchers to the 5th International Conference on Fisheries, Aquatic, and Environmental Sciences (ICFAES) 2023, an international forum dedicated to the exchange of cutting-edge scientific knowledge and advancements. This event represents the culmination of collaborative efforts from researchers across the globe, contributing to the discourse and progress in marine and fisheries. This conference is hosted by Faculty of Marine and Fisheries of Syiah Kuala University and Co-hosted by The Indonesian Association of Oceanologists (ISOI) Banda Aceh Chapter, and The Indonesian Association of Marine Scholars (ISKINDO). We invite submissions for the 5th ICFAES 2023 under the theme of "*Integrated System in Marine and Fisheries Sector to Support Sustainable Development Goals(SDGs)*". Integrated systems in the marine and fisheries sector play a crucial role in harmonizing economic, social, and environmental aspects to promote sustainable development. The 5th ICFAES 2023 is divided into three multidisciplinary topics: Fisheries, Aquatic, and Environmental Science. This conference will provide an excellent forum for sharing knowledge and research findings among academicians, professionals, and governments. Topics of interest cover all theoretical and practical aspects of Fisheries, Aquatic, and Environmental Sciences. The primary goal of ICFAES 2023 is to promote a multidisciplinary understanding of the latest research and innovations through discussion, exchange, and sharing among researchers, practitioners, academicians, and students.

As chairman of the 5th ICFAES, it gives me great pleasure and honor to extend to you all a very warm welcome on behalf of the international conference committee. I wish to take this opportunity to welcome the participants from overseas as the representatives of the students, lecturers, and researchers. A special acknowledgment is extended to our distinguished keynote speakers: Prof. Li Jun He, Dr. Ir. Muhammad Irham, S.Si, M.Si, Dr. Mahmudur Rahman, and Mikhail Solovyev, Ph.D., for graciously contributing their expertise.

I hope this event will catalyze strengthening international cooperation in the transfer of knowledge. In addition, I am most thankful for the ceaseless efforts of staff members of the Marine and Fisheries Faculty of Syiah Kuala University including our Rector, Prof. Dr. Ir. Marwan., IPU, and our Dean, Prof. Dr. Muchlisin ZA, S.Pi, M.Sc. ICFAES is an annual Conference, organized by the Faculty of Marine and Fisheries of Universitas Syiah Kuala. Since our 1st conference in 2018, ICFAES has published 264 papers from Indonesia and abroad in the IOP Conference Series: Earth and Environmental Science, indexed by Scopus. In 2023, 83 papers were submitted to ICFAES 5th, and only 59 will be accepted for publication. The accepted paper will be scheduled to be published for the next available slot in the IOP Conference Series: Earth and Environmental Science.

Finally, we would like to thank all the authors who submitted their papers to ICFAES 2023, the reviewers, the keynote and invited speakers, the moderator, the committee, and our volunteers for all their work in preparing and running the conference. We wish you a very warm welcome and hope that you will have a very productive and enjoyable time at the 5th ICFAES 2023.

Warm regards,

Ilham Zulfahmi, M.Si
Chairman of the 5th ICFAES 2023

Conference Committees
Keynote Speaker

- Dr. Muhammad Irham, S.Si, M.Si (Universitas Syiah Kuala, Indonesia)
- Prof. Li Jun He (East China Normal University, China)
- Dr. Mahmudur Rahman (The University of Newcastle, Australia)
- Mikhail Solovyev, Ph.D (Russian Academy of Science, Russia)

Invited Speakers

- Fajar Kurniawan, S.T., M.A.P., M.MG (Marine and Coastal Resources Management Entity, Indonesia)
- Dr. Agung Setia Batubara (Universitas Negeri Medan, Indonesia)
- Dr. Reniel Cabral ((James Cook University, Australia)
- Wijayanto, M.Sc (Asian Integrated Training Network, Indonesia)
- Ahasan Habib, Ph.D (Universiti Malaysia Terengganu, Malaysia)

Scientific Committee

- Prof. Dr. Ir. Syamsul Rizal (Universitas Syiah Kuala, Indonesia)
- Prof. Li Jun He (East China Normal University, China)
- Kizar Ahmed Sumon, PhD (Bangladesh Agricultural University, Bangladesh)
- Assoc. Prof. Dr. Sukree Hajisamae (Prince of Songkla University, Thailand)
- Dr. Pengfei Li (Guangzou Academy of Science, China)
- Dr. Reniel Cabral (James Cook University, Australia)
- Dr. Mahmudur Rahman (University of Newcastle, Australia)
- Dr. Fayakun Satria (Badan Riset dan Inovasi Nasional, Indonesia)
- Dr. Agung Setia Batubara (Universitas Negeri Medan, Indonesia)
- Mikhail Solovyev, Ph.D (Russian Academy of Science, Russia)
- Dr. Firman M Nur (Badan Riset dan Inovasi Nasional, Indonesia)
- Dr. Nur Fadli, M.Sc. (Universitas Syiah Kuala)

Steering Committee

- Prof. Dr. Marwan, IPU
- Prof. Dr. Taufik Fuadi Abidin, S.Si., M.Tech
- Prof. Dr. Muchlisin Z.A., S.Pi, M.Sc
- Dr. Muhammadar, MP
- Dr. Nur Fadli, M.Sc
- Dr. Ir. Muhammad Irham, M.Sc
- Dr. Zulkarnain, S.Si, M.Si

Executive Committee

- Ilham Zulfahmi, M.Si (Chairman of the Committee)
- Rianjuanda, M.Si (Vice Chairman)
- Cut Nuzlia, M.Sc (Secretary)

- Nurfadillah, S.Si., M.Si (Treasurer and Finance)
- Khairunnisa, S.Pi, M.Sc (Website and Promotion)
- Maria Ulfah, M.Si (Logistic)
- Dr. Irma Dewiyanti, M.Sc (Event)
- Edy Miswar (Information and Technology)
- Said Ali Akbar, M.Si (Publication)
- Rita Susana, S.E. (Household and Documentation)
- Suraiya Nazlia, M.Sc (Accomodation and Transportation)

The 5th International Conference on Fisheries, Aquatic and Environmental Sciences

ICFAES

The 5th ICFAES 2023

Integrated System in Marine and Fisheries Sector to Support Sustainable Development Goals(SDGs)

Banda Aceh, 25th October 2023
Hybrid Conference

Keynote Speakers

Dr. Muhammad Irham
University Syiah Kuala, Indonesia

Dr. Mahmudur Rahman
University of Newcastle, Australia

Mikhail Solovyev, Ph.D
Russian Academy of Science, Russia

Invited Speakers

Dr. Agung Setia Batubara
Universitas Negeri Medan
Indonesia

Dr. Reniel Cabral
James Cook University
Australia

Prof. Li Jun He
East China Normal University
China

Wijayanto, M.Sc
Asian Integrated
Training Network Indonesia

Dr. Fayakun Satria
National Research and Innovation Agency
Indonesia

Ahasan Habib, Ph.D
Universiti Trengganu Malaysia

Scope

Fisheries · Aquatic · Environmental Sciences

Registration Fee	Domestic	Foreigner
Participant (without certificate)	Free	Free
Participant (certificate)	Rp. 75.000	$10
General Presenter (without publication)	Rp.500.000	$50
Student Presenter (without publication)	Rp.500.000	$50
General Presenter (publication)	Rp.2.850.000	$285
Student Presenter (publication)	Rp.2.750.000	$275
Early Bird General Presenter (publication)	Rp.2.700.000	$270
Early Bird Student Presenter (publication)	Rp.2.650.000	$265

indexed in SCOPUS
E3S Web of Conferences

Schedule

Abstract Submission
July 17 ~~August 15~~ August 31

September, 10th
Notification of Acceptance

Early Payment Deadline
September, 15th

September, 20th
General Payment Deadline

Fullpaper Submission Deadline
October, 3rd

CONFERENCE DATE
October, 25th

Registration Now!
www.icfaes.usk.ac.id

Conference Secretariat:
Faculty of Fisheries and Marine Science
Universitas Syiah Kuala, Banda Aceh Indonesia, 23111
Hotline : +62-852-6020-3678 (Cut Nuzlia, M.Sc)/ +62-852 6274 3552 (Ilham Zulfahmi)
Email : icfaes@usk.ac.id

Hosted by
USK Universitas Syiah Kuala · FKP

Co. Host by
PRKP

Statement of Peer review

In submitting conference proceedings to *Web of Conferences*, the editors of the proceedings certify to the Publisher that

1. They adhere to its **Policy on Publishing Integrity** in order to safeguard good scientific practice in publishing.
2. All articles have been subjected to peer review administered by the proceedings editors.
3. Reviews have been conducted by expert referees, who have been requested to provide unbiased and constructive comments aimed, whenever possible, at improving the work.
4. Proceedings editors have taken all reasonable steps to ensure the quality of the materials they publish and their decision to accept or reject a paper for publication has been based only on the merits of the work and the relevance to the journal.

The 5th International Conference on Fisheries, Aquatic, and Environmental Sciences (ICFAES) 2023, 25th October 2023, Banda Aceh, Indonesia

Proceedings editor(s): Dr. Cut Dara Dewi, M.Si

Proceedings editor(s): Dr. Muhammad Nanda, S.Kel

Proceedings editor(s): Said Ali Akbar, M.Si

Proceedings editor(s): Dr. Vicky Prajaputra, M.Si

Proceedings editor(s): Dr. Yudi Haditiar

30 October 2023

Exploring the interplay of chlorophyll-a, sea surface temperature, and sea surface salinity in aceh waters during january and july 2022

Muh. Nur Hidayat[1], *Reza* Wafdan[2], *Marwan* Ramli[3], *Zainal A.* Muchlisin[1,4,6], *Ichsan* Rusydi[4], *Muhammad* Muhammad[5], and *Syamsul* Rizal[1,2,6,*]

[1]Graduate School of Mathematics and Applied Sciences, Universitas Syiah Kuala, Banda Aceh 23111, Indonesia
[2]Department of Marine Sciences, Faculty of Marine and Fisheries, Universitas Syiah Kuala, Banda Aceh, 23111, Indonesia
[3]Department of Mathematics, Universitas Syiah Kuala, Banda Aceh 23111, Indonesia
[4]Department of Aquaculture, Faculty of Marine and Fisheries, Universitas Syiah Kuala, Banda Aceh 23111, Indonesia
[5]Department of Fisheries Resource Utilization, Faculty of Marine and Fisheries, Universitas Syiah Kuala, Banda Aceh, 23111, Indonesia
[6]Research Center for Marine Sciences and Fisheries, Universitas Syiah Kuala, Banda Aceh 23111, Indonesia

Abstract. This study investigates the relationship among chlorophyll-a (Chl-a), sea surface temperature (SST), and sea surface salinity (SSS) in Aceh Waters, Indonesia using data from January and July 2022. Chl-a, SST, and SSS data were retrieved from the Copernicus Marine Environment Monitoring Service (CMEMS) database. Pearson correlation analysis was employed to assess the connections among Chl-a, SST, and SSS within the Aceh Waters region. The findings reveal that all three parameters - Chl-a, SST, and SSS - conform to the seasonal monsoon patterns observed in January and July 2022. The correlation analysis conducted for January revealed the following relationships: a negative correlation between Chl-a and SST (-0.649), an inverse correlation between Chl-a and SSS (-0.215), and a positive correlation between SST and SSS (0.493). Conversely, correlations for July reveal a negative correlation between Chl-a and SST (-0.503), a positive correlation between Chl-a and SSS (0.039), and a negative correlation between SST and SSS (-0.478). Overall, this study elucidates the complex relationship among Chl-a, SST, and SSS in Aceh Waters, which is influenced by seasonal monsoon variations. Understanding this relationship is essential for assessing marine environmental dynamics and their potential impacts on ecosystems and human activities in the region.

* Corresponding author : srizal@usk.ac.id

https://doi.org/10.1051/bioconf/20248701001

1 Introduction

Aceh Waters, situated at the northwestern tip of Sumatra Island, Indonesia, possesses unique aquatic characteristics strongly influenced by seasonal shifts in wind patterns. The active southwest monsoon from June to September and the northeast monsoon from December to February play a crucial role in determining the locations of fishing grounds. These monsoon systems not only impact the region's fishing resources but also affect the abilities of the fishermen in Aceh Waters [1].

Furthermore, Aceh Waters, one of the largest sea regions in Indonesia, is considered to hold immense potential for marine and fisheries resources owing to its distinct aquatic characteristics. Due to its vital role in the regional marine ecosystem, over the years, many research efforts have been directed towards this area [1-7]. Research conducted by Ilhamsyah et al. [8] identified four different types of water masses within this area, including a salt-lens formation measuring 60 miles in diameter and 25 meters in thickness. This salt-lens possesses a core salinity exceeding 35 psu and is located at a depth of 75-100 meters below the thermocline layer in the Northern Weh Island region. This salinity front may serve as an early indicator of upwelling, a phenomenon closely linked to the presence of pelagic fish. This discovery implies that Aceh Waters holds promising potential for the future development of marine resources.

In the pursuit of understanding the marine dynamics of Aceh Waters, it is crucial to investigate the connection between oceanographic factors, such as chlorophyll-a (Chl-a) and sea surface temperature (SST). Chl-a acts as a proxy or indicator of phytoplankton biomass, which forms the foundation of the marine food chain. Additionally, SST and sea surface salinity (SSS) are critical environmental variables that significantly impact capture fisheries. The research conducted by [9] has revealed noteworthy findings concerning Chl-a and SST. It reveals a distinct inverse relationship between Sea Surface Temperature (SST) and the yellowfin tuna (*Thunnus albacares*) catch. This discovery suggests that as sea surface temperatures increase, the yellowfin tuna catch tends to decrease.

Moreover, the study underscores the influence of chlorophyll-a on fish quantity caught, establishing a robust positive correlation between chlorophyll-a levels and the yellowfin tuna catch. This implies that regions with higher chlorophyll-a levels in the ocean are more likely to yield greater yellowfin tuna catches. The influence of chlorophyll-a (Chl-a), sea surface temperature (SST), and sea surface salinity (SSS) on fish aggregation in the Red Sea is significant [10]. Chl-a is an indicator of phytoplankton biomass, which is the base of the marine food chain. Fish are attracted to areas with high Chl-a concentrations because there is more food available. SST and SSS affect the distribution and abundance of marine life, including fish. Some fish species are more tolerant of high salinity than others.

In the Red Sea, fish aggregation is highest during the winter months when Chl-a concentrations are high and SST and SSS are lower [10]. This is likely due to a combination of factors, including strong vertical mixing and upwelling, which bring nutrients to the surface and promote phytoplankton growth. Fish are also attracted to the open waters of the Red Sea, where Chl-a concentrations are higher than in the coastal waters.

Several studies have examined the relationships between Chl-a, SST, and SSS. In the Bay of Bengal (BoB) and the Northern Bay of Bengal (NBoB), researchers Hidayat et al. [11-12] conducted analyses to determine the correlations among Chl-a and SST, Chl-a and SSS, as well as SST and SSS. Research by Torregroza-Espinosa et al. [13] identified spatial-temporal variations and trends and calculated correlations of Chl-a, SST, and SSS at the Magdalena River mouth in the Caribbean Sea. Research by Zhang et al. [14] calculated the correlation between Chl-a and SST in the East China Sea. These studies have revealed connections between Chl-a, SST, and SSS in various research areas. However, this relationship is context-dependent and can be influenced by regional factors, requiring specific regional studies.

Based on the description above, the research questions can be formulated as follows. Firstly, what is the seasonal variability of chlorophyll-a (Chl-a), sea surface temperature (SST), and sea surface salinity (SSS) in Aceh Waters? Secondly, what relationships exist between Chl-a, SST, and SSS within Aceh Waters? This study aims to explore the interplay between three vital oceanographic variables—Chl-a, SST, and SSS—within Aceh Waters. Chl-a serves as a metric for phytoplankton biomass, a cornerstone of the marine food chain. Additionally, SST and SSS are crucial environmental parameters that exert significant influence on fisheries. By comprehending the seasonal fluctuations of these variables and their intricate interactions, we can enhance our comprehension of Aceh Waters' marine dynamics. This knowledge can subsequently contribute to the effective management of marine resources in the region, including fisheries and coastal ecosystems.

The study's specific objectives encompass quantifying the seasonal variations of Chl-a, SST, and SSS within Aceh Waters, identifying the correlations between Chl-a, SST, and SSS within Aceh Waters. Additionally, this study seeks to assess the repercussions of these relationships on marine ecosystems within Aceh Waters. The data used in this analysis were sourced from the Copernicus Marine Environment Monitoring Service (CMEMS) database, a comprehensive global repository of marine environmental data. This database encompasses satellite-derived information on Chl-a, SST, and SSS.

To analyze the data, we employed both descriptive statistics and correlation methods. Descriptive statistics aided in summarizing key aspects of the data, including mean values, medians, and standard deviations. Meanwhile, correlation methods were employed to scrutinize the connections between these oceanographic variables.

2 Materials and Methods

The research was conducted in the waters of Aceh, with coordinates ranging from 3°N to 6.75°N and 93°E to 97.5°E, as depicted in Figure 1. The research employed monthly mean datasets for Chl-a, SST, and SSS for both January and July 2022. These datasets were sourced from the Copernicus Marine Environment Monitoring Service (CMEMS) database [15, 16].

To ensure data quality and reliability, rigorous data quality control procedures were implemented. Any erroneous or inconsistent data points were identified and corrected, preserving the dataset's integrity. Additionally, data preprocessing techniques were applied to standardize and prepare the data for subsequent analysis.

The geographical distribution of sampling points within Aceh Waters was thoughtfully selected to encompass a representative and diverse range of environmental conditions. This strategic selection aimed to capture variations in Chl-a, SST, and SSS across different areas of Aceh Waters, accounting for potential local influences and gradients.

To assess the seasonal variability of Chl-a, SST, and SSS, a variety of statistical analyses were employed. These included boxplot analysis, QQ plot analysis, and correlation matrix analysis. Boxplot analysis was utilized to visualize the statistical distribution of key data parameters such as the lowest value, the value at the 25th percentile, median, the value at the 75th percentile, and the highest value. It also served as a means to identify outliers and compare data groups [17].

Quantile-quantile (QQ) plot analysis, a simple yet powerful method, was used to visualize data distribution [18] and examine theoretical distributions, such as the normal distribution, within the Chl-a, SST, and SSS datasets. If data points on the QQ plot align with a straight line, it indicates that the data approximates a normal distribution. In general, statistical methods like correlation analysis, regression, t-tests, and analysis of variance assume that data follows a normal distribution. However, if the sample size exceeds 100 observations, deviations from normality are not a major concern [19, 20].

The correlation matrix was employed to determine the relationships between Chl-a and SST, Chl-a and SSS, as well as SST and SSS in Aceh Waters. This matrix depicted Pearson correlation coefficients between all pairs of variables. The Pearson correlation coefficient is expressed in Eq. 1 [11]:

$$r = \frac{\sum(x_i-\bar{x})(y_i-\bar{y})}{\sqrt{\sum(x_i-\bar{x})^2\sum(y_i-\bar{y})^2}} \quad (1)$$

where x_i and y_i represent pairs between Chl-a and SST, Chl-a and SSS, as well as SST and SSS, and $\bar{x}$ and $\bar{y}$ denote the averages of the data. Additionally, each subplot off the diagonal contained a scatterplot of a variable pair with a least-squares regression line, the slope of which matched the displayed correlation coefficient. Each diagonal subplot presented the distribution of a variable as a histogram.

The p-value served the purpose of hypothesis testing, indicating whether there was a relationship between the observed phenomena (null hypothesis). If the elements off the diagonal had a p-value smaller than the significance level of 0.05 (α=5%), the corresponding correlation in R was deemed significant. Correlation coefficient values were interpreted as follows: if $0 \leq R \leq 1$, the correlation was considered insignificant or negligible; if $0.1 \leq R \leq 0.39$, it was deemed weak; if $0.4 \leq R \leq 0.69$, it was moderate; if $0.7 \leq R \leq 0.89$, it was strong; and if $0.9 \leq R \leq 1$, it was very strong [21].

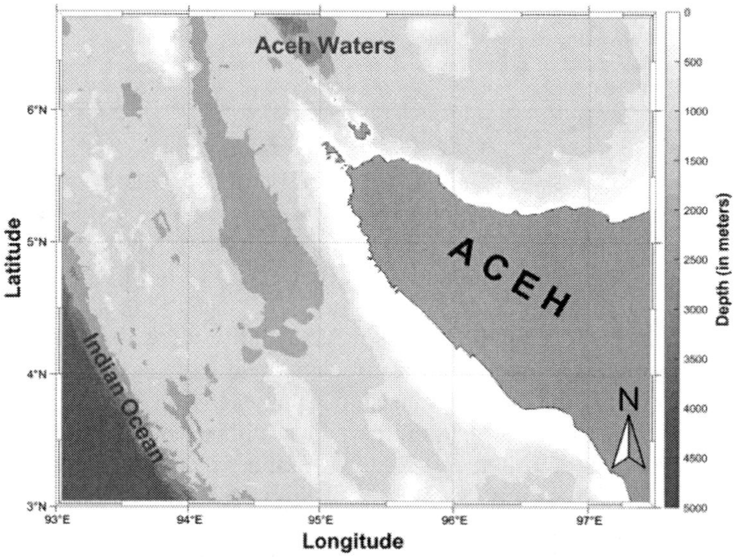

Fig. 1. Bathymetry for Aceh Waters. Colors indicate varying depths from 0 to 5000 meters. Data derived from SRTM15+ [22].

3 Results and Discussion

3.1 Boxplot and QQ plot

This study employed a total of 248 data points for each of Chl-a, SST, and SSS in Aceh Waters. Figure 2 presents Boxplots and QQ Plots for these datasets in both January and July. Figure 2(a) displays the Boxplot for Chl-a in January. Chl-a values in January exhibited a minimum of 0.15 mg/m^3 and a maximum of 1.59 mg/m^3. The median Chl-a value reached 0.4 mg/m^3. The long upper whisker indicates high variability in Chl-a within the range of 0.8

mg/m^3 to 1.59 mg/m^3. In contrast, Chl-a showed uniformity within the range of 0.15 mg/m^3 to 0.8 mg/m^3.

Figure 2(b) illustrates SST in January. SST ranged from a minimum of 28.5°C to a maximum of 30.6°C, with a median SST of 29.4°C. The median line evenly divides the box, indicating symmetrical data. Overall, SST values concentrated within the range of 29.0°C to 29.8°C. Figure 2(c) presents SSS in January. SSS had a minimum of 32.2 PSU, a maximum of 34.5 PSU, and a median of 33.6 PSU. The median SSS closely aligned with the upper quartile, indicating data skewness or uniformity at values >33.6 PSU.

Figure 2(d) displays the Boxplot for Chl-a in July. The minimum Chl-a value in July was 1.13 mg/m3, while the maximum was 0.36 mg/m3, with a median Chl-a value of 0.2 mg/m3. It is also evident that Chl-a in July exhibited outliers above 0.36 mg/m3. Figure 2(e) shows the Boxplot for SST in July. SST ranged from a minimum of 28.9°C to a maximum of 30.3°C, with a median of 29.3°C. SST values in July were concentrated within the range of 29.1°C to 29.6°C. Some outliers were also detected above 30.3°C. Figure 2(f) illustrates the Boxplot for SSS in July. Minimum, maximum, and median SSS values were 32.5 PSU, 33.8 PSU, and 33.2 PSU, respectively. The median, located at the center of the box, along with equal lengths of upper and lower whiskers, indicated a symmetrical Boxplot, suggesting no skewness in the data.

Comparison of the Boxplots for Chl-a, SST, and SSS in January (Figure 2(a-b)) and July (Figure 2(d-e)) reveals that these variables had higher values in January compared to July. This is evident from the range of minimum and maximum values for each variable Chl-a, SST, and SSS.

Figure 2(c) presents the QQ Plot for Chl-a (green dots), SST (dark blue dots), and SSS (light blue dots) in January. The Chl-a data follows a straight line pattern and tends to align with the red line at theoretical quantiles -1 and 2.3. Deviation from the line occurs for values less than -1 (above the line) and greater than 2.3 (below the line). This indicates that data within the central range, between -1 and 2.3, approximates a normal distribution. Values less than -1 indicate longer tails to the left or left skewness (negative). Values greater than 2.3 indicate longer tails to the right or right skewness (positive). SST and SSS data exhibit similar patterns. Theoretical quantile values for SST and SSS from -1 to 1 approximate a normal distribution, but there is deviation in the tails of the distribution, suggesting some skewness in the data.

Figure 2(d) presents the QQ Plot for Chl-a, SST, and SSS in July. It is evident that all Chl-a data points follow a straight line pattern and closely align with the red line, indicating that the dataset conforms to a normal distribution. Conversely, SST and SSS data show slight deviation in the tails of the distribution. For SST, data deviates from the line at values <-1.4 and >1.3, while for SSS, data deviates from the line at values <-2 and >2.2.

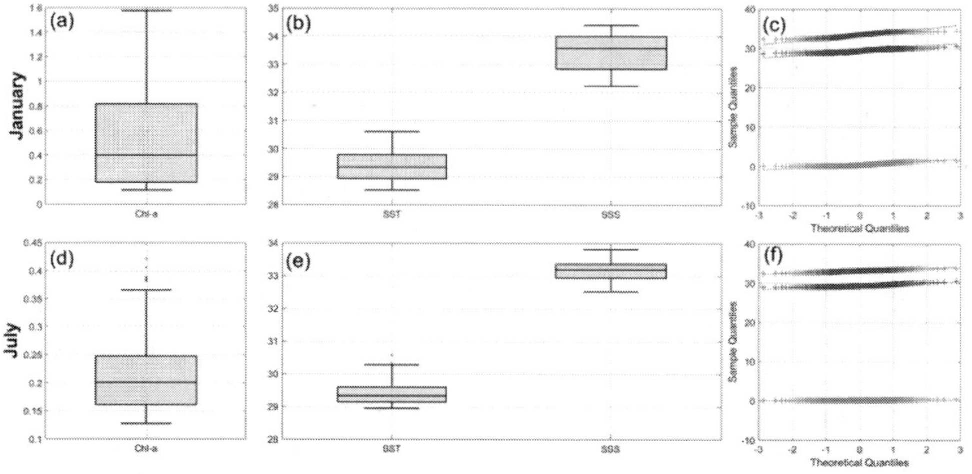

Fig. 2. Boxplots for (a) Chl-a, (b) SST, and SSS for January, and (d) Chl-a, (e) SST, and SSS for July, while (c) QQ plot for January and (f) QQ plot for July. Box plots depict the median and interquartile range (IQR) of data for the 3 variables. Whiskers extend up to 1.5 times the IQR. Red dots represent outliers. Colors. Green in the QQ plot indicates Chl-a, dark blue indicates SST, and light blue indicates SSS.

3.2 Correlation Matrix

Figures 3(a) and 3(b) depict the correlation matrices of Chl-a, SST, and SSS data for the months of January and July, respectively. Figure 3(a) illustrates the correlations between Chl-a and SST, Chl-a and SSS, as well as SST and SSS, along with the distribution of each variable in histograms for January. Each variable pair in January is marked significant, as indicated by the red-colored correlation values. In January, the correlation between Chl-a and SST is moderate and negative, the correlation between Chl-a and SSS is weak and negative, and the correlation between SST and SSS is moderate and positive. Negative correlation indicates an inverse relationship, in this case, when Chl-a is high, SST tends to be low, or when Chl-a is high, SSS tends to be low. Positive correlation indicates a direct relationship, in this case, when SST is high, SSS is also high.

Figure 3(b) illustrates the correlations in July. The Chl-a and SSS pair in July is not significant, as indicated by the black-colored correlation value. Meanwhile, the Chl-a and SST, as well as SST and SSS pairs in July, exhibit moderate and negative correlations, respectively. The non-significant correlation indicates either no or very little linear relationship between Chl-a and SSS in July. On the other hand, the negative correlation suggests an inverse relationship between Chl-a and SST, as well as SST and SSS in July.

Table 1 presents the corresponding correlation coefficient values and p-values for the Chl-a and SST, Chl-a and SSS, as well as SST and SSS pairs. All p-values in January are smaller than the significance level of 0.05, indicating that all variable pairs in January have significant correlations. In contrast, not all p-values in July are smaller than 0.05. Only the Chl-a and SST, as well as SST and SSS pairs, exhibit significant correlations. This indicates that only these pairs have significant correlations in July.

The Pearson correlation coefficient (r) serves as a metric to gauge both the magnitude and direction of a linear relationship between two variables. It falls within the range of -1 to 1, where -1 signifies a perfect negative correlation, 1 indicates a perfect positive correlation, and 0 denotes an absence of any discernible relationship between the movements of the variables [23]. As observed in January, all combinations of variables—Chl-a and SST, Chl-a and SSS, as well as SST and SSS—display significant correlations, supported by p-values below the threshold of 0.05 (refer to Table 1).

Conversely, upon analysis in July, the scenario shifts, revealing that solely the Chl-a and SST, as well as the SST and SSS pairs, exhibit noteworthy correlations. Notably, the Chl-a and SSS pair fails to meet the criteria for significance due to its inability to satisfy the requirement of p-values falling below 0.05, as indicated in Table 1.

Both Chl-a and SST exhibit a negative correlation, evident in both January and July. This indicates that as Chl-a values increase, SST values tend to decrease, and vice versa. Conversely, SST and SSS demonstrate a positive correlation in January but switch to a negative correlation in July. This shift indicates a dynamic relationship between sea surface temperature and sea surface salinity across these two months.

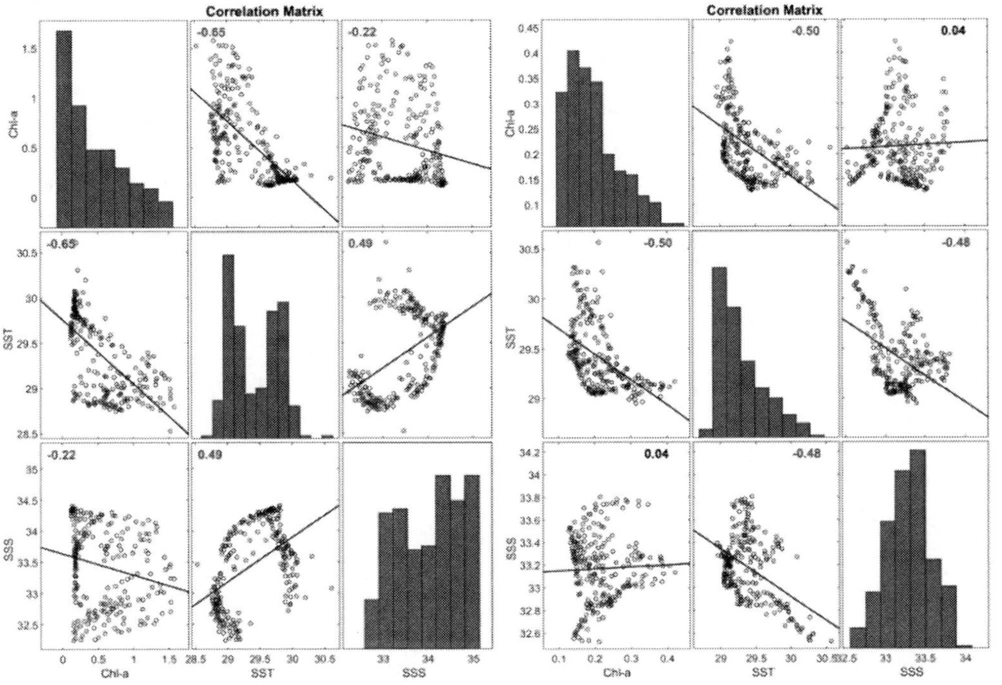

(a) January (b) July

Fig. 3. Correlation matrix for Chl-a and SST, Chl-a and SSS, as well as SST and SSS pairs in (a) January and (b) July.

Table 1. Correlation coefficients (r) and p-values

No	Pairs	January		July	
		r	p-value	r	p-value
1	Chl-a and SST	-6.5×10^{-1}	5.2×10^{-31}	-5.0×10^{-1}	2.4×10^{-17}
2	Chl-a and SSS	-2.2×10^{-1}	6.5×10^{-4}	Not significant	5.4×10^{-1}
3	SST and SSS	4.9×10^{-1}	1.4×10^{-16}	-4.8×10^{-1}	1.4×10^{-15}

4 Conclusions

The analysis of Chl-a, SST, and SSS data in Aceh Waters revealed significant seasonal variability, notably higher values in January compared to July across all variables. Boxplots and QQ Plots provided insights into the distribution and normality of these variables, showing variations in values and patterns between the two months. Correlation analysis highlighted significant relationships among the variables in January, with Chl-a negatively correlated with SST and exhibiting weaker correlations with SSS, while SST and SSS displayed a moderate positive correlation. However, in July, while Chl-a and SST, as well as SST and SSS, remained significantly correlated, the Chl-a and SSS pair did not, indicating differing ecological contexts between the two months. These findings emphasize the dynamic nature of the marine environment in Aceh Waters and the need for comprehensive understanding to support effective marine resource management and ecosystem conservation efforts in the region. Further research is warranted to uncover the underlying mechanisms driving these seasonal shifts in correlations.

Acknowledgments

This research is funded by the Ministry of Education, Culture, Research and Technology of the Republic of Indonesia with the scheme "*Penelitian Pendidikan Magister menuju Doktor untuk Sarjana Unggul* (PMDSU)" for the 2023 fiscal year with the contract number [587/UN11.2.1/PT.01.03/DPRM/2023] and the scheme of "*Penelitian Profesor*" for the 2023 fiscal year with the contract number [9/UN11.2.1/PT.01.03/PNBP/2023]. Author would like to thank to the Ocean Modeling Laboratory, Department of Marine Sciences, Universitas Syiah Kuala, Indonesia, for providing research facilities.

References

1. H.A. Haridhi, M. Nanda, Y. Haditiar, S. Rizal, Ocean Coast. Manag. **154**, 46-54 (2018)
2. M. Ikhwan, Y. Haditiar, R. Wafdan, M. Ramli, Z.A. Muchlisin, S. Rizal, Renew. Sustain. Energy Rev. **159**, 112220 (2022)
3. T. Iskandar, M. Ikhwan, Y. Haditiar, S. Rizal, Songklanakarin J. Sci. Technol. **44**(1) (2022)
4. H.T.A. Rasyidi, Y. Haditiar, M. Ikhwan, R. Wafdan, I. Setiawan, S. Sugianto, S. Rizal, *Effect of wind-driven currents circulation in Sabang Waters*, in IOP Conference Series: Earth and Environmental Science, IOP Publishing **348** 1 012094 (2019)
5. I. Setiawan, Y. Haditiar, M. Syukri, N. Ismail, S. Rizal, Heliyon **9**(6) (2023)
6. I. Setiawan, Y. Haditiar, M. Ikhwan, Z. Nufus, M. Syukri, N. Ismail, S. Rizal, J. Sustain. Sci. Manage. **15**(8), 122-35 (2020)
7. H.A.A. Wahyudi, Y. Haditiar, M. Ikhwan, R. Wafdan, I. Setiawan, M. Muhammad, S Sugianto, S. Rizal, *Numerical study of tides in the Bay of Sabang*, in IOP Conference Series: Earth and Environmental Science, IOP Publishing **348**, 1, 012096 (2019)
8. Y. Ilhamsyah, Y. Koesmaryono, R. Hidayat, I.W. Nurjaya, A.S. Atmadipoera, S. Rizal, Adv. Environ. Sci. **10**(3), 200-208 (2018)
9. A.B. Sambah, A. Muamanah, L.I. Harlyan, T.D. Lelono, F. Iranawati, A. Sartimbul, Aquacult. Aquar. Conserv. Legis. **14**(2), 897-909 (2021)
10. S.T. Daqamseh, A.K. Al-Fugara, B. Pradhan, A. Al-Oraiqat, M. Habib, Sensors, **19**(9), 2069 (2019)
11. M.N. Hidayat, R. Wafdan, M. Ramli, Z.A. Muchlisin, S. Rizal, Glob. J. Environ. Sci. Manage. **9**(3), 389-402 (2023)
12. M.N. Hidayat, R. Wafdan, M. Ikhwan, M. Ramli, S. Rizal, *Analysis of chlorophyll-a, sea surface temperature, and sea surface salinity in the Bay of Bengal*, in IOP Conference Series: Earth and Environmental Science, IOP Publishing **1221**, 1, 012071 (2023)

13. A.C. Torregroza-Espinosa, J.C. Restrepo, J. Escobar, J. Pierini, A. Newton, J. South Am. Earth Sci. **105**, 102978 (2021)
14. C. Ji, Y. Zhang, Q. Cheng, J. Tsou, T. Jiang, X.S. Liang, Int. J. Appl. Earth Obs. Geoinf. **68**, 252-261 (2018)
15. P.Y. Le Traon, A. Reppucci, E.A. Fanjul, L. Aouf, A. Behrens, M. Belmonte, A. Bentamy, A. Lotfi, E. Carval, S.A. Ciliberti, H. Claustre, E. Clementi, G. Coppini, G. Cossarini, A.-M. Alonso-Muñoyerro, A. Delamarche, G. Dibarboure, F. Dinessen, M. Drevillon, Y. Drillet, Y. Faugere, V. Fernández, A. Fleming, M.I. García-Hermosa, M. Sotillo García, G. Garric, F. Gasparin, C. Giordan, M. Gehlen, M.L. Gregoire, S. Guinehut, M. Hamon, C. Harris, F. Hernandez, J.B. Hinkler, J. Hoyer, J. Karvonen, S. Kay, R. King, T. Lavergne, B. Lemieux-Dudon, L. Lima, B. Nardelli Buongiorno, G. Nolan, A. Pascual, J. Pistoia, M.I. Pujol, A.C. Pequignet, E. Peneva, B. Pérez Gómez, L.P. de la Villeon, N. Pinardi, A. Pisano, S. Pouliquen, R. Reid, E. Remy, R. Santoleri, J. Siddorn, J. She, J. Staneva, A. Stoffelen, M. Tonani, L. Vandenbulcke, K. von Schuckmann, G. Volpe, C. Wettre, A. Zacharioudaki, Front. Mar. Sci. **6**, 234 (2019)
16. J.M. Lellouche, E. Greiner, O. Le Galloudec, G. Garric, C. Regnier, M. Drevillon, M. Benkiran, C.E. Testut, R. Bourdalle-Badie, F. Gasparin, O. Hernandez, B. Levier, Y. Drillet, E. Remy, P.Y. Le Traon, Ocean Science, **14**(5), 1093-1126 (2018)
17. K. Hu, Methods Protoc. **3**(4), 64 (2020)
18. H.L. Ford, A.C. Ravelo, P.J. Polissar, Science, **347**(6219), 255-258 (2015)
19. A. Ghasemi, S. Zahediasl, Int J Endocrinol Metab. **10**(2), 486 (2012)
20. P. Mishra, C.M. Pandey, U. Singh, A. Gupta, C. Sahu, A. Keshri, Ann. Card. Anaesth. **22**(1), 67 (2019)
21. P. Schober, C. Boer, L.A. Schwarte, Anesth. Analg. **126**(5), 1763-1768 (2018)
22. B. Tozer, D.T. Sandwell, W.H. Smith, C. Olson, J.R. Beale, P. Wessel, Earth Space Sci. **6**(10), 1847-1864 (2019)
23. C. Andrade, Indian J. Psychol. Med. 41(3), 210-215 (2019)

 https://doi.org/10.1051/bioconf/20248701002

Analysis of sea currents, sea temperature, and sea salinity variations in the malacca strait during january and july 2022 using vertical sections

Muh. Nur Hidayat[1], *Reza* Wafdan[2], *Marwan* Ramli[3], *Zainal A.* Muchlisin[1,4,7], *Sugianto* Sugianto[5], *Makwiyah A.* Chaliluddin[6], *Thaib* Rizwan[6], and *Syamsul* Rizal[1,2,7,*]

[1]Graduate School of Mathematics and Applied Sciences, Universitas Syiah Kuala, Banda Aceh 23111, Indonesia
[2]Department of Marine Sciences, Faculty of Marine and Fisheries, Universitas Syiah Kuala, Banda Aceh, 23111, Indonesia
[3]Department of Mathematics, Universitas Syiah Kuala, Banda Aceh 23111, Indonesia
[4]Department of Aquaculture, Faculty of Marine and Fisheries, Universitas Syiah Kuala, Banda Aceh 23111, Indonesia
[5]Department of Soil Sciences, Faculty of Agriculture, Universitas Syiah Kuala, Banda Aceh, 23111, Indonesia.
[6]Department of Fisheries Resource Utilization, Faculty of Marine and Fisheries, Universitas Syiah Kuala, Banda Aceh, 23111, Indonesia
[7]Research Center for Marine Sciences and Fisheries, Universitas Syiah Kuala, Banda Aceh 23111, Indonesia

Abstract. This study analyzes the sea currents, temperature, and salinity in the Malacca Strait during the northeast (NE) and southwest (SW) monsoons of 2022, represented by monthly average data from the Copernicus Marine Service (CMEMS) data portal for January and July, respectively. Vertical sections are created to visualize the layer structure and variations of sea parameters within the water column. The findings reveal differences in sea currents, temperature, and salinity between the NE and SW monsoons. July (SW monsoon) records consistently warmer temperatures than January (NE monsoon), both at the sea surface and seabed. Surface salinity is higher in July than in January, while seabed salinity is greater in January than in July. In January, the salinity-influenced mixed layer depth (MLD) appears to be thicker than the temperature-influenced MLD. There are no significant differences in the thickness of the temperature- and salinity-influenced MLDs during July. These variations in sea currents, temperature, and salinity at different depths provide a comprehensive understanding of the marine environmental structure and dynamics of the Malacca Strait.

[*] Corresponding author: srizal@usk.ac.id

1 Introduction

The Malacca Strait (MS) is a vital and strategic waterway for many countries because it is a major transportation route for oil and commodities, and it supports a variety of activities, including aquaculture, fishing, shipping, oil drilling, and energy generation [1]. According to Wang et al. [2], the Strait of Malacca is one of the most important maritime transportation routes in the world. However, they are also vulnerable to a variety of risks, including political instability, military conflict, piracy, terrorism, and vessel incidents. These risks can disrupt global trade and have a significant impact on the global economy.

The MS is highly influenced by changes in wind season. The southwest monsoon (SWM) occurs between June and September, while the northeast monsoon (NEM) takes place between December and February [3]. It is known that monsoon winds influence ocean circulation [4,5]. Several studies have been conducted in the MS, such as simulations of currents and tides [1,6,7].

The monsoon in the MS undergoes seasonal changes, shifting from SWM in summer (June to September) to NEM in winter (November to March), with two transitional periods in spring (April to May) and autumn (October) [8]. The monsoon has a major influence on both living and non-living processes in the Strait of Malacca, including nutrient availability and biological productivity [9]. The winter NEM is the most productive season in the Strait of Malacca because of strong mixing of the water column, which brings nutrient-rich deep water to the surface [10]. This nutrient-rich water supports the growth of phytoplankton and other marine organisms, which in turn support the marine food web.

In contrast to the winter NEM, during the summer SWM, the water column is well-stratified, meaning that the layers of water are clearly separated and do not mix easily, preventing the upward flow of nutrients to the surface layer. This results in low biological productivity. The strong SWM pushes the warm surface layer away from the coast, preventing the nutrient-rich deep water from reaching the surface.

The seasonal variation in nutrient availability and biological productivity in the Strait of Malacca has important implications for fisheries and other marine industries. For example, fish catches are typically higher during the winter NEM season when biological productivity is high. It is important to understand the seasonal variation in biophysical processes in the Strait of Malacca in order to manage its marine resources sustainably.

Sea temperature, salinity, and currents are basic parameters that influence the ecological and environmental conditions in the MS. Several studies have investigated these variables. Mansor et al. [11] showed that surface water in the MS in March is fresher and warmer than in August, while at the bottom depth in March it is more saline and cooler than in August. In addition, the mixed layer depth for both months did not exceed about 15 m for temperature and salinity, with a thermocline and halocline layer observed below the mixed layer depth. In the water column, Rizal et al. [3] showed that the temperature pattern in the MS shows a vertical gradient with values of 29°C at the surface and 19°C at the bottom. Meanwhile, salinity in the MS ranges from 31.5 at the surface to 34.5 at the bottom. Regarding ocean currents, Rizal et al. [12] found that for the bottom current, current speed ranges from 0-20 cm/s to the northwest. For the surface and layers 30-50 m, current magnitude is generally larger in February than in August. Whereas for the bottom layer, the current magnitude between February and August is relatively the same.

The definition of mixed layer depth (MLD) is closely tied to the homogeneous nature of the upper ocean layer. This uppermost stratum of the ocean undergoes thorough mixing, resulting in uniform temperature, salinity, and density throughout [13]. This well-mixed layer is formally referred to as the mixed layer itself. The boundary demarcating the lower limit of this mixed layer is known as MLD, and it typically extends to depths of several tens of meters [14].

The mixed layer constitutes the segment of the ocean that experiences direct influence from atmospheric mechanisms. Its significance in controlling air-sea interactions and climate dynamics is substantial. An indispensable parameter in the realms of physical, chemical, and biological oceanography is the MLD, as described by Giunta et al. [15]. MLD exhibits spatiotemporal variations across different time scales, as documented in several studies [16,17]. These fluctuations in MLD have significant ramifications for biological activity and play a pivotal role in shaping the dynamics of oceanic-air interactions [18,19]. Essentially, the MLD serves as the foundation for critical processes, including heat and freshwater exchanges between the ocean and the atmosphere, as well as influencing phenomena such as water mass formation [20], subduction from surface layers to greater depths [21], phytoplankton blooms [22,23], primary production [24], and regulation of seasonal ecosystems [25].

This research focuses on investigating the seasonal variations in sea temperature, sea salinity, and sea currents within the Malacca Strait. To gain deeper insights, we employ vertical sections and collect temperature and salinity profiles at various depths. This approach enables us to conduct a comprehensive comparison of conditions at the sea surface and the sea bottom during January and July, representing the NEM and SWM of the year 2022, respectively. Additionally, we examine sea temperature and sea salinity at a specific location within the Malacca Strait, situated at latitude 4°N and longitude 100°E, to determine the MLD.

Our specific research inquiries are outlined as follows. Firstly, to what extent does the difference between sea surface temperature and sea bottom temperature vary in January and July 2022? Secondly, what are the variations in sea surface salinity compared to sea bottom salinity during January and July 2022? Thirdly, how does the magnitude of sea currents differ between the sea surface and sea bottom in January and July 2022? Fourthly, what is the relationship between the MLD and temperature fluctuations in January and July 2022? Fifthly, how does the MLD change in response to salinity variations during January and July 2022?

2 Materials and Methods

The research was conducted in the Malacca Strait, focusing on vertical data collection at latitude 4°N, as depicted in Figure 1. This study utilized monthly average data of sea temperature, sea salinity, and sea currents, collected for January and July 2022. The data were sourced from the Copernicus Marine Service (CMEMS) data portal [26,27].

This research employed vertical section analysis for sea currents, sea temperature, and sea salinity in the Malacca Strait. Vertical sections are a graphical way to illustrate how a water property changes both vertically and horizontally. This is achieved by contouring data on vertical slices of the sea. Vertical section analysis is a type of spatial analysis along a vertical profile in the sea, spanning from the sea surface to the seabed. Vertical sections enable the identification of patterns, trends, and variations in oceanographic parameters at various depths. The analysis aims to understand the vertical structure and variations of sea currents, sea temperature, and sea salinity in the Malacca Strait in both January and July. Furthermore, station profiles at longitude 100°E and latitude 4°N were derived to provide a clearer view of the conditions of the isotherm and isohaline layers, as well as mixed layer depth (MLD) in January and July.

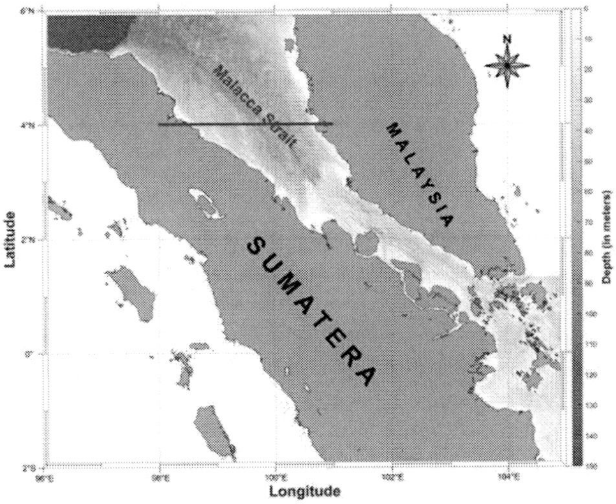

Fig. 1. Research location in the Malacca Strait. The study focuses on the cross-section (red line) at latitude 4°N. Colors indicate varying depths from 0 to 150 meters. Data derived from SRTM15+ [28].

3 Results and Discussion

Figure 2 shows vertical profiles detailing sea temperature, salinity, and currents within the 0 to 65-meter depth range for January and July 2022. Figures 2(a) and 2(d) delineate sea temperature vertical profiles for January and July, respectively, depicting temperature gradients through color variations. In January, temperatures spanned from 28.6°C (at the seabed) to 30.8°C (at the sea surface), while in July, they ranged from 29°C (at the seabed) to 31.4°C (at the sea surface). This indicates that both at the sea surface and at the seabed, the sea temperature in July is higher compared to January. This research correlates with the findings of Amiruddin et al. [29], Haridhi et al. [30,31], and Isa et al. [32,33].

Amiruddin et al. [29] observed that Sea Surface Temperature (SST) during the NEM typically stays between 28 to 29°C, while during the SWM, it elevates to around 30°C. Their investigation is grounded in the comprehensive world ocean database spanning from 1900 to 2005. In a separate study, Haridhi et al. [30,31] identified a marked impact of the monsoon system on SST in Sumatra. They noticed SST reaching its peak during the SWM (August) and hitting its lowest point during the NEM (February) due to prevailing wind patterns.

Meanwhile, Isa et al. [32] reported that SST during the NEM averaged around 28.5°C, whereas during the SWM, it reached approximately 29.9°C. In another study by Isa et al. [33], who analyzed 33 years of satellite-derived AVHRR data, they highlighted cooler temperatures prevailing in the MS and the Andaman Sea from December to February, averaging around 28–29°C, indicating lower SSTs during the NEM. However, temperatures surged to about 29.5°C in March, likely due to increased solar radiation. Comparatively, SST values were higher in July when contrasted with those recorded in January.

Meanwhile, Amiruddin et al. [29] compared vertical sections between the NEM and SWM, noting that seabed temperatures during the NEM were lower compared to those recorded during the SWM. These findings align with our study outcomes. Figure 2(b) and 2(e) present vertical sections depicting sea salinity for January and July, respectively, where color variations represent differences in sea salinity. In January, salinity ranged from 29 PSU (at the sea surface) to 34.2 PSU (at the seabed), while in July, it varied from 31.4 PSU (at the sea surface) to 33.8 PSU (at the seabed). This indicates higher surface salinity in July compared to January, while higher seabed salinity is observed in January than in July.

Amiruddin et al. [29] compared salinity vertical sections during the NEM and SWM, revealing that surface salinity is higher during the latter compared to the former. However, their findings indicated higher seabed salinity values during the NEM than the SWM. This echoes our observations of elevated surface salinity in July compared to January, while January exhibits higher seabed salinity than July. Similarly, Mansor et al. [11] reported analogous results. According to Mansor et al. [11], surface salinity is higher during the SWM (31.5 PSU) than the NEM (30.45 PSU). Conversely, seabed salinity values are higher during the NEM (33.95 PSU) than the SWM (31.75 PSU).

At the sea surface, during the NEM, according to Amiruddin et al. [29], SSS is around 31 PSU. The low salinity during the NEM is caused by the influx of low-salinity water around 30.5 PSU from the west coast of Peninsular Malaysia. During the SWM, the movement of high-salinity water from the Andaman Sea increases SSS to more than 32 PSU in the Strait of Malacca.

Figure 2(c) and 2(f) show vertical cross-sections illustrating sea currents in January and July, respectively. The varying colors represent differences in sea current intensity. During January, sea currents range from 0 to 0.25 m/s, while in July, they fluctuate between 0 and 0.1 m/s. Both surface and seabed sea currents exhibit higher strength in January compared to July. These observations align with the study conducted by Rizal et al. [12]. Rizal et al. [12] demonstrated that the current intensity in February (representing the NEM) was greater than that observed in August (representing the SWM), both at the sea surface and at the seabed.

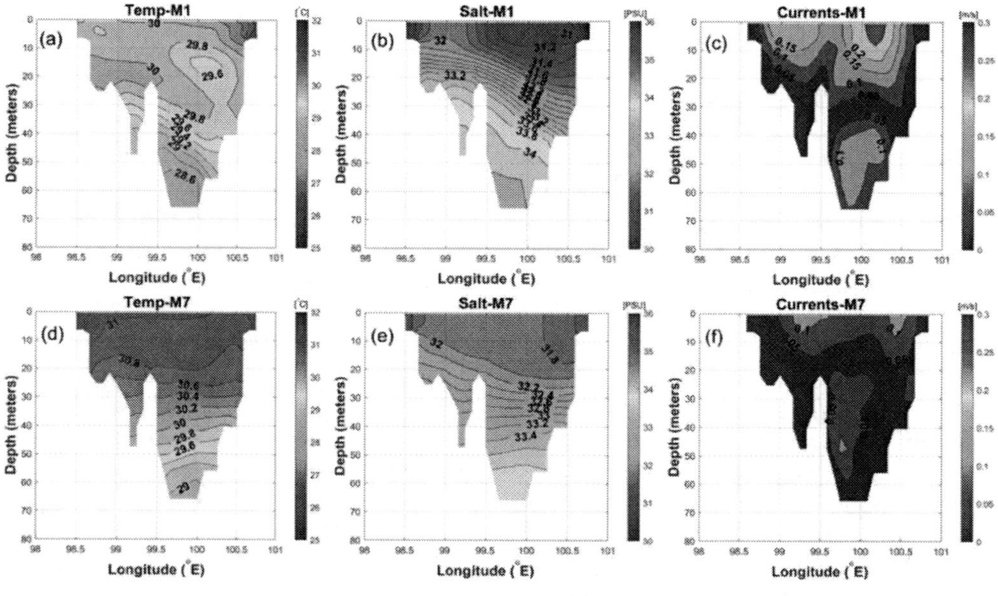

Fig. 2. Vertical sections for (a) Temperature (Temp-M1), (b) Salinity (Salt-M1), and (c)

Currents (Currents-M1) in January and (d) Temperature (Temp-M7), (e) Salinity (Salt-M7), and (f) Currents (Currents-M7) in July 2022 across the Malacca Strait at latitude 4°N. The location is indicated in Fig. 1 by the red line in the Malacca Strait. Colors on the map and the temperature color bar represent temperature differences in °C. Similarly, colors on the map and the salinity and sea currents color bars represent salinity differences in PSU and sea current velocities in m/s, respectively.

Figure 3 is a derivation from Figure 2, delineating temperature and salinity in January and July 2022 at latitude 4°N and longitude 100°E, respectively. In January, the isotherms appear relatively shallow, whereas the isohalines extend to about 10 meters. Within Figure 3(a), the MLD variations influenced by temperature and salinity are evident, with the salinity-induced

MLD being thicker than that caused by temperature. The upper well-mixed layer in the ocean arises due to turbulent mixing processes propelled by surface wind stress, along with heat loss and freshwater influx from the atmosphere [34]. This contributes to the creation of a relatively warm and fresh thin mixed layer, influenced by precipitation and surface heating [35].

Figure 3(b) exhibits isotherm and isohaline layers in July, showcasing similarities extending to roughly 20 meters. Notably, there are no substantial disparities in MLD thickness influenced by temperature and salinity in July. Furthermore, Figure 3(a) and 3(b) illustrate that the isotherm and isohaline layers, as well as the MLD, are deeper and thicker in July than in January. These findings correspond with the results presented by Mansor et al. [11].

In their study, Mansor et al. [11] showed that the MLD concerning temperature in the Malacca Strait measured approximately 10.35 meters in March (representing the NEM). Conversely, during August (representing the SWM), the MLD for temperature in the Strait increased to around 13.6 meters. Likewise, the MLD for salinity in the Malacca Strait during March registered at 10.35 meters, whereas in August, it measured approximately 12.5 meters. Both datasets indicate that the MLD was shallower in March when compared to August.

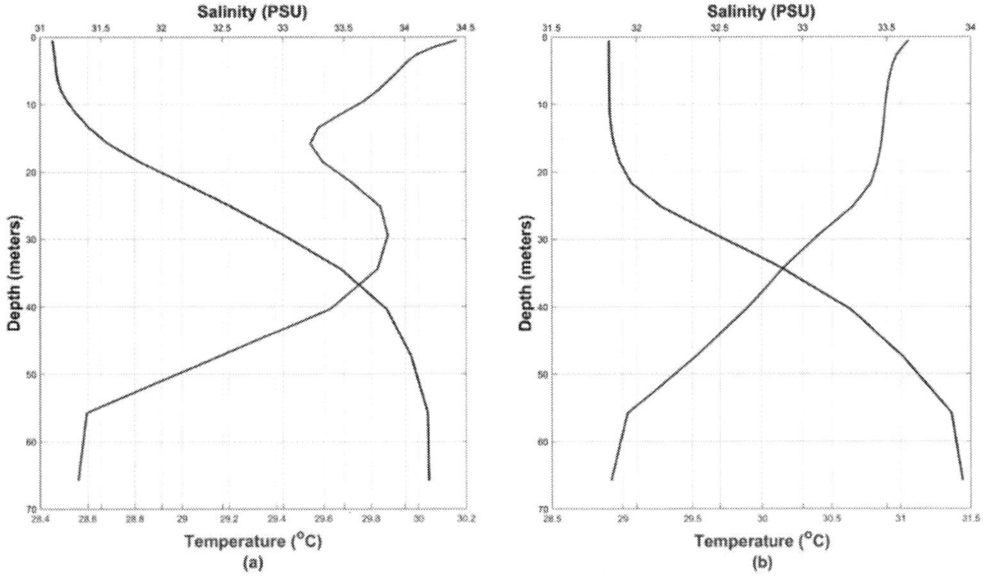

Fig. 3. Temperature profiles (red color) and salinity profiles (blue color) for (a) January and (b) July 2022 at latitude 4°N and longitude 100°E in the Malacca Strait.

4 Conclusion

The Malacca Strait's sea conditions exhibited considerable seasonal changes between January and July 2022. January revealed temperatures ranging from 28.6°C to 30.8°C, while July experienced higher temperatures, ranging between 29°C and 31.4°C, consistently showcasing warmer conditions at both surface and seabed levels compared to January. Salinity levels varied similarly, with January showing a range of 29 PSU to 34.2 PSU, contrasting with July's range of 31.4 PSU to 33.8 PSU. While surface salinity was higher in July, seabed salinity peaked in January. The sea currents also demonstrated noticeable differences between these months, with January recording stronger currents, reaching 0 to 0.25 m/s compared to July's weaker 0 to 0.1 m/s, particularly notable within specific longitudes in January. Figure 2 data collectively emphasizes distinct seasonal trends in temperature, salinity, and currents, providing a foundation for further investigation into the underlying factors driving these fluctuations and their implications on the local marine environment and ecosystems. The study delves into the temporal patterns of ocean thermal and salinity profiles at latitude 4°N and longitude 100°E during January and July 2022. January reveals shallower isotherms and isohalines, extending approximately 10 meters deep, with the salinity-influenced mixed layer depth appearing thicker than the temperature-influenced one. Contrastingly, July depicts isotherm and isohaline layers extending to about 20 meters, showcasing no significant differences in the mixed layer depth influenced by temperature and salinity. The data highlights considerable depth and thickness variations in July compared to January, emphasizing a distinct seasonal shift in the ocean's thermal and salinity structure in this location. These findings are critical for understanding the ocean's response to seasonal variations and their broader impact on the marine environment and ecosystems in the Malacca Strait. Additionally, they offer crucial insights into the implications of climate change on marine ecosystems in this region, warranting further research to comprehend their long-term effects. Studies like those outlined by Giunta and Ward [15] can lay the groundwork for a comprehensive understanding of MLD and its global implications.

Acknowledgment

This research was funded by the Ministry of Education, Culture, Research and Technology of the Republic of Indonesia under the scheme of "*Penelitian Terapan Unggulan Perguruan Tinggi* (PTUPT)" for the fiscal year 2023 with the contract number [549/UN11.2.1/PT.01.03/DPRM/2023] and the "*Penelitian Pendidikan Magister menuju Doktor untuk Sarjana Unggul* (PMDSU)" scheme for the fiscal year 2023 with contract number [587/UN11.2.1/PT.01.03/DPRM/2023]. The author would like to thank the Ocean Modeling Laboratory, Department of Marine Sciences, Universitas Syiah Kuala, Indonesia, for providing research facilities.

References

1. Y. Haditiar, M.R. Putri, N. Ismail, Z.A. Muchlisin, M. Ikhwan, S. Rizal, Heliyon **6**(9) (2020)
2. S. Wang, H. Jia, J. Lu, D. Yang, Reliab. Eng. Syst. Saf. **235**, 109254 (2023)
3. S. Rizal, P. Damm, M.A. Wahid, J. Sündermann, Y. Ilhamsyah, T. Iskandar, Am. J. Environ. Sci. **8**(5), 479-488 (2012)
4. Y. Haditiar, Z. Zulfandi, T. Iskandar, I. Setiawan, S. Sugianto, S. Rizal, *Simulation of wave refraction in the western waters of Aceh, Indonesia*, in IOP Conference Series: Earth and Environmental Science, IOP Publishing **216**, 1, 012014 (2018)
5. H.A. Haridhi, M. Nanda, C.R. Wilson, S. Rizal, Reg. Stud. Mar. Sci. **8**, 114-121 (2016)
6. Y. Haditiar, M.R. Putri, N. Ismail, Z.A. Muchlisin, S. Rizal, Eng. J. **23**(6), 129-143 (2019)
7. Y. Haditiar, S. Rizal, F. Abdullah, *Current simulation in the Malacca Strait and part of South China Sea due to wind*, in 2016 12th International Conference on Mathematics, Statistics, and Their Applications (ICMSA), IEEE 47-50 (2016)
8. M.S. Idris, H.L. Siang, R.M. Amin, M.J. Sidik, J. Mar. Syst. **236**, 103799 (2022)
9. K.K. Liu, S.Y. Chao, P.T. Shaw, G.C. Gong, C.C. Chen, T.Y. Tang, DSR I: Oceanogr. Res. Pap. **49**(8), 1387-1412 (2002)

10. C. Du, Z. Liu, S.J. Kao, M. Dai, Geophys. Res. Lett. **44**(22), 11-510 (2017)
11. K.N.A.A.K. Mansor, N.H. Roseli, F.S.M. Ali, M.F.M. Akhir. J. Coast. Res. **39**(5): 921–932 (2023)
12. S. Rizal, I. Setiawan, T. Iskandar, Y. Ilhamsyah, M.A. Wahid, M. Musman, Sci. Malays. **39**(4), 519-524 (2010)
13. A.B. Kara, P.A. Rochford, H.E. Hurlburt, J. Geophys. Res. Oceans. **105**(C7), 16803–16821 (2000)
14. I. Bessa, A. Makaoui, K. Hilmi, M. Afifi, Model. Earth Syst. Environ. **4**(1), 151–160 (2018)
15. V. Giunta, B. Ward, J. Geophys. Res. Oceans. **127**(4), e2021JC017904 (2022)
16. S. Lim, C.J. Jang, I.S. Oh, J. Park, J. Mar. Syst. 96–97, 1–14 (2012)
17. C.P. Abdulla, M.A. Alsaafani, T.M. Alraddadi, A.M. Albarakati, Ocean Sci. **14**(4), 563–573 (2018)
18. X. Costoya, M. deCastro, M. Gómez-Gesteira, F. Santos, PLoS ONE, **9**(6) (2014)
19. A. Srivastava, S. Dwivedi, A.K. Mishra, Oceanol. **60**(2), 169–186 (2018)
20. J. Yu, B. Gan, Z. Jing, L. Wu, J. Climate. **33**(24), 10419–10436 (2020)
21. J.C. Marshall, R.G. Williams, A.J.G. Nurser, J. Phys. Oceanogr. **23**(7), 1315–1329 (1993)
22. A. Calbet, M.D. Agersted, S. Kaartvedt, M. Møhl, E.F. Møller, S. Enghoff-Poulsen, M.L. Paulsen, I. Solberg, K.W. Tang, K. Tönnesson, D.E. Raitsos, T.G. Nielsen, Sci. Rep. **5**(1), 11240 (2015)
23. O. Schofield, M. Brown, J. Kohut, S. Nardelli, G. Saba, N. Waite, H. Ducklow, Phil. Trans. R. Soc. A. **376**(2122), 20170173 (2018)
24. S. Itoh, I. Yasuda, H. Saito, A. Tsuda, K. Komatsu, J. Mar. Syst. **151**, 1–14 (2015)
25. T. Xue, I. Frenger, A.F. Prowe, Y.S. José, A. Oschlies, Biogeosci. **19**(2), 455-475 (2022)
26. P.Y. Le Traon, A. Reppucci, E.A. Fanjul, L. Aouf, A. Behrens, M. Belmonte, A. Bentamy, L. Bertino, V.E. Brando, M.B. Kreiner, M. Benkiran, T. Carval, S.A. Ciliberti, H. Claustre, E. Clementi, G. Coppini, G. Cossarini, M. De Alfonso Alonso-Muñoyerro, A. Delamarche, G. Dibarboure, F. Dinessen, M. Drevillon, Y. Drillet, Y. Faugere, V. Fernández, A. Fleming, M.I. Garcia-Hermosa, M. García Sotillo, G. Garric, F. Gasparin, C. Giordan, M. Gehlen, M.L. Gregoire, S. Guinehut, M. Hamon, C. Harris, F. Hernandez, J.B. Hinkler, J. Hoyer, J. Karvonen, S. Kay, R. King, T. Lavergne, B. Lemieux-Dudon, L. Lima, C. Mao, M.J. Martin, S. Masina, A. Melet, B. Buongiorno Nardelli, G. Nolan, A. Pascual, J. Pistoia, A. Palazov, J.F. Piolle, M.I. Pujol, A.C. Pequignet, E. Peneva, B. Pérez Gómez, L. Petit de la Villeon, N. Pinardi, A. Pisano, S. Pouliquen, R. Reid, E. Remy, R. Santoleri, J. Siddorn, J. She, J. Staneva, A. Stoffelen, M. Tonani, L. Vandenbulcke, K. von Schuckmann, G. Volpe, C. Wettre, A. Zacharioudaki, Front. Mar. Sci. **6**, 234 (2019)
27. J.M. Lellouche, E. Greiner, O. Le Galloudec, G. Garric, C. Regnier, M. Drevillon, M. Benkiran, C.E. Testut, R. Bourdalle-Badie, F. Gasparin, O. Hernandez, B. Levier, Y. Drillet, E. Remy, P.Y. Le Traon, Ocean Sci. **14**(5), 1093-1126 (2018)
28. B. Tozer, D.T. Sandwell, W.H. Smith, C. Olson, J.R. Beale, P. Wessel, Earth Space Sci. **6**(10), 1847-1864 (2019)
29. A.M. Amiruddin, Z.Z. Ibrahim, S.A. Ismail, Res. J. Environ. Sci. **5**(1), 49–58 (2011)
30. H.A. Haridhi, M. Nanda, C.R. Wilson, S. Rizal, Reg. Stud. Mar. Sci. **8**(2016), 114-121 (2016)

31. H.A. Haridhi, M. Nanda, Y. Haditiar, S. Rizal, Ocean Coast. Manage. **154**, 46–54 (2018)
32. N.S. Isa, M.F. Akhir, I. Khalil, P.H. Kok, N.H. Roseli, J. Sustain. Sci. Manag. **15**(4), 66-77 (2020)
33. N.S. Isa, M.F. Akhir, P.H. Kok, N.R. Daud, I. Khalil, N.H. Roseli, Reg. Stud. Mar. Sci. **39**, 101402 (2020)
34. T. Ezer, J. Geophys. Res. Oceans **105**(C7), 16843–16855 (2000)
35. Y. Ushijima, Y. Yoshikawa, J. Phys. Oceanogr. **49**(7), 1769– 1787 (2019)

Antibacterial Activity of Extract Green Seaweed *Ulva intestinalis* in North Waters of Aceh

Irwan irwan[1*], *Herwina* Herwina[1], *Ramadani* Ramadani[1], Nora Idiawati[2], *Eli* Nurliza[3], and *Nursafiah* Nursafiah[4]

[1]Department of Marine Sciences, Faculty of Marine and Fisheries, Universitas Syiah Kuala, Banda Aceh 23111, Indonesia
[2]Department of Marine Sciences, Faculty of Mathematics and Natural Science, Tanjungpura University, Pontianak 78124, Indonesia
[3]Department of Indonesian Languange Education, Faculty of Teacher Training and Education, Universitas Serambi Mekkah, Banda Aceh, Indonesia
[4]Department of Biology Education, Faculty of Teacher Training and Education, Gunung Leuser University, Kutacane 24664, Indonesia

Abstract. The appearance of bacterial resistance and pathogenic infections makes efforts to find new drugs as natural bacterial agents continue. One of them is through the utilization of marine organisms such as seaweed. Some studies show the content of seaweed bioactive compounds has potential in the pharmacological field. This study aims to determine the type of secondary metabolite compounds and antibacterial activity of ethanol extract of *Ulva intestinalis*. Seaweed extraction using the maceration method for 3 x 24 hours and antibacterial activities were tested by disc diffusion method. The results showed that there were 5 types of secondary metabolite compounds in the extract, namely alkaloids, flavonoids, glycosides, saponins, and steroids. The antibacterial activity test of seaweed extract against *Propionibacterium acnes* and *Aeromonas hydrophila* bacteria in the concentration range of 50% to 100% showed a moderate inhibition zone diameter

1 Introduction

Indonesian waters have high biodiversity, so they are rich in marine products. One of the marine products that has the potential to be developed both for the food and pharmaceutical industries is seaweed [1]. The prospect of seaweed is supported by many bioactive compounds such as alkaloids, carotenoids, flavonoids, terpenes, tocopherols, peptides, polysaccharides and polyphenols [2]. Several studies reported that seaweed bioactive compound components can be antioxidants, antibacterial, anti-inflammatory, antidiabetic, antidiarrheal, antimicrobial, and antibacterial [3, 4, 5].

Several investigations have revealed that seaweed has antibacterial properties. Red seaweed *E. cottoni* and *Sargassum sp.* have antibacterial action against *P. acnes* [6], as has

* Corresponding author: irwan@usk.ac.id

red seaweed *Gracilaria sp.* [7]. Brown seaweed exhibits antibacterial action against *Pseudomonas aeruginosa*, *Enterococcus faecalis*, and *Staphylococcus aureus*, according to [8].

Ulva intestinalis is a species of green seaweed found in Indonesian waters that belongs to the *Ulva sp* group [9]. Ulva intestinalis bioactive substances include antibacterial, anticancer, and anti-larvicidal properties [10]. According [11] methanol, n-hexane, and dichloromethane extracts of *U. intestinalis* have antibacterial activity against *Bacillus cereus*, *Enterococcus faecalis*, *Escherichia coli*, and *Listeria monocytogenes*. U. intestinalis from methanol, dichloromethane, and water extracts have antibacterial activity against *Vibrio algnolyticus*, *V. vulnificus*, and *V. parahaemolyticus*, according [12] even though many exhibit antibacterial activity, research articles on their efficiency in suppressing the growth of gram-positive bacteria (*Propionibacterium acnes*) and gram-negative bacteria (*Aeromonas hydrophila*) are scarce. In truth, both bacteria are harmful and could be detected in people and fish. This research aims to discover the bioactive components and antibacterial activity of an ethanol extract of *U. intestinalis* from northern Aceh waters against gram-positive and gram-negative bacteria.

2 Materials and Methods

2.1 Time and Place

This study was taking place during February and March of 2023. Green Seaweed (*Ulva intestinalis*) were collected at Ulee Lheue Beach in Banda Aceh's Meuraxa District. Extraction activities and phytochemical tests was conducted at Syiah Kuala University's Marine Chemistry and Biotechnology Laboratory, Faculty of Marine and Fisheries, and antibacterial activity tests was conducted at Syiah Kuala University's Microbiology Laboratory, Faculty of Mathematics and Natural Sciences.

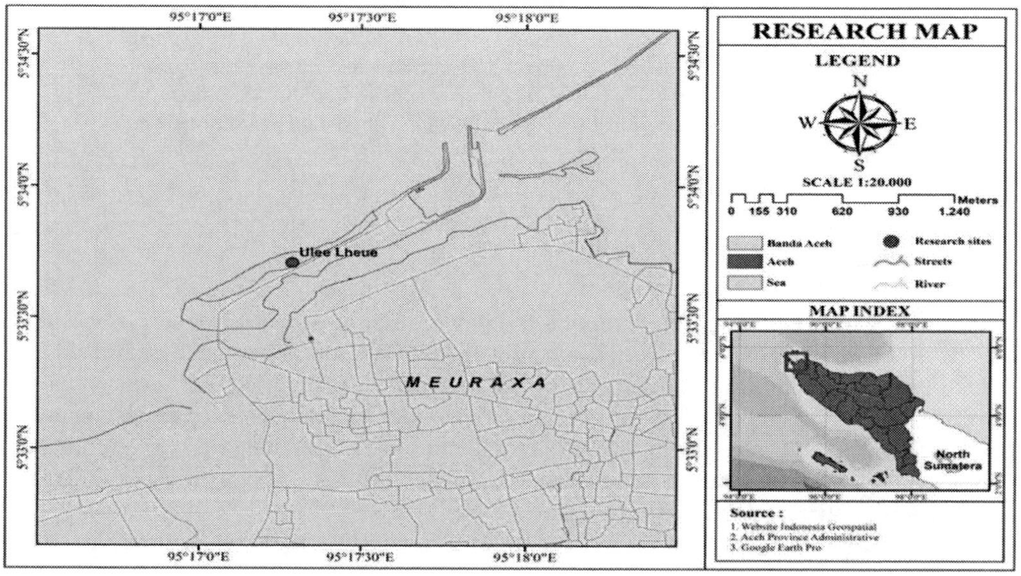

Fig. 1. Map of Sampling Location

2.2 Tools and Material

The tools used are glassware, rotary vacuum evaporator, micropipette, analytical balance, UV-Vis spectrophotometer, incubator, hot plate, autoclave, tube rack, light microscope, needle ose, refrigerator, caliper, Hot place, and aluminum foil.

The materials used green seaweed (*Ulva Intestinalis*), ethanol, aquadest, DMSO, Clindamycin, Muller Hilton agar, nutritional agar, and bacterial tests for *P. acnes* and A. hydrophila.

2.3 Sample Extraction

A total of 100 grams of dried *U.intestinalis* green seaweed samples were macerated in ethanol for 3 x 34 hours at a sample-to-solvent ratio of 1:5. The extraction products were filtered using filter paper, evaporated using a rotary evaporator at 40 ^{0}C to generate a thick extract, and then phytochemical analyses were performed [13].

2.4 Phytochemical Test

Phytochemical examination examined green seaweed *Ulva Intestinalis* for saponins, alkaloids, tannins, phenolic, steroids, terpenoids, and flavonoids. Standardized methodologies were used for phytochemical screening [14].

2.5 Antibacterial Activity

2.5.1. Making agar media

Mixed 2 grammes of powdered nutrient agar (NA) with 100 mL of distilled water in an Erlenmeyer flask. Heat and stir on a hotplate until dissolved. Wrapped in plastic wrap, they were sterilised in an autoclave at 121°C for 15 minutes [15].

2.5.2. Bacteria regeneration

Isolates of *P. acnes* and *A. hydrophila* were inoculated at a concentration of 1 ose into 25 mL Muller Hinton Agar (MBA), then incubated for 6 hours at 28 ^{0}C in an incubator shaker. Fresh bacterial colonies were required for antibacterial action [15].

2.5.2 Antibacterial activity test

The disc method was used in this study to determine antibacterial activity. *Ulva intestinalis* crude extracts were created at concentrations of 25%, 50%, 75%, and 100%. The paper discs were tested for antibacterial activity by soaking them in the crude extract of *U.intestinalis* in each concentration variety and then drying them in the air. Furthermore, the paper disc was placed on the surface of the medium Nutrient agar (NA), which already included the test bacteria, in this example, *P. acnes* (gram-positive bacteria) and *A.Hydrophila* (gram-negative bacteria), and incubated at 37 ^{0}C for 24 hours. A caliper is used to measure the diameter of the produced clear space [16].

3 Results and Discussion

3.1 Phytochemical Screening Test

The results of the phytochemical test showed five bioactive compounds contained in the ethanol extract of *Ulva intestinalis*, namely alkaloids, flavonoids, glycosides, saponins, and terpenoids. The test results can be seen in Table 1 below.

Table 1. Phytochemical screening test of *Ulva intestinalis* extract from Ule lheue Coast

Metabolite secondary	Methods	Results
Alkaloids	Wagner	Positive
Flavonoids	Shinoda	Positive
Glycosides	Keller-Killiani	Positive
Saponins	Froathing	Positive
Terpenoids	Salkowski	Positive

The crude extract of green seaweed (*U.Intestinalis*) showed positive results for the existence of alkaloids and Saponins compounds, as determined by phytochemical testing. According [17] and [18], alkaloid has antibacterial pharmacological effects. Additionally, saponins have antibacterial activity against pathogenic foodborne microorganisms [18].

3.2 Antibacterial Activity

Establish a clear area around the extract-containing paper disc to test antibacterial activity. [19] reported that the diffusion test examined the diameter of the clean zone, which is the extract's antibacterial compound's ability to suppress bacterial growth. The sensitivity of the cultured media organism, incubation circumstances (temperature, duration), and agar dispersion rate (media content and concentration) affect the inhibition zone [20]. Table 2 shows the inhibitory area for each green seaweed extract concentration.

Table 2. Inhibitory zone in various concentration of *U.Intestinalis* extract

Concentration (%)	Average Inhibitory Zone (mm) (n=3)	
	Propionibacterium acnes	***Aeromonas hydrophila***
25	-	-
50	6.10	7.58
75	7.20	7.28
100	8.20	7.17
Clindamycin (+)	17.20	21.18
Ethanol (-)	0	0

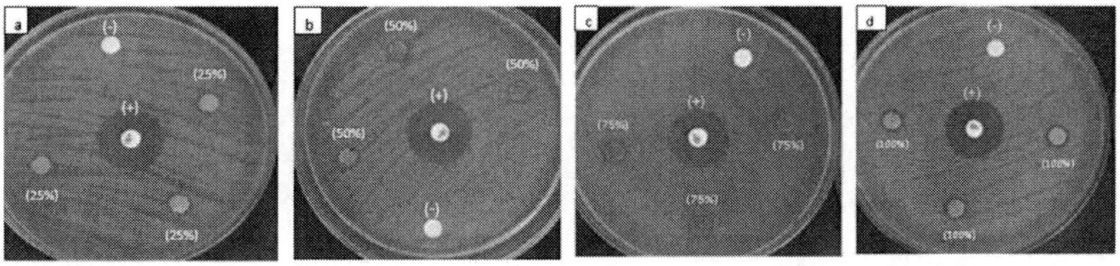

Fig. 2. Diameter of the Inhibition Zone of *Propionibacterium acnes*

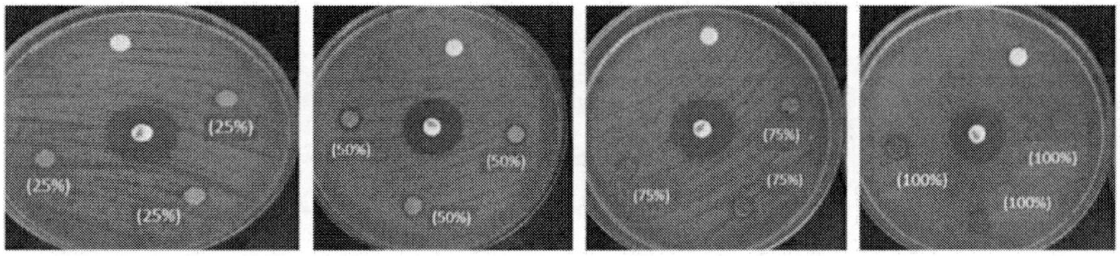

Fig. 3. Diameter of the Inhibition Zone of *Aeromonas hydrophila*

Based on Table 2, the antibacterial test on ethanol extract of *Ulva intestinalis* against *P. acnes* and *A. hydrophila* showed moderate antibacterial activity in the concentration range of 50% to 100%. [21] states that the 5-10 mm inhibition zone range offers a moderate inhibitory response. The 25% concentration, the lowest concentration, cannot inhibit the growth of *P. acnes* and *A. hydrophila* bacteria. It is suspected that the active compound content of *Ulva intestinalis* is not strong enough to denature the bacterial cell wall, and the low concentration of the extract carries little active substance. Meanwhile, at concentrations of 50%, 75%, and 100%, a moderate inhibition zone began to form, characterized by an inhibition zone that enlarged as the concentration of the extract increased. [22] found that the suspension of bacterial turbidity and media agar thickness can impact the inhibition zone of bacteria growth.

In green seaweed extracts, bioactive substances like flavonoids, triterpenoids and saponins can also prevent bacterial growth. According [23], triterpenoid compounds have antibacterial effects because they include extensive polysaccharide groups and sulfates that can inhibit transport through bacteria's thicker cell membranes. Triterpenoids prevent the growth of bacteria by inhibiting DNA-binding enzymes [24].

Flavonoid substances affect energy metabolism and hinder bacterial growth. It also denatures protein associations, breaking membranes and cell walls and poisoning bacteria. Flavonoid chemicals' hydroxyl molecules interact with bacterial DNA, supporting their harmful effect [25, 26]. Saponin chemicals inhibit bacteria by lowering lipid tension by binding to lipopolysaccharides on cell walls, inducing lysis and death [27]. The latest research affirmed that the presence of phenolic compounds influences the antibacterial effectiveness of *Ulva sp* [28, 29].

4 Conclusion

The extract of ethanol from green seaweed (*Ulva intestinalis*) contains triterpenoid, alkaloid, glycoside, flavonoid, and saponin. Antibacterial activity assays against *P.acnes*, and *A.hydrophila* exhibited inhibited zone diameters from all moderate concentrations. While this research shows that *Ulva intestinalis* could have potential as a therapeutic agent

Acknowledgements

The authors would like to thank the Lembaga Penelitian dan Pengabdian (LPPM) Universitas Syiah Kuala that supported this research [392/UN11.2.1/PT.01.03/PNBP/2023].

References

1. T. Handayani. 2019. J.Osean. **1**, 1-14 (2019)
2. C. Jacobsen., A.D. Sørensen, A. D. M., Holdt, S. L. Akoh, D.B Hermund. A. Food. Technology **10**, 541-568 (2019)
3. K. Ponnuchamy, J.A. Jacob. N. Reviews. **5**, 589-600 (2016)
4. S.A. Akbar, M. Hasan, S. Afriani, C. Nuzlia, Biodiversitas Journal of Biological Diversity **24(10)**, 5283 (2023)
5. S. AlNadhari, N.M Al-Enazi, F. Alshehrei. F. Ameen. E. Research. **194**, 110672 (2021)
6. N. Nurjanah, B.E. Aprilia, A. Fransiskayana, M. Rahmawati, T. Nurhayati. J. Pengolahan Hasil Perikanan Indonesia. **21**, 304-316 (2018)
7. V.L. Firdausy, H.P. Sudarno. Activity test of crude extract Gracilaria sp. on Propionibacterium acnes growth inhibition in vitro (Universitas Indonesia, UI Press, 2017)
8. F. Afrin, T. Ahsan, M.N. Mondal, M.G. Rasul, M. Afrin, A.A Silva, C. Yuan, A. K. M. A. Shah.(2023). J. Food. Chemistry. Advances. **3**, 100393 (2023)
9. C. Nufus, A.A. Nurjanah. J. Pengolahan. Hasil. Perikanan. Indonesia. **20**, 620-632 (2017)
10. M. Atlantic, N.E. Caribbean. Ulva intestinalis Linnaeus Synonym. Enteromorpha Intestinalis (Linnaeus) (2014)
11. W. Srikong, N. Bovornreungroj, P. Mittraparparthorn, P. Bovornreungroj. J. Science. Asia. **43**, 88-95 (2017)
12. N. Klongklaew, J. Praiboon, M. Tamtin, P. Srisapoome, P. 2020. J. Marine.Drugs. **18**, 1-23 (2020)
13. R.T. Manalu, T. Teodhora, E.A Sipayung. J. Farmasetis. **3**, 101-106 (2020)
14. W.C. Evans. Trease and Evans' Farmacognosy (2002)
15. S. Hainil, S.F. Sammulia, A. Adella. J. Surya. Medika. **7**, 86-95 (2022)
16. M. Misna, K. Diana, K. (2016). J. Farmasi. Galenika. **2**, 138-144 (2016)
17. Y. Yan, X. Li, C. Zhang, L. Lv, B. Gao, M. Li. (2021). A. Review.Antibiotics, **10**, 318 (2021)
18. Y. Liu, Y. Cui, L. Lu, Y. Gong, W. Han, G. Piao. A. Der. Pharmazie, **353**, 2000120 (2020)
19. E.M. Sari, W.F. Ma'ruf. J. Pengolahan. Bioteknologi. Hasil. Perikanan. **3**, 16-24 (2014)

20. S. Nimah, W.F. Ma'ruf, A. Trianto. 2012. J. Pengolahan. Bioteknologi. Hasil. Perikanan **1,** 9-17 (2012)
21. S. Sudrajat, S. Sadani, S. Sudiastuti, S. J. Tropical. Pharmacy. Chemistry. **1**, 303-311 (2012)
22. P. Zeniusa, M.R Ramadhian. S.H. Nasution, N. Karima. J. Lampung. **8**, 2 (2019)
23. A. E. Farouk, A.H. Faizal, B.H. Ridzwan. J. Biochemistry. Biotechnology. **3**, 64-69 (2007)
24. N. Nurjanah, L. Izzati, A. Abdullah. J. Marine. Sciences. **16,** 119-124 (2011)
25. P.P.A. Bangun. 2021. J. Ilmiah. Farmasi. Attamru. **1,** 1-5 (2021)
26. V.Y. Liling, Y.K. Lengkey, C.N. Sambou, R.R. Palandi. J. Biofarmasetikal. Tropis **3,**112-121 (2020)
27. M. Muflihah. *Analisis Variasi Konsentrasi terhadap Uji Toksisitas Akut Golongan Senyawa Metabolit Sekunder dari Ekstrak Biji Pepaya (Carica papaya L.) pada Larva Udang (Artemia salina Leach)* In Proceeding of Mulawarman Pharmaceutical Conferences, 5-6 June, Samarinda, Indonesia (2015)
28. J.L. McCauley, P.C. Winberg, B.J. Meyer, D. Skropeta. Alga. Res. **35**, 586-594 (2018)
29. S. Mezghani, I. Bourguiba, I. Hfaiedh. M. Amri. Biol. Bull. **225,** 1–7 (2013)

The length-weight relationship of *Magallana gigas* in Krueng Cut, Province of Aceh, Indonesia

Chitra Octavina[1*], *Shella* Afriana[1]

[1]Department of Marine Science, Faculty of Marine and Fisheries, Universitas Syiah Kuala 23111, Indonesia

Abstract. This study aims to determine the length-weight relationship of *Magallana gigas* in Krueng Cut, Banda Aceh City. The research was conducted in October 2020 at Krueng Cut River, Syiah Kuala District, Banda Aceh City. The oyster identification process was carried out at the Laboratory of Marine Biology, Faculty of Marine and Fisheries, Universitas Syiah Kuala. The method used in this study was simple random sampling method. In addition, water quality data were also collected in situ and substrate collected ex situ. The Ostreidae group found in Krueng Cut, Banda Aceh City is the *Magallana gigas* species. The average oyster found in Krueng Cut, Banda Aceh City is small with length growth ranging from 20 to 50.2 mm. The results showed that the growth pattern of *Magallana gigas* was classified as negative allometric at three research stations in Krueng Cut River, Banda Aceh, this was based on the value of $b < 3$. Environmental factors from the three research stations in Krueng Cut, Banda Aceh City are still within the normal limits of *Magallana gigas* life.

1 Introduction

Bivalves are widely distributed throughout coastal waters of Indonesia, especially in various shallow water ecosystems [1]. One of the potential bivalve species that can be developed in order to increase economic income is the oyster species. Oysters belong to the Ostreidae family of mollusc phyla that live in marine and brackish water habitats and usually live attached to mangrove roots, pier poles and various objects of dead coral rocks at the bottom of the waters [2]. Oysters tend to like warm waters and are protected from waves and sloping surfaces with mud, sand, gravel and rock substrates [3]

Oysters are one of the important potential resources in Indonesia, because in general oysters can be utilized for various human needs although only a few types are of high economic value [4]. One of the oyster species that is often consumed by the community is the meat oyster (Ostreidae). This species is a source of protein for local communities and plays an important role in the livelihood of fishing communities [5].

Assessment of the existence of oysters in nature must be done. It is necessary to know how the abundance in nature is still available sufficient for the balance of the ecosystem or even in a state of decline. One form of assessment of the morphometric biota is the

[*] Corresponding author: chitraoctavina@usk.ac.id

relationship between length and weight of the ostreidae family on the coast. So far there is still very little research related to the length-weight relationship of meat oysters (Ostreidae) conducted in Aceh. The research that has been done includes [5; 6; 7]. However, research on length-weight relationships has never been conducted, especially in Krueng Cut, Banda Aceh, so this research is important to contribute to the development of oysters in Aceh in the future, especially for oyster conservation and cultivation. This study aims to determine the length-weight relationship of *Magallana gigas* in Krueng Cut, Banda Aceh City.

2 Material and Methods

2.1 Time and location

This research was conducted from October to November 2020 at Krueng Cut, Banda Aceh City (Figure 1).

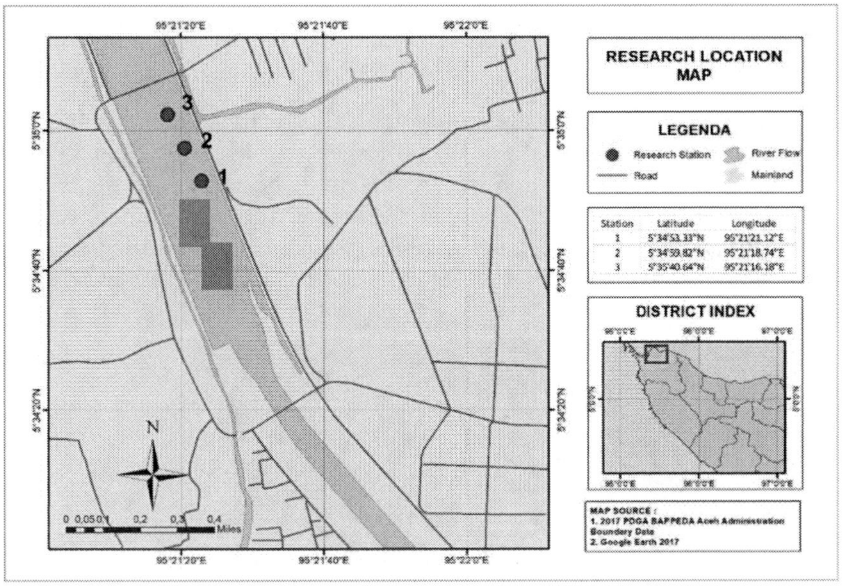

Fig. 1 Location research in Krueng Cut, Banda Aceh

2.2 Material

The tools and materials used in this study consist of Global Positioning System (GPS), Transect, Knife, Plastic samples, 10% formalin, thermometer, pH meter, refractometer, and DO meter.

2.3 Methods

Determination of the station point is determined with a distance of 10 m. When sampling oysters then placed transects in simple random sampling [8] which is done based on the stone throw (10 times each site). The oysters that have been found are put into a container and given 10% formalin. Then after the oyster sample is taken, the water quality parameters are measured in situ, while the substrate sample is inserted in a plastic containers and measured ex situ. Oyster, water and substrate samples are stored in a cool box and taken to the Marine

Biology Laboratory of the Faculty of Marine and Fisheries and the Soil Chemistry Laboratory of the Faculty of Agriculture, Universitas Syiah Kuala for analysis.

Oyster samples that have been preserved with formalin then identified morphologically based on mollusk identification book [9]. Then the sample is measured shell length (SL), shell width (SW), shell thickness (ST), content weight (WM), and total weight (SMW) to obtain morphometric measurements.

2.4 Data Analysis

The data analysis used in this study are 2, namely the analysis of growth patterns using a formula that refers to $W = {}_aL^b$ [10], then the determination of the class interval based on $K = 1 + (3.32 \text{ Log } n)$ [11].

3 Result and Discussion

3.1 Identified of *Magallana* sp.

Based on the results of identification using morphological parts of the outer shell and inside the shell using identification books and the World Register of Marine Science (WoRMS) website, the ostreidae found at the research site are thought to be *Magallana gigas* of pacific oyster (Figure 2). Generally, *Magallana* genus has an elongated shell, the left valve is always larger than the right valve and tends to cup. *Magallana gigas* has a white outer and inner shell is white, the former aductor muscle is colorless, lives in shallow waters with a depth of 3 m [5].

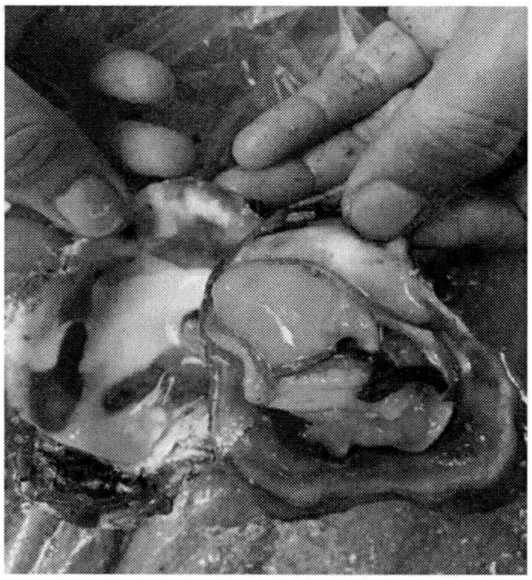

Fig. 2 *Magallana gigas*

The classification of *Magallana gigas* according to WoRMS updated in 2023 kingdom Animalia, phylum Mollusca, class Bivalvia, order Ostreida, family Ostreidae, genus *Magallana*, species *Magallana gigas*.

3.2 Size Frequency Distribution

The number of *Magallana gigas* found in the research location, Krueng Cut, was 172 individuals. The average oyster found in Krueng Cut, Banda Aceh is small with a length ranging from 20 - 50.2 mm (Figure 3).

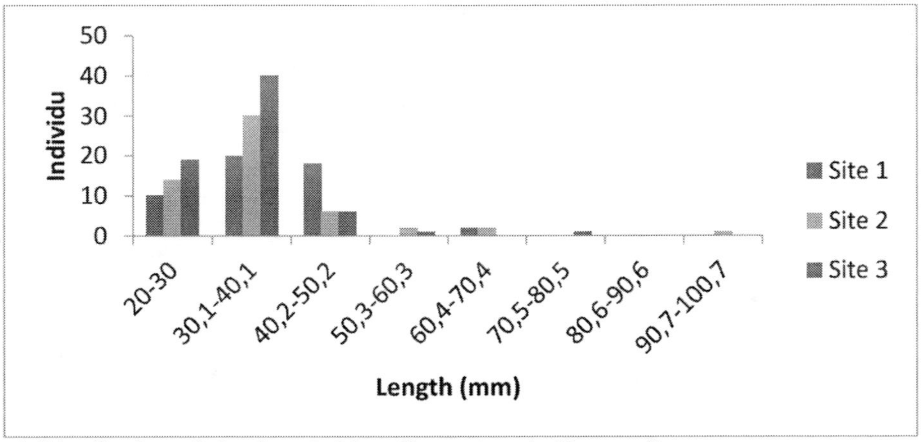

Fig. 3 Distribution of size length *Magallana gigas* in Krueng Cut

The length found at this location is almost similar to the length found by (Octavina et al., 2015). At this size, the results of the study [5] indicate that *Magallana* at that time was in a mature gonadal condition. However, in the study [12] the length of *Magallana* found was much longer, namely 70.24 mm, so it can be concluded that this kerrang can reach a higher length depending on the habitat conditions.

Furthermore, the observation of the weight class interval of *Magallana gigas* in Krueng Cut showed that the highest frequency of the three stations was in the size range of 7.4 - 8.4 g with a total of 28 individuals (Figure 4).

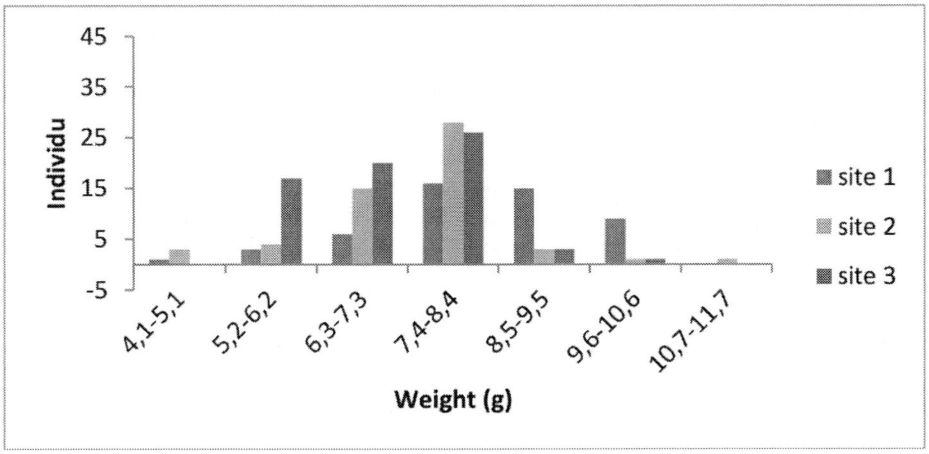

Fig. 4 Distribution of size weight *Magallana gigas* in Krueng Cut

The weight of meat or *Magallana* content in this study was higher in value compared to research [13]. The high value of *Magallana* content weight from this study is thought to be this group in the condition of collecting the entire body mass for spawning preparation so that the accumulation of energy for spawning can be optimal [5].

3.3 Length and Weight Relationship

Analysis of length-weight relationship is an important part of growth to compare the condition or relative health of certain populations or individuals. Growth is a biological process that is formulated as an increase in body length and weight within a certain period. Weight is considered as a function of length, the function value of length is the growth pattern (b value), the greater the b value indicates the better the condition of the waters [14; 6]

The length-weight relationship of *Magallana gigas* found in Krueng Cut is negative allometric (Figure 5). This is due to the indication of the b value criteria <3 [10]. Negative allometric pattern is a pattern of length increase faster than body weight increase.

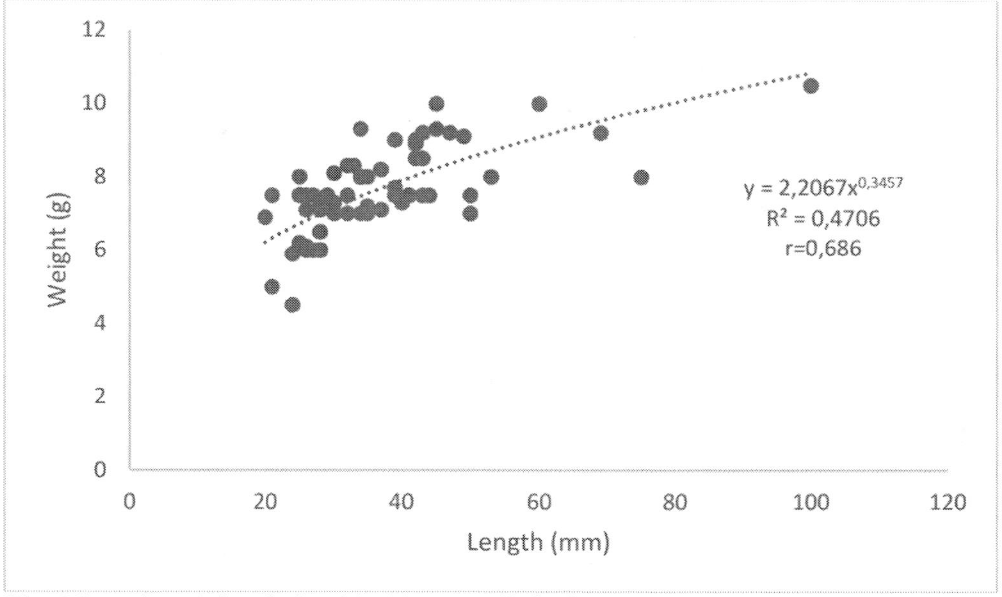

Fig. 5 Length-weight relationship *M. gigas* all site in Krueng Cut

Growth patterns can be seen from the b value in the relationship between body length and weight [16]. The increase or decrease in growth patterns that occur affects the b value and R^2 value found in *M. gigas*. Negative allometric growth in *M. gigas* is thought to be related to its flattened and elongated shape to achieve this shape, the increase in shell length is faster than the increase in weight. The suitability of the effect of body length on the total weight of *M. gigas* oysters can be seen based on the magnitude of the coefficient of determination (R^2).

Length growth is closely related to weight gain. According to [5] growth in shellfish is in the most prominent part, namely the shell, so that growth is an increase in the length of the shell followed by an increase in the body. The relationship between length and weight of meat oysters in Kuala Gigieng has a close correlation. Overall, the oyster growth pattern tends to be thin because the length-weight relationship curve formed tends to be sloping, thus indicating that length growth is faster than weight gain and growth tends to be unbalanced [17; 6; 2; 7].

The high and low value of b does not appear to be influenced by water quality parameters. The results of the analysis of water quality parameters consisting of temperature, salinity, pH, and DO have relatively constant values. The temperature at the three stations ranged from 33^{0}C to 34^{0}C and still supported the growth and reproduction of meat oysters. This is in accordance with the statement of [18] that *Magallana* is able to live in a temperature range of 5^{0}C to 35^{0}C (Optimum 11^{0}C to 34^{0}C) and still survive at temperatures of -50C.

The salinity range in Krueng Cut Banda Aceh is 10-13 ppt. salinity levels at the three stations still support the life of meat oysters. [19] stated that *Magallana* can tolerate a salinity range of 10 ppt to30 ppt (general range 20 ppt to 30 ppt).

The pH measurement value shows that in Kreung Cut, Banda Aceh City tends to be neutral with a range between 8.09 - 8.13. According to [20] meat oysters can live with a pH range of 6.80 to 9.25. However, if it is less or more than the pH range, the oyster will die. Dissolved Oxygen (DO) content at the three stations ranged from 6.96 mg/l to 7.2 mg/l where the highest DO level was located at station I which reached 7.2 mg/l. [21] stipulates that the dissolved oxygen content is >5 ppm for the benefit of marine tourism and marine biota. [5] said that *Magallana* can survive for 5 days in waters containing >1 mg/l dissolved oxygen. Thus it can be concluded that the water quality parameters in Krueng Cut are optimal conditions for the survival of *Magallana gigas* organisms.

4 Conclusion

The growth pattern of *Magallana gigas* tends to form a negative allometric pattern, which is a faster increase in length than weight gain, because the flat shape of the *Magallana gigas* oyster and its life at the bottom of the substrate makes *Magallana gigas* faster to increase body length compared to body weight in order to adapt to its environment. Physico-chemical factors of the aquatic environment are still below the quality standard threshold and support the life of *Magallana gigas* in Krueng Cut, Banda Aceh City.

References

1. A. van der Schatte Olivier, L. Jones, L. Le Vay, M. Christie, J. Wilson, S.K. Malham, Reviews in Aquaculture, **12**, 1 (2020)
2. C. Lodeiros, P. Valentich-Scott, J. Chávez-Villalba, J.M Mazón-Suástegui, J.M Grijalva-Chon. Journal of Shellfish Research, **39,** 2 (2020).
3. C.H. Sepúlveda, M. I. Sotelo-Gonzalez, C.C. Osuna-Martínez, M.G. Frías-Espericueta, R. Sánchez-Cárdenas, M. E. Bergés-Tiznado, A. M. Góngora-Gómez, M. García-Ulloa, Environmental Geochemistry and Health, **45,** 5 (2023)
4. R. Rochmady, S.B.A Omar, L.S Tandipayuk. Agrikan: Jurnal Agribisnis Perikanan, **6,** 1 (2013)
5. C. Octavina, F. Yulianda, M. Krisanti, Z.A Muchlisin. AACL Bioflux, **8,** 5 (2015)
6. L. Kasmini, T.A Barus, M.A Sarong, M.B Mulya Depik, **7,** 1 (2018)
7. R. Rosanawita, I. Dewiyanti, C. Octavina. Jurnal Ilmiah Mahasiswa Kelautan Dan Perikanan Unsyiah, **2**,1 (2017)
8. F. Tankoua, O. Buffet, P. E., Amiard, J. C., Berthet, B., Mouneyrac, C., & Amiard-Triquet, C.. *Ecotoxicology and Environmental Safety*, **88.** (2013)
9. Kozloff, E.N. Marine invertebrate of the Pasific Northwest. Washington Press. London (1987)
10. M. I. Effendie, Metode Biologi Perikanan, (1979)
11. H.A. Sturges, H. A. Journal of the American Statistical Association **21,** 153 (1926)
12. F. Yulianda, M. A Samosir, R Siregar, M. D. Wildan. Habitus Aquatica, **1,** 2 (2021)
13. A. B. R., C. A. S. Mulki, dan J. S. *Journal of Marine Research* (2014)
14. C. Octavina, M. Irsad, A.I. Tareqh, J. Niwatana, M. Ulfah, N. Nurfadillah, S. Agustina, F.M. Nur, N.T.M. Anh. IJMS, **28,** 3 (2023)

15. F.M. Nur, A.S. Batubara, N. Fadli, S. Rizal, M.N Siti-Azizah, Z.A. Muchlisin, European Zoological Journal, **89,** 1 (2022)
16. A. Zabarun, Bahtiar, & Haslianti. Jurnal Manajemen Sumberdaya Perairan, **2**, 1 (2016)
17. N.P. Cavaleiro,C. Lazoski, C.R. Tureck, C.M. R. Melo, V.S. do Amaral, B. J. Lomovasky, T.M. Absher, A. M. Solé-Cava, Journal of Experimental Marine Biology and Ecology, **511.** (2019)
18. S. Nehring, Database of the North European and Baltic Network on Invasive Alien Species **5** (2006)
19. R. J. M. Mann, M. Harding, Southworth. Estuarine, Coastal and Shelf Science. **85** (2009)
20. S. Diederich, Journal of Experimentalmarine Biologi and Ecology, **328,** 2 (2006)
21. Kepmen LH No. 51 Tahun 2004. Tentang baku mutu air laut. (2004)

Food packaging based on biodegradable polymers from seaweeds : a systematic review

Said Ali Akbar[1]*and *Andi* Mustari[2]

[1]Department of Aquaculture, Faculty of Marine and Fisheries, Universitas Syiah Kuala, Banda Aceh 23111, Indonesia
[2]Department of Electrical Network Installation and Maintenance, Akademi Komunitas Negeri Aceh Barat, Aceh Barat 23681, Indonesia

Abstract. As a result of its brief lifespan, food packaging contributes significantly to environmental contamination through the rapid accumulation of plastic in the environment. In order to mitigate these impacts and provide a food packaging solution that is more environmentally sustainable, researchers have created biodegradable and biobased polymers, which are presently being introduced to the market. The current state of research regarding the incorporation of seaweed into food packaging and active packaging is summarized in this study. In order to emphasize the benefits of polysaccharides and draw attention to the constraints of current research, this study also presents a synopsis of the effects of seaweed incorporation on a range of properties, including chemical, physical, thermal, antioxidant, antimicrobial, and mechanical attributes, in addition to the release of active compounds. A multitude of polysaccharides, including those found in seaweed, have the potential to enhance the antibacterial, thermal, and mechanical properties of these polymers, among other attributes. In addition to increasing hydrophilicity and mechanical properties such as tensile strength and elongation at break, they suggest employing it as active packaging. This is feasible as a result of the antioxidant properties of seaweed, which inhibit lipid oxidation and decrease toxic, mutagenic, or carcinogenic free radicals, thereby extending the nutritional value and shelf life of food. Certain varieties of seaweed have exhibited the capacity to impede the proliferation of bacteria classified as gram-positive and gram-negative. Thus, their prospective application as antimicrobial packaging is indicated.

1 Introduction

A crucial component of the contemporary food sector is packaging. Food preservation is essential as it guarantees both the food's safety and integrity [1]. Due to these benefits, packaging is a crucial component of our supply chain; yet, there are drawbacks as well, including the possibility of contaminant residue transfer, issues with cost and energy efficiency, and sustainability [2]. For food packaging, materials like glass, paper, different metals, and plastic are regularly used; however, plastic is the material that is used most frequently [3]. The majority of plastic used in food packaging is thermoplastic plastic, which softens when heated and subsequently cools to its original state [4].

The most widely used thermoplastics are polypropylene, polyethylene terephthalate, low-density polyethylene, and high-density polyethylene [5,6]. These materials typically have a

* Corresponding author: saidaliakbar@usk.ac.id

lifespan of less than a year before being thrown away. Because of its brief lifespan, plastic garbage in the environment accumulates quickly. Every year, currents and wind carry 10–20 million tons of plastic into the ocean, harming the marine environment's ecosystem and fauna in the process. Additionally, it causes microplastic to enter the food chain and include hazardous compounds [7]. Significant progress has been achieved in the search for workable, environmentally acceptable alternatives, such as biodegradable and/or biobased polymers, through a number of research conducted in recent years [8,9].

Natural and fossil fuels are used to create biodegradable polymers like polylactic acid (PLA) and polyhydroxy butyrate (PHB), among others [10]. In a brief amount of time, the microbes may metabolize these biodegradable polymers and then restore them to nature. In contrast, bio-based plastics are entirely composed of bio-based materials such as proteins, polysaccharides, and microbial polyesters that are either created by microorganisms through fermentation or taken from plants or marine organisms [11]. The polysaccharides employed in bio-based polymers are both non-toxic and capable of undergoing biodegradation. These biological materials are also recognized for their biocompatibility, which refers to their ability to function appropriately within a specific application while eliciting an appropriate response from the host. Therefore, these characteristics offer numerous advantages for food packaging, particularly for use in edible coatings and films. Polysaccharides provide a sustainable option for active packaging because of their inherent biological characteristics, such as antibacterial and antioxidant properties [12]. Seaweeds, which are rich in polysaccharides, have great potential as a raw material for active packaging. When combined with biodegradable polymers, they offer a more environmentally friendly alternative to traditional materials [13].

2 Materials and Methods

In this study, we collected the references (papers) from Indonesia in Google Scholar. Recent references published between 2019 until 2023 were retrieved using some keywords "Food Packaging Seaweeds" and “Food Packaging polymers”. The present review offers a comprehensive synthesis of the existing research concerning active packaging and the application of seaweed in the realm of food packaging. Moreover, it provides an exhaustive examination of the ramifications that arise from the incorporation of seaweed into packaging materials, with a particular focus on the implications for the materials' chemical, mechanical, physical, thermal, antioxidant, and antibacterial properties. This article presents a succinct summary of the active compounds that are liberated when seaweed polysaccharides are integrated into food packaging. Furthermore, prospective trends and legal considerations concerning the use of seaweed polysaccharides in food packaging were also addressed in the conversation.

3 Results and Discussion

3.1 Seaweeds

Benthic algae, or seaweeds, are multicellular, large algae that grow faster than land plants. Because seaweeds have higher photosynthetic efficiencies than terrestrial plants, they acquire biomass faster [14]. It contributes significantly to the aquatic biomass, generates almost half of the primary productivity on Earth, and is widely utilized for feed and other purposes. Seaweed is quite abundant, thus during the past ten years, as demand for it has grown, seaweed agriculture has flourished. Consequently, a substantial quantity of seaweed waste is generated from diverse industrial procedures, including the food industry, which imposes strict quality standards on the seaweed [15]. Within a circular economy, these waste materials can be utilized as a third-generation biomass for various purposes such as pharmaceuticals, animal feed, agriculture, and biofuels. Due to the potential severe impact on the marine ecology, it is crucial to recycle the waste material instead of releasing it into the environment. It also implies that algae have a great deal of potential for use as a sustainable energy source and as high-value products [16].

Around the world, seaweed has over 10,000 different species and can grow up to 180 meters below the surface on solid surfaces like granite. It is a highly significant marine resource in terms of its rich biodiversity. Seaweeds can be classified into three classes according to their pigmentation: red (Rhodophyta), brown (Ochrophyta), and green (Chlorophyta). The seaweeds contain a diverse range of bioactive compounds that have multiple applications in the fields of medicine, agriculture, and pharmaceuticals. The several varieties of seaweeds are listed in Table 1, along with corresponding polysaccharides and their molecular structures [17,18].

Table 1. Significant seaweed polysaccharides and their chemical composition.

Seaweed	Polysaccharide	Chemical structure	Sugars	Ref
Brown seaweed	Laminarin	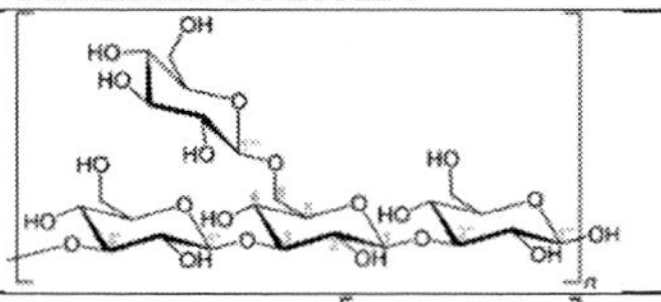	Glucose, guluronate, sulphated fucose, mannuronate, mannitol	[14-16]
	Alginates	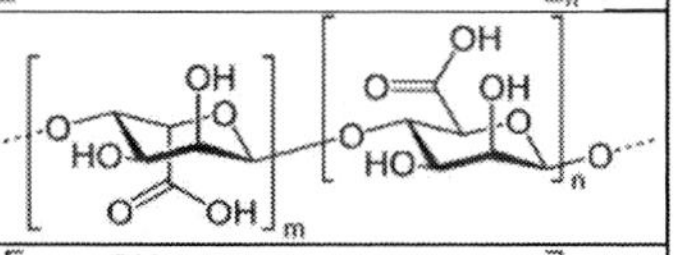		
	Cellulose	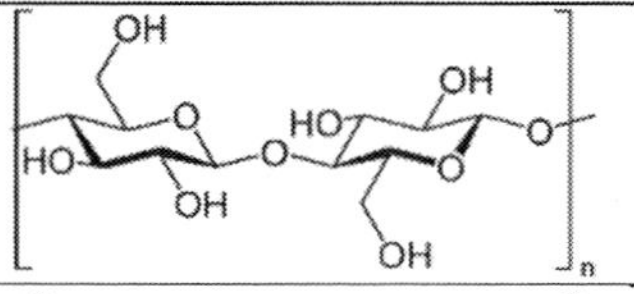		
	Fucoidan	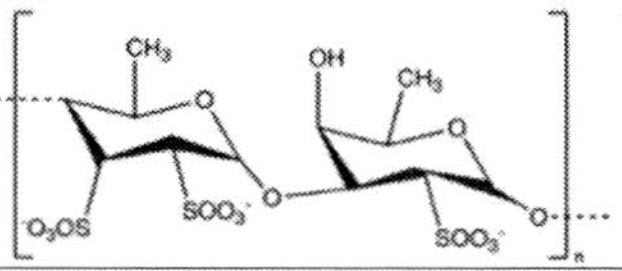		
Red seaweed	Agaropectin	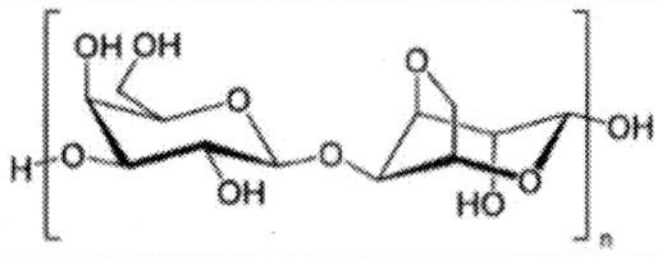	3,6-anhydro-D-galactose, glucose D-galactose, D-fructose,	[17-20]
	Carrageenan	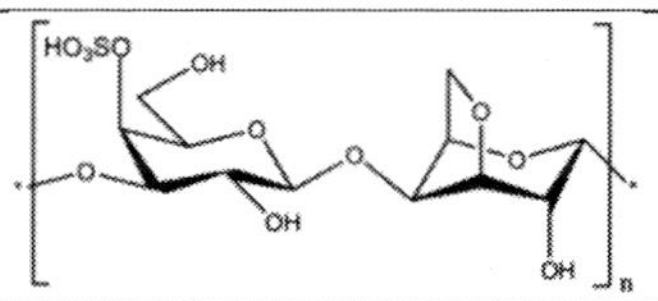		
	Cellulose	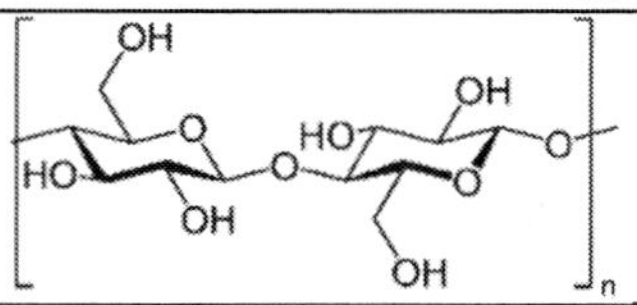		
	Xylans	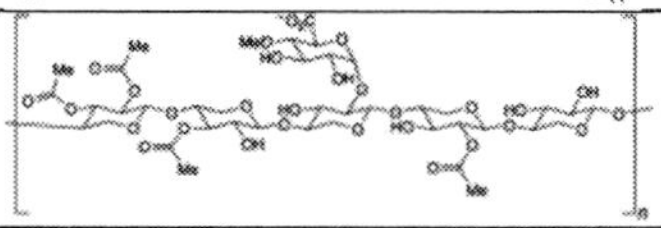		

	Mannans	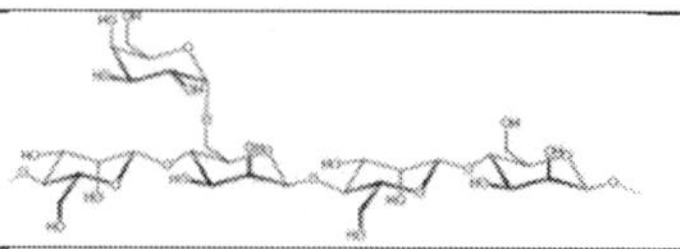		
Green seaweed	Ulvan	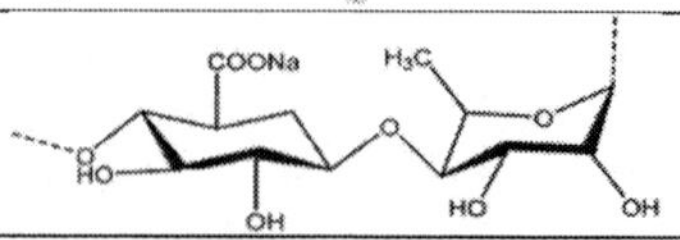	Rhamnose, galactose, Glucose, xylose, uronic acids,	[21-22]
	Starch	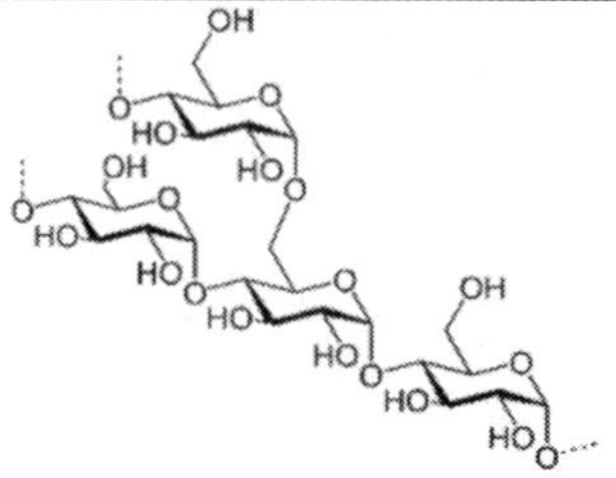		
	Xylopyranose	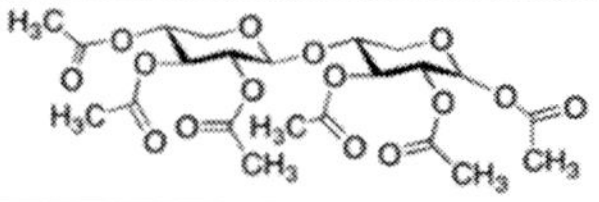		
	Glucopyranose	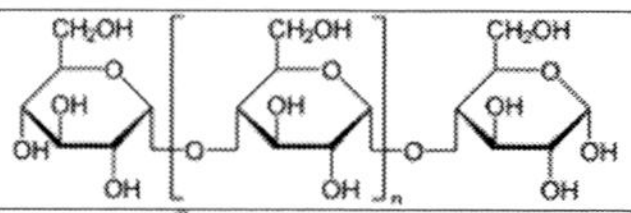		
	Xyloglucan	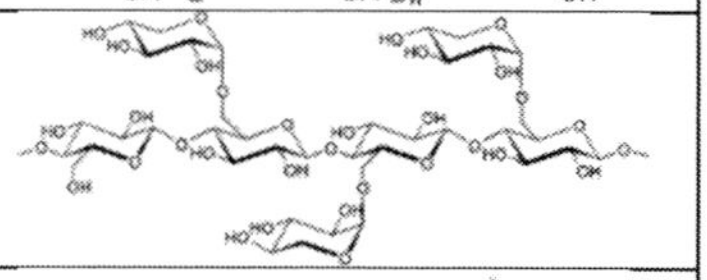		
	Glucuronan	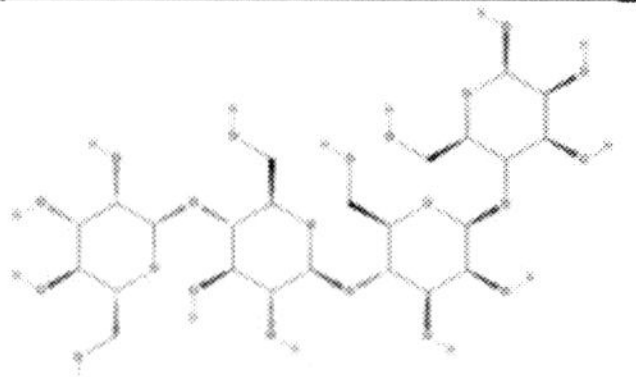		
	Cellulose	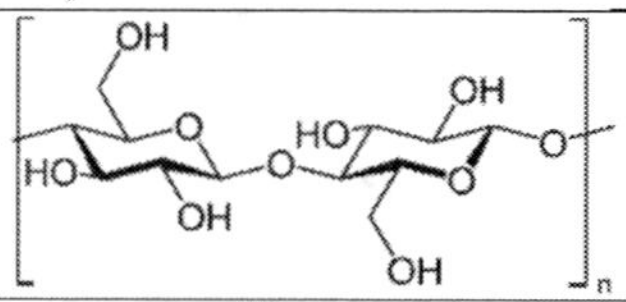		
	Hemicellulose	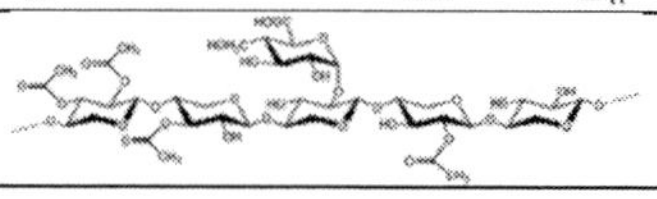		

The red and blue pigments phytocyanin, phycoerythrin, and chlorophyll A differentiate red seaweeds from their brown and green counterparts. Red seaweed is home to a vast array of species due to its varied life cycles and plant morphologies. The species that occur most frequently include Poryphyra capensis, Aeodes orbitosa, and Notogenia stiriata [23]. These seaweeds contain sulphated galactans, which are extensively utilized biopolymers in the food industry, specifically for products like agar and carrageenan. Agar and carrageen are categorized as non-digestible oligosaccharides with immune-modulating, prebiotic, anti-tumor, and antioxidant properties [24]. They are also non-cariogenic in humans. Agarose and agaropectin, the two primary ingredients of agar, have structural and functional similarities to carrageenan. Due to its gelling, emulsifying, and thickening qualities, it can be used commercially in food, medicine, and other uses in addition to being beneficial in science.

Agar's chemical structure is made up of residues of α(1–4)-3,6-anhydro-l-galactose and β 9(1–3)-d-galactose, with tiny amounts of esterified sulphate up to 6% (w/w).

Carrageenans are composed of α (1–4)–3, 6-anhydro-d-galactose and β (1–3)–d-galactose backbones that are sulphated to differing degrees [25]. It is hydrophilic and anionic, and the degree of sulphate ester determines how soluble it is—the more sulphate ester, the more soluble it is. They are separated into three classes based on the degree of sulphation; the Lambda type has the maximum sulphation at 40% (w/w). They have thickening qualities but are unable to form gels like the other carrageenans. Because iota carrageenans have less sulfur than lambda carrageenans, they may gel softly when calcium ions are present. The Kappa family of carrageenans has the lowest sulphate ester content, with a sulphate level of 20% (w/w). The carrageenans are employed to produce cohesive and transparent films due to their ability to form robust gels when exposed to potassium ions [26].

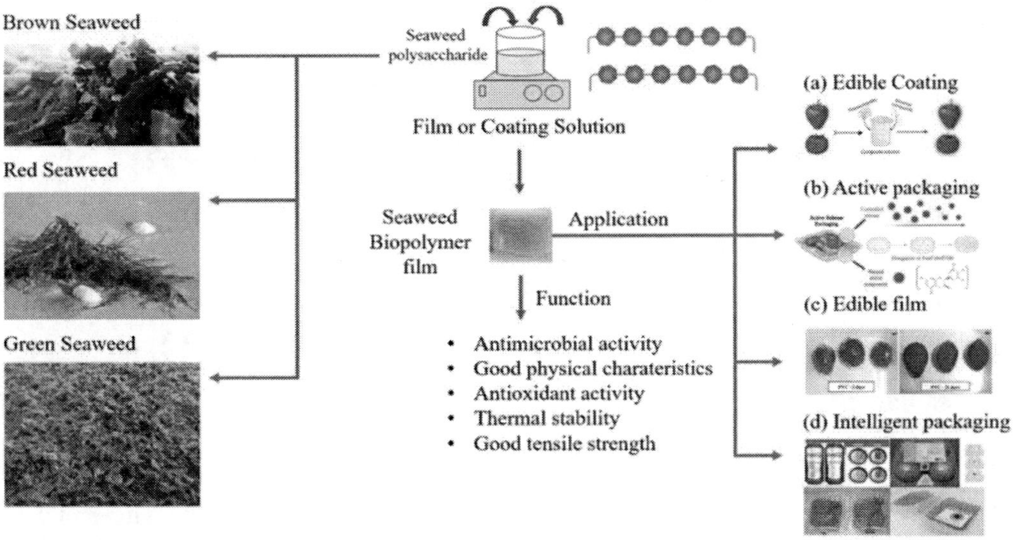

Fig. 1. The process of creating food contact materials using seaweed. (a) The impact of edible coatings on the longevity of fresh strawberries has been studied [29]. (b) Active packaging that is biodegradable and has controlled properties [30] (c) The application of edible coatings on fruits and vegetables [31] (d) Several instances of intelligent bio-based food packaging are provided [32].

Green algae inhabit both saline and freshwater due to the presence of chlorophyll a and b, the pigments responsible for their green hue. Common green seaweeds consist of the Chaetomorpha and Cladophora species, in addition to sea lettuce, which is composed of the Ulva and Monostroma species. Ulvan, which makes up an estimated 36% of the dried weight of green seaweed, is a complex carbohydrate found in its cell walls [27]. There are a multitude of prospective applications for this substance in the domains of agriculture, medicine, and sustenance. The substance's structure is determined by the presence of monosaccharides that are interconnected via α- and β-(1,4) bonds. Among these monosaccharides are rhamnose, glucuronic acid, xylose, and iduronic acid. Additionally, the substance comprises repeating disaccharide units, aldobiuronic acid and ulvanbioses. Similar to various other seaweed polysaccharides, the properties of ulvan are substantially influenced by its eco-physiology, source species, and extraction methods. Recent research has shown that ulvan gels exhibit thermo-reversibility, meaning they can be reversed by changes in temperature. Additionally, variations in pH and ion concentration can affect both the structure of ulvan and the formation of the gel. The bioactivity of ulvan is influenced by its molecular mass. Fractions with higher molecular weight exhibit a more pronounced ability to reduce total serum and LDL-cholesterol levels, whereas polysaccharides with smaller molecular weight demonstrate a greater antioxidant activity and a stronger effect in reducing triglycerides and HDL-cholesterol. In addition, studies have demonstrated that the polysaccharide can effectively inhibit a range of viruses, depending on the specific virus type and dosage administered [28].

The size, species, and general appearance of brown seaweeds vary greatly, but they are often colored brown by a pigment that is produced during photosynthetic processes. Common brown seaweeds include Fucus species, Zonaria species, and kelp like Laminaria pallida. Brown seaweed has three types of polysaccharides: fucoidans (FUC), laminarians, and alginate [16]. Alginates are made up of alternating blocks of α-l guluronic acid and ß-d mannuronic acid and are found in the intracellular matrix and cell walls of brown seaweed. Alginates, which swell in water, are a flexible material that can be utilized as a disintegrating agent in tablets or as a film by virtue of their interaction with divalent and trivalent cations. its properties are dependent on the monomer sequence. The primary polymer in brown seaweed that stores glucose is called laminarin, which has a wide range of documented bioactivities, including antioxidant and anti-tumor properties [19]. However, its primary industrial application is as a ligand for innate immune system pattern recognition receptors [18]. FUC, a complex, heterogeneous sulphated polymer made of l-fucose and sulphate ester groups with trace amounts of other molecules such (acidic-)/monosaccharides, proteins, and acetyl groups, is another polymer that may be isolated from brown seaweed. The polysaccharide's content and properties are significantly influenced by the species, time of year, place of origin, and extraction technique. However, due to its potential anticancer, antiviral, anticoagulant, and anti-inflammatory effects, it is mostly utilized in biomedical-related domains. Seaweed also contains colors, phenolic compounds, fatty acids, sterols, alkaloids, terpenes, and halogenated chemicals [21].

3.2 Active packaging

The term "active packaging" pertains to a system in which the product, container, and surrounding environment work in concert to prolong the product's shelf life, enhance safety measures, retain sensory characteristics, and uphold the product's quality. The product is comprised of constituents that promote the movement of substances between the product and the environment, the packaging, and the product itself. The method by which food contact materials are produced utilizing seaweed (Figure 1). A monolayer system integrates the active component into the polymer, whereas a multilayer system controls the release of the active chemical by enclosing it between the polymer layers [33]. Methods such as chemical addition and temperature control are crucial for this procedure. Several advanced food packaging systems have been developed in recent times, such as ethylene scavengers, moisture management packaging, carbon dioxide generating systems, oxygen scavengers, and flavor and odor absorbent packaging. Table 2 presented the current applications of active packaging.

Table 2. Current applications of active packaging.

Packaging type	Active Compound	Packaging Structure	Improvements	Ref
Ethylene sc avenger	Potassium permangana te	An ethylene scavenger supported by sepiolite, utilizing potassium permanganate as its main component.	Protract the maturation of recently harvested apricots, as well as potentially other fruits sensitive to C_2H_4, thereby extending their ability to be stored after harvest and enhancing	[34]

			their appeal to consumers.	
Oxygen and volatile organoleptic compound scavenger	Silica gel	NS	Decreased spoilage and disease rate. Chlorophyll content loss can be delayed by up to 8 days when stored at temperatures ranging from 29.4 to 33.2 °C and relative humidity levels of 68 to 73%.	[35]
	Vermiculite	Sachet	Delayed softening of the pulp and degradation of vitamin C. The temperature will be maintained at 25 degrees Celsius with a relative humidity of 54% for a duration of 5 days.	[36]
	Halloysite nanotube	LDPE Film	The elongated shelf life of tomatoes and bananas when they are preserved at a temperature of 4 °C and packaged in films.	[37]
	Zeolite-based minerals	LDPE film	The shelf life can be extended up to 20 days at a temperature of 4 °C by creating a balanced atmosphere, resulting in improved sensory quality.	[38]
	TiO_2	Chitosan film	The quality of the tomatoes was compromised and the maturation process was prolonged when they were stored at 25 °C and 50% relative humidity.	[39]
Oxygen scavengers	Potassium sorbate and sodium benzoate and	Chitosan, Pullulan, and Pectin and are utilized as edible coatings.	Strawberries have been found to be efficacious in mitigating weight loss, averting fruit softening, postponing color degradation, and sustaining titratable soluble solids levels.	[40]
	BHT or Sesamol	(1) HDPE/HDPE (BHT)/EVA, (2) HDPE /HDPE (active)/ethylene-vinyl acetate (EVA), (3) LLDPE/HDPE (active)/HDPE	A method that prevents lipid oxidation in linoleic acid and breakfast cereal in an effective manner.	[41]
	Hesperidin Pickering	Gelatin film that has been activated	The encapsulation of hesperidin resulted in	[42]

	emulsion with chitosan nanoparticle stabilization		improved compatibility and enhanced antioxidant activity prior to its incorporation into the film.	
	Carum copticum EOs and *Origanum vulgare* ssp. gracile	Two-dimensional lignocellulose and chitosan/cellulose nanofibers.	The active film displayed notable antioxidant and antimicrobial characteristics, thereby indicating that CNF and LCNFa have the capacity to control the release of bioactive compounds from the films.	[43]
	Iron-based	Utilizing iron-based oxygen scavengers in modified atmosphere packaging	Minimization of microbial and oxidative alterations in chicken thigh meats	[44]
Antimicrobial packaging	Lythrum salicaria extract	Cellulosic nanomaterials	The preservation of the effectiveness of antimicrobial nanoparticles or antimicrobial enzymes during immobilization without any substantial decrease.	[45]
	Tea tree oil (TTO)/β-CD complex	Poly(ethylene oxide) (PEO)	Both the efficacy of the antibacterial agent release mechanism from PEO nanofibers and the antibacterial activity exhibited by PEO nanofibers were enhanced. Escherichia coli demonstrated the most significant antibacterial effectiveness.	[46]
	Copaiba oil	Pectin/Nanoemulsion	The nanoemulsion that was formed demonstrated remarkable compatibility between the oil and the matrix of the film. Antimicrobial efficacy against *Staphylococcus aureus* and *Escherichia coli* has been enhanced.	[47]
	Cumin EO	Sodium caseinate/TiO_2 NPs 3D guar gum	The incorporation of both titanium dioxide (TiO2) and cumin essential oil (EO) led to a synergistic enhancement in the	[48]

			mechanical, barrier, and antimicrobial characteristics of the films.	
	Nisin	Starch/ Halloysite nanotubes 1D	The film experienced a reduction in size and an enhancement in its ability to kill microorganisms in the presence of halloysite.	[49]
	Oregano essential oil in a capsule	The research centers on the application of complex coacervates composed of soluble soybean polysaccharide and acid-soluble soy protein for the purpose of stabilizing Pickering emulsion, which is generated by films of soluble soybean polysaccharide.	The immobilization and improved retention of essential oil (EO) within the film matrix were outcomes of the encapsulation process, which produced films with sustained antimicrobial activity.	[51]
	Cinnamon EO	Zein/Chitosan NPs 3D	The coexistence of CEO (chief executive officer) and CSNPs (chitosan nanoparticles) led to enhanced inhibitory effects on E. coli and S. aureus bacteria. This was a result of the enhanced transportation of essential oil (EO) by loading it into colloidal silica nanoparticles (CSNPs).	[51]

Antimicrobial packaging reduces, prevents, or delays the proliferation of microorganisms through the use of antimicrobial compounds on or within food packaging. This results in a prolonged pre-microbial development period and a deceleration of microbial multiplication [41]. Antimicrobial packaging comprises four distinct categories: antimicrobial coating, antimicrobial sachet or pad, intrinsically antibacterial polymer, and direct incorporation in polymer. Moisture-controlling packaging can be categorized into two distinct types: RH controllers, which eliminate humidity from the air within the packaging, and moisture absorbers, which absorb liquids within the packaging. The former are frequently utilized in the production of dehydrated food items, including candies, snacks, and almonds. Desiccants including calcium oxide, silica gel, and inorganic salt mats are also incorporated. Pads, linens, and blankets comprise moisture absorbers, which are composed of two layers: a microporous polymer, such as polyethylene, is positioned atop a highly absorbent polymer, such as granules, which flow readily [45]. These are utilized for the purpose of controlling the amount of liquid present in packaged goods, fish, poultry, and produce.

Ethyene, a small molecule, acts as a growth-promoting hormone, accelerating the ripening of fruits and vegetables. Consequently, leafy greens experience a reduced postharvest shelf life, increased food softening, and an accelerated rate of chlorophyll breakdown [49]. One possible approach to reduce the level of ethylene around the product (as shown in Table 2) is to use a sachet or blanket containing potassium permanganate. This substance acts as an oxidizing agent and deactivates ethylene when it comes into contact with inert minerals that are immobilized. Amines, aldehydes, and sulphides may arise as a result of protein and other organic compounds breaking down during the preservation of food items like fish and poultry. One can selectively collect these volatile molecules to prevent the mixing of odors from different products during transportation. The packaging, which is resistant to odors, utilizes polyethylene terephthalate, a permeable port, and an odor-absorbing sachet containing nickel and charcoal. To eliminate amines produced during the degradation of protein in fish packing, one can add citric acid or other acidic substances to the polymer film [51].

3.3 Utilizations of seaweed in food packaging

The potential hazards associated with the use of chemicals as active agents in packaging stem from their capacity to migrate from polymers to food. As a result, scientific inquiry is placing an increasing amount of emphasis on the utilization of natural substances rather than chemical active agents. Seaweeds are regarded as prospective basic materials or active constituents on account of their copious polysaccharide content. They enhance the functionality, sustainability, and sensory attributes of the product when blended with any biodegradable polymer, thereby surpassing the performance of conventional materials. Potential applications for seaweeds include the manufacture of sachets, biodegradable plastics, sustainable packaging, consumable packaging, and active packaging [52]. The most recent applications of seaweed polysaccharides as packaging material were detailed in Table 3.

Table 3. Recent application of seaweed polysaccharides on packaging material.

Type Of Application	Polysaccharides	Characteristics of the Packaging Material	Packaging Matrix	Reference
Smart and intelligent packaging	κ-Carrageenan, agar	Enhanced UV–visible light obstruction, pH responsiveness, and physical characteristics. Visual color changes were observed in the films used for the buffer solution (pH 2.0–12.0), ammonia vapor (80 M), and pork spoilage studies.	*Clitoria ternatea Linn*, nano-ZnO, and Nano-TiO_2	[53]
Smart and intelligent packaging	Carrageenan	Enhanced UV protection without diminishing film transparency. A remarkable color change that is pH-responsive across a broad pH range of 2–12. Effectively monitoring the freshness of packaged milk was intelligent film.	Propolis, Shikonin, and Gelatin	[54]
Smart and intelligent packaging	κ-Carrageenan	Enhanced capacity to retain water vapor and block UV–visible light. Enhanced antioxidant and pH-responsive characteristics. Can be utilized for	Mulberry polyphenol extract	[55]

		monitoring the perishability of milk.		
Active packaging	*Fucus vesiculosus*	Enhanced the dimensions, resistance to stretching, and ability to withstand deformation. Prevent the oxidation of lipids in packaged poultry meat for a minimum of 25 days during storage.	Whey protein	[56]
Active packaging	Gelidium sesquipedale	Improved characteristics for the integration and liberation of biologically active extracts.	Nanocellulose, Polyvinyl alcohol,	[57]
Active packaging	Sodium alginate	An elevation in fracture strain. Moisture content, water vapor permeability, and oxygen permeability have all increased. Effective against *S. aureus* and *E. coli*.	Pyrogallic acid, $CaCl_2$, Carboxymethyl cellulose	[58]
Active packaging	Furcellaran	Elevated dimensions, moisture level, and resistance to stretching. Observations were made regarding the variations in color under different pH conditions. Significant levels of antioxidant activity.	Rosemary extract/Gelatin hydrolysate	[59]
Active packaging	Sodium alginate	By means of microencapsulation, lemon grass oil can be released under precise control. The films demonstrated inhibitory effects on the proliferation of *L. monocytogenes* and *E. coli*.	Lemongrass oil	[60]

By incorporating nanocrystalline cellulose, the mechanical properties, thermal stability, tensile strength, and elastic modulus of biodegradable nanocomposites fabricated from alginate were all improved. Furthermore, the research documented a reduction in swelling properties, water vapor permeability, and elongation at break. However, an investigation that examined the incorporation of agar and nanoclay into a film intended for biodegradable food packaging revealed that both the swelling ratio and tensile strength were enhanced [53]. Additionally, its solubility in water, the angle of surface contact between water and the substance, and the rate of water vapor permeation were all diminished. As a consequence of integrating lemongrass essential oil and alginate into the packaging film for fresh cut fruit, weight loss, color alteration, and respiration rate were all reduced. Furthermore, negligible alterations were observed in the physical and sensory attributes, accompanied by minimal quantities of microbial activity [58]. Furthermore, the results obtained from employing a salted alginate film in microwaveable packaging revealed an improvement in heat dispersion and the film functioning as a conductor throughout the heating process. To determine the bactericidal effect of a mixture of carrageenan, locust bean gum, and organically modified nano-clay against L. monocytogenes, an investigation was undertaken. The primary aim was to increase the longevity of culinary products. Furthermore, it resulted in increased tensile strength, delayed thermal degradation, elongation at the point of fracture, and level of aggregation. An additional investigation revealed that the amalgamation of carrageenan and grapefruit seed extract demonstrated enhanced characteristics in terms of water vapor permeability, elongation at break, and UV barrier properties. Additionally, the results revealed a significant reduction in tensile strength, elastic modulus, and water contact angle, in addition to noteworthy antibacterial effectiveness, particularly against gram-positive bacteria [60].

3.4 Impact of seaweed integration on the characteristics of film, including antimicrobial and antioxidant properties

The mechanical, thermal, optical, and chemical properties of other polymers are altered when seaweed is added. To produce a packaging that is both safe and effective, it is necessary to thoroughly assess these effects, which are depending on the seaweed polysaccharides that were added. The type of seaweed polysaccharide being included and the polymer in which it will be incorporated are both important factors in determining the results [57]. Chemical characteristics like hydrophobicity or hydrophilicity, as well as the film's composition with functional groups, are crucial in determining how safe and effective food packaging is. Because of the various components and how they interact, the integration may vary greatly. After adding the agar to the SPS, Fourier transform infrared (FTIR) spectroscopy was used to look for changes in the functional groups in the film (Fig. 2) [58. Little peak shifts to lower wave numbers were observed, but there were no appreciable alterations in the recorded peak as shown by the spectra. Increased intermolecular hydrogen bonding is shown by these, suggesting that the two components are compatible and interact. One possible explanation for the increased moisture absorption of the film upon agar insertion could be its higher hydrophilicity as a sulphated polysaccharide. Overall, these investigations demonstrate that the most frequent modification in chemical characteristics following the seaweed addition was an increase in hydrophilicity. Seaweeds' anionic nature causes greater absorption of moisture, swelling, and a decrease in water contact angles. Nonetheless, the findings demonstrate that variations in crystallinity are influenced by both the seaweed and the way the polymer and polysaccharide interact.

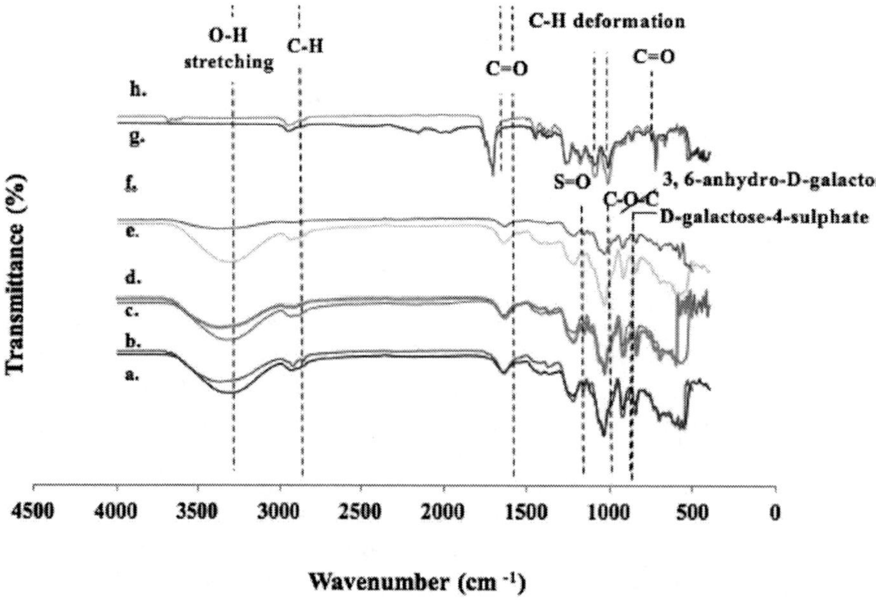

Fig. 2. An analysis was conducted on the Fourier transform infrared (FT-IR) spectra of seaweed-based films, which incorporated additives and conventional mulch film, both prior to and subsequent to their burial in soil [61]. (Copyright Of The MDPI).

It is challenging to speculate on the morphological alterations that may result from the incorporation of components, such as seaweed polysaccharides, because much depends on the interaction between the polymer and the seaweed [62]. Nevertheless, it is critical to examine them due to the fact that the amalgamation of diverse materials might give rise to voids and induce additional modifications that have the potential to impact technical characteristics. The effect of monomer arrangement on the physical properties of alginates has been demonstrated. Alginate's strong affinity for multivalent cations, which causes the formation of gels, and its ability to transition between a sol and gel state irrespective of temperature are its primary physical characteristics. The surface morphology of agar films with different concentrations of agar and sugar palm starch was analyzed using SEM. After agar was added to the mixture, the SEM image revealed a smooth and homogeneous surface devoid of any concentrations or phases of sugar palm starch, according to the study. This finding suggests that the two constituents of the film are harmonious and operate efficiently when combined [63]. On the other hand, breaking structures emerged on the surface as the Agar concentration rose. This could be because to increased filler content in the film matrix and polymer bonding. The morphological alterations brought about by the addition of esterified alginate to the Polyhydroxy butyrate mix were examined using SEM analysis of Polyhydroxy butyrate film samples. The inclusion of polysaccharide caused a modification in the morphological composition of the Polyhydroxy butyrate blend, displaying an exposed, uneven, and dense surface texture. The surface morphology of a film made of sodium alginate and polyvinyl alcohol was analyzed [64]. The analysis of the SEM image demonstrated that the blend film exhibited a uniform texture, suggesting excellent miscibility and the absence of phase separation.

In the context of food packaging, polymer films that possess the appropriate mechanical properties include elongation at break, elastic modulus, and tensile strength. The tests demonstrate that the film can withstand a range of stresses, including those encountered in the preparation, handling, and storage of food, without experiencing any degradation in its integrity [65]. Intermolecular interactions between polymer chains, as well as the integrity and solubility of the constituents, have a substantial impact on the mechanical properties of composite films produced by combining. On the subject of mechanical properties, seaweed polysaccharides have been the subject of numerous investigations. In approximately 50% of the examined studies, the integration of seaweed polysaccharides into the appropriate polymer resulted in an increase in tensile strength. The utilization of flour and alginate in the production of food packaging films with remarkable mechanical properties [66] was observed. Alginate and other agar or starch-based films demonstrate remarkable mechanical properties.

The melting temperature (Tm) and glass transition temperature (Tg) are crucial thermal parameters that reflect the level of intermolecular bonding among polymer chains. They are also required to offer information that must be taken into account during the process and determine whether the polymers are appropriate for use in food packaging [67]. Following the seaweed's addition, collagen with fucoidan added showed better thermal characteristics (Fig. 3). The constituents of the mixture exert a substantial impact on the manner in which the incorporation of seaweed influences the thermal properties. Varying degrees of miscibility, crystallinity, and overall interaction can have significant impacts on the temperature at which crystallization and melting take place.

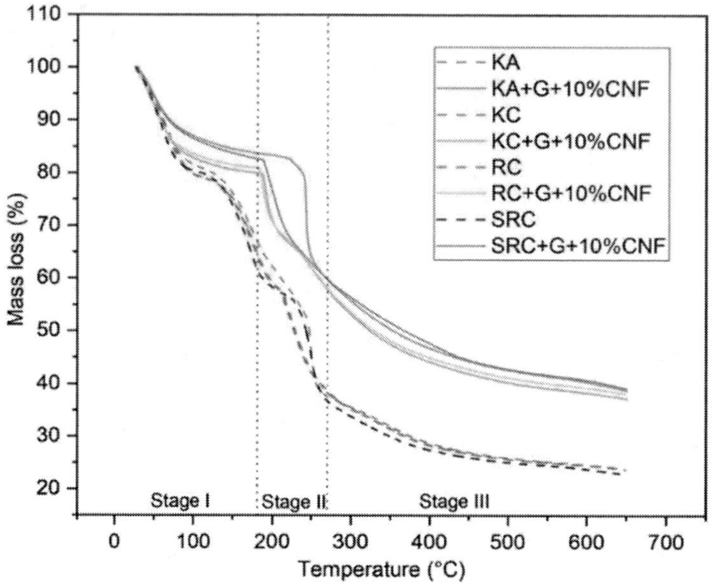

Fig. 3. Thermogravimetric analysis (TGA) was conducted on bioplastic film matrices derived from seaweed, which were loaded with cellulose nanofiber [68] (Copyright Of The MDPI).

Due to the fact that food safety is a major concern in the industry, antimicrobial packaging is indispensable for active packaging. Fundamental to ensuring the sustainability and safety of packaging is the incorporation of natural antimicrobials like chitosan. Considerable research has been devoted to the examination of carvacrol, durian starch, and carrageenan films. The inclusion of purified saccharides, including carrageenan and alginate, in a polymer film exhibited minimal or no inhibitory impact on L. monocytogenes and E. coli bacteria [54]. *S. aureus* was not inhibited by the durian starch/carrageenan control films, as determined by the disk diffusion test. After pre-diffusing the film with 10% carvacrol for 24 and 48 hours, all gram-positive bacteria were completely suppressed. Furthermore, the introduction of nano-clay or nano silver did not result in any antibacterial properties in the carrageenan-based composite film when tested against L. monocytogenes or E. coli. Another investigation [55] found that an agar blend containing both gram-positive and gram-negative food-borne pathogenic bacteria did not demonstrate any antibacterial properties. The addition occurred only after a combination of six unique copper nanoparticles with strong antibacterial properties was introduced.

Lipid oxidation significantly affects food and food product quality. Oxidation produces agents like free radicals, which are cytotoxic, carcinogenic, and mutagenic, and can lead to serious health issues [58]. Foods that have experienced lipid oxidation may also have less nutritional value and a shorter shelf life. Antioxidant packaging is intended to reduce the degree of protein breakdown and lipid oxidation that occurs inside the packaging. Antioxidant-containing active packaging is a viable substitute for conventional packaging, which incorporates or coats the film with antioxidants. The trend toward fewer synthetic additives in packaging has increased interest in natural antioxidants like plant extracts and essential oils because they are safer and have more health advantages [53]. To summarize, the findings suggest that a wide range of polysaccharides possess antioxidant activity to different extents. This information holds potential for the formulation of packaging containing antioxidants. To bolster the antioxidant properties and prevent lipid oxidation while increasing food safety, it is feasible to integrate substances such as mulberry extract and plant essential oils for this purpose. Additional research is necessary to ascertain the antioxidant properties of various extracts and polysaccharides derived from seaweed. This is crucial for the successful application of seaweed as an antioxidant agent in active food packaging [52].

4 Conclusion

Seaweed integrated into natural polymers holds significant potential to revolutionize food packaging in response to increasing consumer awareness of sustainable alternatives. Reducing the use of synthetic packaging could potentially result in a decrease in the level of plastic pollution in our environment. Seaweed possesses antibacterial and antioxidant properties that have the potential to extend the shelf life of food and decrease food waste resulting from decomposition. Further investigation is necessary in order to determine the precise influence of seaweed on the shelf life of particular food items. To optimize safety and mitigate potential hazards linked to seaweed utilization, it is critical to strengthen current regulations concerning the accumulation of toxic metals and other potentially hazardous substances. This will facilitate the broader incorporation of seaweed in food products and packaging materials. To customize the mechanical characteristics of the film for a specific application, a polymer blend or natural plasticizer can be employed. All things considered, it appears to be a viable option for more active and sustainable movies with space for development. Seaweed is being used to create biodegradable packaging that is more practical and significant than existing techniques, which has prompted greater study into process optimization to make seaweed a potential rival in the market. Furthermore, because seaweed has mechanical, antibacterial, antioxidant, and releasing qualities, more thorough research should be done to determine which particular areas will benefit most from seaweed's advantages over conventional packaging materials. A more sustainable approach to food packaging also requires improving the current polysaccharide membranes through the use of additives like lipids, other polymer blends, and a greater range of potentially changed polysaccharides.

Acknowledgment

This study is a part of the university's internal research, which is supported by the Institute for Research and Community Service (LPPM) and the Faculty of Marine and Fisheries, Universitas Syiah Kuala.

References

1. M.S. Firouz, K. Mohi-Alden, M. Omid, Food Res. Int. **141**, 110113 (2021)
2. J. Zheng, L. Tian, S. Bayen, Crit Rev Food Sci **63**, 1 (2021)
3. A. Dey, C.V. Dhumal, P. Sengupta, A. Kumar, N.K. Pramanik, T. Alam, J. Food Sci. Technol **58(9)**, 3251 (2021)
4. A. Nazrin, S.M. Sapuan, M.Y.M. Zuhri, R.A. Ilyas, R.S.F.K.S. Syafiq, S.F.K. Sherwani, Front. Chem. **8**, 213 (2020)
5. A. Surendren, A.K. Mohanty, Q. Liu, M. Misra, Green Chem. **24**, 8606 (2022)
6. J. Promsorn, N. Harnkarnsujarit, Food Control, **142**, 109273 (2022)
7. E.B. Jadhav, M.S. Sankhla, R.A. Bhat, D.S. Bhagat, Environ. Nanotechnol. Monit. Manag **16**, 100608 (2021)
8. F. Wu, M. Misra, A.K. Mohanty, Prog. Polym. Sci, **117**, 101395 (2021)

9. L. Motelica, D. Ficai, A. Ficai, O.C. Oprea, D.A. Kaya, E. Andronescu, Foods, **9(10)**, 1438 (2020)
10. M.A. Sani, M. Azizi-Lalabadi, M. Tavassoli, K. Mohammadi, D.J. McClements, Nanomaterials, **11(5)**, 1331 (2021)
11. R. Priyadarshi, J.W. Rhim, Innov. Food Sci. Emerg. Technol, **62**, 102346. (2020)
12. V.T. Weligama Thuppahige, M.A. Karim, Compr. Rev. Food Sci. Food Saf **21(1)**, 689 (2022)
13. Q. Wang, W. Chen, W. Zhu, D.J. McClements, X. Liu, F. Liu, NPJ Sci. Food **6(1)**, 18 (2022)
14. J. Liu, S. Luthuli, Y. Yang, Y. Cheng, Y. Zhang, M. Wu, J. Choi, H. Tong, Food Sci. Nutr. **8(10)**, 5195 (2020)
15. S.A. Akbar, M. Hasan, S. Afriani, C. Nuzlia, Biodiversitas Journal of Biological Diversity **24(10)**, 5283 (2023)
16. L. Wang, J.Y. Oh, J. Hwang, J.Y. Ko, Y.J. Jeon, B. Ryu, Antioxidants, **8(10)**, 493 (2019)
17. T.H. Nguyen, T.L.P. Nguyen, T.V.A. Tran, A.D. Do, S.M. Kim, Asian Pac. J. Trop. Biomed. **9(12)**, 501 (2019)
18. M. Choulot, I. Michalak, L. Jing, A. Szymczycha-Madeja, M. Wełna, N. Bourgougnon, C. Le Guillard, Algal Res. **75**, 103239 (2023)
19. R. Balti, M.B. Mansour, N. Zayoud, R. Le Balc'h, N. Brodu, A. Arhaliass, A. Masse, Food Biosci., **34**, 100522 (2020)
20. C. Chellapandian, B. Ramkumar, P. Puja, R. Shanmuganathan, A. Pugazhendhi, P. Kumar, Process Biochem, **80**, 58. (2019)
21. R. Ishwarya, B. Vaseeharan, S. Kalyani, B. Banumathi, M. Govindarajan, N.S. Alharbi, S. Kadaikunnan, M.N. Al-anbr, J.M. Khaled, G. Benelli, J. Photochem. Photobiol. B, Biol. **178**, 249 (2018)
22. U. Sonchaeng, P. Wongphan, W. Pan-Utai, Y. Paopun, W. Kansandee, P. Satmalee, M. Tamtin, P. Prapat Kosawatpat, N. Harnkarnsujarit, Polymers, **15(16)**, 3342 (2023)
23. M.O. Akrong, A.K. Anning, G.N.D. Addico, K.A.A. deGraft-Johnson, A. Adu-Gyamfi, M. Ale, A.S. Meyer, Reg. Stud. Mar. Sci, **44**, 101719 (2021)
24. H. Hasby, N. Nurhafidhah, G. Pamungkas, S.A. Akbar. *Rasayan J. Chem*, **13(1)**, 38 (2020)
25. I. Albertos, A.B. Martin-Diana, M. Burón, D. Rico, Food Packag. Shelf Life, **22**, 100382 (2019)
26. B.B. Sedayu, M.J. Cran, S.W. Bigger, Polymers, **12(5)**, 1145 (2020)
27. Y. Shlosberg, G. Schuster, N. Adir, Front. Plant Sci. **13**, 955843 (2022)
28. S. Yalçın, Ö. Karakaş, E.Ş. Okudan, K.S. Başkan, S.D. Çekiç, R. Apak, J. Chromatogr. Sci., **59(4)**, 325 (2021)
29. D. Khodaei, Z. Hamidi-Esfahani, E. Rahmati, NFS J, **23**, 17 (2021)
30. J.R. Westlake, M.W. Tran, Y. Jiang, X. Zhang, A.D. Burrows, M. Xie, ACS Food Sci. Technol. **2(8)**, 1166 (2022)
31. R.G. Cruz-Monterrosa, A.A. Rayas-Amor, R.M. González-Reza, M.L. Zambrano-Zaragoza, J.E. Aguilar-Toalá, A.M. Liceaga, Polysaccharides, **4(2)**, 99 (2023)
32. A. Nešić, G. Cabrera-Barjas, S. Dimitrijević-Branković, S. Davidović, N. Radovanović, C. Delattre, Molecules, **25(1)**, 135 (2019)
33. M.S. Firouz, K. Mohi-Alden, M. Omid, Food Res. Int. **141**, 110113 (2021)
34. R. Alonso-Salinas, J.R. Acosta-Motos, E. Núñez-Delicado, J.A. Gabaldón, S. López-Miranda, Agronomy, **12(3)** 616 (2022)
35. D. Bhattacharjee, R. Dhua, Int J Pure Appl Biosci **5,** 64 (2017)
36. W. de Souza, M. Bezerra, P. Dantas, A. da Curnha, R. Sales, Acta Agron **66**, 331 (2017)

37. C.E. Tas, S. Hendessi, M. Baysal, S. Unal, F.C. Cebeci, Y.Z. Menceloglu, H. Unal, Food Bioproc Tech **10(4),** 789 (2017)
38. B.J. Ahn, K.K. Gaikwad, Y.S. Lee, J Appl Polym Sci. **133(43),** 44138 (2016)
39. P. Kaewklin, U. Siripatrawan, A. Suwanagul, Y.S. Lee, Int J Biol Macromol **112,** 523 (2018)
40. M.Z. Trevino-Garza, S. García, M. del Socorro Flores-Gonzalez, K. Arevalo-Nino, J. Food Sci. **80,** M1823 (2015)
41. X. Zhu, K. Schaich, X. Chen, K. Yam, Packag. Technol. Sci. **26,** 31 (2013)
42. I. Dammak, R.V. Lourenco, P.J. Amaral Sobral, J. Food Eng. **240,** 9 (2019)
43. E. Jahed, M. Alizadeh Khaledabad, M. Rezazad Bari, H. Almasi, Polym. **117,** 70 (2017)
44. D. Lee, H.J. Lee, D.Y. Jung, H.J. Kim, A. Jang, C. Jo, Food Res. Int, **155,** 111046 (2022)
45. S. Mohammadalinejhad, H. Almasi, M. Esmaiili, Mater. Sci. Eng. C Mater. Biol. Appl. **105,** 110115 (2019)
46. H. Cui, M. Bai, L. Lin, Carbohydr. Polym. **179,** 360 (2018)
47. L. Norcino, J. Mendes, C. Natarelli, A. Manrich, J. Oliveira, L. Mattoso, Food Hydrocoll. **106,** 105862 (2020)
48. M. Alizadeh-Sani, J.W. Rhim, M. Azizi-Lalabadi, M. Hemmati-Dinarvand, A. Ehsani, Int. J. Biol. Macromol. **145,** 835 (2020)
49. S.M.M. Meira, G.Zehetmeyer, J.M. Scheibel, J.O. Werner, A. Brandelli, Food Sci. Technol, **68,** 226 (2016)
50. Q.R. Liu, W. Wang, J. Qi, Q. Huang, J. Xiao, Food Hydrocoll. **87**, 165 (2019)
51. N. Vahedikia, F. Garavand, B. Tajeddin, I. Cacciotti, S.M. Jafari, T. Omidi, Z. Zahedi, Colloids Surf. B Biointerfaces **177**, 25 (2019)
52. S. Lomartire, J.C. Marques, A.M. Gonçalves, *Appl. Sci*, **12(6)**, 3123 (2022)
53. L. Liu, J. Zhang, J. Shi, X. Huang, X. Zou, D. Zhang, X. Zhai, Z. Yang, Z. Li, Y. Li, Int. J. Biol. Macromol. **165,** 758 (2020)
54. S. Roy, J.W. Rhim, ACS Appl. Bio Mater. **4,** 770 (2021)
55. Y. Liu, Y. Qin, R. Bai, X. Zhang, L. Yuan, J. Liu, Int. J. Biol. Macromol. **134,** 993 (2019)
56. M.A. Andrade, C.H. Barbosa, V.G.L. Souza, I.M. Coelhoso, J. Reboleira, S. Bernardino, R. Ganhao, S. Mendes, A.L. Fernando, F. Vilarinho, Coatings **11,** 229 (2021)
57. J.P. De Oliveira, G.P. Bruni, M.J. Fabra, E.D. Zavareze, A. Lopez-Rubio, M. Martinez-Sanz, Food Hydrocoll. **89**, 337 (2019)
58. S.A. Akbar, A. Rochliadi, V. Suendo, N. Saidi, L. Lelifajri, A.A. Mardhiah. *Rasayan J. Chem*, **11(4),** 1525 (2018)
59. S. Jancikova, E. Jamroz, P. Kulawik, J. Tkaczewska, D. Dordevic, Int. J. Biol. Macromol. **131,** 19 (2019)
60. R.O. Bustos, F.V. Alberti, S.B. Matiacevich, J. Food Sci. Technol. **53**, 832 2016.
61. M.H. Chong, E.W.N. Jafarzadeh, S. Paridah, M.T. Gopakumar, D. Tajarudin, H.A. Thomas, *Polymers*, ***11*(2)**, 210 (2019)
62. M. Elangovan, P. Anantharaman, M. Kavisri, M. Moovendhan, *Biomass Convers. Biorefin.* **1** (2022)
63. M.I.J. Ibrahim, S.M. Sapuan, E.S. Zainudin, M.Y.M. Zuhri, *J. Mater. Res. Technol.* **9(1)**, 200 (2020)
64. M.P. Sudhakar, D. Magesh Peter, G. Dharani, *Environ. Sci. Pollut. Res*, **28**, 33899 (2021)
65. H. Dysjaland, I. Sone, E. Noriega Fernández, M. Sivertsvik, N. Sharmin, *Molecules*, **27(23)**, 8356 (2022)

66. R. Jumaidin, S.M. Sapuan, M. Jawaid, M.R. Ishak, J. Sahari, *Int. J. Biol. Macromol.* **97**, 606 (2017)
67. G.D. Ulrich, R. Faez, *J Polym Environ*, **30(9)**, 3688 (2022).
68. W.A. Wan Yahaya, N.A.M. Azman, F. Adam, S.D. Subramaniam, K.H. Abd Hamid, M.P. Almajano, *Polymers*, **15(13)**, 2884 (2023)

Microplastics contamination in bivalves off the island in the strait of malacca and its potential health risks

Fuad Miskon[1*], *Intan Noor Munira* Ghazali[2], *Fikriah* Faudzi[2], *Fitri* Yusof[2], *Azaima* Razali[3], *Muhammad Zahir* Ramli[1], *Noor Artika* Hassan[4] and *Norhafezah* Kasmuri[5]

[1]Institute of Oceanography and Maritime Studies (INOCEM), International Islamic University Malaysia, 25200 Kuantan, Pahang, Malaysia
[2]Department of Marine Science, Kulliyyah of Science, International Islamic University Malaysia, 25200 Kuantan, Pahang, Malaysia
[3]Department of Chemistry, Kulliyyah of Science, International Islamic University Malaysia, 25200 Kuantan, Pahang, Malaysia
[4]Department of Community Medicine, Kulliyyah of Medicine, International Islamic University Malaysia, 25200 Kuantan, Pahang, Malaysia
[5]School of Civil Engineering, College of Engineering, Universiti Teknologi Mara, 40450 Shah Alam, Selangor, Malaysia

Abstract. The widespread presence of microplastics in the ocean is a significant threat to marine life and humans. A study was conducted to investigate the extent of microplastic contamination in the coastal waters of Langkawi and Penang, situated on the northern coast of Peninsular Malaysia. Rock oysters (*Saccostrea cucullata*) were utilized as bioindicators due to its availability in all sampling sites to evaluate microplastics,by considering its abundance, types, polymer composition, and potential health risks related to consumption. Soft tissues were digested with 10% KOH, and the resulting microplastics were examined using a stereo microscope and microplastics polymer were identified through ATR-FTIR. Kok Beach and Penarak Beach exhibited notably higher microplastic abundance, mainly in the form of filaments with predominant black and red colours. The most common polymer types were cellulose triacetate (CTA) and polycyclohexanedimethylene terephthalate (PCT). Hazard Quotient values, indicating potential health risks from consuming *S. cucullata*, surpassed a critical threshold at all locations. The study's findings suggest that it serves as a fundamental reference for future research on microplastic contamination in the islands along the northern coast of Peninsular Malaysia.

* Corresponding author : fuadm@iium.edu.my

1 Introduction

The persistent presence of microplastics in the environment can be attributed to the worldwide increase in plastic production [6,54,55,60]. Microplastics, which range in size from 0.001 mm to 5 mm [1,31,58], pose a threat to a variety of marine organisms due to their small dimensions. There is evidence indicating that microplastics have the ability to release contaminants, such as metals and additives, act as carriers for hydrophobic pollutants, and create new habitats for microorganism colonization [27,30,38]. Preliminary examinations of microplastics within organisms are crucial, as they establish a basis for comprehending how these particles are absorbed from the environment into living organisms [4,11,14-16,20,27,34,36,48,62]. Earlier assessments in Malaysia focused on microplastics in surface water, sediment, and organisms like fish, molluscs, and zooplankton. However, challenges arise in comparing data due to variations in locations and digestion methods [15,24,26,32,43,44,59]. Notably, previous studies in Langkawi and Penang concentrated solely on sea surface water, neglecting the pathway of microplastics into marine organisms [32]. Our research involved a comprehensive survey of microplastics in the coastal waters of Langkawi and Penang, with a specific focus on quantifying the diversity of microplastics in rock oysters (*Saccostrea cucullata*), which served as bioindicators. *S. cucullata* was readily available across all sampling sites and could attach to hard substrates. Consequently, it was present throughout all sampling sites. This encompassed evaluating microplastic abundance, shapes, colours, polymer composition, and assessing potential health risks associated with consuming contaminated rock oysters.

2 Methodology

2.1 Sample collection and laboratory preparation

Nine sites along the rocky shore were chosen for investigation (refer to Table 1), considering the presence of *S. cucullata* on rocks and the accessibility of the sites. These sites were distributed across 4 locations in Penang and 5 locations in Langkawi (see Figure 1). A total of 258 individual *S. cucullata*, each approximately the same size (± 4 cm), were randomly gathered using a chisel and hammer [2]. Only samples with intact shells were selected and placed in aluminum foil bags, then sealed in a vacuum container. To maintain freshness, the samples were preserved in ice during transport and stored at -20 °C in the laboratory until further analysis [47].

Table 1. Coordinates of sampling sites.

Site ID	Sampling Site	Coordinate	
		Latitude	Longitude
S1	Teluk Nangka, Miami Beach, Penang	5.4781 °N	100.2673 °E
S2	Pasir Panjang Beach, Penang	5.3006 °N	100.1842 °E
S3	Sungai Batu Beach, Teluk Kumbar, Penang	5.2791 °N	100.2404 °E
S4	Teluk Mat Inca, Teluk Bahang, Penang	5.4637 °N	100.2271 °E
S5	Pasir Tengkorak Beach, Langkawi, Kedah	6.4307 °N	99.7267 °E
S6	Tengah Beach, Langkawi, Kedah	6.2713 °N	99.7309 °E
S7	Penarak Beach, Langkawi, Kedah	6.3044 °N	99.8614 °E
S8	Pasir Hitam Beach, Langkawi, Kedah	6.4357 °N	99.7985 °E
S9	Kok Beach, Langkawi, Kedah	6.3657 °N	99.6754 °E

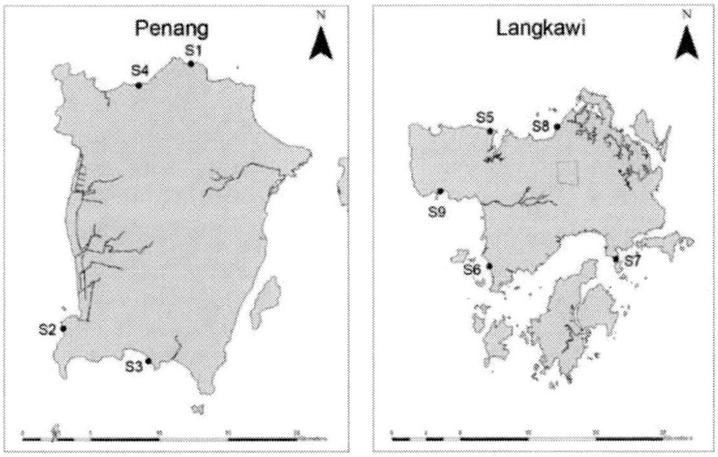

Fig. 1. Sampling sites along Penang and Langkawi.

At each rocky shore site, 30 samples were randomly organized into six groups, each containing five *S. cucullata* individuals for digestion. Within each batch, one procedural blank and one positive blank were included. After thawing at room temperature, soft tissues were separated from the shells using dissecting tools, and the samples were placed in a petri dish [7]. Following this, the samples were subjected to a 24-hour drying process in an oven set at 60 °C [37]. The wet and dry weights of the soft tissue were recorded, and the samples were then moved to a 250 mL Erlenmeyer flask containing a 10% KOH solution [3,4,35]. The flask, covered with aluminium foil, was placed in an oscillator shaker operating at 80 rpm and 60 °C for 24 hours [10]. Density separation of the digested solution was achieved by adding 50 mL of 50% KI for 8 hours [15,23]. The resulting solution underwent filtration using a 47 mm diameter Whatman GF/C filter paper disc with a 1.2 µm pore size and was subsequently dried at 60 °C for 24 hours until a constant dry weight was reached [35]. The filter paper discs were stored in petri dishes, and the dry weights of the filter paper discs before and after Fourier-transform infrared spectroscopy (FTIR) were recorded [40].

2.2 Physical Identification and Polymer Identification of Microplastics

The Whatman GF/C filter paper discs were examined under Nikon SMZ745 stereo microscope featuring a magnification range from 0.67x to 5x [4,7,17]. A microscope camera attached to the eyepiece of the stereo microscope was connected to a computer, and the ImageView software facilitated the capture of presumed microplastic images for size estimation. Validation of polymer characterization was conducted through ATR-FTIR [17]. Presumed microplastics were identified based on their shape, with recorded observations of shape, size, and colour aiding in subsequent polymer identification [40,60]. Particles marked during pre-assumed physical identification underwent polymer characterization using an ATR-FTIR (Perkin Elmer FTIR Spectrometer Frontier) equipped with a diamond crystal sensor and PerkinElmer Spectrum software, covering wavenumbers from 4000 cm-1 to 600 cm-1. The samples were kept in an oven at 60°C until analysis, and prior to library search, all particles underwent baseline correction using OriginLab 2021 to enhance spectrum quality. With a search score exceeding 0.65 and successful polymer characterization, 37 particles were identified as microplastic.

2.3 Quality assurance and quality control

To mitigate atmospheric microplastic contamination, a closed chamber was utilized, and all surfaces underwent acetone wiping before analysis. Glassware and equipment were thoroughly washed with filtered distilled water and shielded with aluminum foil [8,29,38,47]. Chemical solutions used in the experiment were dissolved in filtered distilled water and further filtered using Whatman GF/C filter paper [56]. Each procedural step was accompanied by both negative and positive blanks to track contamination throughout the procedures [42]. No damaged to microplastics in positive blank, and no microplastics detected from negative blank.

2.4 Calculation of Hazard Quotient

The hazard quotient (HQ) serves as a ratio to assess risk in non-carcinogenic scenarios, helping estimate the significance of a particular risk [52]. An HQ below 1 is considered acceptable, while an HQ exceeding 1 indicates an unacceptable condition. The formula used to calculate HQ is derived from the study conducted by Sharif et al. [45].

$$HQ = \frac{EDI}{RfD} \qquad (1)$$

Here, EDI (mg/kg/day) represents the intake of the specific microplastic polymer through shellfish consumption, and RfD (mg/kg/day) stands for the reference dose, which is the estimated safe limit for microplastic polymer intake.

3 Result and discussion

3.1 Biometric data

Table 2 displays the biometric data, including mean and standard deviation, for *S. cucullata* across all sites, covering shell length, shell width, wet weight, and dry weight of soft tissue. To ensure data comparability between sites, *S. cucullata* specimens with similar shell sizes were collected during sampling [57]. Allometric data for bivalve species are recognized to correlate with factors like substratum nature, salinity, temperature, and other chemical characteristics of the water, influencing dimensional relationships [41]. Samples obtained from areas near water sources discharged from land, such as estuaries or riverine zones, exhibited higher wet weight yields. This trend was observed in Teluk Nangka (S1), Sg Batu (S3), and Teluk Mat Inca (S4). In contrast, locations with low yields of wet tissues were observed in isolated areas like Pasir Panjang (S2), Pasir Tengkorak (S5), and Pasir Hitam (S8). Penarak (S7) recorded the lowest wet tissue yield due to a densely populated rock oyster environment attached to the substrate. This led to a competitive environment, resulting in oysters stacking on top of one another and, consequently, reducing the oyster size and soft tissue yield.

Table 2. Biometric data of sample collected in each site.

Station	Shell length (cm)	Shell width (cm)	Wet weight (g)	Dry weight (g)
S1	5.2±0.5	4.6±1.1	2.0885±0.6023	0.7305±0.2872
S2	3.9±0.7	3.7±0.8	1.3349±0.5996	0.5170±0.2408
S3	5.2±0.6	4.9±0.8	3.0500±0.8664	1.1133±0.5720
S4	5.4±0.7	4.8±0.7	2.2390±0.6359	0.8984±0.4844

S5	4.9±0.7	4.1±0.6	1.1942±0.4000	0.3658±0.0845
S6	5.0±0.6	4.2±0.8	1.5890±0.4782	0.6317±0.3122
S7	4.5±0.9	3.6±0.7	0.9929±0.4516	0.4452±0.2496
S8	4.4±0.7	3.7±0.5	1.0947±0.2798	0.3478±0.0959
S9	4.6±0.4	3.9±0.4	1.3707±0.2418	0.4717±0.1586

3.2 Microplastic abundance

The negative blank and positive blank utilized in the experiment underwent filtration using filter paper. The negative blank showed no presence of microplastics on the filter paper, while the positive blank demonstrated that the microplastics remained intact throughout the experiment. These observations confirm the absence of cross-contamination and deterioration during the experimental procedures [35]. In terms of microplastic abundance in S. cucullata, Penarak (S7) exhibited the highest concentration with approximately 0.40 item per wet weight of soft tissues, and Kok Beach (S9) recorded the highest abundance with 0.43 item per individual (see Figure 2). This result aligns with previous studies where microplastic contamination in bivalves ranged from 0.1 to 20 microplastics per gram of wet tissue [12]. The selected sampling sites displayed varying levels of microplastic pollution, with Penarak and Kok Beach experiencing heavy pollution due to fisheries equipment and boating activities, respectively. Our findings indicated that the microplastic concentrations in both sites exceeded those found in bivalves from Vietnam [35] and Europe [37]. However, differences in results could be attributed to the distinct protocols used by Phuong et al. [36]. The identification technique played a crucial role, as visual observation alone was insufficient for characterizing microplastics. Advanced methods such as Raman spectroscopy, capable of detecting samples as small as 1 µm, and µFTIR, capable of identifying microplastics ranging from 3 to 800 µm depending on the working mode, are recommended [35]. Conversely, Pasir Panjang (S2) and Pasir Hitam (S8) exhibited the lowest microplastic abundance, aligning with the biometric data, where both locations had low yields of wet weight tissues.

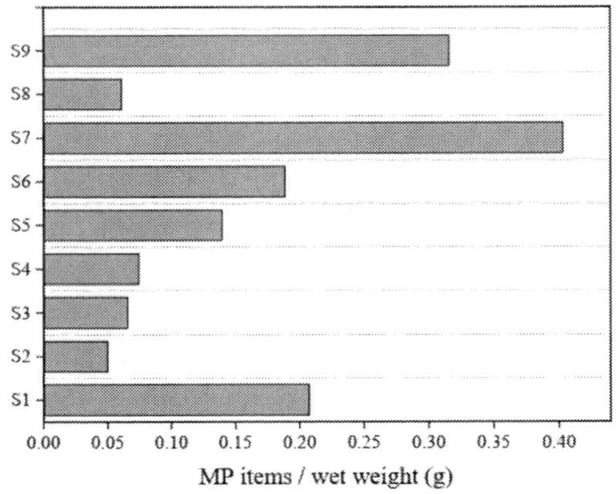

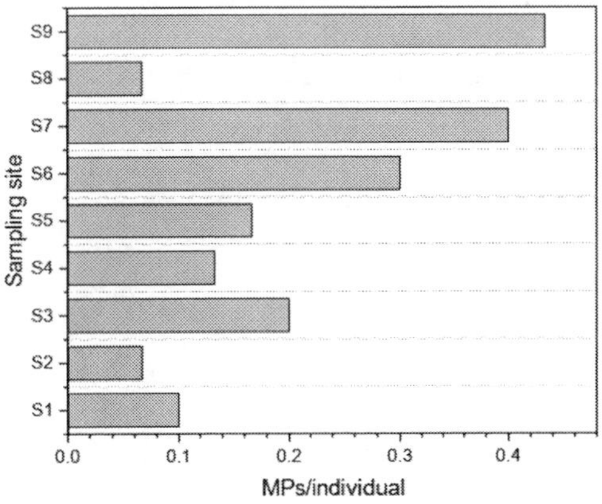

Fig. 2. Abundance of microplastics per wet weight according to sites (left) and microplastics per individual of *S. cucullata* according to sites (right).

3.3 Physical characteristics of microplastics

The predominant shape of microplastics found was filaments with a black colour, followed by fragments exhibiting a red colour (see Figure 3). However, the sources of microplastics based on colour have not been determined, as there were no additional sediment and seawater samples collected for establishing a correlation between the colour of microplastics in soft tissues and the sources of pollution. In a previous microplastic study in the coastal waters of Penang and Langkawi, fragments, film, granules, filaments, and foam were commonly identified in surface water using visual sorting identification [15,32]. Microplastic beads were detected in the soft tissues at sites in Langkawi, a finding not consistent with other studies that did not detect microplastic beads except in fish samples from Skudai, Johor [43], and Northwest Peninsular Malaysia [15]. The distribution of microplastic shapes and colours is depicted in Figure 3.

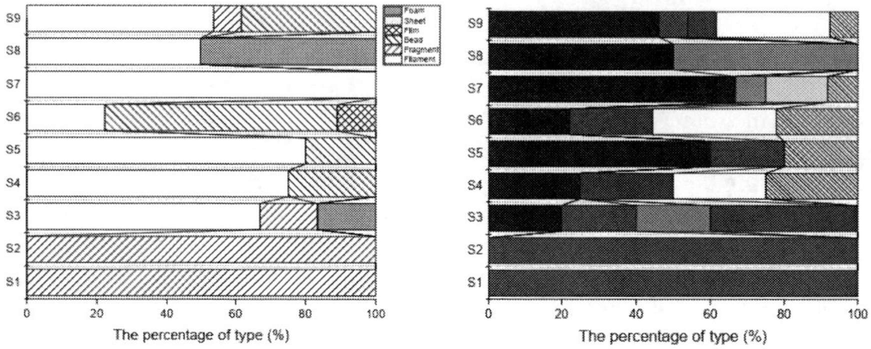

Fig. 3. Distribution of shapes of microplastics (left) and colour of microplastics (right) found according to sites.

The sizes of the identified microplastics ranged from 167.93 µm to 11,268 µm, with most samples falling around 5,000 µm (see Figure 4). Microplastic size is crucial, as plastics of different sizes interact with various levels in the food pyramid [11]. Microplastics can interact at the foundational level of the food pyramid, potentially leading to bioaccumulation or biomagnification [25]. Plastics, irrespective of size, can interact with toxic pollutants such as plasticizers, polybrominated diphenyl ether (PBDE), heavy metals, and polychlorinated biphenyls (PCBs), inducing toxicity, especially at the cellular level [19,21,57]. Moreover, plastics with a size less than 1000 µm can penetrate cell barriers and be transported to other organs through a process known as adherence, ultimately causing oxidative stress and an inflammatory response by cells [12,27].

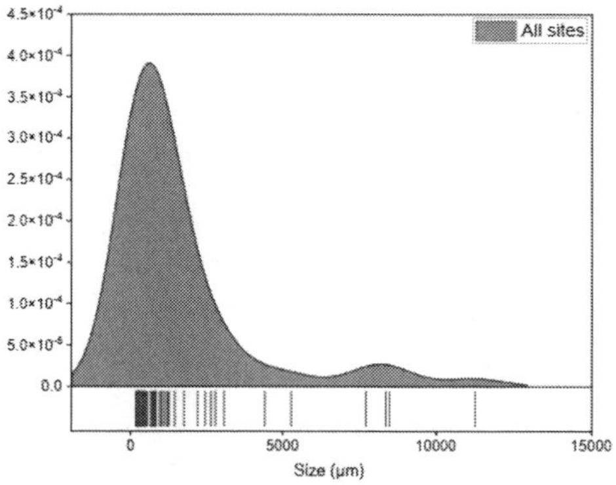

Fig. 4. Distribution of microplastic counts versus size and its kernel density estimation for microplastic in *S. cucullata* for all sampling sites.

3.4 Polymer distribution and chemical composition

A total of 36 microplastic items were successfully identified based on their chemical composition, revealing the presence of six distinct polymers. The most prevalent microplastics were cellulose triacetate (CTA) and polycyclohexanedimethylene terephthalate (PCT) (see Figure 5). The composition of these polymers can serve as a valuable tool for tracing the origin of the plastics. For instance, CTA can originate from natural sources or may be used in the production of cigarette filters, textile fibers, photographic film, surface coatings, and membranes for various separation processes [13]. Polyester polymers like PCT indicate diverse activities in the area, potentially linked to single-use plastic items [49]. Sites with elevated microplastic abundance in both individual and wet tissues, specifically Penarak (S7) and Kok Beach (S9), exhibited the highest polymer identification. The heavy pollution from boat activities in these sites justifies the diverse array of polymers found in *S. cucullata*.

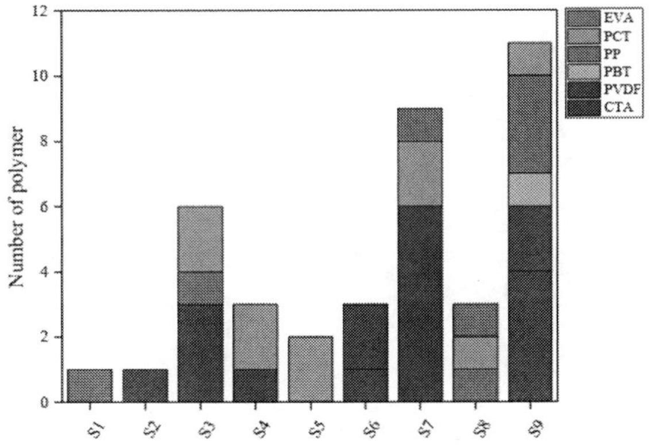

Fig. 5. Distribution of the number of polymers according to sites for microplastic found in *S. cucullata*.

The FTIR spectra of the polymers identified in this study are illustrated in Figure 6. For PVDF, distinctive peaks appear at wavenumbers of 2924 cm-1 (CH2 stretching), 1447 cm-1 (CH2 deformation), 1040 cm-1 (strong C-F stretching), and 855 cm-1 (strong CH bending) [33]. CTA displays peaks at wavenumbers of 3337 cm-1 (-OH stretching of unacetylated cellulose), 2972 cm-1 (C-H stretching of CH2 or CH3), 1738 cm-1 (C=O stretching of the acetyl group), 1435 cm-1 (C-C stretching), 1369 cm-1 (C-C-O), 1218 cm-1 (C-O stretching of the acetyl group), and 1027 cm-1 (C-O-C of the cellulose backbone) [13]. PBT exhibits peaks at wavenumbers of 2965 cm-1 (medium C-H stretching of alkane compound class), 1715 cm-1 (strong C=O stretching of carboxylic acid compound class), 1245 cm-1 (strong C-O stretching of alkyl aryl ether compound class), 1094 cm-1 (strong C-O stretching of aliphatic ether), and 726 cm-1 (aromatic C-H out-of-plane bend) [5]. PP presents peaks at wavenumbers of 2918 cm-1 (C-H stretch), 1457 cm-1 (CH2 bend), 1375 cm-1 (CH3 bend), 1169 cm-1 (CH bend, CH3 rock, C-C stretch), 998 cm-1 (CH3 rock, CH3 bend, CH bend), 975 cm-1 (CH3 rock, C-C stretch), and 843 cm-1 (CH2 rock, C-CH3 stretch) [22]. PCT demonstrates peaks at wavenumbers of 2918 cm-1 (medium C-H stretching of alkane compound class), 1722 cm-1 (strong C=O stretching of carboxylic acid compound class), 1562 cm-1 (medium C=C stretching of cyclic alkene), 1451 cm-1 (C-C stretching), 1167 cm-1 (strong C-O stretching of ester polymer class), and 719 cm-1 (strong C-H bending) [5]. EVA shows peaks at wavenumbers of 2918 cm-1 (C-H stretching), 1736 cm-1 (C=O stretching), 1468 cm-1 (CH2 bending, CH3 bending), 1238 cm-1 (C(=O)-O stretching), and 1018 cm-1 (C-O stretching) [5].

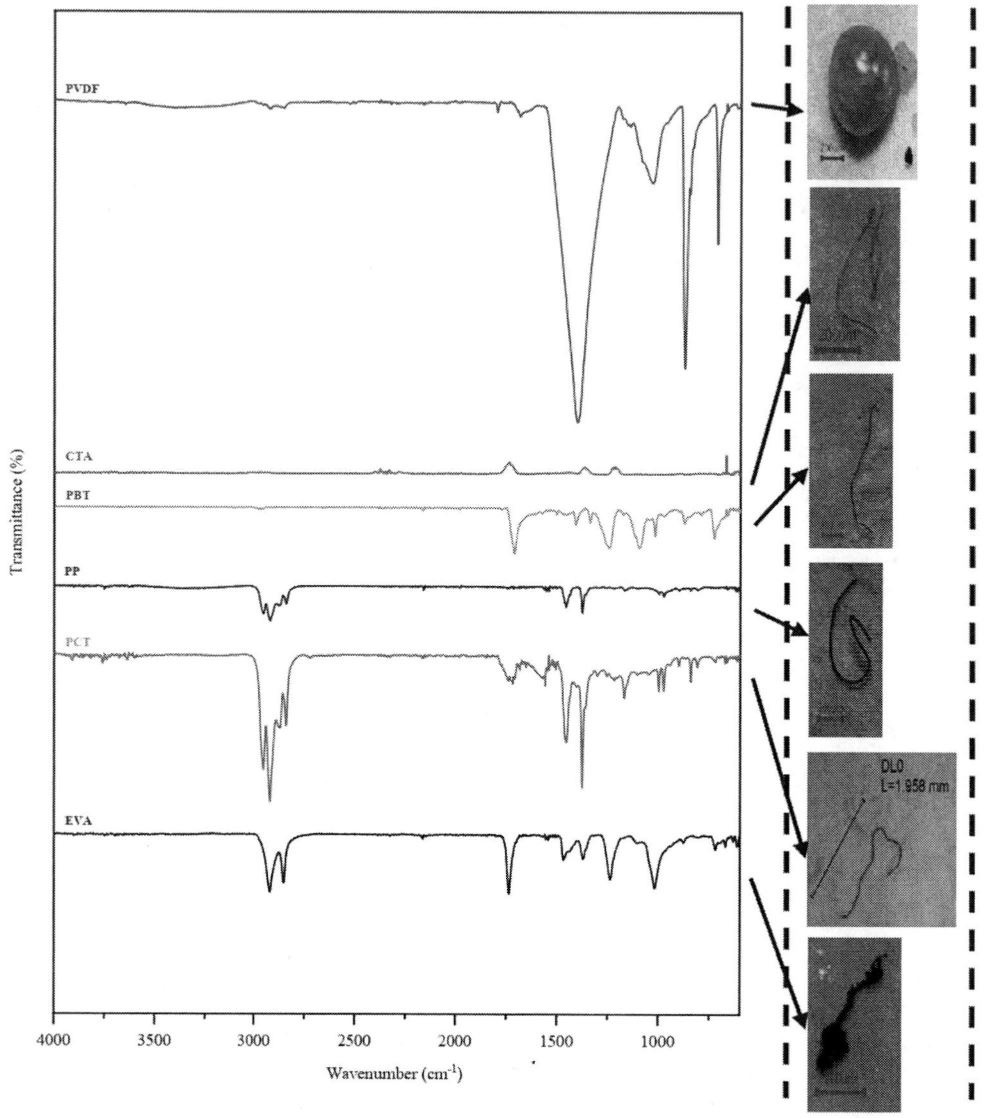

Fig. 6. Compilation of FTIR spectra for microplastic polymers found in Langkawi and Penang.

3.5 Health risk analysis

Exposure to microplastics through shellfish consumption raises potential concerns for human health. Given that the Reference Dose (RfD) value for the polymer remains undetermined, we employed the RfD value of the monomer for the most abundant microplastic found at each site. The methodology, including the calculation for Estimated Daily Intake (EDI) and Hazard Quotient (HQ), has been detailed in the methodology section. Continuous consumption of contaminated shellfish could potentially lead to hepatic and urinary diseases [52,63]. It's crucial to note that the study solely focused on the impact of microplastics and did not consider potential toxicity from other sources. All evaluated sites displayed a Hazard Quotient exceeding 1 (refer to Table 3), indicating a potential risk in consuming oysters from these sites. However, sites without an RfD value for the monomer were labeled as "Not Evaluated" (NE). Table 3 provides information on the dominant pollutant found at each site based on polymer identification in subsection 3.4, along with its Estimated Daily Intake (EDI) projection and Reference Dose (RfD). The Hazard Quotient was calculated for sites with an available RfD monomer value, and any value exceeding 1 predicts potential health effects such as hepatic and urinary issues if consumption continues according to the estimated daily intake (EDI) projection.

Table 3. Hazard quotient (HQ) calculations for all sampling sites.

Site	Pollutant	EDI	RfD	HQ	Health Effects
S1	EVA	0.023	NE	NE	NE
S2	CTA	0.318	0.03	10.611	Hepatic, urinary [50]
S3	CTA	0.116	0.03	3.857	Hepatic, urinary [50]
S4	PCT	0.261	0.10	2.608	Urinary [51]
S5	PCT	1.367	0.10	13.667	Urinary [51]
S6	PVDF	1.679	NE	NE	NE
S7	CTA	4.545	0.03	151.503	Hepatic, urinary [50]
S8	CTA	4.633	0.03	154.437	Hepatic, urinary [50]
S9	CTA	1.307	0.03	43.552	Hepatic, urinary [50]
*EDI(mg/kg/day); RfD(mg/kg/day); NE: Not evaluated, as the value of RfD for the monomer was not available *All the health effects were taken from the IRIS US EPA website from the previous research					

4 Conclusion

This study offers a comprehensive overview of the physical and polymer characteristics of microplastics, along with potential health effects that may arise from the consumption of *S. cucullata* contaminated with microplastics (see Table 4). The gathered data establishes a baseline for the investigation of microplastic contamination in bivalves. This information holds valuable implications for informing public policy and can serve as a predictive tool for the Malaysian fisheries industry in the future.

Table 4. Summary of microplastic data for all sampling sites

Site	Item / w.w.	Item / individual	Shape	Colour	Size range (µm)	Polymer found	HRA
S1	0.2075	0.1000	Fragment	Red	167-1057	EVA	>1
S2	0.0499	0.0667	Fragment	Red	198-423	CTA	>1
S3	0.0656	0.2000	Filament, fragment, foam	Black, orange, brown	270-8485	CTA, PCT, PP	>1
S4	0.0744	0.1333	Filament, bead	Black, brown, white, colorless	164-2608	PCT, CTA	>1
S5	0.1396	0.1667	Filament, bead	Black, brown, colorless	295-1727	PCT	>1
S6	0.1888	0.3000	Filament, bead, film	Black, brown, white, colorless	226-1428	CTA/PVDF, PVDF	>1

S7	0.4029	0.4000	Filament	Black, orange, yellow, colorless	227-2765	CTA, EVA, PCT	>1
S8	0.0609	0.0667	Filament, foam	Black, grey	747-11268	EVA, PP, PCT	>1
S9	0.3161	0.4333	Filament. fragment, bead	Black, red, brown, white, colorless	291-7702	PBT, CTA/PVDF, CTA/PVT, CTA, PP	>1

Acknowledgements

The study presented in the manuscript was fully funded by the Fundamental Research Grant (FRGS19-057-1744/FRGS/1/2016/STG01/UIAM/03/4), Ministry of Higher Education (MOHE), Malaysian Government.

References

1. N. Ajith, S. Arumugam, S. Parthasaranthy, S. Manunupoori & S. Janakiraman Env. Sci. Poll. Res. **27**, 25970 (2020)
2. A. Arkhipkin, E. Boucher, M. Gras, & P. Brickle, J. Mar. Bio. Assoc. UK, ***97***, 735 (2017)
3. B. R. Baechler, E. F. Granek, M. Hunter & K. E. Conn, Lmnlgy Ocn. Let., **5**, 54 (2020)
4. I. L. N. Bråte, R. Hurley, K. Iversen, J. Beyer, K. Thomas, C. C. Steindal, N. W. Green, M. Olsen, & A. Lusher, Env. Poll., **243**, 383 (2018)
5. Z. Chen, J. N. Hay & M. J. Jenkins, Thermochimica Acta **552**, 123 (2013)
6. Y. Cho, W. J. Shim, M. Jang, G. M. Han & S.H. Hong, Env. Poll., **245**, 1107 (2019)
7. G. A, Covernton, B. Collicutt, H. J. Gurney-Smith, C. M. Pearce, J. F. Dower, P. S. Ross & S. E, Dudas, Aqua. Env. Inter., **11**, 357 (2019)
8. K. Davidson & S. E, Dudas, Arc. of Env. Cont. Tox., **71**, 147 (2016)
9. B. De Witte, L. Devriese, K. Bekaert, S. Hoffman, G. Vandermeersch, K. Cooreman & J. Robbens, Mar. Poll. Bul., **85**, 146 (2014)
10. A. Dehaut, A. L. Cassone, L. Frère, L. Hermabessiere, C. Himber, E. Rinnert, G. Rivière, C. Lambert, P. Soudant, A. Huvet, G. Duflos & I. Paul-Pont, Env. Poll., **215**, 223 (2016)
11. N. Digka, C. Tsangaris, M. Torre, A. Anastasopoulou & C. Zeri, Mar. Poll. Bul., **135**, 30 (2018)
12. J. Ding, J. Li, C. Sun, F. Jiang, C. He, M. Zhang, P. Ju & N. X, Ding, Sci. Tot. Env., 739 (2020)
13. P. Fei, L. Liao, B. Cheng & J. Song, Analy. Mtds, **9**, (2017)
14. E. M. Foekema, C. De Gruijter, M. T. Mergia, J. A. Van Franeker, A. J. Murk & A. A, Koelmans, Env. Sci. Tech., **47**, 8818 (2013)
15. Y. H. Foo, S. Ratnam, E. V. Lim, M. Abdullah, V. J. Molenaar, A. T. S. Hwai, S. Zhang, H. Li & N. B. M. Zanuri, PeerJ., **10** (2022)
16. M. Hamaguchi, H. Shimabukuro, H. Usuki & M. Hori, Mar. Bio. Rec., **7** (2014)
17. Y. S. Ibrahim, A. A. Azmi, S. Abdul Shukor, S. Tuan Anuar & S. A. Abdullah, Middle-East J. Sci. Res., **24**, 2129 (2016)
18. Y. S. Ibrahim, R. Rathnam, S. T. Anuar & W. M. W. M Khalik, Malay. J. of Analy. Sci., **21**, 1054 (2017)
19. Y. S. Ibrahim, S. Tuan Anuar, A. A. Azmi, W. M. A. Wan Mohd Khalik, S. Lehata, S. R. Hamzah, D. Ismail, Z. F. Ma, A. Dzulkarnaen, Z. Zakaria, N. Mustaffa, S. E. Tuan Sharif & Y. Y. Lee, J. Gastroent. Hep., **5**, 116 (2021)

20. K. B. Iversen (2018)
21. N. Jaafar, A. Azfaralariff, S. M. Musa, M. Mohamed, A. H. Yusoff & A. M. Lazim, Sci. Tot. Env., 799 (2021)
22. M. R. Jung, F. D. Horgen, S. V. Orski, C. V. Rodriguez K. L. Beers, G. H. Balazs, T. T. Jones, T. M. Work, K. C. Brignac, S. J. Royer, K. D. Hyrenbach, B. A. Jensen & J. M. Lynch, Mar. Poll. Bull., **127**, 704 (2018)
23. A. Karami, A. Golieskardi, Y. Ho, V. Larat & B. Salamatinia, Sci. Rep., **7** (2017)
24. S. Karbalaei, A. Golieskardi, H. Hamzah, S. Abdulwahid, P. Hanachi, T. R. Walker & A. Karami, Mar. Poll. Bul., **148**, 5 (2019)
25. P. J. Kershaw & C. M. Rochman (2016)
26. W. M. A. W. M Khalik, Y. S. Ibrahim, S. Tuan Anuar, S. Govindasamy & N. F. Baharuddin, Mar. Poll. Bul., **135**, 451 (2018)
27. P. Kolandhasamy, L. Su, J. Li, X. Qu, K. Jabeen & H. Shi, Sci. Tot. Env., **610–611**, 635 (2018)
28. L. Krishnakumari, V. R. Nair & K. Govindan, J. Ind. Fis. Assoc., **20**, 21 (1990)
29. J. Li, D. Yang, L. Li, K. Jabeen & H. Shi, Env. Poll., **207**, 190 (2015)
30. Y. Liu, K. Zhang, S. Xu, M. Yan, D. Tao, L. Chen, Y. Wei, C. Wu, G. Liu & P. K. S. Lam, Gon. Res. (2021)
31. F. Murphy, M. Russell, C. Ewins & B. Quinn, Mar. Poll. Bul., **122**, 353 (2017)
32. M. Najihah, M. S. Ismail, C. K. Yap & K. K. Ku Yaacob, J. Bsc. & App. Sci., **16**, 50 (2020)
33. P. Nallasamy & S. Mohan, Ind. J. P. & App. Phys., **43**, 821 (2005)
34. K. L. Ng & J. P. Obbard, Mar. Poll. Bul., **52**, 761 (2006)
35. N. N. Phuong, Q. T. Pham, T. T. Duong, T. P. Q. Le & A. Frederic, Viet. J. Ear. Sci., **41**, 252 (2019)
36. N. N. Phuong, L. Poirier, Q. T. Pham, F. Lagarde & A. Zalouk-Vergnoux, Mar. Poll. Bul., **129**, 664 (2018).
37. N. N. Phuong, A. Zalouk-Vergnoux, A. Kamari, C. Mouneyrac, F. Amiard, L. Poirier & F. Lagarde, Env. Sci. Poll. Res., **25**, 6135 (2018)
38. X. Qu, L. Su, H. Li, M. Liang & H. Shi, Sci. Ttl. Env., **621**, 679 (2018)
39. A. Razali, M. F. Miskon & S. N. Che Nan, Irn. J. Chem. Chem. Eng., (2022)
40. M. Renzi, C. Guerranti & A. Blašković, Mar. Poll. Bul., **131**, 248 (2018)
41. H. U. Riisgård, P. S. Larsen & D. Pleissner, Helgo. Mar. Res., **68**, 193 (2014)
42. M. F. M. Santana, L. G. Ascer, M. R. Custódio, F. T. Moreira & A. Turra, Mar. Poll. Bul., **106**, 183 (2016)
43. S. Sarijan, S. Azman & M. I. Mohd Said, *Microplastics Pollution in Skudai and Tebrau River, Malaysia* in Proceedings of 7th International Graduate Conference of Engineering, Science and Humanities. Universiti Teknologi Malaysia (UTM), (2018)
44. S. Sarijan, S. Azman, M. I. Mohd Said & M. H. Lee, Env. Asia, **12**, 75 (2019)
45. R. Sharif, E. Chong & C. K. Meng, Malay. J. Nut., **22**, 301 (2016)
46. Y. T. Singh, J. Mar. Bio. Assoc. UK, **99**, 385 (2019)
47. L. Su, H. Cai, P. Kolandhasamy, C. Wu, C. M. Rochman & H. Shi, Env. Poll., **234**, 347 (2018)
48. Z. D Taha, R. Md Amin, S. Tuan Anuar, A. A. Abdul Nasser & E. S. Sohaimi, Sci. Ttl. Env., 786 (2021)
49. R. C Thompson, Y. Olson, R. P. Mitchell, A. Davis, S. J. Rowland, A. W. G. John, D. McGonigle & A. E. Russell, Sci., 304, (2004)
50. US EPA, *CAS No. 123-91-1* in Chemical Assessment Summary, (2000)

51. US EPA, *CASRN 120-61-6* in Chemical Assessment Summary; US Env. Pr. Ag., (2002)
52. US EPA, *Human Health Benchmarks* (2018)
53. L. Van Cauwenberghe, M. Claessens, M. B, Vandegehuchte & C. R. Janssen, Env. Poll., **199**, 10 (2015)
54. J. C, Vermaire, C. Pomeroy, S. M. Herczegh, O. Haggart & M. Murphy, FACETS, **2**, 301 (2017)
55. T. Wang, B. Li & D. Wang, Env. Sci. Poll. Res., **28**, 60753 (2021).
56. J. E. Ward, S. Zhao, B. A. Holohan, K. M. Mladinich, T. W. Griffin, J. Wozniak & S. E. Shumway, Env. Sci. Tech., **53**, 8776 (2019)
57. S. Webb, H. Ruffell, I. Marsden, O. Pantos & S. Gaw, Mar. Poll. Bul., **149** (2019)
58. S. L. Wright, R. C. Thompson & T. S. Galloway, Env. Poll., **178**, 483 (2013)
59. T. Y. Hwi, Y. S. Ibrahim & W. M. A. W. K. Khalik, Sains Malaysiana, **49**, 1479 (2020)
60. N. Z. Zahari, P. M. Tuah, M. R. Junaidi & S. A. M. Ali, Water (Switzerland), **14** (2022)
61. A. H. Zainuddin, A. Z. Aris, M. R. M. Zaki, F. M. Yusoff & S. Y. Wee, Mar. Poll. Bul., **174** (2022)
62. J. Zhu, X. Yu, Q. Zhang, Y. Li, S. Tan, D. Li, Z. Yang & J. Wang, Sci. Ttl. Env., **659**, 649 (2019).
63. S.A. Akbar, S. Afriani, C. Nuzlia, S. Nazlia, S. Agustina. (2023). *Depik* **12(3)** 259-273 (2023)

Characterization of Sodium Alginate-Based Edible Film with Addition of Gelatin and Casein

Aef Permadi*[1], *Nur* Hidayah[1], *Wenny* Imelda Simamora[1], *Christianto* Imanuel[2], *Gilina* Wahyu Endah[2], *Siti* Zachro Nurbani[1], *Mohammad* Sayuti[1], *Niken* Dharmayanti[1], and *Rufnia* Ayu Afifah[1]

[1]Fish Processing Technology, Jakarta Technical University of Fisheries, Jakarta.
[2]CV. Sari Mutiara Agar, Malang, Jawa Timur.

Abstract. Edible film is a form of biodegradable packaging that is easily biodegradable and can be consumed at the same time with the product. The purpose of this study is to find characterization of edible films including physical tests (thickness, tensile strength, elongation, elasticity and high temperature resistance), chemical tests (water content and water vapor transmission rate), solubility tests and biodegradability tests on sodium alginate/glicerol-based edible films with the addition of gelatin and casein. This study was conducted using the single factor group random design method with data analysis using SPSS 29 software. The results showed that edible films made from only sodium alginate and glycerol only have advantages in characteristics of elasticity, edible films made from sodium alginate, glycerol and gelatin have advantages in characteristics of elongation, water content, solubility and biodegradability. Edible films made from sodium alginate, glycerol and casein have advantages in characteristics of consumer preferences, thickness, tensile strength, high temperature resistance and water vapor transmission rate.

1 Introduction

In the last two decades, plastic packaging has captured the world's packaging market share replacing cans and glass packaging. Plastic-based goods are synthetic polymer materials that are difficult to degrade in nature. The increase in the use of plastic-based goods is directly proportional to the plastic waste produced, which ultimately leads to the breakdown of the balance of nature [1]. Typically, edible packaging provides a consumable coating or wrapping made of sustainable, biodegradable material around the food, which leads to no waste being produced.

One of the packaging that can be used as an alternative is biodegradable packaging, which is packaging that is easily decomposed by the activity of microorganisms after being used and disposed of into the environment. In addition, biodegradable packaging is able to ensure product safety [2]. Packaging applications with edible materials can be done by coating the product directly with edible solutions that are made first [3]. Edible film is used to maintain the shelf life and quality of food products because it can inhibit the migration of water, oxygen, carbon dioxide, flavor and fat [4]. Edible film is a thin layer that has a function as a packaging material or coating for food ingredients that can also be eaten together with packaged products, and can be degraded by nature biologically [5].

One of the raw materials used in making edible film is sodium alginate. The function of *edible film* on sodium alginate in addition to protecting food products is also the original

* Corresponding author: aef.permadi@politeknikaup.ac.id

appearance of the product can be maintained. Sodium alginate compounds have the potential to form biopolymer film or coating components because sodium alginate has a unique colloidal structure, as a stabilizer, binder, suspension, film forming, gelling and emulsion stability [6]. However, *sodium* alginate-based edibles have weaknesses where their resistance to water is low, has properties that are easily oxidized and barrier properties to water vapor are also low so that it can affect the quality of the film produced and this deficiency will affect the shelf life of food to be less optimal [7]. For that, it is necessary to add other protein ingredients such as gelatin and casein.

Gelatin is a substance obtained from the extraction of cartilage collagen or animal skin such as cows, fish and pigs [8]. In applications in edible film, gelatin has the expected properties in its ability to form films and its high barrier properties. In addition, gelatin-based films can be used as carriers of bioactive additives so that they can achieve active packaging functions and are able to protect food from oxidation [9]. Gelatin is one of the edible film forming materials that has transparent, strong, flexible and able to inhibit the transfer of oxygen and carbon dioxide and film from skin gelatin material has properties that are not compact and sticky, therefore it is necessary to add ingredients that can improve the physical quality of edible film, namely casein protein [10]. Edible film from casein is a technology that makes for food preservation and can be used as an alternative to synthetic polymers. Films and layers made of casein are easily formed due to their exposed secondary structure [11]. The film also contains a barrier against water vapor, oxygen and carbon. The purpose of the study was to find characteristic of sodium alginate/glycerol-based edible films with the addition of gelatin and casein. which is resistant to heat during the processing of sausage products, has biodegradable properties, consumers acceptance, low of water vapor transmission rate, good in elongation, solubility and high tensile strength.

2 Materials and Method

2.1 Materials

The materials used in this study were food grade sodium alginate (Ocean Health), food grade casein (International Healthcare), food grade and halal gelatin (Gelita), food grade glycerol (Wilmar), aquades, food grade silica gel(natural from Zeloid quarry rock), water and soil medium. The equipmentss used in this study were beaker glass (pyrex), measuring cup (pyrex), 1 mL drip pipette (OneMed), hotplate stirrer (HMS-79), petri dish (OneMed), porcelain cup (pyrex), spatula (stainless steel), glass jar, silicon mold 15 x 15 cm, analytical balance accuracy 0.01 gram (Jadever), dehydrator (DHY-D5A), oven (Kirin), aluminum foil, thermometer rod, 0.01 mm accuracy micrometer, distillation tool, gas stove, plastic container, boiling pan (stainless steel), gas stove, Universal Testing Machine and moisture analyzer (Radwag).

2.2 Preparation of Edible film

Edible film made consists of three treatments or formulas, namely: 1) sodium alginate with added glycerol and aquades; 2) sodium alginate with added glycerol, aquades and gelatin; 3) sodium alginate with added glycerol, aquades and casein. The preparation of edible film is carried out by referring to [12], which has been modified, namely sodium alginate weighed as much as 3 g, and put into a 250 mL erlenmeyer with 100 mL aquades added, then added 10% glycerol (b/w) then stirred with a magnetic stirrer and heated with a *hotplate* at 70°C for 15 minutes. After that, the edible solution of film condition is poured on a glass mold measuring 15*15 cm and dried using a 50°C oven for 24 hours. For the the second treatment/formula an addition of 6 grams of gelatin was added and the third treatment/formula added 7 grams of casein. Each treatment/formula is made three times.

2.3 Consumer Preferences

Consumer preference testing of sodium alginate-based edible films was carried out by testing favorability using 30 panelists. Panelists were asked to rate the appearance, texture and aroma of edible film based on a hedonic score sheet on a scale of 1-9 with scale 1 very disliked, scale 5 neutral and scale 9 very fond. Hedonic testing is carried out on each edible film making treatment.

2.4 Characterization of edible films

2.4.1 Thickness of edible film

The thickness test refers to [13]. The film solution is poured with a film mold silicon plate with a size of 15 x 15 cm and flattened with a film leveling silicon plate until it forms a thin and flat sheet of film. Then allowed to dry and put in the oven at 50°C for 24 hours. After being lifted from the oven, the edible film is removed from the mold and placed on a plastic mat and placed in a kelium plastic bag [14]. For silicon plates, square plates with sides measuring 15 cm and 5 cm high are used. Thickness is measured using a digital micrometer (accuracy 0.001 mm) measured at 5 points thus representing the sample and improving accuracy.

2.4.2 Tensile strength

Tensile strength testing was carried out referring to [15]. Edible film is cut according to standard provisions. Both ends of the sample are clamped to the tensile testing machine found in the universal testing machine. The machine is set with a starting distance between the clamps of 10 mm and a speed of 10 mm/min. Next, the initial length is recorded and the ink tip of the logger is placed at position 0 in the graph. The start button is pressed and the tool will pull the sample until it breaks and the tensile strength (Fmax) is recorded by measuring the amount of tensile force required until a sample is disconnected. Then the tensile strength will be obtained by the following equation:

$$\gamma = \frac{Fmax}{A} \times 100\ \%$$

2.4.3 Elongation

The elongation test refers to [15], which is based on the lengthening of the film when the film breaks. The percentage of elongation is measured using the Universal Testing Machine. The measurement of break elongation is carried out in the same way as tensile strength testing. The extension is calculated by the following equation: :

$$Elongation\ (\%) = \frac{strain\ at\ breaks\ length\ (mm)}{initial\ length\ (mm)} \times 100\%$$

2.4.4 Elasticity

Elasticity is a measure of the flexibility of an object or the ability of objects to return to their original shape after being subjected to attractive or compressive forces. Formulated elasticity through a comparison of tensile strength and elongation further expressed in Mega Pascal [16], formulated by the equation:

$$Elasticity(MPa) = \frac{tensile\ strength}{elongation}$$

2.4.5 High temperatures resistance

The high temperatures resistance test is carried out by making a film measuring 5 x 5 cm which will be inserted into a waterbath with a temperature adjusted to the cooking temperature of the sausage, which is 80^{0}C for 40 minutes [17]. Then periodic checks are carried out every five minutes on the film by paying attention to the solubility of the film to water decoction and the resistance of the film to the temperature of heating water.

2.4.6 Water content

The way this method works, is that an empty cup is heated in an oven at 105^{0}C for 30 minutes for drying cup, then cooled in a desiccator for 15 minutes and weighed (W0). Then a sample of 2 grams is put in a dish that has known weight, weighed (W1), then dried in the oven at 105^{0}C for 3 hours to evaporate water inside sample and cooled in a desiccator for 15-30 minutes. Next, the cup and its contents are weighed and dried again for 1 hour, and cooled in a desiccator and finally weighed again (W2).

$$Water\ content\ (\%) = \frac{W1 - W2}{W1 - W0} \times 100$$

2.4.7 Water vapor transmission rate (WVTR)

Water vapor transmission rate testing (ASTM E96-01, 1997), *edible film* cut with a diameter of 4.5 cm (adjusting to the diameter of a beaker glass of 50 mL as a water vapor transmission test medium). The film is placed on top of the beaker glass until it covers the surface of the beaker glass that has been filled with 2 grams of silica gel (0% relative humidity), then tightened with a rubber band. The cup is weighed, then placed into a goblet glass containing aquades (relative humidity 100%, temperature 28^{0}C). The cup is weighed at intervals of 24 hours.

2.4.8 Solubility

Solubility testing of edible film is carried out by cutting edible film and weighing as much as 2 grams then immersed in aquades for 24 hours. The sample is then filtered and then dried at 100^{0}C for 30 minutes. The solubility of *edible film* can be calculated by the equation:

$$Solubility(\%) = \frac{\Delta W}{W0} \times 100$$

2.4.9 Biodegradability

Biodegradability tests are performed to determine the speed of edible films in being degraded by microorganisms in an environment. Biodegradability test using soil burial test method or burial method. Film samples are cut to a weight of 3 x 3 cm. The sample was weighed initially (2 grams) before burial (W1). The samples were buried for 12 days in soil containing compost and cow dung as a source of microorganisms to degrade the film. Samples are observed every 3 days by taking samples and cleaning the soil with tissues, then weighing the final weight (W2). Next, the percent weight loss and biodegradability rate are calculated with the following equation [18].

$$Biodegradability(\%) = \frac{W1 - W2}{W1} \times 100$$

2.5 Data analysis

The research design used was a one-way group randomized design with 3 repeats. The data were analyzed with Kruskal-Wallis and analysis of variance (ANOVA) using SPSS 29 software.

3 Result and Discussion

3.1 Consumer preferences

Consumer preference test is a test to determine the level of liking or acceptability of edible film. The results of hedonic rating tests on several edible attributes of alginate-based films can be seen in Table 1.

Table 1. Characteristic of sodium alginate/glycerol-based edible film with addition of gelatin and casein

Characteristic	**Edible film**			
	Na alginate+Glycerol	**Na alginate+Gelatin +Glycerol**	**Na alginate+Casein+ Glycerol**	**Commercial (control)**
Consumer preferences -Appearance -Texture -Aroma	 8.29 ± 0.42 [a] 8.28 ± 0.34 [a] 8.72 ± 0.34 [a]	 6.42 ± 0.26 [b] 8.69 ± 0.33 [b] 6.28 ± 0.44 [b]	 8.68 ± 0.43 [c] 8.68 ± 0.47 [bc] 8.74 ± 0.41 [ac]	 - - -
Thickness (mm)	0.11 ± 0.01 [c]	0.14 ± 0.01 [d]	0.10 ± 0.01 [b]	0.08 ± 0.01[a]
Tensile strength (MPa)	10.69 ± 0.03 [c]	7.12 ± 0.09 [a]	14.78 ± 0.02 [d]	7.30 ± 0.01 [b]
Elongation (%)	41.98 ± 0.07 [b]	95.11 ± 0.10 [d]	66.64 ± 0.07 [c]	13.04 ± 0.02 [a]
Elasticity (MPa)	0.25 ± 0.00 [c]	0.07 ± 0.00 [a]	0.22 ± 0.00 [b]	0.56 ± 0.00 [d]
High temperature resistance (second)	480.44 ± 3.44 [b]	359.33 ± 3.38 [a]	657.89 ± 7.39 [c]	2400 ± 0.00 [d]
Water content (%)	17.80 ± 0.01 [d]	12.67 ± 0.01 [b]	14.56 ± 0.01 [c]	10.36 ± 0.01 [a]
Water vapor transmission rate (g/m^2.24 hour)	1.00 ± 0.17 [c]	2.04 ± 0.12 [d]	0.64 ± 0.06 [b]	0.00 ± 0.00 [a]
Solubility (%)	33.33 ± 0.38 [c]	51.84 ± 0.18 [d]	23.19 ± 0.34 [b]	2.43 ± 0.03 [a]
Biodegradability (%)	81.68 ± 0.28 [c]	90.95 ± 0.28 [d]	76.78 ± 0.29 [b]	2.67 ± 0.29 [a]

Note: values with different letter superscripts on the line indicate a noticeable difference between treatments

Based on consumer preference of edible film, the preferred formula is edible sodium alginate film with the addition of casein protein. The characteristic appearance of edible film without protein and with the addition of casein is clean, clear but not transparent and slightly shiny for films with the addition of casein. As for the appearance of the film with the addition of clean gelatin, it is not too clear, the color is slightly yellowish to golden and not transparent and not shiny. The yellow to golden color of gelatin materials is caused by several factors, namely the source of raw materials, production processes, acid concentration, extraction temperature, moisture content and ash content [19].

The texture value of edible film is the highest value in the film formula with the addition of gelatin followed by the film with the addition of casein. Gelatin has hydrocolloid properties that can form elastic and flexible films so that gelatin can form a good tissue structure on edible films [20]. Casein has hydrocolloid properties that can form elastic and flexible films and casein has good adhesive properties so that it can form strong edible films so that it can form a good tissue structure on edible films [21].

Aroma is very important for a product because it is a consumer consideration in choosing products and consuming products [22]. Films with the addition of casein produce films that do not have an aroma because the material is made with natural ingredients such as materials that tend not to have a strong odor because they do not use chemical additives that can give a certain aroma or odor to the resulting product [21]. Meanwhile, films with the addition of gelatin produce an unpleasant aroma caused by several factors such as the use of additives in making gelatin, the source of raw materials derived from cowhide and the ash production process when producing gelatin [19].

3.2 Thickness of edible film

The thickness value of the edible film (Table 1), shows that the thickness of the sodium alginate edible film without protein differs markedly from the edible film with the addition of gelatin protein and casein protein. The thickness of edible film in each formula in this study has met Japanese Industrial Standard (0,25 mm). The thickness value of edible film without protein in this study is lower when compared to the study of [12], sodium alginate-based film has a thickness value of 0.14 – 0.71 mm. The factors that cause differences in thickness values are the surface area of printing, the method of heating application.

In the sodium alginate formula with the addition of gelatin has a higher thickness value compared to other formulas. The edible thickness value of sodium alginate film with the addition of gelatin in this study is lower than the research of [23]. Gelatin-based films which have a thickness value of 0.18 mm, this is due to the high concentration of glycerol and gelatin, the higher the film thickness value. According to [23], gelatin and glycerol are hydrophilic which are able to bind water and in the evaporation process which causes a decrease in water content in edible film, which in turn can affect film thickness where the higher the water content, the higher the thickness value on edible film [24].

In the edible formula alginate film with the addition of casein has a higher thickness value [25]'s research, casein-based films have an edible film thickness value of 0.080 – 0.082 mm. But in this study, films with the addition of casein produce the lowest thickness value, this is because according to [26], casein has hydrophobic properties that can repel water so that in the evaporation process which causes a decrease in water content in edible film, which in turn can affect the thickness of the film where the lower the moisture content, the lower the thickness value on the edible film.

3.3 Tensile strength

The tensile strength of the edible film shows that there is a significant difference between the administration of proteins that vary in the sodium alginate-based edible film. In [27] stated that in the 1975 Japanese Industrial Standard (JIS), edible film has a minimum tensile strength value of 0.3 MPa. The tensile strength value of edible film in each formula in this study has met JIS 1975. The tensile strength value determines the resistance of edible film in packaging products, the lower the value, the more fragile the edible film due to its weak molecular bonds [28]. The tensile strength value of edible film without protein in this study has almost the same value when compared to the research of [12], which has tensile strength values ranging from 0.07-0.84 MPa. The factors that cause the tensile strength value are not much different, namely the concentration and ratio of constituent materials and also the method of application and drying that affect the distribution of drying on films that have several similarities. The sodium alginate formula with the addition of gelatin has a lower tensile strength value when compared to other formulas and compared to [28] research which has a tensile strength value ranging from 9,609-465,904 MPa. Meanwhile, the edible formula of sodium alginate film with the addition of casein protein has a higher tensile strength value compared to other formulas and the tensile strength of edible casein film in the study of [29], which has a tensile strength value ranging from 4.47-12.98 MPa.

3.4 Elongation

The elongation value of edible film shows that there is a real difference between the treatment in the use of edible film In [27] stated that in the 1975 Japanese Industrial Standard (JIS), *edible* film has an elongation of at least 70%. The elongation value of *edible film in* each formula in this study, there are films that have not met JIS 1975. This can be caused by several factors such as the collaboration of the basic constituent materials used.

From the results of the elongation test or percent elongation, it is stated that the film formulation with the addition of gelatin is an edible film that meets JIS 1975. This is due to the addition of gelatin to sodium alginate and glycerol. Gelatin has good elastic properties and has the ability to stretch and return to its original shape after being applied pressure or tension [30]. Thus, the addition of gelatin in the edible film formulation can provide high elasticity to the film, so as to increase the elongation of the film as well. The elongation value in this study has almost the same elongation value when compared to the research of [28], which has an elongation value ranging from 8-108%. The factor that influences this is by the factor of using a low glycerol concentration, where the higher the glycerol concentration can reduce the tensile strength of elasticity properties in the gelatin film [31].

In the alginate formula with the addition of casein has an elongation that has not met JIS 1975, but casein has the ability to form a film with good elasticity and casein as the main protein in milk, also has significant elastic properties. So that the addition of casein in the formulation of sodium alginate-based edible film can provide elasticity and tensile strength to the film which can increase the elongation of the film before fracture or break up on film [31]. The elongation value in this study is lower when compared to the research of [29], which has an elongation value ranging from 33.07-80.50%. The factors that affect the elongation value are in the use of constituent concentrations and glycerol, the use of different types of gelatin and drying application methods applied in filmmaking.

3.5 Elasticity

The elasticity value of the edible film shows that there is a real difference between the administration of proteins that vary in the sodium alginate-based edible film and the elasticity of the sodium alginate. Edible film without protein is significantly different from the edible film with the addition of gelatin protein and casein protein. The value of elasticity in the study was obtained from the value of tensile strength divided by the elongation value in this study. If the division results are getting bigger, the elasticity of the film decreases [10], the smaller the value of the elasticity division, the more elastic the edible film is because elastic edible films tend to be easier to apply to food products, easier to form and cut, and more easily broken down by digestive enzymes in the human body. However, in this study, the edible film produced did not meet the elasticity of JIS 1945, because the resulting film has high elasticity and cannot be applied to all food products. Such a film with high elasticity is good for use on fresh produce and not necessarily good on processed products such as sausages.

3.6 High temperature resistance

The high temperature resistance test refers to the cooking temperature of sausages, which is 80^{0}C for 40 minutes [17]. Through the high temperature resistance test, it can evaluate, see that the integrity of the film is not damaged or not released or destroyed when exposed to high temperatures during the boiling process when making the film as a coating against sausages. This is important to ensure that the coating provides protection, improving the quality of sausage products [32].

The value of high temperature resistance in edible film shows that there is a real difference between the administration of proteins that vary in sodium alginate-based edible film. The results of the edible film boiling temperature test in the study stated that the three edible film formulations had not been able to meet the minimum temperature limit for boiling sausages until cooked. Alginate has thermo irreversibility properties, where the alginate gel becomes resistant to heat, which means sodium alginate can be used for high temperature treatment without damage [33]. But not for a long time, so edible film based on alginate is commonly used to coat processed low-temperature food products such as coating on dodol [6] and fresh crabs [34]. Gelatin is one type of protein that is widely obtained from natural collagen found in the skin and bones and in the manufacturing process does not use temperatures that are too high to avoid denaturation in gelatin proteins [35], such as research on horseskin gelatin which can denature at high temperatures and reactions take place exothermic in a short time [20]. So that gelatin-based edible films are commonly used as coatings on fresh fruit products [36]. As for casein which is one type of protein that is widely obtained from milk and is used in the manufacture of various food products, such as cheese, yogurt and ice cream. In the process of making casein, it begins to show a decrease in viscosity at 80°C so that casein can denature at high temperatures. So that casein-based edible film is commonly used as a coating for chicken meat products [37].

3.7 Water content

The results of water content test in the study stated that from among the three edible film formulations, it has not been able to meet the minimum limit of water content. The water content of films without protein in the study had the same value when compared to the research of [38], which had water content of carrageenan-based films ranging from 17.14% to 20.86%. This is because alginate film has hydrophilic properties that can absorb water and form a gel-like structure so that it has a high enough water content in the film [6].

From the results of the water content test, it states that the film formulation with the addition of gelatin is the film that has the lowest water content of the three formulas and has met Japan Industrial Standard. The water content of the film with the addition of gelatin in the study had the same value as the research of [28], which had a gelatin-based water content ranging from 10.46% to 13.88%. Then followed by the film formula with the addition of casein which has a moisture content value that has met Japan Industrial Standard. In general, films composed of protein have low water content, because protein has low water content as well [39].

In this study using the main raw material alginate with the addition of protein and glycerol so as to increase the value of water content in edible film. This is because in making edible films using glycerol plasticizers, where glycerol has an OH group, where the OH group is hydrophilic, which has the ability to bind water. So, the lower the glycerol concentration results in the lower the water bound to the edible film matrix, so that the edible film moisture content is also low [28]. So the value of water content in the film with the addition of gelatin, casein in alginate-based films and the use of glycerol which are both hydrophilic or can bind water, can affect the value of edible film moisture content .

3.8 Water vapor transmission rate (WVTR)

The water vapor transmission value shows that there is a marked difference between protein administration (gelatin and casein) which varies in edible film-based. Water vapor transmission in edible alginate film without protein differs markedly from edible film with the addition of gelatin protein and casein protein. From the results of the water vapor transmission test, it states that the film formulation with the addition of gelatin has met the Japan Industrial Standard and has the highest water vapor transmission value of the three formulas. The water vapor transmission value in the film formula with the addition of gelatin has the same value as [40]'s research, about gelatin-based films has a water vapor transmission value ranging from 0.1 to 0.3 g/m^2.24 hours. Then followed by a film without the addition of protein. This is because the constituent components of the film, namely sodium alginate, gelatin and glycerol are hydrophilic [6], which can increase the rate of water vapor transmission.

The film with the addition of casein has met the Japan Industrial Standard and has the lowest moisture transmission value of the three formulas. The value of water vapor transmission in films with the addition of casein has a slightly higher value when compared to [41], about gelatin-based films has a water vapor transmission value ranging from 0.3 to 0.5 g/m^2.24 hours. This affects the low value of water vapor transmission in films with the addition of casein protein, because casein is hydrophobic, so it can form flexible films with a rather high permeability to water vapor, and low to oxygen and carbon [26]. The increase in the rate of water vapor transmission in edible film can be caused by several factors, such as the film experiencing swelling by water vapor [40], composition of edible film forming materials, such as alginate, gelatin, casein, glycerol; film thickness can affect water vapor transmission, where the thinner the film, the higher the water vapor transmission; humidity of the environment around the film during testing as well may affect the water vapor transmission value.

3.9 Solubility

The solubility value of the edible film shows that there is a real difference between the administration of proteins that vary in the sodium alginate-based edible film and the solubility of the edible film of sodium alginate without protein is significantly different from the edible film with the addition of gelatin protein and casein protein. From the solubility test results, it states that the film formulation with the addition of gelatin has met the Japan Industrial Standard and has the highest solubility value of the three formulas. The solubility value of the film formula with the addition of gelatin has a higher value when compared to the research of [13], on gelatin-based films which have solubility values ranging from 29.10% to 39.90%. Then followed by a film without the addition of protein. The high solubility value of edible sodium alginate film with the addition of gelatin protein is because sodium alginate and gelatin have hydrophilic properties that can absorb water well [23].

The film with the addition of casein has not met the Japan Industrial Standard and has the lowest solubility value of the three formulas. This affects the low solubility value of the film with the addition of casein protein, because casein is hydrophobic, so it can form flexible films with a rather low permeability to water vapor, and low to oxygen and carbon [26]. The factors that affect the solubility value of edible films are different, namely the composition of film materials, including the type and concentration of forming materials such as proteins and polysaccharides can affect the solubility of films in certain solvents; Interaction with glycerol plasticizers and physical properties or physical characteristics of casein such as molecular structure, particle size, and intermolecular bond strength can affect solubility in edible film-forming solutions.

3.10 Biodegradability

The biodegradability value of edible film shows that there is a real difference between the administration of proteins that vary in sodium alginate-based edible films and also the solubility of edible sodium alginate films without protein is significantly different from the biodegradability of edible films with the addition of gelatin protein and casein protein. The biodegradability value of films with the addition of gelatin protein has met SNI 7188.7: 2016 and is the highest value of the three formulations, followed by films without protein addition. This can happen because films composed of sodium alginate and gelatin can degrade naturally in a relatively short time, are non-toxic and biodegradable.

In general, films made from sodium alginate and gelatin are used as basic materials in making biodegradable films [42], so they can be used in various applications in the food and non-food industries that are environmentally friendly. Films with the addition of casein have also met SNI 7188.7: 2016 and have the lowest biodegradability value of the three formulas, because casein is hydrophobic, so it can form flexible films with a rather low permeability to water vapor, and low to oxygen and carbon [26]. However, in edible film research that is made intended as a sausage casing, so that the best biodegradability value has a low value. And in this study the best value for the biodegradability of sausage casings was with the addition of casein.

4 Conclusion

The characteristics of edible film consisting of consumer preferences, thickness, tensile strength, elongation, elasticity, high temperature resistance, water content, vapor transmission rate, solubility and biodegradability, are influenced by the use of sodium alginate, gelatin and casein in the manufacture of edible films. Edible films made from only sodium alginate and glycerol only have advantages in characteristics of elasticity, edible films made from sodium alginate, glycerol and gelatin have advantages in characteristics of elongation, water content, solubility and biodegradability, and edible films made from sodium alginate, glycerol and casein have advantages in characteristics of consumer preferences, thickness, tensile strength, high temperature resistance and water vapor transmission rate. An interesting result is that the use of single sodium alginate or mixed with gelatin or casein is not able to withstand high temperatures when compared to commercial edible films. Another interesting thing is that commercial edible films used in research are more difficult to decompose by microbes than edible films made from sodium alginate or sodium alginate mixed with gelatin or casein.

References

1. R. Silvia Nasution. Journal of Islamic Science and Technology, **1(1)**, 97–104 (2015).
2. M.K. Wahyu. Journal of Research Communication, **5(August)**, 12–42 (2009).
3. Marshall. Journal of Islamic Science and Technology, **1(1)**, 97–104 (2015).
4. C. Winarti. Jurnal Penelitian dan Pengembangan Pertanian, **31(3),** 85-93 (2012).
5. Guilbert. Journal of the American Chemical Society, **123(10)**, 2176–2181 (2013).
6. Y. Nasyiah, I. Darmanto, Wijayanti. Jurnal Pengolahan dan Bioteknologi Hasil Perikanan, **4(4)**, 7–14 (2015).
7. M. Hijriawati , E. Febriana. Jurnal Farmaka, **14(1)**, 8–16 (2016)
8. A. Widyastuti Perdani. Prosiding Pendidikan Boga Busana, **16-1**, (2021).
9. A.W. Perdani. Prosiding Pendidikan Teknik Boga Busana, **16(1)**, 4–7 (2021)
10. M. Fera, Nurkholik. Journal of Food and Life Sciences, **2(1)**, 45–56 (2018).
11. A. Shendurse. Journal of Nutritional Health & Food Engineering, **8(2)**, 219–226. (2018).
12. N.Thea.Puspita Langit, A. Ridlo, S. Subagiyo. Journal of Marine Research, **8(3)**, 314–321 (2019)
13. S. Rahayu Kalaka, A.Silvana Naiu, R. Husain. Jambura Fish Processing Journal, **4(2)**, 64–71 (2022)
14. E. Saravani Pak, S. Najafi. Ghaghelestani, M. Ali Najafi. Journal of Food Science and Technology, **9(59)** (2020)
15. C.Togas, S. Berhimpon, R.I. Montolalu, H.A. Dien, F. Mentang. Jurnal Pengolahan Hasil Perikanan Indonesia, **20**, 468– 477 (2017)
16. W. Setiani, T. Sudiarti, L. Rahmidar. Valensi **3-2**, 100-109 (2013).
17. M. Elvira.Yuniar, D. Nur Azizah. Jurnal Pangan dan Agroindustri, **9(3)**, 139–147 (2021).
18. K. Haryani, M. S. Al Anshar, V. Hermansyah. Prosiding SEMNASTEK 2022, November, 1–10, (2022).
19. M. Nurilmala, A.M. Jacoeb, R.A. Dzaky. Jurnal Pengolahan Hasil Perikanan Indonesia, **20(2)**, 339 (2017)
20. D. Trimelya Gela. Skripsi, 73 (2016)
21. F. Marrudin, A. Ako, Hajrawati, M. Taufik. Persepsi Masyarakat Terhadap Perawatan Ortodontik Yang Dilakukan Oleh Pihak Non Profesional, **53(9)**, (2013).
22. H. Hermanto, A. Susanty. Jurnal Riset Teknologi Industri, **14(2),** 253 (2020).
23. Q.F. Rahmi, E. Wulandari, J. Gumilar. Jurnal Teknologi Hasil Peternakan, **3(1)**, 19 (2022)
24. A.J. Sitompul, E. Zubaidah. Jurnal Pangan Dan Agroindustri, **5(1)**, 13–25 (2017)
25. Ikhwan, A. Deimon. Universitas Brawijaya, (2019).
26. A. Suci Ningsih, E. Dewi, L. Kalsum, E. Margaretty. Seminar Nasional Inovasi dan Aplikasi Teknologi Di Industri 4.0, 190–198 (2019).
27. L. Urip Widodo, S. Neza Wati, N. Made Vivi. Jurnal Teknologi Pangan, **13(1)**, 59–65. (2019).
28. A. Ariani Hendra, A. Rulianto Utomo, E. Setijawati. Jurnal Teknologi Pangan dan Gizi, **14(2)**, 95–100, (2015).
29. Fatma, R. Malaka, M. Taufik. Fakultas Peternakan Universitas Hasanudin, 214 – 219. (2014).
30. H. Ali, A. Baehaki, S. Dwi Lestari. Jurnal Teknologi Hasil Perikanan I, **6(1)**, 26–38 (2017)
31. H. Suryadri, R. Andriani, D. Aditya. Chempublish Journal, **5(2)**, 93–104. (2020).
32. F. Farida, N. Amaliah. Journal of Tropical Agri Food, **1(2)** ,79. (2020)
33. W. Dwi Wardani. Skripsi, Fakultas Pertanian Universitas Sebelas Maret Surakarta,

https://doi.org/10.1051/bioconf/20248701007

(2008).
34. M. Seno Herbowo, P. Har Riyadi, R. Romadhon. Jurnal Pengolahan dan Bioteknologi Hasil Perikanan, **5(3)**, 37–44 (2016).
35. F. Firawati. Prosiding Seminar Nasional Biologi dan Pembelajarannya, 74–82 (2018).
36. Y. Lu, Q. Luo, Y. Chu, N. Tao, S. Deng, L. Wang. National Library Of Medicine, (2022).
37. T. Markoska, D. Daniloski, T. Vasiljevic, T. Huppertz. National Library Of Medicine (2021).
38. A. Rusli, M. Metusalach, M. Tahir. Jurnal Pengolahan Hasil Perikanan Indonesia, **20(2)**, 219 (2017).
39. A. Safari, M. Fadhlillah, S.D. Rachman, N.I. Anggraeni, F.F. Isnanisafitri, S. Ishmayana. Chimical et Natura Acta, **8(1)**, 50 (2020).
40. S. Sunardi, A. Rahman Maulana. Walisongo Journal of Chemistry, **4(1)**, 8–16 (2021).
41. Y. Wahyuni. Thesis, University of Hasanuddin, (2017).
42. Z. A. Nur Hanani, Y.H. Roos, J. P. Kerry. International Journal of Biological Macromolecules, **71**, 94–102 (2014).

A nesting ability of male siamese fighting fish *(Betta sple ndens)* under different level of turbidity derived from palm oil mill effluent

Ilham Zulfahmi[1*], *Agung* Setia Batubara[2], *Adli* Waliul Perdana[3], *Tasya* Alniza Putri[3], *Badratun* Nafis[4], *Furqan* Maghfiriadi[5]

[1]Department of Fisheries Resources Utilization, Faculty of Marine and Fisheries, Syiah Kuala University, Banda Aceh, 23111, Indonesia
[2]Department of Biology, Faculty of Mathematics and Natural Sciences, Universitas Negeri Medan, Kota Medan, 20221, Sumatera Utara, Indonesia.
[3]Department of Aquaculture, Faculty of Marine and Fisheries, Syiah Kuala University, 23111, Indonesia
[4]Department of Biology, Faculty of Mathematics and Natural Sciences, Syiah Kuala University, Banda Aceh, 23111, Indonesia
[5]Center for Aquatic Research and Conservation, Universitas Islam Negeri Ar-Raniry, Banda Aceh23111, Indonesia

Abstract. Contamination of Palm Oil Mill Effluent (POME) into water bodies potentially disrupting the visualization and affects behaviourial changes such nesting ability of male Siamese Fighting Fish (*Betta splendens*). The study examined the effects of turbidity due to POME contamination toward nest area as one of the indicators for the reproductive behavior of the *betta splendens.* The research used Completely randomized Design with different level of turbidity: <1 NTU (control), 20 NTU (treatmen A), 40 NTU (tretment B), 60 NTU (Treatment C). Preference tests were carried out in a 6-L aquarium (20x20x20 cm^3). Nest area was measured each hour for 8 hours. Nest area was captured using camera and measured with imageJ. Result showed there was no significant difference in nest area of *Betta splendens* in different level of turbidity. The range of nest area in sequence until the end of exposure were 0-0.8 cm^2, 0.7-2.5 cm^2, 4.6-6.7 cm^2, 4.5-7.2 cm^2, 7.6-8.7 cm^2, 9.4-11.11 cm^2, 9.7-12.5 cm^2, dan 11.6-14.14 cm^2. Turbidity due to POME collected from the last pond with low pollutant contamination has no impact on nesting ability of betta fish. Nest building behavior is often correlated with hormone concentrations due to pollutant contamination with a certain consentration of pollutant.

* Corresponding author : ilham.zulfahmi@usk.ac.id

1 Introduction

Indonesia is listed as the world's leader in palm oil production. Indonesia's palm oil plantation area reached 11 million hectares (Mha) and produced 31 million tons of palm oil in 2015 [1]. The palm oil waste produced is in the form of solid, liquid and gas waste which has the potential to reduce water quality and cause pollution [2]. POME causes damage to the reproduction of nile tilapia (*Oreochromis niloticus*) such as decreasing the performance of tilapia hormones and resulting in a decrease in gonadsomatic index and spermatocrit [3]. Exposure to POME reduces the concentration of estradiol, testosterone and progesterone hormones in tilapia *(Oreochromis niloticus)* [4]. Exposure to POME also caused histological alteration in the liver tissue of zebrafish (*Dario rerio*), including congestion, hemorrhage and hyperplasi [5]. Several related cases are thought to be the result of high levels of by-products from the palm oil processing process which contain various organic materials in the form of Biological Oxygen Demand (BOD, range: 25000–65714 mg/L), Chemical Oxygen Demand (COD, range: 44300–102696 mg/L), and Total Suspended Solid (TSS, range: 18000–46011 mg/L) as well as various heavy metals [6][7].

High concentrations of TSS in water bodies increase water turbidity. Previous study revealed that POME waste contamination in Bangka Regency has increased the turbidity of the Mabat river from 5.04 Nephelometric Turbidity Units (NTU) under normal conditions to 23.5-65.4 NTU [7]. This increase in turbidity has the potential to block the entry of sunlight which has a negative impact on photosynthetic activity, reduces the availability of dissolved oxygen and disrupts the visual performance of fish [8]. Increased turbidity affects interactions between male and female fish and reduces spawning effectiveness [26].

The Siamese fighting fish (Betta splendens) is widespread in Indonesia from rivers, wetlands and swamps [27][28]. The population of Siamese fighting fish has been decreasing due to various factors such as habitat loss and water pollution [29]. Additionally, the introduction of non-native species and competition for resources have also contributed to the decline in their population. In addition, high levels of turbidity indicate the presence of pollutants that accumulate in fish tissues and cause physiological and behavioral changes, which ultimately affect fish populations [25]. Therefore, although there is no direct relationship between turbidity and Betta fish populations, the impact of turbidity on their behavior and the presence of pollutants in turbid water may affect Betta fish populations indirectly. The decreasing population of Siamese fighting fish is a cause for concern, as these fish are not only popular in the aquarium trade but also play a crucial role in their natural ecosystems [30]. Increased turbidity can affect fish behavior due to visual limitations [5]

In an effort to produce new individuals, male Betta fish will make bubbles and form a foam nest (bubblenest) on the surface of the water. Bubble nests play important rule in guarding eggs during spawning [10]. The results of research [5] reported that turbidity can influence the level of predation on emerald shiner fish (Notropis atherinoides). Fish reared at 40 NTU turbidity showed a reduction in predation due to reduced visual performance. Another study by [6] also reported that increased turbidity due to POME contamination of ~60 NTU had affected the eating behavior of Betta splendens fish as indicated by a reduction in the fish's home range and slow food detection efforts. The turbidity level of 8.7 NTU was caused by the addition of bentonite which caused a decrease in foraging behavior in E. fonticola [9]. The same case was also observed in different types of fish including *Oncorhynchus tshawytscha* and *Oncorhynchus mykiss,* that increasing turbidity caused a decrease in the number of fish that foraged and the amount of prey consumed [22][23][24]. Although no research has directly examined the impact of turbidity on the ability of fish to build nests, information from related studies can provide insight into how environmental conditions at different levels of turbidity can influence the overall behavior and abilities of fish. Therefore, this research aimed to investigated the impact of turbidity due to POME on the ability of Siamese Fighting Fish to build nests.

2 Material and Methods

2.1 POME and Fish Collection

A total of 20 L of POME from the eighth settling pond (last pond) (100 NTU) was collected from Palm oil Factory located in Aceh Province. Until the exposure period, POME was stored at a temperature <4°C to avoid degradation. A total of 25 pairs of Siamese Fighting Fish with uniform colors (predominantly red) were collected from local traders located in Banda Aceh. The total length range of male and female Siamese Fighting Fish is 4-4.5 cm. The weight of male Siamese Fighting Fish ranges from 1.53-1.96 g, while the weight of the fish Female Siamese Fighting Fish range between 1.53-2.10 g. Male Siamese Fighting Fish are characterized by black spots on the dorsal fin area and red vertical lines on the gill covers. Meanwhile, female Siamese Fighting Fish are characterized by their bulging stomach shape accompanied by 2-3 gray vertical lines on the side of the body [11].

Fish acclimatization was carried out individually in transparent containers for one week (Photoperiod 12:12 hours light: dark; DO: 6.2 mg O^2/L; temperature: 25 C; PH: 7.5 and turbidity: <1 NTU). To avoid stressful conditions due to interactions between fish, each side of the acclimatization container was lined with opaque paper. Feed in the form of silk worms was given twice a day (09.00 and 16.00 WIB) ad libitum [6]

2.2 Experimental Design

The research design was a Completely Randomized Design (CRD) with 4 treatments accompanied by 3 repetitions for each treatment. The range of turbidity in the treatment refers to the results of previous research regarding pollution of palm oil liquid waste in Indonesian waters [7]. The details of each treatment are as follows: control (turbidity level <1 NTU), treatment A (turbidity level 20 NTU), treatment B (turbidity level 40 NTU), treatment C (turbidity level 60 NTU). The turbidity value for each treatment is obtained by mixing POME into the treatment container until it reaches the specified turbidity level. Validation of the media turbidity value was carried out using a turbidity meter (Lutron TU-2016, Taiwan).

The experimental container is an aquarium 20 x 20 x 20 cm^3 with a volume of 6 liters of water. Each aquarium is filled with one pair of Siamese Fighting Fish. Bubble attachment media in the form of clear plastic measuring 5x5 cm is placed on the water surface close to one end of the tank for the male to build a bubble nest. Nest area was documented every hour and measured using the ImageJ application version 1.5.3 (Figure 3). All maintenance and experimental stages were approved by the Animal Ethics Committee of Faculty of Science and Technology, Universitas Islam Negeri Ar-Raniry (Animal Ethics approval number: BIO/ET/2020/1970).

Fig. 1. Schematic diagram of aquarium setup during introduction period (20 x 20 x 20 cm^3). The aquarium is given a partition with a size of 8x8 with a volume of 1L of water. The bubble attachment medium is transparent plastic (5x5 cm^2).

3 Result and Discussion

Nest size in each treatment increased with increasing duration. The results of statistical analysis showed that there were no significant differences between treatments from the first hour to the last hour during the exposure period ($p>0.05$). The range of nest sizes built by Siamese Fighting Fish between treatments from the first hour to the eighth hour respectively are 0-0.8 cm^2, 0.7-2.5 cm^2, 4.6-6.7 cm^2, 4.5-7.2 cm^2, 7.6-8.7 cm^2, 9.4-11.11 cm^2, 9.7-12.5 cm^2, and 11.6-14.14 cm^2

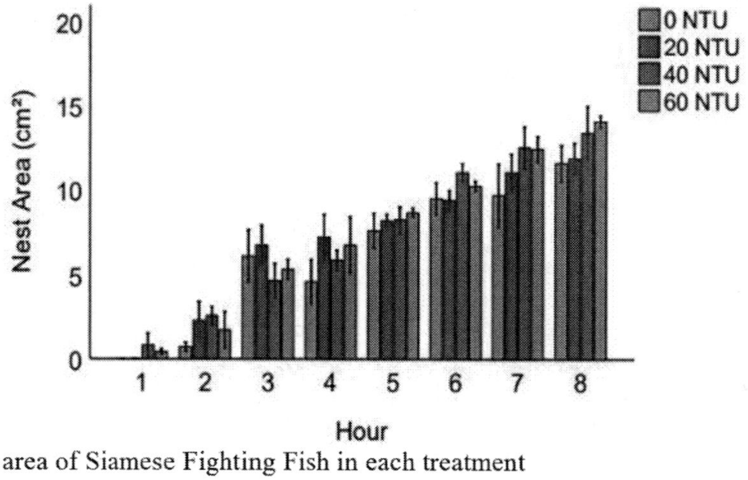

Fig. 2. Nest area of Siamese Fighting Fish in each treatment

A number of studies have reported the impact of turbidity on aquatic organisms [12][13]. Turbidity is closely related to visual function which directly impacts changes in behavior [14]. In natural environments, turbidity has been recognized as a major threat to the diversity of aquatic systems [15]. Male Siamese Fighting Fish have a vital role in reproductive success through nest formation, egg placement and parental care [16]. The behavior of male Betta fish before spawning is usually to release bubbles as a nest for their eggs later. According to [31] Betta fish are bubble nesters, that is, they make foam nests before spawning and put the eggs in them. The bubble substrate is useful so that the eggs do not sink to the bottom of the water, usually these bubbles are placed on aquatic plants [32]. The nest built by the male

Betta fish acts as a container for the eggs and provides a safe environment for the initial growth and development of the offspring [33]. The function of the bubble nest is to protect the embryo and provide more oxygen or nutrients to the embryo. The size of the bubble nest built by male fish can influence the reproductive choices of female fish and the survival rate of fish larvae [17].

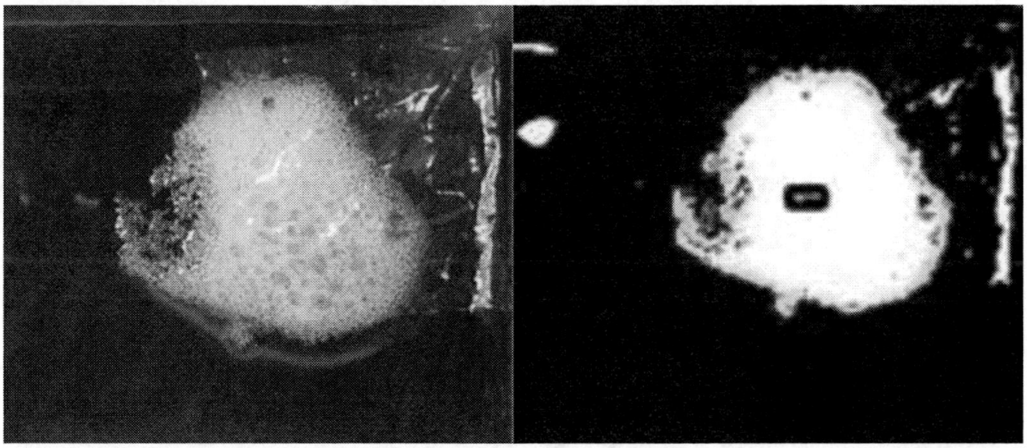

Fig. 3. The description of bubble nest and the measurements methode using ImageJ.

Nest building by male Betta fish involves several stages. These stages include preparing raw materials, making nests, and monitoring. The purpose of making betta fish nests is to maintain the health of the fish, maintain the survival of the fish, and help the development of the fish. The time needed to make a betta fish nest is very short, provided that a nest made in 1 hour can keep fish for 2-3 months. Betta fish nests can be beard-shaped, small, or large, depending on existing needs and guarding conditions [34]. Mucus-coated nest bubbles are formed from air and mucus in the buccal cavity of Siamese fighting fish. This mucus is very thick, it can allow the bubbles to survive longer in the water. Male fish can collect these bubbles to make nests. Betta fish are all unique, so the frequency of making nests is also unique. Some bettas may develop them regularly, while others may develop them less frequently or not at all. Daily, weekly, or monthly are the common time periods for male Betta fish to build a nest. Some nests are large, some are small, and their thickness can also vary. Regular bubble nest building can be a sign that Betta fish are happy and healthy, and suitable for mating [35]

The results of this study indicate that increasing turbidity does not affect the fish's ability to build nests. We suspect that the motivation to build a nest is influenced by internal factors, while turbidity is an external factor that only plays a small role during the nest building period. According to [17] larger male fish build larger bubble nests. This suggests that large nests will be able to accommodate more fish eggs and larvae which can only be built by larger males. However, the number of eggs per clutch did not correlate with male body size or nest bubble area. Habitat arrangement provides better results in building bubble nests in aquariums [18]. [10] reported that Siamese Fighting Fish pairs exposed to fluoxetine did not have bubble nests of consistent and optimal size. Exposure to fluoxetine causes a decrease in locomotion [19], a decrease in androgen hormones and sexual motivation in male fish [20]. As a result, during the spawning period some of the eggs fall to the bottom of the container and take the male's time to collect the eggs and return them to the bubble nest [10].

Fish have varied nest building strategies in turbid conditions through special sensory adaptations. Sensory adaptation is an alternative for fish to navigate and build nests in murky water. One of the sensory adaptations is the electrical signals utilized by fish for passive and active electrolocation and communication in poor visual conditions. Electrical signals have evolved in fish that live in turbid water and tend to be active at night [36]. Active electrolocation involves the production of electrical signals, the discharge of electrical organs by specially modified cells (electrocytes). Electrocytes are formed from muscle fibers or from nerve endings in apteronotids [37][38]. The electric field around the fish is detected by electroreceptors in the body and especially in the head [39][40]. Electroreceptors can detect differences between living and inanimate objects as well as differences in size, shape, and distance by monitoring changes in signal amplitude, resistance, and capacitance [41]. In the context of nest building, fish have special sensory adaptations to navigate and build nests in turbid water by utilizing the lateral line system. This system allows fish to detect changes in water pressure, helping to find suitable nesting sites even in low visibility conditions. All fish have a special mechanical sensor (mechanosensory) lateral line system to detect disturbances in the water. Due to the viscosity and density of water, disturbances created by the presence of potential prey and other moving objects are detected via mechanoreception [21].

4 Conclusion

No significant differences were observed of nest area of Siamese Fighting Fish between treatments from the first hour to the last hour of exposure period ($p>0.05$). The results of this study indicate that increasing turbidity does not affect the fish's ability to build nests. This suggests that large nests will be able to accommodate more fish eggs and larvae that only larger males can build.

Acknowledgments

The study was financed by a Research Grant from Directorate of Research and Community Service, the Ministry of Education Culture Research and Technology Republic of Indonesia under the "Penelitian dasar Kompetitif Nasional (PDKN) 2023" Scheme, Grant No. 062/E5/PG.02.00.PL/2023.

References

1. F. Harahap, S. Leduc, S. Mesfun, D. Khatiwada, F. Kraxner, S.E. Silveira. **12(3),** 420. (2019).
2. E.F. Mauliza. JFU. **1(1)**, 1-8. (2020).
3. M. Muliari, I. Zulfahmi, Y. Akmal, N.W.K. Karja, C. Nisa, K.A. Sumon, M.M. Rahman. *Env. Sci. Poll.* Res, **27**, 30592-30599. (2020).
4. M. Muliari, I. Zulfahmi, Y. Akmal, N.W.K. Karja, C. Nisa, K.A. Sumon. *Adv. Anim. Vet. Sci*, **7(11)**, 1035-1041. (2019).
5. C.L. Nieman, & S.M. Gray. JFB, **95(1),** 186-199. (2019).
6. I. Zulfahmi, A.S. Batubara, A.W. Perdana, A.P. Andalia, D. Nuzulli, M. Hidayat, F.M. Nur, K.A. Sumon, M.M. Rahman. *App. Ani. Behav. Sci*, **257**, 105790. (2022).
7. S. Zahara, U. Umroh, E. Utami. *Akuatik: J. Sumberd. Perair.* **10(1)**, 21–25. (2016).
8. M. Rois, & H. Fresillia. *Tunas Geo*, **6(2),** 116-123. (2017).
9. L.J.S. Becker, E. M. Brooks, C. R. Gabor & K. G. Ostrand. TAMN, **175**, 55–63. (2016).
10. M.N. Forsatkar, M.A. Nematollahi, B.M. Amiri, W. Huang. *Ecotoxicology*, **23**, 1794-1802. (2014).

11. A.R. Axelrod, W.E. Burgess. Nept. City, NJ: TFH Publications. (2004).
12. B.B. Chapman, L.J. Morrell, C.R. Tosh, J. Krause. *Proc. of the Royal Soc. B: Bio. Sci*, **277(1686)**, 1395-1401. (2010).
13. S.M. Leahy, M.I. McCormick, M.D. Mitchell, M.C. Ferrari. *Bio. Letters*, **7(6)**, 811-813. (2011).
14. S.M. Ehlman, D. Martinez, A. Sih. *Behav. Eco. and sociobio*, **72**, 1-9. (2018).
15. S.M. Gray, S. Sabbah, and C. W. Hawryshyn. *Eco. of Fresh. Fish,* **20**, 529-536. (2011).
16. T.M. Montgomery, A.C. Brown, H.K. Gendelman, M. Ota, E.D. Clotfelter. *Env. toxi*, **29(3**), 243-252. (2014).
17. M. Jaroensutasinee, & K. Jaroensutansinee. *J. Fish Bio*, **58(5),** 1311-1319. (2001).
18. M.M. Harlioglu, S. Mise-Yonar. *Inter. J. of Sci. & Tech*, **3(1),** 53-58. (2008).
19. J.G. Kohlert, B.P. Mangan, C. Kodra, L. Drako, E. Long, H. Simpson. *Psych. rep*, **110(1)**, 51-62. (2012).
20. J.A. Mennigen, C.J. Martyniuk, K. Crump, H. Xiong, E. Zhao, J. Popesku, ... & V.L. Trudeau. *Physi. Genom*, **35(3)**, 273-282. (2008).
21. S. Løkkeborg, S.I. Siikavuopio, O.B. Humborstad, A.C. Utne-Palm, & K. Ferter. *Review, F. Bio, and Fish*, **24**, 985-1003, (2014).
22. C. Wellington, D.A. Beauchamp, & J.L. Vogel. *North American J. of Fish. Manag*, **30(3),** 731-741. (2010)
23. K. Huene- mann, J.A Sweka, & K.J. Hartman. TOTAFS, **141(3)**, 742-752, (2012)
24. R.M. Zamor, & G.D. Grossman. TOTAFS, **136(5)**, 1283-1292. (2007)
25. G.R. Scott, & K.A. Sloman. *Aqua. Toxic*, **68(4)**, 369-392. (2004).
26. B.M. Laimeheriwa & M.A.R. Stefanno. *AQUANIPA-JIKP, 5*(**3**), (2023).
27. M.P. Hammer, S.M.N. Skarlatos, E.W. Needham, D.N. Wilson, M.A. Barton, D. Lonza. *Biol Invasions*, **21**, 2269–2279, (2019).
28. B. Panijpan, S. Namkang & L. Parames. *Sci. Asia* **46(4),** (2020).
29. A. Sermwatanakul. *Fish for the People*, *17*(**2**), 42-48, (2019).
30. K. Srikulnath, W. Singchat, N. Laopichienpong, S.F. Ahmad, M. Jehangir, N. Subpayakom, A. Suntronpong, K. Jangtarwan, T. Pongsanarm, T. Panthum, N. Ariyaraphong, J. Camcuan, P. Duengkae, S. Dokkaew, N. Muangmai. *Gen. Genom*, **43**, 91–104, (2021).
31. G. Sanford. *App. Press.* **68,** (1995).
32. G.A.D. Wahjudy. *Skrip. Uni. Air. Sby*. (2016)
33. E.N. Collias, & E.C. Collias. *Prince. Uni. Press*, **(857)**, (2014).
34. Agustiansyah. *J. Pro. Kes. Ind.* **1(2),** (2006).
35. K.L. Shephard. *Bio. Fish.* **4,** 473–485. (1994).
36. P.K. Stoddard, M.R. Markham, V.L. Salazar, & S. Allee. *Physic & behav*, ***90*(1),** 11-20. (2007).
37. M.V.L. Bennett. *Ann of the NY Academ. of Sci*, ***188*(1)**, 242-269. (1971).
38. K.S. Boyle, C. Orphal, & P. Eric. *Proc. of the Roy. Soc. B: Biol. Sci,* (**281**), (2014).
39. Crampton, G.R. William, & D.H. Carl. *Copeia,* 48-60, (2005).
40. Zupanc, K.H. Günther & H. B. Theodore. *Electroreception*, 5-46, (2005).
41. Caputi, A. Angel & B. Ruben. *J. of Comp. Physi,* **(192)**, 587-600, (2006).

Identification of pathogenic bacteria (*Vibrio* sp) in the waters of the industrial area of north aceh regency and lhokseumawe city

Muliari Muliari[1]*, *Mahdaliana* Mahdaliana[2], *Irfannur* Irfannur[3], *Agung* Setia Batubara[4], *Dea* Opiani[5], *Salmarika* Salmarika[1]

[1]Department of Marine Science, Faculty of Agriculture, Malikussaleh University, North Aceh, Indonesia
[2]Departement of Aquaculture, Faculty of Agriculture, Malikussaleh University, North Aceh, Indonesia
[3]Departement of Aquaculture, Faculty of Agriculture, Almuslim Univer*sity, Bireuen, Indonesia
[4]Departement of Biology, Faculty of Mathematics and Natural Sciences, Medan State University, North Sumatra, Indonesia
[5]Student Department of Marine Science, Faculty of Agriculture, Malikussaleh University, North Aceh, Indonesia

Abstract. North Aceh Regency and Lhokseumawe City's waters are mangrove ecosystems, estuaries, fisheries management activities, and industrialization activities. Activities in the industrial areas of North Aceh Regency and Lhokseumawe City have the ability to contaminate waters and affect the chemical and physical composition of the water, which has a direct impact on the abundance of harmful bacteria (*Vibrio* sp). The study looked at the distribution of harmful bacteria (*Vibrio* sp) in water near industrial areas in North Aceh Regency and Lhokseumawe City. The purpose of this study is to investigate the abundance of pathogenic bacteria (*Vibrio* sp.) in the waterways of North Aceh Regency and Lhokseumawe City's industrial areas. This study will also look into the relationship between water quality factors and the abundance of harmful bacteria. The study was carried out in July and August 2023 at four stations placed in marine waters near the industrial area. The research methods used in this research are survey methods and laboratory analysis. Purposive sampling and the ANOVA test were the analysis and sampling techniques employed in this study. The findings demonstrated that there was no discernible variation in *Vibrio parahaemolyticus* between stations 1, 2, 3, and 4. Station 2 saw a strong *Vibrio cholerae* effect ($p > 0.05$), resulting in 87.20 74.36^b CFU/ml. Vibrio total from the four research stations that significantly affected station 2 ($p > 0.05$) were 280.00 167.26^b CFU/ml.

1 Introduction

North Aceh Regency and Lhokseumawe City's waters are near mangrove ecosystems, estuaries, fisheries management activities, and industrialization activities. This coastline area is one of Aceh Province's economic growth areas, having been recognized as a special economic zone (KEK) in which new industries will operate [1]. North Aceh Regency and Lhokseumawe City are Aceh Province regencies/cities with significant potential in the

*Corresponding author: muliari@unimal.ac.id

marine and fisheries sector, including both large and small pelagic fish. Fish is a popular source of animal protein since it is very inexpensive and easy to procure [2].

Bacteria are naturally occurring microbes in marine settings, and some species are pathogenic [3]. Vibrio species found in estuaries and coastal waters pose the greatest threat to human health worldwide [4]. *Vibrio* sp is an agarolytic marine bacterium commonly found in marine and estuarine habitats [5]. The presence of *Vibrio* sp bacteria is significantly impacted by salinity, ammonia, and water temperature [6]. Many bacteria can proliferate and contaminate the fish as a result of poor management [7]. *Vibrio* sp has caused mass mortality in Indian fisheries (Scylla sp.) [8]. *Vibrio* sp bacteria are frequently found in marine waters. *Vibrio* sp bacteria have been detected on contaminated water surfaces as a result of organic waste disposal [9].

The marine industrial area on the coast of North Aceh Regency and Lhokseumawe City is one of the most important places for aquaculture production, such as shrimp cultivation. *Vibrio* sp is a pathogenic bacterium that causes lethal illness outbreaks and is the most dangerous in shrimp farming [10]. This disease can be fatal [11]. Shrimp farming intensification has resulted in water quality issues and the proliferation of dangerous microorganisms [12]. Shrimp are also extremely vulnerable if exposed to harmful microorganisms [13]. A high *Vibrio* sp bacterial density can make shrimp susceptible to vibriosis [14]. *Vibrio* sp can also destroy dyes like trypan blue [15], remove diesel oil efficiently [16], and prevent the establishment of toxic algae [17]. Alkalinity and organic matter are influenced by environmental conditions [18]. Water contaminants are produced as a result of industrialization, climate change, and urbanization [19].

Currently, fisheries production is declining due to illness infections, one of which is caused by bacteria [20]. Water quality can be harmed by pollution from industry, residential activities, and aquaculture [21]. The huge number of industries emerging and developing in Lhokseumawe and North Aceh Regency can have an impact on the water environment [22]. The purpose of this study is to assess the number of pathogenic bacteria (*Vibrio* sp) in the waterways of North Aceh Regency and Lhokseumawe City, as well as the relationship between water quality measures and the abundance of pathogenic bacteria. It is widely recognized that the objective of the program to increase the value of fisheries production was to increase the annual productivity of fish production in Lhokseumawe City [23]. Decades-long human anthropogenic research has altered the aquatic environment significantly, and assessments of non-metallic contamination sources in the Asin Pusong reservoir, Lhokseumawe City, have also been conducted [24]. Consequently, the aforementioned research must also be conducted in the bodies of water encircling other industrial zones in Lhokseumawe City. To proactively mitigate disasters, environmental conservation is of the utmost importance [25]. Bacteria including V. cholera are prevalent in estuaries, wetlands, rivers, and the ocean [26]. The microbes that frequently contaminate seafood are members of the genus Vibrio. The aforementioned bacteria strain has the potential to contaminate seafood, thereby diminishing its export quality [27]. Flows from river estuaries that transport both inorganic and organic detritus into seawater are the source of the elevated concentrations of nitrate and phosphate in coastal [28].

2 Research methods

The study was conducted in July and August of 2023. Water quality assessments in the Lhokesumawe City and North Aceh Regency industrial waterways were being conducted at the same time as water samples were being taken. The research location points were conducted at 4 stations, with 25 location points at each station. Over the course of two months, one sample of the water was taken every week. Following the determination of the coordinates using a Global Positioning System (GPS) instrument, the sampling coordinates were determined.

Water samples of *Vibrio* sp. were collected once at each of the one hundred recorded locations. The measurement data indicates the water quality both at the time the data is collected and at the designated collection site, which is the water column or body of water. The sample is subsequently identified through a series of analytical phases in the laboratory. Establishing agar media is the initial step; planting Vibrio, Vibrio TCBS, and Vibrio TSA make up the subsequent three stages, respectively. Following completion of these procedures, the specimen is positioned within an incubator at a temperature range of 35C to 37C for a duration of 24 hours.

With a purposive sampling technique and the ANOVA test, this study used survey methodologies as well as laboratory analyses. The research site is depicted in Figure 1. Purposive sampling was employed to choose the sampling location in order for the data collected afterwards to be more representative [29]. Sampling is an important step in the analysis process [30].

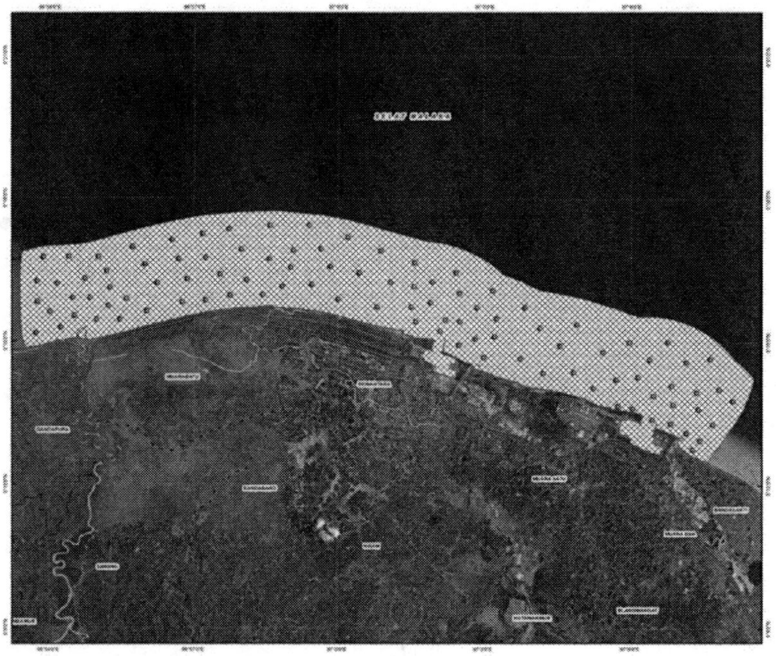

Fig. 1. Location of research sample points

2.1 Statistical data analysis

An ANOVA test was used to evaluate significant variations in water parameters for pathogenic bacteria at the research site. In this study, data was analyzed using SPSS Statistics 20 software to determine the total bacteria in industrial area waters at four locations.

3 Result and discussion

In this research, a study was carried out regarding water quality in industrial areas at each station. The following is data on water quality parameters from 4 research stations located in industrial areas in North Aceh Regency and Lhokseumawe City (Table 1).

Temperature, salinity, brightness, pH, and DO differ dynamically and evenly among the four research locations when compared. At station 1, the temperature ranged from 32.3°C to

20.8°C, with an average of 31.12°C. At station 2, the temperature ranged from 31.9°C to 30.5°C, with an average of 31°C. The average temperature at station 3 was 31.3°C, with a range of 32.3°C and 30.8°C. Additionally, the temperature at station 4 ranged from 32°C to 31°C, with an average of 31.44°C.

Table 1. Data on water quality parameters at the research location

No	Station	Water Quality Data Parameters														
		Temperature			Salinity			Brightness			pH			DO		
		Q	R	X	Q	R	X	Q	R	X	Q	R	X	Q	R	X
1	Station 1	32.3	20.8	31.1	25	20	22	1.1	0.6	0.9	8.1	8.1	8.1	5.5	4.8	5.2
2	Station 2	31.9	30.5	31	25	20	23.4	7	0.5	3.4	8.8	8	8.1	10.9	7.9	8.7
3	Station 3	32.3	30.8	31.3	22	18	20	6	0.8	4.1	8.2	8	8.1	9.8	7.3	7.8
4	Station 4	32	31	31.4	25	16	19.1	6	1	3,4	8.14	8.1	8.1	7.9	6.9	7.4

Information:
T = Height
R = Low
X = Average

At station 1, the salinity parameter ranges from 25 ppt to 20 ppt, with an average of 22 ppt. Station 2 had a salinity range of 25 ppt to 20 ppt, with an average of 23.44 ppt. The average salinity at station 3 was 20.04 ppt, with a range of 22 ppt to 18 ppt. Additionally, with an average of 19.16 ppt, station 4 had the highest 25 ppt and lowest 16 ppt. Table 1 shows the parameters that were measured for this study's pH and DO results.

Temperatures at the four research stations range from 28.2 to 32.2°C, salinity from 18.5 to 24.2, and brightness from 0.7 to 5 meters. According to the research findings of Patty et al., 2020, this is still regarded as normal and beneficial for marine biota if the temperature runs between 28.8 and 30.5°C, the salinity is between 25.0 and 34.0, and the brightness is between 4.5 and 31.5 meters. In addition, the pH and DO from the four sites varied from 8.0 to 8.3 and 8.5 to 6.7 (mg/l), respectively. The seawater quality standard (Minister of Environment Decree No. 51 of 2004) covers the pH range of 7–8.5 and the ideal DO of >5 (m/l), which are the ideal conditions for the flourishing of marine biota. According to Mahbubah et al. [31], the range for water quality is normal. Water security is a crucial concern, according to Yasin et al. [32]. We are all affected by poor water quality, which is one of the most significant indicators of environmental contamination [33].

The statistical analysis revealed that there was no significant difference in *Vibrio parahaemolyticus* between stations 1, 2, 3, and 4. Furthermore, *Vibrio cholerae* exhibited a substantial effect ($p > 0.05$) on station 2, namely 87.20 74.36^b CFU/ml. The total Vibrio from the four research sites that had a significant effect on station 2 ($p > 0.05$) was 280.00 167.26^b CFU/ml. The overall Bacterial Count from the four stations had a significant effect on station 2 ($p > 0.05$), with a difference of 301.20 1595.02^b CFU/ml. The overall plankton concentration from the four locations with a significant effect on station 2 ($p > 0.05$) was 24900.0 13439.71^b CFU/ml. Dinoflagellates from the four stations had a substantial effect on station 2, namely 56.13 30.40^b CFU/ml ($p > 0.05$).

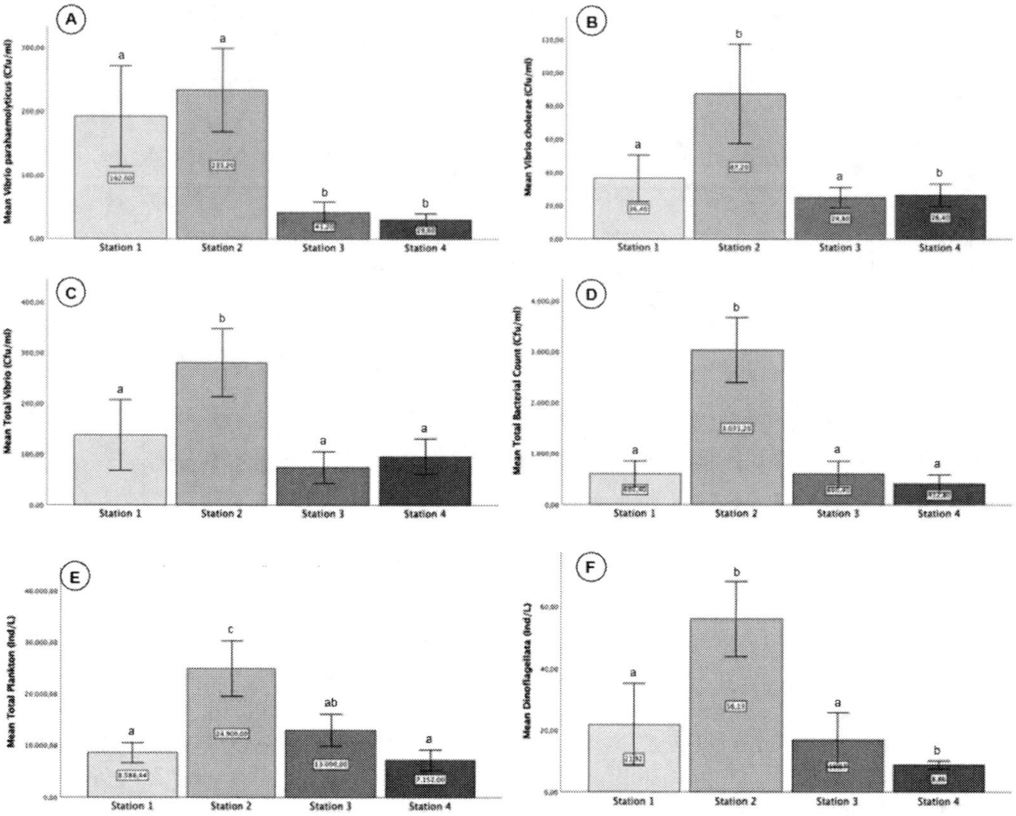

Fig. 2. Graphs (A) *Vibrio parahaemolyticus*, (B) *Vibrio cholerae*, (C) Total Vibrio, (D) Total Bacterial Count, (E) Total Plankton, (F) Dinoflagellates

Vibriosis is a disease caused by the presence of *Vibrio* sp bacteria that has the ability to kill aquatic species [34]. This bacterium thrives in salty environments [35]. Vibriosis illnesses, such as the bacteria *Vibrio parahaemolyticus*, frequently cause large fatalities in grouper fish commodities [36]. N-hexane extract from *R. stylosa* leaves has the potential to be used as an antibacterial agent in tilapia (*Oreochormis niloticus*) against *Vibrio* sp and *A. hydrophila* [37]. *Sargassum* sp. extract was found in the white sea bass diet and was found to alter phagocytic activity when tested with *Vibrio alginolyticus* bacteri [38]. Apart from the primary trigger, *Vibrio* sp dinoflagellates are the cause of white feces sickness [39]. The distribution of phytoplankton is regulated by the physical and chemical characteristics of water, particularly the availability of nutrients [40]. These characteristics include dissolved oxygen, temperature, salinity, pH, suspended solids/SPM, nitrate, and phosphate. Mass shrimp cultivation is susceptible to mortality caused by Vibrio sp., a facultative anaerobic gram-negative bacterium [41]. Plankton abundance is a useful tool for determining the fertility level of the waters in the research area. The presence of garbage contaminates the aquatic ecosystem since it lowers the water's quality and, in particular, the diversity of plankton [42]. The research site is used for brackish water aquaculture, particularly shrimp cultivation, in addition to industrial area activities.

4 Conclusion

The research results of the relative temperature distribution were the highest with a value of 32.3°C and the lowest with a temperature value of 20.8°C. Furthermore, the highest relative seawater salinity is 25 ppt and the lowest is 16 ppt. ANOVA results showed that there was no significant difference in *Vibrio parahaemolyticus* between station 1, station 2, station 3, and station 4. *Vibrio cholerae* had a significant effect on station 2 ($p > 0.05$), namely 87.20 ± 74.36^b CFU/ml. The total Vibrio from the four stations at the research location which had a significant effect on station 2 ($p > 0.05$) was 280.00 ± 167.26^b CFU/ml. The total Bacterial Count from the four stations had a significant effect on station 2 ($p > 0.05$) showing a significant difference, namely 301.20 ± 1595.02^b CFU/ml. The total plankton from the four stations that had a significant effect on station 2 ($p > 0.05$) was 24900.0 ± 13439.71^b CFU/ml. Dinoflagellates from the four stations had a significant effect on station 2 ($p > 0.05$), namely 56.13 ± 30.40^b CFU/ml.

Acknowledgments

We would like to thank the Ministry of Education, Culture, Research, and Technology Indonesia for funding this research through the Fundamental Research-Regular scheme (Number: 132/E5/PG.02.00.PL/2023, 7/UN45.2.1/PT 01.03/VI/2023).

References

1. W. Komarawidjaja, A. Riyadi, Y.S. Garno. Jurnal Teknologi Lingkungan **18**(2) 251-258 (2017)
2. F. Riana, A.S. Junaedi, M. Zainuri. Jurnal Kelautan Tropis. **24**(3) 353-363 (2021)
3. K. Mahardika, I. Mastuti, R. Septory, D. Roza, Z. Zafran, A. Nasukha. Jurnal Riset Akuakultur **16**(1) 49-59 (2021)
4. B. Meena, L. Anburajan, T. Sathish, A.K. Das, N.V. Vinithkumar, R. Kirubagaran, G. Dharani. Marine pollution bulletin **144** 105-116 (2019)
5. V. Letchumanan, W.S. Tan, W.F. Yin, K.G. Chan. Progress In Microbes & Molecular Biology **3**(1) (2020)
6. A. Pariakan, M. Rahim. Journal of Fisheries and Marine Research **5**(3) 547-556 (2021)
7. B. Ihsan. Jurnal Pengolahan Hasil Perikanan Indonesia. **24**(1) 89-96 (2021)
8. S. Saha, A.D. Dash. Uttar Pradesh Journal of Zoology **42**(19) 48-58 (2021)
9. F. A. Fitri, F. Feliatra, D. Yoswati. Asian Journal of Aquatic Sciences **3**(2) 189-192 (2020)
10. L.A. Prastiti, A.H. Verdian, N. Fatimah, K. Fathurohman, A.F. Siburian. Intek Akuakultur **6**(2) 148-157 (2022)
11. I. Fatmala, H. Pranggono, L. Linayati. Jurnal Litbang Kota Pekalongan **16** (2019)
12. B. Widigdo, M. Yuhana, A. Iswantari, C. Madonsa, I.D. Sapitri, Y. Wardiatno, A.A. Hakim, F. Nazar. Jurnal Pengelolaan Sumberdaya Alam dan Lingkungan (Journal of Natural Resources and Environmental Management) **11**(1) 130-140 (2021)
13. M.D. Sani, A.Y. Maharani, M.I. Riandy, R. Joko, K. Susilo, P.A. Wiradana, A. Soegianto. Ecology, Environment and Conservation. **26**(3) 1271-1275 (2020)
14. S.H. Amrullah, K. Mar'iyah, A.A. Afandi. Indigenous Biologi: Jurnal Pendidikan dan Sains Biologi **6**(1) 8-14 (2023)
15. S.D. Khandare, N. Teotia, M. Kumar, P. Diyora, D.R. Chaudhary. Biocatalysis and Agricultural Biotechnology **51** 102802 (2023)

16. H. Zhou, L. Jiang, K.Li, C. Chen, X. Lin, C. Zhang, Q. Xie. Science of The Total Environment **793** 148529 (2021)
17. Y. Wang, S.Li, G. Liu, X.Li, Q. Yang, Y. Xu, Z. Hu, C.Y. Chen, J.S. Chang. Bioresource technology **295** 122246 (2020)
18. C. Handayani, A.D. Pahlewi, A. Listriyana. Zona Laut: Journal of Ocean Science and Technology Innovation 56-61 (2023)
19. C. Zamora-Ledezma, D. Negrete-Bolagay, F. Figueroa, E. Ni, M. Zamora-Ledezma, F. Alexis, V.H. Guerrero. Environmental Technology & Innovation **22** 101504 (2021)
20. S. Haryanto, M. Ansari. Buletin Teknik Litkayasa Akuakultur **18**(2) 153-156 (2020)
21. S.R.P. Maresi, P. Priyanti, E. Yunita. Al-Kauniyah: Jurnal Biologi **8**(2) 113–122 (2015)
22. A.A.F. Harahap, K. Khairunnisa, V. Novalia. Jurnal Glosains **3**(2) 79-86 (2022)
23. S. Sufi. Negotium: Jurnal Ilmu Administrasi Bisnis **3**(2) 96-119 (2020)
24. R. Ezraneti, S. Syahrial, E. Erniati. Jurnal Kelautan Tropis **24**(1) 34-44 (2021)
25. D.A. Rahman, M.B. Abubakar, A. Zulhilmi, M. Muchlis, M. Rizwan. Jurnal Malikussaleh Mengabdi **2**(1) 201-213 (2023)
26. J.F. Mehrabadi, P. Morsali, H.R. Nejad, A.A.I. Fooladi. Journal of infection and public health **5**(3) 263-267 (2012)
27. S.R. Sari, A. Baehaki, S.D. Lestari, E. Arafah. Jurnal Pengolahan Hasil Perikanan Indonesia **23(**3) 542-547 (2020)
28. F. Musafiraand, K. Khairunnisa. Jurnal Kelautan dan Perikanan Indonesia **3**(2) 66-78. (2023)
29. T.F.Z. Aufar, K. Kunarso, L. Maslukah, D.H. Ismunarti, A. Wirasatriya. Indonesian Journal of Oceanography **3**(2) 189-196 (2021)
30. C. Scapuzzi, B. Benedetti, M. Di Carro, E. Chiesa, N. Pussini, E. Magi. Applied Sciences **12**(6) 2951 (2022)
31. N.U. Mahbubah, I. Wirawan, S.O. Madyowati. Jurnal Ilmiah Kelautan dan Perikanan **4**(3) 264-272 (2023)
32. S.N.T.M. Yasin, M.F.M. Yunus, N.B.A. Wahab. Journal of Educational and Learning Studies **3**(1) 14 (2020)
33. I. Yaroshenko, D. Kirsanov, M. Marjanovic, P.A. Lieberzeit, O. Korostynska, A. Mason, I. Frau, A. Legin. Sensors **20**(12) 3432 (2020)
34. K.N. Ambat, I.W. Abida, R. Maherlina. Jurnal Ilmiah Kelautan **3**(3) 66-72 (2022)
35. P. Kusumaningsih. Food Scientia: Journal of Food Science and Technology **2**(1) 99-118 (2022)
36. R.A. Zulfaizi, G.I. Setiabudi, I. Mastuti, K. Mahardika. Jurnal Perikanan Unram **12**(3) 378-388 (2022)
37. S.R. Rahayuningsih, S.S. Patimah, T. Mayanti, M.M. Journal of Marine Research **12**(1) 1-6 (2023)
38. M. Azuwarita, R. Wulandari, A. Zahra. Intek Akuakultur **5**(1) 27-39 (2021)
39. M. Adam, I. M. Widiastuti, A.Y. Yayan, E. Insivitawati, R.F. Pakaya, A. Soegianto, A. Khumaidi. Jurnal Ilmiah Perikanan dan Kelautan **14**(1) (2022)
40. S. Abubakar, N. Akbar, A. Baksir, H. Umasangadji, N. Najamuddin, I. Tahir, R.E. Paembonan, F. Ismail. Jurnal Kelautan: Indonesian Journal of Marine Science and Technology **14**(2) 149-163 (2021)

41. F.R. Mahulauw, A. Lamadi, M. Mulis. Nike: Jurnal Ilmiah Perikanan dan Kelautan **10**(1) 031-039 (2022)
42. S. Safnowandi. Jupe: Jurnal Pendidikan Mandala **4**(5) 195-201 (2019)

Comprehensive analysis of harmful algal blooms in indonesia: from occurrence to impact

Tumpak Sidabutar[1*], *Endang* Sunarwati Srimariana[2], *Hendrik* Cappenberg[1], and *Sam* Wouthuyzen[1]

[1]Research Centre for Oceanography, National Research and Innovation Agency (BRIN). Jl. Pasir Putih I, Jakarta 14430, Indonesia
[2]Faculty of Fisheries and Marine Sciences, IPB University, Jl. Agatis Darmaga Bogor, Bogor 16680, Indonesia

Abstract. The occurrence and frequency of harmful algal blooms have become a significant problem in Indonesia's coastal waters since 1991. This article aims to thoroughly analyze the diversity of algal species involved in these harmful algal blooms (HABs). It highlights that some algae, such as toxic species, can produce dangerous toxins, while others, such as non-toxic algae, remain harmless. This article reviews research studies that describe the occurrence, frequency, and causative types of bloom events in several contaminated coastal regions of Indonesia. The blooms had many consequences for fisheries, the aquatic environment, the economy, and public health. Among the identified species, *Pyrodinium bahamense var. compressum* is known for its high toxicity. The most common taxa contributing to bloom tragedy are *Chaetoceros, Noctiluca,* and *Skeletonema*. The study recognizes that human-induced nutrient enrichment is a major and significant factor in triggering the bloom phenomena. The paper recommends various management strategies and further research initiatives to prevent and reduce the impacts of HABs in Indonesia.

1 Introduction

An algal bloom is a notable rise in the biomass, population size, or abundance of microalgae. It has a detrimental effect on human health and industries, including tourism, aquaculture, and fishing. Algal blooms are becoming one of the most significant issues that require immediate attention and should be brought to the attention of the public, scientists, and the government due to their rising occurrence, amplitude, and impact [1,2].

An international maritime phenomenon, harmful algal blooms (HABs) are also significantly affecting Indonesia's waters. Indonesia's coastal waters have suffered dramatically due to harmful algal blooms. Hazardous algal blooms have recently become more frequent and widely distributed in Indonesian coastal waters. For the past ten years, and even now, algae blooms have been observed in several coastal waters. Additionally, the causing species have been found in certain areas where they have never before created bloom issues. An algal bloom is a seasonal occurrence that occurs naturally in the maritime environment. It is a sign of eutrophication, characterized by increased nutrient concentrations [1]. Seawater becomes discolored due to frequent phytoplankton blooms, especially in coastal waters [2].

* Corresponding author: tumpaksid@gmail.com

Algae can proliferate in water systems, producing toxic algal blooms that endanger humans and marine life. Despite being natural events, there are circumstances in which they could become hazardous. One of these is the overabundance of biomass, which might affect the oxygen content of the water. Algal blooms are usually attributed to anthropogenic activities that provide nutrients to the marine ecosystem [3]. Eutrophication is the primary cause of algal blooms [2,3]. Algal bloom epidemics are closely related to human activities along the coast, such as fishing, tourism, aquaculture, and marine pollution [4]. Algal blooms have been related to eutrophication, industrialization, domestic coastal areas, and global climate change [5].

Their primary causes are twofold: first, anthropogenic forces that cause eutrophication, and second, natural processes like circulation, buoyancy, and river flow. Most people agree that anthropogenic activity is the main factor behind all blooms. When algae meet favorable climatic circumstances, they multiply and create dense concentrations of cells known as blooms [4,5]. Because specific algal blooms can develop toxins, they can disturb microbial ecosystems, cause mass fish fatalities, and contaminate seafood. An algal bloom of this kind is referred to as toxic or hazardous [6].

There are three known hazardous algal blooms. Toxic algae blooms are the first to kill large numbers of fish and other marine life by producing solid poisons. The second bloom is not hazardous, but because of the low oxygen content in the water, it is detrimental to aquatic life. The third is safe for marine life but non-toxic. Few non-toxic blooms have detrimental impacts; most only tint water. Numerous toxic plants can bloom without turning the water a different hue. On the other hand, when the water is clear and the concentration of algae is low, bad things can happen [6,7]. Mass fish fatalities, disturbance of the marine ecosystem, financial losses for fishermen, and poisoning of tourists and customers are only a few of the negative consequences of a hazardous algal bloom [6].

In Indonesia, studies on HABs have been carried out regularly. Nonetheless, many algal bloom occurrences in this nation have not been made public or recorded. Several investigations have documented algal blooms; these studies have identified the primary drivers for algal blooms [7]. The algal bloom producer changes from a non-toxic to a toxic species, and the bloom events happen randomly. An overview of red tide, or HABs, species, and algal blooms in Indonesian coastal waters from 1990 to 2019 is the goal of this page. Most of our conclusions were derived from our observations and our examination of previous reports and documents, including unpublished data. This research aims to present a thorough understanding of the different elements influencing Indonesia's seas and the effects of HABs on marine equilibrium. We will also be able to determine interventions that can assist in raising the standard of the nation's water. We want to use this study to educate the public and policymakers about the several elements influencing Indonesia's waters and how organic pollution affects the ocean's equilibrium.

2 Methods

2.1 Research Location

The tragedy of an algal bloom situation in Indonesian coastal waters was first documented in 1990, and it is still happening today. Nevertheless, many algal bloom incidents in Indonesian seas have not been documented or published. Conversely, algal bloom occurrences in Indonesian waters are becoming more prevalent. We monitored several Indonesian coastal waters, such as Jakarta Bay, Ambon Bay, Kao Bay, and Lampung Bay, where algal blooms were common. Figure 1 displays a map of Indonesia along with the research location. These investigations were carried out by the Research Center for Oceanography (RCO-LIPI) as a component of the Indonesian HABs monitoring program. The majority of area investigations are the result of the HABs' growing experience. The experiments were mainly carried out in the middle of the dry season and during the March–April–May transition period from the wet to the dry season (June-Jul-Aug). The rainy or western season runs from October to March, whereas the dry or eastern season is from April to September.

2.2 Sampling phytoplankton

A cone-shaped phytoplankton net with a squeeze device attached to the end was used to gather concentrated phytoplankton samples. The mouth opening diameter measures 25 cm, the net length measures approximately 125 cm, and the pore size, or mesh size, is 20 m. Ballast was attached to the net's end to aid in vertical pulling down. In the sampling approach, the phytoplankton net is dropped vertically to a depth of 7–10 m, and from that point on, it is continuously and gently pulled to the surface. We put the gathered samples into vials that had been preservative-infused. The samples were promptly preserved in an acidic Lugol solution [8]. Using an Olympus inverted microscope, the preserved phytoplankton samples were taken to the lab for additional quantitative and qualitative investigation (model IX50-S8F2). Phytoplankton identification and cell abundance calculations using the Sedgwick-Rafter Counting Cell [8]. Sources for identifying phytoplankton with [9–11]. using the technique outlined in [8] to count the quantity of phytoplankton cells. Each station's total number of phytoplankton cells was measured per cubic volume.

2.3 Nutrient and oceanographic parameters

Most of the nutrients were dissolved inorganic, including phosphate (PO_4-P) and nitrogen (NO_3-N). At every fixed station, water samples were taken to analyze the nutrients. Using a Kemmerer sampler, surface waters are sampled at a depth of 1.0 m for nutrient analysis. Every fragment was promptly moved to a container cleaned with acid and acidified using 1% v/v HNO3. A spectrophotometer with wavelengths of 690 nm for phosphate and 543 nm for nitrate was used to measure the concentration of each nutrient. Following the procedures [12], Millipore filter paper with a pore size of 0.45 m was used to filter the water sample. The Philips PYE Unicam spectrophotometer model (model PU8600) was utilized. All samples were generally examined as quickly as possible after being collected, especially if the concentration was low. On-site measurements comprised oceanographic characteristics such as salinity, temperature, dissolved oxygen, and water depth. A YSI dissolved oxygen meter (Model 59) was used to directly detect the water's temperature. An ATAGO handheld refractometer determined the salinity (in parts per mille). By submerging the electrode, the pH meter of the TOA model of the HM-IK model measures saltwater's acidity (pH).

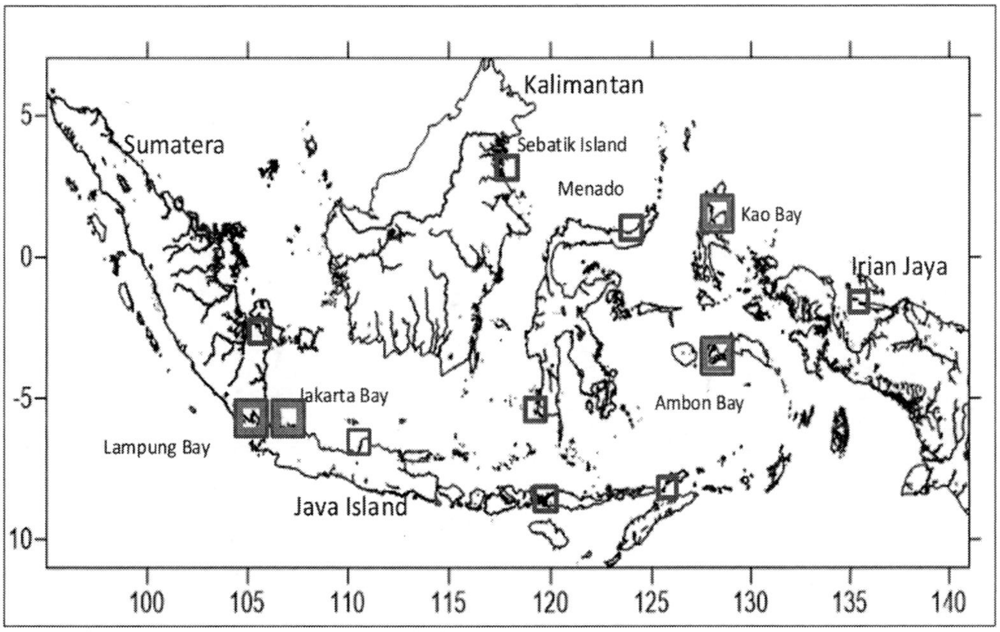

Fig. 1. Diagram depicting algal bloom incidents in the marine regions of Indonesia. Although not every algae outbreak resulted in fish kills, such events have been confirmed in Jakarta, Lampung, Ambon, and Kao bays (red and blue box in the middle). (A red box indicates algae blooms).

3 Results and Discussions

3.1 Harmful Algal Bloom (HABs) in Indonesian waters

The first tragedy of red tide (algal bloom) in Indonesia was documented in 1980 when an outbreak of the toxic dinoflagellate *Pyrodinium bahamense var compressum* caused a red color in Kao Bay. This led to the first recorded report of a dangerous algal bloom in Indonesia's coastal waters. The causative species causing problems in Malaysia, Papua New Guinea, and the Philippines is similar to this toxic species [13,14]. What was notable was the fact that most of the bloom spread near the mangrove regions. There were also many dead fish floating near the shore. The earthquake in eastern Indonesia in 1980 was the cause of the bloom episode in this bay. Dinoflagellate cysts deposited on bottom sediments were likely transported and destroyed by the earthquake. Under unstable conditions, the dinoflagellate forms cysts or resting cysts and is deposited in the bottom sediments over extended periods [15]. After being cleared of sediment and placed in environments with improved salinity, temperature, oxygen levels, and light intensity, the cysts hatch. This deadly species has never previously bloomed, as evidenced by a dormant cyst in depositional mud [15-17]. Some residents near the bay consider the red color of blood in the water to be a superstitious belief or a curse. Luckily, their superstitious beliefs proved extremely useful, as people avoided poisoning by refusing to eat fish or shellfish caught from the water during the episode [13,18].

There are three different types of harmful algal blooms (HABs) in Indonesia. The first cause is an accumulation of toxic algae, leading to many marine life's extinction. Most blooms that are not harmful change the color of the water, but few are harmful. The second type of HAB involves a toxic algal bloom that either paralyzes or poisons shellfish, thereby killing them (diarrhea shellfish poisoning). There are times when toxic plankton blooms do not change the color of the water. The bloom of harmful plankton with loads below the permissible limit is not flagged as an HAB event. For this reason, not every bloom of toxic plankton is recorded as an HAB event. The color of the water seemed to depend on what species were blooming in it. Much of the color comes from the mixture of one or two species in the phytoplankton population during bloom

Table 1. The frequency of discoloration due to algal blooms in Jakarta Bay

Month	**Periods**										**Frequency**
	2004	**2005**	**2006**	**2007**	**2008**	**2009**	**2010**	**2011**	**2013**	**2015**	
Jan	-	-	-	-	-	-	-	-	-	-	-
Feb	-	-	-	-	-	-	-	-	-	-	-
Mar	x	-	-	-	-	-	-	-	x	x	3
Apr	x	x	-	x	x	x	x	x	x	x	9
May	x	x	x	x	x	x	x	x	x	x	10
Jun	x	-	x	-	-	x	x	x	x	x	7
Jul	x	-	-	x	-	-	-	-	x	x	4
Aug	x	-	-	x	-	x	-	-	-	-	2
Sep	x	x	x	x	x	x	x	x	x	x	10
Okt	x	x	x	x	x	x	x	x	x	x	10
Nov	-	-	x	x	x	x	x	x	x	x	8
Dec	-	-	-	-	-	-	-	-	-	-	-

Note: Periods 2004-2007, according to Wouthuyzen *et al.* (2007)

In 2000, algal blooms occurred more frequently and in more locations in Indonesia's coastal waters. In Indonesian waters, diatoms, dinoflagellates, and cyanobacteria are the main species causing blooms. Some are toxic, such as *Pyrodinium bahamense var. compressum* (Pbc). This type of HAB is dinoflagellates, and these organisms are known to produce potent toxins. The most common bloom species include *Noctiluca, Skeletonema, Chaetoceros, Trichodesmium, Thalassiosira*, and *Ceratium furca*. HABs are associated with various conditions in the region, including a reverse monsoon wind, river run-off, rising coastal waters with southwest and northeast winds, and near-shore eutrophication. To date, poisonous or toxic species have been found in four locations: Ambon Bay, Kao Bay, Jakarta Bay, and Lampung Bay. *Pyrodinium* (Pbc), *Alexandrium, Gymnodinium,* and *Pseudonitschia* are poisonous species already known to inhabit these locations. Ambon Bay, Kao Bay, and Lampung Bay are the most dangerous places because of the toxic species *Pyrodinium* (Pbc), which produces saxitoxin. Due to the increasing number of blooms, the occurrence of this species in Lampung Bay is becoming increasingly problematic. The dinoflagellate *Pyrodinium* (Pbc) typically forms a dormant cyst during blooming. This cyst can remain in the sediment for a long time. Therefore, integrated planning, including constant monitoring, is carried out in Lampung Bay, Kao Bay, and Ambon Bay.

3.2 Harmful Algal Bloom (HABs) in Jakarta Bay

Since 2000, the most common algal bloom in Indonesia has been observed in Jakarta Bay. Table 1 shows the frequency of algal bloom-related discoloration in Jakarta Bay between

2004 and 2015. From 2004 to the present, algal blooms have become more frequent and widespread. Algal blooms typically reoccur in a specific bay region near the mouth where thirteen rivers flow into that bay. The algae bloom in Jakarta Bay has worsened and now poses a significant threat to the ecosystem and is affecting the local economy and fisheries. Jakarta Bay is already known for its severe eutrophication. A sign of eutrophication [19,20], characterized by increased nutrient concentrations, especially phosphorus and nitrogen [21,22], is the appearance of algal blooms. Increasing inputs of phosphorus (P) and nitrogen (N) are often the leading cause of coastal eutrophication [23]. Nowadays, red tide or algal blooms are often associated with coastal eutrophication in Jakarta Bay [24-27]. The leading cause of the bloom of the phytoplankton population in Jakarta Bay is the eutrophic condition or nutrient enrichment [21,22].

In Jakarta Bay, algal bloom phenomena occurred mainly from March to June, with sporadic occurrences from September to November (transition from dry to wet season). Algae blooms do not discolor surface water between December and February, the rainy season. The phenomenon of algal bloom in Jakarta Bay is related to the dry season, when sufficient nutrients, especially nitrogen and phosphorus, are in the water after the rainy season. It has also been observed that algal blooms in Jakarta Bay typically occur in March and April, towards the end of the rainy season [30]. During our study, several algal blooms occurred in Jakarta Bay between 2008 and 2015. But not every algae bloom ends with fish mortality. In April and November 2008, and again in November and December 2015, fish died from two algae bloom events. Physical elements that harm fish and other aquatic life have been linked to bloom events. The main driving force for this phytoplankton bloom is eutrophication. But other elements, such as water stratification and currents, can also influence bloom development. The most often tragic fish killing during the bloom in Jakarta Bay was caused by the depletion of oxygen concentration in the water, with the DO concentration in the surface water during the bloom being less than 2.0 ppm and even close to zero at the bottom [30].

3.2 Harmful Algal Blooms (HABs) in Lampung Bay

The first algal bloom disaster was documented in Lampung Bay in 1991 when many farmed shrimp died in brackish water ponds on the east coast. The brackish water pond is the only place in the bay where fish kills occur frequently during the bloom. *Trichodesmium erythraeum* was identified as the most causative species [31]. *Pyrodinium* was found in Lampung Bay in 1999, mainly concentrated in Hurun Bay, and this has been linked to an algal bloom tragedy in the bay [32]. The *Pyrodinium* species can cause shellfish poisoning and lead to paralysis when consumed by humans via infected shellfish. However, there were no reports of fish kills in this bay during the *Pyrodinium* bloom.

Table 2. The primary causative species of algal bloom in Jakarta Bay

No	Causative Species	Potentially	Impact
1	*Skeletonema sp*	Harmful	Discoloration, fish kill
2	*Chaetoceros sp*	Harmful	Discoloration, fish kill
3	*Thalassiosira sp*	Harmless	Discoloration
4	*Trichodesmium sp*	Harmful	Discoloration, fish kill
5	*Noctiluca (green)*	Harmful	Discoloration, fish kill
6	*Ceratium furca*	Harmless	Discoloration
7	*Alexandrium sp*	Toxic	-
8	*Pseudo-nitzchia sp*	Toxic	-
9	*Dinophysis sp*	Toxic	-

10	*Coscinodiscus sp*	Harmless	Discoloration
11	*Leptocylindricus sp*	Harmless	Discoloration
12	*Thalassiotrix sp*	Harmless	Discoloration
13	*Bacteriastrum sp*	Harmless	Discoloration
14	*Pyrodinium* (Pbc)	Toxic	-
15	*Gymnodinium sp*	Toxic	-

In August 2005, Lampung Bay (Hurun Cove) recorded a *Noctiluca* bloom with the highest abundance of 10^5 cells/L. During the bloom period, fish deaths occur in the floating fish culture nets (KJA). An increase in fish mortality in floating net fish farms at these locations has been linked to *Noctiluca* blooms. Table 3 lists the algal bloom periods in Lampung Bay between 2002 and 2013. Between 2002 and 2013, Lampung Bay experienced an increase in algal blooms. Since the 2000s, there have been repeated outbreaks of the phytoplankton population in Lampung Bay. Phytoplankton blooms, which occur throughout the year, have also tended to increase in quantity, frequency, and range. During the algae bloom of 2002-2004, there were no significant fish deaths; only a few dead fish appeared in some floating net cages. However, there were significant fish losses owing to the 2008 *Noctiluca* bloom. Since then, the re-occurrence of *Noctiluca* blooms has led to a decline in fish kills and die-offs in the wild, particularly among fish kept in floating net cages.

Noctiluca was the predominant species causing algal blooms in this bay from 2005 to 2010, as shown in Table 3. This species is the predominant or even dominant species during most blooming phases. Often, the outbreak of the *Trichodesmium* population occurs before or at the same time as the *Noctiluca* bloom. Other causative species, including *Pyrodinium bahamense var. compressum* (Pbc), were detected in water in 2005, 2006 and 2007 without discoloration. *Pyrodinium* blooms often occur in colorless surface water. However, its presence is dangerous as it can produce a highly potent toxin such as saxitoxin. The *Pyrodinium* species was predominantly observed in the pearl mussel and fish breeding area of Hurun Cove as the farming location. *Phaeocystis sp* was the most common causative species of bloom events in 2009, followed by *Dynophysis sp* in 2010 and 2011. *Trichodesmium sp* was widespread from 2002 to 2012, and *Ceratium sp* occurred in 2011. *Pseudonitzchia sp* was widespread from 2009 to 2012, and *Cochlodinium sp* was widespread from 2012 to 2014. A dinoflagellate species called *Alexandrium sp* was discovered in the bay in 2014, which outbreak after the disappearance of *Cochlodinium sp*. Nevertheless, no fish mortalities were reported [33].

Only a few bloom causative species in Lampung Bay are considered dangerous or toxic, including *Pyrodinium sp* and *Cochlodinium sp*. Some causative species are considered harmful or harmless or cause bloom-forming or surface water discoloration. The abundance of *Cochlodinium sp* was laying between 10^5 and 10^6 cells per liter during the bloom period in October 2012, harming many cultured fish in Lampung Bay. The bloom of this dinoflagellate species is the first event documented, recorded, and observed in the sample collected from this bay. So far, before this bloom incident, this species had never been detected, not even in the samples.

Table 3. The episode of causative species of algal bloom in Lampung Bay (2002-2014)

Species	**Occurrence**												
	2002	**2003**		**2005**	**2006**	**2007**	**2008**	**2009**	**2010**	**2011**	**2012**	**2013**	**2014**
Pyrodinium	x	x	-	x	-	-	-	-	x	-	-	-	x
Noctiluca	x	-	x	x	x	x	x	x	x	x	x	x	x
Phaeocystis	-	-	-	-	-	-	-	-	x	x	-	-	-

Dinophysis	-	-	-	-	x	x	-	-	-	-	x	-	-
Trichodesmium	-	-	x	x	-	-	x	-	x	x	x	-	-
Ceratium	-	-	-	-	-	-	-	-	-	x	-	-	-
Prorocentrum	-	-	-	-	-	-	-	-	x	-	x	-	-
Pseudonitzchia	-	-	-	-	-	-	-	-	x	-	x	-	-
Cochlodinium	-	-	-	-	-	-	-	-	-	-	x	x	x
Alexandrim sp	-	-	-	-	-	-	-	-	-	-	-	-	x

Source: Modified from Muawanah [33]

The bloom of *Cochlodinium sp* was characterized by the deep reddish-brown coloring of some of the surface water, giving the impression of an oily suspension. Fish suffer breathing difficulties as the mucous membrane covers their gills, which is believed to cause fish mortality overnight until the early morning hours. Farmed fish such as snapper, grouper, pomfret, and cobia were killed in the bloom incident. When phytoplankton bloom reached 10^6 cells/L in mid-October 2012, fish kills occurred, which increased again at the end of October. As the bloom reached high density, fish kills occurred and worsened water quality. Although the first bloom lasts nearly two weeks, fish kills still occur in floating net cages on the west side of the bay. Similar *Cochlodinium* blooms also occurred in China, Japan, and Korea, causing fish mortalities [34-36].

Cochlodinium polykrikoides was identified as the causative species following the bloom incident [15]. The bloom of *Cochlodinium sp* has been reported in Southeast Asia [37]. This species is widespread, ranging from the waters of the Atlantic through Europe to Asia. This species has never been discovered in Indonesian seas before. The first bloom of *Cochlodinium* was reported in Southeast Asia, and the first bloom of *Cochlodinium* was reported until the 1990s. Subsequently, in the 1990s, South Korea's fishing industry lost approximately US$100 million annually [37]. Since then, this species has bloomed on the coasts of North America, Europe, and Asia. Three subgroups or ribotypes of *Cochlodinum polykrikoides* have been satisfactorily identified as belonging to this species. The Philippine ribotype, the US-Malaysia ribotype, and the Southeast Asian ribotype (the original Japan-Korea group) [15]. In addition, this species was spotted on the Italian Mediterranean coast in the late 1990s. In 2001, a significant bloom was observed in the Black Sea along Tanjung Big Utrish and Odesa (*http://phyto.bss.ibss.org.ua*). The bloom tragedy of this species was also observed in the north of Tarragona in 2011[38].

The phytoplankton responsible for the algal blooms in Lampung Bay included cyanobacteria, dinoflagellates, and diatoms (Table 4). These are usually diatoms, which are potentially dangerous because when they bloom due to a lack of oxygen, they harm fish and other marine life. When they bloom, some harmless species usually coexist with the other species and change the color of the surface water. Dinoflagellates, including *Pyrodinium* (Pbc), *Cochlodinium sp., Dynophysis sp., Alexandrium sp., and Prorocentrum sp.,* are among the most common poisonous species. *Pyrodinium* is the most toxic species belonging to the dinoflagellate. This species is typically found near the mangrove forest on the western side of Lampung Bay, where fish farms are located. In addition, the bloom of *Cochlodinium sp.* was also distributed on the western side of the bay where fish farms exist. Some typical species are also found among the other phytoplankton species, as the other toxic species, such as *Dynophysis sp., Alexamdrium sp., and Prorocentrum sp.*, occur predominately but never discolor the surface.

Table 4. The main species responsible for algal blooms in Lampung Bay.

No	Causative Species	Potentially	Impact
1	*Trichodemium sp*	Harmful	Fish mortality
2	*Thalassiotrix sp*	Harmless	Discoloration
3	*Rhizosolenia sp*	Harmless	Discoloration
4	*Chaetoceros sp*	Harmful	Fish mortality
5	*Skeletonema sp*	Harmful	Fish mortality
6	*Noctiluca sp*	Harmful	Fish mortality
7	*Ceratiun furca*	Harmful	Fish mortality
8	*Pseudo-nitzchia*	Toxic	-
9	*Phaeocystis sp*	Harmless	Discoloration
10	*Dinophysis sp*	Toxic	-
11	*Prorocentrum sp*	Toxic	-
12	*Pyrodinium bah. var. compressum*	Toxic	Discoloration
13	*Cochlodinium polykrikoides*	Toxic	Fish mortality
14	*Alexandrium sp*	Toxic	-

3.3 Harmful Algal Blooms (HABs) in Ambon Bay

An algae bloom occurred in the waters of Ambon in 1994 due to a *Pyrodinium* outbreak, which resulted in toxins when shellfish harvested from the bay were consumed. Shellfish collected from the coast resulted in the deaths of some people, and some should be hospitalized [14]. Since the tragedy of the *Pyrodinium* incident, another bloom belonging to *Trichodesmium* and *Noctiluca* occurred in Ambon Bay [39]. The bloom of *Trichodesmium erythraeum* in Ambon Bay in 1995 was the cause of the fish-killing event. In August, during the rainy season, other patches of blooms colored the surface water. There were no dead fish during this bloom in the bay. The outbreak of green *Noctiluca scintillans* appeared after the *Trichodesmium* bloom. It seems that these two bloom-maker species are causally related. The bloom of green *Noctiluca scintillans* occurred in late August during the rainy season. Bloom formation appeared on the surface during the calm conditions of the bay at low tide. The bloom formed a thick layer of green cells on the surface water. At the bloom time, the phytoplankton population had a total abundance of 10^4 cells/L, with 10^3 cells/L belonging to the *Noctiluc*a species. The first that occurred was *Trichodesmium* blooms and then followed by green *Noctiluca*. Green *Noctiluc*a are often found during the lowest tide period. No fish were killed, but the bloom resulted in declining fish production and baitfish numbers in that bay.

A yellowish-colored *Chaetoceros* bloom appeared in the Inner Ambon Bay in 1996 [39]. A bloom occurred in August, during the rainy season, but did not result in any fish kills. At that time, there were no fish kills in the bay's wilderness because there was no fish culture in the floating net. However, it is rare for multiple fish to be caught in the water during the blooming period. Up to 50% of the total population of phytoplankton species were diatoms, which dominated the bloom. The diatom *Chaetoceros sp.* was approximately 10^4 cells per milliliter. Neither the local fishing business nor the bay's environmental conditions were affected by the bloom. However, the aesthetic appeal of the bay was temporarily affected by the uneven bloom of phytoplankton on the sea surface.

At the beginning of November 1997, during the transition period, an unexpected bloom of *Alexandrium affine* occurred in Ambon Bay [40]. Although this species has occasionally been spotted in samples from this bay, this is the first time it has been seen in an abundance stage. Interestingly, this species thrived during the El Nino event, which resulted in an anomalous climate with higher-than-normal air temperatures and less precipitation. During the *Alexandrium sp* blooms, the inner bay was covered with a million cells per liter, turning the surface water reddish-brown. There were no fish kills during the bloom that could have threatened surrounding fisheries or other biota. Whether this type of dinoflagellate was poisonous or not was still a mystery.

After the 1998 El Niño events, the phytoplankton population tended to increase, in parallel with the rainfall in the area. The *Chaetoceros* outbreak, which was irregularly distributed across the inner and outer bays, was the cause of the bloom phenomena in July 1998. The patchy bloom turned the sea surface to a yellowish color. The number of *Chaetoceros* species exceeds 10^4 cells/L of the total abundance of the phytoplankton population, which reaches 10^5 cells/L. There were no fish deaths or human injuries during the bloom period. Table 5 lists the predominant bloom-forming species in Ambon Bay and their impacts.

Table 5. The primary causative species of HABs in Ambon Bay

Period	Occurrence	Causative species	Potential	Impact
1994	July	*Pyrodinum bahamense (Pbc)*	Toxic	Discoloration, fish kill
1995	August	*Trichodesmium sp, Noctiluca sp*	Harmless	Discoloration
1996	August	*Chaetoceros sp, Skeletonema sp*	Harmless	Discoloration
1997	November	*Alexandrium affine*	Toxic	Discoloration, fish kill
1998	March	*Chaetoceros sp*	Harmless	Discoloration
1999	August	*Noctiluca scintillans*	Harmless	Discoloration
2002	March	*Chaetoceros sp, Skeletonema sp*	Harmless	Discoloration
2003	April	*Chaetoceros sp, Skeletonema sp*	Harmless	Discoloration
2004	June	*Pyrodinium (Pbc)*	Toxic	Discoloration
2005	August	*Noctiluca scintillans*	Harmless	Discoloration
2006	April	*Chaetoceros sp, Skeletonema sp*	Harmless	Discoloration, fish kill
2009	June	*Pyrodinum (Pbc)*	Toxic	Discoloration
2012	June-July	*Pyrodinum (Pbc)*	Toxic	Discoloration

4 Conclusion

The three regions have different types of algae that cause HABs, as well as seasonal and annual changes. HABs are common in Jakarta Bay, Ambon Bay, and Lampung Bay. HABs occur in Jakarta Bay all year round, in the eastern half of the country from August to October and in the western region from April to June. Algal blooms are most commonly caused by diatoms such as *Skeletonema* and *Chaetoceros*. The most toxic species are dinoflagellates, including *Alexandrium sp., Pyrodinium bahamense* var. *compressum*, *Gymnodinium sp., and Cochlodinium sp*. Lampung Bay is often dominated by *Trichodesmium,* a species that produces HABs alongside *Noctiluca scintillans*. *Pyrodinium bahamense*, a poisonous species, is becoming increasingly common in Ambon Bay. In addition, the types of algae blooms triggered differed from year to year. The diatom genera *Thalassiosira, Chaetoceros,* and *Skeletonema* are common and seasonal species that play an essential role. In Indonesian coastal waters, the most common red tide species are *Skeletonema, Chaetoceros, Trichodesmium,* and *the Green Noctiluca.*

Acknowledgment

The authors would like to thank the government and the authority of BRIN. Thanks to the researchers and technicians at my university who support this project.

References

1. Anderson DM. UNESCO. Hydrobiologia **512**, 215-231 (2007)
2. Anderson DM 2008. Harmful Algae. **8**, 39 – 53 (2008)
3. Davidson K, Gowen RJ, Harrison PJ, Fleming LE, Hoagland P, Moschonas G, Jour. Env. Manag. **146**, 206-216 (2014)
4. S.A. Akbar, M. Hasan, S. Afriani, C. Nuzlia, Biodiversitas Journal of Biological Diversity **24(10)**, 5283 (2023)
5. Duarte CM Hydrobiologia **629**, 263-269 (2009)
6. Anderson DM, Gilbert PM, Burkholder JM, Estuaries, **25**, 704–726, (2002)
7. Brown AR Rev. Aquacult. **12**, 1663–168, (2020)
8. Anderson DM, Cembella AD, Hallegraeff GM. Ann. Rev. Mar. Sci. **4**, 143–176 (2012)
9. Sournia A Phytoplankton Manual Monographs on Oceanogr. Methodology UNESCO Paris. (1978)
10. Newell GE, Newell RC *Marine Plankton* A Practical Guide Anchor Press London. (1977)
11. Yamaji IE Illustration of the Marine Plankton of Japan Houkusho Osaka Japan 369 (1966)
12. Thomas CR *Marine phytoplankton*. Academic Press Inc. San Diego 262 (1995)
13. Parson TR, Maita Y, Lalli CM Manual of che. and biol. Pergamon Press Canada, (1984)
14. Wiadnyana NN, Sidabutar T, Matsuoka K, Ochi K, Kodama M, Fukuyo Y. IOC/UNESCO Paris, 53-56, (1996)
15. Wiadnyana NN, Sediadi A, Sidabutar T, and Yusuf SA. Proc. of the IOC/WESTPAC, 3rd Inter. Sci. Symp. 22-26 Nov 1994, Bali, Indonesia: 104-112, (1994).
16. Thoha H, Muawanah, Bayu IMD, Rachman A, Sianturi OR, Sidabutar T, Iwataki M, Takahashi K, Avarre J-C, and Masseret E . Front. Microbiol. **10**:306 (2019)
17. Rachman A, Thoha H, Sianturi OR, Bayu MD, Fitriya N, Sidabutar T, Witasari Y, Wibowo SPA, and Iwataki M. Philippine Jour. of Nat. Sci. 24 Nos.1 & 2, 104-115, (2019)
18. Sidabutar T, and Srimariana ES. IOP Conf. Series: Earth and Env. Sci 718 012091.(2021)
19. Sumadiharga, O. K. Lonawarta **2**: 10-17 (Indonesian) (1977)
20. Damar A, Prismayanti AD, Rudianto BY, Ramli A and Kurniawan F 2021 IOP Conf. Series: Earth and Env. Sci. 744 012009 (2021)
21. Yuliana. *IPB* Graduate School, Ph.D. Dissertation, 17 (2012)
22. Glibert P, Burkholder JM 2011 Chinese Jour. l of Oceano. and Limno. **4** :724-738 (2011)
23. Damar A, Ph.D. Thesis The Faculty of Math. and Nat. Sci. Christian-Albrechts-Univ. Kiel. (2023)
24. Sidabutar T, Srimariana ES, and Wouthuyzen S. IOP Conf. Series: Earth and Env. Sci. 429 012021 (2020)
25. Sidabutar T, Srimariana ES, and Wouthuyzen S. IOP Conf. Series: Earth and Env. Sci. 744 012077 (2021)
26. Damar A, Hesse K J, Colijn F and Yonvitner, Deep-Sea Res. Part II **163.** 72–86, (2019)
27. Damar A, Ervinia A, Kurniawan F and Rudianto, 2021 IOP Conf. Ser. Earth Env.Sci. 744 012010 (2021)
28. Sidabutar T, Dissertation. IPB Graduate School. Bogor Agri. University Bogor 123 (2017)
29. Sidabutar T. 1997. Proc. Sem. Mar. LIPI-UNHAS, Ambon, 240-247. (In Indonesian) (1997)

30. Wouthuyzen Sam, Tan CK, Ishizaka J, Hoang Son TP, Varis Rans, Tarigan S, and Sediadi A 2007 Proc. Symp. ALOS Data Model Sci. Prog. in Kyoto. (2007)
31. Adnan Q. Paper presented at Penang Meeting, Nov. 12 (1992)
32. Widiarti R Proc. JSPS-DGHF, Intl. Symp. on Fish. in Tropical Area, Bogor, 306 (2000)
33. Muawanah, Haryono T, Widiatmoko W. Report of BBPBL (In Indonesia), (2013)
34. Iwataki M, Kawami H, Matsuoka K. Phycol. Research. **55**: 231 (2007)
35. Iwatakia M, Kawami H, Mizushima K, Mikulski CM, Doucette GJ, Relox JR, Anton A, Fukuyo Y, Matsuoka K. Harmful Algae **7**, 271–27 (2008)
36. Iwataki M, Kawami H, and Matsuoka K. Phycol. Res. **55**, 231–239 (2007)
37. Kim CJ, Kim HG, Kim CH, and Oh HM. Harmful Algae **6**, 104–111 (2007)
38. Rene A, Garces E, Camp J. Harmful Algae **25**, 39–46 (2013)
39. Sidabutar T, Wiadnyana NN, and Praseno DP. Proc. of the ASEAN-CANADA Tech. Conf. and Mar. Sci. Penang, Malaysia, 19-28 (1996)
40. Wagey GA, Wiadnyana NN, and Taylor FJR. SEAHAB **4,** 1-2 (1998)

Reproductive performance of Zebra Fish *(Danio rerio)* exposed to palm oil mill efflluent in chronic toxicity

Ilham Zulfahmi[1*], *Sayyid* Afdhal El Rahimi[2], *Khalisa* Huwaina Adani[3], *Ihdina* Alfi Nasution[3], *Badratun* Nafis[4], *Cicin* Sintaria Utami[5]

[1]Department of Fisheries Resources Utilization, Faculty of Marine and Fisheries, Syiah Kuala University, Banda Aceh, 23111, Indonesia.
[2]Department of Marine Science, Faculty of Marine and Fisheries, Syiah Kuala University, Banda Aceh, 23111, Indonesia.
[3]Department of Aquaculture, Faculty of Marine and Fisheries, Syiah Kuala University, 23111, Indonesia
[4]Department of Biology, Faculty of Mathematics and Natural Sciences, Syiah Kuala University, Banda Aceh, 23111, Indonesia
[5]Department of Biology, Faculty of Science and Technology, Universitas Islam Negeri Ar-Raniry, Banda Aceh, 23111, Indonesia

Abstract. Palm Oil Mill Effluent (POME) is potentially harmful to the aquatic environment. POME contains high organic material including COD (Chemical Oxygen Demand), BOD (Biochemical Oxygen Demand) TTS (Total Suspended Solid) and various type of heavy metals. of zebra fish *(Danio rerio)*. Reproductin has an important role in producing new individuals which directly affect the population. Impaired reproductive performance potentially impairs juvenile production optimization. The present study investigated how sub-chronic toxicity of POME impact the reproductive performance used Completely randomized Design (CRD) in three treatments and four replicates based on value of LC50-96 hours (5.156 ml/l): Control (0 ml/L), treatment A 10% POME (0,5 ml/L), treatment B 20 % POME (1 ml/L). The fecundity, relative fecundity, GSI, and egg diameter were analyzed. Data was analyzed with Analysis of Variance (ANOVA) and followed with Least Significance Difference (LSD) test. Results showed that fecundity in treatment A (149 ± 38.70) and treatment B (85± 11.35) were significantly decreased compared to the control (219 ± 42.38) ($P<0.05$). While relative fecundity significantly decreased in treatment B (0.33 ± 0.13) rather than control (0,87 ± 0,14). Significantly decline is also observed on GSI in tretment A (4.79 ± 2.55%) and treatment B (2.55 ± 0.21%) compared to control (6.96 ± 1.70%). While the egg diameter only shows a significantly decline in treatment B (0.57 ± 0.18 mm) compared to control (0.71 ± 0.27 mm).

[*] Corresponding author : ilham.zulfahmi@usk.ac.id

1 Introduction

Indonesia is the largest palm oil producer. Since 1980, palm oil production in the form of CPO (Crude Palm Oil) has continued to increase with an average growth of 11.13% per year [1]. POME contains high concentrations of Biochemical Oxygen Demand (BOD) ranging from 25000-65714 mg/L, Chemical Oxygen Demand (COD) of 44300-102696 mg/L, and suspended solids ranging from 18000 to 46011 mg/L [2]. Previous studies revealed that POME cause significant decrease in plankton diversity and disrupts the physiology and reproduction of fish [3][4].

Fish is a ideal bioindicator for assessing pollution due to its sensitivity [5]. POME reported cause the alteration in reproduction of Nile tilapia (*Oreochromis niloticus*) including decreased reproduction hormonal as well as decreased gonad somatic index (GSI) and spermatocrit [4]. Exposure to POME reduced the concentration of estradiol, testosterone and progesterone in Nile tilapia *(Oreochromis niloticus)* [6]. Zebrafish *(Danio rerio)* embryos exposed to POME resulted in mandibular malformations, head and eye malformations, and yolk sac edema[7]. Exposure to POME has a negative impact on the hatching rate, survival rate, and heart rate of Nile tilapia. Hatching and survival rates decrease significantly with increasing POME concentration.Besides, malformation and heart rate also increased significantly as the concentration of POME increased [4].

Study related to the impact of POME on fish reproduction has previously been reported on Nile tilapia *(Oreochromis niloticus).* Exposure to 3.130 mg/L POME causes a significant decrease in GSI, and disorientation of gonad shape [8]. In addition to, exposure to several heavy metals (Fe, Cu, Pb and Cd) in Nile tilapia *(Oreochromis niloticus)* caused several degenerative impact and a decrease in the number of seminiferous tubules in the testes and shows deformation or changes in normal shape and the presence of severe lymphocyte infiltration in the ovaries [9].

To date, zebrafish have been used in several toxicity tests [10][11][12]. Previous studies reported the changes in reproductive capacity, increased DNA damage, and alteration in gene expression of zebrafish *(Danio rerio)* due to uranium exposure [13]. Exposure to Hg caused various side effects in fish at physiological, histological, biochemical, enzymatic, and genetic levels. Certain fish species are more sensitive to Hg toxicity than other fish species. Therefore, Hg-induced toxicological pathology in fish is influenced by factors such as species, age, environmental conditions, exposure time, and exposure concentration [14]. Exposure to mercury during the early stages of life disrupts the balance of sex hormones and gametogenesis by changing the expression of mRNA for genes involved in the hypothalamic, pituitary, gonadal axes, resulting in lower fecundity values [15]. To best our knowledge, information underlying the impact of POME on fish reproduction is still very limited. Hence, this study aims to investigated the effect of POME exposure in chronic concentration on several of reproductive variable (gonadosomatic index, fecundity, and oocyte diameter) of zebrafish.

2 Material and Methods

2.1 Study Area, POME Collection and Fish Acclimatization

This research was conducted from August to September 2023 at the Zoology Laboratory, Ar-Raniry State Islamic University, Banda Aceh. POME was obtained from the palm oil factory located in Nagan Raya Regency, Aceh Province. The POME was transported to the laboratory and stored in a cool room at < 4 °C to avoid biodegradation. As many as 45 adult female zebrafish *(Danio rerio)* with an average length of 2.7 ± 1.16 cm and weight 0.34 ±

0.26 gr were purchased at fish trader located in Banda Aceh. Adult female zebrafish are characterized by morphological characteristics of larger body size, bulging stomach, reddish urogenital opening, and paler body color [16]. Acclimatization of fist test was done for seven days. During acclimatization, the fish were fed with commercial feed (Wonder Tropical/ Protein content of 40%) twice daily (09.00 and 17.00 Indonesian Western Time) at a satiation level. The feeding was stopped 24 h before the toxicity test.

2.2 Sub Chronic Toxicity Test

The research design was a Completely Randomized Design (CRD) with four treatments followed by three repetitions for each treatment. POME is prepared to the concentration required for the sub-chronic test, namely 10% and 20% of the acute concentration (LC_{50} 96 hours; 5.156 ml/L)[10]. The detail of sub-chronic concentrations of each treatment including control (0 mL/L POME), Treatment A (0.5 mL/L POME), and Treatment B (1 mL/L POME). Preference test was conducted in aquarium (20×15×20 cm) with 3 L of volume. Each aquarium was filled with five zebrafishs. Fecundity, GSI, and egg diameter from each treatment was measured it the end of exposure period. Preference test lasting for 28 days.

The GSI was calculated using the following formula [17]:

$$GSI = 100x\,(\frac{Wg}{Wt})$$

Where, GSI is the gonadosomatic index, Wg is the gonad weight, and Wt is the total body weight. The total fecundity (F) was calculated using the equation as follows [18, 19]:

$$F = \frac{X}{Q} x\, Wg$$

Where F is fecundity, X is the partial number of eggs of the gonad, Wg is the gonad weight and Q is the partial weight of gonad. Relative fecundity (RF) was measured using the equation as follows [20]:

$$RF = \frac{F}{W}$$

Where, RF is relative fecundity (egg/g), W is the body weight and F is the total number of eggs of the gonad. The measurement of oocyte diameter was done on fish with stage IV gonad development. Gonad developmental stage IV of zebrafih was indicated by its morphology referred to [18]. The number of oocytes measured for every treatment consisted of 150 eggs using a microscope equipped with a micrometer (magnification 40X) (Figure 1).

$$D = \sqrt{Dv \times Dh}$$

Where, D is Egg diameter (mm), Dv: Vertikal egg diameter (mm) and Dh: Horizontal egg diameter (mm) [21].

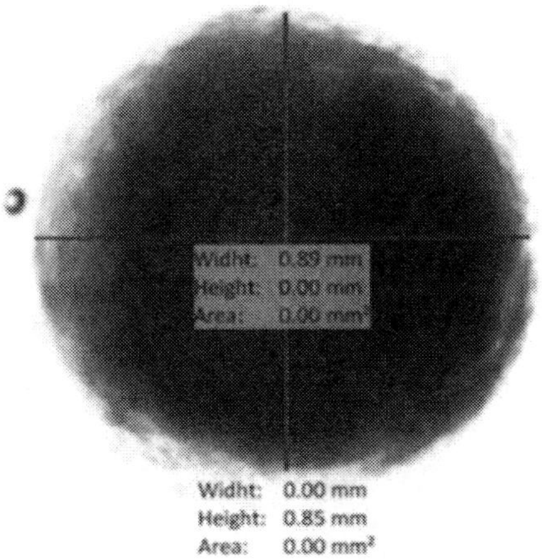

Fig. 1. Guidance in measuring the diameter of zebrafish eggs

3 Result and Discussion

3.1 Fecundity

The fecundity values ranged from 85 to 219 eggs/individual. The highest fecundity value was found in the control (219 eggs ± 42.38), followed by treatment A (149 ± 38.70 eggs), and treatment B (85 ± 11.35 eggs). There was a significant difference infecundity value between control and treatments. The highest relative fecundity was recorded in control (0.87 ± 0.14 eggs), followed by treatment A (0.54 ± 0.25 eggs) and treatment B (0.33 ± 0.13 eggs). There was a significant difference between the control and treatment B, but no significant when compared to treatment A (Table 1).

Table 1. Fecundity and relative fecundity of Zebrafish *(Danio rerio)* between treatment

Paramenters	Control	Treatment A	Treatment B
Fecundity (egg)	219±42,38^{a}	149±38,70^{b}	85±11,35^{b}
Relative fecundity (egg/mg)	0,87± 0,14^{a}	0,54±0,25^{ab}	0,33±0,13 ^{b}

* Different letters on the same row indicate significant differences (P<0.05).

A decrease in fecundity might correlated with gonad development alteration, characterized by increased follicles/germinal cells. Previous research revealed that a decrease in fecundity might correlated with a decrease in the GSI. Exposure to mercury ($Hg^{2+)}$ at a concentration of 15µg/L in the early stages of zebrafish (*Danio rerio*) results in disruption of gonad development which is characterized by an increase in early stages follicles/germ cells and a decrease in late stage follicles/germ cells [22]. In addition, exposure to Carbamazepine (CBZ) and Gemfibrozil (GEM) 0.5 and 10 µg L^{-1} caused oocyte atresia and changes in ovarian histology which was a direct effect on oocyte development which caused a decrease in fecundity of zebrafish [23].

3.2 GSI

The result showed that the highest average GSI value was observed in control (6.96 ± 1.70%), while the lowest value was in treatment B (2.55 ± 0.21%) (Figure 2). There was no significant difference in GSI between the control treatment A (4,797 ± 2,55%). However, a significant difference was observed between the control and treatment B.

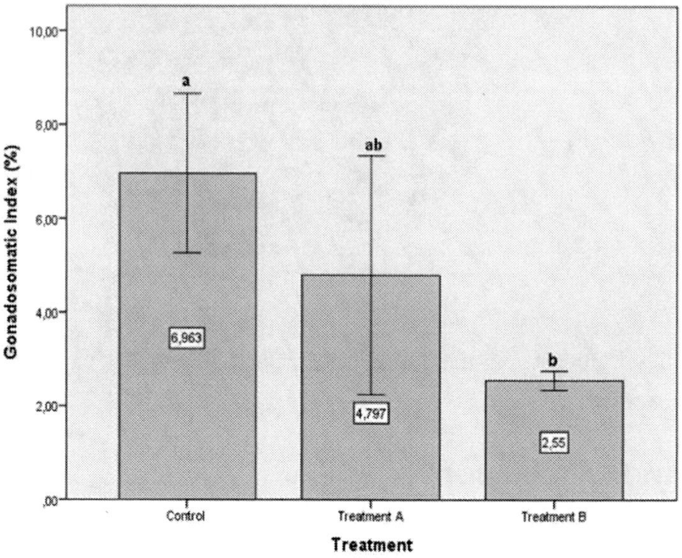

Fig. 2. The value of GSI of Zebrafish *(Danio rerio)* between treatment. (*different letters indicate a significant difference of P<0.05).

The GSI value was decreased as the concentration of POME increases. The low value of GSI was linked to the oocyte's diameter shrinking. Our study showed that the average diameter of oocytes seemed to shrink more when the concentration of POME in exposure media increased. Smaller oocyte weight had been decreased gonad weight so that the GSI values became low. According to [24], the shrinking diameter of oocytes was caused by disturbance in the vitelogenesis process due to pollutant exposure. Such exposure has caused damage to liver tissue so that the process of vitelogenesis becomes disrupted [25]. Fish exposed to mercury (Hg) showed inhibited testicular and ovarian growth and delayed gonad development. Exposure to POME triggered a decline in gonadosomatic index and shrunk the oocyte diameter of female Nile tilapia *(Oreochromis niloticus)* [8]. GSI decreased significantly in zebrafish with exposure to endocrine-disrupting chemicals [10]. Heavy metals are endocrine disruptors that can cause hormonal imbalances that affect various physiological processes, such as reproduction [26]. Cu exposure for 30 days changed steroid hormone levels and endocrine-related gene expression in the HPG (Hypothalamic-Pituitary-Gonadal) in zebrafish (*Danio rerio)* [27]. Flutolanin influences the content of Gonadotropin-releasing hormone (GnRH), regulating reproductive performance through the synthesis of gonadotropin hormones in the HPGL (Hypothalamus-Pituitary-Gonad-Liver) axis, thereby causing a decrease in reproductive capacity in zebrafish (*Danio rerio*) [22].

3.3 Egg Diameter

The result showed that the average value of zebrafish egg diameter in treatments A and B was smaller than in the control treatment. The average value of egg diameter decreases with the increasing concentration of POME (Figure 3). There was no significant difference in egg diameter between the control (0.71 ± 0.1 mm) and treatment A (0.62 ± 0.06 mm). However, a significant difference was observed in egg diameter of control compared to treatment B (0.58 ± 0.08 mm).

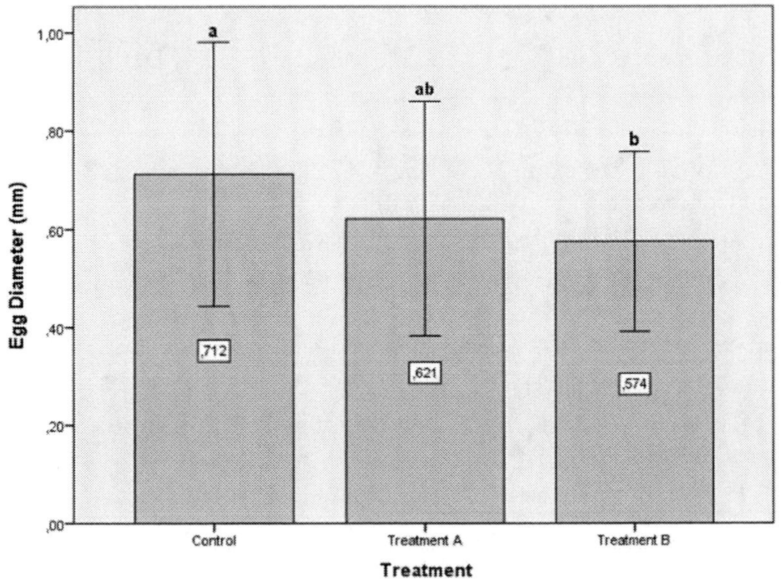

Fig. 3. The value of egg diameter of Zebrafish *(Danio rerio)* between treatment *different letters indicate significant differences of P<0.05

Similar previous research reported that the average oocyte diameter in tilapia *(Oreochromis niloticus)* show a decrease trend when the concentration of POME in the exposure media increase [8]. A decrease in oocyte diameter due to heavy metals has also been reported in tilapia [28]. Water quality plays an important role in fish survival. Inappropriate water quality can have a direct impact on growth, development, metabolism, immune function, behavior, stress, reproduction, and various other disorders.

4 Conclusion

Exposure to POME in sub-chronic concentrations for 28 days caused a significant decrease in reproductive performance which indicated by a decrease in the Gonadosomatix Index (GSI), fecundity, and egg diameter of zebrafish *(Danio rerio)* (P<0.05). Fecundity in treatment A and B were significantly decreased compared to the control. While relative fecundity significantly decreased in treatment B compared to control. Significantly decline is also observed on GSI in tretment A and B compared to control. While, the egg diameter only shows a significantly decline in treatment B compared to control.

Acknowledgments

The study was financed by a Research Grant from the Universitas Syiah Kuala under "Penelitian H-Index 2023" Scheme [Grant No : 122/UN11.2.1/PT.01.03/PNBP/2023].

References

1. Directorate General of Plantations, Ministry of Agriculture, Republic of Indonesia, (2020)
2. J. Wang, Q. Mahmood, J.P. Qiu, Li. Y.S., Y.S. Chang, L.N. Chi, X.D. Li, BioMed. Res. Int. 1–9 (2015)
3. M. Muliari and I. Zulfahmi, JIPK, 6, **2**, 137-146 (2016)
4. M. Muliari, Y. Akmal, I. Zulfahmi, N. W. Karja, C. Nisa, M. Mahyana, R. Humairani. In E3S Web of Conferences. 151. 01022 (2020)
5. C. Tokatli, Environ. Sci. Pollut. Res. **29**, 58. 87231-87244 (2022)
6. M. Muliari, I. Zulfahmi, Y. Akmal, N. W. K. Karja, C. Nisa, K. A. Sumon. Adv. Anim. **7,** 11 (2019)
7. Y. Hashiguchi, M. R. Zakaria, M. Toshinari, M. Z. M. Yusoff, Y. Shirai, M. A Hassan. Environ. Pollut. **277**,116780 (2021)
8. I. Zulfahmi, M. Muliari, Y. Akmal, A.S. Batubara. Egypt. J. Aquat. Res. **44**:327–332 (2018)
9. S. A. Elgaml, T. T Saad, M. Hamed, V. H. Zaki. Int j Fish Aquat **7**, 5, 542-547 (2019)
10. I. Zulfahmi, S. A. El Rahimi, S. D. Suherman, Almunawarah, A. Sardi, K. Helmi, N. Badratun, A. W. Perdana, K. H. Adani, I. A.A. Nasution, K. A. Sumon, M. M. Rahman, Chemosphere **340**, 139788 (2023)
11 I. Y. Wiendarlina, N. Herlina, E. Mareta, J. Ilm. Farm, **12.** 1 (2022)
12. Y. Xiong, X. Chen, F. Li, Z. Chen, Z. Qin. Aquat. Toxicol. **246,** 106143. (2022)
13. P. Shankar, E, J, D. Titus, L. Truong, K. Hayward, L, G. Hudson, R. L. Tanguay. Environ Pollut. **269**:1-23. (2021)
14. P. Morcillo, M. A. Esteban, A. Cuesta, Environ. Sci. **4**, 3 (2017)
15. D. Xie, Q. Chen, S. Gong, J. An, Y. Li, Z. Liu, Y. Shen, J. P. Giesy, Aquat. Toxicol. (2020)
16. N. Khosim, H. Latuconsina, and R. A. Suhada, *J. Sci. Technol.* **3,** 2 (2023)
17. K.M. Amtyaz, M.K. Atiqullah, M. Z. Khan, M.U.A. Hashmi, J. Entomol. Zool. Study **1**, 28–31 (2013)
18. K. Selman, R. A. Wallace, A. Sarka, and X. Qi. J. Morphol. **218**, 2 (1993)
19. F. Duponchelle, P. Cecchi, D. Corbin, J. Nuñez, M. Legendre, Environ. Biol. Fishes **57**, 155–170. (2000)
20. S. C. Baker and R. C. Heidinger. Trans. Illinois State Acad. Sci**. 87,** 145–150. (1994)
21. N, J. Rodriguez, Z. J. Oteme, S. Hem, Aquat. Living Resour. **8**:291-296 (1995)
22. M. Teng, C. Wang, M. Song, X. Chen, J. Zhang, and C. Wang. Environ. Res. (2020)
23. M. Galus, N. Kirischian, S. Higgins, J. Purdy, J. Chow, S. Rangaranjan, H. Li, Aquat. Toxicol. 132-133. (2013)
24. S. S. Mlambo, J. H. Van Vuren, I. E. Barnhoorn, M. S. Bornman. Environ. Toxicol. Pharmacol. 28, 1 (2009)
25. M. G. V. Paraso, J. K. C. Morales, A. A. Clavecillas, and M. S. E. G. Lola. Bull. Environ. Contam. Toxicol. **98,** 638-642 (2017)
26. S. Li, Q. Sun, Q. Wu, W. Gui, G. Zhu, D. Schlenk, Environt Pollut. **249**, 1049-1059 (2019)
27. G, J. Gautam, R. Claube, Turkish J. Fish. Aquat. Sci. **18**:1205-1214. (2018).

28. M. Azab, Mohamed A. Aly-Eldeen, H. M. Khalaf-Allah, and M. M. A. El-Battal1. Egypt. J. Aquat. Biol. Fish. **23**, 3 (2019)

Analysis of total suspended solids in Lampuuk-Lhoknga Beach Waters, Aceh Besar, Indonesia

Ichsan Setiawan[1*], *Yopi* Ilhamsyah[1], *Vieriz* Andriy Fitrada[1], *Muhammad* Haiqal[1], and *Syarifah* Meurah Yuni[2]

[1]Marine Sciences Department, Faculty of Marine and Fisheries Universitas Syiah Kuala Banda Aceh 23111, Indonesia
[2]Mathematics Department, Faculty of Mathematics and Natural Sciences Universitas Syiah Kuala Banda Aceh 23111, Indonesia

Abstract. Research on total suspended solids in the waters of Lampuuk-Lhoknga Beach, Aceh Besar was conducted on March 16, 2023. Lampuuk-Lhoknga Beach is a beach that has high activity that affects the concentration level of total suspended solids which becomes a problem for waters if the concentration level is high. This study aims to analyze and calculate the concentration of total suspended solids in the waters of Lampuuk-Lhoknga Beach. This study used purposive sampling method by taking beach water samples from several predetermined stations. Water samples that have been taken are analyzed using the gravimetric method. Based on the results of the study, the concentration of total suspended solids at 10 stations (the northernmost, Station 1 to the southernmost, Station 10) ranged from 100 mg/l to 1500 mg/l. The lowest concentration result was found at Station 4 near the cliff, which was 100 mg/l, while the highest result was found at Station 5 near tourism activities, which was 1500 mg/l. The high concentration of total suspended solids is thought to be caused by high human activity in all fields ranging from marine tourism, culinary areas to industrial areas.

1 Introduction

Indonesia is located between two continents and two oceans which makes Indonesia a dynamic country, both economically and politically. Indonesia's geographical location flanked by two continents and two oceans also makes Indonesia a strategic area, especially in marine potential. This can be seen from Indonesia's vast sea area [1]. Indonesia has a vast marine area, almost 70% of Indonesia's territory is oceanic, Indonesia also has abundant biological and non-biological resources. Because of this, Indonesia is given the status of being one of the countries in the world that has high marine biodiversity [2].

Lampuuk-Lhoknga waters as one of the coastal areas is an area filled with various human activities, especially in the field of tourism, the impact of these various activities has a direct effect on water quality on the beach, one of which is the increase in Total Suspended Solids (TSS) content. Coastal areas are dynamic, unique, and vulnerable to environmental change.

*Corresponding author: ichsansetiawan@usk.ac.id

Water quality is essential for all living things [3]. Therefore, water resources must be protected so that they can be utilized by humans and other living things properly. Utilization of water resources for the benefit of all living things must be done wisely for the benefit of current and future generations [4].

TSS is one of the water quality parameters used to measure the concentration of solid particles contained in surface water. TSS concentration can provide important information about water quality and can be an indicator of water pollution. Therefore, TSS analysis is very important in water resources management. TSS is a place where heterogeneous reactions will take place, heterogeneous reactions themselves will serve as the starting material for the formation of sediment and can hinder the ability to produce organic substances in a water body [5]. According to Wirasatriya [6] High TSS can cause other things such as reducing the photosynthetic activity of marine plants, both micro and macro, causing the oxygen content released by marine plants to decrease and causing fish to die. Increased TSS content causes turbidity that can inhibit the entry of sunlight into the water column. The lack of sunlight intensity that enters the water causes the high TSS that occurs in Lampuuk-Lhoknga waters to inhibit phytoplankton growth. TSS concentration can indicate the condition of sedimentation in a water body [7]. Waters that have high TSS concentrations tend to experience high sedimentation as well. TSS is suspended solid material with a diameter of > 1 μm that is retained by a millipore sieve with a pore diameter size of 0.45m. TSS consists of mud and fine sand as well as microorganisms.

The main cause of TSS in waters is soil erosion or soil erosion that is carried into water bodies [8]. To measure the amount and distribution of suspended matter in water bodies, TSS analysis is used. The range of TSS can indicate whether there is sedimentation in the water [7]. Waters with high TSS concentrations often have high sedimentation rates. Siswanto [9] claims that areas with significant sedimentation rates are areas such as estuaries and coastlines that have high amounts of TSS. The pattern of sedimentation that takes place is influenced by current patterns in coastal waters and estuaries [10]. The distribution of TSS can be significantly influenced by the tidal range [11]. The distribution of TSS can be used to evaluate the degree of sedimentation in an area.

Lampuuk-Lhoknga Beach is a tourist beach that is one of the factors in the movement of the regional economy. Good beach tourism is good tourism in all respects ranging from management, access, cleanliness, and others. Water quality is included in beach cleanliness. Water quality management and water pollution control which includes chemical parameters (temperature and TSS), and microbiological parameters (total coliform bacteria). Determination of water quality status using the pollution index method compared to PP No. 822/2001 water quality standards.

This study aims to analyze and determine the concentration of TSS in the surface waters of Lampuuk-Lhoknga beach. Later the results of this study are expected to provide information that can be used as a basic source and input material for further research and can determine the quality of water in the Lampuuk-Lhoknga Waters which affects the water so that it inhibits living conditions in the waters will be disrupted.

2 Research methods

This research was conducted in the waters of the Lampuuk Beach and Lhoknga Beach tourist areas, Lhoknga District, Aceh Besar Regency, Aceh Province.

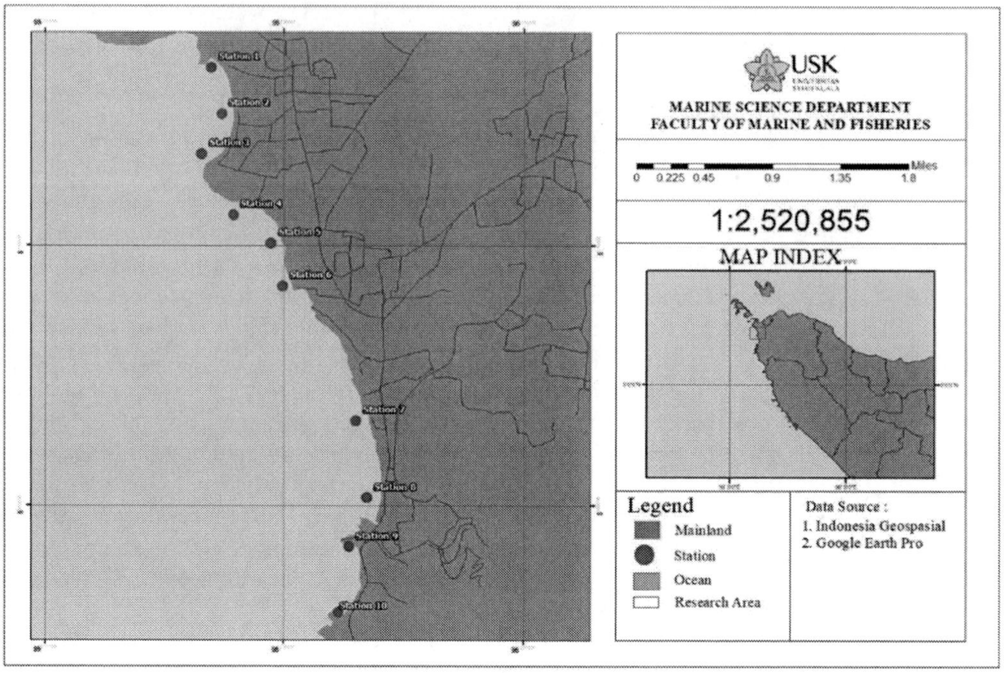

Fig. 1. Research Location Map.

2.1 Location point determination

Determination of sampling station points in the Lampuuk-Lhoknga Beach Waters area (Figure 1). Determined as many as 10 research stations taken along the coastline of Lampuuk Beach and Lhoknga Beach. Determination of the research location point is based on differences in beach management, the number of tourist visit to the beach, and uniqueness of different locations.

2.2 Water sampling and analysis

Water sampling was carried out using a 1 liter sample bottle with a depth of ± 1 meter from the water surface. Furthermore, the filled sample bottle is inserted into the coolbox [2]. Analysis of TSS samples using the Gravimetric method, where samples are filtered using filter paper. The filter paper used is Whatmann filter size 1.25µm, before use the filter paper must first be weighed to get its weight, the water sample must first be homogenized before filtering, the volume of sample water tested is 500 ml at each station. Filter paper that has been used to filter 500 ml of sample water is then heated using an oven at 105°C for 1 hour, then the filter paper is cooled in a desiccator then weighed and recorded the results. To determine the distribution of TSS values, the samples obtained were analyzed using the formula according to the Badan Standarisasi Nasional (2004) SNI 06-6989.3-2004 [12]:

$$TSS\left(\frac{mg}{l}\right) = \frac{(A-B)x1000}{v}$$

Description:
TSS = Total Suspended Solids (mg/l)
A = weight of filter paper + dry residue (mg)
B = weight of filter paper (mg)
V = sample volume (ml)

3 Results and discussions

TSS is the residue of total solids retained in the filter with a particle size of no more than 2μm. Materials such as bacteria, fungi, algae, clay, mud, sulfides, and metal oxides are materials included in TSS. These materials, which function as the earliest sediment-forming materials that can hinder the ability to produce organic substances in a water body, make these materials a place where heterogeneous reactions take place. The increase in TSS is caused by several things such as mangrove logging activities, sand mining, high wave phenomena, tides, industrial waste, and all other human activities.

The amount of TSS concentration in the waters also shows the sedimentation conditions in the waters. Usually, flocculation and filtration lead to TSS removal. TSS contributes to turbidity reduction by increasing light penetration for photosynthesis and air visibility. This means that turbidity values cannot be converted into TSS values. Turbidity is the shape change required to display light. Instead, scattering is produced using suspended particles found within the sample. Turbidity is an optically murky concept. Due to changes in the size and shape of particles and materials, the color and intensity will change. TSS is always measured as dry weight and the drying procedure must be observed to avoid errors caused by retained moisture. This makes TSS one of the most important physical indicators in determining or seeing the condition of waters and coastal environments. In addition, TSS is often used as an early indicator to determine the physical changes that occur in an environment.

Measurement of TSS concentration levels can be done directly or indirectly. Direct measurements are measurements using the gravimetric method or measurement of water content, this method is the most common method used to measure the concentration of TSS. While indirect measurements can be used through worldview 3 satellite imagery or other remote sensing.

The principle of TSS analysis is as follows:

The homogeneous test sample is filtered with filter paper measuring 0.7μm-1.5μm in accordance with the National Standardization Agency, then the residue retained on the filter paper is dried at 103°C to 105°C using an oven for 1 hour, the increase in the weight of the filter paper is the TSS residue in the test volume. In this trial, the author used 500 ml of seawater from each station as the test volume.

Table 1. TSS results.

No	Research Station	TSS Concentration (mg/l)
1	Station 1	280
2	Station 2	600
3	Station 3	500
4	Station 4	100
5	Station 5	1500
6	Station 6	1240
7	Station 7	360
8	Station 8	280
9	Station 9	280
10	Station 10	840

TSS concentration levels at 10 stations are very high values in Table 1, because according to Ministerial Decree. No.51/MENKLH/2004 [13], the quality standard TSS value set for marine tourism is 20 mg/l. At Stations 1, 2, 3, which are located in the Lampuuk beach tourism area ranging from 280 mg/l to 500 mg/l, the thing that causes TSS levels at the sampling location point at this station is high is tourist activities such as culinary areas, restaurant waste, and pollution from motorized vehicles such as boats, jetski, and others. At Stations 5, 6, which are located in the bathing area and the Lhoknga culinary area, are the location points with the highest TSS content values in the range of 1500 mg/l to 1240 mg/l, apart from the high human activity at this location, the flow of the Krueng Raba tributary which empties into the Lhoknga beach which can carry mud, sand and other materials from the mainland to the coastal waters, especially in seasons with high rainfall. At Stations 8 and 9 are the closest collection location points to the industrial area has a concentration of 280 mg / l, this is a high level for industrial areas, because according to the Ministerial Decree. No P.16 / MENLHK / SETJEN / KUM. 1/4/2019 [13] the wastewater quality standard is 50 mg/l. While the lowest TSS levels are found at the point of Station 4 collection location with a TSS concentration of 100 mg/l, this is because at the point of Station 4 collection location is a location that is minimal with human activity because the location is far from the rural road to the beach is the thing that causes this location to be minimal activity. However, the level or concentration of TSS at the Station 4 collection point is still very high from the quality standard set by Ministerial Decree.No.51/MENKLH/2004 which is 20 mg/l, this is because at this location there is still a flow from the river mouth which causes sediment or other materials to be carried into coastal waters.

There are other things that cause high TSS concentration levels such as particles suspended in water, these particles can be inorganic particles such as detritus and organic particles such as living components such as phytoplankton, zooplankton, bacteria, fungi, benthos, and others. Apart from being influenced by materials originating from land, the concentration of TSS in marine waters also comes from air and displacement of suspended sediment from erosion.

The high concentration of TSS in the waters along the Lampuuk-Lhoknga coastline indicates that the water quality in the area is poor. Due to the production of organic compounds in water may be inhibited by suspended particles, which are the first materials to form deposits and the site of heterogeneous chemical reactions. Therefore, the higher the content of TSS in the water, it will block the entry of sunlight into the water, so that it will interfere with the photosynthesis process or photosynthesis does not take place perfectly which will cause a decrease in dissolved oxygen levels released into the water. If dissolved oxygen levels in a body of water are low, it causes many negative impacts on the waters, one of which is that aquatic biota cannot respire so that it can cause death. Ecosystem balance will be disrupted if there is mass death in aquatic biota. Low dissolved oxygen levels will also cause a foul odor in a body of water.

Therefore, the less content of TSS in a water body will have a good impact on the life of the surrounding ecosystem. Not only does it have an impact on the disturbance of marine biota life, increasing high levels of TSS on an ongoing basis also has a direct effect on changing the depth of a body of water for a long time. In addition, the rapid distribution of TSS will continue to have an impact on changes in suspended sediment deposition along the coast, as it will lead to constant sedimentation and land development of land areas resulting from high levels of sedimentation. Based on the results obtained, the high level of TSS concentration in the waters of Lampuuk-Lhoknga Beach makes these waters polluted, this is supported by previous research by Khairunna et al. [14] where TSS shows that the value obtained does not exceed the specified threshold with a range of values of 4 mg/l to 14 mg/l, so this value does not exceed the threshold determined by KEPMEN LH 51 of 2004 with a value of 80 mg/l. The BOD_5 value obtained did not exceed the KEPMEN LH 51 Year 2004 threshold and the range of values obtained from this study from the lowest to the highest was 0.8 mg/l to 2.3 mg/l. While the range of TSS concentrations in the waters of Lampuuk-Lhoknga Beach exceeds the established limit.

4 Conclusion

The concentration of TSS levels at the 10 stations ranged from 100-1500 mg/l with a test volume of 500 ml of water. The lowest concentration result is found at Station 4 which is 100 mg/l, while the highest result is found at Station 5 which is 1500 mg/l. The results of TSS concentration levels at 10 stations are high levels because the value of 10 stations that have exceeded the TSS quality standard based on Kepmen. No.51/MENKLH/2004 which is 20 mg/l for marine tourism and Kepmen. No. P.16/MENLHK/SETJEN/KUM. 1/4/2019 for waste.

Acknowledgments

Authors express gratitude to the Research and Community Service Institute of Universitas Syiah Kuala for financial assistance under Grants for *Lektor Kepala* in 2023, with contract number: 145/UN11.2.1/PT.01.03/PNBP/2023.

References

1. W. Winnarsih, E. Emiyarti, L. O. A. Afu, *Sapa Laut*, ***1***, 2 (2016)
2. T. B. Solihuddin, M. E. Sari, G. Kusumah, *Bulletin Geologi Tata Lingkungan*, ***21***, 3 (2011)
3. W. Pudjiastuti, *Social marketing: strategi jitu mengatasi masalah sosial di Indonesia*, Jakarta (2016)
4. M. S. Tarigan, E. Edward, *Makara Journal of Sciences*, *7*, 3 (2010)
5. H. Helfinaslis, S. Sultan, R. Rubiman, *ILMU KELAUTAN: Indonesian Journal of Marine Sciences*, ***17***, 3 (2012)
6. A. Wirasatriya, *Buletin Oseanografi Marina*, *1*, 1 (2011)
7. A. D. Siswanto, *Studi konsentrasi total suspended solid (tss) di sepanjang pantai Kwanyar Bangkalan*, Seminar Nasional Teknologi Ilmu Kelautan Surabaya, ITS, Surabaya (2009)
8. H. Effendi, *Telaah kualitas air bagi pengelolaan sumber daya dan lingkungan peraturan*. FPIK-IPB, Bogor (2000)
9. A. D. Siswanto, *Kajian laju sedimentasi dan perubahan garis pantai di perairan Delta Bodri, Kabupaten Kendal*, Skripsi, FPIK-UNDIP, Semarang (2004)
10. A. Solikhin, *Kajian morfologi dan arus di perairan Bodi, Kendal*, Skripsi, FPIK-UNDIP, Semarang (2004)
11. E. Sulistyorini, *Fluktuasi total suspended solid (tss) berdasarkan karakteristik pasang surut di perairan Delta Bodri Kabupaten Kendal*, Skripsi, FPIK-UNDIP, Semarang (2004)
12. Badan Standardisasi Nasional, *metode pengujian kualitas fisika air. air dan limbah bagian 3: cara uji padatan tersuspensi solid (total suspended solid, tss) secara gravimetri*, SNI 06-6989.3-2004 (2004)
13. Keputusan Menteri Negara Lingkungan Hidup Nomor 51 Tahun 2004, *tentang baku mutu air laut*, Jakarta
14. N. Khairunna, S. Agustina, I. Setiawan, I. Irwan, M. Ramadhaniaty, R. Sakinah, S. Keumala, K. Ondara, *JKPI*, **1**, 3 (2021)

Estimation of seagrass blue carbon stocks in Ahmad Rhang Manyang and Ujung Pancu Waters, Aceh Besar District

Khairunnisa[1,3*], *Wilda* Ulya[1,2], *Haekal* Azief Haridhi[1,3], and *Irma* Dewiyanti[1,3]

[1]Department of Marine Sciences, Faculty of Marine and Fisheries, Universitas Syiah Kuala, Banda Aceh, 23311, Indonesia
[2]Ocean Diving Club, Faculty of Marine and Fisheries, Universitas Syiah Kuala, Banda Aceh, 23311, Indonesia
[3]Research Centre for Marine Sciences and Fisheries, Universitas Syiah Kuala, Banda Aceh, 23111, Indonesia

Abstract. The blue carbon was concepted by the United Nations Environment Program (UNEP) in 2009, refers to the ability of marine and coastal ecosystems such as mangrove, seagrass, and tidal marshes to capture and store carbon dioxide through photosynthesis. This study aims to analyze the estimation of seagrass blue carbon stock and sequestration in Ahmad Rhang Manyang and Ujung Pancu. Data were collected using the plot transect method to assess cover, density, biomass and use model calculations called Seagrass Carbon Converter (SCC) to assess seagrass carbon stocks. Three species of seagrass were found at four research stations *are Halodule pinifolia, Halophila ovalis, Halophila minor*. The result showed that seagrass cover in Ahmad Rhang Manyang waters was 27.95% in the poor category and in Ujung Pancu waters was 39.4% in the medium category.. The seagrass carbon stock in Ahmad Rhang Manyang 123.93 gram C/m^2 and in Ujung Pancu was 231.43 gram C/m^2. The seagrass carbon absorption value in Ahmad Rhang Manyang was 4.16 TC/Ha/year (tons of carbon per hectare per year) and Ujung Pancu is 15.5 TC/Ha/year.

1 Introduction

The blue carbon was concepted by the United Nations Environment Program (UNEP) in 2009, refers to the ability of marine and coastal ecosystems such as mangrove, seagrass, and tidal marshes to capture and store carbon dioxide in their biomass and sediment through photosynthesis [1]. These ecosystems can take in carbon faster than the terrestrial ecosystems [2].

* Corresponding author: knisa@usk.ac.id

Seagrass captures and stores carbon dioxide from the atmosphere through diffusion system or the process of transferring CO_2 from the atmosphere to the sea which is then used for the photosynthesis process [3,4]. The great carbon sequestration potential by seagrasses plays the important role in reducing carbon dioxide levels in the atmosphere, thereby offering a viable way for climate change mitigation (5,6).

Research on seagrass blue carbon has been carried out in several regions of Indonesia, where the carbon storage ranges from 0.52 – 65.98 tons C/ha [7-9]. However, there has been limited research on this topic within Aceh waters. Therefore, conducting this research is important to collect data inventory, which can serve as a foundational resources to take the action to conserve and preserve seagrass ecosytem.

Several place in Aceh Waters such as Pulo Aceh Island, Matahari Island, Lamteng Bay, Ujung Pancu and Ahmad Rhang Manyang Bay covered with seagrass beds. The research locations in Ahmad Rhang Manyang waters and Ujung Pancu waters were chosen based on consideration of the closest access that could be reached by researchers. This study aims to analyze the estimation of seegrass blue carbon stock and sequestration in those areas.

2 Material and Method

2.1 Study Area

This research was conducted in May and June 2023 at Ahmad Rhang Manyang waters and Ujung Pancu waters, both are located in Aceh Besar Regency (Fig 1). The laboratory work was carried out at the Marine Chemistry and Fisheries Biotechnology Laboratory, Faculty of Marine Affairs and Fisheries, Syiah Kuala University.

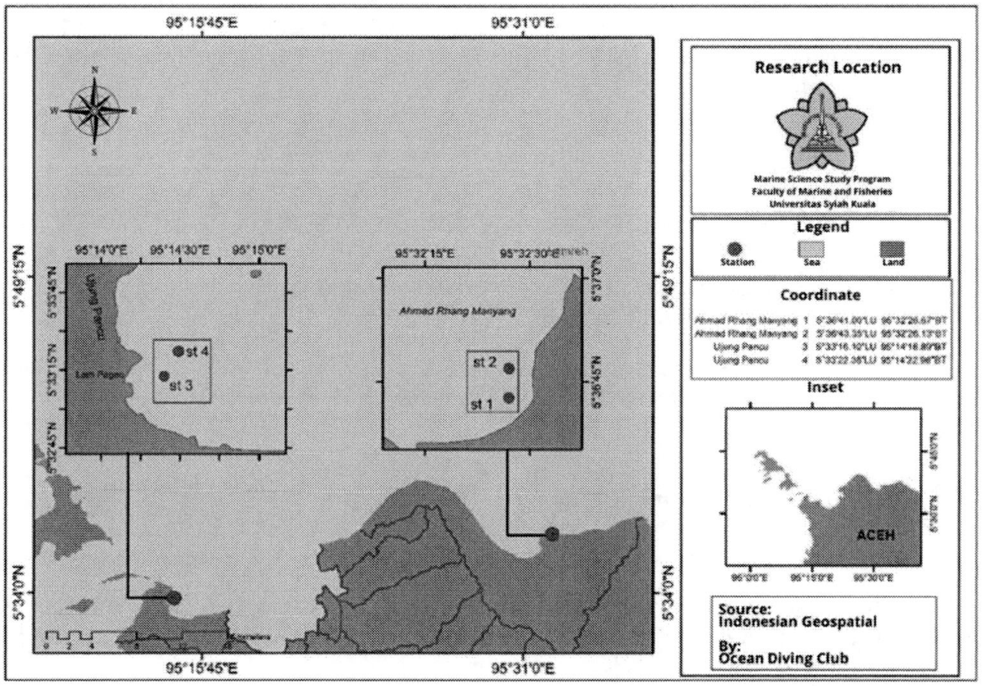

Fig. 1. Study Area

2.2 Data Collection

2.2.1 Seagrass Coverage and Density

Seagrass data collection was carried out using the Seagrass Net parallel transect plot method (Fig 2) [10]. The transect was used based on the consideration of the seagrass meadows area at the research location where the seagrass grows horizontally along the coastline, with the length of the transect lines adjusted. The sampling point was taken at 5 meters after the first vegetation was found from the shoreline [11]. Visual observations of seagrass cover were carried out on 50cm x 50cm plots beside the 50m transect line.

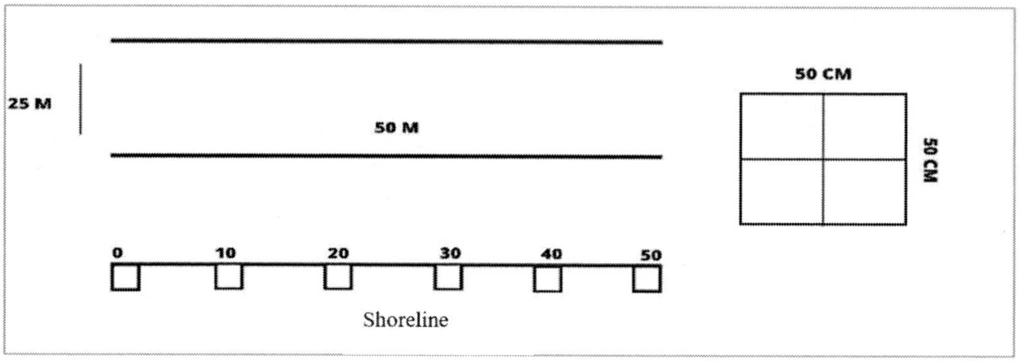

Fig. 2. Sampling Illustration

2.2.3 Seagrass Biomass and Carbon

Seagrass samplesfor biomass (3 – 5 shoots) were taken to the laboratory and separated into above ground parts (leaf blades and leaf midribs) and below ground parts (roots and rhizomes). The samples that have been separated were dried using an oven at 60°C for 48 hours and weighed until reached constant dry weight [12]. After that, the biomass that has been obtained is used to find the carbon value through the equations.

2.3 Data Analysis

2.3.1 Seagrass Coverage

Seagrass coverage is the percentage of seagrass canopy that covers the subtrate area. The calculation for percentage used the equation [13]:

$$Seagrass\ coverage\ (\%) = \frac{Total\ coverage\ in\ the\ plot}{4} \quad (1)$$

$$Average\ seagrass\ cover(\% = \frac{the\ total\ value\ of\ seagrass\ cover\ for\ all\ transects}{Total\ of\ quadrant\ transect} \quad (2)$$

After the seagrass cover for each location is obtained, the seagrass health status can be determined based on Decree of Minister of Environment No. 200 of 2004 which is classified into 3 categorizes (Table 1).

Table 1. Seagrass Condition Categorizes

Seagrass cover (%)	Condition	
≥ 60 %	Good	Rich/Healthy
30 % - 59,9 %	Moderate	Less rich/Less Healthy
≤29,9 %	Bad	Poor

2.3.2 Seagrass Density

The density of seagrass for each species was calculated by dividing the number of seagrass shoots by the plot quadrant area of 0.25 m^2 using the following equation[14]:

$$D = \frac{N}{A} \quad (3)$$

where D is density (shoots/m2), N = number of shoot, A = plot area

2.3.3 Seagrass Biomass

Biomass sampling was carried out on seagrass shoots which were then separated into two parts consisting of above-ground biomass (AGB) and below-ground biomass (BGB). The calculation for biomass used the equation[12]:

$$AGB{=}AGB\ xDS \quad (4)$$

$$BGB{=}BGB\ xDS \quad (5)$$

2.3.4 Carbon Stock and Sequestration

Seagrass carbon stock is the amount of carbon that can be absorbed and stored in seagrass body tissues. In this research, carbon storage estimation were carried out using the Seagrass Carbon Converter (SCC) developed by LIPI using an approach to biomass content, cover and seagrass density in one area, the formula used is as follows [15]:

$$\text{SSC} = 6.874866 + 0.003865[Density] + 0.297085[Biomass] + 0.010671[Coverage] \quad (6)$$

$$SC = 1.74315 + 0.08654[\text{Coverage}] \quad (7)$$

Where:
SSC : Standing Stock Carbon
SC : Carbon sequestration

3 Result and Discussion

There were two seagrass species found in Ahmad Rhang Manyang, namely *Halodule pinifolia and Halophila ovalis* and two species in Ujung Pancu *namely Halodule pinifolia* and *Halophila minor* (Table 2). The two research locations are categorized as associated seagrass meadows.

Table 2. Seagrass Distribution

No	Species	Ahmad Rhang Manyang		Ujung Pancu	
		St 1	St 2	St 3	St 4
1	*Halodule pinifolia*	+	+	+	+
2	*Halophila ovalis*	+	+	-	-
3	*Halophila minor*	-	-	+	+

The assessment of seagrass condition is determined through the canopy percent cover and its classification is aligned with the categories stipulated in Decree of Minister of Environment no. 200 of 2004. Seagrass condition in Ahmad Rhang Manyang was categorized as damaged/poor with a value of 27.95%, while in Ujung Pancu was categorized as moderate/unhealthy with a value of 39.40% (Fig 3).

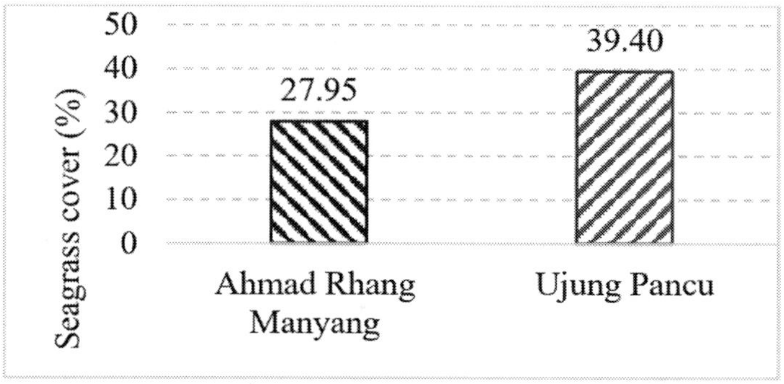

Fig. 3. Seagrass Cover in Ahmad Rhang Manyang and Ujung Pancu

The seagrass density can be seen in table 3. Based on this data, the lowest seagrass density was obtained at station 2 with 368.44 individuals/m^2 in the Ahmad Rhang Manyang. The highest seagrass density was obtained at station 3 with 846.11 individuals/m^2 in Ujung Pancu waters. Both waters are dominated by *Halodule pinifolia.*

Table 3. Seagrass Density

Seagrass Species	Density (ind/m^2)			
	Ahmad Rhang Manyang		Ujung Pancu	
	St 1	St 2	St 3	St 4
Halodule pinifolia	516.67	296.44	737.33	702.89
Halophila ovalis	22.67	70	-	-
Halophila minor	-	-	105.78	-
Total	540.33	368.44	846.11	706.89

Ujung Pancu waters exhibited higher seagrass cover and density than in Ahmad Rhang Manyang waters. This discrepancy can be attributed to various factors including fishing activities, seagrass extent, and substrate composition. The waters of Ahmad Rhang Manyang possess a more limited expanse of seagrass due to their proximity to the coral reef system, consequently influencing the average values of cover and density. In contract, Ujung Pancung predominantly features a muddy subtrate type, impacting nutrient absorption and subsequently influencing segrass growth. Generally, sediments that contain mud are generally richer in organic material than sandy sediments [16].

The results of biomass analysis show that seagrass in Ujung Pancu exhibits higher below ground biomass at 423.13 grams C/m^2 and above ground biomass with a value of 321.28 grams C/m^2, in contrast to the seagrass in Ahmad Rhang Manyang where the below ground biomass at of 227.44 grams C/m^2 and 159.7 grams C/m^2 for above ground biomass (Fig 4).

The largest seagrass carbon stock were obtained from seagrass beds in Ujung Pancu waters at 231.43 grams C/m^2 and the smallest seagrass carbon stocks were obtained from seagrass beds in Ahmad Rhang Manyang at 123.93 grams C/m^2 (Fig. 5).

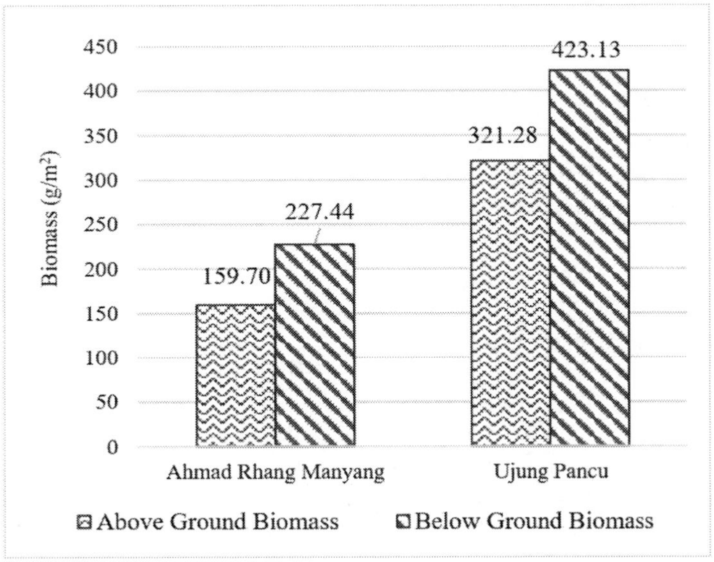

Fig. 4. Seagrass Biomass

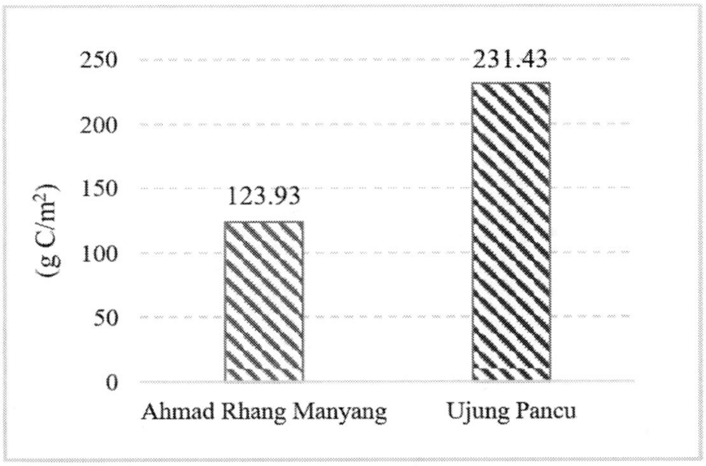

Fig. 5. Seagrass Carbon Stock

Carbon sequestration is the ability of seagrass to store carbon over a long period of time. Observation results show that carbon stock in tonnes per hectare each year in Ahmad Rhang Manyang waters at 4.16 tonnes of carbon per hectare per year (TC/ha/year) and in waters Ujung Pancu at 5.15 tons of carbon per hectare per year (TC/ha/year) (Fig 6). These values show overall that seagrass in Ujung Pancu waters store more carbon each year than Ahmad Rhang Manyang waters.

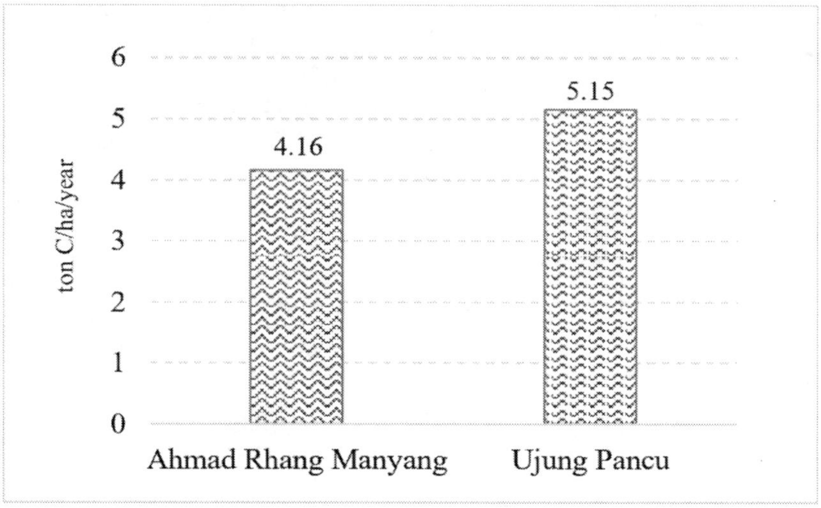

Fig. 6. Seagrass Carbon Sequestration

Seagrass in Ujung Pancu exhibits higher carbon stock and sequestration than segrass in Ahmad Rhang Manyang. A large biomass value influences the amount of carbon stock and sequestraion in seagrass stands. Seagrass with a large biomass will store large carbon stocks [17]. The related topic has also been studied at Megiat Beach, Nusa Dua Bali where the total above ground carbon were lower than the below ground carbon [18]. The higher carbon stock in the above ground part likely attributed to its placement within the substrate, rendering it less susceptible to direct impacts from fluctuations in water conditions such as temperature and sunlight exposure [19]. Also, the smaller substrate size has a positive effect on the system for absorbing nutrients and organic materials.

4 Conclusion

There are two seagrass species found in Ahmad Rhang Manyang waters, namely *Halodule pinifolia and Halophila ovalis* and also two species in Ujung Pancu waters *namely Halodule pinifolia* and *Halophila minor.* Seagrass condition in Ahmad Rhang Manyang waters is categorized as damaged/poor with a value of 27.95%, while in Ujung Pancu waters is categorized as moderate/unhealthy with a value of 39.40%. The largest seagrass carbon stock were obtained from seagrass beds in Ujung Pancu waters at 231.43 grams C/m^2 and the smallest seagrass carbon stocks were obtained from seagrass beds in Ahmad Rhang Manyang at 123.93 grams C/m^2

References

1. J.W. Fourqurean, C.M. Duarte, H. Kennedy, N. Marbà, M. Holmer, M.A. Mateo, E.T. Apostolaki, G.A. Kendrick, D. Krause-Jensen, K.J. McGlathery, O. Serrano. Nat. Geo. **5**, 505 (2012)
2. Jang and Awiati. Selisik. **9**, 36 (2023)
3. A. Rustam, T.L. Kepel, R.N. Afiati, H.L Salim, M. Astrid, A. Daulat, P. Mangindaan, N. Sudirman. Segara. **10**, 107 (2014)
4. S.C. Maberly, A.W. Stott, B. Gontero. Front. Plant Sci. **13** (2022)
5. M. Stankovic, R.A. Rappe, F. Carly, F. D. Galon, M.D. Fortes, M.S. Hossain, W. Kiswara, C.V. Luong, P. Minh-Thu, A.K. Mishra, T. Noiraksar, N. Nurdin, J. Panyawai, E. Rattanachot, M.Rozaimi, U.S. Htun, A.Prathep. Sci. Total Environ. **783** (2021)

6. G. Gao, J. Beardall, P. Jin, L. Gao, S. Xie, K. Gao. J. Appl. Ecol. **59**, 1686 (2022)
7. R. Hartati, I. Pratikto, T.N. Pratiwi. Buloma **6**, 74 (2017)
8. Ganefiani, Suryanti, N. Latifah. IJFST. **14,** 115 (2019)
9. I.K.S. Negara, I.W.G. Karang, I.N.G Putra. JMRT. **3**, 82 (2020)
10. F.T. Short, L.J. McKenzie, R.G. Coles, K.P. Vidler, J.L. Gaeckle. *SeagrassNet Manual for Scientific Monitoring of Seagrass Habitat, Worldwide edition*. (University of New Hampshire Publication, 2006).
11. A. Rustam, N.S. Adi, A. Daulat, W. Kiswara, D.S. Yusup, S.A. Rappe *Pedoman Pengukuran Karbon di Ekosistem Padang Lamun* (ITB Press, 2019).
12. S. Rahmawati, U.E. Hernawan, K. McMahon, B. Prayudha, H.B. Prayitno, A.J. Wahyudi, M. Vanderklift, M. 2019. Blue Carbon in Seagrass Ecosystem (UGM Press, Yogyakarta, 2019)
13. S. Rahmawati, A. Irawan, I.H. Supriyadi, M.H. Azkab. *Panduan Monitoring Padang Lamun* (LIPI, Jakarta, 2014)
14. M.H. Azkab. Oseana. **14,** 1 (1999)
15. A.J. Wahyudi, S. Rahmawati, A. Irawan, H. Hadiyanto, B. Prayudha, M. Hafizt, A. Afdal, N.S. Adi, A. Rustam, U.E. Hernawan, Y.P. Rahayu, M.Y. Iswari, I.H. Supriyadi, T. Solihudin, R.N.A. Ati, T.L. Kepel, M.A. Kusumaningtyas, A. Daulat, H.L. Salim, W. Kiswara. Ocean Sci. J. **55**, 85 (2020)
16. N. El-Radaideh, A.A. Al-Taani, T. Al-Momani, K. Tarawneh, A. Batayneh, A. Taani. *Lake and Reservoir Manag.* **2014**, 32 (2014)
17. Khairunnisa, I. Setyobudiandi, and M. Boer. *JITKT* **10**, 639 (2018)
18. I.K.V.S. Rahadiarta, I.D.N.N. Putra, Y. Suteja. 2018. JMAS. **5**, 1 (2018)
19. S. Supriadi, R.F. Kaswadji, D.G. Bengen, M. Hutomo. IJMS. **19**, 1 (2014)

Microplastic in Beach Sediment of Nasi Island, Aceh Besar Regency, Indonesia

Sri Agustina[1,2*], *Diva* Cesarylla Darmawan[1], *Ichsan* Setiawan[1], *Syahrul* Purnawan[1], *Sofyatuddin* Karina[3], *Muhammad* Irham[1], and *Maria* Ulfah[1]

[1]Marine Science Department, Faculty of Marine Science and Fisheries, Universitas Syiah Kuala, Banda Aceh, Aceh, Indonesia, 23111.
[2]Research Centre of Marine Science and Fisheries, Universitas Syiah Kuala, Banda Aceh, Aceh, Indonesia, 23111.
[2]Research Centre of Marine Science and Fisheries, Universitas Syiah Kuala, Banda Aceh, Aceh, Indonesia, 23111.
[3]Laboratory of Marine Chemistry and Fisheries Biotechnology, Faculty of Marine Science and Fisheries, Universitas Syiah Kuala, Banda Aceh, Aceh, Indonesia, 23111.

Abstract. Microplastic might be a small size of plastic that is directly released into the environment or the result of larger in size of plastic degradations. The presence and effect of microplastic have been reported globally. However, the existence of microplastic trapped in beach sediment of Nasi Island is limited. This research was conducted to analyse the type of sediment, abundance and type of microplastic polymer found in beach sediments in Nasi Island, Aceh Besar Regency, Aceh Province, Indonesia. Sampling was carried out in April 2022 at Nipah Beach and Alue Riyeung Beach in Nasi Island, Aceh Besar Regency. Sediment type was analysed by dry sieve method, using a stratified sieve shaker while the observation of microplastic and its polymer were identified by microscope and Fourier Transform Infrared Spectroscopy (FTIR), respectively. The results showed that the types of sediment at both beaches belonged to the sand type. There were three types of microplastics found, namely fragments, films, and fibres. The highest abundance of microplastics was found in a 0.063mm filter at Nipah Beach with an amount of 2651 Particles/Kg while the lowest was found in a 0.125mm filter with an amount of 8 Particles/Kg at Alue Riyeung Beach. As the polymer types, it is found that fragment type microplastics are likely to have PETE (Polyethylene Terephthalate) polymers, film type microplastics are likely to have PS (Polystyrene) polymer types and Fibre microplastics are likely to have LDPE (Low-Density Polyethylene) polymers.

1 Background

Pollution of plastic have dominated the marine ecosystem which is up to 75% of the pollutant accumulates in rivers, beaches, sea surfaces and seabed [1][2]. Plastic bags, fishing equipment, food and drink containers are the most common observed categories of plastics

* Corresponding author: sri_agustina@usk.ac.id

covering on beaches with percentage more than 80% wash up on beaches [3]. Its nature, which is difficult to degrade (non-biodegradable), makes this waste categorized as the largest contributor to pollution and causes damage to the natural balance [4]. Due to its properties which is difficult to degrade, the use of plastic in large quantities and continuously will certainly have a significant impact on human health and also the environment.

Plastic debris thrown into the sea will gradually degrade into small particles that might be consumed by marine organism accidentally. Particles with a size of <5 mm are defined as microplastics. Microplastics can float or sink because the density of microplastics is lighter than seawater's, such as polypropylene as the basic material for plastic bottles which will float and spread widely in the ocean. Other microplastics such as methyl methacrylate which is an acrylic base material that is denser than sea water and is likely to accumulate on the seabed, this means that large amounts of microplastics can eventually accumulate in the deep sea and then settle in sediment and ultimately disrupt the marine food chain [5].

Nasi Island, which is administratively included in the Aceh Besar Regency, has a good ecosystem because it is supported by its position between the Andaman Sea, the Indian Ocean and the Strait of Malacca. Nasi Island is a fishing area for fishermen and also for recreation activities. High fishing activities also have an impact on increasing debris release in the waters. Research on marine debris in Aceh is still relatively limited, especially regarding data on the abundance of microplastics. Therefore, this research aims to analyse the presence of microplastics in two waters in Aceh, namely Alue Riyeung Beach and Nipah Beach, Nasi Island, Aceh Besar.

2 Material and method

This research was carried out on Nasi Island, Aceh Besar Regency in April 2022 (Figure 1). Samples were taken at 2 stations, namely Nipah Beach and Alue Riyeung Beach with 3 repetitions at each station. There were 3 analyses carried out for each sample, namely sediment classification, identification of microplastic types, and determination of microplastic polymers.

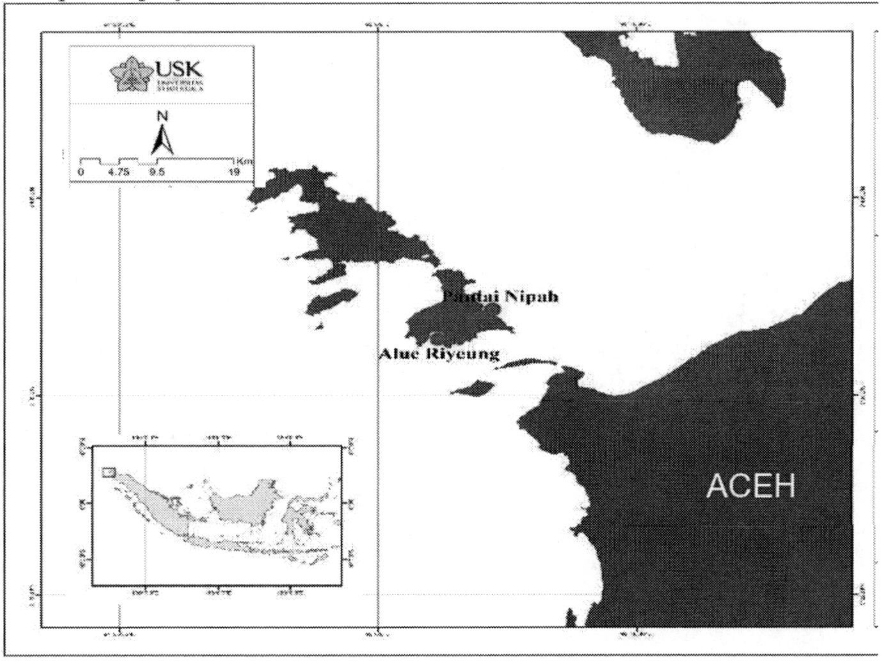

Fig. 1. Study Area of Nasi Island, Aceh Besar Regency. Red dots represent the sampling stations.

2.1 Sediment classification

A total of 200g of sediment samples were sieved using a sieve shaker for 10 minutes with stratified sieve of 2mm, 1mm, 0.5mm, 0.25mm, 0.125mm, 0.063mm and 0.038mm separation mesh sizes. The sediment samples left at each level was weighed to calculate the percentage of each fraction following the formula on equation 1. Then, the sediment classification referred to the Folk Triangle [6].

$$Percentage\ of\ weight = \frac{fraction\ weight}{Total\ weight\ of\ sample} x100 \quad (1)$$

2.2 Identification of microplastic type

The sieved sample of each fraction was added to a 1:1 H_2O_2 solution to destruct the organic matters. After 30 minutes of heating at 90°C, the sample is removed and cooled to room temperature. Density separation was carried out using saturated NaCl solution for 24 hours. Samples were filtered using 0.45 µm cellulose nitrate filter paper to collect all the floating matters. The identification of microplastic by a binocular microscope with a lens magnification of 4x/0.10 was carried out to see the presence of microplastics in filtered samples [7]. Microplastic abundance was calculated using the following formula on equation 2.

$$Microplastic\ abundance = \frac{Number\ of\ microplastic}{Total\ weight\ of\ sediment} \quad (2)$$

2.3 Determination of microplastic polymer

The spectroscopic method was used to obtain the chemical structure of microplastics from sediment samples. Sediment samples that have been identified using a microscope are then combined into the similar type, namely fragment, fibre and film. Polymer analysis is conducted by the FTIR method based on the wave number value at each spectrum peak.

3 Results and discussion

3.1 Sediment classification

Sediment at both locations (Nipah Beach and Alue Riyeung Beach) on Nasi Island is classified into 2 types of sediment, namely sand and gravel (Table 1). Both beaches tend to have the same type, namely sand. There is 1 station on Nipah Beach which shows a different type, namely the gravel sand fraction, and the table above also shows that there is no type of sand that is very fine or classified as mud on both beaches where sediment samples were taken.

Table 1. Sediment classification in Nasi Island

Sampling points	Percentage of sediment fraction			Type of sediment
	Gravel	Sand	Mud	
AR1	0	100	0	Sand
AR2	0	100	0	Sand
AR3	0	100	0	Sand
NP1	22.69	77.31	0	Gravelly sand
NP2	3.80	96.20	0	Sand
NP3	0	100	0	Sand

Note: AR1= sample from Alue Riyeung Beach at point 1; AR2= sample from Alue Riyeung Beach at point 2; AR3= sample from Alue Riyeung Beach at point 3; NP1= sample from Nipah Beach at point 1; NP2= sample from Nipah Beach at point 2; NP3= sample from Nipah Beach at point 3.

3.2 Microplastic identification

Figure 2 shows the abundance of microplastics in different sediment sizes. The highest abundance was found in filters with a size of 0.125mm on Nipah beach with a total of 2651 particles/Kg, while at the same size Alue Riyeung beach only had 172 particles/Kg. Meanwhile, the lowest abundance was found in the 4mm filter at Alue Riyeung beach, where there was no abundance of microplastics at that sediment size, and at the same size, Nipah beach had an abundance of 470 particles/Kg.

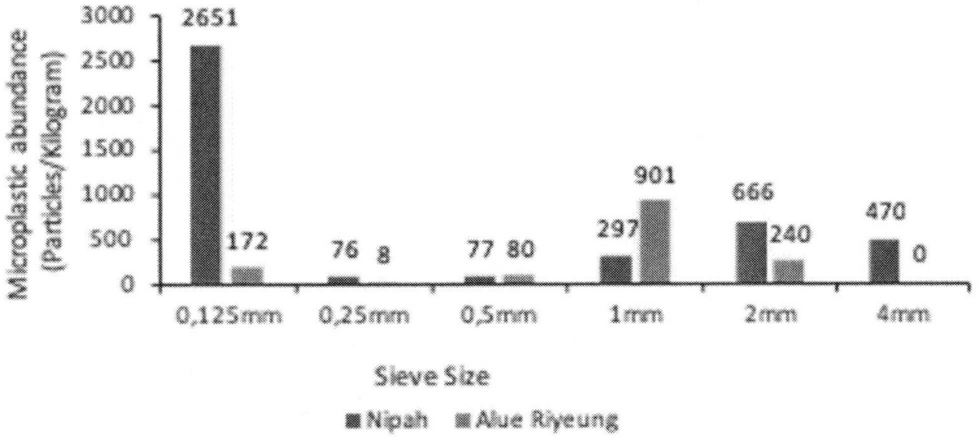

Fig. 2. Microplastic abundance in different sieve size.

3.3 Determination of microplastic polymer

Polymer type analysis is carried out by interpreting data produced by FTIR waves. The microplastics that have been identified are then separated according to type for the FTIR process. In this research, the results of the analysis show the possibility that fragment type microplastics have a PETE (Polyethylene Terephthalate) type polymer, film type microplastics have a PS (Polystyrene) type polymer, and fibre type microplastics have an LDPE (Low-Density Polyethylene) type polymer (Table 2).

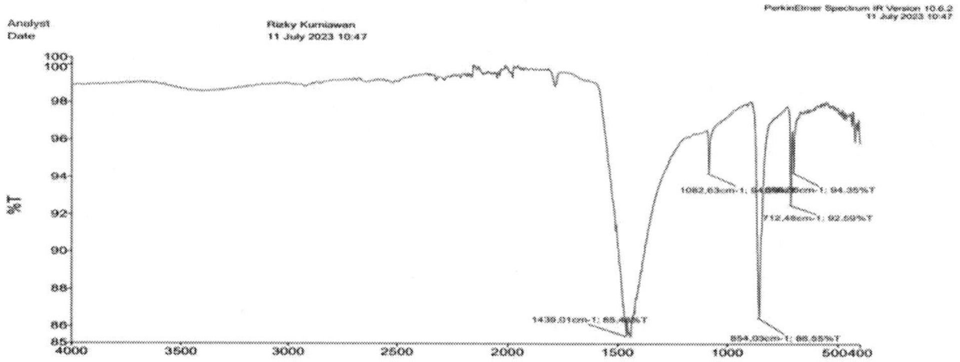

(a)

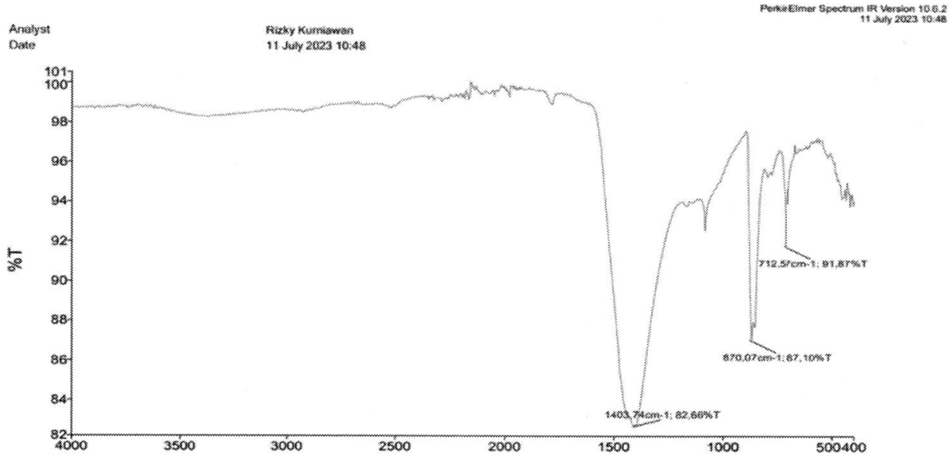

(b)

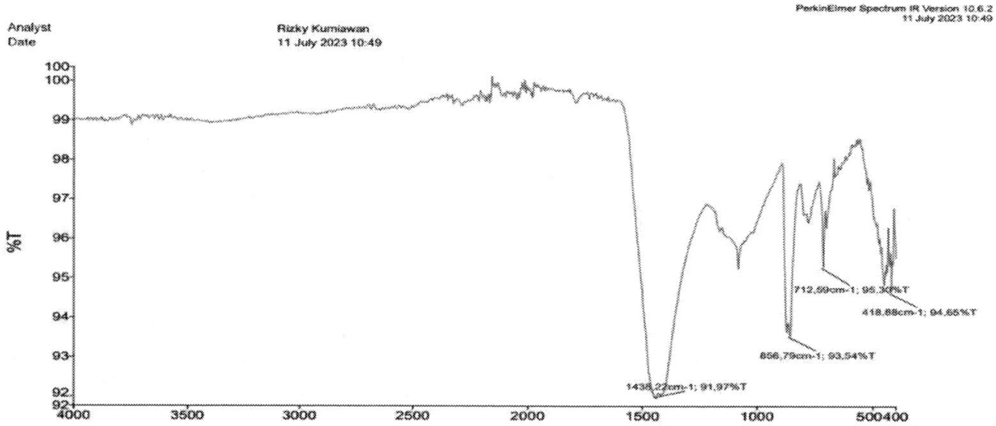

(c)

Fig. 3. The FTIR Spectrum of Microplastic Type: (a) Fragment; (b) Film; (c) Fibre.

 https://doi.org/10.1051/bioconf/20248702007

3.4 Discussion

In this study, 3 types of microplastics were found in the samples, namely Fiber, Fragment and Film with the amounts listed in table 4.2. Microplastics on Nipah beach were dominated by the film type, where the total number of film type microplastic particles was found to be 221 particles, followed by the fragment type with a total of 77 particles and then the fiber type with a total of 43 particles. In contrast to Nipah beach, on Alue Riyeung beach, fiber type microplastics occupy the first position with a total of 53 particles, followed by film type microplastics with a total of 20 particles and fragment type with a total of 2 particles. The types of microplastics that commonly enter the waters are fragment, fiber and film microplastics [9]. These three types are also found on both beaches. Microplastic fragments are another result of cutting plastic products with very strong synthetic polymers. Fragments come from plastic consumer products. The origin of these fragments can be cosmetic bottles, fishing lines, industrial raw materials, polymer plastic fragments that can be decomposed by oxidation.

Table 2. Absorption bands of microplastic in Nasi Island.

Microplastic type	Absorption bands (cm^{-1}) {8]	Absorption bands (cm^{-1}) (this study)	Assignment	Polymer
Fragment	1440 1094 840 717	1439 1082 854 712	CH_2 Bend C-O stretch CH_2 rock, C-CH_3 stretch CH_2 rock	PETE
Film	1409 840 717	1403 870 712	Aromatic ring stretch CH_2 rock, C-CH_3 stretch CH_2 rock	PS
Fibre	1440 840 717	1438 857 712	CH_2 Bend CH_2 rock, C-CH_3 stretch CH_2 rock	LDPE

Fiber is a type of microplastic that has a thin, elongated shape and size. Fiber type microplastics come from synthetic materials such as thread, fishing line and fishing nets [10]. Film-type microplastics have the characteristics of being shaped like sheets or plastic fragments, generally this type is used as material for making plastic bags or plastic packaging [11]. Film has a lower density than other types of microplastics so it is easier to transport. Microplastic particles in the form of fragments are easier to find because they have a low mass so they float on the surface of the water and the fragments are also formed by plastic fragments that flow in low temperature waters and are exposed to ultraviolet light.

The highest abundance of microplastics based on sediment level was produced by Nipah beach which was found on a 0.125 mm sieve with a total of 2651 particles/kg, in contrast to Alue Riyeung beach on the same sieve size, which only had an abundance of 172 particles/kg. Meanwhile, the lowest was produced by Alue Riyeung beach with a filter size of 4mm which did not contain any microplastics. According to {12], a higher percentage of mud (soft sediment) can cause microplastics to be trapped very tightly. Meanwhile, sand sediments have gaps that are less dense and allow microplastics to not easily be retained due to the influence of strong wave energy in the coastal area.

Analysis of polymer types obtained 3 polymer results from the 3 types of microplastics found. Fragment type microplastics may contain PETE/PET (Polyethylene Terephthalate) type polymers. PET is a thermoplastic resin from the polyester group. PETE is widely produced in the chemical industry and is used in synthetic fibers, beverage bottles and food containers, thermoforming applications, and combined with glass fibers in engineering resins. The increasing amount of PET use causes the amount of PET waste to quickly increase as well. Even though PET is not a harmless material, PET is a type of plastic that is non-degradable so it can cause environmental problems [13]. The results of further analysis show that film type microplastics likely contain PS (Polystyrene) type. Polystyrene is a polymer with styrene monomer, a liquid hydrocarbon made commercially from petroleum. At room temperature, polystyrene is usually a solid thermoplastic, melting at higher temperatures. Styrene is classified as an aromatic compound. Fiber type microplastics show the possibility of containing LDPE (Low-Density Polyethylene) polymers. This type of plastic can be said to be the parent plastic because it can still be recycled into other types of plastic. LDPE has a density value smaller than pure water, around 0.920 g/cm^3 [14,15].

4 Conclusion

It can be concluded that Nipah and Alue Riyeung beaches have the same type of sediment, namely sand, but there is one part of Alue Riyeung beach that shows gravelly sand. The results of the analysis of the presence of microplastics on both beaches stated that there were 3 types of microplastics on both beaches, namely Fiber, Fragment and Film. The abundance of microplastics based on sediment size shows that the highest abundance is found in filters with a size of 0.125 mm on Nipah beach with the majority being film types and the lowest is found on filters with a size of 4mm on Alue Riyeung beach with a total of 0. Fragment type microplastics are likely to have PETE (Polyethylene) type polymers. Terephthalate), film type microplastics may have a PS (Polystyrene) type polymer and fiber type microplastics may have an LDPE (Low-Density Polyethylene) type polymer.

Acknowledgments

This research was funded by UNIVERSITAS SYIAH KUALA, with contract number: 147/UN11/SPK/PNBP/2022 Dated February 11th, 2022 (on behalf of Sofyatuddin Karina).

References

1. S. Agustina, S.P. Nuraini, S. Purnawan, E.E.W. Siregar. Depik Jurnal Ilmu-Ilmu Perairan, Pesisir dan Perikanan, **9**, 1, (2020)
2. M. A. Browne, P. Crump, S.J. Niven, E.L. Teuten, A. Tonkin, T. Galloway, R.Thompson. Environmental Science and Technology **45**, 21 (2011)
3. M. Thiel, I. A. Hinojosa, L. Miranda, J.F. Pantoja, M.M. Rivadeneira, N. Vásquez. Marine Pollution Bulletin, **71** (2013)
4. H. Hiwari, N. Purba, Y.N. Ihsan, L. P. S. Yuliadi, P.G. Mulyan, P. G. Pros Sem Nas Masy Biodiv Indon. **5**, 2 (2019)
5. N. Seltenrich. (2015). Environ Health Perspect **123**, (2015).
6. S. Purnawan, I. Setiawan, Marwantim. Depik, **1**, 1 (2012)
7. G. Peng, B. Zhu, D. Yang, L. Su, H. Shi, D. Li. Environmental Pollution, **30** (2017)
8. M. R. Jung, F. D. Horgen, S.V. Orski, V. Rodriguez, K. L. Beers, G. H. Balazs, T. T. Jones, T. M. Work, K. C. Brignac, S. Royer, K. D. Hyrenbach, B. A. Jensen, J. M. Lynch. Marine Pollution. **127**, (2018).
9. A. A. Layn, Emiyarti. Ira. Sapa Laut, **5**, 2 (2020)
10. S. Yudhantari, G. Hendrawan, P.R. Pusphita. Journal of Marine Research and Tecnology. (2019).

11. F. M. Septian, N. Purba, M. U. K. Agung, L. P. S. Yuliadi, L. F. Akuan, P. G. Mulyani, Jurnal Geomaritim Indonesia, **1**, 1 (2018)
12. D. L. Watters, M. M. Yoklavich, M. S. Love. Marine Pollution Bulletin. **60**, 1 (2010)
13. R. Siddique. Berlin: Springer Verlog Berlin Heidberg. 93-120. (2008)
14. D. S. Achilias. Thessaloniki . Greece. (2012).
15. S.A. Akbar, S. Afriani, C. Nuzlia, S. Nazlia, S. Agustina, Depik, **12(3)**, 259-273 (2023)

Carbon stock estimation of mangrove ecosystem in the Kuta Raja Subdistrict, Banda Aceh, Aceh Province

Irma Dewiyanti[1,2*], *Khairina* Khairina[1], *Sayyid Afdhal* El-Rahimi[1]

[1]Marine Science Department, Faculty of Marine and Fisheries, Universitas Syiah Kuala, Banda Aceh 23111, Indonesia
[2]Marine biology laboratory, Faculty of Marine and Fisheries, Universitas Syiah Kuala, Banda Aceh 23111, Indonesia

Abstract. Mangroves play a crucial role in maintaining the environmental balance through their ability to absorb and store carbon. This study aimed to analyze carbon storage in mangrove ecosystems, including aboveground biomass, belowground biomass, deadwood biomass, and organic soil material. The sampling method used in this research was purposive sampling to select three observation stations. The results of this research indicate that the biomass of mangrove tree stems in the Kuta Raja Subdistrict, Banda Aceh, was 117.9 tons/ha, which was the highest value compared to the roots at 47.2 tons/ha, and deadwood at 2.2 tons/ha, as part of carbon storage in mangrove plants. The carbon stock at the research site was highest in mangrove tree at 55.43 tons/ha, followed by mangrove root at 22.17 tons/ha, and deadwood at 1.04 tons/ha, and organic soil material at 2.7 tons/ha. The total carbon stock of the mangrove ecosystem was 81.37 tons/ha.

1 Introduction

Mangrove ecosystems can protect coastal environment from severe impact of natural climates [1], high productivity through nutrients recycle, and reduce pollution and erosion [2]. Another advantage of mangroves is that they can reduce carbon through a sequestration mechanism or absorb carbon from the atmosphere, which is then stored as biomass and soil organic matter [3;4]. Mangroves can reduce carbon through a mechanism of sequestration or absorption of carbon from the atmosphere, which is then stored in mangrove parts such as plants, litter, and soil organic matter to prevent an increase in carbon dioxide [5;6]. This absorption occurs through photosynthesis, and plants absorb carbon in the form of CO_2 [7]. Global warming is an increase in atmospheric temperature caused by greenhouse gases such as carbon dioxide (CO_2), methane (CH_4), nitrous oxide (N2O), perfluorocarbon (CFCS), and sulphur hexafluoride (HFCS) [8]. Global warming causes extreme weather, increasing the Earth's temperature so that it can melt polar ice. Carbon dioxide has a greater influence than other greenhouse gases [9].

Mangrove biomass is high and is considered to contribute significantly to global carbon fluxes because it can store large amounts of carbon [10]. Mangrove as biofilter can maintain balance in aquatic environments [11]. The city of Banda Aceh has an area of

* Corresponding author: irmadewiyanti@usk.ac.id

mangroves that remained after the tsunami disaster in 2004, spread over 4 sub-districts and has a carbon potential of 78.73 tonnes/ha [12]. Estimating carbon stores in mangrove forests is a forest-conservation effort [13]. It can be used as a strategy by managers to carry out mangrove forest conservation efforts. Several studies have been conducted on carbon estimation, such as that conducted by Dewiyanti, et al [7] and Rahmah, et al [12] in Banda Aceh, Aceh Province.

Kuta Raja is one of the sub-districts in Banda Aceh which has a mangrove ecosystem covering an area of 47.9 ha in 2015 after the tsunami disaster [14]. It is located in the community and on roads crossed by many vehicles. The existence of this ecosystem is very good for environmental balance; therefore, it must be maintained. The sustainability of mangrove forests can be implemented in three ways: protection, preservation, and utilization. The aim of this study was to analyze carbon storage in the mangrove ecosystem, including upper surface biomass, lower surface biomass, dead wood, and soil organic matter.

2 Methodology

2.1 Location and time of research

The research was conducted from May until August 2023 in the mangrove forest. There were three stations established using purposive sampling method, namely Gampong Pande as station 1, Gampong Jawa as station 2, and station 3 was in Gampong Peulanggahan, Banda Aceh, Aceh Province, Indonesia. The soil samples were analysed in Soil and Plant Research Laboratory, Faculty of Agriculture, Universitas Syiah Kuala, included drying, and the bulk density.

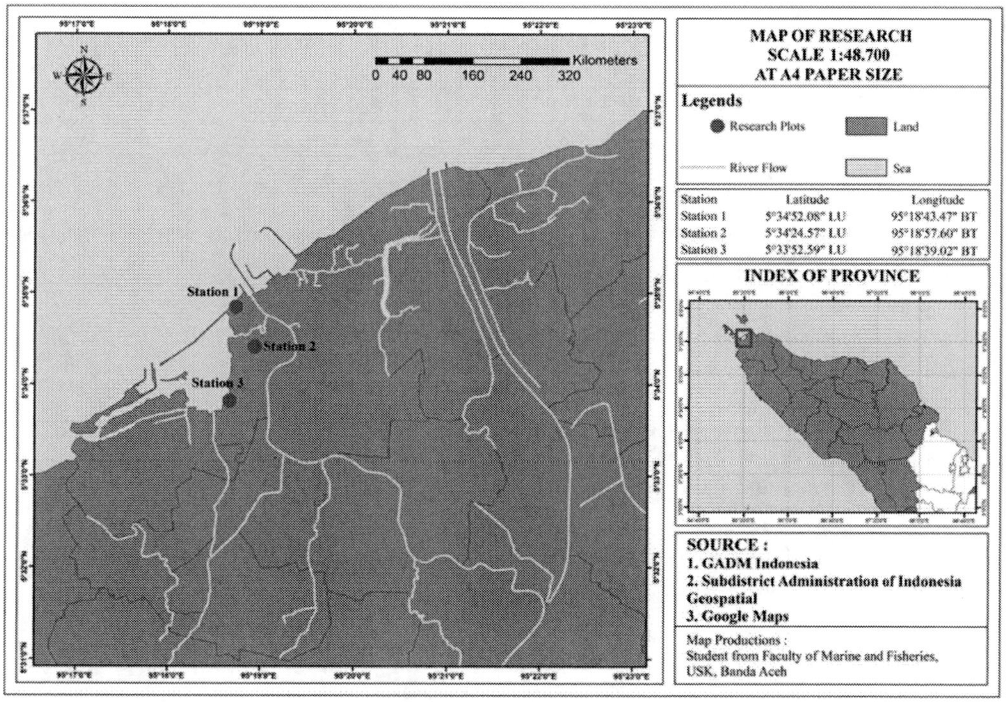

Fig. 1. Map of research locations

2.2 Tools and materials

The main tools used in this study were ropes for transects, measuring tape, sacks, Global Positioning System (GPS), and digital scales. Meanwhile, the material used was soil from the mangrove ecosystem as a sample.

2.3 Research methods

2.3.1 Determining research locations

Research location determine by considering the mangrove dominance, density, and diameter in 3 station. The station was a square with dimensions of 10 m × 10 m. Based on the Indonesian National Standards [15], measuring and calculating carbon reserves in mangrove forests was carried out using field measurements by calculating carbon reserves from the carbon pool or carbon storage section, which consists of surface biomass, namely trees, subsurface biomass, namely roots, dead wood, and soil organic matter. .

2.3.2 Tree biomass data

Tree biomass measurements were carried out by identifying the type of tree, measuring the diameter at breast height (DBH), and recording the data. The tree biomass was calculated using an allometric equation.

Table 1. Allometric Equations

Type	Biomass	Source
Rhizophora apiculate	B = 0.043 D2.63	Amira, 2008
Rhizophora mucronata	B = 0.1466 D2.3136	Dharmawan, 2010
Avicennia alba	B= 79211D2.470895	Tue et al., 2014
Rhizophora stylosa	B = 0.0139D2.1072	Clough & Scott 1989
Avicennia marina	B = 0.1848D2.3524	Dharmawan & Siregar, 2008
Sonneratia alba	B =0.258D2.287	Kusmana et al., 2018
Xylocarpus Granatum	B = 0, 1832D2,21	Tarlan, 2008

B=Biomass and D=Diameter

Mangrove biomass carbon was calculated based on the Indonesian National Standard [15] as follows:

$$Cb = B \text{ x } \% \text{ C organic}$$

Where:
Cb = carbon content of biomass (kg)
B = total biomass (kg)
%C organic = percentage value of carbon content, equal to 0.47.

The carbon content of the biomass per hectare was calculated using the following equation:

$$C_n = \frac{C_x}{1000} \times \frac{10000}{l_{plot}}$$

Where:
C_n= carbon content per hectare in each carbon pool of each plot (ton/ha)
C_x = carbon content in each carbon pool in each plot (kg)
l_{plot}= Plot area of each carbon pool (m^2)

2.3.3 Dead Tree Biomass Data Collection

Dead trees were measured by measuring DBH and determining the integrity of the dead tree. The dead tree biomass was calculated using the allometric equation and multiplied by a correction factor for the level of dead tree integrity.

(a) b) (c) (d)

Fig.2. Level of tree integritySource: Indonesian National Standard 7724:2011

Caption:
a: living tree
b: dead tree (correction factor 0.9)
c: dead tree without leaves and twigs (correction factor 0.8)
d: dead tree without leaves, branches, or twigs (correction factor: 0.7)

2.3.4 Root biomass measurement

Root biomass measurements are calculated using the following formula:

$$B_{bp} = NAP \text{ X } B_{ap}$$

Information:
B_{bp} = biomass below ground level (kg)
B_{ap} = biomass above ground /mangrove tree (kg)
NAP = root–shoot value, equal to 0.40. Based on Indonesian National Standard 7724 [15], the shoot root value for the
mangrove ecosystem is 0.40.

2.3.5 Measurement of soil carbon content

Measurement of soil organic carbon content in the mangrove ecosystem was measured using five soil sample points (four corners of the station and one in the middle of the station). Soil samples were taken at a depth of 21 cm. The samples were weighed by wet weight and then taken to the Soil and Plant Research Laboratory, Faculty of Agriculture, Universitas Syiah Kuala to be dried, then weighed by dry weight and the "bulk density" value was calculated (weight of dry soil mass per unit volume) [16]. The calculation of soil carbon adopted the loss on ignition (LOI) method, which involves weighing the weight of the sample that is lost after burning. The samples were dried in an oven at 60°C for 48 hours [17].

Soil carbon calculations based on the Indonesian National Standard [15] are as follows:

$$C_t = Kd \times \rho \times \% \text{ C organic}$$

Where:
C_t = soil carbon content (g/ cm^2)
Kd = sampling depth (cm)
ρ = bulk density (g/cm^3)
% C organic = percentage value of carbon content, equal to 0.47.

The calculation of soil organic carbon content per hectare was calculated using the following equation:

$$C_{soil} = Ct \times 100$$

Where:
C_{soil} = soil organic carbon content per hectare (tons/ha)
C_t = soil carbon content (g/cm^2)
100 the conversion factor from g/cm^2 to tonnes/ha

3 Results and discussion

3.1 Composition of mangrove vegetation

Four species of mangroves were found at three research location, there were *Rhizophora mucronata*, *R. stylosa, R. apiculata*, and *Xylocarpus granatum*. *R. apiculata* was found at stations 1 and 2, the number of trees were 3 trees/100 m^2 and 44 trees/100 m^2, respectively. Furthermore, *R. mucronata* was found at station 3 as many as 12 trees/100 m^2. At station I, three mangrove species were found, there were *R. apiculata* with a number vegetation of 3 trees/100 m^2, *R. stylosa* (11 trees/100 m^2), and *X. granatum* with 2 trees/100 m^2 (Table 2). In general, the research location was dominated by *Rhizhopora* sp., this species has the ability to adapt in the varied coastal environment condition. According to Kolinug et al. [18] and Srikanth et al. [19], *Rhizophora* sp. can adapt to high salinity and has supporting roots that can absorb air in muddy substrates. *Rhizophora* is mangrove type that can live in various types of substrate and higher ability to survive than other mangrove type [20], and *Rhizophora* from Rhizophoraceae family had wide distribution and tends to adapt well in the environmental condition of mangrove community [21]. Moreover, Canizares et al. [22] reported that Rhizophoraceae is most important and widespread mangrove species in the Pacific.

3.2 Mangrove vegetation biomass

Biomass is expressed as the amount of organic compounds present in the plants. Determination of biomass uses a destructive method which is carried out without damaging the plants, namely by using the allometric equation. At the research location, *R. mucronata* has a higher tree biomass than other species found at the research location with a total biomass of 194.45 tons/ha because it has the highest average diameter value. Table 2 shows the biomass of each species mangrove vegetation. *R. mucronata* has a higher biomass than the other mangrove types [23]. Mangrove plant biomass is influenced by mangrove type, tree diameter, and tree density [24]. *R. mucronata* also had the largest average diameter compared with the other types found at the research site.

Table 2. Average diameter and total biomass of mangrove species

Station	Location	Species	Average diameter (cm)	Tree (stem) biomass (tons/ha)	Root biomass (ton/ha)	Dead Tree Biomass (tons/ha)
I	Gampong Pande	*R. apiculata*	5.45	19.08	7.63	3.69
		R. stylosa	6.63	4.12	1.65	0
		X. granatum	5.8	1.84	0.74	0
2	Gampong Jawa	*R. apiculata*	9.1	80.58	32.23	2.97
3	Gampong Pelanggahan	*R. mucronata*	10.7	194.45	77.8	0

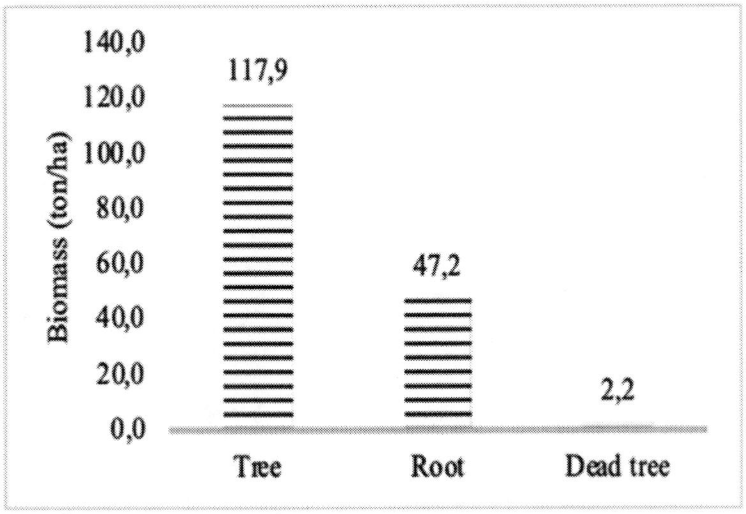

Fig. 3. Biomass of mangrove vegetation

The root biomass of mangrove plants is directly proportional to the tree biomass. The greater the value of tree biomass, the greater the root biomass of mangrove plants [14]. At the research location, the highest biomass was in mangrove tree (stem), and followed by root, and dead tree, the value were 117.9 tons/ha, 47.2 tons/ha, and 2.2 tonnes/ha, respectively. Dead tree biomass at the research location was obtained with a correction factor of 0.7 according to the condition of the tree, which only left a few branches [15]. The total biomass at the research location was 167.3 tons/ha. Rahmah et al. [12] reported that the mangrove ecosystem in coastal area, Banda Aceh, Aceh Province Indonesia has net production value because it has a biomass tree vegetation was ranged between 62.9–398.8 tons/ha.

3.3 Mangrove vegetation carbon stock

The carbon stock of mangrove plants was higher in station 3 at tree level as much as 91.4 tons/ha than station 1 (37.0 tons/ha) and station 2 (37.87), it means that this station stored the most carbon in plants compared to other locations due to the high vegetation biomass at this location. According to Indonesian National Standard [15], 47% of tree biomass is in the form of carbon. Therefore, the greater the biomass value, the higher the carbon stock stored [25]. Nedhisa and Tjahjaningrum [26] argued that most of the biomass from plants is carbon; therefore, biomass greatly influences the carbon stock that can be stored by mangrove plants. Biomass stored in mangrove vegetation of Kerala, southwest coast of India is 117.11 ± 1.02 tons/ha [27]. Furthermore, the mangrove ecosystem in the Anambas Islands Marine Tourism Park, Indonesia has a biomass of 574. According to Indonesian National Standard [15], the carbon stock of mangrove plants consists of carbon stock stored in the tree trunk as the top surface (above ground), roots as part of the lower surface (below ground) apart from soil carbon, and dead trees found in the ecosystem [28]. Mangrove tree trunks store the most carbon compared to other parts. This is supported by research conducted by Rahmah et al. [12], where tree trunk estimated carbon stock of 270.12 tons/ha. Carbon stock has positively correlation with mangrove density and size, where the higher the mangrove density, the more carbon stock tends and the larger tree size will produce higher biomass and carbon stores [29].

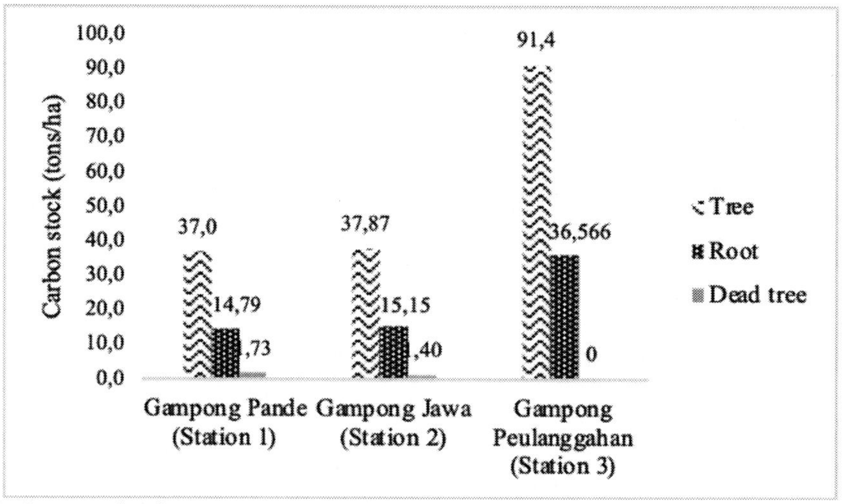

Fig. 4. Carbon stock of mangrove vegetation in each station

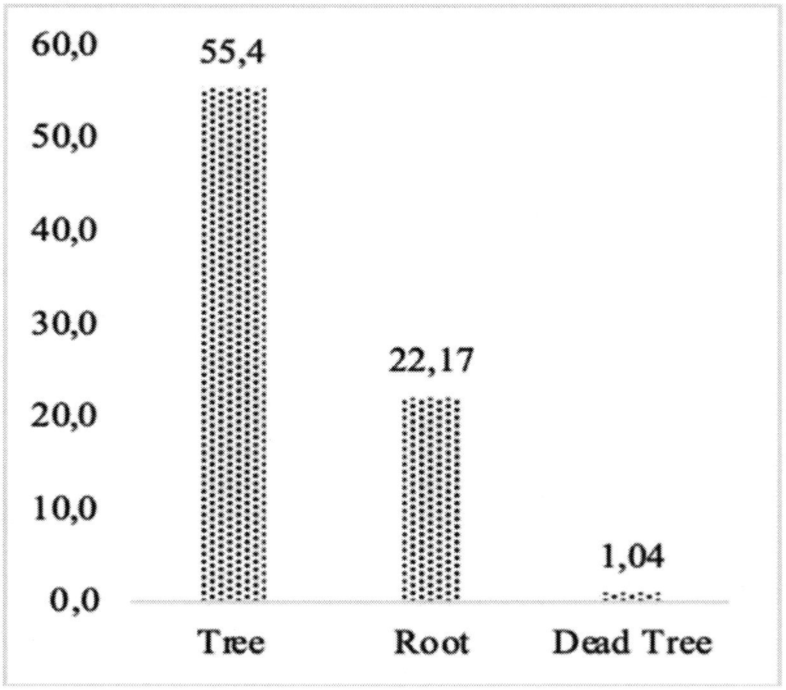

Fig. 5. Carbon stock of mangrove vegetation in each component of mangrove

3.4 Mangrove soil carbon stock

The soil carbon stock in station 1, 2, and 3 were 2.18 tons/ha, 4.22 tons/ha, and 4.54 tons/ha, respectively. In total, the soil carbon stock at the research location was 10.94 tons/ha. The soil carbon stock at the study area is less than the result reported by Rahmah et al. [12] in Banda Aceh, Aceh Province which has a soil carbon stock of 14.84 tons/ha at a soil depth of 15-30 cm. Low soil carbon stock in study area assumed because the location is close to the sea and influenced by tides, therefore, the composition of sand is higher than dust and silt. The sand fraction has a limited capacity to stabilize organic compounds compared to silt and clay soil [30] due to the greater porosity of sand. Several studies show that changes in organic carbon in soil are related to sand fraction, and it have the greatest influence on the Carbon cycle [30; 31]. The soil content of the dust fraction is positively correlated with organic matter, high organic carbon because the soil content of the dust fraction is also high [32]. C-organic at the research location was classified as very low, with an average of 0.99%, organic C is low with a range of <1.00%, low 1.00-2.00%, medium 2.01-3.00%, and high >5.00%, respectively [33].

Total carbon stock in study area included above ground carbon stock (tree), and below ground carbon stock consisted by root, dead tree, and soil organic was 89.58 tons/ha (Table 3). The total above and below ground carbon stock were 55.43 tons/ha and 34.15 tons/ha. The result of this study in line with result reported by Rahmah et al. [12], mangrove carbon stock in Banda Aceh is 89.24 tons/ha. Harefa et al., [34] reported that mangrove carbon stock ini Deli Serdang, Indonesia more than 58.87 tons/ha dominated by *Rhizophora apiculata*. Some biology parameters of mangrove vegetation can influence the carbon stock, for example stem diameter, species density, and diversity of tree species are parameters that affect the biomass and carbon content an ecosystem [35].

Table 3. Total carbon stock in study area

Stock Carbon (tons/ha)				Total Carbon Stock (tons/ha)
Tree	Root	Dead tree	Soil	
55.43	22.17	1.04	10.94	89.58

4 Conclusion

Carbon stock can be determined by above and below ground biomass. The above ground carbon stock is higher than the below ground in mangrove ecosystem Kuta Raja Subdistrict, Banda Aceh, Aceh Province. In term of tree component, the greatest biomass and carbon stock were found in tree (stem) compared to root and dead wood. Whereas the lowest carbon stock was in dead wood. Soil organic matter as below ground stores 10.94 ton/ha of carbon. The total carbon stock of the mangrove ecosystem is 89.57 tons/ha, which includes tree (stems), roots, dead wood and soil.

References

1. M.W. Beck, N. Heck, S. Narayan, P. Menendez, B.G. Reguero, S. Bitterwolf, S. Torres-Ortega, G.M. Lange, K. Pfliegner, V.P. McNulty, Ecosyst. Serv **56**, 101440 (2022)
2. C.E. Lovelock, E. Barbier, C.M. Duarte, PLoS Biol **20**, e3001836 (2022)
3. K.A.R.S. Perera, M.D. Amarasinghe, Geoderma **347**, 80-89 (2019)
4. P. Taillardat, D.A. Friess, M.Lupascu, Biol. Lett **14,** 20180251 (2018)
5. V. Maulidia, A.A. Akbar, Jumiati, Arifin, A. Sulastri, Journal of Wetlands Environment Management **10**, 1 (2022)
6. M. Chatting, I. Al-Maslamani, M. Walton, M.W. Skov, H. Kennedy, Y.S. Husrevoglu, L. Le Vay, Front. Mar. Sci **9**, 781876 (2022)
7. I. Dewiyanti, M. Martunis, S. Agustina, IOP Conf. Ser.: Earth Environ. Sci **348**, 012119 (2019)
8. J. Samidjo, Y. Suharso, Journal Od Ivet University **24**, 2 (2017)
9. H.A. Rahman, Int. j. Malay-Nusantara stud **1**, 2 (2018)
10. J. Hutchison, A. Manica, R. Swetnam, A. Balmford, M. Spalding, Conserv. Lett **7**, 3 (2014).
11. E. Hilmi, L.K. Sari, T.N. Cahyo, C. Kusmana, E. Suhendang, Biotropia **26**, 3 (2019)
12. F. Rahmah, H. Basri, Sufardi, Jurnal Manajemen Sumberdaya Lahan **4**, 1 (2015)
13. N.S. Suhaili, J.L.J. Fei, F.W. Sha'ari, M.I. Idris, S.M. Hatta, J. Kodoh, N.A. Besar, Biodiversitas **21**, 12 (2020)
14. S. Saputra, S. Sugianto, D. Djufri, Jurnal Edukasi dan Sains Biologi **5**, 1 (2016)
15. Standar Nasional Indonesia. Pengukuran dan penghitungan cadangan karbon (ground based forest carbon accounting), (2011)
16. J. Howard, S. Hoyt, K. Isensee, E. Pidgeon, M. Telszewski, Coastal Blue Carbon: Methods for assessing carbon stocks and emissions factors in mangroves, tidal salt marshes, and seagrass meadows. International Union for Conservation of Nature. Arlington, Virginia, USA, (2014).
17. S. Suryono, N. Soenardjo, E. Wibowo, R. Ario, E.F. Rozy, Buletin Oseanografi Marina **7**, 1 (2018)
18. K.H. Kolinug, M.A. Langi, S.P. Ratag, W. Nurmawan, *In Cocos.* **5**, 4 (2014)
19. S. Srikanth, S.K.Y. Lum, Z. Chen, Trees **30**, 2 (2015)
20. R.Y. Putri, A. Bashri, Jurnal Riset Biologi dan Aplikasinya **5**, 2 (2023)
21. R. Utina, A.S. Katili, N. Lapolo, T. Dangkua, Biodiversitas **20**, 3 (2019)

22. L.P. Canizares, R.A. Seronay, AACL Bioflux **9**, 3 (2016)
23. N.M. Heriyanto, and E. Subiandono, Jurnal Penelitian Hutan dan Konservasi Alam **9**, 1 (2012).
24. D. Rachmawati, I. Setyobudiandi, E. Hilmi, Omni-Akuatika **10**, 2 (2014)
25. W.C. Adinugroho, K. Sidiyasa, Jurnal penelitian Hutan dan Konservasi alam **3**, 1 (2006)
26. P.I. Nedhisa, I.T. Tjahjaningrum, Jurnal Sains **8**, 2 (2020)
27. K.M. Harishma, S. Sandeep, V.B. Sreekumar, Ecological Processes **9**, 31 (2020)
28. R.R.K. Sinaga, F. Kurniawan, S. Roni, D.Y.W. Laia, W. Andrito, J.R. Hidayati, IOP Conf. Ser.: Earth Environ. Sci **1148,** 012003 (2023)
29. R. Mardliyah, R. Ario, R. Pribadi, Journal of Marine Research **8**, (2019)
30. J.L. Yost, A.E. Hartemink, Advances in Agronomy **158**, 221-268 (2019)
31. P.L. Woomer, J. Tungani, G. Odhiambo, F.M. Mwaura, African Crop Science Conference Proceedings **7**, 479-484 (2005)
32. K.V.S. Andrade, F.S.R. Holanda, T.O. Santos, M.B.S. Santana, R.N. Araújo Filho, Floresta e Ambiente **25**, 2 (2018)
33. T.R. Fitriana, Biodivesitas **7**, 1 (2006)
34. M.S. Harefa, Z. Nasution, M.B. Mulya, A. Maksum, Biodiversitas **23**, 2 (2022)
35. M.F. Hikmatyar, T.M. Ishak, A.P. Pamungkas, S. Soffie, A. Rijaludin, Al-Kauniyah Jurnal Biologi 8, 1 (2015)

Estimation of carbon stocks in mangrove sediments in Ruyung Village, Mesjid Raya District, Aceh Besar Regency

Harum Farahisah[1*], *Putri* Khairani Amalia Siregar[1]

[1]Marine Science Department, Faculty of Marine and Fisheries, Universitas Syiah Kuala Banda Aceh 23111, Indonesia

Abstract. Exploration related to blue carbon has increased recently. As one of the blue carbon ecosystems, the mangrove ecosystem is one of the coastal ecosystems that can absorb and store carbon. This research aims to calculate mangrove density and estimate carbon stocks in sediments in the mangrove ecosystem in Ruyung Village, Mesjid Raya District, Aceh Besar Regency. Mangrove density was calculated at three stations within 10 x 10 m transects. Sediment samples were also taken compositely, and carbon analysis was carried out in the sediment using the LOI (loss on ignition) method. There were two species of mangroves found at the research location, namely *Rhizophora apiculata* and *Avicennia officionalis*. The research results show that the average mangrove density at the research location is 0.20 ind/m^2. The average soil density and sediment carbon stock at all research stations is 1.24 g/cm^3 and 271.66 MgC/ha respectively. The mangrove ecosystem, both biomass and surrounding sediment, can store carbon, which is one of the ways to mitigate global warming.

1 Introduction

Global warming occurs due to increasing carbon emissions into the atmosphere and will ultimately result in climate change. Apart from the ecological functions that we generally know, namely as feeding, nursery and spawning ground, mangroves play an important role in mitigating the effects of greenhouse gases generated by anthropogenic activities such as deforestation, agriculture, and industrial processes [1]. Mangrove forests have the same role as other forests, namely to absorb and sequester carbon dioxide (CO_2) so that they can help prevent climate change [2]. Various studies have confirmed that mangroves have a faster carbon sequestering capacity than other ecosystems such as grasslands or tropical rainforests [3]. Mangrove stored carbon in the biomass and in the soil/sediment below. Soils/sediment in mangrove rich of carbon that extend from 0,5 m to 3 m in depth and accomodate 49%-98% of the carbon stored at the mangrove ecosystem[4,5].

* Corresponding author: harumfarahisah@usk.ac.id

The mangrove forest area in this research is in Ruyung Village. Ruyung Village is one of the villages in Mesjid Raya District, Aceh Besar Regency. This mangrove area is located between human activities such as community settlements and shrimp ponds. Research related to carbon stocks in mangrove sediments has been carried out several times, including research in Ayah District Kebumen Regency [6], Baturapa Village Bolaang Mongondow District [7] and the Ngurah Rai Grand Forest Park, Bali [8]. In Aceh Besar district, there was research related to carbon uptake in mangrove vegetation (not in the sediment below) specifically in Beureunut village [9]. However, research regarding the estimation of carbon stocks in mangrove sediments in Mesjid Raya District, Aceh Besar Regency has not yet been carried out, so this research aims to analyze carbon stocks in mangrove sediments in this area.

2 Methods

This research was carried out in the mangrove forest of Ruyung Village, Mesjid Raya District, Aceh Besar Regency in March 2023 (Fig. 1).

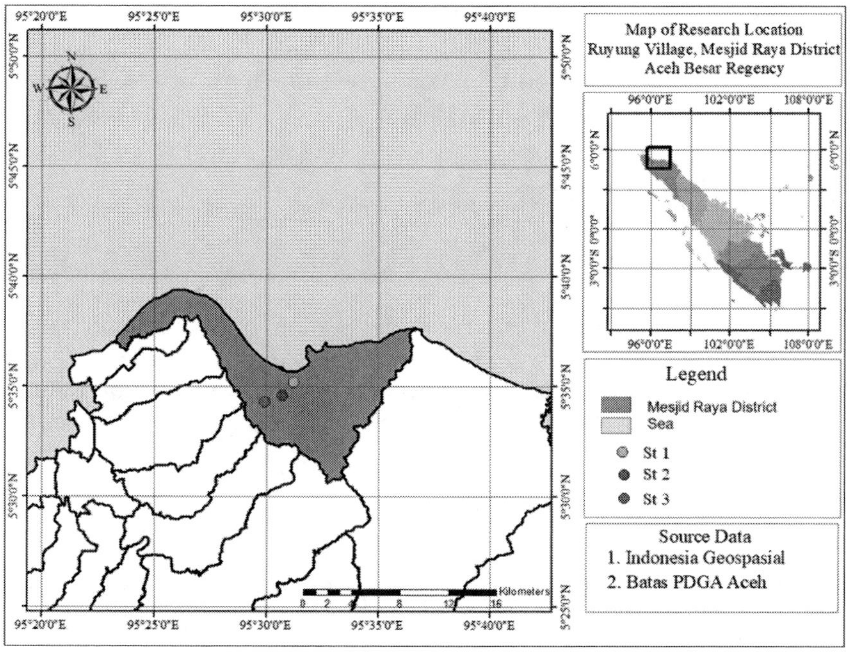

Fig. 1. Map of research location.

There were three research stations in the mangrove ecosystem of Ruyung Village. Each station is 10 meters from the water flow. Each station has a 10 x 10 m plot to calculate mangrove density [8]. Sediment samples were taken 3 times in the plot using a modified pipe with a diameter of 5 cm and a length of 50 cm. The analyzed samples were taken from a depth of 20-30 cm from the sediment surface in each plot at the station. Apart from sediment samples, data on water quality and sediment texture were also analyzed.

Analysis of sediment carbon content was carried out using the LOI (Loss on Ignition) method [10]. Sediment samples were taken and dried in an oven at 60°C for 48 hours. After drying, the sample was ground, and 3 grams was taken to be burned in a muffle furnace at 450°C for 4 hours and the final weight of the sample was recorded. Calculations and data analysis include soil density, dry ash, % C, and total soil carbon content using the following formula:

a) Soil density/ Bulk density

$$\text{Soil density (BD)} = \frac{oven-dry\ masa\ (g)}{sampel\ volume\ (cm^{3})} \quad (1)$$

b). Lost on Ignition

$$\%\,BO = \left(\frac{Wo-Wt)}{Wo} x100\right) \quad (2)$$

% BO = Percentage value of sediment organic matter lost in the combustion process; Wo: Size Initial weight (3 grams); Wt = Final weight after burning (grams).

c). Carbon percentation

$$\%\,C = 1/1{,}724 \text{ x } \%BO \quad (3)$$

d). Total Soil carbon

$$\text{Soil C (Mg/ha)} = \text{BD x SDI x}\%C \quad (4)$$

Soil C= Estimation of carbon stocks; BD: Soil Density (g/cm^3) SDI = Sample depth interval (cm); % C = Percent carbon content of organic sediment material.

3 Results and discussion

3.1 Mangrove Density

The mangrove species found at this research location consist of two species, namely *Rhizopora apiculata* and *Avicennia officinalis*. At station 1 dominated by *Rhizopora apiculata*, at station 2 they are dominated by *Avicennia officinalis* and there are several dead mangroves found, and at station 3 only one type of mangrove was found, namely *Rhizopora apiculate* (Table 1).

Tabel 1. Mangrove density

Station	Mangrove (individu)		$Di(Ind/m^2)$
	Rhizopora apiculata	*Avicennia officinalis*	
1	23	2	0,25
2	5	10	0,15
3	35	-	0,35

The highest mangrove density is at station 3 with a value of 0.35 ind/m^2, followed by station 1 (0.25 ind/m^2 and station 2 (0.15 ind/m^2). The average mangrove density in Ruyung village is 0, 25 ind/m^2. When compared in Banda Aceh and Aceh Besar [11], the types of mangroves found are more abundant with higher densities, namely *Rhizophora mucronata, Rhizophora apiculata, Bruguiera parviflora, Bruguiera cylindrica*, and *Avicenna lanata* with an average of The density at tree level was 0.82 ind/m^2. This is thought to be because the sampling stations were in two locations, namely Banda Aceh and Aceh Besar, so the type and density of mangroves were also greater.

3.2 Sediment texture and water quality parameters

The water quality parameters observed were pH, salinity and temperature. Based on Table 2, it is known that each station is in the range 7 which is neutral pH. According to [12], a pH value in the range of 7-8 is still in the range of good conditions for mangrove vegetation. pH also has a small effect on mangrove growth but is closely related to decomposer activity. If the pH is low, it will make decomposer activity very low so that the process of changing organic material into inorganic will be slow. The higher the pH, the activity of organisms in decomposing organic matter will increase.

Table 2. Sediment texture and water quality parameters

Station	pH	Salinity (ppt)	Temperature (°C)	Texture
Station 1	7,6	31	30	Sandy loam
Station 2	7,6	31	30	Sandy loam
Station 3	7,8	30	29	Sandy clay loam

The salinity and temperature values at the research station range between 30-31 ppt and 29-30 °C. Based on the decision of the Minister of Environment and Forestry No. 51 of 2004, the optimal temperature and salinity in the mangrove ecosystem are 28 – 32 °C and ~34‰ respectively, so it is still within sea water quality standards for mangroves. The types of sediment textures obtained in the research were sandy loam and sandy clay loam sediments. The difference between these two types of sediment is that the sandy loam sediment type has a rather clear rough feel, forms rather hard balls, and is easily crushed, while the sandy clay loam sediment type feels smooth with a few parts that are a bit rough, somewhat sticky, can be formed when pressed. and can be rolled easily.

3.3 Soil density/ Bulk Density

Soil density is the weight of soil in oven/dried conditions per unit volume. From the results of the analysis that has been carried out, it is known that the soil density value at each station has a different value. The sample volume is obtained from the results of measuring the height of the sample interval and the diameter of the corer. Soil density values at each station can be seen in Figure 2.

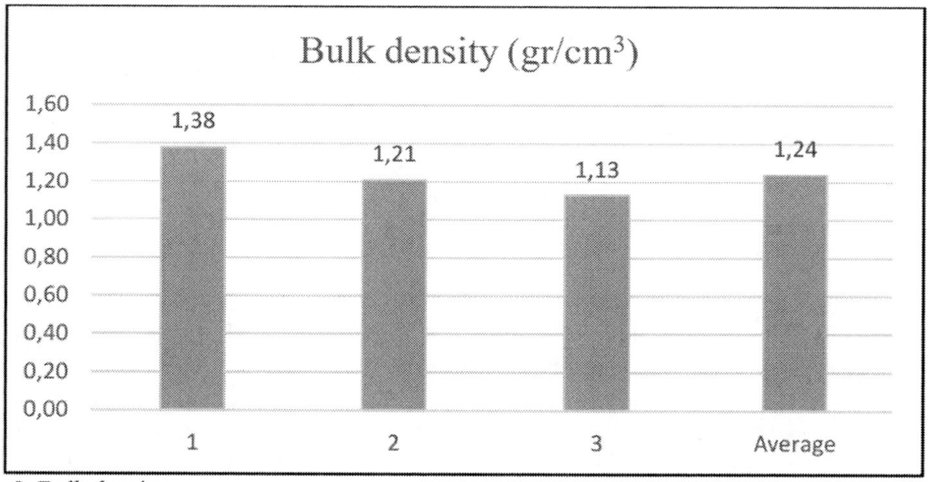

Fig. 2. Bulk density

The highest soil density value was at station 1, namely 1.38 g/cm³, the lowest value was at station 2, namely 28.4 g/cm³. The average value of soil density obtained from this research is 33.8 g/cm³. The results of this research are greater than the soil density value obtained from the results of research on carbon stocks in mangrove sediments carried out by [7] namely 0.39 g/cm³ and from the results of stock research soil carbon carried out by [13] was 0.58 g/cm³. This is thought to be due to the texture of the sediment at the research location being sandy loam and sandy clay so that it has a higher density.

3.4 Percentage of carbon organic sediment

Organic carbon in sediment is one of the factors that make up organic compounds in waters. According to [14], organic carbon is the main factor for improving soil quality and for storing carbon. The organic carbon percentage value is obtained from the results of the LOI (Loss on Ignation) method to determine organic material and its conversion to organic carbon. The average value of organic carbon percentage for all stations can be seen in Figure 3.

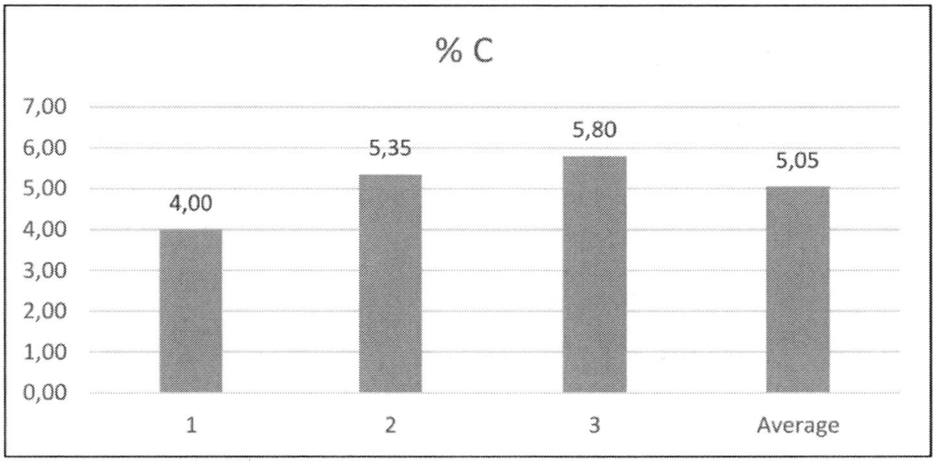

Fig. 3 Carbon organic percentation

Based on Figure 3, the lowest organic carbon percentage value is at station 1 and the highest is at station 3. This difference in carbon percentage occurs because the mangrove density level at station 3 is greater than at the other stations. The average percentage of sediment organic carbon at all stations is 5.05%C. This organic carbon percentage value is smaller than the organic carbon percentage value in the Sediment at Northern of Bunaken National Park Coast [15] where the organic carbon content at a depth of 0-30 cm ranges from 6.77-13.78%C.

3.5 Total carbon sediment

The highest soil carbon is at station 3, namely 347.67 MgC/ha and the lowest is at station 1, 177.70 MgC/ha. The average value of overall carbon content is 271.66 mgC/ha (Figure 4). The average carbon stock in this study is greater than the results of research [13] which analyzed the carbon stock in Karimun Java mangrove sediments, namely 72.75 MgC/ha and 213.12 MgC/ha at a depth of 15-30 cm. However, the carbon stock in Ruyung village is lower than the average carbon stock in Baturapa village which is 398.82 MgC/ha [7]. According to [16] there are several factors that influence carbon stores in the soil, including environmental factors such as land use and soil physico-chemical factors such as temperature, pH, pores, soil texture, soil density and others.

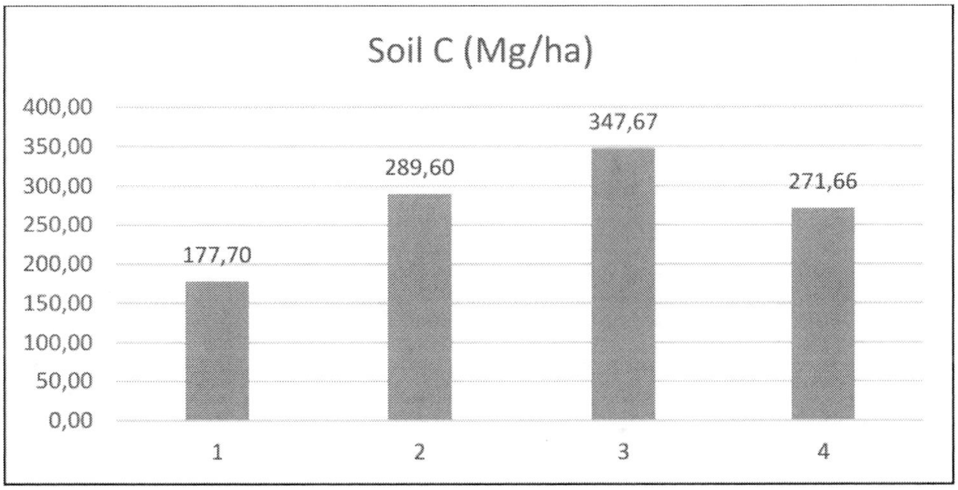

Fig. 4 Total carbon soil

4 Conclusion

The mangrove species found at the research location were *Rhizopora apiculata* and *Avicennia officinalis*, with an average density of 0.20 ind/m^2. The soil density and carbon percentage were 1.24 g/cm^3 and 5.05 %C, respectively. The total carbon content in the mangrove sediments of Ruyung village was 271.66 MgC/ha. From this research, it can be concluded that sediment in mangrove ecosystem can store carbon.

References

1. H. Akram, S. Hussain, P. Mazumdar, K. O. Chua, T. E. Butt, and J. A. Harikrishna, Forests **14**, 1698 (2023)
2. D. O. Suman, *Mangrove Management: Challenges and Guidelines* (Elsevier B.V., 2019)
3. J.-J. Zhu and B. Yan, Sci. Total Environ. **822**, 153438 (2022)
4. D. C. Donato, J. B. Kauffman, D. Murdiyarso, S. Kurnianto, and M. Stidham, (2012)
5. T. Alsumaiti and S. Shahid, J. Agric. Crop Res. **7**, 1 (2019)
6. M. A. Pangestika, N. Soenardjo, and R. Pramesti, J. Mar. Res. **12**, 89 (2023)
7. A. Marbun, A. P. Rumengan, J. N. W. Schaduw, C. P. Paruntu, P. A. Angmalisang, and V. E. Manopo, J. Pesisir Dan Laut Trop. **8**, 20 (2020)
8. C. R. Irawati, I. N. Merit, and I. M. Sudarma, ECOTROPHIC J. Ilmu Lingkung. (Journal Environ. Sci. **15**, 154 (2021)
9. C. Akbar, Y. Arsepta, I. Dewiyanti, and Bahri Samsul, J. La'ot Ilmu Kelaut. **1**, 63 (2019)
10. J. Howard, S. Hoyt, K. Isensee, E. PIdgeon, and M. Telszewski, *Coastal Blue Carbon: Methods for Assessing Carbon Stocks and Emissions Factors in Mangroves, Tidal Salt Marshes, and Seagrasses* (Conservation International, Intergovernmental Oceanographic Commission of UNESCO, International Union for Conservation of Nature, Arlington, Virginia, USA., 2014)
11. H. Farahisah, N. Azizah, I. Dewiyanti, and M. Ulfah, 23111 (2022)
12. E. Poedjirahajoe, D. Marsono, and F. Wardhani, J. Ilmu Kehutan. **11**, 29 (2017)
13. N. Hickmah, L. Maslukah, S. Y. Wulandari, D. N. Sugianto, and A. Wirasatriya,

Indones. J. Oceanogr. **3**, 419 (2021)
14. S. Supriyadi, D. Jurusan, B. Pertanian, and P. Unijoyo, (2023)
15. R. J. Verisandria, J. N. W. Schaduw, C. F. A. Sondak, M. Ompi, A. Rumengan, J. Rangan, P. Studi, I. Kelautan, F. Perikanan, U. Sam, P. Studi, M. Sumberdaya, F. Perikanan, and U. S. Ratulangi, **1**, (2018)
16. I. G. Agung, I. Mahasani, N. Widagti, I. W. Gede, and A. Karang, **1**, 14 (2015)

Effect of different bait towards eel (*Monopterus Spp* and *Anguilla Spp*) catching results in Kampung Baru, Bakongan Subdistrict, South Aceh

Nanda Rizki Purnama[1,4], *Alfi* Syahri Ramadhan[1], *Edy* Miswar[1,3,4*], *Nasruddin* Nasruddin[2]

[1]Departement of Fisheries Resource Utilization, Faculty of Marine and Fisheries, Universitas Syiah Kuala, Banda Aceh, Indonesia
[2]Karawang Maritime and Fisheries Polytechnic, Karawang, West Java, Indonesia
[3]Center for Environmental and Natural Resources Research (PPLH-SDA), Universitas Syiah Kuala, Banda Aceh 23111, Indonesia
[4]Faculty of Marine and Fisheries, Universitas Syiah Kuala, Banda Aceh 23111, Indonesia

Abstract. Kampung Baru is a village in South Aceh Regency. The fishing gear used by most fisherman is a traditional fishing gear. The traditional fishing gear usually used is trap. It is usually used by fishermen to catch eel fish. This study aims to find out the most effective type of bait to catch eel by using trap fishing gear in Kampung Baru Village, Bakongan subdistrict, South Aceh. The types of bait that are tested are congealed, earthworms, chicken intestines and shrimp paste at sampling locations that are usually used by local fishermen. The study used experimental fishing testing methods conducted from April to May 2023. Data analysis used is to determine the influence of different bait by using a complete randomized design. Research results show that the most catch uses a type of earthworm bait 20 eels. Data analysis using Analysis of Variance averages 2.05 eel per day. The smallest real difference test results show that each treatment has a different value and significant. So, it can be concluded that the effect of using different bait has an effect on the catch of eel fish (*Monopterus spp* and *Anguilla spp*), and the most effective of bait is earthworms.

1 Introduction

South Aceh is one of the regencies in Aceh Province which consists of 18 sub-districts, 43 settlements and 260 villages with its capital, Tapakuan [3]. Bakongan District is one of the districts of South Aceh Regency. The population is predominantly live-eyed as fishermen, farmers and traders. Kampung Baru Village is one of them. Most of the villagers of Kampung Baru work as fishermen. The fishermen of Kampung Baru Village still use traditional fishing gear. The traditional fishing gear usually used is trap. Trap is a passive trap that makes it difficult or impossible for prey to escape again. This fishing gear is well known among fishermen. This bubu is usually used by fishermen in Kampung Baru village to catch eel fish.

* Corresponding author: edymiswar@usk.ac.id

Eels are fierce predators in swamp and rice field environments. The food is small fish, worms and crustaceans. Its habit is to nest in muddy holes and wait for passing prey. Eels hide under rocks and mud during the day [11]. This happens because eels cannot tolerate direct sunlight. Eels are animals that actively search for food at night [12].

Bait is one of the important factors affecting the success of eels caught [13]. The type and size of the bait used must provide stimulation for the target fish to approach and eat the bait. The rivers and canals of the village of Kampung Baru, Bakongan District, are full of rice paddies, and fishermen use them as bait to catch eels. Additionally, the use of bait in the form of earthworms (*Lumbricus rubells*) is also often used to catch eels [1].

The fishermen of Kampung Baru Village usually only use earthworm bait and golden eggplant. For other kinds of bait fishermen have never tried. Sintia [8] also concluded that the use of golden bait, worms and trash fish in trap in Pemayung subdistrict resulted in the same number, weight and length of eel caught. So that the bait is suitable for farming areas.

Fishermen use traps as a fishing gear for catching eels. This trap is made from woven wire resembling a cylindrical tube with a diameter of 10 - 15 cm and a length of 100 - 120 cm, with a conical shape towards the tip [14]. At the front, a valve is installed as a trapping so that eels that enter the fishing gear cannot get out [4]. Some of the traps are covered using black plastic and on the back a lid is attached using an old drink bottle. The purpose of covering the trap with black plastic is to follow the characteristics of eel fish which actively search for food at night [12].

Traps is a fishing gear commonly known among fishermen. Traps are often also called traps and guiding barriers. This fishing gear is often used by fishermen in Kampung Baru village to catch eels. This trap is made from woven bamboo, woven rattan and woven wire [15]. According to Sudirman and Mallawa [16], traps are a type of trap that is installed permanently in the water for a certain period of time which makes it easier for fish to enter but difficult to get out. In operation, it can use bait or without bait. The use of bait in a fishing gear operation serves to invite or stimulate fish so that the operating system will be more effective [17].

Based on the initial survey, the problem of bait to catch this eel still does not have a solution point. At certain times of the season these gulls and earthworms are difficult to find. As a result, many of the fishermen do not carry out the fishing process because the main bait that is commonly used does not exist. So that in this case, a study needs toward on the effect of different bait on the catching eel. The bait tested on this assessment are rice, earthworms, chicken intestines and shrimp paste.

2 Research Methods

Research was conducted in April–May 2023. The research site was conducted in Kampung Baru Village, Bakongan District, South Aceh Regency. The research location map is as follows:

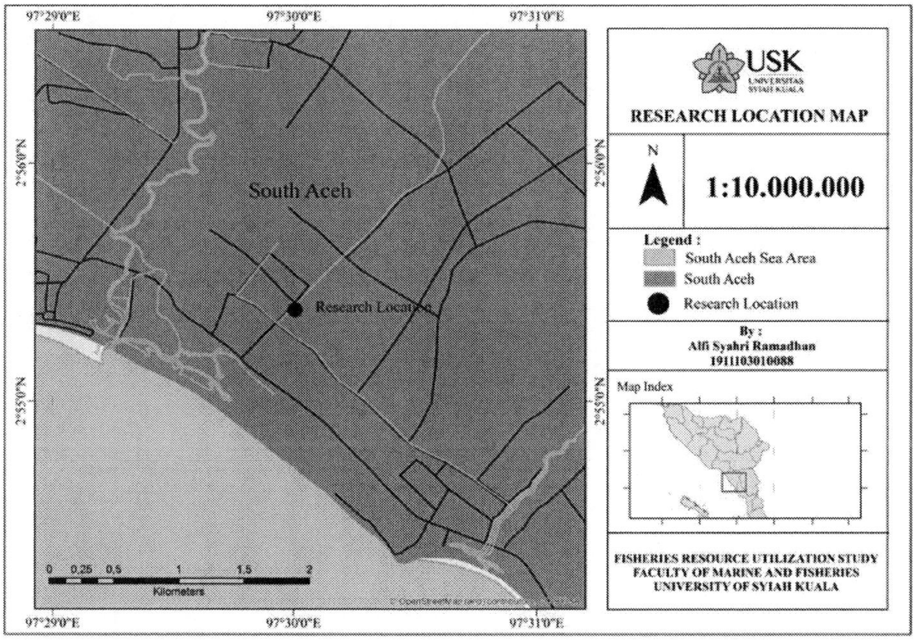

Fig. 1. Caption of the Figure 1. Below the figure.

2.1 Sampling Procedure

The types of bait tested were earthworms, golden snails, shrimp paste and chicken intestines. The bait was tested using trap fishing gear. Traps are installed along oil palm plantation ditches which are usually used by fishermen to catch eels using traps. At this location, 8 units of trap fishing gear were placed, each trap fishing gear used different bait and was installed randomly. The distance for installing eel traps was 5 – 10 meters. Sampling was carried out 4 times a week, so the total sampling for 1 month 5 days was 20 times.

2.2 Data Analysis

Analysis of the data to find out the construction description of this eel-baking device using descriptive analysis, and the influence of the difference in the bait type of bubu-baking device in Kampung Baru Village using the Complete Random Design (RAL) model. The RAL equation [6] is as follows:

$$Yij = \mu i + \tau i + \varepsilon ij \qquad (1)$$

Explanation:

i = 1, 2, … , t and
j = 1, 2, … , r
Yij = Observations on the i-th treatment and the j-th trial.
μ = General flatness
τi = Effect of i-th treatment
εij = Random effects on the i-th treatment and the j-th trial.

The effect of treatment on variability is observed using either the Analysis of Variance (ANOVA) test method or the analysis of a variety print using the SPSS program. If the results of variety analysis show a distinctly different effect between each treatment, it will be conducted with the BNT Test.

The hypotheses tested in this study are as follows:
H_0: There is no effect of different types of bait on the catch of fish
H_1: There is an effect of different types of bait on the catch of fish

The test decision rules according to [6] are as follows:
a. If F-count < F-table > (0.05) then H < reject H>
b. If F-count < F-table > (0.05) then H < reject H>

3 Results and Discussion

3.1 Results

Based on research results that the catch showed eel fish were caught with all kinds of bait tested. During the study conducted, the number of caught fish was 43 consisting of 41 *Monopterus albus* and 2 *Anguila spp*. The range of fish obtained is from 15 g to 270 g with a total weight of 3,740 g. The catch can be seen in Table 1 as follows:

Table 1. Eel Fish Caught

No	Day/Date	The bait				Heavy (g)
		Snail	Earthworm	Chicken intestines	Shrimp paste	
1.	Sunday 30/4/2023	0	0	2	0	230
2.	Monday 01/5/2023	0	0	0	0	0
3.	Tuesday 02/5/2023	0	0	0	0	0
4.	Wednesday 03/5/2023	0	0	0	0	0
5.	Sunday 07/5/2023	0	2	0	1	440
6.	Monday 08/5/2023	1	0	0	0	80
7.	Tuesday 09/5/2023	0	0	1	0	90
8.	Wednesday 10/5/2023	1	0	0	0	100
9.	Sunday 14/5/2023	1	0	1	0	180
10.	Monday 15/5/2023	1	0	0	0	80
11.	Tuesday 16/5/2023	0	2	0	0	180
12.	Wednesday 17/5/2023	0	3	1	0	205
13.	Sunday 21/5/2023	1	1	0	0	75
14.	Monday 22/5/2023	1	3	1	0	395
15.	Tuesday 23/5/2023	1	3	0	0	570
16.	Wednesday 24/5/2023	0	0	1	0	160
17.	Sunday 28/5/2023	0	3	0	1	330
18.	Monday 29/5/2023	2	0	0	0	85
19.	Tuesday 30/5/2023	1	2	1	0	235
20.	Wednesday 31/5/2023	0	1	1	0	205
Total		10	20	9	2	3.635

Research has shown that eel is widely caught in traps that use earthworm bait, snail and chicken intestines. Meanwhile, in trap using shrimp paste bait, less fish caught eel. The most common catch is in ground worm bait, with 20 eels at the same location or an average of one day.

Table 2. Anova test number of catches (tails)

Source of Variation	*SS*	*def.*	*MS*	*F*	*P-value*	*F crit*
Between Groups	8.2375	3	2.746	4.561		2.72494392
Within Groups	45.75	76	0.602			
Total	53.9875	79				

Therefore, further tests with the Real Differential Test (BNT) need to know the effects of different bait. The results of the BNT test are shown in Table 3 as follows.

Table 3. BNT Test

Treatment	Average	Average BNT
Earthworm	1	1.4887
Snail	0.5	0.9887
Chicken intestines	0.45	0.9387
Shrimp paste	0.1	0.5887

The smallest real difference (BNT) test results show that each treatment has a different value and significance.

3.2 Discussion

The eel fishing process in this research uses trap fishing gear. A trap is a fishing tool in the form of a trap and guiding barriers. This trap fishing tool is made from woven bamboo, woven rattan and woven wire [15]. This trap is installed permanently in the water for a certain period of time and in operation it can use bait or without bait [16]. The process of operating trap fishing equipment is carried out in the afternoon because eels are nocturnal fish. This statement is also in accordance with the opinion of Saparinto and Cahyo [12] who state that eels are animals that actively search for food at night. The process of removing/checking the bubu is carried out in the morning.

This research was done for 30 days, on research there are several types of bait used as a test for fish capture by using trap. It is seen from the results of many fish caught in every treatment is not only based on using the use of trap but large influence is caused by water conditions and bait used. The bait factor is very influenced by trap catch. The bait used (type and size) must provide stimulation for fish and other catch targets to approach and eat bait [2]. These factors have closely related relations with aspects of the capture target [7]. Water conditions play an important role in the success of fishing operations, this is because water is a medium for soaking traps [5].

The highest eating activity of eels occurs at night, this is proven by the fact that eels caught in the morning have full stomachs, while eels caught in the afternoon tend to have empty stomachs [18]. Researched the eating habits of two species of eel fish *A. bicolor* and *A. marmorata* in Segara Anakan, Cilacap, and found that eel fish are nocturnal and carnivorous, their main diet is crustaceans [19].

The most interested bait fish on this research is land worms. This relates to the prey that eels are most interested in. The most interesting prey are worms, small fish and insects [20]. Tsukamoto and Kuroki [10] also added that the land worms have a typical smell so liked by eels. This bait is a common bait used by fishermen in the fishing process.

Apart from land worms, bait of chicken intestines also obtained almost the same catch. Each bait has distinctive smell and quickly rotting because many contain water. According to Purwanto et.al [4] more water content on the bait and faster also smell on the bait. This causes on both baits didn't give a maximum catch result. Meanwhile, the bait is very much different from the third bait. It was caused because the bait with the time of trap, this bait will be late in water so that bait will be not effective. Furthermore, the result of environmental pollution and capture of eels that continue to reduce the population of eels [9].

The number of eels caught as much 41 tails. Land worms are the most interesting bait on the Sampling location. The total amount of capture is 20 eels with an average result of the capture of land worms are one eel/day fish bait. Apart from earthworm bait, golden snail bait also resulted in a catch of 10 eels with an average catch of 0.5 eel/day. For example, the eels of a catch during research are 9 eels with an average of 0.45 eel/day. Meanwhile, traps using shrimp paste bait had the lowest of catches, namely 2 eels with an average catch of 0.1 eel/day. Apart from the main catch, there is also a by-catch obtained using traps, namely 2 eels.

Judging from the catches obtained, it is known that there is an influence of differences in the type of bait used to catch eel using trap fishing gear, using analysis of variance (ANOVA). Based on Table 2, it can be seen that the F-calculated treatment value = 4.651 is greater than the F-table at the 5% level. So, it can be concluded that the different of bait used as treatment have a significant effect on fish catches using trap fishing gear. To find out which treatment and group has the most influence, it is continued with the Least Significant Difference (BNT) test. Based on the Least Significant Difference (BNT) test, there are differences in each treatment. Judging from the average value of using different types of earthworm bait, it is very significant, golden snails with chicken intestines are not significantly different and earthworm bait with shrimp paste is very significantly different.

High catches were found in the third, fourth and fifth weeks. Meanwhile, in the first week and second week there were fewer catches because they were influenced by natural factors like flooding. Floods result in an abundance of food availability so that when the process of operating the trap fishing equipment is carried out there is less catch. Seasonal variations cause changes in food availability, temperature, feeding activity and spawning activity [21]. If we look at the catch, many eels are caught in rivers that are overgrown with aquatic plants and are quite muddy. Eels really like aquatic environments where water plants grow and are muddy because these places are rich in prey [12].

Hook rods fishing gear is the most widely used fishing gear for catching eels, and is a selective eel fishing gear [22]. In contrast to Baru Village, South Aceh, fishermen use traps to catch eels and this is an effective fishing tool because the catch does not hurt like fishing rods and can live longer in the trap.

4 Conclusion

The use of different types of bait had an effect on the catch of eel, and that the most effective bait for catching eel using traps was earthworms. Judging from the average value of using different earthworm bait, it is very significant, golden snails with chicken intestines are not significantly different and earthworm bait with shrimp paste is very significantly different.

Reference

1. I. Aldita, A.D.P. Fitri, *Journal of Fisheries Resources Utilization Management and Technology*, **3(3)** 88-95 (2014)
2. H. Boesono, A. Dian, E.Y. Susanto, Journal of Fisheries Resources Utilization Management and Technology **1**, 138-147 (2012)
3. Dinas Kelautan dan Perikanan Aceh Selatan. Laporan Tahunan. Aceh Selatan: DKP Aceh Selatan (2019)
4. A.A. Purwanto, A.D.P. Fitri, B.A. Wibowo, Journal of Fisheries Resources Utilization Management and Technology. **3(3)** 72-81 (2013)
5. Rakhmadevi CC. Waktu Perendaman dan Periode Bulan Pengaruh Terhadap Kepiting Bakau Hasil Tangkapan Bubu di Muara Sungai Radak Pontianak. Program Studi Pemanfaatan Sumberdaya Perikanan. Departemen Pemanfaatan Sumberdaya Perikanan, Fakultas Perikanan dan Ilmu Kelautan, Institut Pertanian Bogor. Bogor. 83 hlm. (2004)
6. A. Raupong, Bahan ajaran mata kuliah rancangan percobaan. Program Studi Statiska Jurusan Matematika Fak. Matematika & Ilmu Pengetahuan Alam Universitas Hasanuddin. (2011)
7. M. Riyanto, Respon Penciuman Ikan Kerapu Macan (Epinephelus fuscoguttatus) Terhadap Umpan Buatan. Sekolah Pascasarjana, Institut Pertanian Bogor. Bogor. 137 pp (2008)
8. Y. Sintia, Pengaruh umpan pada alat tangkap bubu terhadap hasil tangkapan belut monopterus albus di kecamatan pemayung. Pemanfaatan sumberdaya perikanan, Bogor, Intitute Pertanian Bogor. (2022)
9. P. Susatyo, N. Setyaningrum, E.T. Winarni, T. Chasanah, Atang. The Journal of Tropical Life Science, **8(2)** 177-186 (2018)
10. K. Tsukamoto, M. Kuroki. *Eels and humans*. Springer. Japan. (2014)
11. M.F. Miah, M.N. Naser, M.K. Ahmed, *Journal of Global Biosciences*, **4(3)**, 1780-1794. (2015)
12. C. Saparinto, Panduan Lengkap Belut. Penerbit Penebar Swadaya, Jakarta (2012)
13. D. Indrawati, Pengujian Umpan Buatan (Arginin dan Leusin) terhadap ikan Kerapu Macan pada Skala Laboratorium. Bogor; Departemen Pemanfaatan Sumberdaya Perikanan, Fakultas Perikanan Dan Ilmu Kelautan, Institut Pertanian Bogor. Pp 14 (2010)
14. B. Burnawi, Buletin Teknik Litkayasa Sumber Daya dan Penangkapan, **6(1)** 7-9. (2016)
15. Baskoro, M. S., Yusfiandayani, R. Metode Penangkapan Ikan. PT Penerbit IPB Press. pp 23-24. (2019)
16. A.M. Sudirman, Teknik penangkapan ikan. PT. Rineka cipta. Jakarta. (2004)
17. W. Ardha, Pengaruh Perbedaan Jenis Umpan dan Jenis Mata Pancing Layang-layang terhadap hasil tangkapan di Perairan Sendang Biru Malang. Skripsi. Universitas Diponegoro. Semarang. Halaman 67-68. (2010)
18. F.W. Tesch, The Eel. 3rd ED. Blackwell publishing company. UNESCO/WHO/UNEP. Water Quality Assessment. Edited by Chapman, D. Chapman and Hall Ltd. London. (2023)
19. E. Susatyo, I. Yuwono, P. Sulistyo, Sukardi. Study on feeding behaviour of eels and the larvae occurrence in Sagara Anakan. Proceeding of Tropical Eel Fishery Resource, BPPT, Ministry of Research and Technology (2003)

20. F.M. Juniarta, T.Q. Dewi, Budidaya Belut di Berbagai Wadah. Penerbit Penebar Swadaya. Jakarta. pp 12-13 (2016)
21. L. Febriani, Studi makanan dan pertumbuhan ikan bilih (*Mystacoleucus padangensis*) di Danau Singkarak, Sumatera Barat. Bogor. Intitute Pertanian Bogor (2010)
22. N.R., Purnama, R.M. Aprilla, I. Zulfahmi, S.A. Akbar,. Analysis of leading capture fisheries commodities in Pidie district, Aceh, Indonesia. In *IOP Conference Series: Earth and Environmental Science,* IOP Publishing (1221(1) 012042 (2023

Seaweed-based biosorbent for the removal of organic and inorganic contaminants from water: a systematic review

Said Ali Akbar[1,*] and *Khairunnisa* Khairunnisa[2]

[1]Department of Aquaculture, Faculty of Marine and Fisheries, Universitas Syiah Kuala, Banda Aceh 23111, Indonesia
[2]Department of Marine Science, Faculty of Marine and Fisheries, Universitas Syiah Kuala, Banda Aceh 23111, Indonesia

Abstract. Inadequately treated or untreated wastewater contributes substantially to the discharge of undesirable and hazardous substances into aquatic environments. Serious concerns are raised when certain pollutants become persistent and bioaccumulative upon release into the environment. Despite the existence of alternative wastewater treatment technologies, adsorption has consistently demonstrated its efficacy in the treatment of wastewater originating from diverse industrial sources. Adsorption is selected as the optimal method due to its numerous benefits, which include greater efficacy, reduced cost, and convenient accessibility in comparison to alternative treatments. Biosorption using naturally occurring seaweeds can, however, remove contaminants from a variety of sources, including heavy metals, nitrogen, phosphorus, and phenolic compounds, and dyes from the paper, textile, and printing industries. Biosorption utilizing seaweed has surfaces as a feasible substitute for existing technologies in the effective elimination of these pollutants from effluent on account of its environmentally sustainable nature, readily available resources, and cost-effectiveness. An economical adsorbent known as seaweed is examined in this article in order to remove contaminants from effluent. In an extensive table, the application of seaweed in effluent treatment is detailed. The majority of studies, according to published research, have utilized simulated wastewater; biosorption using seaweed to remediate actual wastewater has received less attention.

1 Introduction

Algae, which range in form from unicellular microalgae to multicellular macroalgae, are a diverse group of photosynthetic organisms that inhabit both freshwater and marine environments [1]. Seaweeds, or macroalgae, are organisms that exhibit rapid growth and bear a resemblance to plants; certain species have the capacity to attain a maximum length of 60 meters [2]. Seaweeds are utilized in a multitude of industrial processes on account of their diverse intrinsic properties. These processes include bioremediation and ecosystem balancing to prevent eutrophication; seaweed extract is employed as a biobased fertilizer for crops [4], anaerobic digestion generates energy-rich biogas [5], and seaweeds are even utilized as an edible fresh food [6]. Seaweeds, especially in Asia, are commonly utilized for the production of industrially significant phycocolloids [7] or ingested raw. Additionally, seaweed provides

* Corresponding author: saidaliakbar@usk.ac.id

a natural polymer that is integrated into traditional plastic formulations in order to produce biodegradable plastics [8]. A summary of the industrial and domestic uses of various species of seaweeds is provided in the following section (Table 1).

In recent times, there has been an increasing scholarly focus on the potential adsorption-based treatment of effluent by seaweeds. Primarily due to the presence of sulphated polysaccharides in their fibril matrix and intercellular spaces, macroalgae have a remarkable capacity to bind trace metals and other pollutants. Indeed, potent ion exchangers are the hydroxyl, sulphate, and carboxyl groups present in polysaccharides; thus, they serve as critical sites for metal cation complexation [9]. The cell wall chemical compositions mentioned above exhibit significant variation across various species of seaweeds, with the most suitable composition contingent upon its intended function.

Table 1. Several applications of seaweed in industrial and domestic

Type	Seaweed	Applications	References
Brown seaweed	*Sargassum sp.*	Animal feed, supplements for poultry feed: environmental implications and feasibility	[10]
Brown seaweed	*Eisenia bicyclis*	Phlorotannins inhibit adiposity and disorders associated with obesity, the effects of Eisenia bicyclis stems' in vitro antioxidant and antiglycation properties on the gastrointestinal microbiota of rodents fed a high-sucrose, low-fibre diet are investigated	[11]
Red seaweed	*Kappaphycus alvarezii*	The progress made in the development of biodegradable plastics by incorporating natural polymers into conventional plastic formulations	[12]
Red seaweed	*Gracilaria gracilis*	As a source of multiple products utilized in the biotechnology, nutraceutical, and pharmaceutical industries	[13]
Red seaweed	*Gracilaria fisheri*	In rodents with colonic inflammation, oligosaccharides derived from Gracilaria fisheri ameliorate gastrointestinal dysmotility and gut dysbiosis when used as a flavor supplement or as a source of savory flavor for seafood products.	[14]
Green seaweed	*Arabidopsis thaliana*	Biochar production for the purposes of carbon sequestration and soil improvement	[15]
Brown seaweed	*Saccharina latissima*	Seaweed can be utilized as a circular nutrient (N and P) management system to mitigate eutrophication in aquatic ecosystems. Resource for the production of biobased fertilizers.	[16]

Type	Seaweed	Applications	References
		Potential for bioextraction and blue growth in Danish Saccharina latissima aquaculture—A model for eco-industrial production systems that combat climate change and marine eutrophication	
Brown seaweed	*Laminaria hyperborea*	Bioenergy (methane) potential of seaweed as a promising seaweed bioenergy option	[17]
Brown seaweed	*Saccharina japonica*	Bio-oil production, which can be used as a fuel for green electricity generation, and hydrothermal liquefaction-based combined heat, hydrogen, and power production: A comprehensive feasibility study	[18]

2 Materials and Methods

In this study, we collected the references (papers) from Indonesia in Google Scholar. Recent references published between 2019 until 2023 were retrieved using some keywords " Seaweed Biosorbent". The present status of research on Seaweed Application in Wastewater Treatment.

3 Results and Discussion

3.1 Seaweeds

Benthic algae, or seaweeds, are multicellular, large algae that grow faster than land plants. Because seaweeds have higher photosynthetic efficiencies than terrestrial plants, they acquire biomass faster [19]. It contributes significantly to the aquatic biomass, generates almost half of the primary productivity on Earth, and is widely utilized for feed and other purposes. Seaweed is quite abundant, thus during the past ten years, as demand for it has grown, seaweed agriculture has flourished. This results in a large amount of waste seaweed from various industrial processes, such as the food industry, where the seaweed needs to meet stringent quality requirements [20]. In a circular economy, these wastes can be used as a third generation of biomass for pharmaceutical, feed, agricultural, and biofuel applications. Since releasing these materials into the environment could gravely affect the marine ecology, it is imperative to reuse the waste material. It also implies that algae have a great deal of potential for use as a sustainable energy source and as high-value products [21].

Around the world, seaweed has over 10,000 different species and can grow up to 180 meters below the surface on solid surfaces like granite. It is one of the most important living resources in the ocean in terms of marine biodiversity. Seaweeds can be categorized into three classes based on the type of pigmentation: red (Rhodophyta), brown (Ochrophyta), and green (Chlorophyta). These seaweeds include a variety of bioactive compounds that have numerous uses in the medicinal, agricultural, and pharmaceutical sectors. The several varieties of seaweeds are listed in Table 1, along with corresponding polysaccharides and their molecular structures [22,23].

Table 1. Important seaweed polysaccharides and their chemical structure.

Seaweed	Polysaccharide	Chemical structure	Sugars	Ref
Brown seaweed (*Phaeophyta*)	Laminarin	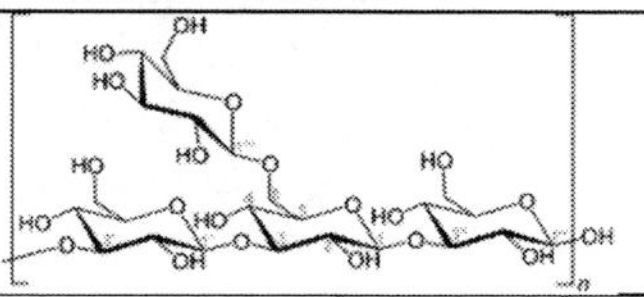	Glucose, guluronate, sulphated fucose, mannuronate, mannitol	[19-12]
	Alginates	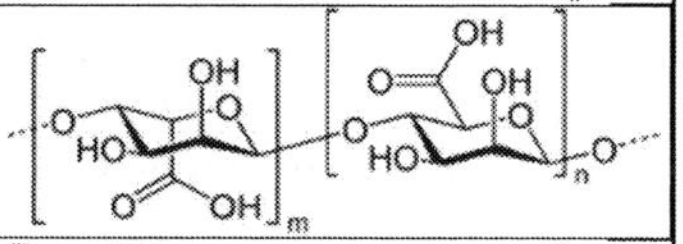		
	Cellulose	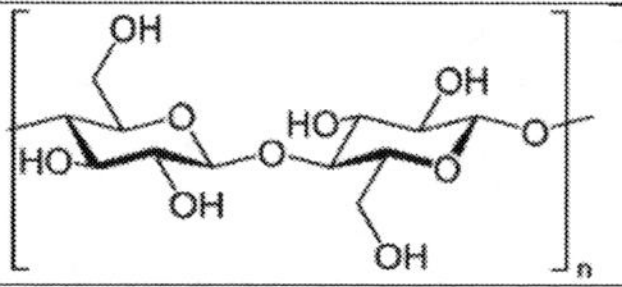		
	Fucoidan	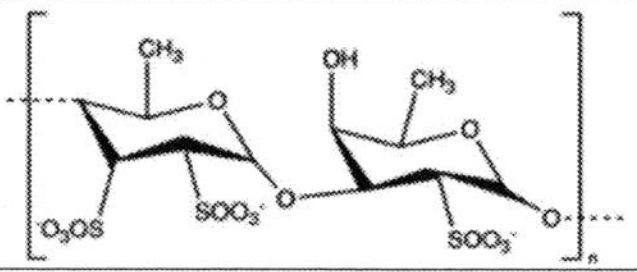		
Red seaweed (*Rhodophyta*)	Agaropectin	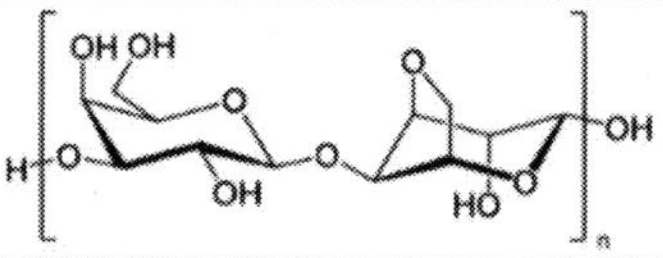	D-galactose, D-fructose, 3,6-anhydro-D-galactose, glucose	[22-25]
	Carrageenan	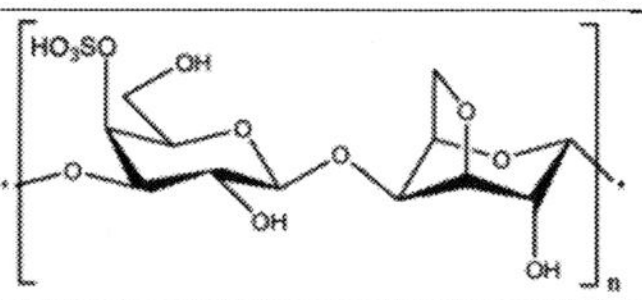		
	Cellulose	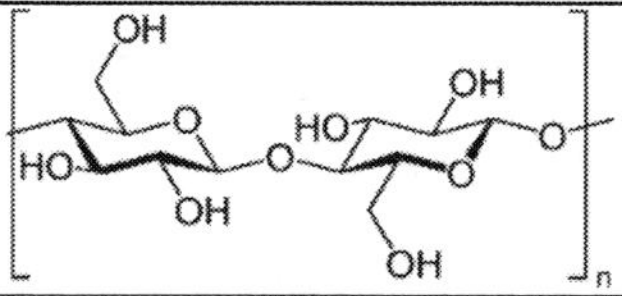		
	Xylans	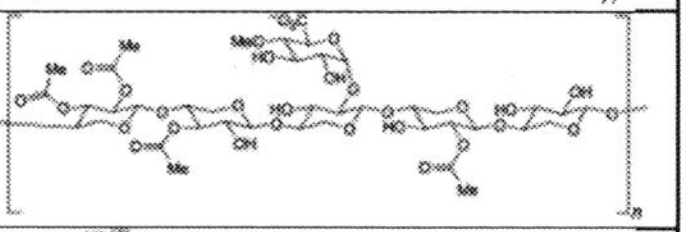		
	Mannans	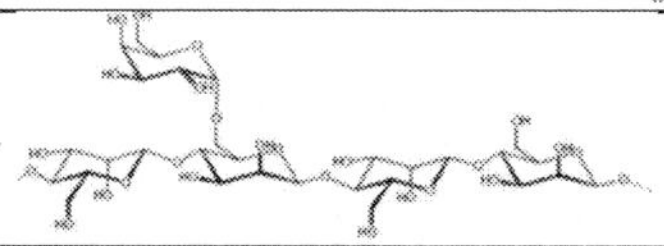		
Green seaweed (*Chlorophyta*)	Ulvan	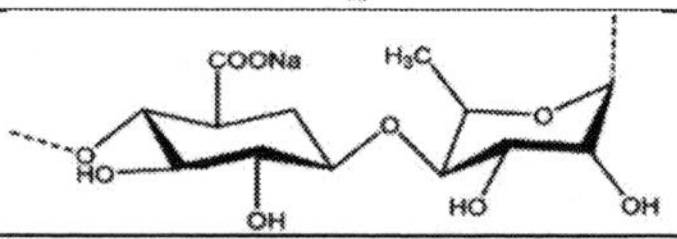	Glucose, xylose, uronic acids, rhamnose, galactose	[26-27]
	Starch	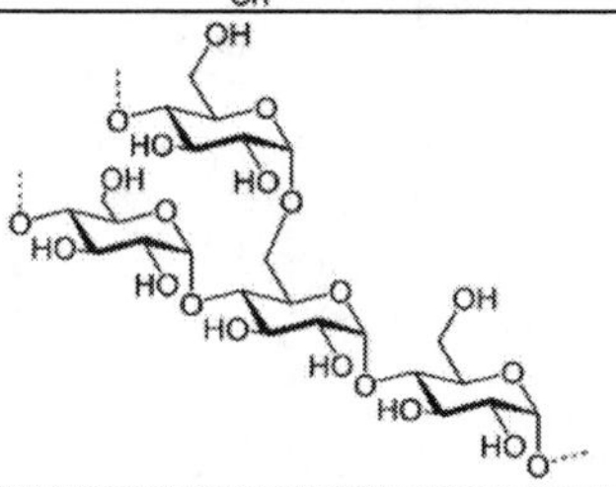		

	Xylopyranose	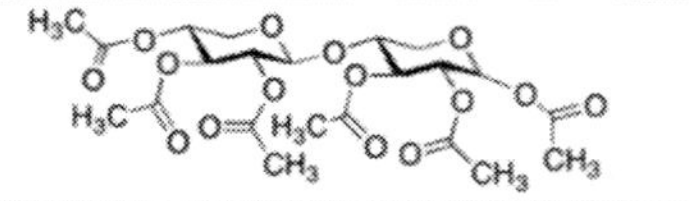		
	Glucopyranose	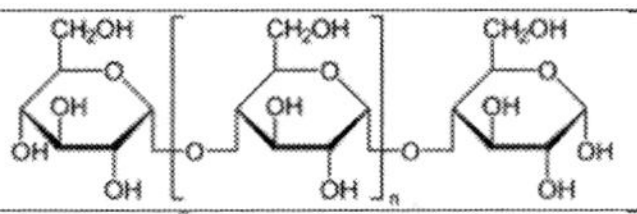		
	Xyloglucan	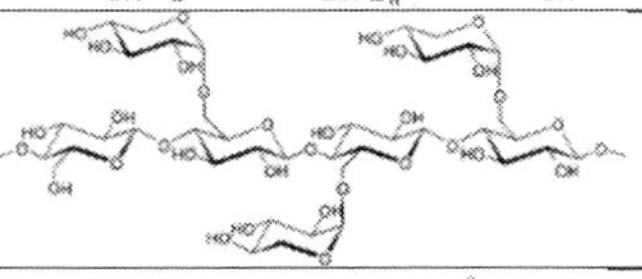		
	Glucuronan	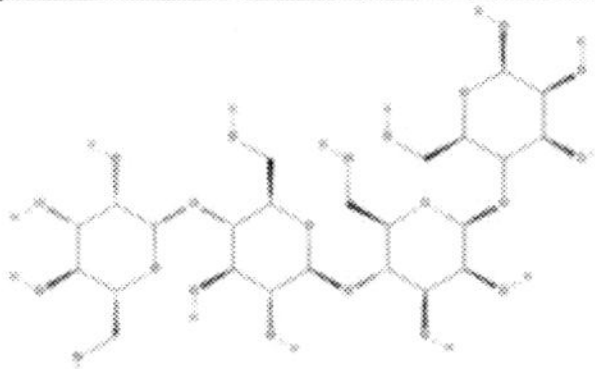		
	Cellulose	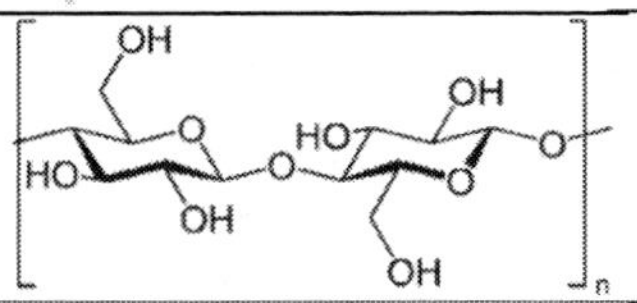		
	Hemicellulose	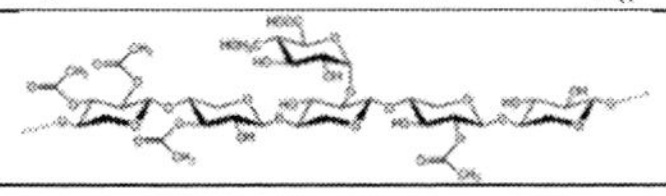		

3.2 Seaweed Application in Wastewater Treatment

Seaweed, apart from its diverse industrial applications, has been extensively studied and implemented as an adsorbent in effluent treatment to substitute functional activated carbon. Wastewater is produced as a byproduct of any given process or activity. It may originate in households, factories, manufacturing sectors, petrochemical industries, aquaculture, agriculture, or landfills. The presence of both organic and inorganic contamination in these wastewaters is widespread. Organic pollution is defined as the existence of substantial amounts of organic compounds; the same holds true for inorganic pollution. These organic compounds are derived from various sources, including domestic sewage, urban runoff, effluents from agriculture and aquaculture, treatment facilities, and industrial effluents from processes such as paper and fiber manufacturing and food processing. Common organic pollutants include substances such as pesticides, fertilizers, hydrocarbons, phenolic compounds, plasticizers, biphenyls, lubricants, greases, detergents, and pharmaceuticals. Chemicals, dyes, benzene, toluene, ethylbenzene, and p-xylene (BTEX) are all classified as organic pollutants [28]. Typical sources of inorganic toxic pollutants include heavy metal ions, arsenides, and fluorides, which originate from industrial sectors including agriculture, paint manufacturing, and others. As a result of the stringent discharge standards established by environmental regulations, wastewater treatment has gained international attention. Numerous investigations are presently underway to examine various methodologies for treating distinct categories of effluent, and adsorption onto macroalgae is not a novel concept. It is concerned with the elimination of heavy metals, phenols, chemical oxygen demand (COD), and biological oxygen demand (BOD), among other things. The removal of carbon fixation, lipid production, total organic carbon (TOC), turbidity, and COD and BOD from wastewater using macroalgae has been the subject of very few studies, with the majority of research concentrating on the removal of heavy metals, phenols, and dyes [29].

3.2.1 Removal of Heavy Metals

In addition to its primary application in effluent purification, seaweed reduces or eliminates toxic heavy metals. The elimination of heavy metals from effluent has emerged as a substantial environmental concern in recent times, primarily due to the detrimental effects that these metals have on living organisms and the environment, particularly when they exceed regulatory limits. Certain heavy metals, even at minute concentrations, are carcinogenic and deleterious [30]. They are capable of readily accumulating in living organisms and do not biodegrade. The accumulation of heavy metals in soil and groundwater is an increasing matter of concern. Soil parent material (lithogenic source) and various anthropogenic sources contribute to the presence of heavy metals in wastewater. These sources include metal smelters, the paint industry, fertilizers, agricultural processes, leather tanning, electroplating, alloy and battery manufacturing, as well as the disposal of other industrial waste materials.

The heavy metals that cause concern on a global scale are chromium (Cr), nickel (Ni), copper (Cu), arsenic (As), cadmium (Cd), mercury (Hg), and lead (Pb) [31]. Pb exhibits severe toxicity towards the reproductive system, nervous system, and kidneys, while Hg damages structures and functions as a neurotoxin by impeding the enzymatic processes involved in regular neurotransmission. Exposure to As in the environment may also result in noncancerous health complications, including the development of tumors unrelated to cancer. As it is absorbed from soils by plants, transported through water bodies into the environment, and accumulated in numerous types of food crops and aquatic plants, it poses a threat to human health. Research has established that rice could potentially serve as the principal source of inorganic As. Human activities, including the production of cement and construction materials, welding alloys, foundries, steel and alloy manufacturing, the electroplating industry, lamps, mines, urban and industrial waste incineration, coal ash, tanneries, fertilizers, and wood preservatives, contribute to environmental exposure to Cd [32].

Nickel toxicity towards plants may induce modifications in root, stem, and leaf development, in addition to germination. Furthermore, it disrupts vital plant physiological processes, including photosynthesis, water balance, and mineral nutrition. Additionally, plant metabolic processes are impacted by nickel's capacity to produce reactive oxygen species, which induce oxidative stress. Studies have also demonstrated that nickel exposure in animals causes inflammation of the lungs and liver and spleen, as well as cardiac toxicity. Chromium, being one of the hazardous elements that finds extensive application in industry, specifically in coatings and metal platings as corrosion inhibitors, ultimately infiltrates water bodies through effluents discharged from various sectors including tanneries, textiles, electroplating, mining, printing, photography, and pharmaceuticals. Excessive chromium concentration results in oxidative injury to cells and genotoxicity [33].

Additional heavy metals coexist in nature, and each heavy metal presents a distinct hazard to both the environment and human health when it surpasses permissible thresholds. Reliable methodologies are imperative in order to effectively eliminate heavy metals from effluent and aquatic environments. Considerable effort has been dedicated to the efficient elimination of heavy metals from effluent. To date, a wide range of treatment approaches have been implemented to decontaminate heavy metals. These include electrodialysis, chemical precipitation, reverse osmosis, ion exchange, ultrafiltration, nanofiltration, coagulation, flocculation, and flotation [34]. However, these approaches have their own drawbacks in comparison to alternative approaches. Adsorption, due to its cost-effectiveness, versatility, simplicity, and efficacy, has emerged as the favored method for the removal of heavy metals. Aside from this knowledge, the primary obstacles to the widespread use of activated carbon, a common adsorbent, are its exorbitant cost and restricted reusability. From this standpoint, biosorption has emerged as a potentially effective technique characterized by its minimal cost, high efficiency even at minute quantities, absence of supplementary nutrient needs, user-friendly operation, and absence of adverse environmental impacts. Seaweed's natural ecosystem-balancing function, in conjunction with the aforementioned characteristics, has prompted an abundance of research into the adsorbent performances and efficiencies of various seaweed varieties in the removal of heavy metals from diverse wastewaters.

Each of the three seaweed groups—green, brown, and red—has been extensively implemented in a multitude of effluent treatment processes. Simulated and actual effluent are evaluated in order to determine the adsorption capacity of seaweed. In conclusion, there is a scarcity of research that specifically examines adsorption using actual wastewater, whereas a substantial number of studies have employed aqueous solutions and simulated wastewater. Pb, Zn, Cd, Ni, and Fe were obviously the primary elements targeted in the majority of the interventions [35]. Research was also conducted on ion exchange theory, in which the sequestration of heavy metals present in effluent is the predominant mechanism. Additionally, it was observed that the quantity of ions bound to biomass was comparable to the quantity of metals displaced from the biomass. Indeed, the inherent ability of seaweed to exchange cations was adequately elucidated by the functional groups, including carboxylic and sulfonic groups, that were present on the outer surfaces of every red, brown, and green seaweed. The aforementioned variations in cell wall chemical compositions among distinct species of seaweeds account for the significant disparity in metal biosorption affinity between brown, green, and red seaweeds. This discrepancy is primarily due to differences in the cell wall matrix.

3.2.2 Removal of Dyes

Adsorption treatments are preferred treatment methods for the elimination of dissolved organic pollutants, such as pigments originating from various industrial sectors including textile, paper, pulp, food coloring, cosmetics, and carpet manufacturing. The dyes utilized in these sectors are reportedly in significant quantities [36]. These dyes not only have the potential to cause cancer but are also detrimental to aquatic organisms and mammals. The majority of textile industry-consumed dyes are resistant to aerobic digestion and do not degrade biologically. Malachite green and methylene blue are the two most frequently encountered dyes in wastewater. An assortment of inexpensive adsorbents were investigated and implemented to eliminate dyes, concurrently with the pursuit of more cost-effective and readily accessible substances [37]. Methylene blue removal agents include activated carbons (AC), including those derived from karanj fruit hulls and olives, as well as waste-based ACs, including those derived from cashew nut shells, Ficus carica bast fiber, Daucus carota plant, and Salix psammophila plant. Adsorbents derived from waste materials, including potato peel-based Limonia acidissima ACs, Daucus carota plant-based ACs, are employed for the purpose of eliminating malachite green dye. The prevalent dyes that cause pollution in effluent are detailed in Table 4, which includes malachite green and methylene blue. Seaweed has been employed as an adsorbent in the preponderance of removal processes; to this day, numerous ongoing investigations utilize natural seaweeds [38]. The adsorption of dyes by seaweed can be primarily attributed to the biosorption process-involved active functional groups, including hydroxyl, carboxyl, carbonyl, amine, and sulfate. Research on kinetic reactions revealed that the data exhibited a strong agreement with the pseudo-second-order kinetic model, suggesting that the chemisorption mechanism was utilized in the process of dye removal from effluent [39].

3.2.3 Removal of Nitrogen and Phosphorus

Due to anthropogenic sources, the presence of excessive inorganic nutrients, such as nitrogen and phosphorus, causes eutrophication in water bodies, which eventually results in an increase in the frequency of detrimental algal blooms and hypoxia. In the majority of freshwater and marine ecosystems, eutrophication has emerged as the most significant water quality concern due to its detrimental effects on coral reef health and community collapse, heightened frequency of fish mortality, and diminished water transparency. Ultimately, these nutrients may also be utilized to promote the proliferation of economically valuable aquatic vegetation, such as seaweeds [40]. The function of seaweed aquaculture in harmonizing ecosystems is not novel. Its most essential characteristic is seaweed's capacity to store a substantial amount of nitrogen in its tissues. The bioextraction of inorganic nutrients from seaweed has garnered significant interest from both academics and authorities.

Experimental kinetic data suit the pseudo-first order model satisfactorily, according to studies, which conclude that the biosorption process involves both extracellular and intracellular transfer [41]. Extracellular transfer encompasses both chemical bonds (chemisorption) and the attraction of phosphate by active sites on the biosorbent surface (physisorption). In contrast, intracellular transfer involves both biotransformation and intracellular accumulation. Additionally, the data provide corroborating evidence that the pseudo-second-order model is the most suitable [42]. CheMIsorption is described by a pseudo-second-order equation that incorporates valency forces resulting from the sharing or exchange of electrons between the biosorbent and adsorbate. Therefore, it can be concluded that the process of phosphate biosorption onto seaweed may involve multiple steps, including chemisorptions [43].

4 Conclusion

The considerable variety of documented applications of seaweeds has garnered the interest of numerous researchers and emphasized the significance of seaweeds during this era. The aforementioned discussion on the various functions and applications of seaweeds will ensure a consistent supply and cultivation of superior raw seaweed materials. It is anticipated that sectors dependent on seaweeds may contribute to socioeconomic development. Undoubtedly, there are numerous opportunities that remain to be discovered in these singular yet resourceful species, such as seaweeds, which are among the most intriguing and intricate forms of life. With this article, an overview of seaweeds in the wastewater treatment industry was intended to be presented. Clearly, biosorption utilizing seaweed is a prospective technique due to its use of a naturally occurring raw material, its higher efficacy, and the low cost of treatment investment. Significantly fewer studies employed actual effluent for remediation purposes, with the majority employing simulations. As a result, it is suggested that forthcoming research places greater emphasis on the purification of actual effluent. The utilization of synthetic or simulated seaweed-based wastewater treatment methods may encounter certain limitations due to their failure to replicate the intricate characteristics of natural wastewater. Furthermore, it is possible that the outcomes derived from simulated wastewater lack practical applicability when it comes to actual wastewater treatment facilities.

Acknowledgment

This study is a part of the university's internal research, which is supported by the Institute for Research and Community Service (LPPM) and the Faculty of Marine and Fisheries, Universitas Syiah Kuala.

References

1. H. P. Manzi, R.A. Abou-Shanab, B.H. Jeon, J. Wang, E.S,*Trends in Plant Science.* **(4)** 596–611 (2021)
2. Y. Ji, K. Gao, *Advances in marine biology*, **(*88*)** 91-136 (2021)
3. E. Trigueros, M.T. Sanz, P. Alonso-Riaño, S. Beltrán, C. Ramos, R. Melgosa, *Journal of Applied Phycology*, **(*33*)** 1181-1194 (2021)
4. D. Dhingra, K. Kandiannan, *Agricultural Engineering Today*, **45(3)** 19-25 (2021)
5. N. Thakur, E.S. Salama, M. Sharma, P. Sharma, D. Sharma, X. Li, ***(18)*** 100120. (2022)
6. E.M. Cabral, M. Oliveira, J.R. Mondala, J. Curtin, B.K. Tiwari, M. Garcia-Vaquero, *Marine Drugs*, **19(4)**, 211 (2021)
7. S. Lomartire, A.M. Gonçalves, *Marine Drugs*, **21(7)**, 384 (2023)
8. J.S.A.R. Junior, G. Turan, *Bulletin of Biotechnology*, **3(1)**, 21-26 (2022)
9. R. Katiyar, A.K. Patel, T.B. Nguyen, R.R. Singhania, C.W. Chen, C.D. Dong, *Bioresource Technology*, **(328)** 124829 (2021)
10. H. Al-Khalaifah, S. Uddin, *Sustainability*, **14(14)**, 8968 (2022)
11. Y. Midorikawa, G. Lee, S. Suzuki, H. Takahashi, T. Kuda, In vitro antioxidant and antiglycation properties of Eisenia bicyclis stems and their effects on the gut microbiota of mice fed a high-sucrose and low-fibre diet. *Journal of Applied Phycology*, **34(6)**, 3237-3246 (2022)
12. M.P. Sudhakar, D. Magesh Peter, G. Dharani, *Environmental Science and Pollution Research*, **28**, 33899-33913 (2021)
13. M.H.Z. Kochesfehani, S. Ataei Jaliseh, M.Z. Kochesfehani, *Journal of Aquatic Physiology and Biotechnology*, **8(4)**, 115-130 (2021)
14. K. Sakena, S. Peerakietkhajorn, B. Siringoringo, P. Muangnil, S. Wichienchot, P. Khuituan, *Journal of Functional Foods*, **71**, 104021 (2020)
15. F. Kenneth, C.F. Joniver, W. Meredith, J.M. Adams, *European Journal of Phycology*, *58*(3), 284-299 (2023).
16. X. Zhang, T. Boderskov, A. Bruhn, M. Thomsen, *Algal Research*, **64**, 102686 (2022)
17. D.A. Smale, A. Pessarrodona, N. King, M.T. Burrows, A. Yunnie, T. Vance, P. Moore, *Scientific reports*, **10(1)**, 12161 (2020)
18. H. Niaz, B. Brigljevic, Y.B. Park, H.C. Woo, J.J. Liu, *ACS Sustainable Chemistry & Engineering*, **8(22)**, 8305-8317 (2020)
19. J. Liu, S. Luthuli, Y. Yang, Y. Cheng, Y. Zhang, M. Wu, J. Choi, H. Tong, Food Sci. Nutr. **8(10)**, 5195 (2020)
20. S.A. Akbar, M. Hasan, S. Afriani, C. Nuzlia, Biodiversitas Journal of Biological Diversity **24(10)**, 5283 (2023)
21. L. Wang, J.Y. Oh, J. Hwang, J.Y. Ko, Y.J. Jeon, B. Ryu, Antioxidants, **8(10)**, 493 (2019)
22. T.H. Nguyen, T.L.P. Nguyen, T.V.A. Tran, A.D. Do, S.M. Kim, Asian Pac. J. Trop. Biomed. **9(12)**, 501 (2019)
23. M. Choulot, I. Michalak, L. Jing, A. Szymczycha-Madeja, M. Wełna, N. Bourgougnon, C. Le Guillard, Algal Res. **75**, 103239 (2023)
24. R. Balti, M.B. Mansour, N. Zayoud, R. Le Balc'h, N. Brodu, A. Arhaliass, A. Masse, Food Biosci, **34**, 100522 (2020)
25. C. Chellapandian, B. Ramkumar, P. Puja, R. Shanmuganathan, A. Pugazhendhi, P. Kumar, Process Biochem, **80**, 58. (2019)
26. R. Ishwarya, B. Vaseeharan, S. Kalyani, B. Banumathi, M. Govindarajan, N.S. Alharbi, S. Kadaikunnan, M.N. Al-anbr, J.M. Khaled, G. Benelli, J. Photochem. Photobiol. B, Biol. **178**, 249 (2018)
27. U. Sonchaeng, P. Wongphan, W. Pan-Utai, Y. Paopun, W. Kansandee, P. Satmalee, M. Tamtin, P. Prapat Kosawatpat, N. Harnkarnsujarit, Polymers, **15(16)**, 3342 (2023)
28. H. Znad, M.R. Awual, S. Martini, S, Molecules, **27(4)**, 1275 (2022)
29. E.H. Foday, B. Bo, X. Xu, Sustainability, **13(21)**, 12311 (2021).
30. A.S. El Shehawy, A. Elsayed, O.A. El-Shehaby, E.M. Ali, The Egyptian Journal of Aquatic Research. **49(4)** 513-519 (2023)
31. A.E.F. Abomohra, M.E. El-Hefnawy, Q. Wang, J. Huang, L. Li, J. Tang, S. Mohammed, Journal of Cleaner Production, **316**, 128341 (2021)

32. H. Luo, S. Xie, Y. Yang, Algal Research, **65**, 102749 (2022)
33. S. Agarwal, M.F. Albeshr, S. Mahboobb, U. Atique, P. Pramanick, A. Mitra, Journal of King Saud University-Science, **34(5)**, 102078(2022)
34. N.Mingu, S.A. Aziz, E.Y.Stidi, M.H.A. Majid, J. Idris, M.S. Sarjadi, Study of Composition and Surface Morphology of Seaweed as Biosorbent: A Review. In Journal of Physics: Conference Series, IOP Publishing. **2314(1)** 012029 (2022).
35. D.A.E.A. Ahmed, S.F. Gheda, G.A. Ismail, Environmental Science and Pollution Research, **28**, 12831-12846 (2021).
36. M. Yadav, S. Thakore, R. Jadeja, Environmental Chemistry and Ecotoxicology, **4**, 67-77 (2022).
37. A.T. Mansour, A.E. Alprol, M. Ashour, K.M. Ramadan, A.H. Alhajji, K.M. Abualnaja, Gels, **8(5)**, 310 (2022)
38. A.T. Mansour, A.E. Alprol, K.M. Abualnaja, H.S. El-Beltagi, K.M. Ramadan, M. Ashour, Polymers, **14(7)**, 1375 (2022)
39. M.I. Al-Zaban, N.K. Alharbi, F.M. Albarakaty, S. Alharthi, S.H. Hassan, M.A. Fawzy, Sustainability, **14(9)**, 5285 (2022).
40. H. Zhang, L. Song, X. Chen, P. Li, Molecules, **26(11)**, 3390 (2021).
41. B.Shahar, L. Guttman, Aquaculture, **532**, 736011 (2021)
42. E. Antunes, A.K. Vuppaladadiyam, R. Kumar, V.S. Vuppaladadiyam, A. Sarmah, M.A. Islam, T. Dada, Cleaner and Circular Bioeconomy, **1**, 100005 (2022).
43. G. Luo, Reviews in Aquaculture, **15(3)**, 1058-1082 (2023).

Pollution status of Aneuk Laot lake Sabang based on pollution index and saprobic index

Nurfadillah Nurfadillah[1,2,4*], *Seri* Maulidawati[1], *Riztania* Anggraini[1], *Cut* Nanda Defira[3], and *Adli* Waliul Perdana[1]

[1]Departement of Aquaculture, Faculty of Marine and Fisheries, Universitas Syiah Kuala, 23111 Indonesia
[2]Graduate School of Mathematics and Applied Sciences, Universitas Syiah Kuala, Banda Aceh 23111, Indonesia
[3]Biology Department, Faculty of Mathematics and Natural Sciences Universitas Syiah Kuala Banda Aceh 23111, Indonesia
[4]Research Center for Marine Sciences and Fisheries, Universitas Syiah Kuala Banda Aceh 23111, Indonesia

Abstract. Pollution that occurs in lake waters needs special attention from various parties in the management of the lake in the future. The activities occurring around the lake result in an increased inflow of pollution into the lake. The aim of this study is to assess the pollution condition of Lake Aneuk Laot in Sabang by utilizing indicators such as the pollution index, CCME WQI (Canadian Council of Ministers of the Environment Water Quality Index), and saprobic index. The investigation took place in both September 2019 and June 2021, employing the stratified random sampling method with four designated observation stations for the sampling process. Parameter measurements analyzed in the pollution index include temperature, depth, current, TDS, TSS, BOD, COD, DO, phosphate, nitrate, ammonia, sulfide, iron, lead, oil and fat, detergent, pH, e-coliform, and parameters used in the saprobic index include phytoplankton data. Based on the analysis of the Pollution Index and CCME WQI it is determined that the pollution status of Lake Aneuk Laot is heavily polluted for Class I, moderately polluted for Classes 1 and 2, and falls under the good category for Class 4. The saprobic index results show the beta-mesosaprobic category with a result of 2.3 (moderately loaded).

* Corresponding author: nurfadillah@usk.ac.id

1 Introduction

The Lake Aneuk Laot (DAL) is a lake that formed from the remains of a dormant volcano's activity and gradually filled with rainwater trapped within it, resulting in the formation of the lake. The lake's surface elevation remains relatively constant, at 25 meters above sea level, and during the rainy season, the water level can rise, although it rarely leads to flooding. According to the findings of [1] Lake Aneuk Laot (DAL) has no surface inlet, and similarly, there is no outlet for the lake. However, it is suspected that the lake has an underground outflow in the form of seepage and water flow through crevices in the lake's bedrock. Lake Aneuk Laot (DAL) holds a crucial role, in line with [2] as a small island like Pulau Weh relies on Lake Aneuk Laot (DAL) as a vital freshwater resource to support all aspects of life for the island's residents.

The waters of Lake Aneuk Laot (DAL) serve as a vital water source for the inhabitants of Sabang. Various activities such as agriculture, livestock farming, the cooling of PLN engines, and recreational facilities around the lake can have negative impacts. These activities are suspected of contributing waste to the water, resulting in a change in the function of Lake Aneuk Laot (DAL), which, in turn, poses a threat to aquatic organisms, especially plankton, which serves as the primary producer in the water. Low-quality water, rendering it unsuitable for various applications, leads to numerous health and environmental repercussions and additionally decreases the overall availability of clean water. The pollution of both surface and groundwater is emerging as a major menace to the accessibility of fresh water resources [3]. This aligns with [4], which asserts that increased human activities in aquatic environments tend to lead to pollution and can disrupt the sustainability of a body of water.

Prior to this, the waters of Lake Aneuk Laot (DAL) were also studied by [5] who noted that the lake's waters were already starting to be polluted due to human waste activities, as evidenced by the high levels of *E. coli*. Similar results were obtained in the research of [6], which found that the microbiological quality of river water and distribution pipes in Aceh Besar Regency and Banda Aceh contained high levels of coliform and fecal coliform. Furthermore, heavy metal pollution is also a dangerous concern. Heavy metals are one of the most commonly encountered environmental pollutants in water, and they can have negative effects on both humans who use the water and the organisms within it [7]. Examining pollution is crucial for establishing the safety of water for various uses. Consequently, numerous studies have explored water pollution, employing indices like the Storet index across different regions. For instance, research conducted by [8] focused on assessing water quality status using methods such as IP, Storet, and CCME WQI. Furthermore, the study conducted by [9] involved the assessment of water quality pollution status, employing the Storet method and various pollution indices.

[10] also undertook a study to assess the water quality status in the Wonokromo River, employing the Storet method and pollution index. The utilization of water quality indices serves the purpose of classifying and comprehending the patterns in changes to water quality status, thereby facilitating the suitable utilization of a water body in accordance with its intended purpose. Widely adopted methods for water quality indices in Indonesia, as stipulated by the Minister of Environment's Decision No. 115 of 2003 concerning Guidelines for Determining Water Quality Status, include the Storet and IP (Index Pollution) methods. Additionally, the CCME WQI (Canadian Council of Ministers of the Environment Water Quality Index), a method developed in Canada, is also employed [11]. Moreover, the saprobic index is another method that can be used to evaluate the degree of water pollution.

The saprobic index is used as a tool to measure water quality and evaluate the impact of organic pollution on aquatic ecosystems. A higher saprobic index value indicates lower water quality because a high value suggests a dominance of organisms tolerant to organic pollution, with sensitive organisms becoming rare or absent. Conversely, a lower value indicates better water quality with a higher presence of sensitive organisms.

One of the biotic components that can serve as a reference for assessing aquatic life i plankton, which plays a role as a bioindicator in aquatic environments. According to [12] the presence of plankton in a body of water is crucial within the aquatic ecosystem, particularly phytoplankton, as they are autotrophic, meaning they can convert inorganic nutrients into organic matter required by living organisms through the process of photosynthesis. Additionally, the discovery of plankton in a body of water contributes to the inventory of plankton biodiversity data, which may not have been fully revealed. Therefore, it is important to have scientific information about this research study to understand the changes in the conditions of Lake Aneuk Laot (DAL) in accordance with its designated class, Based on [13] concerning Environmental Protection and Management. Hence, the objective of this study is to assess the pollution status of Lake Aneuk Laot (DAL) using indicators such as the pollution index (IP), CCME WQI, and saprobic indices.

2 Methodology

2.1 Sampling Area

The investigation took place in the waters of Lake Aneuk Laot (DAL), Sabang City, Aceh Province, during September 2019 and June 2021. Water quality sample analysis was conducted at the Standardization Research Institute in Banda Aceh, while plankton identification was carried out at the Marine Biology Laboratory of the Faculty of Marine and Fisheries at Syiah Kuala University. The research location is illustrated on Map 1.

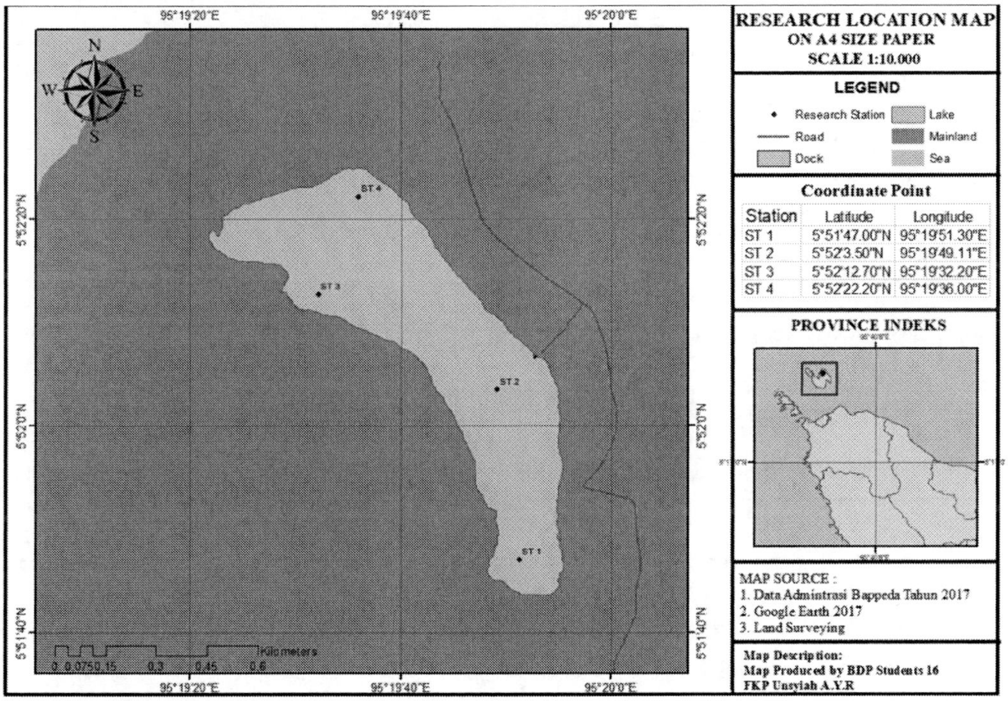

Fig. 1. Location of the research

The method for determining observation stations using the stratified random sampling method. This research was conducted at 4 (four) stations located in the waters of Lake Aneuk Laot, Sabang City, based on the differences in activities along the lake's shoreline, namely: Station 1 (considering the presence of protected forests around the lake and community activities); Station 2 (dock area and recreational activities); Station 3 (PDAM water intake area) and Station 4 (residential area).

2.2 Research procedur

The sampling procedure was conducted quantitatively, employing plankton filtering equipment with plankton net no. 25. Water samples were collected using a Van Dorn water sampler. Plankton observation was executed using the census method, and identification was carried out with the aid of the plankton book [14], along with identification guides for marine diatoms and dinoflagellates [15], and marine phytoplankton [15]. The measured parameters encompassed physical, chemical, and biological factors, including temperature, visibility, Dissolved Oxygen, pH, nitrate-nitrogen (NO3-N), and total phosphate.

2.3 Data Analysis

2.3.1. The pollution index method

The pollution index is examined through two quality indices. The first is the average index (IR), representing the overall pollution level of all parameters in a single observation. The second is the maximum index (IM), indicating the specific parameter that predominantly contributes to a decline in water quality in a given observation [16]. The Pollution Index (IP) for designation j is calculated using the following formula:

$$IPj = \sqrt{\frac{(\frac{Ci}{Lij})^2\ maksimum + (\frac{Ci}{Lij})^2\ rata-rata}{2}} \quad (1)$$

Note:
IPj : Pollution Index for designation j
Ci : Concentration of parameter test result
Lij : Concentration of the parameter according to the quality standard of water designation j
j : Class of designation
(Ci/Lij)m : Maximum Ci/Lij value
(Ci/Lij)r : Average Ci/Lij value

which classifies water quality into four classes, as listed in Table 1.

Tabel 1. Pollution Index Categories

No	Score IP	Description
1.	0 – 1.0	Meets water quality standards
2.	1.1 – 5.0	Slight pollution
3.	5.1 – 10	Moderate pollution
4.	>10	Heavy pollution

Source: [17]

2.3.2. Canadian Council of Ministers of The Environment Water Quality Index (CCME WQI)

The CCME index is one of several water quality indices created by the Canadian Council of Ministers of the Environment [18]. As outlined in [19], the method developed by the CCME comprises three elements.

1. F1 (Scope), expresses the percentage of variables that do not meet the quality standard, at least for one time period (failed variables) relative to the number of variables measured:

$$F1 = [\frac{number\ of\ failed\ variables}{total\ number\ of\ variables} \tag{2}$$

2. F2 (Frequency), expresses the percentage of tests for each parameter that do not meet quality standards (failed tests).

$$F2 = \left[\frac{number\ of\ failed\ tests}{total\ number\ of\ tests}\right] \times 100 \tag{3}$$

3. F3 (Amplitude), expresses the number of times a failed test value does not meet the quality standard. F3 is calculated in three steps, namely:

a) The amount of time that the respective concentration is greater or less than the minimum quality standard. This is called "excursion". If the test value is more than the quality standard:

$$excursion\ i = \left[\frac{failed\ test\ value}{objective}\right] - 1 \tag{4}$$

b) If the test value is less than the quality standard:

$$excursion\ i = \left[\frac{objective}{failed\ test\ value}\right] - 1 \tag{5}$$

c) total excursion value in each measurement

$$F3 = [\frac{nse}{0{,}01\ nse\ +0{,}01}] \tag{6}$$

Note: nse = *normalised sum of the ecursions*

Then the CCME water quality index is calculated by the formula:

$$\text{CCME WQI} = 100 - (\frac{\sqrt{F1^2+F2^2+F3^2}}{1{,}732}) \tag{7}$$

Determination of water quality categories based on the CCME WQI method as follows [20]

Table 2. Water quality category based on CCME WQI method

Value	Status	Category
0 – 44	Poor	Water quality is consistently under threat and often compromised, with conditions typically deviating from natural and desirable levels.
45 – 64	Lack	Water quality is frequently jeopardized and compromised, with conditions commonly deviating from natural and desired levels.
65 – 79	Moderate	Water quality is generally safeguarded, but occasionally, it faces threats and disturbances, leading to conditions that may deviate from natural or desired levels.

80 – 94	Good	Water quality is safeguarded under the assumption that the level of threats and disturbances is minimal, resulting in conditions that rarely deviate from natural or desired levels.
95 – 100	Very good	Ensuring water quality protection suggests that, in the absence of threats and disturbances, water conditions closely mirror their initial or natural state. This index value can be achieved when all measurements of quality standards consistently align over time.

2.3.3. Saprobik Index

The saprobic index is a metric employed to quantify the number of species classified under the organic pollution status in water bodies. This index relies on the presence of organisms in water to assess its status [21]. The Saprobic Index (SI) is determined through the application of the subsequent formula.

$$SI = \frac{1C+3D+1B-3A}{1A+1B+1C+1D} \tag{8}$$

Note:

SI: Saprobic Index

A: Cyanophyceae organism group

B: Dinophyceae organism group

C: Chlorophyceae organism group

D: Bacillariophyceae organism group

Based on the dominant saprobic organisms in a water body, pollution levels can be divided into four categories, as shown in the following table 3.

Table 3. Relationship between saprobic coefficient (SI) and pollution levels [22]

Saprobity level	Saprobiotic index (SI)	Degree of organic pollution
Oligosaprobic	1.0–<1.5	Not to lightly loaded
Oligosaprobic–beta-mesosaprobic	1.5–<1.8	Lowly loaded
Beta-mesosaprobic	1.8–<2.3	Moderately loaded
Beta-mesosaprobic–alfa-mesosaprobic	2.3–<2.7	Critically loaded
Alpha-mesosaprobic	2.7–<3.2	Heavily polluted
Alpha-mesosaprobic–polysaprobic	3.2–<3.5	Very heavily polluted
Polysaprobic	3.5–4.0	Excessively polluted

Source: [22]

3 Result

The pollution index is among the techniques employed to assess the quality status of a water body. As outlined in [13] regarding Environmental Protection and Management, water quality classification comprises four classes. Class I designates water suitable for drinking water sources. Class II is designated for recreational water facilities, freshwater fish farming, livestock, and agriculture. Class III is earmarked for freshwater fish farming, livestock, and agriculture. The outcomes of the pollution index analysis are detailed in Table 4.

Table 4. Pollution status of Lake Aneuk Laot based on the pollution index

No	Class	Pollution Index	Status
1.	Class I	5.9	Moderate pollution
2.	Class II	2.8	Slight pollution
3.	Class III	2.4	Slight pollution
4.	Class IV	0.9	Meets water quality standards

The CCME WQI is a method that compares the results of water quality analysis values to the reference. The results of the CCME index analysis are presented in Table 5.

Table 5. Pollution status of lake aneuk laot based on CCME index

No	Water quality class	Score CCME Index	Status
1.	Class I	32.9	Poor
2.	Class II	48.7	Lack
3.	Class III	71.3	Moderate
4.	Class IV	89,7	Good

Saprobic index value obtained from the results of the analysis based on the relationship between the water saprobity coefficient (X) with the level of water pollution [22] associated with phytoplankton communities classified the occurrence of organic and inorganic material pollution is moderately loaded or beta-mesosaprobic with saprobic value 2.3. Saprobic index values are presented in Table 6.

Tabel 6. Saprobic index values of Aneuk Laot lake

Code	Class phytoplankton	Number of species
A	Cyanophyceae	5
B	Dinophyceae	2
C	Chlorophyceae	4
D	Bacillariophyceae	13
SI (Saprobic Index)		2.3

4 Discussion

The status of Aneuk Laot Lake is included in moderately polluted conditions in class I, lightly polluted in classes II and III, and good water conditions in class IV. Based on the analysis of water quality standards using the pollution index, several parameters are indicated that do not meet water quality standards, namely DO, BOD, COD, TSS,and total phosphate. Thus, the waters of Aneuk Laot Lake are less suitable for use as clean water for consumption, household use, recreational infrastructure / facilities as well as fisheries, agriculture and livestock. The average value of TSS value in Aneuk Laot Lake waters is 102.25 mg/L. These results have exceeded the quality limit for class III water classification for lake waters based on [13] which is 100 mg/L. Lake waters that are utilized by the community as the main source of clean water if they contain high levels of TSS will have an impact on the health of living things if consumed continuously. In addition, excessive TSS levels in lake waters will interfere with the photosynthesis process and result in the death of aquatic organisms [23].

The average value of total phosphate content in Aneuk Laot Lake Waters is 0.3 mg/L. These results have exceeded the quality limit of lake waters based on Government Regulation No. 22 of 2021 for water classification class I - III which ranges from 0.01 - 0.1 mg/L. According to [24] one of the causes of high phosphate levels in waters can come from domestic waste containing detergents [24]. Excessive levels of total phosphate in waters are not only harmful to these waters, but also harmful to living things if consumed for a long time. According to [25] waters with high phosphate levels will trigger rapid growth of aquatic plants. In addition, if these waters are utilized by living things as a source of drinking water continuously, it will have an impact on digestive problems [26].

The average COD content in Aneuk Laot Lake for classes I-IV surpasses the threshold established by Government Regulation No. 22 of 2021, which is 80 mg/L. Elevated COD levels are expected to be inversely proportional to the DO content in the waters. High COD can disrupt the life of aquatic biota and can even cause death in fish and aquatic plants [27]. According to the findings from the research cited in [28], the lake area within the Jayapura City expo arena is categorized as moderately polluted, as determined by the pollution index. This is because seven parameters exceed the quality standard values, namely temperature, BOD, COD, sulfide, lead, copper and phenol. Pollution that occurs in the Lake expo arena area is thought to come from domestic waste and erosion. The research conducted by [29] in Kandung Suli Lake, Kapuas Hulu Regency, indicates that several parameters, including TSS, BOD, and COD have surpassed the quality standard threshold. The discovery of food remnants that settle to the bottom of the waters and the activities of the community around the location are thought to be a source of contributing to the increasing concentration of TSS, BOD and COD in the waters.

Water quality research using the pollution index was also conducted by [30] based on the quality status assessment, Danau Sunter is classified as lightly polluted for class III water quality standards. The test results of the BOD and temperature parameters in Lake Sunter do not meet the quality standard criteria. The increased BOD concentration can be attributed to the deposition and accumulation of organic matter at the bottom of the water, resulting from tourism and household activities. The accumulation of organic matter at the water's bottom can intensify the decomposition process by microbial organisms, subsequently diminishing the dissolved oxygen content.

Based on the analysis of water quality standards in Aneuk Laot Lake using CCME WQI, there are several parameters that exceed the predetermined quality standard values, namely TSS, BOD, COD, DO, total nitrogen and total phosphate. COD and total phosphate parameters are the parameters with the highest concentration values in Aneuk Laot Lake waters. The same research results were conducted by [31], in Lake Batur Bangli Regency, it is known that the COD, total phosphate and dissolved residue parameters have exceeded the quality standard limits based on Bali Governor Regulation No. 16 of 2016. The entry of large amounts of organic compound discharges sourced from community activities, agriculture and fish farming is indicated as the cause of increasing concentrations of COD, total phosphate and dissolved residues in the waters.

The findings from the study conducted by [32] indicate that using the CCME index on Lake Toba, Sipinggan Village is classified as waters with fair to very poor conditions for class I. This is attributed to several parameters that fail to meet quality standards, specifically TSS, BOD, and coliform. The highest TSS concentration obtained a value of 75.3 mg/L, this has exceeded the predetermined quality standard threshold. Waste from agricultural activities in the form of organic and inorganic materials and solids cause TSS concentrations in waters to increase.

Saprobic index (IS) is an index used to calculate the number of species belonging to the status of organic pollution in waters, this index uses the presence of organisms present in the waters to determine the status of waters [21]. Aneuk Laot Lake (DAL) waters have a moderate level of organic and inorganic pollution (beta-mesosaprobic), meaning that there are organic pollutants that have undergone decomposition. The same research was also conducted by [33] the saprobic index in Lake Matano showed that the status of waters with moderate to very light pollution levels with organic and inorganic pollutants dominated by mesosaprobic conditions. This condition is relatively similar to the research conducted by [34], the saprobic index in Lake Toba has a plankton community showing a level of pollution that is classified as very mild to mild with a small load of organic and inorganic pollution that takes place in the mesosaprobic /oligosaprobic phase.

5 Conclusion

The analysis of the Index Pollution and CCME WQI, it is determined that the pollution status of Lake Aneuk Laot is heavily polluted for Class I, moderately polluted for Classes 1 and 2, and falls under the good category for Class 4. The saprobic index results show the beta-mesosaprobic category with a result of 2.3 (moderately loaded).

References

1. N.A. Yusifa, S. Yanto, Sairiyyah dan I. Muhammad. *Jurnal Aceh Physic Social* **8,** 2: 47-54 (2019).
2. C.B. H. Edyanto. *Jurnal Teknik Lingkungan.* **1**:115-124 (2006).
3. B.K. Mishra, P. Kumar, C. Saraswat, S. Chakraborty, A. Gautam. *Water* , ***13,*** 490 (2021).
4. B. Suharto, L. Dewi, A. N. Mustaqiman, T.R.A.K. Marjo. Indonesian Journal of Urban and Environmental Technology, 2, 2 (2019).
5. T. Widiyanto. *Jurnal Limnotek Perairan Darat Tropis*, **24** (2): 83-92 (2017).
6. D.A.I. Sutapa and T. Widiyanto, T. *Jurnal Limnotek Perairan Darat Indonesia. Pusat Penelitian Limnologi LIPI.* **21**(2):135-144 (2014).
7. S.A. Akbar, H.K. Rahayu, Lantanida Journal, 11(1), 51-66 (2023)
8. Y. Romdania , A. Herison, G.E. Susilo. *Jurnal Spatial.* (2018)
9. C. Hermawan. *Jurnal Rekayasa*. **7**(2). (2017).
10. Purnamasari, D. Eva. 2017. Penentuan status mutu air Kali Wonokromo dengan metode storet dn indeks pencemar. *Thesis.* Institut Teknologi Sepuluh November.
11. S.A. Akbar, S. Afriani, C. Nuzlia, S. Nazlia, S. Agustina, Depik, 12(3), 259-273 (2023)
12. P. Wirabumi, Sudarsono, Suhartini. *Jurnal Prodi Biologi.* **6** (3):174-184 (2017).
13. Peraturan Pemerintah Republik Indonesia Nomor 22 Tahun 2021. Tentang Penyelenggaraan Pengelolaan dan Perlindungan Lingkungan Hidup.
14. I.M. Shuthers. D. Rissik. *Plankton. A Guide to Their Ecology and Monitoring for Water Quality.* CSIRO Publish, Melbourne, PP.577 (2009).
15. G. Hassle, E. Syvertsein, K. Streidinger, K. Tangen and C. Tomas. *Marine diatom: identifying marine diatoms and dinoflagellates*. Academic Press. P 598 (1996)
16. Marganingrum, D.Rosmini, A. Sabar, Pradono. *Pusat Penelitian Geoteknologi Lembaga Ilmu Pengetahuan Indonesia,* **23**(1):41-52. (2013).
17. Keputusan Menteri Lingkungan Hidup Nomor 115 Tahun 2003. Tentang Pedoman Penentuan Status Mutu Air.
18. Canadian Council of Ministers of the Environment. Canadian water quality guidelines for the protection of aquatic life: ccme water quality index 1.0, user's manual. CCME. Winnipeg, Canada. (2001).
19. A. Lumb, D. Halliwell, T. Sharma. *Environment Monitoring and Assessment,* **113:**411-429. (2006).
20. A. Lumb, T.C Sharma, J.F. Bibeault. *Water Quality*, **11**(24). (2011).

21. F. Ramadhan, A.F. Rijaluddin, Mardiyansyah. *Jurnal Biologi,* **9**(2):95-102. (2016).
22. M.V. Junqueira, G. Friedrich, P.R. Pereira de Araujo. *Environ Monit Assess*, **163:** 545 – 554 (2010).
23. Suyata, Irmanto, D. Kartika, S. Nurhandayani. *Prosiding Seminar Nasional dan Call for Papers,* 58-65 (2020).
24. A.W. Tungka, Haeruddin, A. Churun. *Journal Of Fisheris Science And Technology,* **12**(1):40-46. (2016).
25. C. Patricia, W. Astono., D.I. Hendrawan. *Trisakti Open Journal Systems*, **1**: 179-185. (2018).
26. Z. Ismail. *International Journal of Water Resources And Environmental Engineering*, **3**(7):132-138. (2011).
27. D. Ramayanti, U. Amna. *Jurnal kimia sains dan terapan*. **1**(1):16-21. (2019).
28. A. Prasetia, A.F. Walukow. *Jurnal Dinamika Lingkungan Indonesia*, **8**(1):42-47. (2021).
29. L. Muthifah, Nurhayati, K.P.Utomo. *Jurnal Teknologi Lingkungan Lahan Basah*, **6**(1):1-10. (2018).
30. A.A. Saputro, Sunaryo, R. Fahdiran. *E- Journal Prosiding Seminar Nasional Fisika*, **9**:125-140. (2020).
31. N.M.H. Sukmawati, A.E. Pratiwi, N.W. Rusni. *Jurnal Lingkungan dan Pembangunan,* **3**(2):53-60. (2019).
32. Tampubolon, Y.C. 2020. Analisis kualitas air tingkat pencemaran di Danau Toba Desa Sipinggan Kabupaten Samosir Sumatera Utara. Skripsi, Universitas Sumatera Utara.
33. A.A Sentosa, A.H. Dimas, S. Hendra. *LIMNOTEK.* **24**(2):61-73. (2017)
34. E.P. Sagala, *LIMNOTEK.* **20**(2):151-158. (2013).

Analysis of the levels of pollution in the Ujong Pancu Beach on Physical-Chemical Quality

Ichsan Rusydi[1*], *Irwan* Irwan[2], *Sayyid* Afdhal El Rahimi[2], *Said* Ali Akbar[1], *Karla* Amelia[3], *Muhammad* Muhammad[2], *Heru* Qowiyul Rahman[2], *Imam* Safir Alwan Nurza[4]

[1] Department of Aquaculture, Faculty of Marine and Fisheries, Syiah Kuala University, Indonesia;
[2] Department of Marine Science, Faculty of Marine and Fisheries, Syiah Kuala University, Indonesia,
[3] Department of Fisheries, Faculty of Agriculture, Gajah Mada University, Indonesia;
[4] Research Center for Elephant Conservation and Forest Biodiversity, Syiah Kuala University, Indonesia

Abstract. Checking and monitoring water quality pollution is an important thing to do. In general, tourist areas are more susceptible to water pollution. The Peukan Bada subdistrict area has Ujung Pancu beach with a length of ± 2 km and is a marine area that plays an important role for the surrounding community. The development of human activities that occur around coastal areas will have an impact on water pollution. The aim of this research is to determine the level of pollution that occurs in the waters of Ujung Pancu Beach using the Pollution Index method based on physicochemical measurements. Water quality values are measured based on water quality standards for marine tourism purposes in the Minister of Environment Decree No. 22 of 2021. Based on the results of the analysis carried out, the result was that the level of pollution obtained was moderate pollution. The average value obtained at high tide is around 8.09 and at low tide around 8.47. Overall, the water quality parameters of Ujung Pancu beach need to be considered so that they are maintained so that they do not have a negative impact on tourist visits and marine tourism activities.

1 Introduction

Ujung Pancu beach is one of the coastal places in Aceh Besar District, with a coastline length of ± 2 km. It is situated in the Lam Pageu Village, Peukan Bada district, Aceh. With a land area of 3.60 km2, Gampong Lam Pageu is situated on the seashore. There is a sea region at Ujung Pancu that is significant to the local population. The Ministry of Maritime Affairs and Fisheries recognized Ujong Pancu as a Regional Marine Conservation Area (KKPD) in 2020. However, because there are a lot of fish in these waterways, the locals use them for fishing. In addition, this location is becoming more and more popular with tourists. A contributing element to the watery ecosystem in this region is the presence of coral reefs surrounding Ujung Pancu, which is close to Pulau Tuan, a small island [1].

The ports, populations, tourism, and industrialization of coastal cities are all factors that contribute to the pollution of their coastal waters. Activities that can produce waste, either directly or indirectly, often disrupt the ecosystem of coastal waters. In addition to causing economic and ecological losses like decreased biological productivity of waters, fish and other marine biota mortality, and damage or decrease in aesthetic value, pollution can have a

* Corresponding author: ichsanrusydi@usk.ac.id

negative impact on people's health and even cause death in urban areas. those who consume the marine biota that is present in coastal rivers [2].

The Ujung Pancu water area faces a number of issues, including marine pollution brought on by waste from nearby communities and shrimp ponds. The seawater surrounding the area may become a different color as a result of this. Organic waste, which is primarily leftover food, can upset the equilibrium of coastal ecosystems by accumulating organic elements that can lead to an increase in blooming algae, which can disturb fish communities in the waters. The primary waste product of ponds is ammonia [3], which is typically present in estuaries in large quantities. Ammonia can cause eutrophication and even algal blooms in the sea or estuary, which makes it dangerous along with other nitrogen compounds like nitrates and nitrites. Disease can also be spread by waste that has not been treated before being disposed of, particularly waste from ponds where cultivated shrimp are sick. Furthermore, water from the sea or other natural sources is what most fish farmers use for cultivation. Temperature, salinity, pH, dissolved oxygen, turbidity or brightness, and nitrate are physicochemical parameters that are frequently used as indicators that measure the quality and pollution status of water [4]. In order to conduct this research, the quality of the waters in Ujung Pancu was assessed in July 2023 at high tide and low tide as required with the Maritime Tourism.

2 Method

2.1 Study Area

The location of the research was carried out in the waters of Lam Pageu Village (Figure 1) in Peukan Bada, Aceh Besar in three stations. The oceanographic parameter analysis was carried out at the Environmental Quality Testing Engineering Laboratory, Chemical Engineering Department, Engineering Faculty, University of Syiah Kuala and the pollution index method analysis was conducted at the Chemical Laboratory of the Faculty of Marine and Fisheries of the University of Kuala Shia on July 10, 2023.

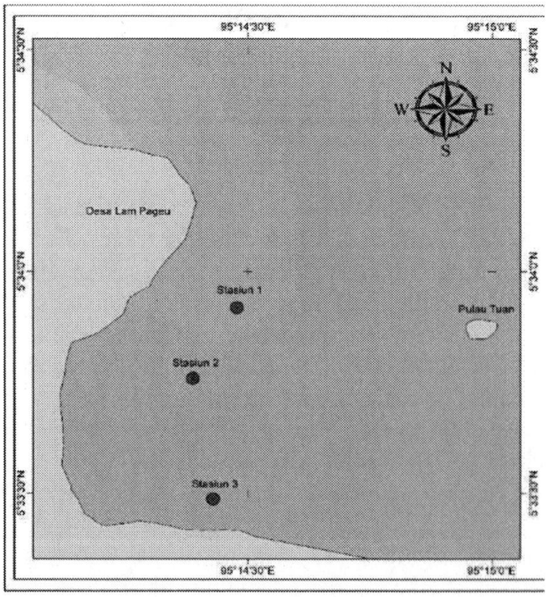

Fig. 1. Sampling sites located in waters of Lam Pageu Village

2.2 Samples collection and processing

The research conducted from 20 to 29 June 2023. The method used in this study is a descriptive method by collecting data from observations carried out in the field with the aim of obtaining a picture of the situation and conditions. Researchers focus on a case to be observed and analyzed accurately in the form of a group or an individual. The stages of research to be carried out are: (1) Preparation of equipment and transportation for water sampling, (2) Prepare equipment for recording the data obtained, (3) Sampling of water data, (4) Testing sample water quality.

Measurement of test parameters using the pollution index of the physical and chemical parameters based on Government Regulation [5]. Water samples were taken to test physical and chemical parameters twice at high tide and low tide. Water samples were taken using a water sampler with a depth of 1 meter. Then put it in a sterile bottle and store it in the coolbox. Water quality parameters such as salinity, temperature, brightness, pH, and DO are carried out *in-situ*. Meanwhile, nitrite, ammonia and phosphate were analyzed at the Chemistry Laboratory, Faculty of Teacher Training and Education, Syiah Kuala University.

3. Results and Discussion

3.1 Test results

The quality of the water used for maritime tourism should ideally meet the standard values both physically and chemically. Water quality values are measured on the basis of the water quality standards. If the measurement results for each parameter not yet fulfilled the quality standards, then the water can be categorized as polluted. The following are the results of the analysis of the physicochemical parameters of the waters of Lam Pageu village at high and low tide, presented in Table 1 below:

Table 1. The analysis of the physicochemical parameters of the waters of Lam Pageu village

Parameters			High tide			Low tide		
Concentration	Unit	BM	St. 1	St. 2	St. 3	St. 1	St. 2	St. 3
Salinity	$^0/_{00}$	33-34	33	30	31	28	32	33
Temperature	^0C	28-32	27.5	28	28	34	33	34
Brightness	m	>6	7.25	6	6.4	6.25	5.25	5.25
pH	-	7-8.5	9.8	9.5	9.4	9.5	9.4	10.4
Oxygen Dissolved (DO)	mg/l	>5	5	5.8	5.5	5,6	5.8	5.3
Nitrite	mg/l	0.06	3.53	3.46	2.58	5.84	3.91	5.72
Ammonia	mg/l	0.3	2.1	1.9	1.54	2.49	2.39	1.44
Phosphate	mg/l	0.015	0.22	0.62	0.72	0.82	1.04	1.06
IP			8.32	8.23	7.90	8.69	8.63	8.31
Category			T. S	T. S	T. S	T. S	T. S	T. S

Nb: BM (Quality standards), St.1 fishing spot, St.2 Tourist spot. St.3 Dock, and TS (Moderately Polluted)

Based on the Table 1 of results from the analysis of measurements of physical and chemical parameters, it was found that the characteristics of the level of seawater pollution at high tide and low tide using the Water Pollution Index method were classified as Moderately Polluted which occurred in all locations. So, the Moderately Polluted category at low tide is found at the fishing location with the highest value of 8.69. The lowest pollution category at high tide is at the dock location with a value of 7.90.

3.2 Discussion

The quality of the sea water used for sea tours and other activities should ideally meet standards, both physically and chemically. Quality values of marine waters that exceed the maximum threshold for their distribution will be classified as polluted waters. As for the influence of the rise in pollution in the waters of Lam Pageu Village, the very high values at the time of the recession have been reduced in the water. Lam Pageu Village is usually used by the community as a tourist destination and fishing in the surrounding waters, activities can be done in the sea slopes that are dominated by water both on the surface and in water. The sea has an extraordinary attraction to mankind [6]. This is reflected in the fact that most of the world's large population lives along the coast. The development of maritime tourism in this era is a business that forms an important component of the extensive tourism industry.

The pollution index method (IP) is an index used in determining the magnitude of the level of contamination according to the parameter of water quality status authorized by the government [7] [8]. The pollution index method has advantages in terms of determining the water quality status [11].

The research according to the Water Pollution Index method was carried out by comparing the level of values between high and low tide of Lam Pageu village in 2023. The water quality pollutant parameters reviewed in this study are then compared to the seawater quality standards. When the quality of the water exceeds the maximum threshold allowed in the standard standard of quality, then the water quality is said to decrease according to its provisions so that it belongs to contaminated water [9].

3.3 Water quality analysis

The quality of the sea water used for maritime tourism activities must meet standards, both physically and chemically. Seawater quality values that do not reach the minimum and significant differences will be classified as polluted waters. As for the results obtained from institutional and laboratory measurements of some physiological and chemical parameters in Lam Pageu coastal waters at the time of high tide presented in Table 1. Salinity affects the life of coral animals due to the presence of osmosis pressure on living tissues. The optimum salinity for coral life ranges between 30–33 ppt, because the presence of coral is fairly low in high rainfall areas [10].

Water temperature is crucial in the existence of biodiversity that is a tourist attraction. The Table 1 indicates that the temperature concentration at the time of planting ranges from 27.5 to 29 °C, where the temperature is the optimal temperature to allow the coral reefs to grow and thrive [11]. The brightness at the time of installation in Lam Pageu ranges from 6-7.25 m with an average depth of 8 to 9 meters. A water that has a high brightness value in normal (bright) weather, giving an indication or indication of the low level of dissolved and suspended particles in the water [12].

The level of dissolved oxygen in the chart in the village of Lam Pageu is high enough that marine biota activity can oxidize organic material quite well. Oxygen is used by aquatic organisms in the process of respiration and decomposition of organic and inorganic substances [13]. While the pH range in Lam Pageu at the time of planting is in the range of

9.4 – 9.8, the value at pH 9 causes a worrying impact on the variety of biosphere that is a tourist attraction. And other parameters that exceed the threshold limit of bacteria such as ammonia (1,54-2,1 mg/L), nitrites (2,58-3,53 mg/ L), and phosphate (0,22-0,72 mg/ l). As for the results obtained from institutional and laboratory measurements of some physical and chemical parameters in the coastal waters of the Ujung Pancu at the time of recession presented in Table 1. On the brightness parameter, depth determines how far the penetration of light is. Brightness is >6 which means the depth at the time of recession must be more than 6 meters to be counted. At the time the average depth in Lam Pageu ranges from 7 to 8 meters.

Ammonia is abundantly found in liquid waste, both domestic waste, agricultural waste, and waste from factories, especially nitrogen fertilizer plants [14]. Liquid waste from ammonia plants contains up to 1000 mg/L ammonium. Therefore, the presence of ammonia in wastewater is severely limited. European countries limit the ammonia content in wastewater to a maximum of 0.5 mg/L, while American countries limit it to 0.77mg/L [15].

The high ammonia presence would disrupt the existence of organisms around the waters. Under the conditions of abundance of oxygen dissolved ammonia will undergo nitrification in water and become nitrites and subsequently with the help of bacteria nitrite undergoes change and into nitrates [16]. With the presence of nitrates and phosphates sufficiently high will give the effect of the occurrence of eutrophication which can stimulate the growth of phytoplankton quickly and result in algae blooming. Nitrate levels above 5 mg/L describe pollution derived from human activity [17].

From community activities which tend to throw industrial waste and household waste into the sea, this also has an impact on the biota of green mussels and oysters due to the high levels of cadmium and lead metal taken from the bottom of the waters, while green mussels in cultivation containers on the surface and water column have safe limit for consumption, because the ability to absorb metal is very high [18].

In water, nitrites (NO_2) can usually be found in very small amounts than nitrates, because they are unstable with the presence of oxygen. Nitrite is an intermediate between ammonia and nitrate (nitrification). Nitrites in water consumed or used in everyday life can be hazardous to health. In humans, excessive consumption of nitrites can cause the oxygen binding process to be interrupted by blood hemoglobin, which subsequently forms hemoglobins that are incapable of binding. Water pollution concentration of nitrites in water commonly used by the public should not exceed 0.06 mg/L [19].

The Natural sources of phosphate in waters are the weathering of mineral rocks and the decomposition of organic materials. Anthropogenic sources of phosphate are industrial waste, domestic waste, drift from fertilizers, and the destruction of organic materials and phosphate minerals [20]. Excessive levels of phosphate in surface water can lead to eutrophication. Eutrophication is the result of a blooming algae in waters 7 so that it can interfere with other biota of those waters. Because there are no gaps in the sunlight so that the organisms below it will die, the dead organisms will be decomposed by bacteria and need a lot of oxygen, so the oxyge levels in the water will be reduced and the water biota in it will be disrupted. [21].

For this reason, carrying capacity is needed to overcome the environmental impact of shrimp ponds to realize sustainable aquaculture industrialization and maintain marine ecosystems which have great potential for the future, Then it is necessary to carry out an environmental impact analysis (AMDAL) if the development of both intensive and semi-intensive pond areas is carried out, so that waste in the waters can be controlled, by reducing the recommended shrimp stocking density, so that the feed provided is also in accordance with pond waste that leaves the environment [22].

5. Conclusions

Based on the results of research on the quality of waters in Lam Pageu at high tide and low tide, it shows moderate pollution with several parameters that have passed the quality standard thresholds such as nitrite, ammonia, phosphate and pH. and the highest level of pollution is when low tide.

References

1. E. Rudi, Asses Res Coral Reefs and Com Percep of Conserv Areas Waters in Ujong Pancu, Aceh Besar. Biospecies **6** 2 (2013)
2. M.F. Samawi, Sys Design Contr Pollu Makassar City Beach Waters (IPB, Bogor, 2007)
3. W.G Zulfikar, Pond Shrimp Poten Pollute Envi. Jala.Tech (2020). https://app.jala.tech/kabar_udang/tambak-udang-berpotensi-mencemari-lingkungan.
4. Najamuddin, Var Season of Ambient Phy-Chem Para Waters Estuary Jeneberang, South Sulawesi. Proceedings of the National Seminar on Maritime Affairs and Small Island Resources **2** 1 (2017)
5. [KMLH] Decree of the State Minister for the Environment Number: 22 of 2021 concerning Sea Water Quality Standards for Marine Biota (2021)
6. M. Fadli, Determ Water Quality Status with the Index Method Sea Water Poll in the East of Sabang City. Syiah Kuala University: Banda Aceh (2023)
7. G.P. Rahayu, R.W Sayekti, M. Sholichin, J of Water Eng, **1** 2 (2021)
8. S.A. Akbar, S. Afriani, C. Nuzlia, S. Nazlia, S. Agustina, Depik, 12(3), 259-273 (2023)
9. R.Y. Tallar, J. Suen. Int Soil Water Conserv Res. **3** 3 (2015)
10. D.R.I Lubis, Analysis quality waters with method index pollution on the beach end kareung Lampuuk, Aceh Besar (Syiah Kuala University, Banda Aceh, 2019)
11. R Dahuri, Marine biodiversity (PT. Gramedia Pustaka Utama, Jakarta, 2003)
12. G. Sudiono, Analysis management reef corals in conser areas sea area island Randayan and surrounding areas Regency Stuffed West Kalimantan Province (Diponegoro University, Semarang, 2008)
13. M. Simanjuntak, Journal Knowledge Oceanography, **12** 2 (2007)
14. E.P. Bonin, E.J. Biddinger, G.G. Botte, J of Power Sources, **182** 284-290 (2008)
15. M. Jorgensen, L. J. Philips, Discourse Analysis as Theory and Method (SAGE Publications Ltd, 2002)
16. A. Sumantri, M.R. Cordova. J Penge SDA dan Ling, **1(2)** 127-134 (2011)
17. H Effendi, Review Water Quality for Manage Resour and Envi Waters (Kanisius, Yogyakarta, 2003)
18. I Rusydi, D Savira, D F Putra, I Dewiyanti, C N Defira, Analysis of Pb and Cd Heavy Metal Contents in Green Mussel (Pernaviridis L.) in Alue Naga Waters, Banda Aceh, ICFAES 2021 IOP Conf. Series: Earth and Environmental Science **869** (2021) 012074 IOP Publishing.
19. A.C. Karlina, A.M, Supriatna, V. Amalia, Analysis of Nitrite Levels in Surface Water and Ground Water Samples in Regency Areas Cilacap Using the Spectrophotometer Method Uv-Vis (Department of Chemistry, Faculty of Science and Technology, UIN Sunan Gunung Jati, Bandung, 2022)
20. S.A. Akbar, H.K. Rahayu, Lantanida Journal, 11(1), 51-66 (2023)
21. M. Indrawan, B. Richard, Prick, J. Supriatna, Biology Conservation-Edition Revision (Indonesian Obor Foundation, Jakarta, 2017)
22. Ichsan, M. Mahmudi, H. Nuddin.Daya Dukung Lingkungan Perairan Berbasis Blue Economy Terhadap Produksi Budidaya Tambak Udang Vanname Di Wilayah Pesisir

Probolinggo. Prosiding Seminar Nasional, Peran Ekonomi Kreatif dalam Pembangunan Nasional yang Berkelanjutan, 79-92 (2013).

Understanding sea wave height conditions in sumatra waters

Yudi Haditiar[1,2*], *Muhammad* Ikhwan[3], *Muhammad* Nanda[1,2], *Haekal Azief* Haridhi[1,2,4]

[1]Department of Marine Science, Faculty of Marine and Fisheries, Universitas Syiah Kuala, Banda Aceh, 23111, Indonesia
[2]Research Center for Marine Sciences and Fisheries, Universitas Syiah Kuala, Banda Aceh 23111, Indonesia
[3]Department of Mathematics, Faculty of Mathematics and Natural Sciences, Universitas Syiah Kuala, Banda Aceh, 23111, Indonesia
[4]Tsunami and Disaster Mitigation Research Center (TDMRC), Universitas Syiah Kuala, Banda Aceh 23111, Indonesia

Abstract. Ocean waves play a crucial role in influencing a wide range of marine and fisheries activities. In this research study, we aim to analyze the climatological conditions of sea wave heights in Sumatran waters over a span of 83 years (1940-2022). We utilize three-dimensional Copernicus numerical model simulations to examine seasonal patterns of sea wave heights and assess the likelihood of extreme wave events. Our findings reveal that high waves, reaching heights of 4-5 meters, are predominantly observed in the western and southern zones off the coast of Sumatra, particularly in the vicinity of smaller islands. Conversely, in enclosed or semi-enclosed waters, such as the lee side of Sumatra and the Malacca Strait, sea wave heights tend to be relatively lower. Seasonally, our study indicates that extreme wave heights are more likely to occur during the west season as opposed to the east season. This research provides valuable insights into the dynamic ocean wave conditions in Sumatran waters, which can have significant implications for various sectors reliant on the ocean environment.

1 Introduction

Waves have a significant impact on various marine processes, as well as offshore and nearshore activities. Surface waves, through momentum and heat fluxes, play a crucial role in influencing the ocean-atmosphere interface [1-2]. Additionally, above the sea surface, waves can influence each other's wind stress [1-3].

Climatic variations in wind speed and direction are recognized as significant drivers of annual and seasonal fluctuations in wave heights within the waters of Indonesia [4-5]. Notably, the highest wave heights and energy levels are observed during the west monsoon, while they are at their lowest during the east monsoon. It has been estimated that the potential wave energy production could reach 16.83 MW in Indonesian waters [6] and 67.29 kW/m

* Corresponding author: yudi@usk.ac.id

along the western coast of Sumatra [7]. These findings underscore the promise of harnessing waves as a viable renewable energy source for both Indonesia and Sumatra.

Surface waves are recognized for their significant impact on the mixing of the upper ocean, often referred to as the Oceanic boundary layer. This influence stems from four primary processes: wave breaking, the Coriolis-Stokes forces [8], Langmuir circulation [9], and mixing attributed to unbreaking waves [1-2]. Experiments conducted by Wu et al. [1] have demonstrated that incorporating these four wave-related processes results in a more faithful representation of sea surface temperature dynamics. Wave breaking, in particular, plays a crucial role in the formation and evolution of the Oceanic boundary layer, a pivotal component in the interconnected ocean-atmosphere system.

Increased climate change activities such as the El Niño-Southern Oscillation (ENSO) in the South China Sea and the Indian Ocean Dipole (IOD) in the Indian Ocean induce significant wave height variations or extreme wave phenomena. This storm wave activity can cause disruption to marine activities such as navigation, fishing, engineering and coastal disasters [10].

Waves in Indonesian waters have been studied from several sides [11]. Wu et al. [12] looked at the influence of swell on atmospheric mixing and wind stress. Rachmayani et al [4,13] analyzed wave characteristics with a spectrum model. Nadya et al. [5] identified wave energy in Javanese waters (Pangandaran Beach) using the Shore Protection Manual (SPM) method. Dogan et al. [3] observed global propagation of air pressure waves due to volcanic activity. Rizal and Ningsih [7,11], calculated the potential wave energy in the waters of West Sumatra using numerical simulations of the wave spectrum. Ningsih et al. [10] studied wave climate characteristics and effects of tropical cyclones on high wave occurrences in Indonesian waters with descriptive and inferential statistics. So far, the influence of swell and local winds on the characteristics of significant wave heights in Sumatran waters has not been studied much.

The interaction between swell and wind-generated waves can give rise to unpredictable high waves, posing significant dangers for mariners. This study aims to investigate the characteristics of surface waves in the waters off western Sumatra by analyzing data from a significant wave height (SWH) model. SWH represents the average height of the highest one-third of ocean or sea waves, encompassing those generated by both wind and swell. It quantifies the vertical distance between wave crests and troughs. In this research, we examine patterns and variations in SWH over time, with a particular focus on the peak periods of the western and eastern monsoons. Additionally, our study identifies a high probability of extreme waves with significant SWH occurring in Sumatran waters. We hope that this research will offer valuable insights into the dynamic conditions of ocean waves in Sumatran waters, carrying significant implications for various sectors that rely on the region, including fisheries, navigation, and the overall marine environment.

2 Research methods

2.1 Research sites

Sumatran waters exhibit a complex array of geographical characteristics. These include active tectonic zones like the Sumatran Fault Zone (SFZ) and the Andaman zone, vast expanses of deep waters such as the Indian Ocean and Andaman Sea, with depths reaching up to 4000 meters. Additionally, there are shallower waters, as seen in the Malacca Strait, with depths of approximately 50 meters. The region also features several small islands, such as Aceh Island, Simeulue, Nias, Siberut, and various other small islands (as shown in Figure 1). The Sumatran Fault Zone (SFZ) extends for about 1900 kilometers and is characterized by numerous segmented segments. It comprises approximately 20 main segments, each varying in length from 60 to 200 kilometers [14]. Given its unique geographical location and the presence of these islands, Sumatran waters experience spatial variations in wind intensity. Consequently, the northern part of Sumatra, located above the equator, encounters stronger trade winds or northeastern monsoons. In contrast, the southern part, dominated by the southwest monsoon, experiences relatively unobstructed wind circulation from the Indian Ocean.

2.2 Research Data

As per ERA5 data, ocean waves are a combination of waves that exhibit diverse characteristics, including height, length, and direction, which are detailed in the two-dimensional wave spectrum. The Significant Wave Height (SWH) is defined as the result of the amalgamation of wind-driven waves influenced by local winds and swell generated by distant winds occurring at various times and locations. This SWH metric is calculated as four times the square root of the integral taken across all directions and frequencies within the wave spectrum.

The research is based on hourly instantaneous Significant Wave Height (SWH) data with a spatial resolution of $\Delta x = \Delta y = 0.5$ degrees obtained from ERA5 (the 5th major atmospheric reanalysis produced by European Centre for Medium-Range Weather Forecasts (ECMWF)), spanning from 1940 to 2022. ERA5 is a fifth-generation reanalysis dataset developed by the Copernicus Climate Change Service (C3S) at ECMWF. This dataset provides historical global climate and weather information, with data available from 1940 onward. ERA5 employs data assimilation techniques, which involve combining model data with real-world observations to create a comprehensive analysis of the atmospheric state. This approach significantly enhances the quality of historical climate and weather data. Furthermore, ERA5 includes an ensemble dataset with ten members, which serves to estimate uncertainty and identify regions that are particularly sensitive to changes in atmospheric conditions. This ensemble approach adds an extra layer of robustness to the dataset, making it a valuable resource for climate and weather research.

2.3 Data analysis

Hourly Significant Wave Height (SWH) data spanning 83 years from 1940 to 2023 were averaged to derive seasonal averages. These seasons are categorized as the northeast season (December-February), the first transition season (March-May), the southwest season (June-August), and the second transition season (September-November). Subsequently, time-series data for SWH in three distinct areas were analyzed. These areas include those located in front of the island (A1-A6), behind the outermost island (B1-B3), and within the Malacca Strait (C1-C3). This analysis was conducted to gain insights into the patterns and impacts of SWH that have formed over the seasons. The seasonal SWH pattern was approximated using a Fourier series equation.

$$y = a_0 + \sum_{i=1}^{n} \quad a_i \, cos\, cos\, (iwx) \, + b_i \, sin\, sin\, (iwx) \tag{1}$$

In Eq. (1), *a*0 represents a constant term (intercept) in the data and is associated with the $i = 0$ cosine term. *w* stands for the fundamental frequency of the signal, while *n* represents the number of terms or harmonics in the series, and n = 2.

In the realm of probability theory and statistics, the Weibull distribution is a continuous probability distribution. Its primary purpose is to represent a wide spectrum of random variables, particularly those relating to the duration between events or the time until failure. This distribution finds practical application in scenarios such as modeling the highest one-day sea level. In this study, extreme wave heights were analyzed using the Weibull distribution. Weibull distribution memiliki dua parameter pembentuk yaitu scale (a) and shape (b). Probability Density Function (PDF) Weibull distribution (Eq. 2) and Cumulative Distribution Function (Eq. 3) of the Weibull distribution are determined as follows:

$$f(a,b) = \{\frac{b}{a}\left(\frac{x}{a}\right)^{b-1} e^{-\left(\frac{x}{a}\right)^{b}} \; 0 \tag{2}$$

$$F(a,b) = \int_{0}^{x} b a^{-b} t^{b-1} e^{-\left(\frac{t}{a}\right)^{b}} dt = 1 - e^{-\left(\frac{x}{a}\right)^{b}} \tag{3}$$

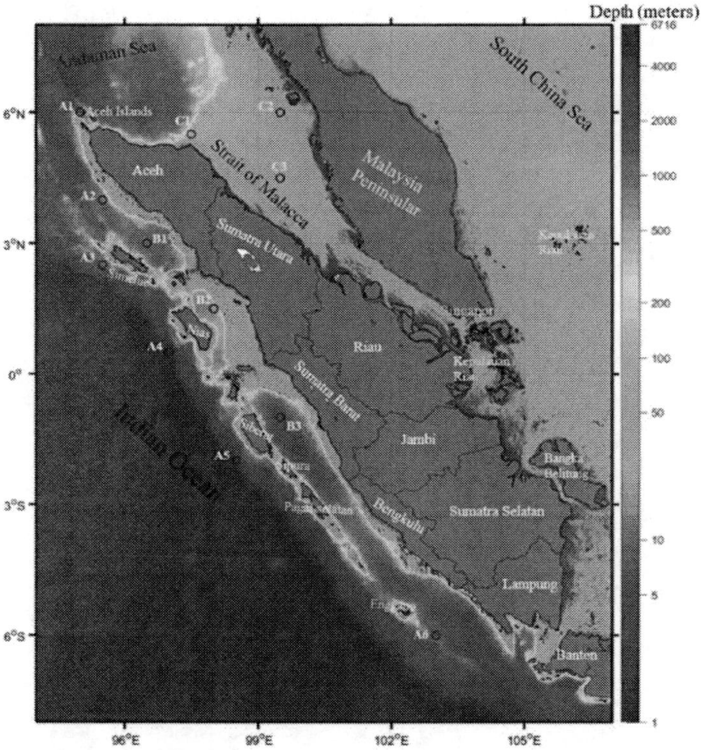

Fig. 1. Bathymetry (meter) (SRTM30)

3 Results and discussion

Figures 2 to 5 illustrate the seasonal SWH climatology in Sumatran waters over an 83-year period from 1940 to 2023. During different seasons, SWH values are as follows: 1.93 meters for the northeast season (December-February), 2.03 meters for the first transition season (March-May), 2.63 meters for the southwest season (June-August), and 2.18 meters for the second transition season (September-November). The peak in SWH occurs during the southwest monsoon, which encompasses June to August (referred to as JJA). This peak is primarily driven by the intensified westerly wind circulation originating from the Indian

Ocean, with the westward winds being relatively unobstructed as they cross the Indian Ocean. High SWH values are observed off the coast of the outer islands of Sumatra, including Simelue, Nias, Siberut, Sipura, and Enggano. As the waves approach these islands, they experience reduction and diffraction, resulting in relatively lower wave heights on the lee side of the islands. Notably, in the South China Sea, high SWH is observed, but this phenomenon is limited to the northeast season (December-February). The SWH patterns in Sumatran waters are intricately linked to the coastline profile and water depth. In open and deep waters such as the Indian Ocean, SWH tends to be relatively high. Conversely, in semi-enclosed waters like the Malacca Strait, SWH is relatively lower. The southwest monsoon, particularly during JJA, induces greater variations in wave height in the waters surrounding Sumatra. This is reflected in a higher standard deviation of SWH (reaching 0.98 meters) during the southwest monsoon compared to other seasons, such as the northeast monsoon (DJF), where the standard deviation only reaches 0.69 meters.

Obtaining information about extreme sea wave heights is critical for effective disaster mitigation. Extreme wave height is determined by considering both the maximum Significant Wave Height (SWH) and the SWH percentile at 0.93 over an 83-year period. The distribution of extreme waves in the waters surrounding Sumatra is illustrated in Fig. 9 and Fig. 10. The analysis results indicate that the maximum wave height in the sea off Sumatra, including the Indian Ocean, can reach heights of up to 4.5 meters. Notably, outer islands like Aceh Island, Simeleu, and Enggano are at a heightened risk of experiencing the impact of these high waves. This information is invaluable for disaster mitigation efforts, particularly in vulnerable coastal areas.

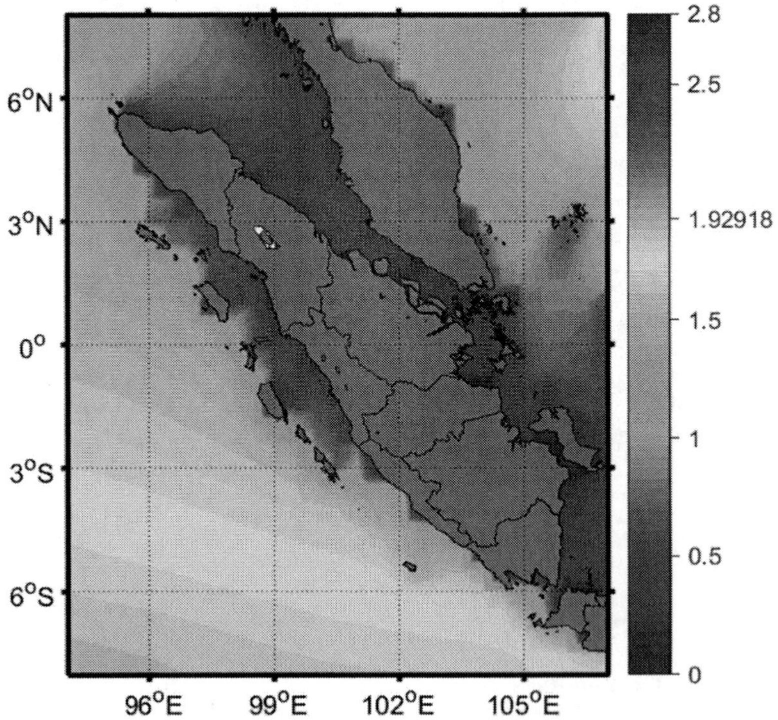

Fig. 2. SWH during DJF (meter)

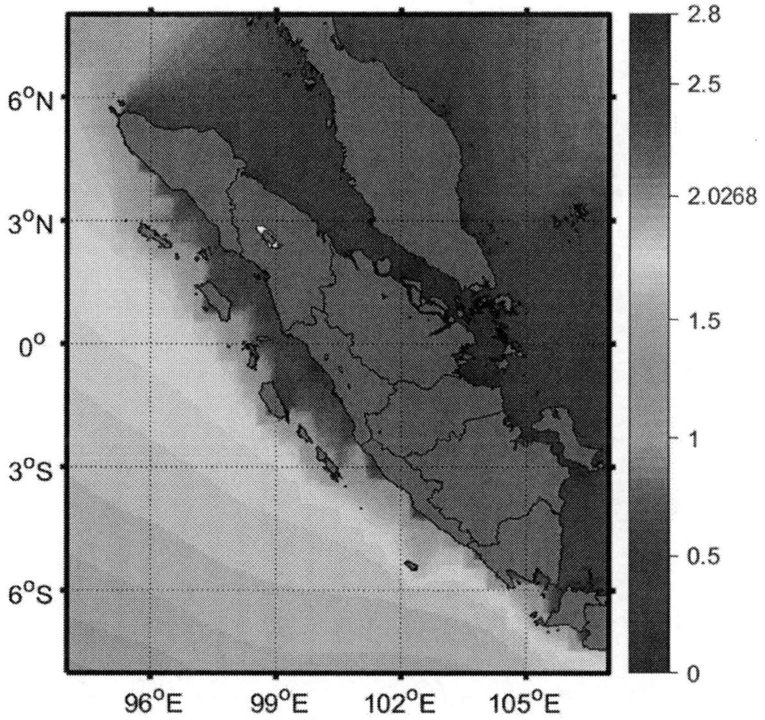

Fig. 3. SWH during MAM (meter)

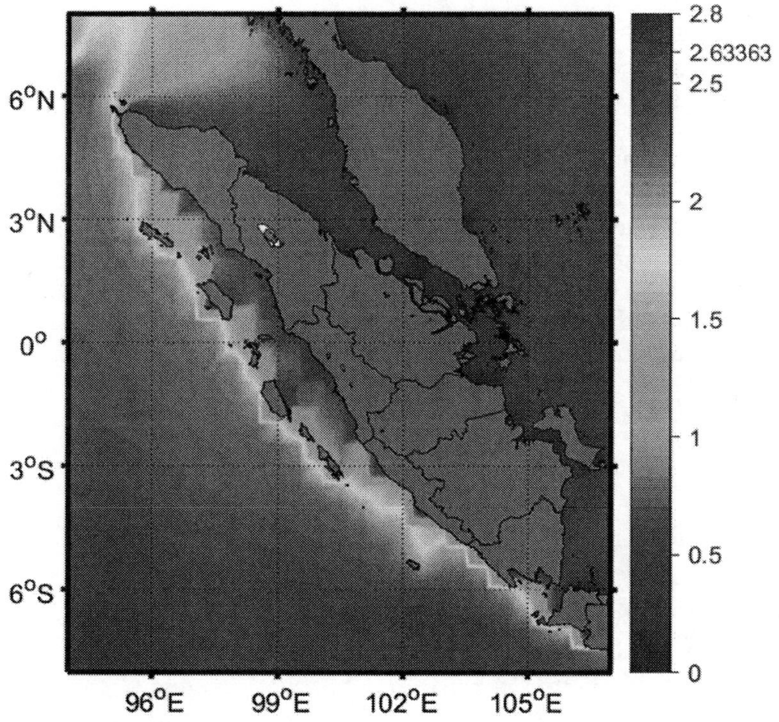

Fig. 4. SWH during JJA (meter)

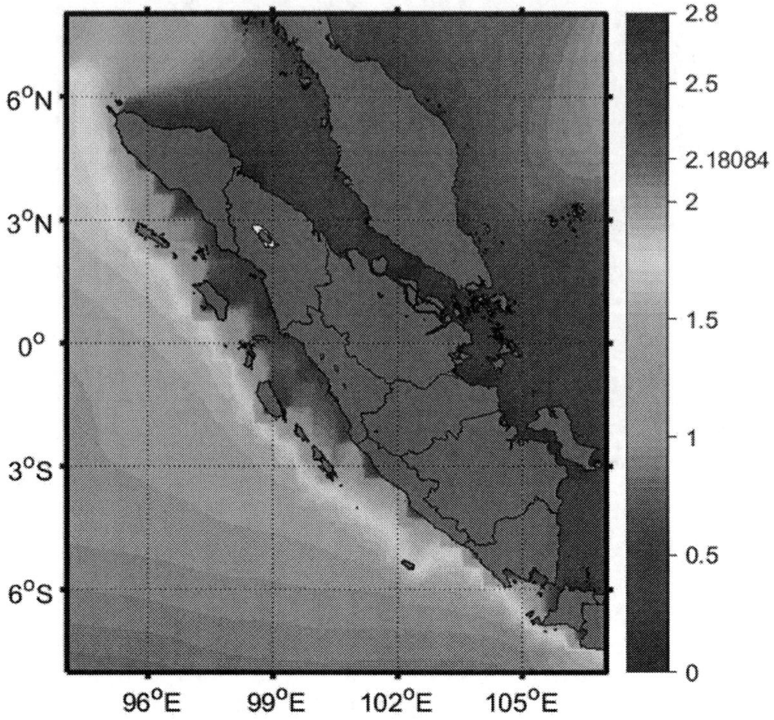

Fig. 5. SWH during SON (meter)

Seasonal signals of Significant Wave Height (SWH) at stations located both in front of and behind the Sumatra islands, as well as within the Malacca Strait, are presented in Fig. 6 to Fig. 8. These signals are derived from the analysis of hourly SWH data spanning from 1940 to 2022. In general, high waves are observed during July, with the peak occurring in August, and these conditions are primarily associated with the southwest monsoon. Notably, the outer island regions exhibit a relatively greater seasonal amplitude (as shown in Fig. 6) compared to areas behind the islands (as seen in Fig. 7). Behind the islands, particularly, low waves are prevalent during February and March. In contrast, the Malacca Strait experiences low waves in April and May (Fig. 8). During January and February, waves are slightly higher in the northern part of the Malacca Strait, owing to the influence of the peak of the northeast monsoon [15]. East winds favor the formation of high sea waves in the northern of the Malacca Strait and Sumatra Waters. However, it's important to emphasize that, on the whole, the highest waves still occur during the southwest monsoon, notably in August. This is because, in August, the winds are strong enough, and the extreme sea levels increase due to the already heightened wind conditions [16-17].

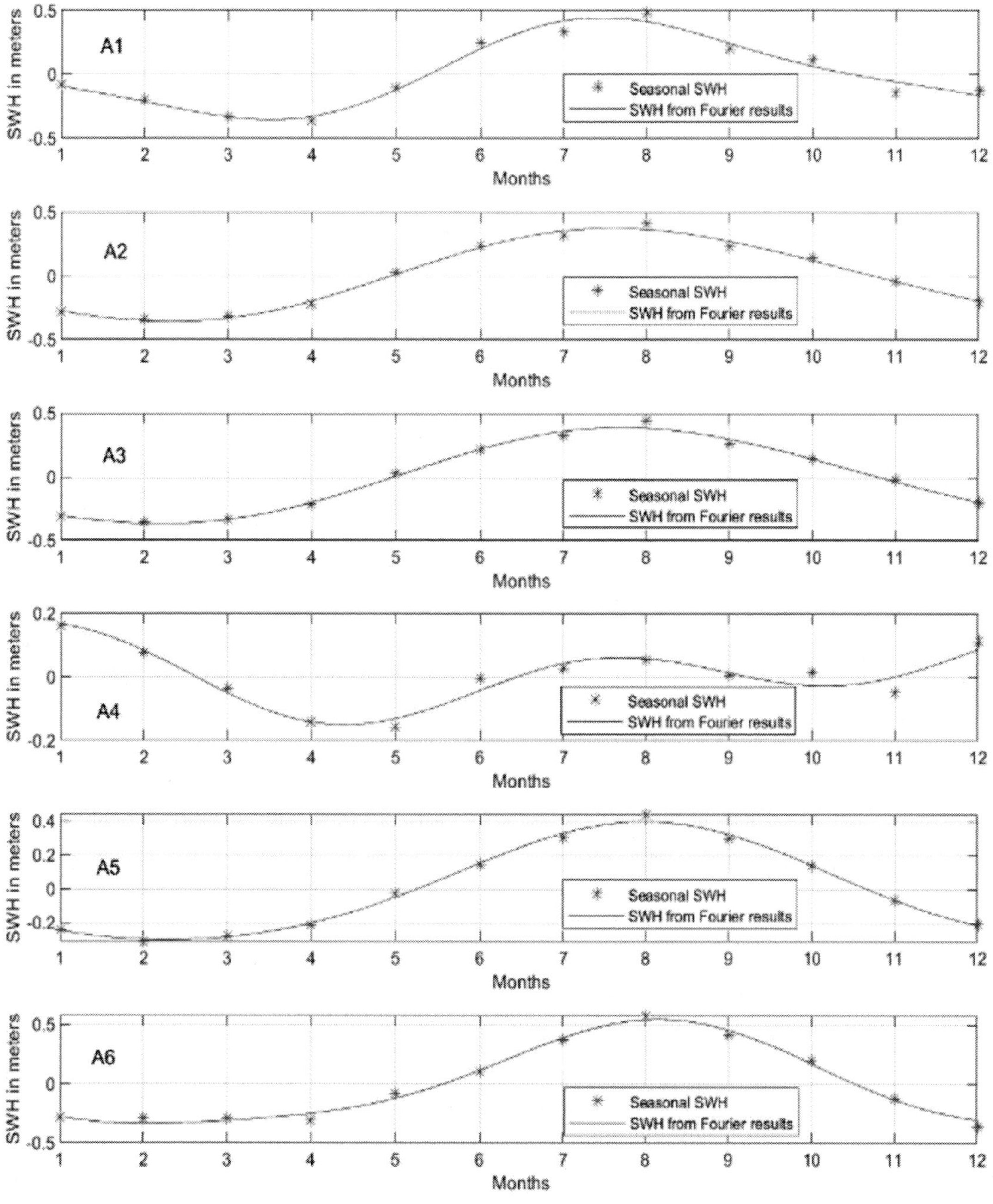

Fig. 6. SWH seasonal signal on the wind side (the island area facing the waves) (see stations A1-A6 in Figure 1).

https://doi.org/10.1051/bioconf/20248702014

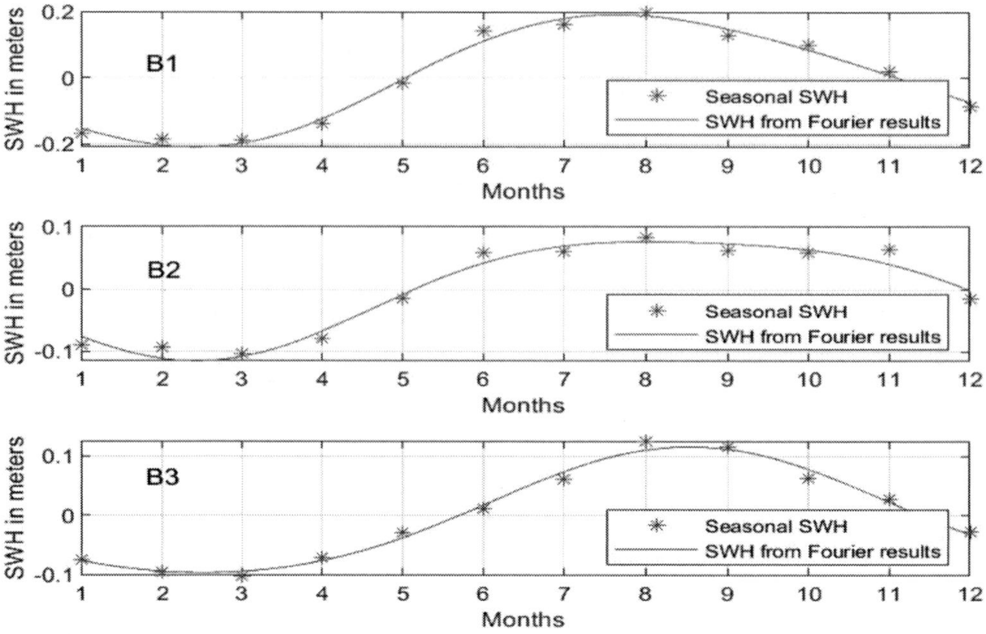

Fig. 7. SWH seasonal signal on the lee side (the area behind the island) (see stations B1-B3 in Fig. 1)

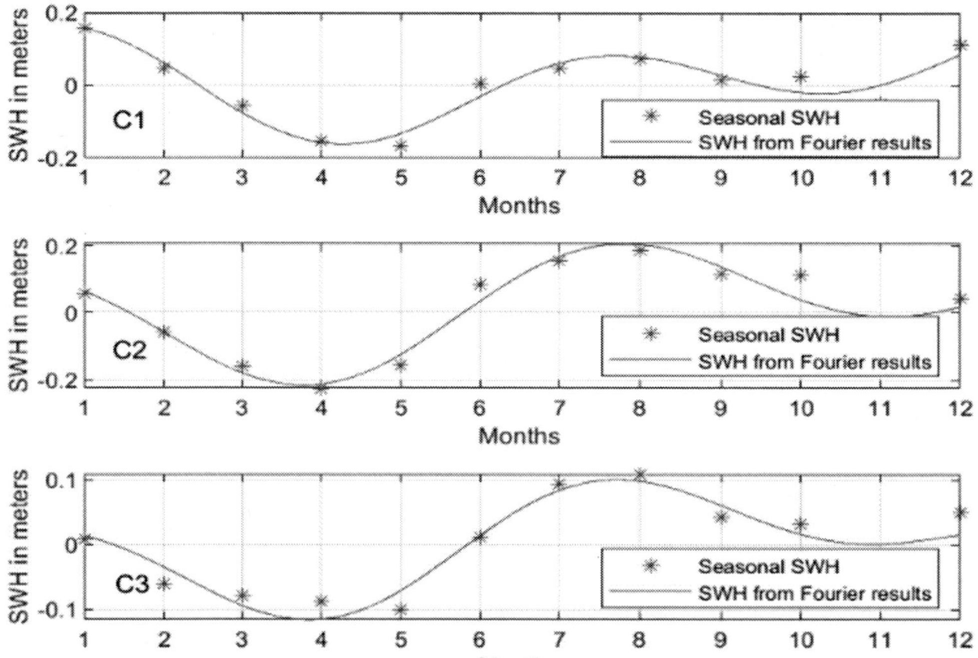

Fig. 8. SWH seasonal signal in the Malacca Strait and the east of Aceh (see stations C1-C6 in Fig. 1).

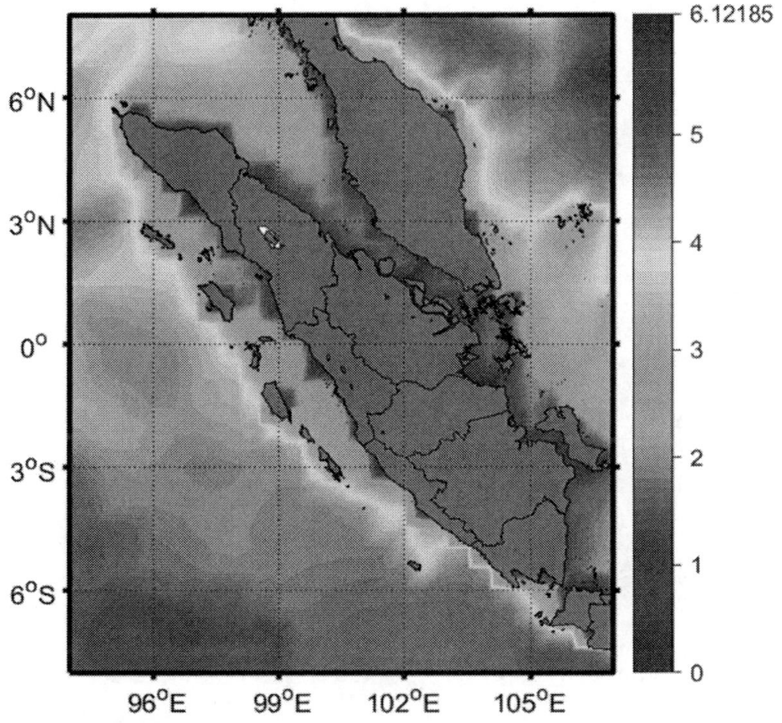

Fig. 9. Maximum SWH (meter)

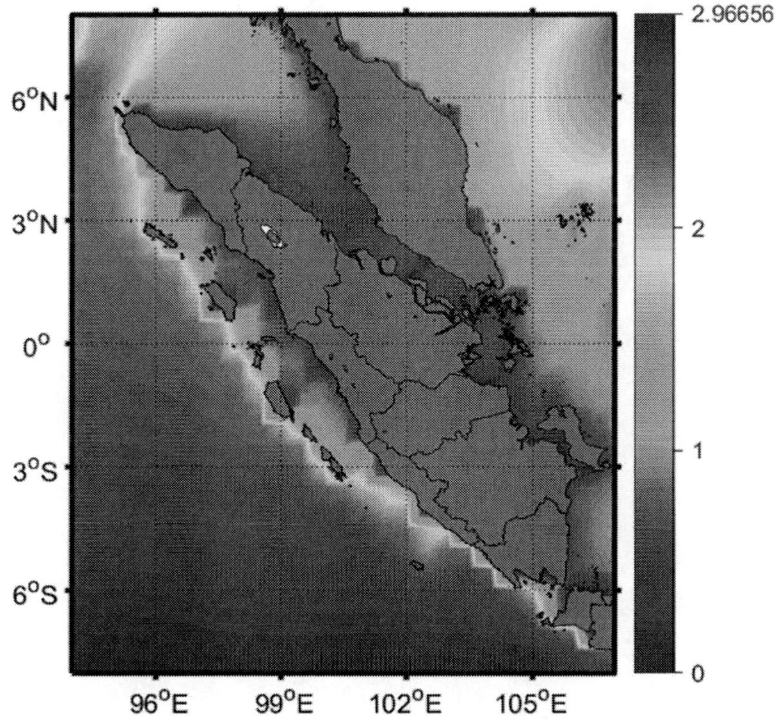

Fig. 10. Percentile 0.93 of SWH maximum (meter)

https://doi.org/10.1051/bioconf/20248702014

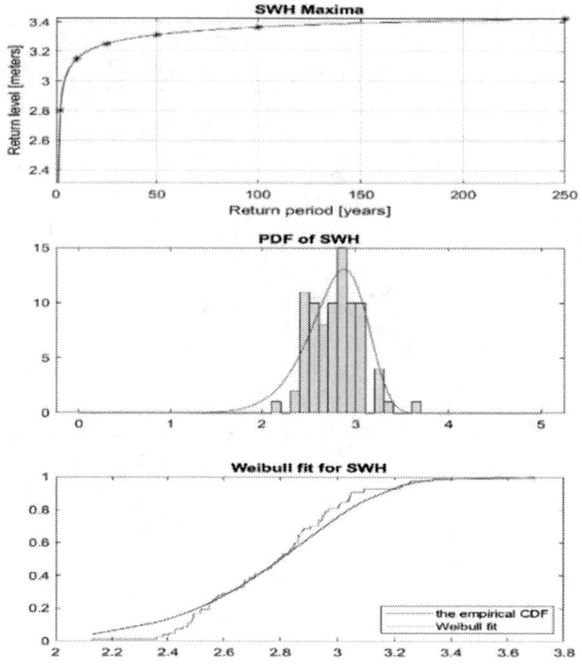

A1

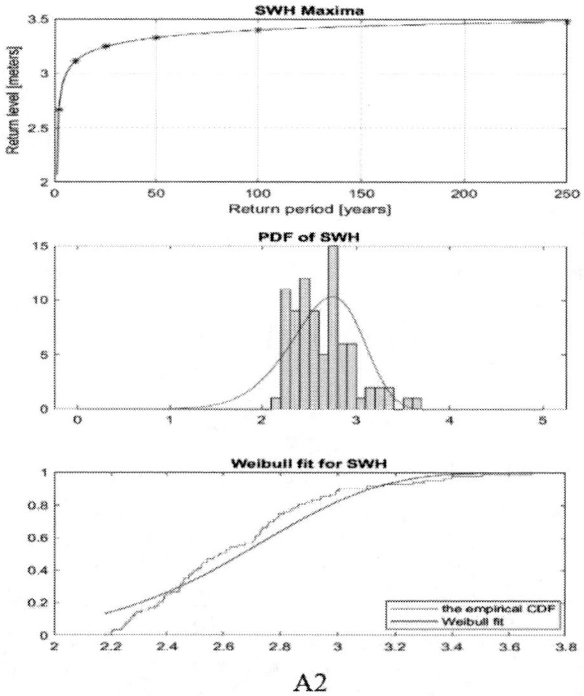

A2

https://doi.org/10.1051/bioconf/20248702014

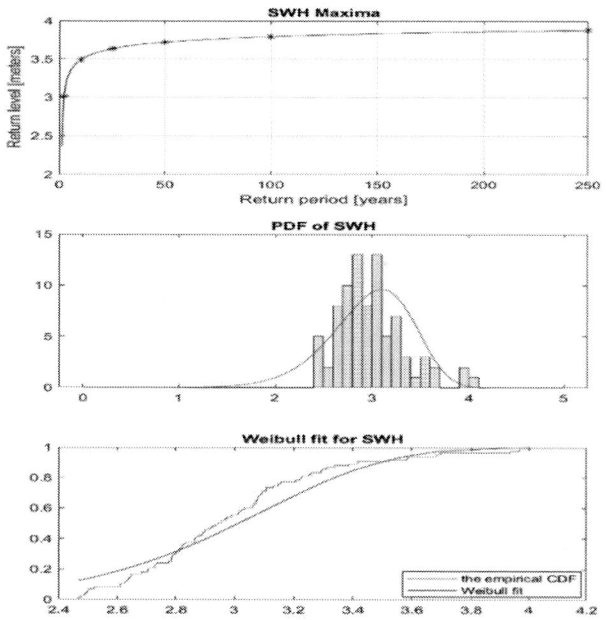

A3

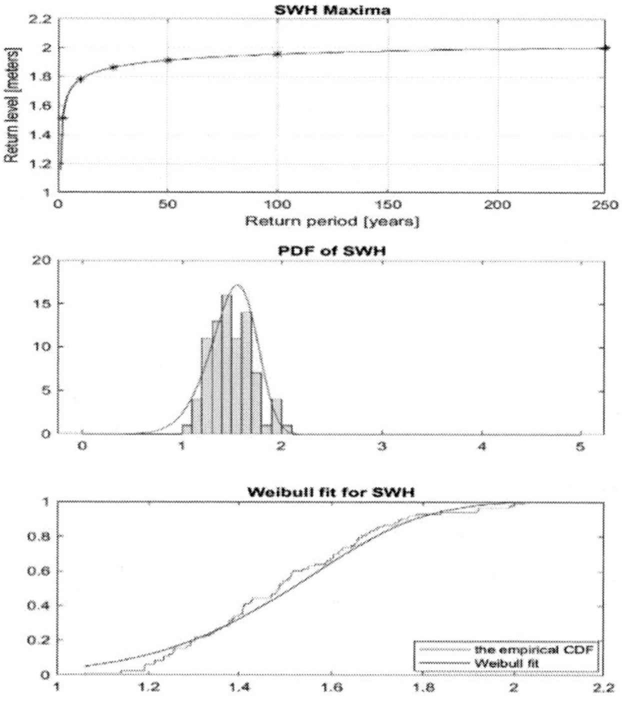

A4

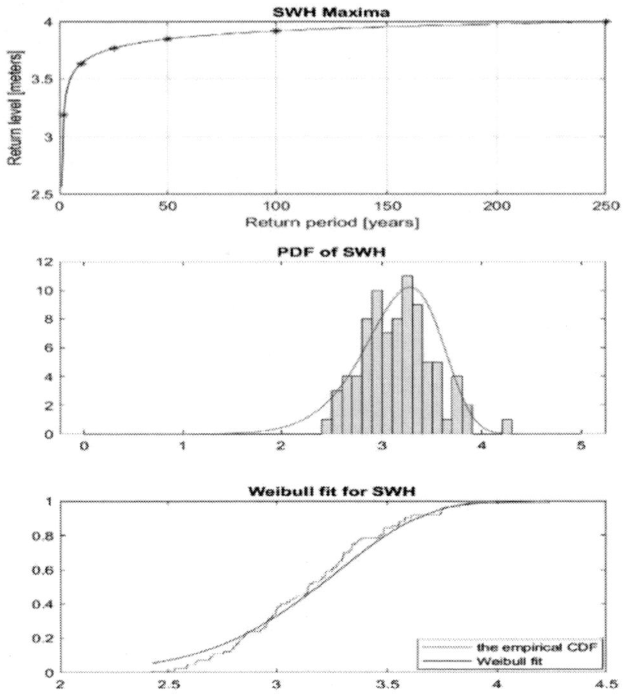

A5

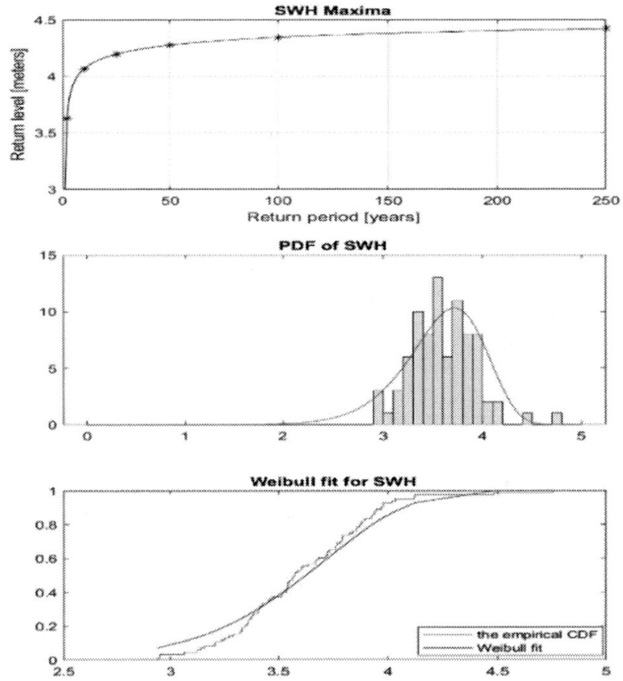

A6

Fig. 11. Probability of extreme waves on the wind side (the island area facing the waves) (see stations A1-A6 in Fig. 1).

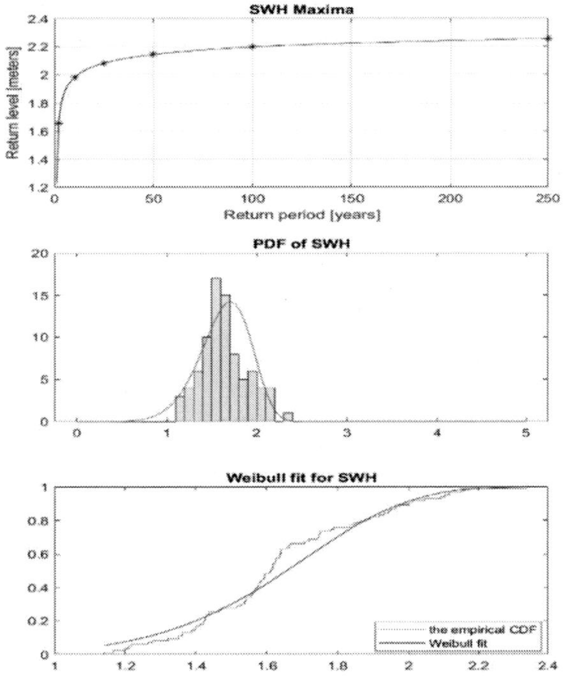
SWH Maxima
Return level [meters]
Return period [years]
PDF of SWH
Weibull fit for SWH
the empirical CDF
Weibull fit

1B

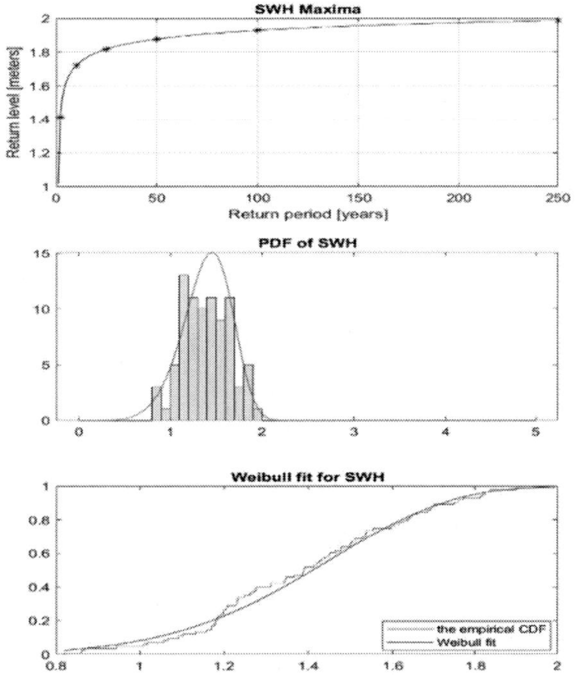
SWH Maxima
Return level [meters]
Return period [years]
PDF of SWH
Weibull fit for SWH
the empirical CDF
Weibull fit

2B

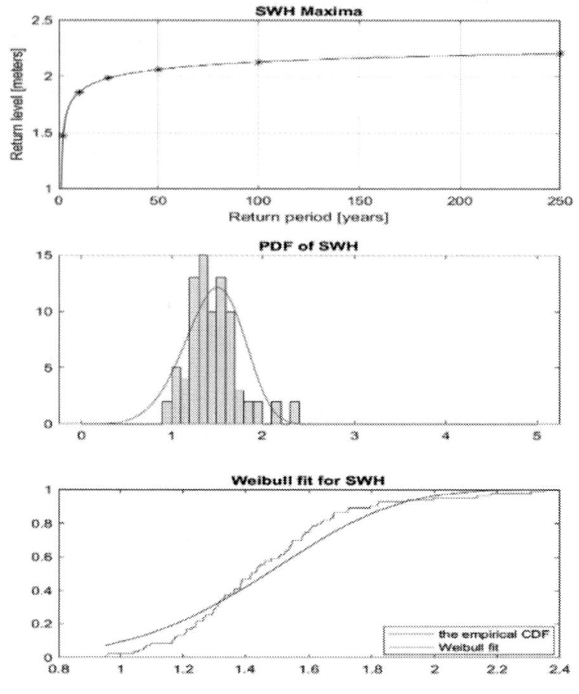

3B

Fig. 12. Probability of extreme waves on the lee side (the area behind the island) (see stations B1-B3 in Fig. 1).

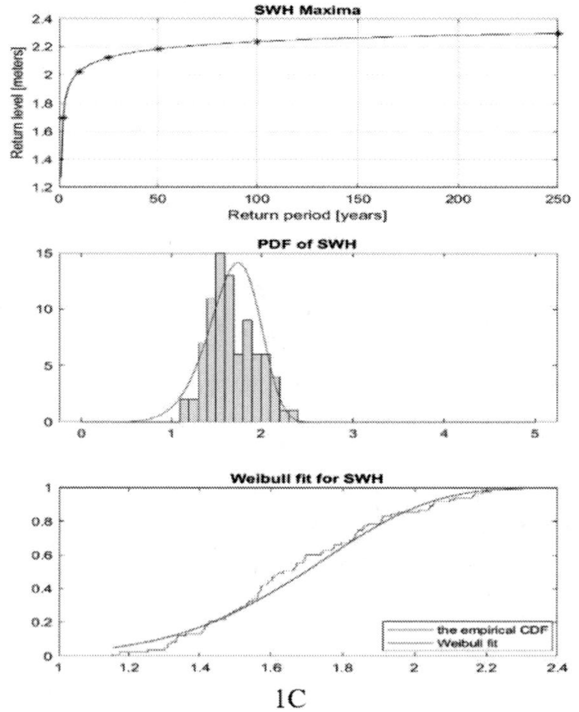

1C

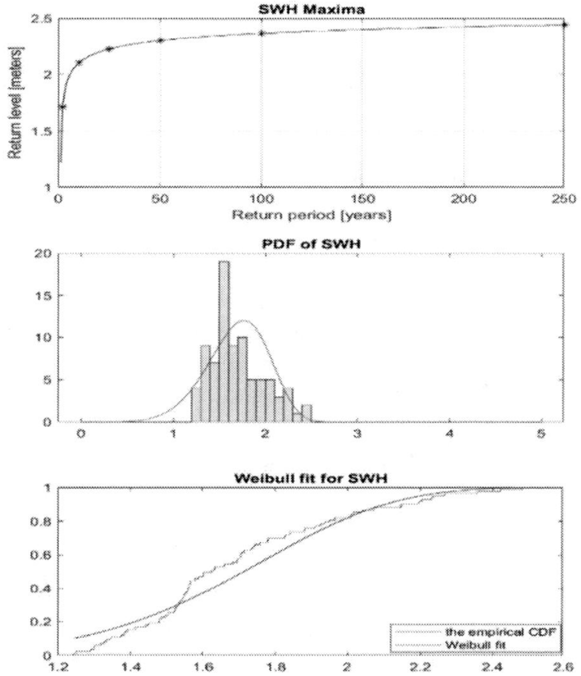

2C

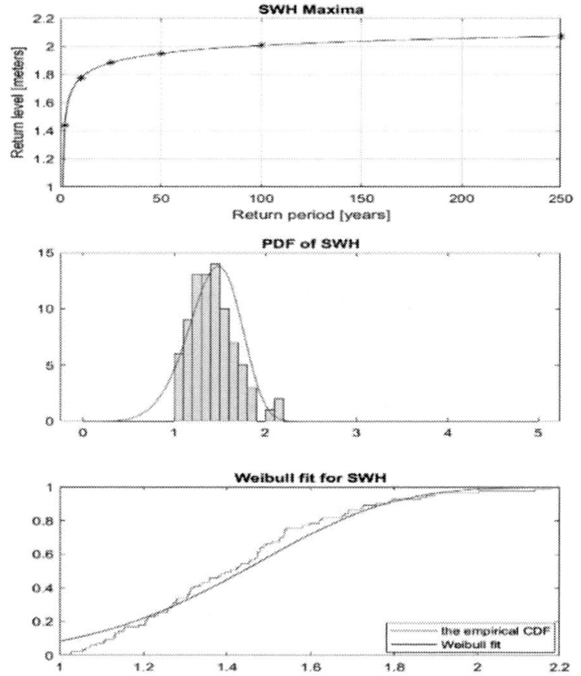

3C

Fig. 13. Probability of extreme waves in the Malacca Strait and the east of Aceh (see stations C1-C6 in Fig. 1).

4 Conclusions

We conducted an analysis of significant wave heights in the waters surrounding Sumatra using data from the ERA5 reanalysis model. The data we used for analysis has an hourly resolution, which is particularly well-suited for studying wave characteristics resulting from both wind and swell. Our analysis revealed that wave height is significantly influenced by both the season and geographic location. During the southwest monsoon, characterized by prevailing west and southwest winds, high waves are observed in most of Sumatra's waters, including the Malacca Strait. The peak in wave height occurs in August, particularly in the Indian Ocean and the outer islands. In the northern waters of the Malacca Strait and Aceh waters, high waves also form in January, indicating the influence of the northeast monsoon. These areas, especially in front of Sumatra's outer islands, are particularly susceptible to sea wave-related disasters. As waves approach the shore, they induce Stokes drift, a phenomenon involving the net transport of water particles in the direction of wave propagation. This process can contribute to the development of nearshore currents. However, a comprehensive study of the impact of waves on nearshore currents requires a coastal hydrodynamic model with higher resolution. Moreover, it is crucial to conduct further analysis to understand trends in wave height and their correlation with climate changes, such as the Indian Ocean Dipole (IOD) and the El Niño–Southern Oscillation (ENSO). Additionally, research on combined waves and ocean currents, particularly using high-resolution models, is essential. This will provide valuable insights into the broader implications of variations in wave height and their connection to larger climatic phenomena.

Acknowledgement

The authors express their gratitude to the Marine Fisheries Research Center (PRKP) and Tsunami and Disaster Mitigation Research Center (TDMRC) of Universitas Syiah Kuala for their invaluable support in facilitating this research. Additionally, sincere thanks are extended to LPDP Indonesia for providing funding to support Innovative Product Research (RISPRO).

References

1. L. Wu, A. Rutgersson, E. Sahlée, and X. G. Larsén, Tellus A: Dyn. Meteorol. Oceanogr. **67**, 1 (2015)
2. L. Wu, A. Rutgersson, E. Sahlée, J. Geophys. Res. Oceans **120**, 12 (2015)
3. G. G. Dogan, A. C. Yalciner, A. Annunziato, B. Yalciner, O. Necmioglu, Ocean Eng. **267**, (2023)
4. R. Rachmayani, N. S. Ningsih, H. Ramadhan, S. Nurfitri, *Analysis of Ocean Wave characteristic in western Indonesian seas using wave spectrum model*, in *MATEC Web of Conferences*, 26-27 September 2017, Bandung, Indonesia (2018)
5. F. H. Nadya, M.L. Syamsuddin, Subiyanto,Q.W. Sari, WSN **170** (2022)
6. I. Alifdini, N. A. Iskandar, A. W. Nugraha, D. N. Sugianto, A. Wirasatriya, A. B. Widodo, Ocean Eng. **165** (2018).
7. A. M. Rizal, N. S. Ningsih, J. Ocean Eng. Mar. Energy **6**, 2 (2020).
8. G. Broström, K. H. Christensen, M. Drivdal, J. E. Weber, Ocean Dyn. **64**, 7 (2014)
9. W. Tsai, G. Lu, J. Fluid Mech. **969** (2023)
10. N. S. Ningsih, A. Azhari, T. M. Al-Khan, Ocean Coast Manag **243** (2023).
11. A. M. Rizal, N. S. Ningsih, Ocean Eng. **251** (2022).

12. L. Wu, A. Rutgersson, E. Sahlée, X. Guo Larsén, J. Geophys. Res. Oceans **121**, 7 (2016)
13. R. Rachmayani, N. S. Ningsih, I. Ardiansyah, J. Ocean Eng. Mar. Energy **9**, 1 (2022)
14. D.H. Natawidjaja, W. Triyoso, J. Earthq. Tsunami **1**, 1 (2007)
15. Y. Haditiar, M. R. Putri, N. Ismail, Z. A. Muchlisin, M. Ikhwan, S. Rizal, Heliyon **6**, 9 (2020)
16. M. Bernardino, M. Gonçalves, R. M. Campos, and C. G. Soares, Ocean Eng. **275** (2023)
17. M. Hieronymus, C. Dieterich, H. Andersson, R. Hordoir, Theor. App. Mech. Lett. **8**, 6 (2018)

Application of ground-penetrating radar for shallow subsurface investigation at the coastal area of lhok village, lhoong, aceh besar regency, aceh

Muhammad Nanda[1, 2, 3, 4], *Nazli* Ismail[3, 4, 5], *Muna* Mahlia[1], *Farhadila* Amril[1], *Badrul* Munir[6], *Amir* Asyqari[3,5], *Husnul* Khatimah[5], and *Yudi* Haditiar[1,2*]

[1]Department of Marine Sciences, Faculty of Marine and Fisheries, Universitas Syiah Kuala, Banda Aceh, Aceh 23111, Indonesia
[2]Research Center for Marine Sciences and Fisheries, Universitas Syiah Kuala, Banda Aceh, Indonesia
[3]Research Center Ethnosciences, Universitas Syiah Kuala, Banda Aceh, Aceh 23111, Indonesia
[4]Tsunami and Disaster Mitigation Research Center, Universitas Syiah Kuala, Banda Aceh, Indonesia,
[5]Physics Department, Universitas Syiah Kuala, Banda Aceh, Aceh 23111, Indonesia
[6]Geophysical Engineering Department, Universitas Syiah Kuala, Banda Aceh, Aceh 23111, Indonesia

Abstract. This research aims to identify subsurface layers in the coastal area of Lhok Village, Lhoong, Aceh Besar Regency, Aceh. Ground Penetrating Radar (GPR) was the primary method to extract information concerning concealed geological structures beneath the Earth's surface, including in coastal areas. GPR is a non-destructive geophysical technique that utilizes electromagnetic waves to explore and map subsurface layers. However, data collection involved field surveys utilizing GPR equipment directed at the ground's surface. During the survey, radar data was processed and analysed to pinpoint various subsurface level. The obtained information contributes to a deeper understanding of the region's geological characteristics and subsurface layers. Additionally, it aids in assessing geological risks, facilitating construction planning, and conserving natural resources in the area. Here, the outcomes of this study may also serve as a foundation for further research in geology and geophysics within the Aceh Coastal region.

1 Introduction

The coastal area represents a dynamic and complex environment with interrelated geology. This transitional zone between the sea and the land continuously adapts to the movements of seawater, encompassing currents, waves, and tides [1]. These marine processes carry and deposit sediments comprising fragments of rocks, minerals, and organic matter from diverse sources [2]. Various factors, including carriers such as water, wind, and human activities, influence transport processes for sediments [3].

* Corresponding author: yudi@usk.ac.id

Here, the sediment deposition process in coastal areas involves continuous erosion, transport, deposition, resuspension and redeposition [4]. Physical, chemical, and biological factors influence these processes, shaping the characteristics of sediment layers in the region. The accumulation of sediments is an ongoing phenomenon, offering valuable insights into the coastal area's geological history and environmental dynamics [5].

The Ground Penetrating Radar (GPR) is an effective method for investigating Shallow Subsurface and sediment layers. This method applied electromagnetic waves ranging from 1 to 1000 MHz for mapping the subsurface structures [6]. GPR offers high-resolution capabilities for distinguishing dielectric properties of materials and shallow subsurface geological formations [7]. However, it comes with a limited penetration depth; as the frequency increases, the wavelength decreases, resulting in higher resolution. A few studies used GPR to investigate the subsurface structure, such as [1–3]

The research used the GPR method to analyze the sediment layers and investigate the shallow subsurface in the coastal area of Lhok Village, Lhoong, Aceh Besar Regency. Furthermore, this area predominantly comprises sand, gravel, mud, and other materials from the Lamno Limestone formation [11] (Fig 1b). Besides, this area has been hit by the 2004 tsunami. The tsunami has left sediments in the area they passed, eroded and covered by coastal sediments due to tides and climate change.

The investigation in this region is important for comprehending sediment layers and subsurface structures. The analysis of GPR data is needed to yield information regarding sediment thickness, identification of sediment layer boundaries, and vertical sediment structure modelling. This information will aid in understanding sediment deposition processes and patterns in the coastal areas of Lhok Village, Lhoong, Aceh Besar Regency.

2 Methods

The acquisition data was collected in the coastal areas of Lhok Village, Lhoong, Aceh Besar Regency, Aceh. A GPR instrument with Series IDS Opera Duo dual-frequency antenna (250 and 700 MHz) [12] was used to perform the survey. Furthermore, the radar antenna is shielded, enabling simultaneous penetration at shallow and deep depths [13]. Further theoretical basis of the GPR method can be found in [7], [14]. The transmitter transmits radar waves into the target medium. If there are variations in the electrical properties different from the surrounding medium, the waves are reflected to the surface and received by the receiver [13], [15].

2.1 Data Acquisition

In this study, data acquisition used the radar reflection profiling system to characterize shallow subsurface layers within the research area. Data was acquired along a 50-meter profile extending from the cave entrance (northwest side) towards the coastline (southeast most) (Fig.1c). Dual frequencies were used to observe variations in the resulting target distribution. However, based on geomorphological observation, this profile's surface predominantly comprises topsoil, sand, karst and vegetation (Fig 1b & 1c). The GPR measurements produced images of subsurface layers with raw data images. Additional correction data and processing steps are required to eliminate and reduce the influence of noise data or errors that occurred during the measurement process.

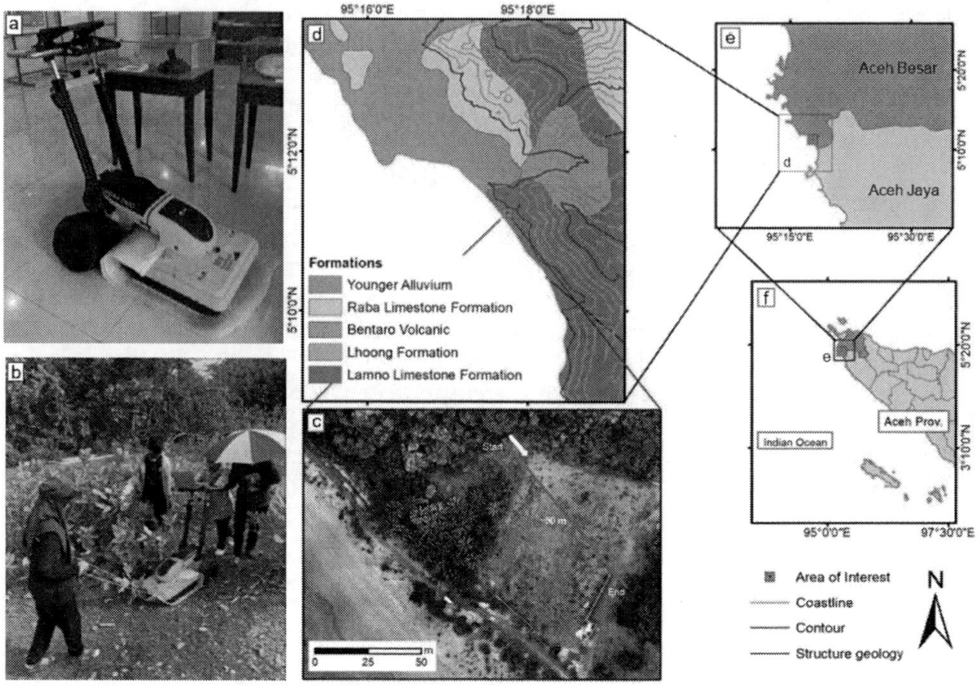

Fig. 1. (a) The GPR Unit used (b) Field view of research site with data acquisition process (c) Profile sketch of GPR data acquisition (d) The study area overlaid by geological maps (modified based on [11]) (e - f) zoom out area study area

2.2 Data Processing

The GPR data was processed in two steps - filtering and visualization. The filtering stage involves separating interference in the measurement data to display the signal from each subsurface layer. In general, this process consists of the following steps in Fig 2:

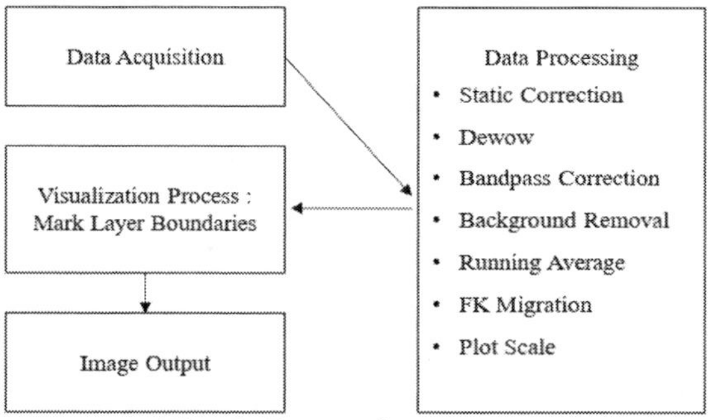

Fig. 2. Research process steps

a) Static correction: Correcting the two-way travel time of the wave to a horizontal reference plane above the ground interface. These corrections are essential for obtaining accurate subsurface images and improving profile interpretation [15 - 16]
b) Substract-mean (Dewow): A noise removal process that effectively eliminates interfering radar signals from the data [18]
c) Bandpass correction: Filtering signals in the frequency domain eliminates unwanted noise, thereby improving the signal-to-noise ratio. Bandpass filters permit the passage of

frequencies within a specific range while diminishing frequencies outside that range [19], [20]

d) Background removal: This filter eliminates repetitive noise throughout the entire data profile [21].
e) Running average: To smooth the data display and make it more organized [22]
f) Fk migration: The filter combines multiple reflected signals to enhance the visibility of the data signal [18 - 20]
g) Plot scale: Adjusting the colour scale and sharpness of the radar plot makes the radar data easier to interpret

The data from the previous step is marked to distinguish individual soil layers during the visualization process. The boundaries separating these layers help clarify the different layers. The thickness of each sediment can be estimated using the layer boundary.

3 Result and Discussion

The data from acquisition in the field is shown on radargram as a time function (two-way travel time) and distance. The data has flat shapes as the waveform (**Fig. 3**). On the other hand, if the wave passes through a different material or medium, the shape of the radargram will be associated and change for not flat longer anymore.

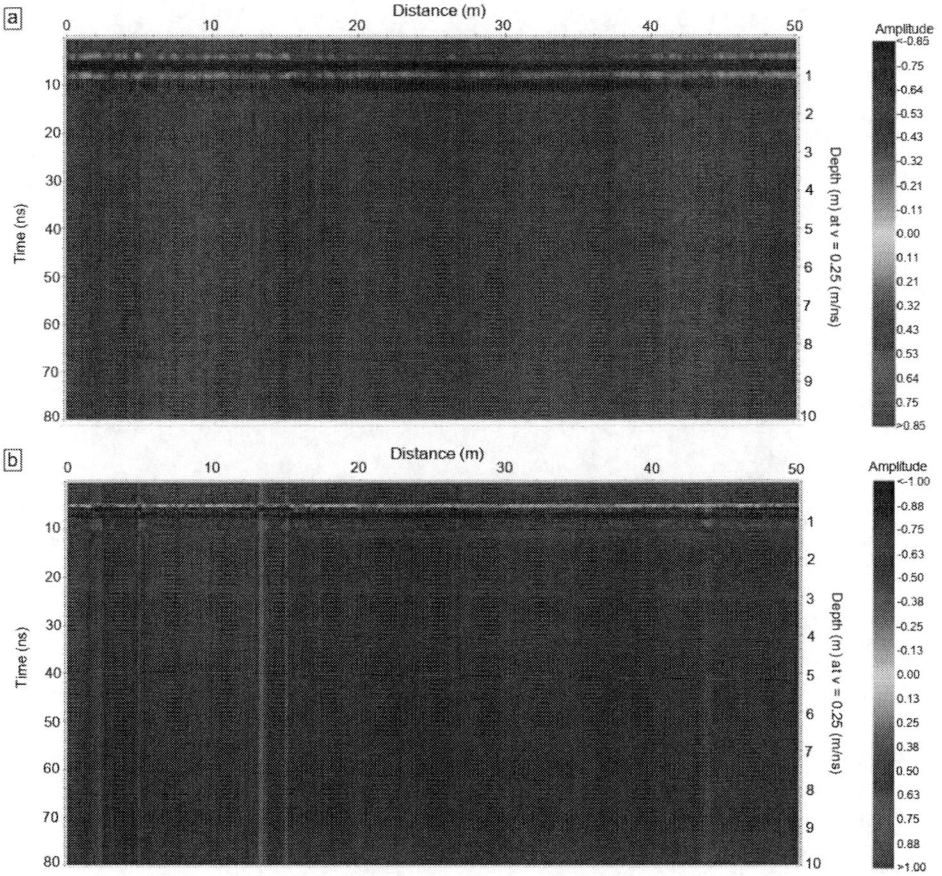

Fig. 3. Raw data images obtained from GPR data acquisition at frequencies of (b) 250 MHz and (b) 700 MHz, respectively.

https://doi.org/10.1051/bioconf/20248702015

Based on data processing, shallow subsurface layers were obtained for several materials/rocks. The interpretation of GPR radargrams showed the distribution of shallow subsurface layers and the presence of three subsurface layers at 250 MHz frequency and 700 MHz frequency, respectively. In **Fig 4**, shallow subsurface layers shown were marked by dashed black lines. However, the interpreted radargram display was based on the research area's geological conditions. Every change or difference in rock types can also be observed from the amplitude contrast (wave velocity is directly proportional to wave amplitude)

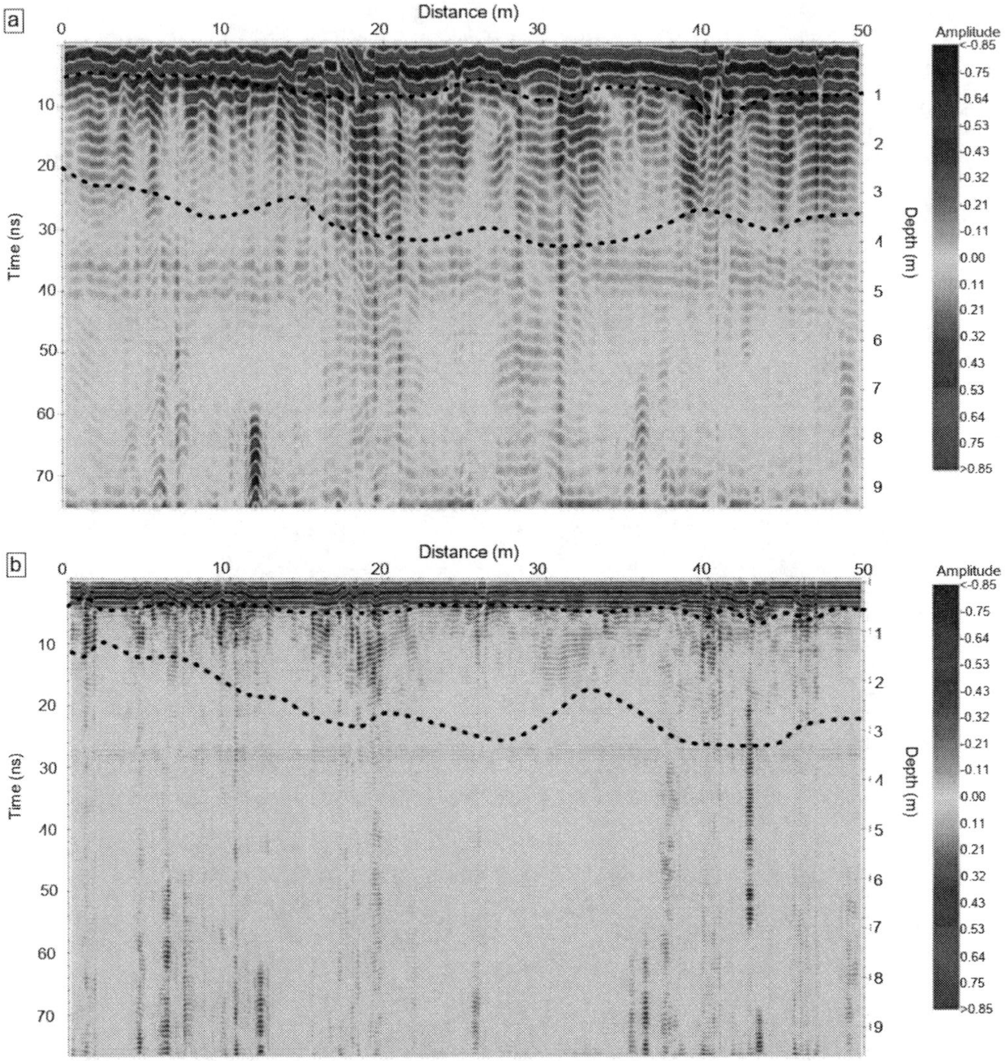

Fig. 4. Interpreted data radargram GPR from (a) 250 MHz and (b) 700 MHz, respectively.

The interpretation of GPR radargrams showed the distribution of shallow subsurface layers. In Fig 3, the results of the radargram interpreted indicate three subsurface layers. The uppermost layer is a pile of topsoil with a depth of 0-1 m. The second layer is sand with a 1.0 - 3.2 m depth along the profile. Moreover, the thickest is fine wet sand that overlaid the rough wet sand below. Finally, the interpretation results indicate that the radargram at the frequency of 250 MHz is better to show the deeper depths but could be clearer at the shallow depth [23]. Therefore, this situation is in contrast at a frequency of 700 MHz.

In the study of coastal sedimentation, it is recommended to use the frequency of 700 MHz with a good penetration ability to see the sedimentation process. In these antennas, the contact of these deposits shows an increase in the signal amplitude [24]. The amplitude analysis of a particular environment and the findings indicated that low amplitude values were correlated with high compaction values. This relationship suggested that high compaction accelerated the decay of signals [25]. The results of the identification of radargram show that bedding in coastal areas, e.g., sand material, is on the surface and gravel on the subsurface. Around the coast, we also found a hill or dune (sediment dominated by sand with good sorting) or lagoon (fine deposits and organic material). In addition, sedimentary bedding in this region is also influenced by tidal currents, which can be a small sediment transport medium, such as fine sand and clay.

4 Conclusions

The interpretation of GPR radargrams has provided valuable insights into the distribution of shallow subsurface layers in the study area. The radargrams revealed the presence of three subsurface layers at different frequencies, with the uppermost layer consisting of topsoil followed by sand and fine wet sand. The result also highlighted the importance of selecting the appropriate frequency for GPR surveys, with the 250 MHz frequency proving effective for deeper depths. In contrast, the 700 MHz frequency offered better clarity for shallow depths. These findings significantly affect coastal sedimentation studies, particularly in identifying sedimentary bedding, such as sand and gravel. The influence of tidal currents on sediment transport, including fine sand and clay, was also observed. Moreover, these results contribute to our understanding of the subsurface characteristics and sediment dynamics in coastal areas, providing valuable information for various geoscience and environmental applications.

References

1. A. McLachlan and O. Defeo, The ecology of sandy shores. Academic press, (2017).
2. S. D. Kalev and G. S. Toor, "The composition of soils and sediments," in Green Chemistry, Elsevier, pp. 339–357, (2018).
3. J. Syvitski et al., "Earth's sediment cycle during the Anthropocene," Nat. Rev. Earth Environ., **3**, 3, pp. 179–196, (2022).
4. S. Song et al., "A global assessment of the mixed layer in coastal sediments and implications for carbon storage," Nat. Commun., **13**, 1, p. 4903, (2022).
5. C. J. Hein and A. D. Ashton, "Long-term shoreline morphodynamics: processes and preservation of environmental signals," Sandy Beach Morphodynamics, pp. 487–531, (2020).
6. A. M. E. Mohamed, I. El-Hussain, A. Deif, S. A. S. Araffa, K. Mansour, and G. Al-Rawas, "Integrated ground penetrating radar, electrical resistivity tomography and multichannel analysis of surface waves for detecting near-surface caverns at Duqm area, Sultanate of Oman," European Association of Geoscientists & Engineers, (2019).
7. J. . Reynolds, An Introduction to Applied and Environmental Geophysics, 2nd ed. UK: Chichester: John Wiley and Sons Ltd, (2011).
8. A. D. Switzer, C. Gouramanis, C. S. Bristow, and A. R. Simms, "Ground-penetrating radar (GPR) in coastal hazard studies," in Geological Records of Tsunamis and Other Extreme Waves, Elsevier, pp. 143–168, (2020).
9. S. Sonkamble and S. Chandra, "GPR for earth and environmental applications: Case studies from India," J. Appl. Geophys., **193**, p. 104422, (2021).

10. I. Z. Roslan, N. Z. Mohamad, N. K. Nasir, A. H. M. Yusof, and N. A. Ismail, "Correlation of ground penetrating radar and 2-D resistivity imaging methods towards shallow subsurface layer at coastal area," Phys. Chem. Earth, Parts A/B/C, **128**, p. 103294, (2022).
11. J. . Bennett et al., "Geology map of the Banda Aceh Quadrangle,Sumatera," Geol. Res. Dev. Centre, Bandung, p. Explanatory note, **19** pp., (1981).
12. G. S. IDS, "Manual Opera Duo," hexagon, www.idsgeoradar.com, (2023).
13. A. P. Annan, "Electromagnetic principles of ground penetrating radar," Gr. penetrating radar theory Appl., **1**, pp. 1–37, (2009).
14. J. Rey, J. Martínez, P. Vera, N. Ruiz, F. Cañadas, and V. Montiel, "Ground-penetrating radar method used for the characterisation of ornamental stone quarries," Constr. Build. Mater., **77**, pp. 439–447, (2015).
15. A. P. Annan, "GPR methods for hydrogeological studies," in Hydrogeophysics, Springer, pp. 185–213, (2005).
16. P. Kearey, M. Brooks, and I. Hill, An introduction to geophysical exploration, vol. 4. John Wiley & Sons, (2002).
17. D. Zhang, R. Zhong, J. C. Li, and F. Zeng, "Topographic correction of GPR Profiles based on Laser data," in IOP Conference Series: Earth and Environmental Science, **17**, no. 1, p. 12251, (2014).
18. S. C. Fisher, R. R. Stewart, and H. M. Jol, "Processing ground penetrating radar (GPR) data," in CREWES Research Report, **4**, p. 11, (1992).
19. X. Xie, C. Zeng, and Z. Wang, "GPR signal enhancement using band-pass and K–L filtering: a case study for the evaluation of grout in a shielded tunnel," J. Geophys. Eng., **10**, 3, p. 34003, (2013).
20. A. A. Adepelumi, O. Fayemi, and J. Akindulureni, "Geophysical mapping of subsurface stratigraphy beneath a river bed using ground penetrating radar: Lagos Nigeria case study," Univers. J. Geosci., **1,** 1, pp. 10–19, (2013).
21. J.-H. Kim, S.-J. Cho, and M.-J. Yi, "Removal of ringing noise in GPR data by signal processing," Geosci. J., **11**, pp. 75–81, (2007).
22. M. Szymczyk and P. Szymczyk, "Preprocessing of GPR data," Image Process. Commun., **18**, 2–3, p. 83, (2013).
23. M. Rasol et al., "Background clutter amplitude and frequency of GPR signals to analyse water content in sedimentary deposits: Urban infrastructure environment," Environ. Res., **212**, p. 113189, (2022)
24. C. G. Leandro, E. G. Barboza, F. Caron, and F. A. N. De Jesus, "GPR trace analysis for coastal depositional environments of southern Brazil," J. Appl. Geophys., **162**, pp. 1–12, (2019).
25. P. Wang, Z. Hu, Y. Zhao, and X. Li, "Experimental study of soil compaction effects on GPR signals," J. Appl. Geophys., **126**, pp. 128–137, (2016).

Microplastic study on (*litopenaeus vannamei*) cultivation enterprises on the east coast of aceh

Muhammadar Abdullah Abbas[1,2], *Furqan* Ramadhan[2], *Sri* Agustina[1]

[1]Department of Marine Science, Faculty of Marine and Fisheries, Universitas Syiah Kuala, Banda Aceh 23111, Indonesia
[2]Department of Aquaculture, Faculty of Marine and Fisheries, Universitas Syiah Kuala, Banda Aceh 23111, Indonesia

Abstract. Microplastics are small plastics measuring <5mm which, if they enter the water, can contaminate aquatic biota such as shrimp. The research aims to identify the number, shape and color of microplastics in shrimp bodies in vanname shrimp cultivation businesses on the East Coast of Aceh. Shrimp samples were taken from 3 stations, namely Pidie, Pidie Jaya and Bireuen. The samples were measured for weight and length, then prepared using H_2O_2 and H_2SO_4 in a ratio of 3:1 20 mL then incubated, heated, then filtered and identified using a Zeiss microscope with 40x magnification. The results obtained from shrimp body samples contained 2 types of microplastics, namely fiber and film. The most dominant microplastics at the three stations were 13 particles in the Pidie location, 9 particles in the Pidie Jaya location and 5 microplastic particles in the Bireuen location. The presence of microplastics that have entered the body of biota in the vannamei shrimp cultivation business is an early warning. So it is necessary to develop research on other consumption biota to determine the presence of microplastics and their impact if consumed by the public.

1 Introduction

Plastic is a material that is widely used for various human needs because it is easy to shape/mould, is durable and has low production costs [1,2]. However, as the production costs and practicality of plastic materials become increasingly cheap, their use becomes increasingly uncontrolled and often enters the environment uncontrollably [3]. The presence of plastic waste in the environment, as a result of exposure to UV rays, increased temperatures, and mechanical abrasion, can cause the plastic waste to break down into small plastic pieces known as microplastics [4].

The presence of microplastics in the aquaculture environment can come from various sources, both internal sources resulting from the aquaculture activity itself and external sources from outside the aquaculture activity. Internal sources of microplastics in aquaculture activities cannot be separated from the use of plastic equipment in all aquaculture activities from seeding to the harvest process. Externally, microplastics also have a significant influence on the presence and distribution of microplastics in the aquaculture environment. Using water from various external sources (rivers, lakes, sea, groundwater) can potentially provide MP pollutant input to the aquaculture environment. Some of these plastic equipment, for example, are used in maintenance containers, nets, fishing gear, water pipes and wiring [5]. The use of tools with plastic material allows plastic fragments to be released and enter the fish rearing media as microplastics.

Data on the presence of microplastics in aquaculture in Indonesian waters, especially in the coastal areas of the east coast of Aceh, is still very minimal, while [6] stated that the level of plastic pollution in Indonesia was very high. Apart from that, research regarding microplastic contamination in aquaculture on the east coast of Aceh is also not yet available. because most research focuses on marine areas, without paying attention to research for aquaculture [7]. Meanwhile, the presence of microplastics in aquaculture has an impact on reducing water quality and threatens aquaculture safety, because aquaculture is a source of income from cultivation in coastal areas on the east coast. It is hoped that the research will be able to confirm microplastic contamination in vaname shrimp on the east coast of Aceh. It is feared that this will have a negative impact on the condition of biota that consume microplastics that accumulate in sediment in waters so that it can cause both physical and chemical damage to internal organs and disrupt the digestive tract system [8].

2 Methods

This study will be held in July - August 2023. Sampling is done at the Vanname Shrimp Cultivation Business at the East Coast of Aceh. This research was carried out in June - August 2023. Samples were taken at a vannamei shrimp cultivation business on the east coast of Aceh. Then the sample identification continued at the Fish Nutrition and Histology Laboratory, Faculty of Marine Affairs and Fisheries, Syiah Kuala University, Darussalam, Banda Aceh, Aceh.

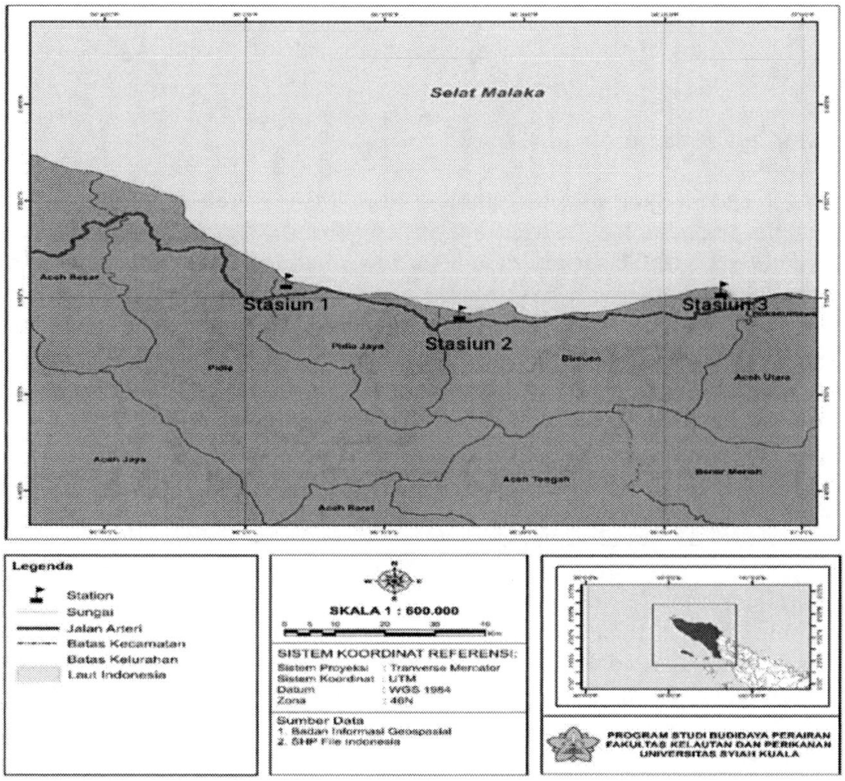

Fig. 1. Research location map

- Station 1 kembang tanjong
- Station 2 panteraja
- Station 3 samalanga

2.1 Sampling technique

This research was carried out in several stages including determining the location by observation, taking samples using random sampling. The work procedure is carried out in 3 stages, namely, 1. Taking samples of vannamei shrimp; 2. Preparation and storage of shrimp samples; 3. Identify microplastic samples from vannamei shrimp samples

2.2 Sampling

Sampling was carried out using a random sampling method, where samples of biota caught at cultivation businesses were taken randomly at 3 points in one location on the east coast of Aceh. Samples were taken from catches in different cultivation businesses, 5 individuals each randomly from each of several vannamei shrimp cultivation sites.

2.3 Sample Preparation

The shrimp sample preparation method used in this research is the method used in [9]. Sample preparation is carried out starting from measuring and recording the weight and length of the shrimp sample using a ruler and scale. Shrimp samples are placed in sample bottles that have been labeled with the name of each sample. The samples that have been taken are then stored in a cool box, then a microplastic analysis test is carried out at the Fish Nutrition and Histology Laboratory, Faculty of Marine Affairs and Fisheries, Syiah Kuala University. Shrimp samples were weighed using a digital scale with an accuracy of 1 gram for further analysis.

2.4 Separation of samples and microplastics

This separation of samples and microplastics refers to research by [10]. Shrimp sample preparation is carried out by starting from measuring the length and weight of the shrimp sample. Separated from the exoskeleton, legs and head of the shrimp. The shrimp meat is then crushed using a mortar and pestle. The finely ground samples were then added with a solution of 30% H_2SO_2 and 30% H_2O_2 as much as 2 times the weight of the shrimp sample. Shrimp samples were incubated for 24 hours. After that, heat the sample jar using a water bath over low heat at around 40-60 °C for 1 hour. After the shrimp sample is crushed, then filter the sample using Whatman filter paper. The sample is then identified using a Zeiss microscope.

2.5 Microplastic Identification

The microplastic samples were ready and placed in a petri dish. Then the type of microplastic in the sample was identified using a Zeiss microscope. Identification of myroplastics using a Zeiss microscope connected to a computer so it is easier to see the sample and with 40x magnification. Microplastic samples are identified based on shape. Based on shape, microplastics are divided into 3 groups, namely fiber, fragments and films. Microplastic samples that have been collected from shrimp meat are cleaned using distilled water, dried and stored in aluminum foil to prevent contamination

3 Result

3.1 Microplastic abundance

The results of the research showed that the highest number of microplastic particles was found in shrimp at the Pidie location, namely 13 particles/g shrimp weight and the least was found in shrimp at the Bireuen location, namely 5 particles/gram shrimp weight. Meanwhile, more fiber type microplastics were found in shrimp, namely 20 particles from all locations , while the least film type microplastics were found in shrimp, namely 3 particles from all these locations.

 https://doi.org/10.1051/bioconf/20248702016

Table 1. Number of Microplastics in Shrimp Samples

Location name	Shrimp sample (tail)	Total average weight of shrimp samples (grams)	Number of microplastics (particle)			Total particles	Microplastic abundance (particles/gram)
			Fiber	Film	fragment		
Pidie	5	8,11	10	2	1	13	1.6
Pidie Jaya	5	9.6	6	0	3	9	0.9
Bireuen	5	7	4	1	0	5	0.7

3.2 Presence and Forms of Microplastics

The presence and form of microplastics found in Vaname shrimp from the 3 locations consisted of 3 types, namely fiber, fragment and film. The most microplastic particles were found in the samples located in Pidie, namely 13 particles, for the particles located in Pidie Jaya, 9 particles were found and the fewest particles were found in the Bireuen location, namely 5. This is because in the Pidie, Pidie Jaya and Bireuen locations, microplastics were found in the body. shrimp because the area is located close to residential areas, fishing activities and remaining pond activities. As stated by Yizheng Li *et al,* . [9] that aquaculture activities are the main cause of the high abundance of microplastics in ponds, for example shrimp fishing and degradation of aquaculture equipment so that pond residues and community activities can cause the entry of microplastic particles through the air or water, as according to [10] that there is a relationship between human activities, tourism and fishermen's activities on the distribution of microplastics in the air and in the waters.

The dominant forms found in the three locations were fiber with a total of 20 particles, film with a total of 4 particles and fragments with a total of 3 particles. In accordance with research by Yizheng Li *et al,* [9], The results of research on the identification of microplastics in cultivation ponds show that fiber is the most dominant microplastic , followed by fragments and film . Similarly, research by [11], identification of microplastics in shrimp samples at all four stations shows that the type of microplastic is the most abundant. found is fiber. Fiber type microplastics come from plastics which are often contaminated with waste, fishing equipment and pond equipment [12].

The most dominant type of microplastic fiber was found in these three locations. The cultivation ponds of these three locations supply water from the sea, because the disposal of household waste into river estuaries which is released into the sea can cause the spread of microplastics [13]. Therefore, many fiber-type microplastics were found in the bodies of vannamei shrimp in this study.

Fragment type microplastics are thought to come from materials such as plastic bottles, paralon pipes and other tools made from plastic produced from aquaculture activities, as well as small-sized waste that can easily move from place to place resulting from the activities of local residents through the air [14]. Fragment type microplastics also come from the fragmentation of large plastics caused by exposure to UV light which causes them to degrade [15]. Fragments have different densities that can make them float, float and sink in waters. Therefore, this type of microplastic can contaminate aquatic biota, both those that move freely and also those that live at the bottom of the water [16].

Very few film-form microplastics were found in these three locations. Film-formed microplastics come from thin plastic fragments and the presence of microplastics in the body of vannamei shrimp is caused by several factors. One of the factors is the presence of waste left over from pond activities, pond equipment used in pond activities and feed storage areas. The film forms in the form of sheets that are thinner than fragments, originating from the fragmentation of plastic bags and other packaging [17].

Based on the results of this research, the most microplastics were found from three research locations, namely Pidie. This is caused by the cultivation system using a semi-intensive system, according [9] Microplastics are found at a higher rate in soil pond sediments compared to the water surface. According to research[18], the high level of microplastics in sediment samples obtained from land ponds is caused by two factors. The first factor is influenced by high anthropogenic activity and the second is within the pond itself. It is impossible for microplastics to enter sediment into earthen ponds and over time the number of microplastics will continue to increase and tend to accumulate at the bottom of the substrate. The cultivation environmental activities at the Pidie location are not well maintained, because there is burning of rubbish near the pond, which allows microplastic pollution through the air. This is in accordance with the statement [13] that microplastics can also enter the aquaculture environment through air contamination. Due to their very small size, microplastics can easily be carried by the wind and float in the air. Meanwhile, relatively few microplastics were found at the Pidie Jaya and Bireun cultivation locations, because the cultivation locations were quite well maintained. The Pidie Jaya and Bireun cultivation locations use an intensive system using HDPE tarpaulin which has many advantages, namely high impermeability, flexibility, strength, anti-tearing and resistance to high temperatures [19]. Therefore, using HDPE tarpaulin is very appropriate to reduce the presence of microplastics. In intensive system ponds in the Pidie Jaya and Bireuen locations, the presence of microplastics is due to the aquaculture activities themselves and plastic equipment such as the use of ropes, cables and pipes. The large number of microplastics in aquaculture ponds is due to the weak hydrodynamics of the pond environment and poor water exchange quality. In the end, the water exchange in this cultivation comes from seawater exchange [9].

This shows that there is a relationship between the microplastic content in shrimp. Microplastics can absorb toxins produced from chemicals in waters and the surrounding environment and can be transferred into the food chain indirectly [4]. This can have a negative impact on the food chain sequentially. Shrimp that consume microplastics for a long period of time will experience death and stunt growth because the plastic particles cannot be digested in the shrimp's body [20]. Due to the size, chemical composition and physical properties of microplastics, this can affect shrimp and water quality and can have an impact on cultivation activities. Another thing that can affect human health is the accumulation of microplastics in ponds consumed by shrimp at that location and will enter the human body if humans consume contaminated biota [21].

Identification of microplastics in vannamei shrimp with the type of microplastic obtained using a Zeiss microscope with 100x magnification and producing images like the following

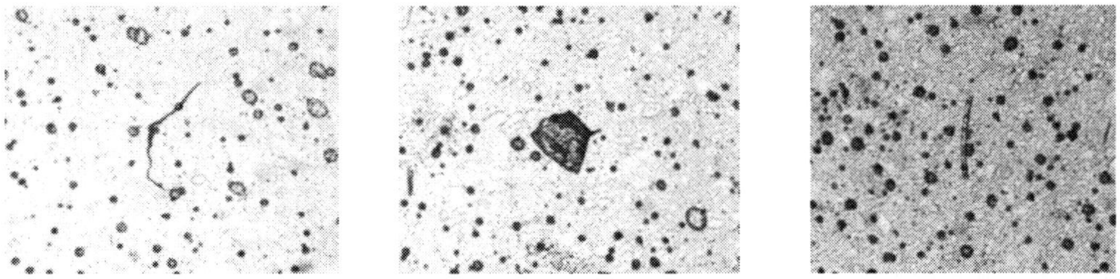

Fig. 2. Microplastic photo sample and kind (a) fiber (b) fragment (c) film

This classification is obtained from matching research results with the characteristics of each microplastic. Several research sources from [17-14] show that the form of microplastic fiber can be recognized by its shape which tends to be elongated and resembles a thread . Microplastic film form It has the form of thin sheets of plastic , this type is also in the form of plastic fragments which are not like sheets and fibers . Fiber and film can be sourced from various types of plastic, most of which are the result of use from domestic activities. The shape of microplastics in general can also be influenced by how long the microplastics are in the water and how long the fragmentation process takes.

3.3 Microplastic Color

The research results showed that there were 4 types of microplastic colors in vannamei shrimp , namely red , black, yellow and clear . The overall color of microplastics is most often found in colors black namely with a total of 18 particles at all sampling locations , while the lowest overall microplastic color was found in the clear color, namely 1 particle at all sampling locations.

Table 2. Number of microplastic colors

Location	Number of colors of microplastics (Particles)				Amount (Particles/gram)
	Red	Black	Clear	Yellow	
Pidie	3	8	1	1	13
Pidie Jaya	2	6	0	1	9
Bireuen	1	4	0	0	5
Total	6	18	1	2	27

The most dominant colors of microplastics in this study found in the three locations were black and red, black and red which came from shrimp fishing equipment such as nets and tools left over from pond activities such as the use of shrimp fishing nets, ropes and cables as per Yizhenli's research. [9], Most microplastics are identified as PP, and there are two main sources, namely fishing equipment such as fishing nets and woven feed bags which can be directly broken down and entered the pond with the feed. Activity waste from pools and clothing threads from pool activities can also enter microplastics into the water. Black and red colors are also artificial colors from anthropogenic results [22] and also colors that have been degraded by sunlight (UV). Fiber type microplastics are dominated by black and red. The color of microplastics, which are yellow and clear, is the least dominant in the three locations. The yellow color also comes from *polystyrene* or food packaging or *Styfoam boxes* , which indicates the length of time they stay on the surface of the sea, land and water, as well as the level of weathering. The yellow color indicates it has been in the water longer and has been oxidized. The color of film type microplastics is dominated by clear and fragment type microplastics are dominated by yellow. The clear color found in these microplastics is thought to have come from plastic bags and the yellow microplastics came from broken styrofoam boxes that were not managed properly and also indicates the length of time the microplastics have been degraded by UV light over a long period of time. [23]

The color of microplastics is used as an indicator for identifying the source, type of contamination, level of degradation and time period. According to [24], microplastic color can be used to identify the source and type of contamination in shrimp. This happens because the color of microplastics can increase the chance of contamination of aquatic biota because their characteristics resemble the natural prey of shrimp [25], the color of microplastic particles is a contamination sector because they can attract shrimp to prey on them due to the color similarity to their natural prey.

The results obtained from shrimp body samples contained 3 types of microplastics, namely fiber and film. The most dominant microplastics at the three stations were 13 particles in the Pidie location, 9 particles in the Pidie Jaya location and 5 microplastic particles in the Bireuen location. The presence of microplastics that have entered the body of biota in the vannamei shrimp cultivation business is an early warning [11]. So it is necessary to develop research on other consumption biota to determine the presence of microplastics and their impact if consumed by the public.

D.Conclusion

The abundance of microplastics from the three locations was 27 microplastic particles and the highest microplastics were found at the Pidie location. The presence of microplastics in vannamei shrimp proves that shrimp can be a means for microplastics to enter the human body through the food chain. The forms of microplastics found in the three research locations were Fiber, Fragment and Film, the highest type of fragment found in the three locations.

E. References

1. P. Azizah, P., A. Ridlo and C. A. Suryono. **9(3)** 326-332 (2020)
2. M. Browne, A.; Galloway, T. S.; Thompson, R. C. **3** 559–566 (2007)
3. M. Cole, M., P. Lindeque, C. Halsband dan T. S. Galloway. **62(12)** 2588-2597 (2011)
4. J. Jambeck, J.R., Geyer, R., Wilcox, C., Siegler, T.R., Perryman, M., Andrady, A., Narayan, R., Law, **347** 768–771 (2015)
5. C. Rochman, C. M., C. Brookson, J. Bikker, N. Djuric, A. Earn, K. Bucci, S. Athey, A. Huntington, H. McIlwraith dan K. Munno. **38(4)** 703-711. (2019)
6. Victoria. Microplastic Contamination in Fresh Waters. ResearchGate. Chemical Engineering, Bandung Institute of Technology. (2026)
7. E. Wicaksono, E. A. 77-91 (2022)

8. S. Wright, R.C. Thompson, T. S. Galloway. The physical impacts of microplastics on marine organisms: a review. Environmental pollution, **178** 483-492 (2013).
9. D. Yona, A. Aisyahdi, and M. A. As'adi. 2020. Identification and Comparison of the Abundance of Plastic Waste Based on Size in Sediment on Several Beaches in Pasuruan Regency. East Java, **8(2)** 375-383 (2020)
10. A. Zhou, Y. Zhang, S. Xie, Y. Chen, X. Li, J. Wang, J. Zou, J., Rev. Aquac. **13**, 719–733. (2021)
11. S.A. Akbar, S. Afriani, C. Nuzlia, S. Nazlia, S. Agustina. (2023). *Depik* **12(3)** 259-273 (2023)
12. L. Stephanie, C. WrightaRichard, Thompsonb, S. Tamara, Gallowaya. Environmantental pollution, **178** (13) 483-492. (2013)
13. J. Boucher, D. Friot, Gland, Switzerland: IUCN. 2017.
14. E.A. Wicaksono, Torani Journal of Fisheries and Marine Science: 77-91 2022.
15. A.L. Andrady, Marine PollutionBulletin, **62(8)** 1596–1605 (2011).
16. D. Adyasari, M.A. Pratama, N.A. Teguh, A. Sabdaningsih, M.A. Kusumaningtyas, N. Dimova, Mar. Pollut. Bull. **171**, 112689. 2021.
17. V.L. Hidalgo-Ruz, R.C. Gutow, M. Thompson, M. Thiel. Environmental science & technology, **46(6)**: 3060-3075 (2012)
18. Annisa, Fuel Oil from Waste K rese k bags , blue ocean publisher , member of I K API, 1 4 15, Jambi (2021).
19. M.A. Browne, T.S. Galloway, R.C. Thompson,. Integr. Environ.Assess. Manage, **3** 559–566 (2007)
20. C.M. Rochman, C. Brookson, J. Bikker, N. Djuric, A. Earn, K. Bucci, S. Athey, A. Huntington, H. McIlwraith, K. Munno. Environmental toxicology and chemistry , **38(4)** 703-711 (2019)
21. A. Deni,. Information on Shrimp Cultivation Technology, Solutions for Increasing Shrimp Production, Central Jakarta, Amafrad Press (2014)
22. A.R. Hastuti, F. Yulianda, Y. Wardiatno. International Journal of Bonorowo Wetlands , **4(2)** 94-107 (2014)
23. S.Rahmawati, Identification of the Presence and Form of Microplastics in Sediment and Fish in the Code River, DI Yogyakarta 2020.
24. S.L. Wright, J. Ulke, A. Font, K.L.A. Chan, F.J. Kelly, Environ. Int. **136**, 105411 (2020).
25. S.A. Akbar, R.R. Fazli, R. R, Meta-Analysis: The effectiveness of artificial diets to increase on survival rate of vannamei shrimp (Litopenaeus vannamei). In *IOP Conference Series: Earth and Environmental Science,* IOP Publishing. **1221(1)** 012066 (2023)

Analysis of types, forms and abundance of microplastics in the mangrove forest area of pusong island, langsa city

Cici Maulida[1], *Muhammad* Irham[2,3,4*], *Saiful* Saiful[5]

[1]Master's Study Program in Integrated Coastal Resources Management, Universitas Syiah Kuala
[2]Marine Science Study Program, Faculty of Marine and Fisheries Science, Universitas Syiah Kuala
[3]Center for Environmental and Natural Resources Research, Universitas Syiah Kuala
[4]Geographic Information Systems Laboratory, Faculty of Marine and Fisheries Science, Universitas Syiah Kuala
[5]Department of Chemistry, Faculty of Mathematics and Natural Sciences, Universitas Syiah Kuala

Abstract. Pollution of mangrove waters by microplastic particles is coastal plastic waste pollution which has become a major water pollution problem in developing countries. However, this pollution problem, both qualitatively and quantitatively, has not yet been discussed comprehensively. Therefore, it is necessary to analyse the level of microplastic presence and identify the types of microplastics of each size found in the water in the mangrove forest area of Langsa City-Aceh. The aim of this research is to analyse the presence, shape and dominance of microplastic particles in the mangrove waters of Langsa City. The research method was carried out by taking water and river sediment samples based on SNI 03-7016-2004 and the samples were the result of a combination of places (integrated samples). Separation of microplastics is carried out by filtering, destroying organic compounds, separating based on specific gravity and filtering using vacuum. Identification of the type, shape and abundance of microplastics is carried out using a microscope. The analysis results show that microplastics in water are generally of the fragment and fibre type with a density of between 7 - 13 particles per Liter with an average size of 200 μm.

1 Introduction

Pusong Island is a coastal area located in Gampong Telaga Tujuh, West Langsa District, Langsa City, Aceh Province. Pusong Island is six miles from the port of Kuala Langsa. Access to this island requires a ferry boat about 30 minutes from Kuala Langsa port. Nearly 80% of the livelihood of Pusong Island residents is as fishermen who use this water access, the rest work as traders, salted fish processors, ferry service providers and others.

* Corresponding author : irham@usk.ac.id

As an estuary area that is still influenced by river flows, Pusong Island is thought to carry a lot of plastic waste which is generally produced by the community [1]. Around 80% of plastic waste comes from land, while 20% comes from the sea [2]. This is exacerbated by the input level being greater than the degradation time, especially for plastic waste which dominates ¾ of marine waste and 10% of which is newly produced waste [3], [4]. The distribution of microplastics is also influenced by wind movements, the movement of marine biota, currents and the characteristics of the polymers that make up the plastic [5]. Coastal areas, especially industrial areas, are hotspots for microplastic accumulation [6], [7].

Microplastic waste can be found on the surface, columns and bottom of waters (substrate), beaches, rivers and estuaries [8]. Microplastics found in sea water come from river flows, as the main route for microplastics from terrestrial sources [9]. Microplastics can also come from community activities around rivers and coasts [10]. Microplastics have an impact on the environment [11] because of their size, microplastics are susceptible to being consumed by organisms in waters by accidentally swallowing them and mistaking microplastic particles for food [7]. Microplastics cause various negative impacts on fish that consume them, such as internal organ infections, blockages, carcinogenesis, endocrine disruption, physical damage (resulting from accumulation, translocation, shape, egesting, population level effects, transfer to the food chain), changes in lipid metabolism and behavioural changes as well as toxicity [6], [12].

Several reports of fishing nets being abandoned/untied and becoming ghost fishing, causing fish to die due to entanglement and it is feared that the consumption of microplastics by fish will cause bioaccumulation and biomagnification in the food chain[13]. In addition, microplastics not only impact the marine environment and organisms, but can also impact humans as final consumers through transfer to the food chain [6]. The presence of microplastic pollutants in aquatic biota consumed by humans can pose a food safety risk, because it has an impact on health [14]. Exposure to microplastics can cause inflammation, oxidative stress, cell poisoning, while the chemicals in them cause digestive and reproductive disorders and trigger an immune response. Inhaled microplastics will cause respiratory problems and inflammation [15].

Coastal areas are the most vulnerable areas to microplastics, this is because coastal areas are dynamic areas and are influenced by tides, so the presence of microplastics will spread throughout the mangrove forest area [16]. Mangrove roots and other natural processes will increase the mechanism for trapping microplastics [17].

Microplastic pollution has been found on the sea surface, bottom waters, water column, beaches, rivers, estuaries, bays and open seas in polar and equatorial regions, including the Arctic and Antarctic seas [8]. As much as 330 million metric tons of plastic waste is produced globally per year [18]. According to [19], the volume of input of plastic waste from the East Asian continent into the surrounding seas is the largest in the world. The presence of microplastics is also found in Indonesia with an estimate of 0.48 to 1.29 million metric tons per year and this is considered the second largest contributor of plastic waste to the sea in the world after China [20].

2 Material and Methods

The material in this research is water from the waters of the mangrove forest area of Pusong Island in Langsa City, Aceh Province. The method used in sampling is the purposive sampling method. Purposive sampling was taken at four representative stations (**Figure 1**), namely water areas and mangrove forest areas closest to residential areas, water inflow from coastal waters to the estuary, mangrove forest areas with low levels of mangrove density and areas with high levels of mangrove forests. the mangrove density is high. Water samples were taken at the top of the surface, but before taking the water samples, environmental quality measurements were carried out at the four stations, namely measuring temperature, salinity, pH and ocean currents.

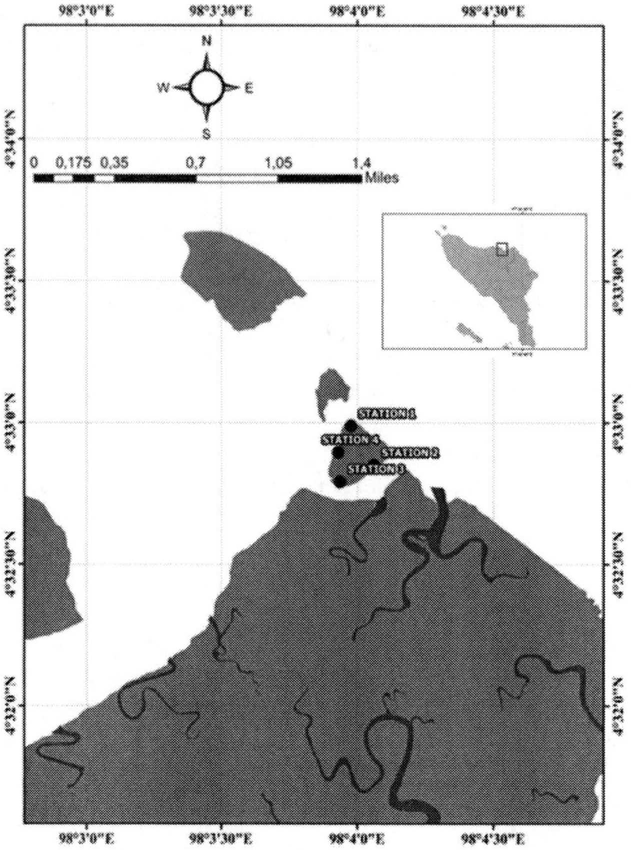

Fig. 1. Map of study area and the location of sampling

Fish acclimatization was carried out individually in transparent containers for one week (Photoperiod 12:12 hours light: dark; DO: 6.2 mg O^2/L; temperature: 25 C; PH: 7.5 and turbidity: <1 NTU). To avoid stressful conditions due to interactions between fish, each side of the acclimatization container was lined with opaque paper. Feed in the form of silk worms was given twice a day (09.00 and 16.00 WIB) ad libitum [6]

2.1 Water Sampling Method

Water samples were taken using a water sampler, then filtered using a plankton net with a diameter of 30 cm and the mesh size is 200 μm. The water samples that have been taken were then put into a 600 ml sample bottle and stored in a storage box to avoid contamination. Next, the separation was carried out using a floatation tool, where the process of separating water from microplastics was carried out in stages by adding 200 ml of sample water every time (3 repetitions) which is then added with filtered sea water to the limit of the floatation tool. Next, the samples were stirred evenly for 5 minutes and let sit for 10 minutes. The sample at the top of the floatation device were poured into a 100 ml sample bottle with every 50 ml of sample water which were separated into a different sample bottle and were labelled. To the 50 ml water sample of microplastics, a 23 ml of 70% ethanol solution was added and to the remaining 50 ml water sample of microplastics no ethanol solution was added to see the differences between the two samples. Ethanol helps to discolour organisms and brightens the colour of plastic [21].

2.2 Extraction of Microplastic Samples

The extraction process was carried out at the Marine Chemistry Laboratory, Faculty of Marine and Fisheries of Syiah Kuala University. On a sample of 50 ml of water, the 20 of 30% H_2O_2 was added using a dropper pipette and stirred with an iron spatula to homogenize the sample. Next, place the sample on a hot plate stirrer at 30 °C for 10 minutes with 150 ml of saturated NaCl solution added for density separation. The density separation process was carried out for 20 minutes and the sample was poured into an Erlenmeyer flask then covered with aluminium foil to leave for 12 hours. The results of this sample extraction were then taken as much as 20 ml for each water sample. The sample that has been extracted was then poured into a funnel with filter paper at the top to filter the sample and then dried for 48 hours (2 days). After the sample was visibly dry, the filter paper on a petri dish was palced and put it in the incubator machine to completely remove the water content on the filter paper at a temperature of 40 °C for 8 hours. After the extraction process is complete, observations was made using a stereo microscope [22].

2.3 Identify Type, Colour and Size of Microplastics

Identification of micro plastics was done by classifying six types of microplastic, namely films, fragments, fibers, foam, pellets and granules [23]. Identification of microplastic types was also carried out based on type, color and size using a stereo microscope (Olympus SZ61) with 10x-40x magnification with an attached software camera [24]. Microplastic particles were counted in a "zigzag" pattern from left to right. The data was processed using Microsoft Excel and displayed in the form of tables and images which were analyzed descriptively [22].

2.4 Calculating the Abundance of Microplastics in Water

Calculation of microplastic abundance can be calculated using the equation:

$$K = \frac{n}{v}$$

Information:
K = Abundance of microplastics (particles/m³)
n = Number of Microplastics (Particles)
v = Sample volume (m³)

3 Result and Discussion

Before collecting microplastic data, environmental parameters are first measured. The results of measuring these parameters can be seen in **Table 1**. The location of station 1 is the point closest and directly facing the settlement of the Pusong Island population, of course this will not have a big influence on the current if seen from the results of the environmental parameter measurements that have been carried out. This is different from station 2, which is the entry point for free coastal waters into the estuary. This point is the entry point for fishermen who want to go to sea to find fish. Currents and salinity also greatly influence this station point, this is proven by the strong currents and high waves that make the condition of the mangrove ecosystem at station 2 not very fertile and relatively low. Meanwhile, at station 3, even though it is adjacent to station 2, the condition of the mangrove ecosystem in this area is fertile because of the good and sufficient supply of nutrients. Many fishermen use this location as a place-to-place crab traps (bubu) which proves that the influence of the current is not that big. Fishermen in this area can maintain their fishing gear. The salinity and temperature measurement data shows that it is suitable for the habitat of mangrove crabs. At station 4, the measurement results show that the condition of the mangrove ecosystem is lower but better at station 4 compared to station 2. However, the condition of the water environment is almost the same as station 1, namely the current, temperature and salinity are relatively low.

Table 1. Results of environmental parameter measurements at sampling stations

Parameters	Station 1	Station 2	Station 3	Station 4
PH	7.1	7.4	7.2	7
Température	31.3^{0}C	35.6^{0}C	31.1^{0}C	35.5^{0}C
Salinity	32	33	32	33
Curent speed	28.25 m/s	7.70 m/s	18.22 m/s	25.18 m/s
Direction	North-East	East	North-East	North-East

3.1 Microplastic Abundance

The highest abundance of microplastics was found in water samples without ethanol, namely 0.275 particles/m³ and the lowest abundance was found in water samples that had added ethanol solution, namely 0.174 particles/m³ (**Table 2**). The high concentration value of microplastics on the surface of the water can be caused by the low mass of microplastics and microplastics being released in seawater with a high salt content, making the microplastics continue to float [25].

Table 2. Abundance of Microplastics at Each Station

Sample	Type of Microplastic	St 1	St 2	St 3	St 4	Average (particle/m³)
Water + Ethanol	Fiber/Line	0,08	0,06	0,08	0,1	0,08
	Fragment	0,08	0,08	0,08	0,1	0,085
	Film	0	0,02	0,02	0	0,01
	Foam	0	0	0	0	0
	Pellet	0	0	0	0	0
Total		**0,16**	**0,16**	**0,18**	**0,2**	**0,175**
Water - Ethanol	Fiber/Line	0,14	0,06	0,12	0,1	0,105
	Fragment	0,26	0,06	0,2	0,06	0,145
	Film	0,02	0	0,02	0,04	0,02
	Foam	0	0	0,02	0	0,005
	Pellet	0	0	0	0	0
Total		**0,42**	**0,12**	**0,36**	**0,2**	**0,275**

As shown in **Table 2**, there were differences in abundance between water samples that had ethanol added and those that had not been added. This proves that the ethanol solution influences the microplastic samples. Kim *et al.*,[26] said that if there are different methods for adding ethanol solution, it can remove several types of microplastic pollutants.

Based on the analysis results, the highest abundance of microplastics was found at station 1 at 0.35 particles/m³ (**Figure 2**), where this station is the point closest to people's homes. According to [27] that high population density can influence the distribution and concentration of the abundance of microplastics found. One of the factors causing the dominance of microplastics in the environment is household activities [4]. Waste from human activities can be leftover clothing threads that come from washing clothing fabrics and plastic ropes that have undergone a degradation process [28]. Furthermore, the second highest abundance of microplastics was at station 2, with a total abundance of 0.28 particles/m³ (**Figure 2**). The high abundance of microplastics at station 2 could be caused by it being a place where sea waters enter the estuary. Apart from that, Station 2 is also a route for Pusong Island fishermen who want to catch fish in the sea. Based on research by [29], the source of microplastics can come from the fragmentation of plastic that enters the environment, either through river flows, run off, tides, carried by the wind, or from the sea such as passing ships. as well as fishing gear

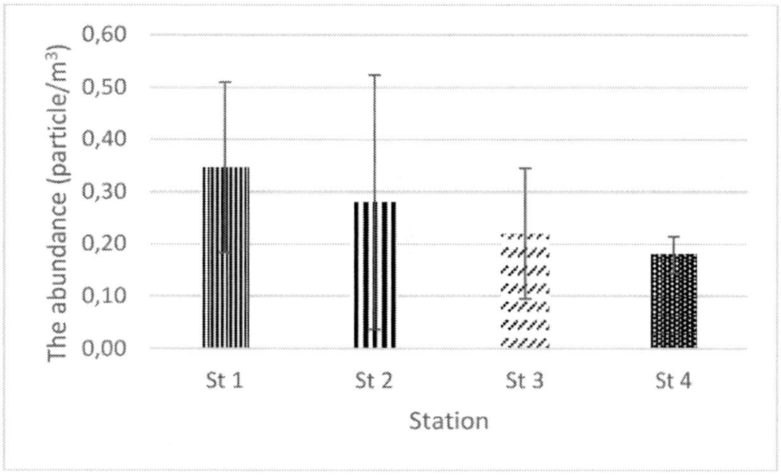

Fig. 2. The abundance of microplastics in Pusong island

The abundance of microplastics at station 2 automatically influences the abundance of microplastics next to it, namely station 3, which has the third highest abundance of microplastics at 0.22 particles/m³ (**Figure 2**). This could be caused by sea currents and winds which carry microplastic particles to this station. According to [11] that repeated movements of currents and gravity can be a factor causing the movement of microplastics in the water column to other places so that the amount of microplastics tends to be different at each station.

At station 3 there is also a river flowing into the sea estuary. River water flow is one of the main routes for microplastics to enter the marine environment [30]. This could be the main factor in the high abundance of microplastics at this station. Meanwhile, the lowest abundance of microplastics was at station 4, namely 0.18 particles/m³. Station 4 is between station 3 and station 1, each of which is only 10 meters away. It is thought that currents will transport microplastic particles in the water column to move them to other places [29].

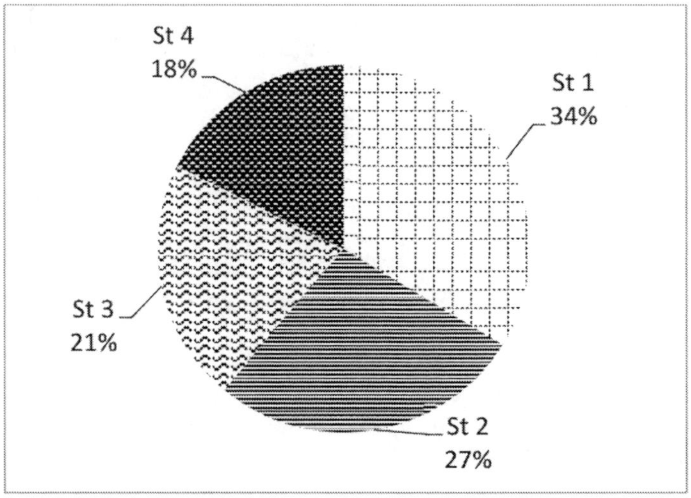

Fig. 3. Percentage of Microplastic Abundance on Pusong Island

Based on the analysis results in **Figure 3**, the highest percentage value of microplastic abundance is at station 1, namely 34%, while the lowest percentage abundance is at station 4, namely 18%. Therefore, the results of this analysis are in accordance with what has been found by several previous researchers in areas with the same characteristics as the findings in this study.

3.2 Type of Microplastic

According to [31] the types of microplastics commonly found on water surfaces are fragments, fibres and films. Based on observations of the water on Pusong Island (**Figure 4**), the highest percentage density of microplastics is the fragment with the value of 49%. Furthermore, the second highest is the fibre at 41% and followed by the film at 7%. Meanwhile, the lowest density percentage is the type of foam, at 2% and pellets are only 1%. Fragment type microplastics are produced from waste such as bottles, jars, and small pieces from Parallon pipes. The high abundance value of microplastic fragment types can be caused by the dominance of the amount of waste on riverbanks, such as plastic bottles or other household plastic waste. The process of fragmentation and size reduction of macro-sized polypropylene plastic waste will occur if the plastic waste flows in river flows and becomes fragmented microplastic waste.

The distribution of fibre-type microplastics can be influenced by fishing activities, fishing nets or abandoned nets which then undergo a monofilament degradation process [10]. According to [28] fibre type microplastics can originate from waste resulting from human activities which enter river flows until they settle and accumulate in sediment. Waste from human activities can be leftover clothing threads that come from washing clothing fabrics and plastic ropes that have undergone a degradation process [28].

As a result of observations of microplastics in water samples on Pusong Island, five forms of microplastics were found, including fragments, fibres, films, foam and pellets (Figure 5). Fragment type microplastics have a rough surface and jagged edges. Fragmented microplastics were the most frequently found in water samples at all observation stations. This has an impact on the high use of solid textured plastic items such as household utensils and beverage bottles. Apart from that, the fibre-shaped microplastics identified in this study were the second most frequently found. Fiber-shaped microplastics were found at all observation stations. Microplastic fibre comes from fishing activities and household activities [2], of course this is in line with the facts on the ground that Pusong Island has a high population density with the majority of the population working as fishermen. Meanwhile, film-shaped microplastics were also found at all observation stations, although not too many, but this indicates that there are plastic bags and other plastic packaging that have experienced degradation and fragmentation due to waste that is not managed properly [32]. Meanwhile, microplastics in the form of foam and pellets were only found at a few stations, and the number found was the lowest compared to the other three forms of microplastics, this was due to their lower or lighter density so they tended to be easily carried away by currents or easily transported [29].

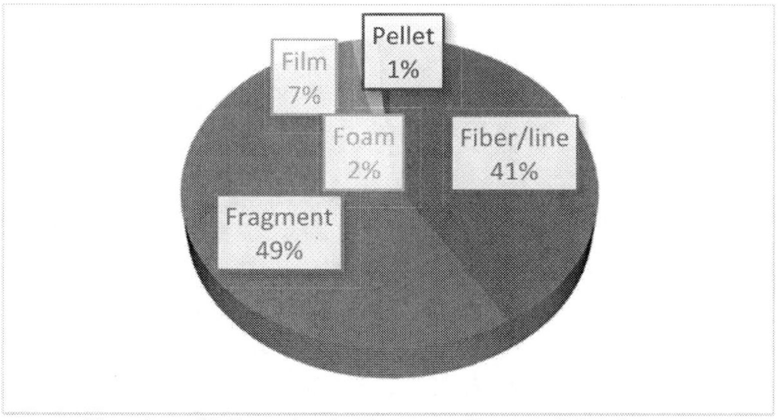

Fig. 4. Percentage Density of Microplastic Types on Pusong Island

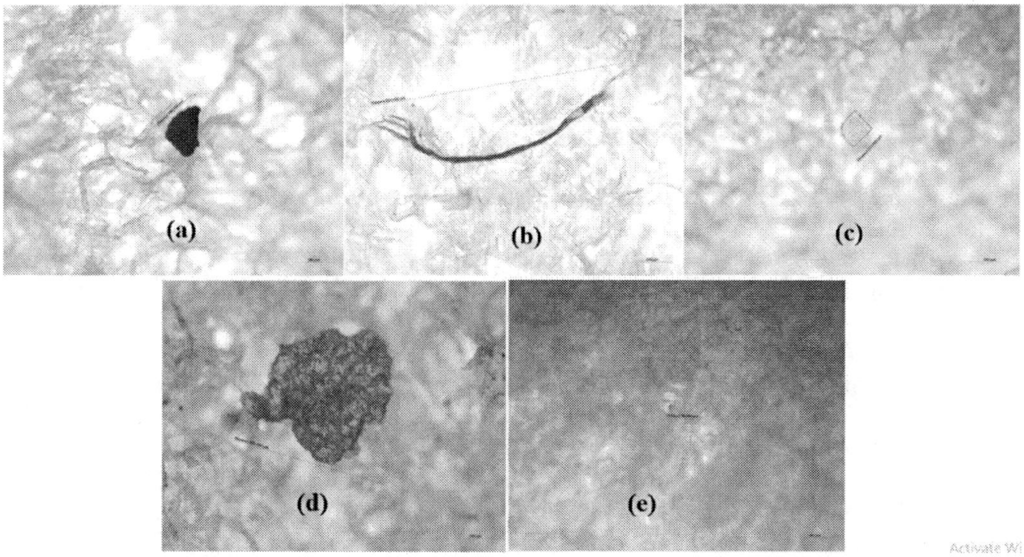

Fig. 5. Forms of microplastics found on Pusong Island (a). Fragments, (b). Fiber, (c). Film, (d). Foam and (e). Pellets

3.3 Type of Microplastic

The colours of microplastics found at all stations were white, black, red, yellow, blue and brown (Figure 6). The most dominant colours of microplastics were black and brown with a total of 41 particles each, while the colour that was least found was green with a total of 5 particles. The differences in the colour of microplastics vary greatly due to the length of time the microplastics are exposed to sunlight so that the polymer undergoes oxidation which results in colour changes in the microplastics themselves [33].

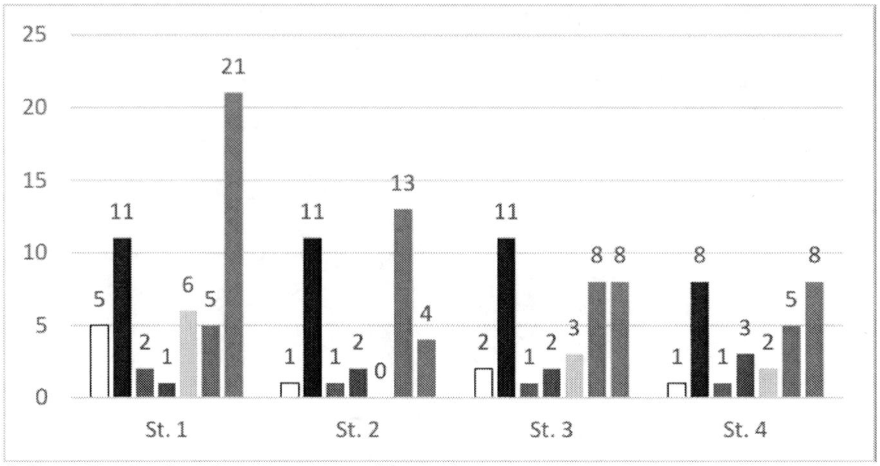

Fig. 6. Microplastic abundance separated by color

As shown in **Figure 6**, not only black, brown and green can be found at all observation stations, but blue, white and red are also found. Yellow is not found at station 2, however, it is found at the other three stations. As for station 1, the colour most found is brown, while at station 2 it is blue. This is different from station 3 which is dominated by black and then station 4 has the two colours that are most found, namely black and brown. According to [34] observing the colour of microplastics will be useful to find out the source of the pollutant. Fragment type microplastics have various colours such as red, green, blue, orange and brown [35]. Fiber is shaped like thread with various colours such as blue, red and black [10]. Fiber comes from fishing lines and net fishing equipment and can also come from clothing that is no longer used and experiences friction so that it breaks down into very small plastic particles, or fabric fibres left over from washing clothes that are carried by currents into the water [36]. The red, blue, and black colours are believed to come from clothing threads and wastewater from washing. The microplastic film type is black and white (grey) like plastic sheets, while the foam type is white and green with a rubbery texture, and the pellets are black, orange, and grey. The presence of colour or transparency makes it possible for marine biota to accidentally eat it [37]. The white or transparent colour is thought to come from plastic bags and Styrofoam that are thrown away carelessly. Red and blue are artificial colours from anthropogenic results or colours that have been degraded by sunlight (UV), while black can be indicated by the large number of contaminants absorbed in microplastics and other organic particles. If the colour of the microplastics found is still dark, it indicates that the microplastics have not undergone significant discolouring.

3.4 The Size of Microplastic

Based on the results of observations of microplastics using a stereo microscope, the sizes of the microplastics found in each type of sample, both sediment and seawater samples, have different sizes.

Microplastics are composed of irregular fibres to round and long fibres [10]. Based on **Table 3,** microplastics in the form of fragments have various sizes in water samples. The smallest fragment size was 17.6 μm and the largest fragment size was found at 611 μm. Of all the sizes of microplastics found, fibre was the largest in all water samples, compared to the sizes of other microplastics. The fibre shape is always elongated, even reaching a size of 1755.5 µm in water without ethanol and the smallest size, namely 31.3 µm in water samples with ethanol added, this result is almost the same as that found by [38]. Meanwhile, for film-form microplastics, the smallest size found was 71.5 µm in seawater samples without ethanol and the largest size was 197 µm found in sediment samples. Meanwhile, the size found in the foam form was 577 µm and the pellet form was 23.4 µm in the water sample without ethanol.

Table 3. Size of microplastics in water samples on Pusong Island

Sample	Type of Microplastics	Size (µm)
Water with Ethanol	Fragment	33,2 - 611 µm
	fibber	31,3 - 828,4 µm
	film	197 µm
Water without Ethanol	Fragment	17,6 - 416,5 µm
	fibber	63,3 - 1755,5 µm
	film	71,5 - 136 µm
	foam	577 µm
	Pellet	23,4 µm

The size differences in microplastics are influenced by the length of the microplastic fragmentation process in waters. The microplastic fragmentation process can also be influenced by ultraviolet radiation and strong mechanical forces from waves and sea currents [11]. Small-sized microplastics show a long fragmentation process, until the plastic particles are broken down into smaller sizes (Hebner and Maurer-Jones, 2020). Many smaller microplastics are buried in deeper layers of sand or sediment before being transported by ocean currents [39].

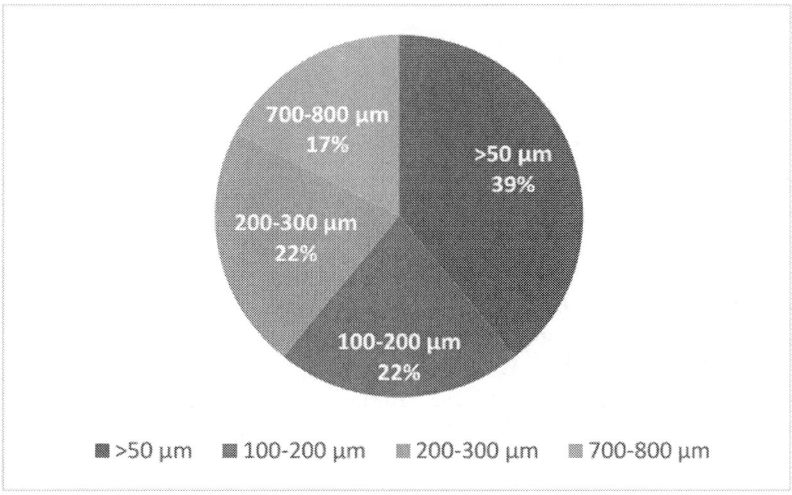

Fig. 7. Percentage of microplastic size in seawater samples added with ethanol

In seawater samples added with ethanol (**Figure 7**), it was found that the most found microplastic size was >50 µm in size at 39%, while the smallest size found was 700-800 µm in size, namely only 17%. In contrast to the size obtained in sediment samples, the size of seawater samples with ethanol added was larger, even reaching 700-800 µm, although only 17%.

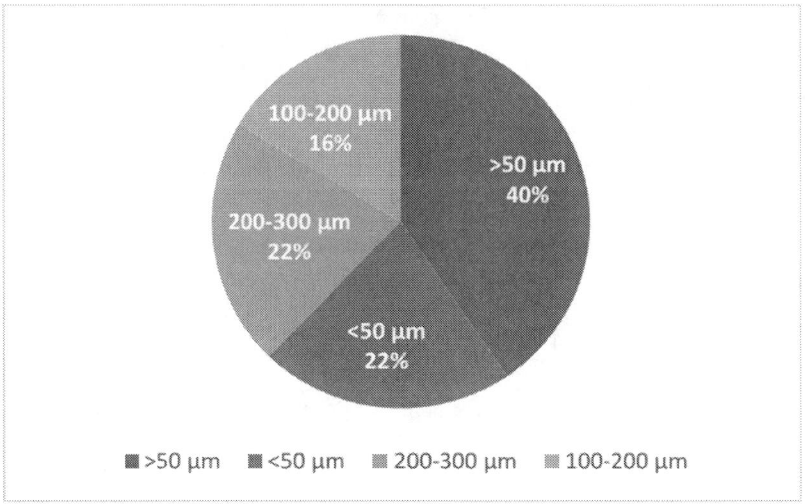

Fig. 8. Percentage of Microplastic Size in Seawater samples without Ethanol

Meanwhile, in seawater samples without ethanol (**Figure 8**), the largest percentage of microplastics was found in sizes <50 µm, namely 40%, while the smallest size found was 100-200 µm, which was only 16%. Of the three samples, both sediment samples and seawater samples, the most dominant size was found to be the same. According to [11,40], the time for plastic to fragment is largely determined by the physical conditions of the environment. The presence of microplastics in an aquatic ecosystem will have the same influence on the surrounding waters [9].

4 Conclusion

The highest abundance of microplastics found on Pusong Island was 0.35 particles/m^3 at station one, while the lowest abundance was at station 4 with a total of 0.18 particles/m^3. The types of microplastics found on Pusong Island consisted of 5 types, namely fragments (49%), fibber (41%), film (7%), foam (2%) and pellets (1%). Meanwhile, the colours of microplastics found at all stations were white, black, red, yellow, blue and brown. Meanwhile, for the size of the microplastics found based on observations using a stereo microscope, the smallest fragment size was 17.6 µm and the largest fragment size was 611 µm, while for the type of fibber that is always elongated, the largest size was around 1755.5 µm. and the smallest size is 31.3 µm. Meanwhile, for the film type the smallest size was found to be around 71.5 µm and the largest size reached 197 µm and for the size of foam type microplastics it was only found with a size of 577 µm and the same thing with the pellet type was found at 23 .4 µm.

References

[1] T. Mahyuddin, *PERSPEKTIF*, **6(2)** 195–204 (2013)
[2] D. A. Ambarsari and M. Anggiani, *Oseana*, **47** 20–28, (2022)
[3] F. Galgani, G. Hanke, S. Werner, and L. De Vrees, "Marine litter within the European Marine Strategy Framework Directive," *ICES Journal of Marine Science*, vol. 70, no. 6, pp. 1055–1064, Sep. (2013)
[4] I. Sari Dewi, A. Aditya Budiarsa, and I. Ramadhan Ritonga, *Depik*, **4(3)** 2015)
[5] V. Hidalgo-Ruz, L. Gutow, R. C. Thompson, and M. Thiel, *Environmental Science &*

Technology, **46(6)** 3060–3075 (2012)
[6] S. L. Wright, R. C. Thompson, and T. S. Galloway, *Environmental Pollution*, **178** 483–492 (2013)
[7] M. Bergmann, L. Gutow, and M. Klages, Eds., *Marine Anthropogenic Litter*. Cham: Springer International Publishing, 2015.
[8] C. A. Peters, P. A. Thomas, K. B. Rieper, and S. P. Bratton, *Marine Pollution Bulletin*, **124(1)** (2017)
[9] A. Nuryanti, "Identifikasi kelimpahan Particle Suspected as Microplastic (PSM) pada sedimen kawasan konservasi hutan Mangrove Wonorejo, Kecamatan Rungkut, Kota Surabaya," 2022.
[10] Q. N. Laila, P. W. Purnomo, and O. E. Jati, **4(1)**, 28–35 (2020)
[11] O. B. Laksono, J. Suprijanto, and A. Ridlo, *Journal of Marine Research*, **10(2)** 158–164 (2021)
[12] B. Jovanović, *Integrated Environmental Assessment and Management*, **13(3)** 510–515, (2017),
[13] J. A. C. C. Nunes, C. L. S. Sampaio, F. Barros, and A. O. H. C. Leduc, *Marine Pollution Bulletin*, **129(2)** 802–805 (2018).
[14] P. C. Hollman, H. Bouwmeester, and R. J. B. Peters, Occurrence And Potential Health Risks," (2013).
[15] O. W. Prakoso, "Karakterisasi Sifat Tarik Komposit Laminat Hibrid Sisal-E-Glass/Polypropylene," 2017.
[16] C. S. L. Simamora, W. Warsidah, and S. I. Nurdiansyah, *Jurnal Laut Khatulistiwa*, **2(3)** 96 (2020).
[17] A. Yunanto, N. Fitriah, and N. Widagti, *JFMR-Journal of Fisheries and Marine Research*, 5(2) (2021)
[18] R. Akhtar, M. Y. Sirwal, K. Hussain, M. A. Dar, M. Shahnawaz, and Z. Daochen, in *Impact of Plastic Waste on the Marine Biota*, Singapore: Springer Nature Singapore, 239–256. (2022)
[19] J. R. Jambeck *et al.*, *Science*, **347(6223)** 768–771, (2015)
[20] L. Noviantini, "Perlindungan Lingkungan Laut dari Limbah Plastik menurut Hukum Internasional dan Hukum Nasional Indonesia," 2022.
[21] F. Rahmadhani, "Identifikasi dan analisis kandungan mikroplastik pada ikan pelagis dan demersal serta sedimen dan air laut di perairan Pulau Mandangin Kabupaten Sampang," 2019.
[22] H. Lestari S, A. Dhamar Syakti, and F. Idris, "Kandungan Mikroplastik Pada Siput Goggong (Strombus sp.) Di Perairan Bintan Bagian," Universitas Maritime Raja Ali, 2023.
[23] L. M. L. Nollet and K. S. Siddiqi, Eds., *Analysis of Nanoplastics and Microplastics in Food*. First edition. | Boca Raton : CRC Press, 2020. | Series: Food analysis and properties: CRC Press, 2020.
[24] M. Lee and H. Kim, *Nanomaterials*, **12(5)** 851 (2022).
[25] M. Mardiyana and A. Kristiningsih, *Jurnal Pengendalian Pencemaran Lingkungan (JPPL)*, **2(1)**, 29–36, (2020)
[26] S. W. Kim, W. R. Waldman, T.-Y. Kim, and M. C. Rillig, *Environmental Science & Technology*, **54(21)**, 13868–13878, (2020).
[27] F. C. Alam and M. Rachmawati, *Jurnal Presipitasi : Media Komunikasi dan Pengembangan Teknik Lingkungan*, **17(3)**, 344–352, (2020).
[28] M. S. Mauludy, A. Yunanto, and D. Yona, *Jurnal Perikanan Universitas Gadjah Mada*, **21(2)** 73, (2019).
[29] W. C. Ayuningtyas, *JFMR-Journal of Fisheries and Marine Research*, **3(1)** 41–45, (2019).
[30] F. Wulan Sari, *Jurnal Jeumpa*, **8(2)** 558–564, (2021).
[31] A. R. A. Lima, M. Barletta, and M. F. Costa, *Estuarine, Coastal and Shelf Science*, **165** 213–225, (2015).
[32] M. Claessens, L. Van Cauwenberghe, M. B. Vandegehuchte, and C. R. Janssen, *Marine Pollution Bulletin*, **70(1–2)** 227–233, (2013).
[33] M. A. Browne, T. S. Galloway, and R. C. Thompson, *Environmental Science & Technology*, **44(9)** 3404–3409 (2010)
[34] A. D. Syakti, *Marine Pollution Bulletin*, **122(1–2)** 217–225 (2017)
[35] M. Kazour, S. Terki, K. Rabhi, S. Jemaa, G. Khalaf, and R. Amara, *Marine Pollution Bulletin*, **146** 608–618 (2019)
[36] K. H. Hanif, J. Suprijanto, and I. Pratikto, *Journal of Marine Research*, **10(1)**, 1–6, (2021)
[37] L. Tosetto, J. E. Williamson, and C. Brown, **123** 159–167 (2017)
[38] V. T. K. Khuyen, D. V. Le, A. R. Fischer, and C. Dornack, *Global Challenges*, **5(11)** (2021)
[39] I. Permatasari, I. Dewiyanti, S. Purnawan, S. M. Yuni, M. Irham, and I. Setiawan, "The correlation between mangrove density and suspended sediment transport in Lamreh Estuary,

Mesjid Raya Subdistrict, Aceh Besar, Indonesia," *IOP Conference Series: Earth and Environmental Science*, vol. 216, p. 012004, Dec. 2018, doi: 10.1088/1755-1315/216/1/012004.
[40] S.A. Akbar, S. Afriani, C. Nuzlia, S. Nazlia, S. Agustina. (2023). *Depik* **12(3)** 259-273 (2023)

A numerical model approach of wave dynamic

Muhammad Irham[1,2,3,*], *Yopi Ilhamsyah*[1], *Edy Miswar*[1,2,3,4]
[1] Department of Marine Science, Faculty of Marine and Fisheries, Universitas Syiah Kuala University, 2311, Indonesia
[2] Center for Environmental and Natural Resources Research, Universitas Syiah Kuala
[3] Geographic Information Systems Laboratory, Faculty of Marine and Fisheries Science, Universitas Syiah Kuala
[4] Department of Fisheries Resources Utilization, Faculty of Marine and Fisheries, Universitas Syiah Kuala, Banda Aceh, 2311, Indonesia

Abstract. The numerical model of wave field distribution using time dependent of wave combination from refraction-diffraction equation on a gentle slope was carried out in Modeling Laboratory of Marine and Fisheries Faculty of Syiah Kuala University. The research of numerical modeling was conducted to understand how the impact of the presence of coastal structures on the coast toward the wave dynamic approach. In the theory, the greater of wave angle, the greater of wave height distribution will be behind the breakwater, thus allowing large sediment transport to make changes to the coast. Therefore, this research is important because the knowledge of the relationship between waves and coastal structure is still limited while the dynamics of the coast have a distinctive significance in the regions. The study aims to determine the wave phenomena that interact with the structure of the coastal building so that the results of this study can be used and utilized as substantial study for future coastal planning. The method used in this study is the application of wave equations that may be solved by numerical model. The model results show that the greater the angle of incidence of waves on the southwest coast, the greater the wave height produced when it hits coastal structures. However, this characteristic does not occur in the west coast region. This is caused by the topographic nature of coastal areas and different coastal characteristics.

1 Introduction

Coastal areas have a significant role in human life, such as the use of marine reserves as a source of food, medicine, transportation, and recreation [1]–[4]. Coastal waters are also vulnerable to various human activities through extensive tourism activities, fishing, ports, [4] and industrial estate development [5]. As a result, industrial development which is directly proportional to the increase in pollutants from time to time causes physical and ecological changes in coastal areas and their surroundings [2]. It is important to study the interactions between the coast and the ocean to understand these changes. A beach usually has significant

* Corresponding author: iham@usk.ac.id

wave characteristics due to non-linear interactions in shallow water depths [6]. Waves trigger the transport, dispersion, and vertical mixing of seawater such as fronts and particle transport [7]. Strong wave-induced currents can cause erosion of seafloor sediments and maintain the shape of the sediment in suspended particles [8], [9]. However, waves and currents vary across the coasts of the world depending on their interactions with coastal geometry, bedtime bathymetry, and wave superposition [10].

The coastal area is unique because of the geographical location of the ocean [11]. In addition, this area is also strongly influenced by the topography of the area around the coast [12]. The topography of the mountainous areas bordering the sea is quite significantly different from the coasts in the lowland valleys and basins [13]. These factors are characteristics that affect the nature of wave propagation and its interaction with coastal waters [14]. The wave propagation interacts with coastal topography having unique wave dynamics [15]. The presence of shoaling is caused by the bathymetric profile of the beach towards the coastal area causing diffraction and refraction of the distribution of waves which causes the supply of energy through divergence due to the influence of forces acting on the coastal topography.

Studies on wave propagation in coastal areas are still lacking, especially in the coastal areas of Aceh. There have been several studies of waves on the coast of Aceh which have been reported in several articles. [11] reported that there were coastal waves that triggered longshore currents around Lhok Nga beach. [16]–[18] reported that seasons determine the flow and transportation of suspended sediments in the northern waters of Aceh. [15] reported that in the waters of West Aceh the influence of wind on currents is quite small, especially during the transition of seasons. Coastal waters have a rough bottom topography and a varied coastline. This condition causes the influence of waves on current circulation to be relatively complex compared to open water [19]. Considering that studies on the propagation of waves are still of minimal importance while an understanding of waves can be useful for environmental mitigation, especially for waters that are busy with industrial activities and waste [13], this study can be used as a solution to answer this challenge.

In line with efforts to utilize marine resources, while still paying attention to environmental sustainability and carrying capacity, knowledge, and technology and the meaning of marine research need to be increased. The fact that research on the sea in the last few decades has been carried out more to support the utilization and conservation of the sea and coastal areas. To fulfil this goal, several studies in coastal areas have focused more on the ecological role of bio-physical-chemical properties where the three subjects are interrelated. On the marine physics side, the interaction of hydro-oceanography with the coast is one of the most important studies to be carried out, especially regarding the interaction and propagation of waves along the coast to see the nature and characteristics of the dynamics that occur. Therefore, the study of wave propagation on the coast is an important focus in this study to find out how wave propagation and wave trajectories in this area react to aspects of coastal geomorphology and hydro-oceanographic dynamics.

2 Material and Methods

2.1 Numerical Method

The study used numerical method based on equations of shoaling and refraction of wave, which have been established by [11], [15], while the input data is obtained from field measurements made during high and low tide.

The numerical wave equations solution will be solved by discretizing the equation of wave (12) which comprises the coefficients of shoaling and refraction from (13), (14), and

(15). Based on wave equation, the wave height can also be calculated by applying the equation below [19].

$$H = H_0 K_s K_r \quad (1)$$

In which H and H_0 are wave and initial wave height respectively, K_s shoaling coefficient, and K_r is refraction coefficient.

2.2 Discretization of Equations

The Applying the central difference approach, the equation (13) can be obtained as:

$$\beta^{n+1} = \frac{(p_t \Delta t - 2)\beta^{n-1} + (4 - 2q_t \Delta t^2)\beta^n}{(2 + p_t)} \quad (18)$$

with

$$p_t = -\frac{\partial c^n}{\partial x} \cos\theta^n + \frac{\partial c^n}{\partial y} Sin\theta^n \quad (19)$$

$$q_t = c^n \left(\frac{\partial^2 c^n}{\partial x^2} \sin^2\theta^n - \frac{\partial^2 c^n}{\partial x \partial y} \sin 2\theta^n + \frac{\partial^2 c^n}{\partial y^2} \cos^2\theta^n \right) \quad (20)$$

While the governing equation used as numerical approach is the wave form Equation (12):

$$\theta^{n+1} - \theta^n = \frac{\Delta s}{c^n} \left(\sin\theta^n \frac{dc^n}{dx} - \cos\theta^n \frac{dc^n}{dy} \right) \quad (21)$$

where $$\frac{dc}{dx} = \frac{c_{i+1,j} - c_{i-1,j}}{\Delta x_i + \Delta x_{i+1}} \text{ and } \frac{dc}{dy} = \frac{c_{i,j+1} - c_{i,j-1}}{\Delta y_j + \Delta y_{j+1}} \quad (22)$$

Here, the tangential component of the wave path in equation (10 and 11) will use the position of the new wave trajectory point which is discretized as:

$$x^{n+1} = x^n + \Delta s \cos\left(\frac{\theta^{n+1} + \theta^n}{s}\right) ds \quad (23)$$

$$y^{n+1} = x^n + \Delta s \sin\left(\frac{\theta^{n+1} + \theta^n}{s}\right) ds \quad (24)$$

where $\Delta s = \Delta t \,.\, c^n$.
with point of x^n, y^n, x^{n+1}, and y^{n+1} *do* not have to have a grid value.

To attain the distribution of the wave trajectory multiply (23 and 24) by interpolating the points in a grid (**Figure 3**). The interpolation formulation used is

$$c^n = c_{i,j}(\zeta - 1)(\eta - 1) - c_{i+1,j}(\zeta)(\eta - 1) - c_{i+1,j+1}(\zeta)(\eta) + c_{i,j+1}(\zeta - 1)(\eta). \quad (25)$$

Where x^n, y^n, x^{n+1}, and y^{n+1} are the positions of x and y at n and $n+1$ time step respectively, β^{n-1}, β^n and β^{n+1} are orthogonal coefficients at β at n-1, n, and $n+1$ time step, θ^n and θ^{n+1} are wave angle at n and $n+1$ time step, i, $i+1$, i-1, and j, $j+1$, j-1 are distance position on grid x and y at distance step, Δx, Δy, Δt, and Δs are the grid spaces of x, y, t, and ζ and η the elevation factor toward x and y grid called staggered mesh scheme (SMS) as shown in **Figure 1** below.

https://doi.org/10.1051/bioconf/20248702018

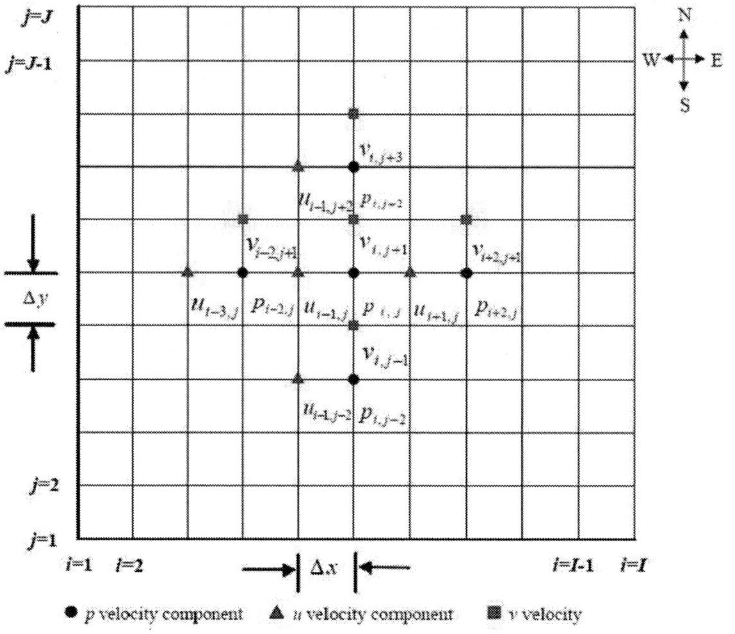

Fig. 1. Staggered Mesh Scheme.

2.3 Model Desaign

The numerical model of ocean waves is built on a gentle slope equation, the input of wave parameters such as incoming wave height (Ho), period (T), wave angle of arrival and the bathymetry. While the program results are in the form of wavelength (L), phase speed (C), wave group speed (Cg), wave number (k), refraction coefficient (Kr), trajectory position (x_x, x_y, y_x, y_y), elevation (ζ), and wave height (H).

In the wave field distribution model is done by using equation (9), where in the simulation, the effect of a wave break is calculated by its energy dissipation factor (fD). The fD energy dissipation factor is first to set zero (0) then given a value when the wave enters the break area, the size depends on the breaking wave coefficient. When waves reach the beach and coastal structure, the reflectivity is applied as the reflection coefficient (Kr). The higher the Kr, the higher the reflected wave in the coastal boundary and coastal structure.

3 Result and Discussion

3.1 Bathymetry Digitization Results

Based on the results of the bathymetric analysis carried out, the deepest depth data obtained in the study in Gampong Suak Puntong, Kuala Pesisir sub-district, Nagan Raya Regency is 18 meters, which is located between Suak Puntong Village, Nagan Raya Regency and Peunagan Cut Ujong Village, West Aceh Regency. The Suak Puntong area is located at the top of the bathymetry image, while at the bottom right is the land of the Suak Puntong gampong.

In the case of this research, a bathymetric survey needs to be carried out to obtain depth data and visualization of the contour or topography of the bottom of the waters. The bathymetric survey includes the process of depicting the bottom of the waters starting from measurement, processing, to visualization of the bottom of the waters.

3.2 Wave Propagation in Suak Puntong

The wave distribution model modeled in this study is the wave model in Suak Puntong by carrying out 2 scenarios of different angles. The wave propagation model was carried out with 2 different scenarios, namely from the direction of the incident wave angle of 180°, 202.5°.

The wave distribution model modeled in this case study is the wave distribution model in the Nagan Raya case study. Other parameters used are the direction of the incident wave ($\alpha 0$) = 180° in meteorological coordinates, period T = 15 seconds, height of the incident wave (H0) = 1.7 m. The results obtained show that the maximum wave height reached 3.5 m which occurred around the jetty.

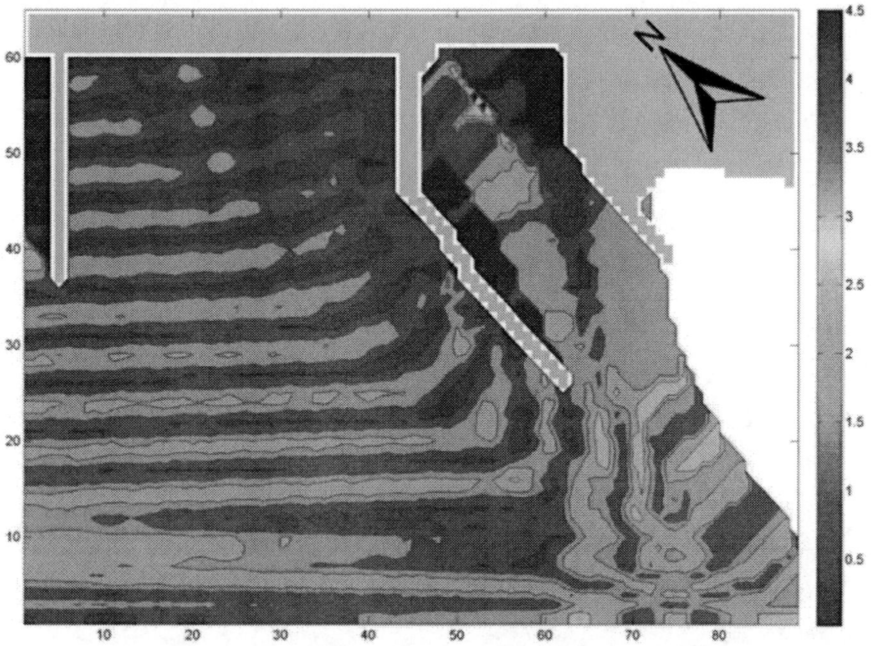

Fig. 2. Wave Propagation from an Incident Wave Angle of 180°

The results obtained from the simulation of the wave ray propagation model in this case study were given a value of Kd (diffraction coefficient) of 0.5, and produced the image shown in **Figure 2** and wave ray profile shown in **Figure 3**. Another result shown for comparison is by looking at the wave height thinning beam profile. The point taken is at x = 50 by looking at x = 50 which experiences a deflection of waves towards the south due to the jetty and around the jetty the wave rays gather or accumulate so that the resulting wave energy is high.

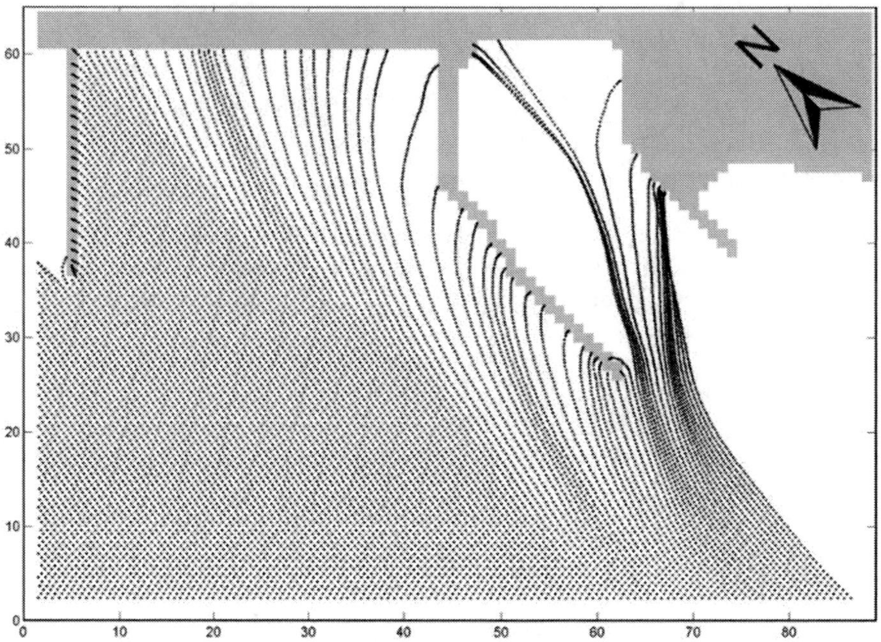

Fig. 3. Wave rays from an incident wave angle of 180°

The wave distribution model modeled in this case study is the wave distribution model in the Nagan Raya case study. Another parameter used is the direction of the wave arrival (α0) = 202.5° in meteorological coordinates, period T = 15 seconds. The results obtained show that the maximum wave height reaches 4 m which occurs around the jetty.

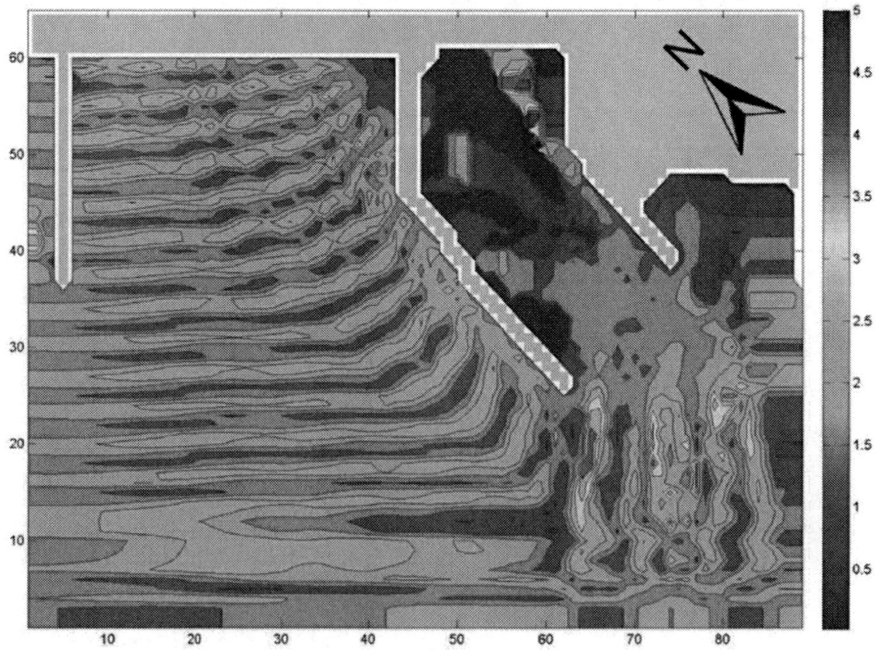

Fig. 4. Wave Propagation from an Incident Wave Angle of 202.5°

The results obtained from the simulation of the wave ray propagation model in this case study were given a value of Kd (diffraction coefficient) of 0.5, and produced the image shown in **Figure 4** and wave ray profile shown in **Figure 5**. Another result shown for comparison is by looking at the wave height thinning beam profile. The point taken is at x = 50 by looking at x = 50 which experiences a deflection of waves towards the west due to the presence of a harbor and around the harbor the wave rays separate so that the wave energy produced is low.

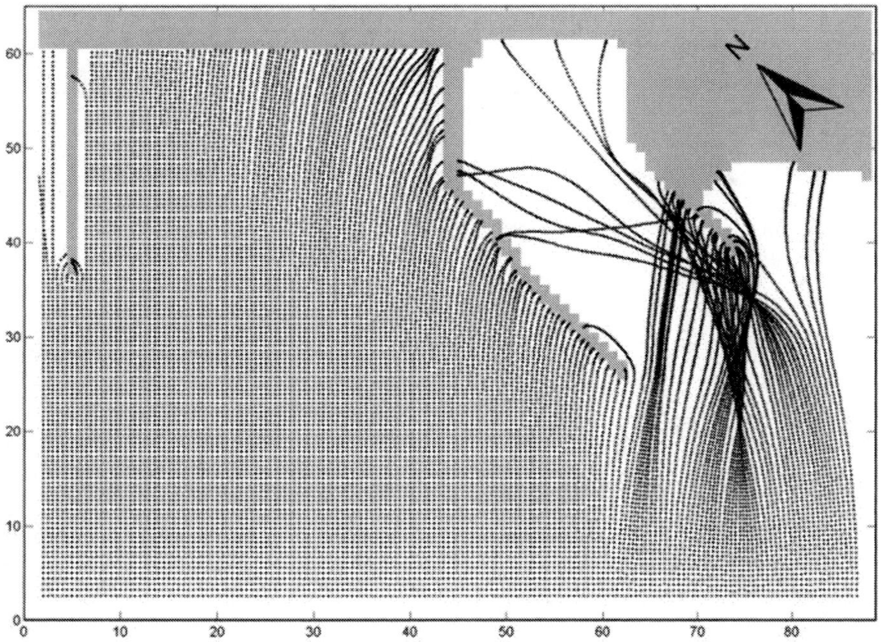

Fig. 5. Wave Rays from an Incident Wave Angle of 202.5°

The results of the wave height propagation pattern show forms of refraction-diffraction equations. The highest waves occur in converging wave rays because they are close to the direction of the wave's incident angle and have a wave barrier in the form of a jetty. This undergoes a diffraction process so that the wave height is very high. This wave model is designed with an incident wave height of 2 m. This incoming wave height is taken from the average significant wave height from WaveWatch III (WW3) Global Wave Model data. Then the wave period value is also obtained from the average of the data. The wave height and period also indicate the stability of the waves in these waters.

In scenario I, wave height (m) with input of incident wave height (H0) = 1.7 m, Tp = 15 seconds, direction of the incident wave angle ($\alpha 0$) = 180° (south). Where in this scenario a refraction process occurs because the wave bends towards the right. The wave rays in this scenario are directed away from each other and indicate a divergent zone. The wave height at an angle α = 180° shows that the highest wave profile with a height of up to 3.5 m occurs around the jetty. Scenario II, wave height (m) with input wave height (H0) = 1.7 m, Tp = 15 seconds, direction of the angle of incidence of the wave

3.3 Flow Model Result

The results of the flow distribution modeled in this study are currents caused by waves where these currents are greatly influenced by the size of the wave and the direction of the wave arrival. The direction of wave arrival is taken from the extreme direction of arrival towards the harbor at high tide and in the west season. The results of data collection in the field show that the offshore incoming wave height is 1.7 meters with a period of 15 seconds from the southwest.

Judging from the current distribution conditions in the area around the port, it can be concluded that the current rotation is strong in the estuary area before entering the port pool, the current is concentrated and rotates at two points in the port pool (**Figure 6**). This occurs as a result of the depth (bathymetry) in the harbor pool being unequal (in one area being deep and in another area being shallow) thus causing instability in the current distribution.

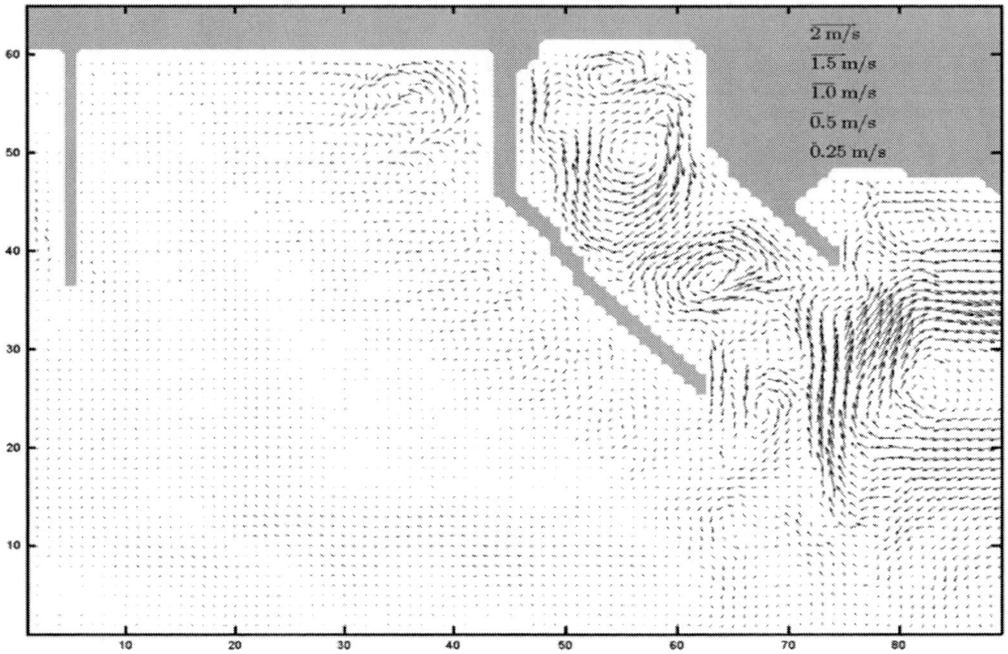

Fig. 6. Distribution of currents caused by waves in the study area.

4 Conclusion

Based on the results of research on wave propagation dynamics models in case studies in Gampong Suak Puntong, Kuala Pesisir District, Nagan Raya Regency, it shows that in scenario I, with the direction of the wave incident angle (α0) = 180°, H0 = 1.7 m, Tp = 15 seconds. The bending waves separate from each other, which indicates a divergence zone and undergo a refraction process where the maximum wave height produced reaches 3.5 m. scenario II, with the direction of the incident wave angle (α0) = 202.5°, H0 = 1.7 m, Tp = 15 seconds. The deflected waves overlap, which indicates a convergence zone and undergo a diffraction process where the maximum wave height produced reaches 4 m. So, in the two wave scenarios taken, there are wave refraction and diffraction phenomena.

References

[1] P. Nova, A. Pimenta-Martins, J. Laranjeira Silva, A. M. Silva, A. M. Gomes, and A. C. Freitas, *Crit. Rev. Food Sci. Nutr* **60(21)** 3680–3692 (2020)
[2] M. Vikas and G. S. Dwarakish, *Aquat. Procedia*, **4** 381–388 (2015)
[3] T. R. Walker *et al.*, "Environmental Effects of Marine Transportation," in *World Seas: An Environmental Evaluation*, Elsevier, 2019, pp. 505–530.

[4] N. I. Kadagi, N. Wambiji, S. T. Fennessy, M. S. Allen, and R. N. M. Ahrens, *Mar. Policy*, **124** 104351 (2021)
[5] S. Wang, B. Lu, and K. Yin, *Mar. Policy*, **130** 104553 (2021)
[6] S. C. Mohapatra, R. B. Fonseca, and C. Guedes Soares, *J. Coast. Res* **344** 928–938 (2018)
[7] D. Justić *et al.*, *Estuaries and Coasts* **45(3)** 621–657 (2022)
[8] H. Chen, X. Liu, and Q.-P. Zou, *Adv. Water Resour* **123** 160–172 (2019)
[9] A. Mahmoodi, M. A. Lashteh Neshaei, A. Mansouri, and M. Shafai Bejestan, *J. Mar. Sci. Eng.*, **8(4)** 284 (2020)
[10] R. Almar *et al.*, **12(1)** 3775 (2021)
[11] M. Irham and I. Setiawan, *Omni-Akuatika*, **13(1)** (2017)
[12] S.-I. Kang, W.-H. Ryang, and S.-S. Chun, *J. Korean earth Sci. Soc* **36(6)** 533–542 (2015)
[13] E. R. Gomes, R. P. Mulligan, K. L. Brodie, and J. E. McNinch, *Coast. Eng* **116** 180–194 (2016)
[14] V. A. Squire, *Philos. Trans. R. Soc. A Math. Phys. Eng. Sci.*, **376(2129)** 20170342, (2018)
[15] I. Setiawan and M. Irham, *Civ. Eng. Dimens.*, **20(1)** 30–34 2018.
[16] M. Irham, E. Miswar, Y. Ilhamsyah, and I. Setiawan, *IOP Conf. Ser. Mater. Sci. Eng.*, **352** 012043, (2018)
[17] I. Setiawan, S. M. Yuni, S. Purnawan, Y. Ilhamsyah, and R. Wafdan, *IOP Conf. Ser. Earth Environ. Sci* **284(1)** 012026 (2019)
[18] I. Setiawan *et al.*, *IOP Conf. Ser. Earth Environ. Sci* **176** 012016 (2018)
[19] H. Zhang, W. Cheng, X. Qiu, X. Feng, and W. Gong, *Cont. Shelf Res* **142** 32–49 (2017)

An ecosystem-based fisheries assessment for some coral fishes in the coastal Seram Laut Island, Maluku

Tirtadanu[1*], *Puput Fitri* Rachmawati[2], Prihatiningsih[1], *Reny* Puspasari[1], *Masayu Rahmia Anwar* Putri[2], *Muhammad* Taufik[1], *Karsono* Wagiyo[1], *Andrias S.* Samusamu[2], Mujiyanto[2], and *Ofri* Johan[2]

[1]Research Center for Fishery, National Research and Innovation Agency (BRIN), Address Jalan Raya Bogor km. 47, Nanggewer Mekar, Cibinong 16915, Indonesia
[2]Research Center for Conservation of Marine Resources and Inland Waters, National Research and Innovation Agency (BRIN), Address Jalan Raya Bogor km. 47, Nanggewer Mekar, Cibinong 16915, Indonesia

Abstract. As the Seram Laut Island (SLI) community relies on coral fish resources, changes in coral fish populations will impact their income and nutritional needs. This research aimed to investigate the SLI ecosystem's status using risk score indicators and the management implications for the sustainability of the coral fish fishery ecosystem. This research used the tier 2 approach semiquantitative ecosystem-based fishery assessment (EBFA) method based on a risk score of 0 to 3, with four objectives: sustainability, habitat, biodiversity, and socio-economy. Several high-risk scores in the red zone were found in the fishing efforts indicator of *Lethrinus obsoletus*, *Siganus canaliculatus*, and *Cephalopolis miniata*, catch per unit effort (cpue) indicator of *S. canaliculatus*, maturity proportion, and fishing gear impact of *L. obsoletus* and *S. canaliculatus* and habitat quality indicator of *S. canaliculatus*. Coral fish's species risk index (SRI) ranged from 1.28 to 1.64 (yellow zone), with *L. obsoletus* having the highest SRI. The coastal of SLI's ecosystem risk index was in the limited range (yellow zone). For ecosystem sustainability, proper management should be developed with a focus on the indicators in the high-risk state (red zone) by spatial planning for coral and habitat conservation, as well as policies based on local wisdom.

* Corresponding author: tirtadanu91@gmail.com

1 Introduction

Coral fish fisheries have significantly contributed to the primary source of income and nutrition for communities in Seram Laut Island (SLI), Maluku. Due to the firm reliance of the SLI community on coral fish resources, changes in coral fish populations will significantly impact their needs. Various factors, including overfishing, destructive fishing gear, and pollution, can affect the coral fish population. Various fish species depend on the coral reef ecosystem as their permanent habitat, nursery ground, and foraging habitat. Hence degradation of their habitat will reduce biodiversity and production [1]. Furthermore, bottom gillnets are a common type of fishing gear in Seram seas, and its fishing methods are linked to habitat destruction due to net snagging on corals, such as *Acropora* sp. [2].

Coral fish are highly vulnerable to population declines due to their high economic values, limited distribution, and reliance on habitat. Several studies have reported declines in coral fish populations due to coral reef degradation and over-exploitation, including in Papua New Guinea and Solomon waters [3-4]. Communities in SLI are highly dependent on coral fish production, so a decrease in its population can cause various problems, including economic decline, lack of nutritional food, and the collapse of the fisheries industry. Therefore, an ecosystem-based fisheries assessment (EBFA) is essential to develop proper management for ecosystem sustainability. This research integrated four objectives, including stock sustainability, habitat, biodiversity, and socioeconomic, in formulating ecosystem-based fisheries management in the coastal SLI.

Ecosystem-based management is essential for adequately managing some coral fish since habitat conditions, fisheries, and socioeconomic factors strongly influence their population. Some countries, including the United States, Australia, and Korea Waters, have performed ecosystem-based fisheries assessments to manage their fisheries [5-7]. Many fishers in SLI have captured the coral fish for over thirty years, but there is still a lack of data and studies on its fisheries. Due to the high dependence of the community on coral fish production, Ecosystem-based assessment for coral fish fisheries in the coastal SLI is essential to develop proper management.

Research that integrates aspects of fisheries with their habitats will provide a good understanding of appropriate management decisions [8]. It is challenging to undertake quantitative studies due to the paucity of time-series data on fisheries production, effort, and habitat conditions. For this reason, this research was conducted to develop a management model combining different fishery aspects using limited data. This work is crucial for monitoring fisheries resources in SLI and implementing management recommendations to prevent fish resource declines, which lead to financial losses, the loss of supplies of food source, and the loss of livelihoods for fishers. This research aimed to investigate the ecosystem's status using risk score indicators and the management implications for the sustainability of the coral fish fishery ecosystem in the SLI coast.

2 Data and methods

2.1 Data collection

The data was collected by filling out questionnaires to 36 fishers on the coast of Seram laut Island (SLI) (Figure 1). Fishers used gillnet and handline, with the three key species including orange-striped emperor fish (*Lethrinus obsoletus*), white-spotted spine fish (*Siganus canaliculatus*), and coral grouper (*Cephalopolis miniata*). *L. obsoletus* and *S. canaliculatus*

 https://doi.org/10.1051/bioconf/20248703001

were the target species of the gillnet, whereas *C. miniata* was the target of the handline. The mesh size of the gillnet was 2.5 inches, and the hook size of the handline was 10-12.

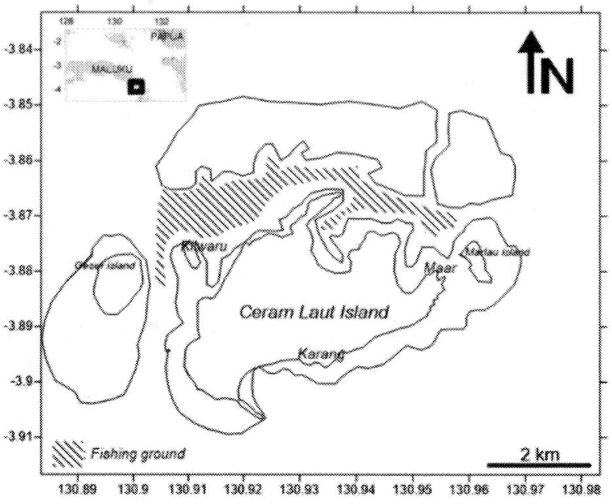

Fig. 1. The scope area for an ecosystem-based fisheries assessment for some coral fishes in the coastal of Seram Laut Island, Maluku.

2.2 Data analysis

The risk score ranges from 0 to 3, with details 0-0.99 being in the safety zone (green zone), 1-1.99 being the limited range (yellow zone), and 2-3 being the high-risk zone (red zone), applied in this study as part of a tier 2 approach semi-quantitative ecosystem-based fisheries assessment (EBFA) [7]. This study used four objectives with 12 indicators (Table 1). The four objectives include sustainability, biodiversity, habitat, and socio-economics.

The Objectives risk index was calculated based on the equation [7]:

$$\mathrm{ORI} = \frac{\sum_{i=0}^{n} I_i W_i}{\sum_{i=1}^{n} W_i} \quad (1)$$

Where ORI is the objectives risk index, I_i is the score of indicator I, Wi is the weighting factor of indicator I, and n is the number of indicators.

The species risk index was estimated for *L. obsoletus*, *S. canaliculatus*, and *C. miniata*, following the equation [7]:

$$\mathrm{SRI} = \lambda_s ORI_s + \lambda_B ORI_B + \lambda_H ORI_H + \lambda_E ORI_E \quad (2)$$

Where SRI is species risk index, λ_S, λ_B, λ_H and λ_E are the the weighting factor of sustainability, biodiversity, habitat quality and socio-economic and the ORI is objectives risk index.

The fisheries risk index (FRI) was determined for the gillnet and handline fisheries using each species' catch per unit effort. To measure the ecosystem risk index (ERI), the FRI was weighted with each fishery's total catch per unit effort (gillnet and long line). The fisheries risk index and ecosystem risk index are calculated based on the equation [7]:

$$FRI = \frac{\sum cpue(a)_i SRI_i}{cpue_i} \quad (3)$$

$$ERI = \frac{\sum cpue(b)_i FRI_i}{\sum Ci} \quad (4)$$

Where cpue (a) and cpue(b) are the catch per unit effort in each species and each fishery (gill net and hand line), respectively. FRI is the fisheries risk index, and ERI is the ecosystem risk index.

Table 3. Objectives, indicators and questionnaires of an ecosystem-based fisheries assessment for some coral fishes in the coastal Seram Laut Island, Maluku.

Objectives	indicators	Questionnaires
Sustainability	1. Cpue	1.1. Trend of cpue
	2. Fishing efforts	2.1. Trend of fishing efforts
	3. Reproduction habitat	3.1. Coral health
	4. Captured size	4.1. Proportion of immature
Biodiversity	5. Bycatch	5.1. Proportion of bycatch
	6. Diversity	6.1. Fishing gear impact on biodiversity
		6.2. Composition change
Habitat	7. Fishing gear	7.1. Fishing gear impact on habitat
	8. Habitat condition	8.1. Habitat damage
	9. Pollution rate	9.1. Oil accident
		9.2. Plastic pollution
		9.3. Household pollution
Socio-economics	10. Income	10.1. Proportion of income and expenses
		10.2. Income trend
	11. Job satisfaction	11.1. Fisher's satisfaction
	12. IUU Fishing	12.1. IUU occurred

3 Results and discussion

The proportion of indicators in *L. obsoletus* included 8.3% in the green zone, 66.7% in the yellow zone, and 25% in the red zone. The proportion of indicators in *S. canaliculatus* included 16.67% in the green zone, 50% in the yellow zone, and 33.3% in the red zone. The proportion of indicators in *C. miniata* included 8.3% in the green zone, 83.3% in the yellow zone, and 8.3% in the red zone (Table 2). Several indicators on *S. canaliculatus* showed a declining catch rate, the rising fishing pressure, destructive fishing gear, and degraded habitat conditions. Habitat destruction will disrupt the life of various commercial fish species in coastal Seram Laut Island (SLI). Furthermore, the capture of *S. canaliculatus* and *L. obsoletus* often causes damage to coral reefs due to net snagging. Habitat degradation and high fishing pressure can be the main factors of declining catch rates. Seagrasses and coral reefs play a vital role as habitats for various fish species. Seagrass ecosystems are nursery grounds for juvenile *S. canaliculatus* and will migrate to coral reef ecosystems during the adult phase [9-10]. Habitat degradation in coral reef ecosystems will lead to an increase in juvenile natural mortality, which will decrease adult fish populations. In addition, habitat degradation will also reduce small fish populations, which are prey for many commercially important fish species [11]. Habitat conservation is essential to managing coral fisheries in the coastal SLI. High fishing pressure will also lead to overfishing, decreasing fish biomass and biodiversity [12] This condition will reduce the fish catch rate, leading to economic losses. Therefore, fishing quotas can be an alternative to maintaining the sustainability of commercial fisheries in the coastal SLI.

The maturity proportion indicator of *L. obsoletus* was found in the high-risk (red zone). Additionally, observational findings revealed that *L. obsoletus* captured by the gillnet in SLI are small immature fish. According to the evidence, *L. obsoletus* nursery ground is located in the fishing area of coastal SLI. When gonad maturity has been achieved, several *Lethrinidae* fish species migrate to offshore to spawn [13]. *Lethrinidae* larvae preferred the habitat associated with seagrass and coral reefs [14]. Fishing activities in the nursery region could hinder recruitment, endangering the sustainability of *L. obsoletus*. Protecting nursery habitats can be an alternative investment for the sustainability of fisheries in SLI waters.

Among all the objectives risk index (ORI), only the sustainability ORI of *L. obsoletus* was high risk (red zone), with an ORI of 2.06 (Figure 2), caused by rising fishing pressure and a small-size capture. Estimates of sustainability's objectives risk index in the high zone were also found in the round scad fishery caught by Purse Seine in Red Sea waters with a sustainability ORI of 2.03 [15]. The operation of non-selective and destructive fishing gear can cause the high ORI of various commercially important fish. Gillnet operations in SLI are carried out in shallow waters with several habitat variations with characteristics in seagrass ecosystems, coral reefs, and sandy substrates. Mature adult *S. canaliculatus* were still found in the fisher's catch. In contrast, adult *L. obsoletus* was not found in the gillnet fishing area, indicating that *S. canaliculatus* adults have more limited movements than *L. obsoletus*. A previous study has also found that emperor fish species *Lethrinus nebulosus* have long spawning migrations compared to other coral fish species [13]. *L. obsoletus* has a relatively long lifespan of about 13 years, with gonad maturity reached at 3-4 years [16]. Fishing operations in the nursery ground areas can inhibit migration and recruitment of emperor fish. Fishing quotas for *L. obsoletus* from gillnet gear can be an alternative to maintaining coral fish' sustainability in coastal SLI.

Table 2. Risk score for all the indicators of EBFA on the orange-striped emperor (*Lethrinus obsoletus*), white-spotted spine foot (*Siganus canaliculatus*) and coral grouper (*Cephalopolis miniata*) in the coastal Seram Laut Island, Maluku.

Objectives	Indicators	Weight	*L. obsoletus*	*S. canaliculatus*	*C. miniata*
Sustainability	Catch per unit effort	3	1.81	2.17	1.97
	Fishing efforts	2	2.17	2.33	2.27
	Reproduction habitat	2	1.86	1.67	1.17
	Maturity proportion	2	2.55	0.77	1.60
Biodiversity	Bycatch	2	0.67	1.17	1.27
	Diversity	2	1.47	1.75	1.10
Habitat quality	Fishing gear impacts	3	2.11	2	1.18
	Habitat quality	2	1.56	2.17	1.97
	Pollution rate	2	1.19	1.06	1.43
Socio-economic	Income	3	1.56	1.83	1.97
	Job satisfaction	2	1.08	0.17	1.87
	IUU Fishing	2	1.36	0.83	0.53

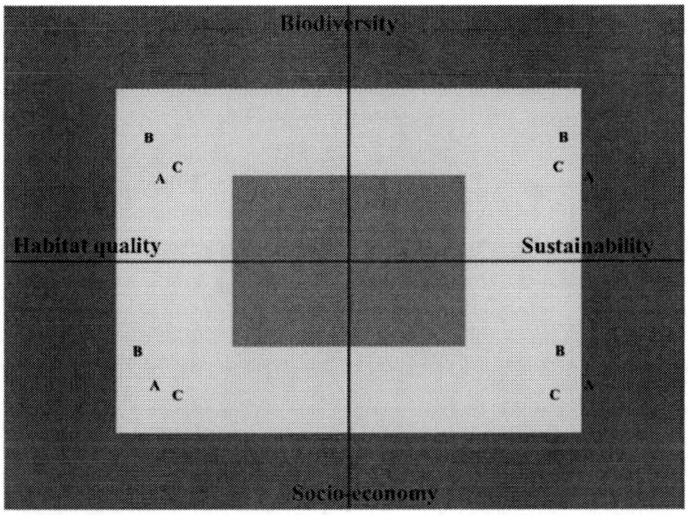

Fig. 2. Objective risk index for (A) orange-striped emperor (*Lethrinus obsoletus*), (B) white-spotted spinefoot (*Siganus canaliculatus*) and (C) coral grouper (*Cephalopolis miniata*) in the coastal Seram Laut Island, Maluku.

Species risk index (SRI) was 1.64 for *L. obsoletus*, 1.54 for *S. canaliculatus* and 1.55 for *C. miniata* (Table 3). The estimated species risk index has also been reported for several commercial species caught by trammel nets in Korean waters, ranging from 1.20 to 1.64 in 2000 [17]. *L. obsoletus* has the highest SRI due to the operation of gillnet gear, which tends to damage coral reefs, and none of the mature adult fish was found in the gillnet fishing area, while *C. miniata* has a lower SRI due to the hand line being operated in the deeper areas and having a less environmental impact than gillnets. This condition is also indicated by the gill nets' fishery risk index (FRI), which is higher than the hand line (Figure 3). Previous research has also found that hand line has less environmental impact than gill net [2]. Therefore, fishing with hand line is more recommended than gillnet for coastal SLI fisheries.

To decide what crucial management has to be done, examining each indication's risk score is essential. According to the findings of the EBFA study, the coast Seram Laut Island ecosystem risk index was in the limited range or yellow zone with a risk score of 1.60 (Figure 4). Management can focus on several indicators with risk scores in the red zone, including the declining catch rate of *S. canaliculatus*, the large number of small immature *L. obsoletus*, degradation of coral reef habitat due to the fishing operation and pollution. Some of these factors may raise the risk of the ecosystem risk index, push the risk into the danger zone and habitat losses, causing fish resource loss, which is vital to the SLI communities.

The conservation of habitat requires proper spatial planning as an investment in the long-term viability of fisheries. Implementing the Seram Laut Island fisheries policy required local community supervision and supporting local wisdom. In addition to habitat conservation, the government needs to provide adequate fish landing facilities so fishers can land and report their catches. Afterward, management based on catch quotas can be implemented for high-risk species. Total Allowable catch can be a management option to help fish stocks recover [18-20]. Specific IUU fishing issues connected to fish bombs need to be addressed strongly. Strict legislation and a close partnership between the local community and political institutions are necessary to strengthen local wisdom [21]. Ecosystem-based management is the key to achieving satisfaction from various parties, both related to the sustainability of fish resources, habitat and the economic sustainability of the community on Seram Laut Island.

Fishing communities have implemented local wisdom-based management for fisheries sustainability in several waters in Indonesia [22]. A local wisdom-based system known as

 https://doi.org/10.1051/bioconf/20248703001

"sasi" can be applied to fisheries in SLI. The management system includes the closure in some areas and spawning periods for several high-risk species. The government can strengthen these management tools by supporting research on spawning areas and seasons, fishing quotas, and optimal catch sizes. Implementing such management is expected to provide fisheries sustainability so that the community's food source and economic needs can continue to be met. To execute management that all stakeholders can agree upon, socialization of the significance of ecosystem sustainability and collaboration between the government, local communities, fishing industry, and researchers are required.

Table 3. Objectives risk index (ORI), Species risk index (SRI), Fisheries risk index (FRI) and Ecosystem risk index (ERI) for some coral fishes in the coastal Seram Laut Island, Maluku.

Fishery	Species	Objectives	ORI	SRI	FRI	ERI
Gillnet	*L. obsoletus*	Sustainability	2.06	1.64	1.62	1.60
		Biodiversity	1.07			
		Habitat	1.69			
		Socio-economy	1.37			
	S. canaliculatus	Sustainability	1.78	1.54		
		Biodiversity	1.46			
		Habitat	1.78			
		Socio-economy	1.07			
Hand Line	*C. miniata*	Sustainability	1.77	1.55	1.55	
		Biodiversity	1.18			
		Habitat	1.48			
		Socio-economy	1.53			

Fig. 3. Fisheries risk index of gill net fisheries (FRI_{GN}) and hand line fisheries (FRI_{HL}) for some coral fishes in the coastal Seram Laut Island, Maluku.

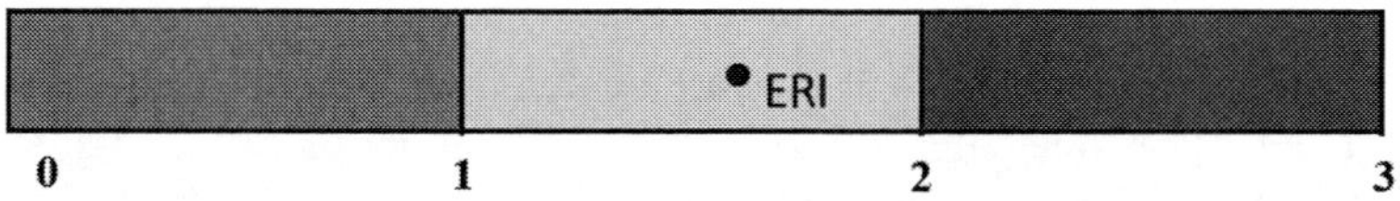

Fig. 4. Ecosystem risk index (ERI) of the coastal Seram Laut Island, Maluku.

4 Conclusion

Some high-risk scores in the red zone were found in the cpue indicator of *S. canaliculatus*, fishing efforts indicator of *L. obsoletus*, *S. canaliculatus*, and C. *miniata*, maturity proportion indicator of *L. obsoletus*, fishing gear impacts indicator of *L. obsoletus* and *S. canaliculatus* and habitat quality indicator of *S. canaliculatus*. Management can focus on reducing the risk of some of those indicators with ecosystem-based management, including the spatial planning for coral reef and habitat conservation, as well as policies based on local wisdom for the sustainability of habitat, fish resources, and the community's economy in Seram Laut Island.

Acknowledgements

The authors thank the Global Environment Facility (GEF-6) and the Ministry of Marine Affairs and Fisheries Republic of Indonesia for supporting this research financially.

References

1. D.J. Coker, S.K. Wilson, M.S. Pratchett Rev Fish Biol Fisheries 1-38 (2013)
2. A. Tupamahu, Haruna, Barbara, B.G. Hutubessy, S.R. Siahainenia, A.C. Nanlohy, K. Hehanusa 2021 AGRIKAN Jurnal Agribisnis Perikanan **14** 44-54 (2021)
3. G.P. Jones, M.I. McCormick, M. Srinivasan, J.V. Eagle *PNAS* **101** 8251-8253 (2004)
4. A.T. Hughes, R.J. Hamilton, J.H. Choat, K.L. Rhodes Plos One **15** 1-17 (2020)
5. NOAA Exploration of ecosystem-based fishery management in the United States A report from the NOAA Science Advisory Board. National Oceanic and Atmospheric Administration, US. 111 pp. (2014)
6. P. Herrón, G.A. Castellanos-Galindo, M. Stäbler, J.M. Díaz, M. Wolff Front. Mar. Sci. **6** 1-17 (2019)
7. C.I. Zhang, S. Kim, D. Gunderson, R. Marasco, J. B. Lee, H. W. Park, J.H. Lee *Fish. Res.* **100** 26-41 (2009)
8. C.M. Roberts Conservation Biology **9** 988-995 (1995)
9. H. Latuconsina, R. Affandi, M.M. Kamal, N.A. Butet Jurnal Ilmu dan Teknologi Kelautan Tropias **12** 89-106 (2020)
10. Suardi, B. Wiryawan, A.A. Taurusman, J. Santoso, M. Riyanto AACL Bioflux **9** 1220-1232 (2016)
11. S.K. Wilson, R. Fisher, M.S. Pratchett, N.A.J. Graham, N.K. Dulvy, R.A. Turner, A. Cakacaka, N.V.C. Polunin Ecological Applications **20** 442-451 (2010)
12. T.D. Eddy, V.W.Y. Lam, G. Reygondeau, A.M. Cisneros-Montemayor, K. Greer, M.L.D. Palomares, J.F. Bruno, Y. Ota, W.W.L. Cheung One Earth **4** 1278-1285 (2021)
13. R.C. Babcock, R.D. Pillans, W.A. Rochester Marine and Freshwater Research **68** 1422-1437 (2017)
14. Y. Nakamura, T. Shibuno, D. Lecchini, Y. Watanabe Aquatic biology **6** 61-65 (2009)
15. L.S. Alsolami, M. Abdelaty, C.I. Zhang Fisheries Research **227** 1-8 (2020)
16. B.M. Taylor, Z.S. Oyafuso, M.S. Trianni Ichtyological Research **64** 423-432 (2017)
17. I. Yeon, M.Y. Song, M.H. Sohn, H.J. Hwang, Y.J. Im, D.H. Kim, H.W. Park, C.I. Zhang Fisheries Research **112** 179-188 (2011)
18. L.E. Dee, S.S. Horii, D.J. Thornhill Biological Conservation **169** 225-237 (2014)
19. T.A. Militz, J. Kinch, D.S. Schoeman, P.C. Southgate Marine Policy **90** 160-167 (2018).

20. Tirtadanu, U. Chodrijah, K. Amri IOP Conf. Series: Earth and Environmental Science **1250** 1-7 (2023)
21. J.M. Monteiro Jurnal Dinamika Hukum **17** 83-88 (2017)
22. D.P Keliat, A. Amirudin, Y. Luqman E3S Web of Conferences **317** 1-12 (2021)

Standardization of ingredient and frying time for squid stick home industry

Daniel H Ndahawali[1], *Fidel* Ticoalu[1], *Itje D* Wewengkang[1], *Nova M* Tumanduk[1], *Fitroh D* Hariyoto[1*], and *Agusta P B L* Soeharso[1*]

[1]Department of Fishery Product Processing Technique, Marine and Fishery Polytechnic of Bitung, Bitung City, North Sulawesi Province, Indonesia 95526

Abstract. Squid sticks are a typically value-added product processed by the Fisherman's Wives Group in Motto village, Bitung City, Indonesia. Standardization can be used as a reference in processing and consumer protection. This study aims to obtain a standard proportion of salt and optimal frying time through hedonic tests, and obtain information on the chemical content of squid stick products. Based on the respondent's responses, it was found that the standard recipe based on the total weight ingredients was 54.69% wheat flour, 19.53% squid, 11.72% garlic, 7.81% sugar cane, 1.95 % sea salt, 1.95% butter, 1.95% chicken eggs, 0.20% ginger and 0.20% emulsifier. While the processing standard, namely frying at 120 °C for 60 seconds using cooking oil. The chemical content of the product is 12.53% protein, 14.89% fat, 5.84% water content, and 65.60% carbohydrates.

1 Introduction

Indonesia is the largest archipelago in the world with 62% of the area is ocean, wich making Indonesia as a country with high marine resource potential. The potential of marine resources in the form of fishery products in Indonesia is not only fish commodities, but also a variety of other commodities such as squid, cuttlefish, and octopus that belongs to cephalopods. With their adaptability, cephalopods seize any chance to thrive in optimal environments [1,2]. As a part of the cephalopods, there are around 400 species of squid, all of which are marine and may be found in all sea depths from the Arctic to the Antarctic [3]. Squid has been recognized as a popular food in today's society because of high protein content [4]. Due to its exceptional quality and delectable flavor, squid is a highly prized commodity with substantial economic value. Capture fisheries provide the potential for extensive harvesting of this resource throughout almost all waters within the Indonesian Fisheries Management Area (WPPNRI) [5]. All parts of the squid, including the mantle, tentacles, ovaries, ink, and visceral organs, can be use as food ingredients. The mantle alone can be consumed raw, boiled, or transformed into various food products like surimi and dried-seasoned squid. Furthermore, the skin serves as a source for cosmetic products. Moreover, melanin-free ink, serine protease inhibitors extracted from the ovaries, and chitooligosaccharide derived from the pen are valuable

* Corresponding author: fitroh.hariyoto@kkp.go.id ; agusta.soeharso@kkp.go.id

functional ingredients that improve the texture and properties of surimi, both dark and white-fleshed varieties [6].

Squid can also be found and is one of the non-fish commodities in the North Sulawesi region, especially in Bitung City. Motto Village, which is one of the villages in Bitung City, precisely in the Lembeh Island area, is a fishing village with squid as the main catch, to be utilized directly or as a product by the fishermen's wives group [7]. One of the efforts to utilize the perishable nature of squid is by processing it into dried products. Dried squid's popularity in China as a versatile snack, side dish, or refreshment is mirrored in its price, which can reach a staggering 497 RMB per kilogram, demonstrating its significant market demand [8]. The dry product in the form of squid sticks which in the production process have been successfully commercialized in Motto Village, become a source of additional income for fisherman's families [7]. The fishermen's wives process squid as a value-added product by making it into squid sticks because squid is classified as a food commodity that is prone to deterioration and eventually rot if not stored frozen.

Several studies conducted on processed dried squid in the form of squid-laver snacks [9] and squid crackers [10] showed the addition of 0.51% salt in squid-laver snacks and the standard proportion of salt in processed squid crackers has not been found. These two studies illustrate that the addition of salt plays an important role in the processing of squid sticks and is also a response to the organoleptic acceptance of processed squid products. Squid sticks produced by a group of fishermen's wives in the Motto Village showed good acceptance assessed through sensory testing involving 30 panelists [11]. Furthermore, as a squid-proecessing product, the making process of squid stick carried out by fishermen's wives is still traditional, which is only done by frying as usual using oil in a simple way, without using automated equipment. The production process is carried out manually by workers and there is no clear standard for determining the amount of ingredients and the production process. [11]. The production process, which is still traditional and yet to be standardized, makes the products produced more vary in shape, color, texture and size [12].

Standardization is necessary to ensure that the raw materials used and the processing process are the same, even if they are processed at different times. Traditional standardization methods hinder the possibility of improvement because they consider achieving standards as the primary goal as if they were the highest and optimal level of performance. The lean method views standardization as the basis for improvement, and is expected to develop standards continuously. Standardization requires information on processing procedures and the composition of ingredients in processed products. There is no information regarding the standardization of the squid stick production process in the fishermen's wives in Motto Village, so the purpose of this study is to obtain the optimal standard proportion of salt and frying time, and obtain information on the chemical content of squid stick products for commercial and regulations.

2 Materials and Methods

2.1 Squid stick production

The process of making squid stick products was carried out at the Marine and Fisheries Polytechnic of Bitung, at Processing Laboratory, based on the initial recipe from the previous research [11]. The process of making squid stick products can be seen in Figure 1, starting from the preparation of 500 grams of raw squid type *Sthenoteuthis oualaniensis* which has previously been cleaned and milled. Then prepare the addition of other ingredients, namely 1400 grams of wheat flour, 300 grams of garlic, 200 grams of sugar cane, 50 grams of butter,

50 grams of chicken eggs, 5 grams of ginger, 5 grams of emulsifier and a certain amount of salt as a variable. The ingredient formula as shown in table 1.

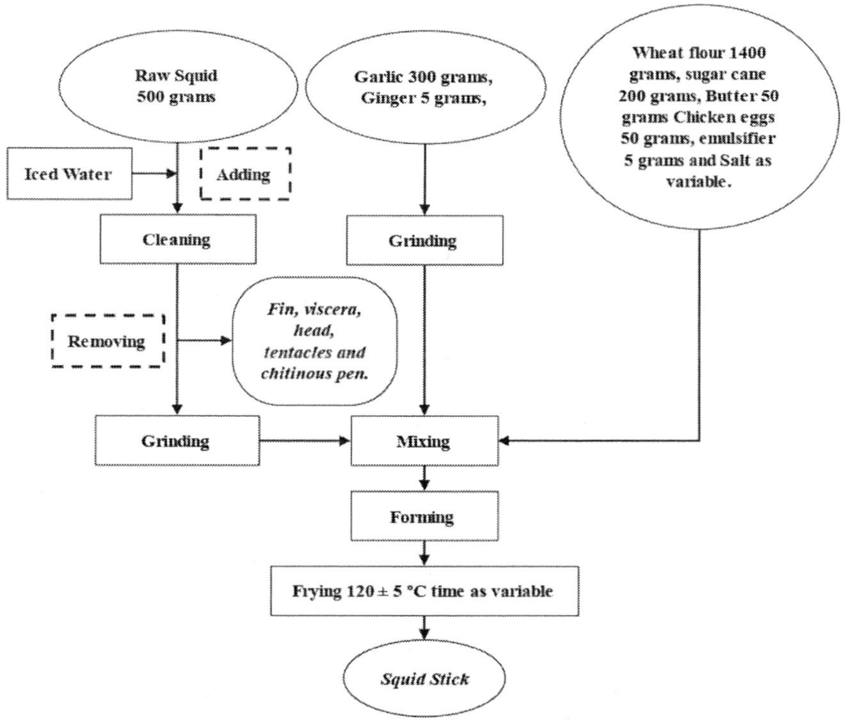

Fig. 1. Procedure for making squid stick products

Table 1. Trial formula for making squid sticks.

Ingredient	Formula (grams)
Wheat flour	1400
Squid	500
Garlic	300
Sugar cane	200
Salt	as variable
Butter	50
Chicken Egg	50
Ginger	5
Emulsifier	5

Garlic and ginger are grined and mixed with finely ground squid, flour, sugar, butter, emulsifier, and salt according to the formula. Product variations were divided into three formulas, with two different frying times to obtain six samples (S1-S6). The difference in squid stick formulation lies in the addition of salt percentage in three formulas (10%, 20%, 30%) based on total weight of squid raw materials used. The next process is dough molding and frying using palm oil at a temperature of 120 ± 5 °C in a predetermined time of 45 and 60 seconds. The fried product then removed from the frying pan and drained. Table 2 shows the experimental design of the process.

Table 2. Trial formula design.

Formula	Salt (grams)	Frying time (second)
Formula 1	50	60
Formula 2	50	45
Formula 3	100	60
Formula 4	100	45
Formula 5	150	60
Formula 6	150	45

2.2 Sensory analysis

The hedonic method was used to test the acceptability of squid sticks through panelist assessment based on Indonesian Regulations for organoleptic and sensory testing Instructions (SNI 012346-2006) [13]. The liking level of squid sticks was analyzed under laboratory conditions (in individual cubicles), by presenting 5-7 cm samples to 30 panelists including healty student aged 17- 22. Samples were randomly coded with three digits, and the sample order was presented in 6 units on white transparent plastic. Acceptance level was determined in terms of appearance, texture, flavor, and overall rating. The scale used is a number from 1 to 6, where 1 = very dislike, 2 = dislike, 3 = normal, 4 = somewhat like, 5 = like, and 6 = very like. The sample with the highest acceptance was in the first position (1st rank), while the least affection was in the last (sixth) position (6th rank).

2.3 Chemical analysis

Proximate analysis was conducted to determine the content value of chemical properties adapted from research on proximate test results on dried squid preparations in the form of squid crackers [10]. The squid stick formulation tested was the one that ranked first (very liked) based on the panelists' affection with the method of the test can be seen on table 3. The purpose of this test was to determine the effect of formula composition and time variation determined as standardized squid stick formulation parameters. Proximate testing was conducted at Saraswati Indo Genetech Laboratory Service, Bogor, Indonesia.

Table 3. Chemical analysis parameter references.

Parameter	Method	References
Moisture content	Oven	SNI 01-2891-1992
Ash content	Gravimetry	SNI 3751:2018
Fat	Soxhlet	SNI 01-2891-1992
Protein	Kjeldahl	AOAC 960.52
Carbohydrate	By difference	SNI 01-2891-1992

2.4 Statitical analysis

Data processing of the results of the panelists' favorability assessment was carried out using statistical analysis to determine which formula had the best acceptability using the hedonic method; with Friedmant-test criteria used to determine the effect of the variable proportion of salt and frying time on the expansion ratio and specific volume of snacks ($p < 0.05$) with SPSS v.26 software.

3 Results and Discussion

3.1 Squid stick characteristic

Traditional processing technologies for water-based products involve the use and increase of temperature, which aims to reduce the water content of the product, resulting in a rough texture [14]. The production process of squid sticks with raw squid and wheat flour at high temperature will result in denaturation of squid protein, evaporation of water content in the ingredients, and gelatinization on the surface of starch components in the flour, causing high crispness value in products processed using the palm oil frying method [15].

All squid stick samples (S1-S6) had a brownish yellow to brown color, which was influenced by the length of frying time. Color is a parameter that directly affects the processing process through heating. Squid sticks fried for 60 seconds visually had a browner color. In the same formula, different frying times will affect the color appearance of food products [15]. Foods processed using oil-frying methods have a distinctive aroma, and the increase in TBARs (Thiobarbituric Acid Reactive substances) levels in foods using deep-frying processing methods is related to the production of pyrazine aromatic compounds. The moisture content change in the frying process affect the maillard reaction and caramelization process, resulting in discoloration of squid sticks during heat treatment from the degree of water loss and oil absorption.

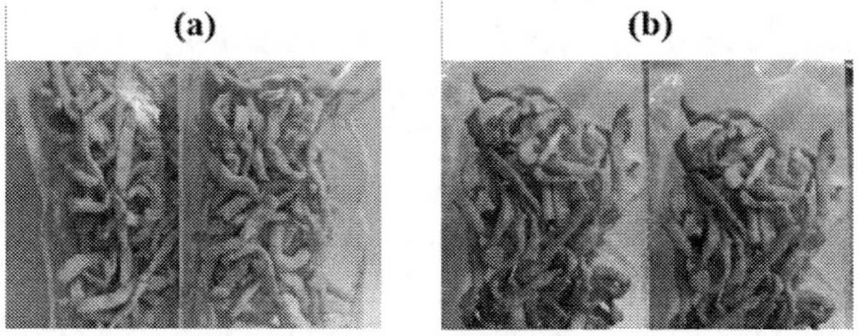

Fig. 2. Squid stick appearance with different time of frying time (a) 45 second (b) 60 second

The six squid stick samples in this study had different crunchy textures. Squid sticks fried for 60 seconds were crispier than those fried for 45 seconds, seen organoleptically through texture analysis. The difference in crispiness is caused by the length of time frying will affect the final moisture content of the product so that it feels crispier in a longer time [16], 9]. The proportion of added salt also affects product acceptance, where squid sticks with a higher salt concentration will have less free water content, affecting the texture of the product [18], 11]. The higher salt content will be absorbed and accumulated more, resulting in a salt crystal fouling effect on appearance, a moderately high saltiness, and a hard texture [20].

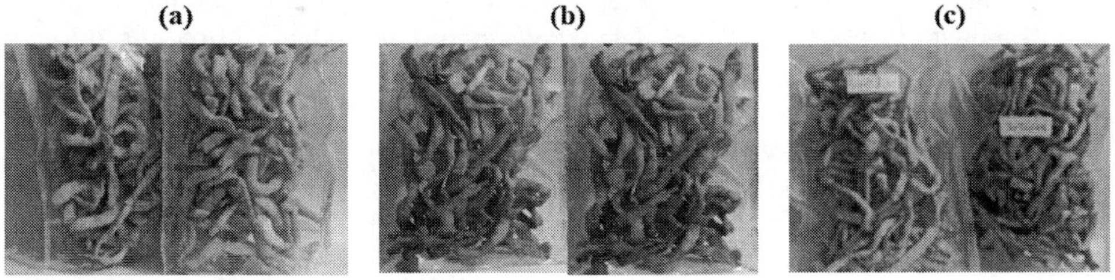

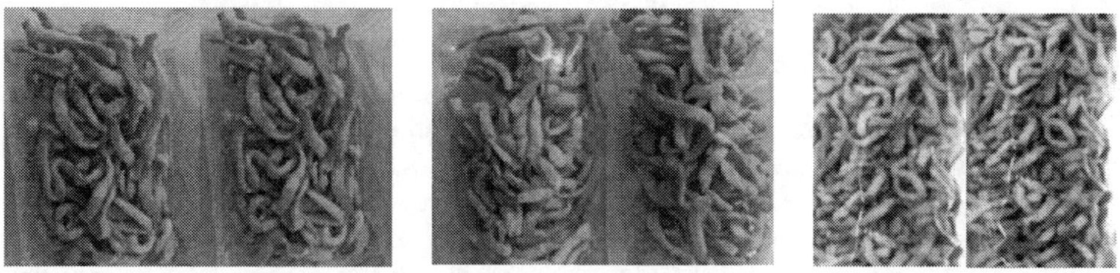

Fig. 3. Appearance of squid stick. (a) Salt 10 % - frying time 45 seconds; (b) Salt 10% - frying time 60 seconds; (c) Salt 20 % - frying time 45 seconds; (d) Salt 20% - 60 seconds; (e) Salt 30% - 45 seconds; and (f) Salt 30% - 60 seconds

3.2 Hedonic test

The results of the sensory assessment using the hedonic test showed that the most preferred sample was the addition of 10% salt with a frying time of 60 seconds. These results are in line with the results of dried salted skipjack tuna [21]. The more salt added will cause the product to get drier and the less water content so that it brings out a more dominant salty taste. The dominance of saltiness will eliminate the taste of other ingredients such as ginger and onions, even the taste of squid raw materials will be camouflaged, thereby reducing consumer preference scores. These results are then used as recipe standards and process standards. Table 4 shows the Friedman test results from the panelists' affections assessment of squid stick products with different variations in salt percentage.

Table 4. Friedman test results in various treatments of squid stick samples.

Formula	Salt (%)	Frying time (second)	Rating value
1	10	60	$1{,}77 \pm 1{,}073^{a}$
2	10	45	$2{,}27 \pm 1{,}311^{b}$
3	20	60	$3{,}33 \pm 1{,}422^{c}$
4	20	45	$3{,}40 \pm 1{,}133^{d}$
5	30	45	$4{,}77 \pm 0{,}971^{e}$
6	30	60	$5{,}50 \pm 0{,}731^{f}$

Sensorially, salt can increase flavoring strength due to interactions with amino acids, flavorings, and MSG, reduce bitterness that occurs upon component interactions, and improve product acceptability in chips, crackers, and biscuits. The addition of salt to flavors also affects the product characteristics of crackers [22]. The study was related to the interaction of salt with other ingredients, such as protein and starch. The addition of salt can cause an increase in protein solubility so that the protein gel becomes more dispersed in the dough [23], 16]. As a result, the dough rises more evenly. The addition of salt to the dough can also affect the colligative properties of the dough which causes the starch gelatinization temperature to be higher. The addition of salt also shifts the optimum moisture content value for developing cracker-like products, i.e. starchy snacks. The optimum moisture content of starchy snacks with salt is lower than that without salt [25]. This shift in optimum swelling moisture content is related to the effect of salt on the boiling line and iso-critical viscosity line through changes in the glass transition of starch components [26].

The frying temperature applied to squid stick products was 120 ± 5 °C. The choice of frying temperature was based on a literature study [11] also from trial and error. Product size and frying temperature affect the color of the resulting product [27]. Based on the results of these studies, it is known that the percentage/amount of salt concentration, length of frying time, and temperature can affect the level of consumer preference for dried squid processed products. Through this study, the most preferred salt content was 10% of the squid raw material with a frying time of 60 seconds at a temperature of 120 ± 5 ºC. The use of 10% salt based on the weight of the squid raw materials used, after being calculated and presented on the basis of the total weight of all raw materials, can be seen in table 5, as the standardized formula.

Table 5. Standard formula for making squid sticks.

Ingredient	Formula (gram)	Formula (%)
Wheat fluor	1.400	54.69
Squid	500	19.53
Garlic	300	11.72
Sugar cane	200	7.81
Salt	50	1.95
Butter	50	1.95
Chicken Egg	50	1.95
Ginger	5	0.20
Emulsifier	5	0.20

3.3 Chemical characteristics of squid stick

In the production process, the part of the squid used as the main ingredient for making squid sticks is the mantle. The muscle weight (mantle and arms) of pelagic squid of the genus Sthenoteuthis reaches 60-65% of the body weight [28]. The exact value indicates the characteristics of active fish, such as bonito and trout (65 % and 57%, respectively). The chemical composition characteristics of squid mantle, fins, arms and tentacles include high protein content (about 17-22% of fresh weight), low lipid content (1-3%) and shallow glycogen content (0.1%). The chemical composition of squid differs from one species to another. Squid from *Loligo vulgaris* species have 1.4% fat, 18.5% protein and 1.5% ash in their mantle [29]. In comparison, the tentacles contained 1.4% fat, 16.1% protein, and 1.6% ash levels.

Based on the analysis results (Table 6), the protein value was lower than that reported by previous studies on squid [21]. This decrease is due to the added ingredients, which are minimal protein components such as wheat flour, at a ratio of three times that of squid. This led to a decrease in the total % of squid. This decrease also leads to affordability as the main raw material of squid is more expensive than other ingredients.

Table 6. Standard formula for making squid sticks.

Parameter	% (w/w)
Carbohydrate	65.60
Fat	14.89
Protein	12.53
Moisture	5.84
Ash	1.14

The moisture content of squid stick products in this study was 5.84% so that it can be categorized into dry products. In addition, when viewed from the characteristics of dry processed products, squid stick products can be classified into the type of processed crackers in accordance with the Indonesian National Standard Agency (SNI No. 8272:2016) concerning Fish, Shrimp, and Mollusk Crackers. The SNI regulation states that the moisture content in cracker products is a maximum of 12%, the protein content in grade III mollusk crackers is at least 2%, grade II is at least 5%, and above 8% for grade I crackers. The results of the protein content test on squid sticks show that as much as 12.53% was protein, so squid stick products are included in class I mollusk crackers, which according to Indonesian regulations can be widely marketed throughout Indonesia. The protein content in squid sticks is thought to be contributed by eggs, wheat flour, and squid raw materials. More protein variations will increase the nutritional value of squid sticks due to the different sources of squid stick protein, such as wheat flour and chicken eggs, compared to the squid itself.

4 Conclusion

Based on the panelists' favorability rating, the standard formula was obtained, namely 54.69% wheat flour, 19.53% squid, 11.72% garlic, 7.81% sugar cane, 1.95% salt, 1.95% butter, 1.95% chicken eggs, 0.20% ginger, and 0.20% emulsifier (from the total weight of the ingredients). While the processing standard, which is frying with a temperature of 120 °C for 60 seconds using cooking oil. The chemical content of squid stick products is 12.53% protein, 14.89% fat, 5.84% moisture content, and 65.60% carbohydrate. In accordance with SNI 8272:2016 concerning Fish, Shrimp, and Mollusk Crackers, the protein content of squid stick categorized as grade I mollusk cracker which can be widely marketed throughout Indonesia.

Acknowledgments

This research was funded by Marine and Fisheries Polytechnic of Bitung (APBN) through the Budget Implementation List (DIPA) of Marine and Fisheries Polytechnic of Bitung for the fiscal year 2023.

References

1. P. G. K. Rodhouse, Deep. Res. Part II Top. Stud. Oceanogr. **95**, 129 (2013)
2. P. G. K. Rodhouse, G. J. Pierce, O. C. Nichols, W. H. H. Sauer, A. I. Arkhipkin, V. V. Laptikhovsky, M. R. Lipiński, J. E. Ramos, M. Gras, H. Kidokoro, K. Sadayasu, J. Pereira, E. Lefkaditou, C. Pita, M. Gasalla, M. Haimovici, M. Sakai, and N. Downey, in *Adv. Mar. Biol.* (2014), pp. 99–233
3. M. Khomeni, *Studi Pelunakan Daging Cumi-cumi (Loligo pealii) dengan Penambahan Filtrat Enzim Bromelin Kasar* (Makassar, 2018)
4. B. Basri, K. S. Harahap, and A. Yelofeva, Aurelia J. **2**, 111 (2021)
5. A. Putridwiaditin, M. Patanda, and R. F Telussa, J. Ilm. Satya Minabahari **8**, (2023)
6. A. Singh, A. Mittal, and S. Benjakul, Food Rev. Int. **38**, (2022)
7. F. D. Hariyoto, A. P. B. L. Soeharso, D. H. Ndahawali, I. D. Wewengkang, and F. J. B. Ticoalu, J. Abdi Insa. **10**, 1484 (2023)
8. R. Shi, M. Huang, J. Wang, C. He, X. Ying, X. Xiong, and X. Xiong, Food Addit. Contam. - Part A Chem. Anal. Control. Expo. Risk Assess. **37**, (2020)

9. C. C. Zhao, G. H. Jiang, and J. B. Eun, J. Food Qual. **2017**, (2017)
10. J. Jumiati, D. Ratnasari, and A. Sudianto, J. Ilm. Perikan. Dan Kelaut. **11**, (2019)
11. I. D. Wewengkang, F. J. B. Ticoalu, D. H. Ndahawali, A. P. B. L. Soeharso, and F. D. Hariyoto, J. Bluefin Fish. **4**, (2022)
12. M. Maidana-Eletti, Deakin Law Rev. **19**, (2014)
13. [BSN] Badan Standarisasi Nasional, *Standar Nasional Indonesia (SNI) Nomor 012346-2006 Tentang Petunjuk Pengujian Organoleptik Dan Sensorik* (Indonesia, n.d.), p. 2006
14. Z. Cui, T. Manoli, T. Nikitchina, and H. Mo, J. FOOD Sci. Technol. **14**, (2020)
15. H. Ning, H. Qiu, J. Miao, Y. Qu, and K. Lai, Appl. Food Res. **2**, (2022)
16. G. O. Fetuga, T. E. Ajayi, and O. R. Karim, African J. Root Tuber Crop. **11**, (2014)
17. K. N. Van Koerten, M. A. I. Schutyser, D. Somsen, and R. M. Boom, J. Food Eng. **146**, (2015)
18. A. Bunger, P. Moyano, and V. Rioseco, Food Res. Int. **36**, (2003)
19. Jamaluddin, H. Syam, and A. Mustarin, in *J. Phys. Conf. Ser.* (2019)
20. D. H. Ndahawali, F. Ticoalu, I. D. Wewengkang, A. P. Soeharso, and F. D. Hariyoto, J. BLUEFIN Fish. **5**, 36 (2023)
21. M. S. Kapoh, J. Dewi, A. S. Wibawa, Y. H. Sipahutar, J. Sirait, *Penambahan Kadar Garam Terhadap Mutu Sensori, Kadar Air, dan Kadar Garam Produk Terpilih Ikan Asin Cakalang (Katsuwonus pelamis)*, in Prosiding Simposium Nasional IX Kelautan dan Perikanan UNHAS **9** (2022)
22. C. S. Cheow and S. Y. Yu, J. Food Process. Preserv. **21**, 161 (1997)
23. N. Pakpahan and N. Nelinda, J. Teknol. Pengolah. Pertan. (2019)
24. S. Maisont and W. Narkrugsa, Kasetsart J. - Nat. Sci. **44**, 251 (2010)
25. R. G. M. Van Der Sman and J. Broeze, Food Funct. **5**, (2014)
26. A. Farahnaky, I. A. Farhat, J. R. Mitchell, and S. E. Hill, Food Hydrocoll. **23**, (2009)
27. Sugito, Hermanto, and Arfah, J. Agroindustri **3**, 83 (2020)
28. G. E. Shulman, M. V. Chesalin, G. I. Abolmasova, T. V. Yuneva, and A. Kideys, in *Bull. Mar. Sci.* (2002)
29. S. Atayeter and H. Ercoşkun, J. Food Sci. Technol. **48**, (2011)

The activity of *Pometia pinnata* leaf extract against pathogenic bacteria in fish

Jarod Setiaji[1*], *Valentio* Febian Prokoso[1], *Heriyanto*[1], *Hisra* Melati[1], *Tengku* Said Raza'i[2], *Suryanti*[1] and Eva Ayuni[1]

[1]Department of Aquaculture, Faculty of Agriculture, Universitas Islam Riau, Pekanbaru 28284, Indonesia.
[2]Marine and Fisheries Faculty, Universitas Maritim Raja Ali Haji, Tanjungpinang, 19100, Indonesia

Abstract. The use of antibiotics to cure bacterial diseases has a negative impact on fish and the environment, so it is necessary to look for natural products that are safe for treating fish diseases. This study aims to determine the activity of *P. pinnata* extract as an antibacterial towards *V. alginolyticus, P. aeruginosa, A. salmonicida, E. ictaluri, A. hydrophila*, and *E. tarda* bacteria. *P. pinnata* extract was obtained by maceration using ethanol solvent. This extract was tested for its phytochemical content, characterized by FT-IR and tested for its antibacterial activity by the agar diffusion method. The results of the phytochemical test of *P. pinnata* extract produced terpenoids compounds flavonoids, saponins and phenolic. The results of the FT-IR characterization of *P. pinnata* extract contained O-H, C-H, C=O and C-O groups. The results of the inhibition test showed that *P. pinnata* extract was able to inhibit the growth of *V. alginolyticus* bacteria by 13.0 mm to 14.7 mm. *P. aeruginosa* 14.5 mm to 17.1 mm. *A. salmonicida* by 12.4 mm to 14.9 mm. *E. ictaluri* by 13.5 mm to 15.5 mm. *A. hydrophila* 13.9 mm to 17.1 mm. *E. tarda* 13.6 mm to 15.5 mm. In conclusion, *P. pinnata* extract can be used to inhibit the growth of pathogenic bacteria in fish.

1 Introduction

Bacteria *Vibrio alginolyticus, Pseudomonas aeruginosa, Aeromonas salmonicida, Edwardsiella ictaluri Aeromonas hydrophila*, and *Edwardsiella tarda* are infect many farmed fish. This pathogenic bacterial infection can cause fish death. *V. alginolyticus* is an opportunistic bacterium, including pathogenic bacteria that can threaten the development of aquaculture globally [1]. In fish, this bacteria can lead to several diseases such as ulcers, septicemia and exophthalmia [2]. *P. aeruginosa* bacteria cause many diseases in aquaculture fish [3]. *P. aeruginosa* causes gill necrosis, hemorrhagic septicemia, abdominal distension, congested kidney dan friable liver [4].

A. salmonicida bacteria cause many diseases in salmon [5]. *A. salmonicida* bacteria can also cause disease in various types of fish [6]. It can cause bleeding, ulcer, hemorrhage,

* Corresponding author: jr.setiaji@agr.uir.ac.id

thereby causing death in the fish [7]. *E. ictaluri* bacteria, is a pathogenic bacteria that causes many diseases in catfish [8].

E. tarda is a pathogenic bacteria that attacks many freshwater and marine fish. The consequences of this bacterial attack cause significant losses in aquaculture activity [9, 10]. *E. tarda* bacteria is a pathogen that causes many diseases in fish. Large deaths in *P. olivaceus* fish cultivation are caused by the bacteria E. tarda [11]. *A. hydrophila* is a pathogenic bacterium and is a Gram-negative bacterium. This bacteria can cause MAS (motile aeromonas septicemia) disease [12, 13]. The result of A. hydrophila infection is bleeding in the fish's body, especially in the chest, fins and stomach [14].

Antibiotics are often used to fight fish diseases that caused by bacterial. However, frequent use of antibiotics can inflict bacterial resistance [15] and can harm fish, especially consumption fish [16]. Natural products made from antibacterial plants such as the *Pometia pinnata* plant are sought after as an alternative to antibiotics.

P. pinnata is a plant that has a tap root and many branches, a shady plant with a tree height reaching 20-40 m. The leaves are green and have a pointed tip. This plant is often found in Indonesia and usually called matoa [17]. The *P. pinnata* plant has efficacy in the pharmaceutical and cosmetic fields [18]. It can be used as a medicinal ingredient, because it contains chemical compounds. *P. pinnata* leaf extract contains glycosides, alkaloids, steroids, flavonoids, saponins, tannins [19, 20] and terpenoid group compounds [21]. Other studies have proven the activity of α-glucosidase inhibitors from the ethanol extract of matoa stem bark as an antihyperglycemic agent [22, 23]. Methanol, ethylacetate and n-hexane extracts of *P. pinnata* seeds are antidiabetic [24].

The benefits of the *P. pinnata* plant include diarrhea, dysentery, antiviral, diabetes, antioxidant [25]. *P. pinnata* leaf extract contains saponin compounds which have antimicrobial activity against Staphylococcus epidermidis [26]. Ethanol extract of *P. pinnata* leaves has antifungal activity against Trichophyton mentagrophytes [27]. *P. pinnata* leaf extract contains flavonoid compounds which have strong antioxidant activity [28, 22]. Ethyl acetate extract of the endophytic fungus *P. pinnata* leaves has antimicrobial against *P. aeruginosa, Candida albi*cans and *S. aureus* [29]. The purpose of this research is to find out the activity of *P. pinnata* extract as an antibacterial towards *E. tarda, V. alginolyticus, P. aeruginosa, A. salmonicida, E. ictaluri, A. hydrophila* bacteria.

2 Material and Method

2.1 Extraction of *Pometia pinnata* leaf

P. pinnata leaves were obtained from the experimental garden of the Faculty of Agriculture at the Universitas Islam Riau Pekanbaru, Indonesia. The research was conducted in May-June 2023 and 3 Kg of leaf samples were collected. These leaves were dried in an oven at 40°C for 48 hours. Then, they were finely ground in a blender. The obtained *P. pinnata* powder was macerated in 90% ethanol solvent for 36 hours. Subsequently, to obtain the extract, the macerated phytrate is evaporated at 50°C using a rotary evaporator [30].

2.2 Phytochemical test and Fourier-transform infrared spectroscopy (FT-IR)

Chemical analysis was performed on the *P. pinnata* extract to detect the presence of different compounds, which include alkaloids, flavonoids, phenolics, saponins, and terpenoids. In the alkaloid test, 250 μL of Mayer reagent was added to 10 mg of the *P. pinnata* extract, a white precipitate was formed. When Dragendorff reagent was added, it turned orange, indicating a positive result. For the flavonoid test, 0.05 g of magnesium and

10 drops of 37% HCl were mixed with 10 mg of *P. pinnata* extract and stirred for 1 minute. A positive result was indicated by a colour change to red.

The phenolic test, 10 mg of *P. pinnata* extract was mixed with 500 μL of 5% $FeCl_3$, and a positive result was indicated by colour change to blue. To test for saponins, 10 mg of *P. pinnata* extract was shaken with 5 mL of distilled water for 1 minute, followed by the addition of 150 μL of 1N HCl and another 1-minute of shaking, resulting in a positive result due to the formation of the results have been shown foam. For the terpenoid test, mixing 10 mg of *P. pinnata* extract with 10 drops of CH_3COOH and 3 drops of H_2SO_4, gave a positive result indicated by the formation of a red colour [30].

To further determine the functional groups present in the *P. pinnata* extracts, FT-IR spectroscopy analysis was performed. 1 mg of each extract was ground, added to KBr, and mixed vigorously until homogeneous. The infrared absorption of the resulting mixture was then measured in the wavelength range of 4500–450 cm, allowing identification of the functional groups.

2.3 Inhibitory activity of *P. pinnata* extract

P. pinnata extract was tested on the bacteria *V. alginolyticus, E. ictalurid, P. aeruginosa, A. salmonicida, E. tarda, A. hydrophila* using the agar diffusion method. 1 mL of pathogenic bacterial inoculum (0.08-0.1 at 600nm) was mixed evenly in 15 mL of NA medium and poured into a petri dish. After the NA media had solidified, Oxytetracycline paper was used as a positive control, as a negative control a paper disk was used that was treated with 30 mL of methanol. The concentrations of *P. pinnata* extract tested were 100 mg mL^{-1}, 200 mg mL^{-1}, 300 mg mL^{-1}, 400 mg mL^{-1}, 500 mg mL^{-1}. From this concentration, 30 μL was taken and dropped on a paper disc, then incubated for 1 days at 30°C. Observation of the inhibitory power of the extract was the area around the paper disc that was not growing with pathogenic bacteria [30].

2.4 Data Analysis

The collected data was tabulated and descriptive analysis was carried out by evaluating the amount of inhibitory power formed as an antibacterial indicator.

3 Results and Discussion

3.1 Phytochemical test and functional groups

The purpose of this test is to find out the compounds that have an active role in antimicrobial activity. Compounds in the *P. pinnata* extract were identified by monitoring the colour changes that occurred after being given the reagent. This phytochemical test on the *P. pinnata* extract was carried out to confirm the chemical compounds in the *P. pinnata* extract, then identify the main chemical compound groups that have an active role for antibacterial (Table 1).

Table 1. Result Phytochemical Test

No.	Compound	Results	Test
1	Terpenoids	Formed colour-red	+
2	Flavonoids	Formed colour-pink	+
3	Saponins	Foamy	+

4	Phenolic	Formed colour blue-black	+
5	Alkaloids	No orange precipitate is formedias	-

There was no noticeable color change or precipitate formation when P. pinnata extracts were exposed to Mayer's and Dragendroff's reagents. Conversely, samples containing alkaloids were altered and turned white or orange due to interaction with these conventional reagents, as documented in reference [31].

When *P. pinnata* extracts were exposed to Lieber-Burchard reagent, the colour changed to red. This colour change to red was caused by acid oxidation and was promoted by the presence of H_2SO_4. During this process, in these compounds electrons are separated from hydrogen, which expanded the conjugation of the compound and was visually reflected in the red colour [32]. The presence of terpenoids can be seen from the red color that forms in samples from the oxygenation and dehydrogenation processes.

The compounds contained in *P. pinnata* extract are categorized by changing the color of the extract, to peach color. Furthermore, the red color formed is caused by benzopyrylium salt. The reduction of polyhydroxy groups from flavanols produces a red color due to the existence of the Mg element in HCl [33].

Foam formation occurred when *P. pinnata* extract was mixed with HCl. The positive extract contained saponin, when the extract formed foam around 5 minutes after adding the reagent [34]. Height of the foam produced in the extract after adding 1 drop of 1N HCl varied between 1 to 5 cm.

P. pinnata extract was positive for containing phenolics, this was known by observing the color change in the extract to dark blue when reacted with ferric chloride. The colour formed occurs due to a reduction process involving phosphomolybdic acid and phosphotungstate with the reagent, resulting in the formation of a blue colour. The blue color formed indicates the presence of phenolic content. This happens because phenolic ions reduce heteropoly acid, which forms molybdenum tungsten, so the color becomes dark [35]. The dark blue color is caused by the formation of sugar bonds in the extract.

FT-IR spectrum results are assumed that at a length of 3420 cm^{-1} there is an O-H group, where flavonoid, phenolic, terpenoid and saponin compounds have C bonds with O-H. The 2891 cm^{-1} spectrum shows C-H. The spectrum of 1753 cm^{-1} shows the C=O group possessed by flavonoids. The spectrum of 1075 cm^{-1} shows the C-O group found in structure of flavonoids, phenolics and saponins [36].

3.2 Inhibitory Activity

Inhibition test results showed that *P. pinnata* was able to inhibit the growth of A. *V. alginolyticus* bacteria by 13.0 mm to 14.7 mm. On *P. aeruginosa* by 14.5 mm to 17.1 mm. On *A. salmonicida* by 12.4 mm to 14.9 mm. On *E. ictaluri* by 13.5 mm to 15.5 mm. On *E. tarda* by 13.6 mm to 15.5 mm. On *A. hydrophila* 13.9 mm to 17.1 mm (Table 2). The inhibitory ability of *P. pinnata* extract on pathogenic bacteria is classified as strong [37].

Table 2. Inhibition activity of *P. pinnata* extract against pathogen bacteria

P. pinnata **Extract concentration ($mg.mL^{-1}$)**	**Inhibition zone diameter (mm)**					
	Vibrio alginolyticus	***Pseudomonas aeruginosa***	***Aeromonas salmonicida***	***Edwardsiella ictaluri***	***Edwardsiella tarda***	***Aeromonas hydrophila***
100	13.0 ± 0.14	14.5 ± 0.07	12.4 ± 0.28	13.5 ± 0.57	13.6 ± 0.07	13.9 ± 0.49
200	14.0 ± 0.14	15.3 ± 0.35	13.1 ± 0.28	14.3 ± 0.42	14.5 ± 0.07	15.2 ± 0.07

300	14.5 ± 0.28	15.8 ± 0.07	13.7 ± 0.42	14.4 ± 0.28	15.0 ± 0.14	15.7 ± 0.21
400	14.6 ± 0.28	16.8 ± 0.42	14.6 ± 0.07	14.8 ± 0.07	15.2 ± 0.07	16.3 ± 0.78
500	14.7 ± 0.42	17.1 ± 0.85	14.9 ± 0.64	15.5 ± 0.42	15.5 ± 0.14	17.1 ± 0.92

The most notable result from testing the inhibitory effect of *P. pinnata* extract inhibitory activity was significant activity against *P. aeruginosa*. The amount of inhibitory power for each bacteria varies, this is because the reaction of pathogenic bacteria to *P. pinnata* extract is not the same. The concentration and type of antimicrobial substance are very influential in inhibiting microbial growth. The size of the inhibition zone formed is directly proportional to the increase in the concentration of the extract given [38]. The presence of a clear zone in the culture media for pathogenic bacteria indicates that the *P. pinnata* extract contains antibacterial compounds like terpenoids, saponins, phenolics, and flavonoids.

Terpenoids are antibacterial, because terpenoids have lipophilic properties which can damage cell membranes [39]. Terpenoids have the ability to interact with porins, namely transmembrane proteins found in the membrane layer of bacterial cells. Damage to porins is caused by strong polymer bonds from these interactions. Consequently, the bacterial cell wall permeability system is disrupted, and the supply of nutrients to the cell is reduced, so that bacterial growth stops.

The terpenoids present in this fraction possess antibacterial properties due to their ability to inhibit lipid production and alter the structure of cell membranes by impeding the synthesis of ergosterol [40, 41]. The antimicrobial efficacy of terpenoid compounds stems from their lipophilic characteristics. Terpenoids can disintegrate and disrupt the lipid component within the bacterial plasma membrane, leading to an imbalance in cell membrane permeability. This disruption results in the disturbance of ion homeostasis within bacterial cells [42].

Flavonoids have demonstrated the ability to form complex compounds with characteristics that can damage microbial cells. The interaction of flavonoids with proteins in bacterial cells can damage the components inside the cells. Additionally, this compound can direct energy transfer into the bacterial cytoplasm, thus hindering bacterial mobility [43]. Flavonoid compounds inhibit bacterial growth by binding to the cell surface, causing damage to cell walls and enzyme activity. The hydroxyl group and beta ring elements in flavonoids are involved in antibacterial mechanisms. Flavonoids can influence porin activity, biofilm formation, permeability, thereby disrupting the cytoplasmic membrane and inhibiting the formation of nucleic acids. [44].

Flavonoid compounds, when acting as antibacterial agents, employ several mechanisms to disrupt bacterial functions, including damaging cell walls and interfering with metabolic processes. These mechanisms encompass three primary categories: inhibition of nucleic acid synthesis, disruption of cytoplasmic membrane function, and interference with energy metabolism [45]. Furthermore, flavonoid compounds can also destroy the permeability of bacterial cell.

Saponins exhibit antimicrobial properties. As an antibacterial, saponin damages the cell membrane permeability system, so that cells lose proteins and enzymes [31, 46]. If saponin interacts with the sterol membrane, it can destroy the cell wall structure, resulting in the detached of important materials of the cell. The inhibitory mechanism for saponins in bacteria is to damage the permeability of bacterial cell membranes. Apart from that, saponins also have hydrophilic and lipophilic properties and can bind to the lipid components of bacterial cells, causing lysis and rupture of bacterial cells [46]. Consequently, when this interaction takes place, the cell wall becomes susceptible to breaking or undergoing lysis, allowing antibacterial substances to enter the cell easily and disrupt its metabolism, ultimately leading to bacterial death [47].

The hydroxyl groups contained in phenolics show strong antibacterial characteristics. Hydroxyl groups damage protein structures and prevent nutrients from entering bacterial cells. Additionally, hydroxyl groups dissolve fats in cell walls, thereby disrupting bacterial cell membranes [48].

The hydroxyl groups in phenolic compounds can impair the permeability from the cytoplasmic membrane system, which disrupts the transport of essential organic ions into cells, ultimately leading to the death of bacterial cells. Phenolic compounds also possess biological functions such as anti-carcinogenic, anti-inflammatory, and antioxidant properties. Numerous phenolic compounds have been demonstrated to be effective in inhibiting a variety of pathogenic bacteria [49]. Phenolics cause the flagella to break in bacteria, causing damage to the membrane and cell walls, resulting in macromolecules leaving the cell [50].

4 Conclusion

P. pinnata leaf extract contains terpenoid, flavonoid, saponin and phenolic compounds. P. pinnata leaf extract is effective in inhibiting the accretion of pathogenic bacterial in fish like *V. alginolyticus, E. ictaluri, P. aeruginosa, A. salmonicida, A. hydrophila, E. tarda.*

Acknowledgments

The authors would like to thank all the colleagues who gave valuable support to this study.

References

1. Y. Chen, S. Cai, J. Jian, Fish. Shellfish Immunol. **86**, 875- 881 (2019)
2. Y.D. Wang, Y.H. Wang, C.F. Hui, J.Y. Chen, Fish. Shellfish. Immunol. **52**, 31- 43 (2016)
3. C.F. Souza, M.D. Baldissera, S.N. Descovi, C.C. Zeppenfeld, M. Camila, C.M. Verdi, R.C.V. Santos, A.S. da Silva, B. Baldisserotto, Microbiol. Pathog. **129**, 271-276 (2019)
4. Y. Hilliam, S. Kaye, C. Winstanley, J. Med. Microbiol **69**, 3–13 (2020)
5. X. Li, K. Fan, Y. Liu, Y. Liu, P. Liu, J. Fish. Dis. **44,** 7: 961-977 (2021)
6. Z. Xu, P. Jin, X. Zhou, Y. Zhang, Q. Wang, X. Liu, S. Shao, Q. Liu, Appl. Env. Microbiol. **87,** 21: 1-15 (2021)
7. K. Valderrama, M. Soto-Dávila, C. Segovia, I. Vásquez, My Dang, J.J. Santander, J. Fish. Dis. **42**, 11: 1601-1608 (2019)
8. S. Kalindamar, H. Abdelhamed, A.O. Kordon, L.M. Pinchuk, A. Karsi, Front. Vet. Sci. **8**, 681609: 1-13 (2021)
9. Y. Sun, J. Zhang, L. Qin, C. Yan, X. Zhang, D. Liu, PLoS One. **12**, 3: e0172783 (2017)
10. M.Y. Davies, M.G. de Oliveira, P.V.M. Cunha, S.L. Franco, S.S.L. Santos, Z.L. Moreno Vet. Quart. **38**, 99-105 (2018)
11. X.L Guan, B.C. Zhang, L. Sun, Devlop. Comp. Immunol. **103**, 103531 (2020)
12. D. Zhang, G.S. Moreira, C. Shoemaker, J.C. Newton, D.H. Xu, FEMS Microbiol. Lett. **363,** 9: 1-5 (2016)
13. G. Chaix, F. Roger, T. Berthe, B. Lamy, J.E. Bilak, R. Lafite, F.J. Leray, F. Petit, Front. Microbiol. **8**, 1259 (2017)
14. H. Abdelhamed, I. Ibrahim, W. Baumgartner, M.L Lawrence, A. Karsi, Front. Microbiol **8,** 1519 (2017)

15. M.G. Rasul, B.C. Majumdar, The Saudi J. Life Sci. **2,** 3: 81–88 (2017)
16. S. Rodriguez-Mozaz, I. Vaz-Moreira, S.V.D. Giustina, M. Llorca, D. Barcelo, S. Schubert, T.U. Berendonk, I. Michael-Kordatou, D. Fatta-Kassinos, J.L. Martinez, C. Elpers, I. Henriques, T. Jaeger, T. Schwartz, E. Paulshus, K. O'Sullivan, K.M.M. Parnanen, M. Virta, T.T Do, F. Walsh, C.M. Manaia, Envi. Int. **140**, 105733: 1-11 (2020)
17. W.M. Sidoretno, U. Abdurrab, J. Indo. Nat. Research Pharm. 3, 1: 16-25 (2018).
18. W. Rahmah, H. Hamzah, S. Hajar, S.S. Ressandy, E.M. Putri, J. Farmasi Sains Praktis, 7, 1: 59-66 (2021).
19. S. Sutomo, N. Hasanah, A. Arnida, A. Sriyono, J. Pharm. 8, 1: 101-110 (2021).
20. Y.C. Pamangin, R.D. Pratiwi, S. Dirgantara, J. Kimia. **4,** 1: 52-62 (2020).
21. F. Maryam, B. Taebe, D.P. Toding, J. Man. Pharm. Indo, **6,** 1: 1-12 (2020)
22. E. Tahalele, Sutriningsih, Indo. Natural Resear. Pharma. J. **3,** 2: 44-55 (2018)
23. S.P. Mataputun, J.A. Rorong, J. Pontoh, J. MIPA Unsrat. **2,** 2: 119-123 (2013)
24. M. Sukiman, J.A. Margaretha, C. Irawan, Hanafi, L. Sulistiawaty, The Pharma Innov. J. **7,** 5: 10-12 (2018).
25. L. Wulandari, A.S. Nugraha, A.U. Himmah, J. Kefarm. Indo. **11,** 2: 132–141 (2021)
26. R. Rossalinda, F. Wijayanti, D. Iskandar, J. Sains Terap. Kimia. **3,** 1: 1-8 (2021).
27. W. M. Sidoretno, M. Gustari, Photon: J. Sain Kesehat. **11,** 2: 137-148. (2021)
28. M.O. Hehakaya, H.J. Edy, J.P. Siampa, Pharmacon, **11,** 4: 1778-1785 (2022)
29. A. Fajrina, D.D.A. Bakhtra, A.E. Mawarni, J. Farmasi Higea. **12,** 1:81-89 (2020)
30. J. Setiaji, J. Hariwitonang, T.I. Johan, Heriyanto, IOP Conf. **1118**, 012006: 1-6 (2022)
31. Khadijah, A.M. Jayali, S. Umar, I. Sasmita, J. Kimia Mulawarman. **15,** 1: 11-18 (2017)
32. Y.I. Masadi, T. Lestari, I.K. Dewi, J. Kebid. Kesehat. Tradi. **3,** 1: 32- 40 (2018)
33. A.I. Aribowo, C.F Lubis, L.M. Urbaningrum, N.D. Rahmawati, S. Anggraini, J. Health Sains. **2,** 6: 751-757 (2021)
34. M.L.F. Kumalasari. F. Andiarna, Indo. J. Health Scien. **4,** 1: 39-44 (2020)
35. M. Tahir, A. Muflihunna, Syfrianti, 2017 J. Fitofar. Indones. **4,** 1: 215-218 (2017)
36. Hasby, Nurhafidhah, G. Pamungkas, S.A. Akbar, Rasayan J. Chem **13(1)** 38-43 (2020)
37. W.W. Davis, T.R. Stout, Applied and Env. Microbiol. **22,** 4: 666-670 (2009)
38. R.P.A. Rahmah, M. Bahar, Y. Harjono, Biogenesis. **5,** 1: 34-41 (2017)
39. P. Xiu, R. Liu, D. Zhang, C. Sun, Applied and Env. Microbiol. **83,** 12: 1-14 (2017)
40. F. Kuswandani, M.H. Satari, A.M. Maskoen, J. Dent. Indones. **26,** 3: 119-125 (2019).
41. N. Marfu'ah, S. Luthfiana, I. Ichwanuddin, Pharm. J. Islamic Pharmacy, **5,** 2: 1-10 (2021).
42. B. Ergüden, Lett. Applied. Microbiol. **73**, 4: 438-445 (2021)
43. D.F. Manik, T. Hertiani, H. Anshory, J. Khazanah. **6,** 2: 1-11 (2016)
44. I. Gorniak, R. Bartoszewski, J. Kroliczewski, Phytochem. Rev. **18,** 2: 241–272 (2019
45. A.N. Panche, A.D. Diwan, S.R. Chandra, J. Nutri. Scien. **5,** e47: 1-15 (2016)
46. L. Ravi, V. Manasvi, L.B. Praveena, Asian J. Pharmac. Clini. Res. **9,** 2: 344-347 (2016)
47. F. Nurzaman, J. Djajadisastra, B. Elya, J. Kefarmasian Indo. **8,** 2: 85-93. (2018)
48. F. Hardiansi, D. Afriliana, A. Munteira, E.D. Wijayanti, Pharmacy Med. J. **3,** 1: 16-22 (2020)
49. A. Siregar, M.A. Mutia, A. Napiah, Pharma. J. Islamic Pharmacy. **6,** 1: 1-8 (2022)

50. Zhang G, Yang Y, Memon FU, Hao K, Xu B, Wang S, Wang Y, Wu E, Chen X, Xiong W and Si H, Front. Microbiol. **12,** 738896: 1-14 (2021)

The effect of citric and acetic acid treatment on gelatin production from catfish skin

Nami Lestari[12*], *Lamhot P.* Manalu[1], *Taufik* Hidayat[1], *Lukman* Junaidi[1], *Eddy Sapto* Hartanto[1], Rienoviar[1], *Suroto Hadi* Saputra[3], *Armen* Zulham[4], and *Dheni Mita* Mala[5]

[1]Research Centre for Agroindustry, Research Organization for Agriculture and Food, National Research and Innovation Agency (BRIN), 16911 Cibinong, Indonesia
[2]School of Postgraduate Studies, Universitas Djuanda, 16720 Bogor, Indonesia
[3]Research Center for Biomass and Bioproduct, Research Organization for Life Sciences and Environment, National Research and Innovation Agency (BRIN), 16911 Cibinong, Indonesia
[4]Research Center for Cooperative, Corporation, and Peoples Economy, Research Organization for Governance, Economy, and Community Welfare, National Research & Innovation Agency, 12710 Jakarta, Indonesia
[5]Centre for Agro-based Industry (BBSPJIA), Ministry of Industry, 16122 Bogor, Indonesia

Abstract. The demand for halal gelatin is increasing, but the supply is decreasing. One solution to this problem is to produce halal gelatin from fishing waste, particularly catfish skin. A study was conducted to determine the impact of adding acid to the gelatin extraction process. Citric and acetic acid were used at concentrations of 0.05%, 0.075%, and 0.1%. The yield of gelatin varied from 14.97% to 24.1% depending on the concentration of acid used. The highest yield was obtained using a concentrated solution of acetic acid. Gelatin extracted using a citric acid concentration of 0.075% had the lowest water content. The pH of the solution ranged from 4 to 5, and the viscosity ranged from 10 to 14 MPa. The resulting gelatin exhibited a strength of between 200 and 259 blooms. The production process that used acetic acid on immersion was the most efficient because it resulted in high quantities of gelatin, low water content and viscosity, and good gel strength.

1 Introduction

Gelatin is a protein that is derived from the hydrolysis of collagen found in animal bones or skins [1, 2, 3]. It is a peptide mixture with protein. Gelatin is widely used in the food and non-food industries [3-4] due to the characteristics of the hydrocolloid, which play a vital role in the properties of the products. According to Said [3], the characteristics of gelatin are influenced by the properties of the collagen that make it up. Collagen is a derivative of high molecular weight fibrous proteins [2], with long polypeptide chains that form about 50 to 1000 amino acid chains. These chains are mainly composed of the amino acids glycine, proline, and hydroxyproline [3, 5].

The global market demand for standard food gelatin products was $1638 million in 2021, according to Straits Research [6]. This demand is predicted to rise to $2389 million in 2030,

* Corresponding author: namilestari65@gmail.com

with a cumulative annual growth rate (CAGR) of 4.3% during the period 2022-2030. Although there is an increase in global gelatin market demand, the demand for halal gelatin products required by halal industries, such as the food, cosmetics, and pharmaceutical industries [4], is also growing. The State of the Global Islamic Economy (SGIE) report of 2022 suggests that Muslims around the world spent $2 trillion on food, medicine, and cosmetics in 2021. This spending reflects a year-to-year growth of 8.9% and is expected to reach $2.8 trillion by 2025 with a CAGR of 7.5% [7]. The Global Islamic Economy Indicator report, which covers 81 countries, shows that Indonesia is the fourth-largest country in demand for halal products, after Malaysia, Saudi Arabia, and the United Arab Emirates [7]. Indonesia, as the world's largest Islamic country [8], needs many halal products to be consumed, including gelatin that is much needed in the Indonesian halal industry. However, Indonesia still imports gelatin from countries such as the United States, Spain, Japan, and China. The US is the world's largest gelatin producer, with a 31.26% market share, followed by Spain with a market share of 12.72%. The demand for gelatin-containing foods and beverages in Indonesia, such as ice cream, yogurt, and chocolate, as well as cosmetics and pharmaceuticals [4, 9], is increasing rapidly, which is why Indonesia needs to produce more halal gelatin.

Indonesia is blessed with abundant biological resources that can be utilized as raw materials for the production of halal gelatin. These include cowskin, goatskin, camelskin, and fishskin. Biopolymers of partial hydrolysis nature that produce biodegradable packaging films with excellent mechanical properties, high quality, good biocompatibility, non-toxicity, and excellent film-forming properties can be derived from fish gelatin [10]. Given the varied and widely distributed fishery products throughout Indonesia, catfish is a potential raw material for gelatin production, according to Indonesian local wisdom. The Director-General of Marine Competitiveness and Fisheries of the Ministry of Marine Affairs and Fisheries (KKP) has dismissed the idea that catfish (*Pangasius sp*) is not a fish that is in great demand in the world market. On the contrary, catfish is one of the most sought-after fish in the global market, with a projected Indonesian production of 380,000 tons by 2022, most of which is absorbed by the domestic market. This strengthens the potential for using catfish as a raw material for gelatin production [4, 10]. The skin part of the catfish can be used to produce gelatin [11].

Gelatin from the skin of the *Pangasianodon gigas* fish contains 89.1 g of protein/100 g and 0.75 g of fat/100 g, with high amounts of proline and hydroxyproline amino acids (211 residues per 1000 residues) [12]. The amino acid composition of catfish gelatin is slightly different from bovine gelatin, but the gelatin from catfish skin has greater gelatine flowering strength, viscosity, foam capacity, and foam stability compared to cow skin gelatin. SDS-PAGE gelatin of catfish leather also shows a high band intensity on the main protein components, particularly the α-, β-, and γ-components, which are similar to standard type I or type A collagen. Furthermore, it was found that catfish leather gelatin has a high gel strength (276 ± 5 g) and a different amino acid composition compared to pig leather gelatin [11].

The success of gelatin extraction from catfish skin depends on the type of solution used during the extraction process [11-12]. The other research has also expressed a similar view, stating that the method of gelatin isolation affects the success of the process [13]. It also has identified acetic acid and citric acid as the most commonly used solvents in the pre-treatment of raw materials [13]. The type of solvent used in pre-treatment determines the type of gelatin produced, with acid pre-treatment producing A-type gelatin, while base pre-treatment produces B-type gelatin [3, 13, 14, 15]. Moreover, the previous researchers have noted that the type and concentration of acid solvent used in pre-treatment affect the characteristics of the resulting gelatin products [16-17].

Based on the description provided above, the problem that this study aims to solve is about the effects of citric acid and acetic acid treatment during the extraction process of gelatine from catfish skin, which can result in high yield and good quality gelatin type A. The objective of this study is to identify the effects of using different types of acids and concentrations during the extraction process, which can help researchers and gelatin business industry professionals to determine the appropriate pre-treatment method to obtain high-quality and high-yield gelatin products from catfish skin.

2 Materials and methods

2.1 Materials

The study primarily uses the skins of catfish (*Pangasianodon sp*) obtained from small and medium-sized industries (SMEs) located in Pasuruan, East Java (as shown in Figure 1) as the main raw material. The chemicals used in the study include sodium hydroxide (NaOH), citric acid ($C_6H_8O_7$), acetic acid (CH_3COOH), aquades (H_2O), filing fabrics, and other laboratory analysis materials obtained from the Process Laboratory of the Grand Chamber of Standardization and Agro-Industrial Services (BBSPJIA) located in Cikaret, Bogor District, West Java, Indonesia.

Fig. 1. The skins of catfish.

2.2 Preparation of gelatin from catfish skin

The process of extracting gelatin from the catfish skin involves modified techniques described in the other research [16-18]. The catfish skin is cleaned by continuously flowing water to remove any residual meat, blood, adipose tissue, abrasions, and extraneous substances. The cleaned catfish skins are stored in a freezer at a temperature of -18°C before use.

The next step is to remove the non-collagen components from the skin by immersing it in a solution of 0.01 N sodium hydroxide for an hour, followed by rinsing with aquades until the pH level reaches 7. The skin is then immersed in solutions containing acetic acid and citric acid at different concentrations for 12 hours. The substance is filtered, and then cleansed using aquades until the pH level reaches 7. The extraction process is conducted using aquades in an 80°C water bath for four hours with a ratio of 1:3 between the catfish skin and aquades. The liquid gelatin produced is filtered using a specialized fabric designed for filtration purposes. The resulting liquid gelatin is then dried in an Excalibur dehydrator at a temperature of 55°C for 20 hours to produce a dry gelatin sheet. The parameters of the extracted gelatin product are determined by analyzing dried gelatin sheets, taking into account the treatment involving acetate and citric acid concentrations. Table 1 presents the treatment code utilized in this investigation.

Table 1. Experimental code of research.

Treatment code	Acid type	Acid concentration (%)
A1	Acetic Acid	0,100
A2	Acetic Acid	0,075
A3	Acetic Acid	0,050
B1	Citric Acid	0,100
B2	Citric Acid	0,075
B3	Citric Acid	0,050

2.3 Gelatin characterization

Gelatin products are characterized based on their physical and chemical properties. Physical properties are determined through visual analysis (color), yield calculation, viscosity and gel strength measurement. On the other hand, chemical properties are determined through water content, ash content, protein content, pH, and metal resin (Cd, Hg, Pb, and As) analysis. Gelatin product characterization is conducted in the Process Laboratory of the Grand Chamber of Standardization and Agro-Industrial Services (BBSPJIA) in Cikaret, Bogor District, West Java, Indonesia, Agroindustrial and Biomedical Laboratory (LAPTIAB) in Serpong District, South Tangerang City, Banten, Indonesia, and Saraswanti Laboratory Indo Genetech (SIG) in West Bogor district, City of Bogor, West Jawa, Indonesia. The results of the characterization will be compared with the Indonesian National Standard (SNI) on Fish Gelatin (SNI 8622:2018) [19] and the British Standards Institution on methods for sampling and testing gelatine (physical and chemical methods) (BS 757:1975) [20].

The yield of gelatin produced from the skin of catfish is calculated by comparing the dry weight of gelatin to the wet weight of the skin. This calculation is based on the characterization of the physical properties of the gelatin product and can be expressed using equation 1.

$$\%\, yield = \frac{Dry\ weight\ of\ gelatin\ (g)}{Wet\ weight\ of\ catfish\ skins\ (g)} \quad (1)$$

The gel strength and viscosity of gelatin products were determined using the Rheoner RE 3305 device and Brookfield Syncro-Lectric Viscometer, respectively, following the Official Procedure of the Gelatin Manufacturers Institute of America as described by the British Standard Institution [21]. For gel strength determination, a 5 mm cylinder probe of size 400 x 0.01 mm was used with a speed of 0.5 mm/s, a sensitivity of 0.2 V. For viscosity measurement, a gelatin sample with a concentration of 6.67% (w/w) was prepared in fine water (7 g of gelatin in 105 ml of fine water) and its viscosity was measured at a temperature of 60 °C and a sliding speed of 60 rpm using a spindle.

The chemical properties of gelatin products are characterized by measuring their acidity (pH) at room temperature using a pH meter [21] according to the standards set by the British Standard Institution. Water levels, ash levels, and protein levels are determined using the standard method of the Official Methods of Analysis of the Association of Official Analytical Chemists (AOAC) [22]. The presence of heavy metal resins in gelatin products is identified by following the guidelines of SNI 7387:2009, which specifies the maximum limit of heavy metal resin permissible in foodstuffs [23].

3 Results and discussions

3.1 Physical and chemical properties of gelatin

The results of the study indicated that varying the concentration of acetic acid and citric acid during the immersion process did not have an impact on the color of the gelatin produced. All treatments produced a transparent and yellowish gelatin color, as shown in Figure 2. This meets the standard set by SNI 8622:2018 [19], which requires the gelatin to be colored up to pale yellow. Similar results were observed when compared to the standard set in BSi 757-1975 [20], which also specifies a pale yellow color.

Fig. 2. The gelatin of catfish skin.

Characterization of the physical and chemical properties of gelatin consists of yield, pH value, water content, viscosity, and gel strength, where the analysis results are shown in Figure 3.

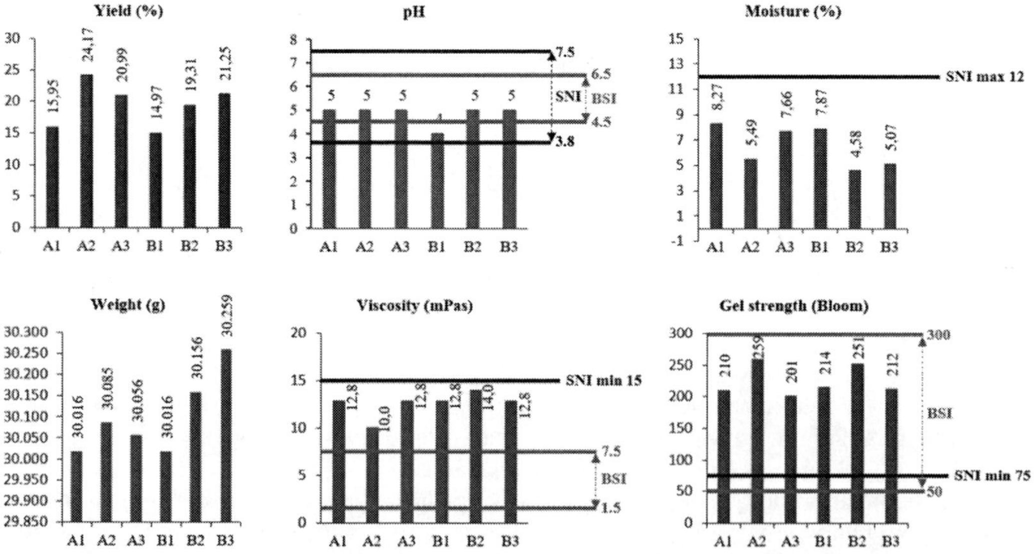

Note: A1 is the symbol for treatment with the addition of 0.100% acetic acid; A2 is the symbol for treatment with the addition of 0.075% acetic acid; A3 is the symbol for treatment with the addition of 0.050% acetic acid; B1 is the symbol for treatment with the addition of 0.100% citric acid; B2 is the symbol for treatment with the addition of 0.075% citric acid; and B3 is the symbol for treatment with the addition of 0.050% citric acid

Fig. 3. Physical and chemical properties of gelatin.

 https://doi.org/10.1051/bioconf/20248703004

According to Figure 3 the concentration levels of acetate and citric acid used during the immersion process have an impact on the yield of gelatin. The results indicate that adding 0.075% acetic acid during the immersion process produces higher gelatin yields compared to adding 0.050% acetic acid. This suggests that a higher concentration of acetic acid modifies more collagen and helps to transform it into gelatin. However, increasing the concentration of acetic acid to 0.1% does not increase the yield of gelatin. In a previous study, adding 0.05 M of acetate acid to the skin of catfish produced a gelatin yield of 69% [15]. This is greater than the yield produced in this study using 0.075% acetic acid. When citric acid is used instead of acetic acid, the yield of gelatin decreases as the concentration of acid increases. This result contradicts the findings of the other research, which found that higher acid concentrations lead to higher yields during the immersion process [24]. Increasing the concentration of acid during the immersion process increases the number of hydrogen ions (H^+) [25]. Increasing the acid concentration speeds up the hydrolysis process, which converts collagen molecules into gelatin more quickly. The faster the hydrolysis, the more collagen molecules are converted, increasing the yield.

According to the study, an increase in H+ ions speeds up the rate of collagen hydrolysis, resulting in more triple helix being fragmented into α, ß, and γ chains. As a result, the collagen is turned into gelatin more efficiently. Figure 3 shows that the pH of gelatin remains unaffected by the difference in acetate and citric acid concentrations used during the immersion process. The resulting gelatin had a pH of 5 for all treatments except for the one with 0.1% citric acid immersion, which dropped to 4. This shows that there is a possibility that washing material (catfish skin) after immersion with 0.1% imperfect citric acid causes the gelatin pH to be still low. This could be due to the acetic and citric acid remaining in the gelatin after rinsing. It was found that the type and concentration of acid used can affect the acidity of the produced gelatin [24]. This supports the findings of our study. However, the pH values for all treatments fall within the range of gelatin-sensitive values specified in BSI 757-1975 [20].

Based on the water level identification study, the concentration of acetic acid and citric acid employed during the immersion process has a direct impact on the water level in the resulting gelatin, as shown in Figure 3. The treatments that included 0.075% and 0.05% citric acid and 0.075% acetate acid resulted in the lowest water content in the gelatin, with values of 4.58%, 5.07%, and 5.49%, respectively. The other three alternative therapies, on the other hand, resulted in much higher water content, ranging from 7.66% to 8.27%. It is worth noting that the water content of all treatments stayed below the maximum amount of gelatin water defined by the SNI of Fish Gelatin (SNI 8622:2018) [19].

Adjusting the concentration of acetate and citric acid during the immersion process does not significantly affect gelatin viscosity, as observed in Figure 3. When 0.075% acetic acid and 0.075% citric acid were added during immersion, viscosity measurements differed. The observed viscosity values for acetic acid were 10 mPa and 14 mPa for citric acid. It was previously discovered that increasing the concentration of acid used in the immersion process increases the viscosity of the resulting gelatin products [24]. This increase in viscosity value can be attributed to the interconnection of polypeptide chains, which in turn increases the molecular weight [26]. The viscosity identification analysis showed that the viscosity values of gelatin for all treatments were within the range of 10 to 12.8 mPa, which exceeded the vulnerability limits specified in BSi 757-1075. The specified range in the BSi 757-1075 is 1.5 to 7.5 mPa. This indicates that the acid concentration provided during the immersion process is excessively high, leading to an increased value of gelatin viscosity.

The results of the gel strength identification reveal that the concentration of acetate and citric acid used in the immersion process affects the strength of the gel produced (as shown in Figure 3). Treatments that included 0.075% acetic acid and 0.075% citric acid during immersion showed the highest gel strength values, at 259 and 251 blooms respectively. It is suspected that the increased acid concentration used in the immersion process improves the gel formation process by breaking down the polymer chain of amino acids. This is supported by Said [3]. The gel strength of all treatments was found to be within the vulnerability value of gel strength, which is a minimum of 75 bloom according to SNI Standard [19], and above 50 – 300 bloom according to BSI Standard [20].

Based on the treatments depicted in the study, it can be concluded that the two most effective treatments were the ones that incorporated 0.075% acetic acid (A2) and 0.075% citric acid (B2). Additional testing will be conducted on both treatments to determine the chemical properties of the most optimal gelatin products.

3.2 Chemical properties of gelatin A2 and B2

According to the results presented in Table 2, both Gelatin A2 and Gelatin B2 meet the standards outlined in SNI 8622:2018 [19]. Neither of the gelatin products contained heavy metal resin, indicating that the catfish skins used in manufacturing were free from heavy metallic resin prior to processing. Additionally, there was no evidence of migration or contamination from the production equipment during manufacturing. Typically, metallic debris in food items arises from contaminated raw materials with heavy metals during the cultivation process [27].

After analyzing the ash and protein content of two gelatin products, A2 and B2, it was observed that B2 had a lower ash value but a higher protein level compared to A2. Research conducted related to application of acid, indicates that an acidic solution can effectively break down the polymer chain of amino acids under optimal conditions, resulting in optimal protein extraction [3]. The other research also suggest that the concentration of the acidic solution used during the immersion process plays a significant role in breaking down the amino acid polymer chain at the right threshold [24]. Specifically, under the same process conditions, a 0.075% concentration of citric acid was found to be more effective than a 0.075% concentration of acetate acid.

Table 2. Chemical properties of gelatin A2 and B2.

Parameter	Gelatin A2	Gelatin B2	SNI 8622:2018
Ash Content (%)	0.68	0.37	Max 3
Protein Content (%)	88.92	93.22	-
Heavy Metal Contamination:			
- Mercury (Hg) (mg/kg)	Not detected	Not detected	Max 0.1
- Timbal (Pb) (mg/kg)	Not detected	Not detected	Max 1.5
- Cadmium (Cd) (mg/kg)	Not detected	Not detected	Max 0.5
- Arsenic (As) (mg/kg)	Not detected	Not detected	Max 0.5

4 Conclusion

The study found that soaking gelatin in different concentrations of acetic acid and citric acid did not affect the color of the gelatin produced. All treatments resulted in transparent, light yellow gelatin. The extraction process yielded gelatin ranging from 14.97 to 24.1%, with the highest yield obtained when using a higher concentration of acetic acid. The use of 0.075% citric acid resulted in the least amount of water in the extracted gelatin. The pH of the gelatin ranged from 4 to 5, with viscosities between 10 and 14 MPa. The gelatin produced had a strength range of 200-259 blooms. The best treatment for gelatin production was the use of a higher concentration of acetic acid, which resulted in high yield, low water content and viscosity, and high gel strength.

References

1. H.W. Ockerman, C.L. Hansen, Animal by-product processing & utilization (CRC Press, Boca Raton, Florida, America, 1999)
2. M. C. Gómez-Guillén, M. Pérez-Mateos, J. Gómez-Estaca, E. López-Caballero, B. Giménez, P. Montero, Trends Food Sci. Technol., **20**, 1 (2009)
3. M.I. Said, *Role and function of gelatin in the development of the food and non-food industry: A review,* in Proceedings of the 2nd International Conference of Animal Science and Technology (ICAST) 5-6 November 2019, Makassar, Indonesia (2019)
4. S. Nirwandar, Int. J. Halal Res., **2**, 1 (2020)
5. A.A. Karim, R. Bhat, Trends Food Sci. Technol., **19**, 12 (2008)
6. Straits Research, Food Grade Gelatin Market, Straits Research, 2023, [Online] (accessed Sep. 30, 2023)
7. DinarStandard, State of the Global Islamic Economy Report 2022, DinarStandard, 2022, [Online] (accessed Sep. 30, 2023)
8. World Population Review, Muslim Population by Country 2023, *World Population Review*, 2023, [Online] (accessed Sep. 30, 2023)
9. Jangkar Global Groups, Data Impor Gelatin Indonesia, Jangkar Global Groups, 2023, [Online] (accessed Sep. 30, 2023)
10. P. Tongnuanchan, S. Benjakul, T. Prodpran, K. Nilsuwan, Food Hydrocoll., **48** (2015)
11. H.Y. Liu, D. Li, S.D. Guo, LWT, **41**, 3 (2008)
12. A. Jongjareonrak, S. Rawdkuen, M. Chaijan, S. Benjakul, K. Osako, M. Tanaka, LWT, **43**, 1 (2010)
13. J. Alipal, N.A.S. Mohd Pu'ad, T.C. Lee, N.H.M Nayan, N. Sahari, H. Basri, M.I. Idris, H.Z. Abdullah, Materials Today: Proc., **42** (2019)
14. I. Zuraida, B.F. Pamungkas, AACL Bioflux, **13**, 5 (2020)
15. S.F. See, M. Ghassem, S. Mamot, A.S. Babji, J. Food Sci. Technol., **52**, 2 (2015)
16. S. Winarti, U. Sarofa, M.Y.R. Prihardi, Food Res., **5**, 4 (2021)
17. A. Hasan Zuhudi, A.H. Nurdiana, Res. Prog. Mech. Manuf. Eng., **4**, 1 (2023)
18. M. Nurilmala, M.T. Nasirullah, T. Nurhayati, N. Darmawan, J. Perikan. Univ. Gadjah Mada, **23**, 1 (2021)
19. DPBM, SNI produk perikanan nonpangan, (Direktorat Pengolahan dan Bina Mutu, Direktorat Jenderal Penguatan Daya Saing Produk Kelautan dan Perikanan, Kementerian Kelautan dan Perikanan, Republik Indonesia, Jakarta, 2019)
20. BSi, BS 757:1975: Methods for sampling and testing gelatine (physical and chemical methods), 1975, [Online] (accessed Sep. 30, 2023)
21. GMIA, Standard testing methods for edible gelatin, (Official Procedure of the Gelatin Manufacturers Institute of America, Inc., 2019)
22. AOAC, Official methods of analysis of AOAC international - 20th Edition, (Association of Official Analytical. Chemist. AOAC International. Virginia USA, 2016)
23. SNI, SNI 7387:2009 tentang batas maksimum cemaran logam berat dalam pangan, (Badan Standarisasi Nasional, Indonesia, 2009)
24. H. Muyasyaroh and A. A. Jaziri, *Effect of different acetic acid concentration on physicochemical characteristics of gelatin from starry trigger fish skin (Abalistes stellaris)*, in Proceedings of the International Conference on Sustainable Aquatic Resources (ICoSAR) 27-28 August 2019, Malang, Indonesia (2019)
25. I. Kołodziejska, E. Skierka, M. Sadowska, W. Kołodziejski, C. Niecikowska, Food

Chem. **107**, 2 (2008)
26. I. Jaswir, H.A. Monsur, H.M. Salleh, African J. Biotechnol.**10**, 81 (2011)
27. S. Elena Cristina, T. Lucia Carmen, Food. **12**, 18 (2023)

Lobster (*Panulirus* spp) catch in simeulue waters, aceh province: composition and results

Irfannur Irfannur[1], *Muliari* Muliari[2], *Syifa* Saputra[3,*], *Yusrizal* Akmal[1], *Agung* Setia Batubara[4], *Asmaul* Husna[1], *Imamshadiqin* Imamshadiqin[2]

[1]Department of Aquaculture, Faculty of Agriculture, Almuslim University, Bireuen, Indonesia.
[2]Department of Marine Science, Faculty of Agriculture, Malikussaleh University, North Aceh, Indonesia
[3]Department of Forestry, Faculty of Agriculture, Almuslim University, Bireuen, Indonesia
[4]Department of Biology, Faculty of Mathematics and Natural Sciences, Medan State University, North Sumatra, Indonesia

Abstract. Aceh Province's main export is lobster. This commodity commands a high economic price in Simeulue Regency, which boasts exceptional commodities in the fisheries industry, particularly lobster. The purpose of this study is to describe the composition of the catch and the productivity of the catch of six different varieties of lobster in the Simeulue water. The research methods employed include field surveys to collect data directly from fishermen. The study was carried out in Simeulue waters in numerous areas in 2023, including Busung, Kampung Air, Teupah Selatan, and Alafan. The types of samples taken were six species consisting of *Panulirus homarus, P. longipes, P. penicillatus, P. versicolor, P. ornatus* and *P. polyphagus*. The research results showed that the catch comprised lobster species such as *Panulirus homarus, P. longipes, P. penicillatus, P. versicolor, P. ornatus*, and *P. polyphagus*. The catch composition was dominated by *Panulirus penicillatus* (44%) and *P. versicolor* (23%). The length and weight growth of the six varieties of lobster is allometric, which is vital for sustainable fisheries management. The research findings also show oscillations in the supply of Simeulue lobster shipments, with peak delivery in 2022 (246,157 ind) and a fall in 2023 (57,654 ind). It is critical to underline the need for sustainable management in maintaining a balance between usage and conservation of fishing resources.

1 Introduction

Simeulue waters in Aceh Province are among the places with the greatest potential for marine resources and key products in the local fishing industry is lobster (*Panulirus* spp.). In addition to having a substantial economic impact, lobster fishing has an ecological effect on the equilibrium of the marine ecosystem. Unrestrained lobster fishing practices have been shown to endanger lobster populations' viability and harm maritime environments [1]. The productivity and quantity of lobsters in Simeulue seas can be impacted by elements like climate change, overfishing, and the degradation of natural habitats.

Lobster is a leading commodity in Aceh Province, particularly in Simeulue Regency. This commodity has a considerable economic value, making it a key pillar in the area's fishing sector [2]. According to statistical statistics, the overall production of lobster catches in Simeulue Regency from 2011 to 2016 was 2.4 tons per month, and continued to decline from

* Corresponding author : syifa.mpbiounsyiah@gmail.com

January 2016 to July 2016 [3]. The existence of lobster as a major fisheries item in Simeulue Regency had a substantial impact on the local economy and enhanced the income of fishermen. One of these is a *Panulirus*-genus crayfish found in Aceh province [4]. In addition to offering financial gains, lobstering has social effects and helps to sustain the culture of coastal communities that have long depended on marine resources.

Lobster fisheries resources have a high economic value and are a popular food commodity [5]. This fishery commodity has a high economic value and is intended to be sold and consumed both domestically and internationally [6]. Because the economic value of lobster is quite high, ranging from IDR 200,000 to IDR 400,000 per kg [7], it can be ensured that the economic benefits generated from this superior commodity are balanced with the protection and preservation of marine resources, which are critical for the future of the Simeulue Regency area. *Panulirus penicillatus, P. homarus, P. versicolor, P. longipes*, and *P. ornatus* are among the species caught off Badung Beach in Bali [8]. *Panulirus versicolor* is one of the primary species of lobster that is extremely vulnerable to fishing, as is the case in West Papua [9]. *Panulirus versicolor* is one of the five types of lobster that are dominantly caught in the waters of South Sulawesi [10]. The catch per fishing effort (CPUE) has decreased as a result of the high intensity of *Panulirus homarus* catches in Gunung Kidul waters [11]. Meanwhile, catch intensity dominates the *P penicillatus* species in Aceh Jaya waters, despite its bigger size compared to other species [4].

Intensive fishing is brought on by the increased demand for lobsters [12]. The average size of lobsters caught getting smaller as a result of a lack of control over fishing intensity. The economic worth of lobsters is decreased due to their decreasing size [13]. Fishing intensification that is not based on sustainable resource management principles, such as fishing using explosives, potash, or other chemicals, will harm habitats and ecosystems [14]. According to Junaidi et al. (2010) [15], such use will negatively impact other biological elements such as ratio imbalance, species extinction, and lobster populations. To manage lobsters according to species and distribution areas, a study on distribution based on species is needed due to the availability of lobsters in the Simeulue seas. Therefore, given that fishing is frequently carried out both locally and for export marketing, there is currently little research regarding the identification and catches of lobsters in Aceh Province. Therefore, more research is required to better understand biological elements and sustainable fishing. This study intends to define the composition of the catchand the productivity of the catch in six types of lobster in Simeulue waters.

2 Research Methods

The research methods employed include field surveys to collect data directly from fishermen and analyze population parameters such as weight and spread. The study was carried out in the waters surrounding the Simeulue Islands. The lobster fishing location is around 2 to 4 kilometers from the shoreline. The fishing fleet includes nets (gill nets), as well as Fishing vessel, with the size less than 5 gross tonnage. The sampling locations were determined at several areas, including Busung, Kampung Air, Teupah Selatan, and Alafan, as shown in (Figure 1) below.

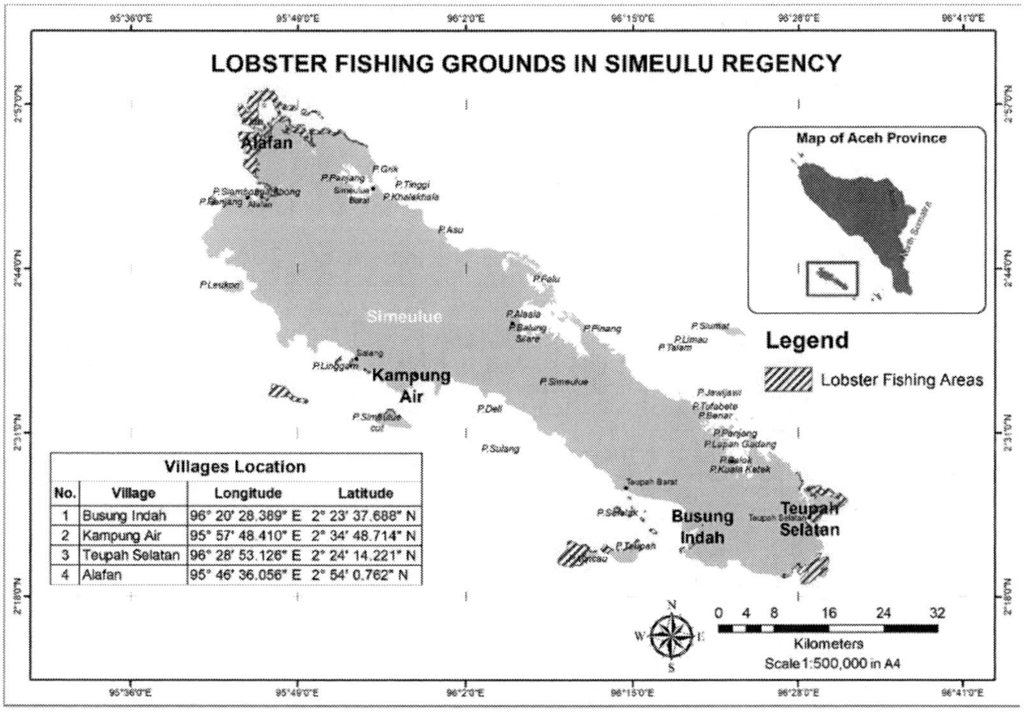

Fig. 1. Map of lobster fishing locations in Simeulue Regency waters

2.1 Composition and Productivity of the Catch

The composition and productivity of lobster catches might differ depending on lobster species, habitat, fishing methods, and other environmental factors. Because of their high economic value, lobsters are a valuable item in the fishing business. The catch composition was examined descriptively and tabulatively. For all lobster catches, the catch composition is displayed in the form of a catch composition graph. referring to Irfannur et al (2017) [4], data on catch composition was collected for 17 days at the research site.

The productivity of lobster fishing is measured using gillnet fishing gear. Catch per Unit Effort (CPUE) is a method for calculating the amount of lobster caught per unit of fishing effort. This analysis was carried out to determine the abundance and level of utilization based on the division between catch (Catch) and fishing effort (Effort) with the equation according to [16]:

$$CPUE=\frac{C}{F} \qquad (1)$$

Information: *CPUE* = Result catch per Unit (kg/trip)

Catch (C) = Total catch (kg)

Effort (F) = Total fishing effort (trip)

By estimating CPUE, we can compare lobster abundance under various situations or observe variations over time. When CPUE rises, either the lobster population or fishing productivity may have increased. On the other hand, a drop in CPUE can be a sign of a population fall or a decline in the efficiency of fishing gear.

3 Results and Discussion

Based on the data collected, lobster catches in Simeulue waters showed variations during the research period. The number of samples taken to see the composition of the catch during the study was 190 of all types of lobsters measured. The types of samples taken were six species consisting of *Panulirus homarus, P. longipes, P. penicillatus, P. versicolor, P. ornatus*, and *P. polyphagus*, as presented in Table 1 below.

Table 1. Composition of types and number and weight of lobsters

No	Type	Number	Type	Weight	Percentage (%)	
					Amount	Weight
1	*Panulirus homarus*	17	*P. homarus*	11.74	8.95	4.59
2	*P. longipes*	28	*P. longipes*	37.01	14.74	14.48
3	*P. penicillatus*	84	*P. penicillatus*	150.08	44.21	58.70
4	*P. versicolor*	45	*P. versicolor*	50.09	23.68	19.59
5	*P. ornatus*	9	*P. ornatus*	4.01	4.74	1.57
6	*P. polyphagus*	7	*P. polyphagus*	2.74	3.68	1.07
Total		190		255.67	100.00	100.00

The growth trend of the five types of lobster caught in Aceh Jaya waters, namely *P. penicillatus, P. homarus, P. longipes, P. versicolor*, and *P. ornatus*, is negative allometric. This means that some portions of the lobster's body grow more slowly than the rest of its body. Negative allometry in lobsters can have a big impact on fisheries management since it influences how lobster size fluctuates over time and how minimum catch sizes should be regulated to keep populations sustainable.

The composition of the catch collected during the research, with the *Panulirus penicillatus* type accounting for 58,70%, followed by the *P. versicolor* type accounting for 19,59%, *P. homarus* accounting for 4,59%, *P. longipes* accounting for 14,48%, *P. ornatus* accounting for 1,57%, and *P. polyphagus* accounting for 1,07%. According to the catch composition, the species *Panulirus penicillatus*, *P. versicolor*, and *P. longipes* account for the majority of the catch and may be the primary objectives for catching lobsters in Simeulue seas. The catches reported are consistent with previous study from Karimunjawa seas [17] and Canggu Beach, Bali [8], which both recorded the presence of similar types of lobsters. References to lobster biological points from research conducted in diverse locales can provide useful information into fisheries management and conservation activities. Knowing the composition of catches from different areas might help discover trends and compare lobster populations in different regions.

The main source of income for the coastal communities in Simeulue Regency is lobster fishing. Due to Indonesia's high lobster catch rates, lobster is a valuable product for the capture fisheries industry [14]. Simeulue waters are a conservation area, particularly for the conservation of fishery resources, which largely concentrates on the social aspect [18]. The significance of this region for fisheries resource protection is intimately tied to many societal factors that have an impact on the long-term viability of fisheries resource management. Therefore, it's crucial to establish sustainable fishing methods that prevent overfishing, such as catch limits, the use of environmentally friendly fishing equipment, and thorough population monitoring.

According to the research findings (Figure 2), the maximum possible catch occurred in February 2023, totaling 5,982 Kg, while the lowest occurred in May 2023, totaling 2,358 kg.

The results demonstrate that the quantity of lobsters caught varies from month to month. The number of lobsters taken in February 2023 reached 5,982 kg, showing that the season or specific conditions in that month enable active fishing activity and significant catches. The number of species caught could be influenced by variables such as the lobster breeding season or changes in migratory patterns. Meanwhile, the quantity of lobsters collected in May 2023 dropped drastically to 2,358 kg. This reduction could be due to a variety of variables such as seasonality or less favorable environmental conditions for fishing, or it could be due to human factors such as harsher fishing rules during that month. However, special attention must be devoted to the management and sustainability of fisheries resources to ensure that the Simeulue people may continue to rely on fisheries as a source of income. Data show that fluctuations in fish catch numbers can have a significant impact on the economic well-being of Simeulue villages that rely on fishing.

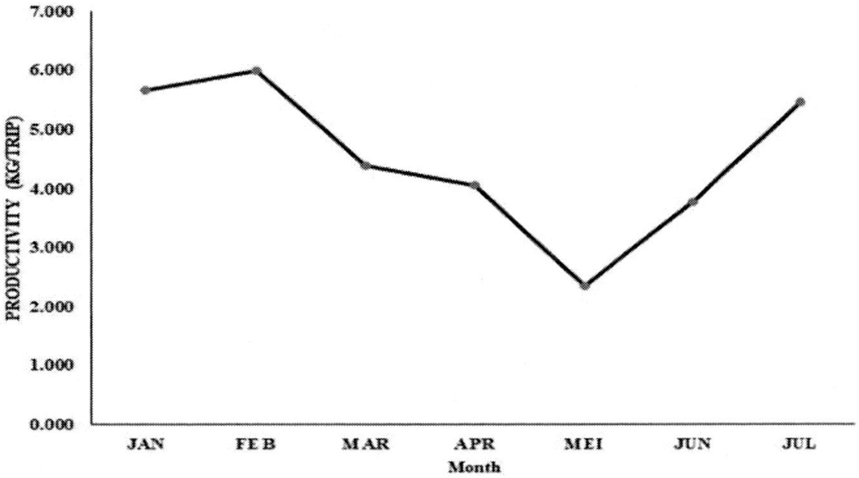

Fig. 2. Productivity of lobster fishing per month in Simeulu waters at four research locations

The catch productivity has decreased significantly when compared to the number of lobsters delivered outside Simeulue Regency (Figure 3). According to data from the Fish Quarantine Agency for Quality and Safety Control of Aceh Fishery Products, KKP Simeulue Regency Working Area, fluctuations in the number of individual lobsters sent show a seasonal pattern, with peak deliveries occurring in 2022 at 246,157 ind, and a decline in 2023 at 57,654 ind. This has a significant impact on the productivity of fishing catches. As a result, the number of lobsters delivered is a crucial indicator of fishing activity, as well as demonstrating the economic importance of the fisheries sector in this region.

https://doi.org/10.1051/bioconf/20248703005

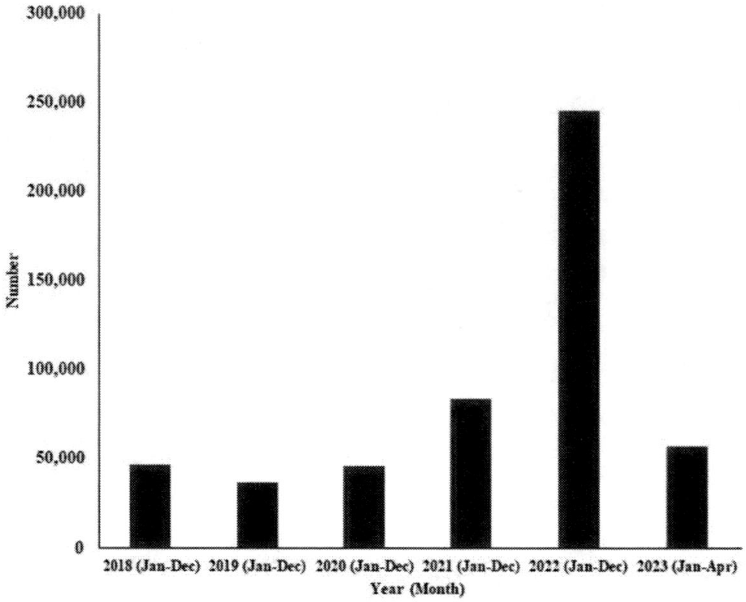

Fig. 3. Number (ind) of lobster shipping supplies per year in Simeulue Regency (Source: Fish quarantine agency for quality control and safety of Aceh fishery products, KKP Simeulue Regency Working Area)

4 Conclusion

Maintaining the sustainability of marine resources and marine ecosystems requires research on the composition and production of lobster catches in the Simeulue water. *Panulirus homarus, P. longipes, P. penicillatus, P. versicolor, P. ornatus*, and *P. polyphagus* are some of the lobsters that have been caught in Simeulue waters. The result showed that *Panulirus penicillatus* made up the majority (44%), followed by *P. versicolor* (23%), *P. homarus* (9%), *P. longipes* (15%), *P. ornatus* (5%), and *P. polyphagus* (as much as 4%). The biggest catch output in Simeulue waters was in February 2023, totaling 5,982 Kg, while the lowest occurred in May 2023, totaling 2,358 Kg. Meanwhile, the greatest delivery supply was in 2022, with 246,157 ind, and as much as 57,654 ind in 2023. Sustainable management practices can be applied to maintain a balance between the utilization and conservation of marine resources in the Semeulue Seas with a fuller understanding of lobster species, catch patterns, and impacts.

References

1. A.S. Samusamu, & Lukijanto. E3S Web Conf **74** (2018)
2. Edwarsyah. Jurnal Ilmu Dan Teknologi Kelautan Tropis **9**(2) 795–804 (2017)
3. A. Zulham, & Z. Nasution. Jurnal Kebijakan Sosial Ekonomi Kelautan Dan Perikanan **6**(2) 153 (2016)
4. Irfannur, R.I. Wahju, & M. Riyanto. ALBACORE Jurnal Penelitian Perikanan Laut **1**(2) 211–223 (2017)
5. S. Tomi. RJOAS: Russian Journal of Agricultural and Socio-Economic Sciences **90**(6) 51–59 (2019)
6. E. Anggraini, P. Arifin, & A. Rahman. Aquatic **4**(1) 1–1 (2021)

7. N. Maisyaroh, Ismail, & H. Boesono. Journal of Fisheries Resources Utilization Management and Technology **3**(3) 3 (2014)
8. M.R. Asvin, I.W. Restu, & R. Ekawaty. Current Trends in Aquatic Sci **2**(1) 108–114 (2019)
9. Tirtadanu, A. Suman, U. Chodrijah, B. Kang, & C.I. Zhang. Ocean and Coastal Management **211** 105780 (2021)
10. Musbir, Sudirman, & M. Palo. Jurnal IPTEKS Pemanfaatan Sumberdaya Perikanan **1**(2) 95–102 (2014)
11. A. Damora, Y. Wardiatno, & L. Adrianto. Marine Fisheries: Jurnal Teknologi dan Manajemen Perikanan Laut **9**(1) 11–23 (2018)
12. T. Tirtadanu, H.N. Yusuf, & C.I. Zhang. Indonesian Fisheries Research Journal **28**(1) 23–31 (2022)
13. B. Pranata, V. Sabariah, & S. Suhaemi. Jurnal Sumberdaya Akuatik Indopasifik **1**(1) 1 (2017)
14. A. Setyanto, N.A. Rachman, & E.S. Yulianto. Jurnal Perikanan Univ Gadjah Mada **20**(2) 49 (2019)
15. M. Junaidi, N. Cokrowati, & Z. Abidin. Jurnal Kelautan **3**(1) 29–35 (2010)
16. P. Sparre, & C.S. Venema. (1999)
17. T. Ernawati, A. Priatna, & F. Satria. Indonesian Fisheries Research Journal **25**(2) 91 (2019)
18. N.R. Ramadhanty, J.F. Setiawan, S. Aini, A. Putra, & P. Arisandi. American Academic Scientific Research Journal for Engineering Technology and Sciences **85**(1) 41–59 (2022)

Diversity of potential and endemic freshwater fishes in Aceh Tamiang

Agus Putra Abdul Samad[1]*, *Zidni Ilman* Navia[2], and *Afrah* Junita[3]

[1]Samudra University, Department of Aquaculture, Aceh, Republic of Indonesia
[2]Samudra University, Department of Biology, Aceh, Republic of Indonesia
[3]Samudra University, Department of Accounting, Aceh, Republic of Indonesia

Abstract. The inventory of fish species is required in developing a fisheries control plan, knowing the fish distribution and the potential of fishes as a basis for water spatial planning. This research aims to evaluate the fish diversity of freshwater fishes in Aceh Tamiang. An explorative study was conducted to investigate freshwater fishes in rivers, reservoirs, and other lotic and lentic ecosystems. Specimens were catched using fishing net, gill net and fish trap. The sampling locations were spread over 10 villages and data were collected from July to August 2023. The water quality parameters observed were temperature, pH, DO and brightness. This study reported a total of 26 species belonging to 11 families including: Cyprinidae, Osphronemidae, Cichlidae, Channidae, Synbranchidae, Clariidae, Zenachopteridae, Bagridae, Anabantidae, Nandidae, and Eleotridae. Whereas Cyprinidae and Osphronemidae are the most dominant families found. Morphometric of body weight varied from 0.6 to 156 g and total length 4.6 to 23.6 cm. The factor condition ranged of 0.26 to 7.95. This study showed the diversities of fishes in Aceh Tamiang which may be important for fisheries management. Furthermore, it serves as a basis for future efforts in domesticating freshwater fish in Aceh.

1 Introduction

Indonesia is well-known as one of the centers of fish biodiversity in the world. Nowadays, Indonesia ranks the second largest diversity of fish species in the world after Brazil [1-2]. It is reported, approximately 4,782 species have been identified in Indonesia, of which 1,243 are freshwater fish species [3]. Various types of freshwater fish are known to have an important role as a source of animal protein for society, especially people living in rural areas [4-5]. Moreover, fish also contains nutrients that fulfill the humans need such as fat, essential amino acids, omega-3 and DHA, vitamins and minerals [6-8].

Biodiversity plays a role in stabilizing ecosystems in waters [9]. Aquatic ecosystems such as: rivers, reservoirs, lakes, and swamps may function as fish producers for community [10]. Rivers are also zoning as a breeding place for fish [11]. The characteristics of aquatic habitats are important in determining the diversity and abundance of fish species. In general, habitat is influenced by the differences in river slope, the presence of forests or plants along the watershed [12] which is associ*ated with the presence of animals that inhabit it [13]. High species diversity demands there is also high-quality habitat. However, in certain cases, the characteristics of a habitat can force the resident biota to adapt in order to survive [14]. Research on the diversity of fish species is considered very important in efforts to maximize

* Corresponding author: agus.putra.samad@gmail.com

the fisheries potential in each region

Aceh Province is one of the regions in Indonesia which has a fresh water area of around 45,000 ha and is a habitat for various aquatic biota including wild fish. The previous study [15] have found 114 species of freshwater fish in Aceh waters, and 26 of them are new fish species. The discovery of these new species shows that there are still hundreds of fish species that have not been discovered [2], and the number of fish types is increasing along with increased exploration activities. Freshwater fish genetic resources also facing serious threats by various human activities. This is caused by fishing activities that are carried out through destructive fishing using materials, tools or methods that damage fish resources and the environment [16], such as the use of toxic materials and cyanide [17], explosives [18-19], electric shocks and others tools that are not environmentally friendly [20].

Although freshwater fishes are the most affected by human activities worldwide [21-22], conservation of freshwater fish diversity has received less attention than other vertebrates [23]. Biodiversity is central to biodiversity conservation [24], and the accurate description, measurement, and assessment of biodiversity has therefore been a long-standing concern in ecology and conservation biology [25]. It is feared that limited information regarding the potential of freshwater fish, especially those that live wild in nature, will reduce public attention to these fish and, in the long term, will encourage the extinction of various types of freshwater fish in nature before they are fully utilized. Indeed, effective conservation and sustainable use of genetic resources need to be carried out in an effort to support sustainable food security for current and future generations.

Tamiang regency is located on the Eastern part of Aceh, Indonesia. Its rivers belong to two major basins (right junction river and left junction river), and flowing into the Tamiang river, of which uses and utilizes by the community surrounding as a necessity for daily life and economy both to earn a living and for daily needs [26]. Moreover, [27] mentioned that Aceh Tamiang is one of the regions with the richest aquatic biodiversity and is also an important producer of freshwater fisheries in Aceh. The rivers in Tamiang have long been used by the local community as a place for fishing activities to fulfill daily dishes. However, it is feared that excessive fishing will result in a drastic reduction in fish populations or could even lead to the extinction of certain fish. Therefore, conservation, domestication and cultivation activities can be carried out to prevent the extinction of fish in this area.

Thus, studies on the diversity of fish species is considered very important in efforts to maximize the fisheries potential in each region. However, until now there is no information about the diversity of fish species in Aceh Tamiang district. Indeed, this research aims to determine the diversity and abundance of fish species in the Aceh Tamiang. It is hoped that the results of this research can support the interests of fish conservation and become a database of fish types found in Aceh.

2 Materials and methods of research

2.1. Time and Place

This research was conducted in Aceh Tamiang district. The sampled areas are spread in 10 stations in 4 sub-districts: Bendahara, Seruway, Karang Baru, and Tenggulun (Fig. 1). Sampling stations is carried out after a location survey and interviews with fishermen and local fisheries service officers. Samples were caught using gill net, fishing poles and traps. These locations were selected based on the presence of a main river with geographical characteristics, topography, distribution of typical wild fish, temperature, organic matter content and water flow speed. The data was collection from July to August 2023. Fish specimen identification was analyses in Laboratory of the Aquaculture Department Universitas Samudra

https://doi.org/10.1051/bioconf/20248703006

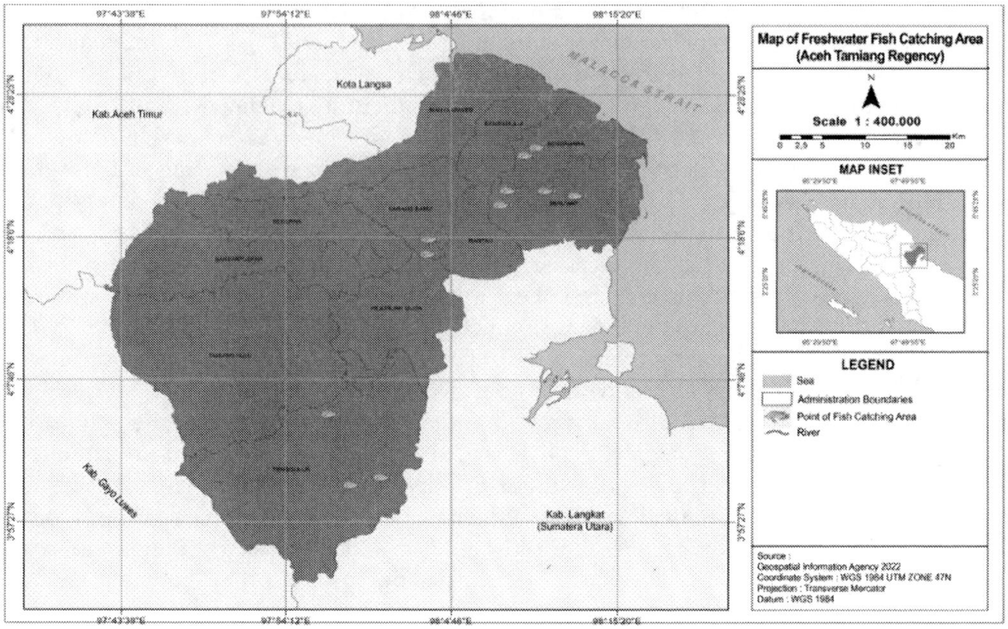

Fig. 1. Map of the study area in Aceh Tamiang, Indonesia

2.2. Data Collection

Exploration of typical fish of Aceh was carried out using the roaming method [28]. All species found at the research location were collected, the local name, habitat and morphological characteristics were recorded [29]. The species was then identified in the laboratory referring to [30-32]. The origin and location of species were determined based on Langaeni et al [33], while the determination of endemic species is guided by the method of Gubiani et al and Zawadzki et al [34-35].The distribution pattern was created based on the coordinates of the fish species found using GPS program [36].

2.3. Research Parameters

Relative abundance of each type of fish is done by calculating the percentage: $Kr = \frac{ni}{N} x100\%$ [27], Kr = Relative Abundance; ni = Number of species i; N = Total number of all species. Meanwhile, to determine whether there is dominance or not, the Simpson Dominance Index: C = Simpson's dominance index; ŋi = Number of individuals of 1st species; N = Number of all species [37]. The dominance index value ranges from 0-1 with the following categories: Index 1 = very high species dominance. Index 0 = nothing dominates. Condition factor calculations are based on the length and weight of the fish. Condition factors are calculated using the formula $K = \frac{Wt}{TL^3}\ x\ 100$ [38], K = condition factor; Wt = final weight (g); TL = average length (cm)

3 Results

3.1. Fish Condition Factor

The fish condition factor describes the plumpness of the fish which is expressed based on length and weight data. The use of condition factor values is important for determining the quality of available fish meat [27]. The results of the fish condition factors are listed in Table 1.

Table 1. Data of fish species and condition factors in Aceh Tamiang

No	Fish Family	Scientific Name	Local Name	W	L	CF
1	Channidae	*Channa striata*	Gabus	54	18,4	0.87
2	Cichlidae	*Tilapia mossambica*	Mujair	63	15.5	1.69
3	Osphronemidae	*Osphronemus gouramy*	Gurami	35.5	12.9	1.65
4	Osphronemidae	*Trichogaster trichopterus*	Sepat	16	10.3	1.46
5	Osphronemidae	*Trichogaster sp*	Sepat	8.9	9.5	1.04
6	Osphronamidae	*Betta splendens*	Cupang	0.6	4.6	0.62
7	Bagridae	*Pterygoplichthys pardalis*	Suik	12	10	1.20
8	Cyprinidae	*Barbonymus sp*	Sirip kuning	25	13.5	1.02
9	Cyprinidae	*Barbonymus sp*	Pangkal hitam	11	10.3	1.01
10	Cyprinidae	*Barbonymus sp*	Tawes silver	11	9.3	1.37
11	Cyprinidae	*Barbonymus sp*	Garis hitam	10	10	1.00
12	Cyprinidae	*Barbonymus sp*	Sirip bening	29	13.7	1.13
13	Cyprinidae	*Barbonymus sp*	Ekor hitam	32	12.5	1.64
14	Cyprinidae	*Barbonymus sp*	Tawar merah	18	11	1.35
15	Cyprinidae	*Barbonymus sp*	Mata merah	9	9	1.23
16	Cyprinidae	*Mystacoleucus padangensis*	Bilis	2.2	7.8	0.46
17	Cyprinidae	Un-identified	Bidadari	12	8.1	2.26
18	Cyprinidae	*Rasbora argyrotaenia*	Aur aur	0.7	4.8	0.63
19	Anabantidae	*Anabas testudineus*	Betok	61.6	14.8	1.90
20	Clariidae	*Clarias sp*	Lele hitam	92	10.5	7.95
21	Clariidae	*Clarias batrachus*	Lele sawah	91	11	6.84
22	Eleotridae	*Oxyeleotris marmorata*	Betutu	12	9.5	1.40
23	Synbranchidae	*Monopterus albus*	Belut	53.9	23.6	0.41
24	Cichlidae	*Tilapia sp*	Nila sungai	156	20	1.95
25	Nandidae	*Pristolepis grootii*	Keper	68.6	14.8	2.12
26	Zenarchopteridae	*Dermogenys orientalis*	Julung-julung	11	16.2	0.26

W= Weight, L= Length, CF= Condition factor

The use of condition factor values has important implications in determining the quality of available fish meat [38]. Table 1 shows that the condition factor of Aceh Tamiang fish during the study period ranged from 0.26 to 7.95. The highest condition factor was found in Clarias sp (7.95) with a length of 10.5 cm and a weight of 92 g. on the other hand, the lowest condition factor value was found in *Dermogenys orientalis*, namely 0.26 with a length of 16.2 cm and a weight of 11 g. The condition factor value indicates that the fish found are thin. [39] stated that if the fish condition factor value is 1 to 3, then the condition of the fish is in a thin body condition (flat), however if the fish reaches a value of 3 up to 4 means the fish has a fat body. The data reported 92.3% of species found in Aceh Tamiang is below 3, which indicated that mostly fishes are in thin condition. Whereas 7.7% is in fat body condition.

Moreover, [9] mentioned that the low value of the condition factor is caused by the high population density in the environment, which resulting in stunted fish growth. [13] also confirm that value of the condition factor depends on many things such as: the number of organisms in nature, the condition of the organisms, the abundance of food, the condition of aquatic environment, and the development of the gonads.

3.2. Fish Relative Abundance

This study showed that the highest fish relative abundance in Aceh Tamiang is found in Cyprinidae family (46.12%), following by Osphronemidae and Cichlidae: 17.55% and 10,25%, respectively. While the lowest relative abundance value is seen Eleotridae (1.09%). The data as shown in Table 2.

Table 2. Relative abundance of fish

No	Fish Family	Relative Abundance (%)
1	Cyprinidae	46.12
2	Osphronemidae	17.55
3	Cichlidae	10.25
4	Channidae	5.28
5	Synbranchidae	4.66
6	Clariidae	4.19
7	Zenarchopteridae	3.57
8	Bagridae	3.42
9	Anabantidae	2.02
10	Nandidae	1.86
11	Eleotridae	1.09

This present study found that Cyprinidae family (*Barbonymus sp*, *Mystacoleucus padangensis, Rasbora argyrotaenia,* Un-identified) reach the highest relative abundance of species. The abundance of this species is expected due to the condition of the aquatic habitat and the quality of the water which is suitable to its living environment. The number of cyprinidae fishes caught was 297 fish. On the other hand, the lowest relative abundance was seen in Eleotridae family (*Oxyeleotris marmorata*), with only 7 fish was catched. [40] stated that cyprinidae family is able to live in fresh water, including shallow waterways, ponds and lakes. Those fishes have also demonstrated the ability to adapt and reproduce rapidly. [9] added that in general the highest abundance value is caused by the ability to utilize the potential of existing resources to fulfill the need and be able to adapt well. Moreover, studies on freshwater fish have been conducted previously [41-42], however, surveys need to be carried out regularly to complete the information about fish species and to obtain information about the health status of fish in a particular ecosystem. In fact, this study discovered an unidentified species belonging to the Cyprinidae family.

3.3. Diversity of Fish

This present study catches 644 fishes which divided into 26 species belonging to 11 families. The dominant species in Aceh Tamiang district belong to the families Cyprinidae (297 fish) and Osphronemidae (113 fish). This is followed by Ciclidae, Clariidae, Channidae, Synbranchidae. Whereas the smallest number of catched spesies is in Eleotridae family (7 spesies) as shown in Fig.2. However, among the 26 species, 1 species belonging to the family Cyprinidae has not been identified.

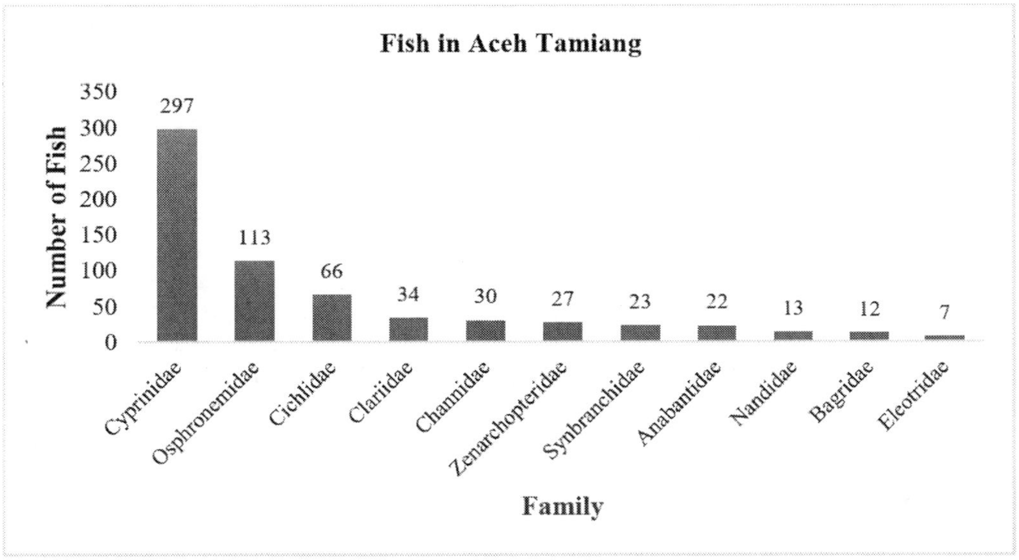

Fig. 2. Number of freshwater fish in Aceh Tamiang

Fish species diversity represents the full extent of ecological adaptation and species evolution in a particular environment. Therefore, it is understandable that fish diversity may vary from place to place [43]. In this study, we found that the distribution differed among locations. At this time, the species caught in Seruway sub-district (lake region) were different from those caught in Tenggulun sub-districts (river region). [44] mentioned that the size of the river channel and its location within the watershed have a significant influence on the composition of fish species. However, human activities can also influence its diversity [45]. To survive and reproduce successfully, fish must be able to adapt to environmental conditions. A community that has low species diversity is influenced by a low uniformity index and also the dominance of one species [46]. Fish species abundance and structure of fish communities in rivers depend on habitat conditions, increasing water volume in the river, the presence of predators, and competition for food. A stable environment also characterized by balanced conditions and contains a diversity of life without any dominant species [27].

Fish diversity is a community character that shows the level of species diversity in a particular community. Diversity type is determined by two properties of the community that are not related to each other, namely the number of types and the uniformity of the abundance of individuals of each type [40]. Species diversity is not only assessed by the number of types, but also based on the distribution of individuals within their species. Type diversity can be considered low if the distribution of individuals is not evenly distributed within each type, even though it has a large number of types. There are two types of ways to analyzing species diversity: based on the curve of the number of species and diversity indices [43]

3.4. Morphometric analysis

The morphometric measurement of the caught fish was described by morphological analysis. The total number of morphological traits measured in this study was 21 indicators (Table 3). The *Tilapia sp* has the heaviest weight (156 g), however the longest length seen in *Monopterus albus* (23.6 cm). Meanwhile, the *Betta sp.* showed the lightest weight (0.6 g), and also the shortest (4.6 cm) among 26 samples.

Morphometric measurements are measurements that can be measured from the fish's body, for example total length, standart length, head length, and height or shape of the fish

[4, 44]. In the present study, 21 indicators were applied to each sample including Total Length, Standard Length, Head Length, Head Height, Body Height, Tail Base Height, Eye Diameter, Distance Between Eyes, Body Width, Length Before Dorsal Fin, Length Before Ventral Fin, Length Before Anal Fin, Dorsal Fin Base Length, Anal Fin Base Length, Ventral Fin Base Length, Pectoral Fin Base Length, Top Tail Fin Length, Middle Tail Fin Length, Low Tail Fin Length.

Information about the relationship between length and weight of a species is absolute needed in estimating fisheries potential [38] because this activity can provide information regarding the composition of organisms available in nature, age gonad maturity, mortality, life cycle, and growth and reproduction of a organisms that live in waters. Apart from that [14] stated that the process of fish growth involves many factors such as: water quality conditions, food availability and continuity, seed stocking size, gender, and the number of other fish that utilize the same resources.

The morphometric characteristics measured in this study were: 4.6 to 23.6 cm for shortest and longest length, respectively and 0.6 to 156 g for the lightest and the weightes, respectively. Whereas the body height ranged from 0.5 to 7 cm. [38] found that fish have body parts that can be measured or counted. However, fish growth process involves some factors such as water quality conditions, availability and continuity of feed, and number of other fish that utilize the same resources [23].

Table 3. Morphometric data of fish found in Aceh Tamiang

No	Scientific Name	MORFOMETRIK (cm)																				
		W (gr)	TL	SL	HL	HH	BH	TBH	ED	DBE	BW	LBDF	LBVF	LBAF	DFBL1	DFBL2	AFBL	VFBL	PFBL	TTFL	MTFL	LTFL
1	*Channa striata*	54	18.4	15	4.6	2.7	2.6	1.9	0.5	2.7	2	5.1	1.5	8.9	10.2	0	6.6	2.2	2.7	3	5.6	3
2	*Tilapia mossambica*	63	15.5	11.8	4	4.5	5	2.1	0.8	0.6	2	5	5	9	10	0	5	3.5	5	4.8	4.8	4.8
3	*Osphronemus gouramy*	36.5	13.9	10.2	3.3	4	5	1.3	0.9	1.1	1.6	6.2	3.3	5.3	4.4	0	4.8	4.9	0.9	2.5	2.7	2.5
4	*Trichogaster trichopterus*	16	10.3	8.9	2.5	2.8	3.2	1	0.5	1.2	1.6	4.9	2.6	3	2.5	0	4.4	8.2	2	1.9	1.5	1.9
5	*Trichogaster sp*	7.9	8.5	6.4	2.0	1.9	2.6	1.3	0.4	0.5	1.2	3.6	1.9	2.6	1.4	0	4.1	6.1	2.0	2.2	1.5	2.2
6	*Betta sp.*	0.6	4.6	3.6	0.6	0.5	0.5	0.4	0.1	0.1	0.3	2.8	1.5	3.4	0.3	0	1	1.3	0.9	1	1	1
7	*Pterygoplichthys pardalis*	18.9	13.9	10.7	2.8	1.8	3	1.4	0.6	0.4	1.5	4.2	5.5	8.2	1.6	2.8	1.6	1.7	2.1	2.9	1.1	2.2
8	*Barbonymus sp*	25	13.5	10.9	2.7	2.8	3.8	1.2	0.8	1.6	1.9	4.7	5.2	8.1	2.5	0	2	2.4	3.1	2.9	1	3.3
9	*Barbonymus sp*	11	10.3	7.5	1.5	2	2.3	1	0.5	0.4	0.6	4	3.7	6.4	1.5	0	0.5	1.5	1.2	2.9	1	3.1
10	*Barbonymus sp*	11	9.3	7.2	1.7	2	2.3	0.9	0.5	0.7	0.5	4.2	3.5	6.1	2	0	0.5	1.8	1.5	3	0.9	3.1
11	*Barbonymus sp*	10	10	7.5	1.7	1.5	2.5	0.9	0.5	0.7	0.8	3.5	3.8	7.2	1.8	0	0.8	1.7	1.6	2.6	1	2.8
12	*Barbonymus sp*	29	13.7	10.8	2.3	2.8	3.8	1.3	0.5	1.7	2	4.8	5.6	8.2	2.3	0	2.1	2.4	3.2	2.8	1	3.2
13	*Barbonymus sp*	32	12.5	9.3	3	2.5	4.5	1.8	1	1.8	2.5	5	4.5	8.8	3	0	2	2.8	3.2	3.2	1	3.4

14	*Barbonymus sp*	18	11	8.5	1.8	1.5	3.3	1.3	0.5	0.8	0.7	4	4.2	7.8	1.4	0	0.8	2	1.8	3	1.5	3.2
15	*Barbonymus sp*	9	9	7	1.5	1.7	2.5	0.9	0.5	1	0.7	3.5	3.8	6.8	1.3	0	0.8	1.5	1.3	2.2	0.8	2.5
16	*Mystacoleucus padangensis*	2.2	7.8	3.6	0.6	0.5	0.5	0.4	0.1	0.1	0.3	2.8	1.5	3.4	0.3	0	1	1.3	0.9	1	1	1
17	Un-identified	12	8.1	6.4	1.8	1.8	3	0.7	0.8	0.8	1.5	2.5	2.7	5.5	10	0	3	1.5	2	1.8	0.8	1.8
18	*Rasbora argyrotaenia*	0.7	4.8	3.6	0.6	0.5	0.5	0.4	0.1	0.1	0.3	2.8	1.5	3.4	0.3	0	1	1.3	0.9	1	1	1
19	*Anabas testudineus*	61.6	14.8	12.5	4.1	2.7	2.2	1.9	0.7	0.5	4.6	4.3	5.2	11.8	7.6	0	5.1	1.4	3.1	2.9	3.2	2.3
20	*Clarias sp*	92	10.5	9	2	1	2	1.2	0.5	1	1	3.5	4.3	4.5	6.1	0	4.5	1	1	1	2	1
21	*Clarias batrachus*	91	11	9.5	2.5	1	2	1.5	0.3	1.5	1	3.5	4.5	5.5	6.5	0	5	1.5	1.5	1	2	1
22	*Oxyeleotris marmorata*	12	9.5	7.5	2	1.5	2.5	1	0.4	0.7	1	3.5	3	5.5	0.3	0.7	2	1.3	1	1.5	2	1.5
23	*Monopterus albus*	53.9	23.6	0	1.6	0.9	0.9	0	0.1	1	1.8	0	0	0	0	0	0	0	0	0	0	0
24	*Tilapia sp*	156	20	16.5	5.5	4	7	2.5	1.5	2.5	3	7	7.5	12	10	0	3.5	5.5	3.5	4	4	4
25	*Pristolepis grootii*	18	12.1	9.8	2.5	1.5	2.3	1.5	0.4	0.8	0.9	3.5	5.4	6.9	1.1	2.3	1.9	1.3	1.9	2.4	0.9	2.4
26	*Dermogenys orientalis*	11	16.2	14	2	0.5	1	0.2	0.3	0.8	0.5	0	0	0	0	0	0	0.5	0	0.3	0.5	0.3

Note: TL (Total Length); SL (Standard Length); HL (Head Length); HH (Head Height); BH (Body Height); TBH (Tail Base Height); ED (Eye Diameter); DBE (Distance Between Eyes); BW (Body Width); LBDF (Length Before Dorsal Fin); LBVF (Length Before Ventral Fin); LBAF (Length Before Anal Fin); DFBL (Dorsal Fin Base Length); AFBL (Anal Fin Base Length); VFBL (Ventral Fin Base Length); PFBL (Pectoral Fin Base Length); TTFL (Top Tail Fin Length); MTFL (Middle Tail Fin Length); LTFL (Low Tail Fin Length)

4 Conclusions

In summary, Aceh Tamiang has 26 species and 11 families (644 fishes) of native freshwater fish. Through morphological examination, *Monopterus albus* and *Betta sp.* have the longest and shortest TL of 23.6 cm and 4.6 cm, respectively. Whereas, *Clarias sp.* and *Dermogenys orientalis* had the highest and lowest CF values of 7.95 and 0.26, respectively. Knowing the diversity and condition factor are important for fisheries management and research on fish population dynamics. In addition, it serves as a basis for future efforts in domesticating freshwater fish.

Acknowledgments

Authors expresses sincere gratitude to all parties who have aided in the completion of this research.

References

1. M.M. Helentina, J. Khairul. Ilmu Alam Lingk. **11**, 7 (2020)
2. S.S. Lenny. J. Peny. Perik. Kel. **11**, 14 (2017)
3. S. Suharsono. Lembaga Ilmu Peng. Indo. (2014)
4. H.L. Ohee. J. Biol. Papua. **9**, 8 (2017)
5. R.K. Hadiaty, S. Wirjoatmodjo. J. Iktiol. Indo. **2**, 6 (2002)
6. E. Weichselbaum, S. Coe, J. Buttriss, S. Stanner. Nutri. Bull. **38**, 49 (2013)
7. M. Glover-Amengor, M.A. Ottah Atikpo, L.D. Abbey, L. Hagan, J. Ayin, J. Toppe. World Rural Observ. **4**, 5 (2012)
8. A.G.T. Tacon, M. Metian. Rev. in Fish Sci. **21**, 6 (2013)
9. S. Wahyuni, S. Sulistiono, R. Affandi. J. Bumi Lest. **14**, 1 (2014)
10. F. Fahrurazi, Y. Riza, E. Ernadi. *J. Kes. Masy.* **5**, 1 (2018)
11. D. Raubenheimer, S.J. Simpson. In *Curr. Op. Insect Sci.* **27**, 1 (2018)
12. A. Ali. J. Bumi Lest. **13**, 2 (2013)
13. S. Purwati, M. Masitah, S. Budiarti, Y. Aprilia. J. Ilmiah BioSmart. **7**, 1 (2021)
14. A. Latief, A. Putra, A.B. Suwardi, B. Baihaqi. Acta Aquatica. **4**, 2 (2020)
15. Z.A. Muchlisin, M.N. Siti Azizah. Intern. J. Zool. Res..**5**, 17 (2009)
16. KKP. 2019. Kepmen-KP. (2019)
17. A.W. Rahim. J. Mar. Fish. **8**, 10 (2017)
18. H. Somun. J. Ilmu Hukum Leg. Opinion. **3**,10 (2014)
19. A. Mustofa, D. Daniel, F. Firmasyah, N. Nubandika. J. Ecol. Fish. Soc. Stat. 20 (2017)
20. N. Nurdin, A. Grydehoj. J. Mar. Island Cult. 5 (2014)
21. P.A. Hastings, H. J. Walker, Jr., R.G. Grantly. Fishes a Guide to Their Diversity. (2014)
22. Kurniawan, Andi. Ekologi Sistem Akuatik. Malang: UB Press. (2018)
23. E.A. Widjaja, Y. Rahayuningsih, J.S. Rahajoe, R. Ubaidillah, I. Maryanto, E.B. Walujo, G. Semiadi. LIPI Press (2014)
24. M. Krokosek, J.D. Olden. Cambridge University (2016)
25. M. Kottelat, J.W. Anthony, N.K. Sri, N.K, W. Soetikno. Periplus Editios. (1993)
26. A.P.A. Samad, A. Junita, B. Baihaqi. In proceeding IOP Conference Series: Earth and Environmental Sci. **1221**, 1 (2023)
27. K. Aprilia, A.S. Agus Putra, R. Rosmaiti. J. Perik. Unram. **13**, 8 (2023)
28. R.L. Delariva, J.C. Silva. Check List. **9**, 5 (2013)
29. C.A. Lucena, H.G. Soares. Zootaxa. **4071**, 24 (2016)
30. W.J. Graca, C.S. Pavanelli. Neotrop. Ichthyol. **6**, 3 (2008)

31. L.F.C. Tencatt, M.R. Britto, C.S. Pavanelli. Neotrop. Ichthyol. **12**, 7 (2014)
32. F.F.Roxo, G.S.C.Silca, L.E. Ochoa, C. Oliveira. Zookeys. **534**, 31 (2015)
33. F. Langaeni, R.M.C.Castro, O.T. Oyakawa, O.A.Shibata, C.S. Pavanelli, L. Casatti. Biota. Neotrop. **7**, 16 (2007)
34. E.A. Gubiani, A.J. Holzbach, G. Baumgartner, L.B.R. Neto, F. Bergmann. Check List. **2**, 5 (2006)
35. C.H. Zawadzki, F.F. Roxo, W.J. Graca. Ichthyol. Explor. Freshwaters. **26**, 10 (2016)
36. D. Viana, C.H. Zawadzki, E.F. Oliveira, H.F. Vogel, W.J. Graca. Biota Neotrop. **13**, 8 (2013)
37. M. Hedja, P. Pysek, Jarosik. J.Ecol. **97**, 10 (2009)
38. A.P.A. Samad, N. Fazillah, R. Humairani, N.F. Hua. J. Biosciences. **29**, 6 (2022)
39. A.P. AS, A. Junita, M. Jamil. Depik **12**, 1 (2023)
40. R. Supanji, W. Adi, E. Utami. J. Sumberdaya Per. **12**, 2 (2018)
41. D. Samitra, Z.F. Rozi. J. Biota. **4**, 1 (2018)
42. T. Sukomono, D.D. Solihin, M.F. Rahardjo, R. Affandi. J. Iktiol. Indo. **13**, 2 (2013)
43. L.S. Syafei. J. Peny. Perik. Kel. **11**, 1 (2017)
44. M.D. Febriani, D. Bhagawati, S. Suryaningsih. J. Ilmiah Biol. Unsoed. **1**, 2 (2019).
45. M. Sirait, F. Rahmatia, P. Pattulloh. Indo. J. Mar.Sci. Technol. **11**, 1 (2018)
46. A. Efendi, A. Astuti, P.I., N.T. Rahayu. J. Pen. Hum. **18**, 2 (2017)

Biometrics of the gastrointestinal tract of *Tor tambra* and *Tor soro* from Aceh, Indonesia

Yusrizal Akmal[1,2], *Rossy* Azhar[2], *Muchammad* Yunus[3*], *Hani* Plumeriastuti[4], *Yeni* Dhamayanti[5], *Agung Setia* Batubara[6]

[1]Sains Veteriner Doctoral Program, Faculty of Veterinary Medicine, Universitas Airlangga, Surabaya. Indonesia
[2]Department of Aquaculture, Faculty of Agriculture, Universitas Almuslim, Bireuen, Indonesia
[3]Department of Veterinary Science, Faculty of Veterinary Medicine, Universitas Airlangga, Surabaya. Indonesia
[4]Department of Veterinary Parasitology, Faculty of Veterinary Medicine, Universitas Airlangga, Surabaya Indonesia
[5]Department of Veterinary Anatomy, Faculty of Veterinary Medicine, Universitas Airlangga, Surabaya. Indonesia
[6]Department of Biology, Faculty of Mathematics and Natural Sciences, Universitas Negeri Medan, Indonesia

Abstract. The gastrointestinal tract of native fish shows structural and functional diversity, which is related to different dietary requirements, feeding habits, phylogeny, age, body shape, and sex. There is some research that discusses the digestive tract of fish, but research on the digestive tract of *Tor tambra* and *Tor soro* is still limited. The purpose of this study was to describe the biometrics of the stomach and intestines of *T. tambra* and *T. soro* harvested from Leuser Ecosystem Area. This study included sample collection, sample preparation, gastrointestinal biometrics, and data analysis. The observation parameters include GaSI (Gastro Somatic Index), ISI (Intestine Somatic Index), RGL (Relative Gut Lengths), and RIL (Relative Intestine Lengths). The measurement data were tested using analysis of variance (ANOVA). The results showed that the highest GaSI, ISI, RGL and RIL values were in *T. tambra* with respective values of 0.50±0.24%, 2.46±0.58%, 44.35±5.22%, and 340.37±33.99%, while the lowest values were found in *T. soro*, whereas 0.22±0.07%, 1.57±0.55%, 32.03±14.99%, and 259.62±65.74%, respectively. The results of the statistical analysis showed that there were significant differences ($P<0.05$) in the GaSI, ISI, and RIL values, while RGL no significant differences ($P>0.05$) between the two species.

1 Introduction

The digestive tract of teleosts has attracted considerable interest in its morphological structure and function related to taxonomy and feeding habits [1]. One of the main factors, according [2][3], affecting fish survival is adequate nutrition which depends on effective ingestion,

* Corresponding author: muchammad-y@fkh.unair.ac.id

digestion, and assimilation of nutrients as well as the physiological competence of the fish to digest and transform ingested material. Digestion is an important process that directly affects the availability of nutrients needed for the body's biological activities; and its efficiency depends on the type and function of digestive enzymes [4]. Feeding habits are influenced by several factors, such as appetite and digestive tract structure, but are also related to the external properties of food, such as protein source and palatability [5]. Fish feeding activities and traits include herbivores that feed on plants, carnivores that consume animal component and omnivores that consume both plants and animals [6]. Diverse feeding habits and habitats affect the digestive tract in terms of structure, morphological, and physiological under various conditions [7].

The gastrointestinal tract in fish shows structural and functional diversity [8][9], which is associated with different dietary requirements, feeding habits, phylogeny, body mass, and shape, and even sex [10][11]. The main differences, observed at macroscopic and microscopic levels, include the shape, size, and structure of the stomach and intestines, the level of anatomical characteristics as well as the histological features of specific parts [7]. Many studies have examined the mechanisms of food demand, digestion and absorption in fish. In addition, it has been reported that the morphology of the digestive tract changes with different feeding habits and environments [9][12]. The shape of the cranium in each species influenced by the individual's genetics and environmental backgrounds, such as feeding habits and water quality [13]. Gill filament morphology can provide information related to the mechanical digestion process [14]. Although, there is a lot of literature discussing the digestive tract in fish, research on the digestive tract in *Tor tambra* and *Tor soro* fish is still limited. The aim of this study was to describe the gastric and intestinal biometrics of *T. tambra* and *T. soro* harvested from Leuser Ecosystem Areas. This information can provide a comparative basis for digestion, absorption and diet studies, as well as contribute to the development of fish farming of *T. tambra* and *T. soro*, which have diverse dietary needs and habitats.

2 Methods

2.1 Sample collection

T. tambra and *T. soro* were collected from river waters in Manis River, Bunin Village, Serbajadi District, East Aceh Regency, sampling in the upstream part of the river which is the habitat of *Tor* fish (97° 36' 12.391" E - 4° 29' 19.262" N). Biometric data analysis was carried out at the Aquaculture Integrated Laboratory, Almuslim University, Bireun Regency, Aceh Province, Indonesia.

2.2 Sample preparation

Fish sample in this study were obtained through direct capture or from fisherman totaling 5 fish of each species. Fish samples had an adult size with a minimum weight of 2 kg and a length of 30-50 cm. The digestive organs of the fish were separated using surgical tools, then placed in a container. Measurement of digestive organs using calipers with an accuracy of 0.01 mm starting from the stomach to the anus. Each organ was weighed using a 0.01 g digital scale. Measurement of the weight and length of the fish's digestive system includes the fish's body, stomach and intestines. Each digestive system data was documented using a digital camera (Canon EOS 400D Japan).

2.3 Digestive tract biometrics

The biometric parameters of the digestive tract in this study are GaSI (Gastro Somatic Index) and RGL (Relative Gut Lenghts) based on the formula of Bhatnagar & Karamchandani [15], ISI (Intestine Somatic Index) and RIL (Relative Intestine Lenghts) based on the formula of Wu et al [16].

$$\text{GaSI} = \frac{\text{fish stomach weight(g)}}{\text{fish body weight (g)}} \times 100$$

$$\text{ISI} = \frac{\text{fish intestine weight (g)}}{\text{fish body weight (g)}} \times 100$$

$$\text{RGL} = \frac{\text{length of fish stomach(cm)}}{\text{total length of fish(cm)}} \times 100$$

$$\text{RIL} = \frac{\text{length of fish intestine (cm)}}{\text{total length of fish (cm)}} \times 100$$

2.4 Data analysis

Biometric data analysis for each parameter was conducted using a one-way variance test. The significantly different criteria used at the 95% confidence level ($p<0.05$). Statistical analysis was conducted with SPPS ver. 22 software.

3 Results and discussion

Biometric studies of fish digestive systems can determine water environmental pollution and potential dangers in an aquatic habitat [17]. The highest GaSI value is found in *T. tambra* reach of 0.50±0.24%, while the lowest value is found in *T. soro* reach of 0.22±0.07%. The results of statistical analysis showed significant differences in GaSI values between the two species ($p<0.05$). The highest ISI value was found in *T. tambra* while the lowest value was found in *T. soro*, which reach of 2.46±0.58% and 1.57±0.55%, respectively. The results of statistical analysis showed significant differences in ISI values between the two species ($p<0.05$) (Table 1). Some species of fish are omnivorous when small and then adopt herbivorous eating habits as their size increases [17]. RGL and GaSI values can indicate that herbivorous fish can change their feeding intensity according to their sexual cycle [18]. Catfish in nature which are carnivorous have a GaSI reaching 36.53% higher than cultivated catfish (29.29%) [17][19].

Table 1. Digestive tract biometric parameters of *Tor tambra* and *Tor soro*, while stars symbol "*" indicate significant differences.

Parameters	*Tor tambra*	*Tor soro*
GaSI	0.50 ± 0.24*	0.22 ± 0.07*
ISI	2.46 ± 0.58*	1.57 ± 0.55*
RGL	44.35 ± 5.22	32.03 ± 14.99
RIL	340.37 ± 33.99*	259.62 ± 65.74*

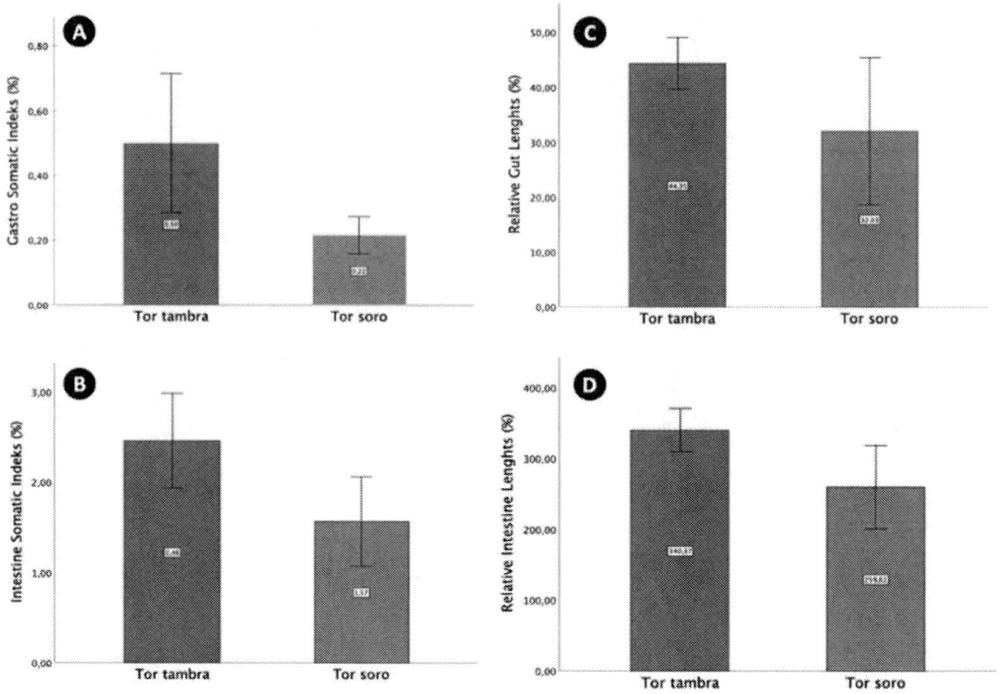

Fig. 1. Digestive tract biometrics of *Tor tambra* and *Tor soro* fish. Description: (A) Gastro Somatic Index, (B) Intestine Somatic Index, (C) Relative Gut Lenghts, (D) Relative Intestine Lenghts.

The highest RGL value was found in *T. tambra* reach of 44.35±5.22% while the lowest value was found in *T. soro* 32.03±14.99%. Statistical analysis showed that there was no significant difference in RGL values between the two species ($p<0.05$). The highest RIL value was found in *T. tambra* (340.37±33.99%) while the lowest value was found in *T. soro* (259.62±65.74%) (Fig. 1). The results of statistical analysis showed significant differences in ISI values between the two species ($p<0.05$). *T. tambra* and *T. soro* feeding habits are omnivorous. *T. tambra* predominantly feed herbivorous-omnivorous food sources, while *T. soro* is predominantly carnivorous-omnivorous. RGL and GaSI values can indicate carnivorous-omnivorous and herbivorous-omnivorous feeding habits [20]. It is clear that the RGL ratio of herbivorous and carnivorous fish decreases gradually, forming a continuum from herbivorous to carnivorous fish [18]. This can be expressed as the angle formed by the constant of the ratio of body length to gut length for each species and in the long run it is possible to know the feeding habits of any species just by plotting its RLG against body length [21]. The research results of Muchlisin et al. [22] strengthens the conclusion that *T. tambra* has an omnivore feeding habit with herbivorous tendencies.

4 Conclusion

The results showed that the highest GaSI, ISI, RGL, and RIL value were found in *T. tambra* while the lowest value were found in *T. soro* which each value of 0.50±0.24%, 2.46±0.58%, 44.35±5.22%, and 340.37±33.99% for *T. tambra* and 0.22±0.07%, 1.57±0.55%, 32.03±14.99%, and 259.62±65.74% for *T. soro*. The results of statistical analysis showed significant differences in the values of GaSI, ISI and RIL while the value of RGL had no significant difference between the two species ($p<0.05$).

Acknowledgements

We would like to thank the Ministry of Education, Culture, Research, and Technology for funding the Doctoral Dissertation Research Grant in 2023 (Contract Number: 114/E5/PG.02.00.PL/2023; 1271/UN3.LPPM/PT.01.03/2023). Our gratitude goes to all parties who have contributed to this research.

References

1. H.A. Al-Abdulhadi, Egypt. J. Aquatic Res. **13**, 387–97 (2005)
2. M. Furne, M. García-Gallego, M.C. Hidalgo, A.E. Morales, A. Domezain, J. Domezain, A. Sanz, Comp. Biochem. Physiol. **149**, 420–425 (2008)
3. J.P. Lazo, R. Mendoza, G.J. Holt, C. Aguilera, C.R. Arnold, Aquaculture **265**, 194–205 (2007)
4. S. Nazemroaya, R. Yazdanparast, M.A. Nematollahi, H. Farahmand, Q. Mirzadeh, Aquaculture **448**, 545–551 (2015)
5. Y. Shen, H. Li, J. Zhao, S. Tang, Y. Zhao, Y. Bi, X. Chen, Aquaculture **538**, 736546 (2021)
6. D.H. Evans, J.B. Claiborne, The physiology of fishes (CRC press, 2006)
7. K. Wołczuk, M. Ostrowski, A. Ostrowska, T. Napiórkowska, Zoology **128**, 38-45 (2018)
8. A.O. Diaz, A.M. Garcia, C.V. Devincenti, A.L. Goldemberg, Anat. Histol. Embryol. **32**, 341–342 (2003)
9. M.L. Dos Santos, F.P. Arantes, K.B. Santiago, J.E. Dos Santos, An. Acad. Bras. Cienc. **87**, 867–878 (2015)
10. S.B. Khojasteh, F. Sheikhzadeh, D. Mohammadnejad, A. Azami, World Appl. Sci. J. **6**, 1525-1531 (2009)
11. R.D. Day, I.R. Tibbets, S.M. Secor, J. Comp. Physiol. B **184**, 497–512 (2014)
12. J.M. Wilson, L.F.C. Castro, Fish Physiol. **30**, 1–55 (2010)
13. Y. Akmal, Y. Dhamayanti, E. Paujiah, Biodiversitas. **21**, 442-450 (2020)
14. S. Almunadiya, M. Muarif, M. Sabar, Y. Akmal, I. Irfannur, M. Muliari, A. Ruzanna, In IOP Conf. Ser.: Earth Environ. Sci. **1221**, 012004 (2023)
15. G.K. Bhatnagar, S.J. Karamchandani, J. Inland Fish. Soc. India **2**, 30-50 (1970)
16. P. Wu, L. Feng, S.Y. Kuang, Y. Liu, J. Jiang, K. Hu, X.Q. Zhou, Aquaculture **317**, 107-116 (2011)
17. P.O. Isibor, O.O. Ogochukwu, I.E. Valerie, T.O. Samson, P. Akinduti, O. Yemisi, Comparative analysis of the trophic status, length-weight relationship, gastro-somatic index and bioaccumulation of trace metals in wild and captive *Clarias gariepinus* (Springer Nature, Switzerland, 2023)
18. B.M. Kurbah, R.N. Bhuyan, J. Appl. Nat. Sci. **10**, 1187-1191 (2018)
19. I. Zulfahmi, Y. Yusrizal, A. I. Burhanuddin, Y. Dhamayanti, E. Paujiah, K.A. Sumon, F. M. Nur, Iraqi J. Vet. Sci. **36**, 549-554 (2022)
20. S. Koundal, R. Dhanze, A. Koundal, I. Sharma, J. Envir. Biosci. **27**, 11-18 (2013)
21. M. Dasgupta, Ind. J. Fish. **51**, 381-384 (2004)
22. Z.A. Muchlisin, A.S. Batubara, M.N.S. Azizah, M. Adlim, A. Hendri, N. Fadli, A.A. Muhammadar, S. Sugianto, Biodiv. J. Biol. Div. **16**, 89-94 (2015)

Traditional fishing gear for sarasin's goby (*Mugilogobius sarasinorum*), endemic to lake Poso, Sulawesi, Indonesia

Kurniawan P. Bandjolu[1,5]*, *Muh.* Herjayanto[2,5], *Ginanjar* Pratama[2], *Lady* D. Khartiono[3,5], *Abdul* Gani[3,5], and *Samliok* Ndobe[4]

[1] Environmental Research and Conservation Division, Mosintuwu Institute, 94663 Poso, Indonesia.
[2] Department of Fisheries Science, Faculty of Agriculture, University of Sultan Ageng Tirtayasa, 42124 Serang, Indonesia.
[3] Aquaculture Study Program, Faculty of Fisheries, Muhammadiyah Luwuk University, 94712 Banggai Regency, Central Sulawesi, Indonesia.
[4] Aquaculture Study Program, Department of Fisheries and Marine, Faculty of Animal Husbandry and Fisheries, Tadulako University, 94118 Palu, Indonesia.
[5] Tim Ekspedisi Riset Akuatika (ERA), Indonesia.

Abstract. Lake Poso is one of the world's ancient lakes and is home to several endemic fish. One of these is the Sarasin's goby *Mugilogobius sarasinorum*, local name *bungu masiwu*. For a long time, the community around Lake Poso, predominantly of the Pamona ethnic group, have used a traditional fishing gear called *inanco*. The objective of this study was to describe *inanco* fishing gear and evaluate the catch composition. The research location in Bo'e Village River (inlet Lake Poso). Research was conducted in March-June 2020. The result showed *inanco* is made from *Arenga pinnata* leaves, with 20 leaves used for each unit. The *inanco* is weighted with stones so the leaves won't drift away. The fish will hide between the leaves in the *inanco* structure, and can be harvested on the following day. The *inanco* is highly selective with minimal environmental impact and has potential as a conservation and sustainable fishing gear.

1 Introduction

Ancient lakes are evolutionary "hotspots" for ichthyofauna, as limited founder populations millions of years ago have since adjusted to changing habitat conditions and ecological niches [1]. In Southeast Asia four lakes or lake systems are considered as ancient lakes: Lake Inlé in Burma, Lake Lanao in Mindanao, Philippines, and Lake Poso and the Malili lakes on Sulawesi, Indonesia [2]. Lake Poso, a freshwater tectonic lake formed around 2 million years ago, is famous as one of the ancient lakes in Indonesia; it has an area of 323 km^2 and a maximum depth of around 450 m [3,4,5]. Geological, geographical and ecological isolation have caused the evolution of unique aquatic biota in Lake Poso [4]. The endemic fauna include several endemic fish species including *Adrianichthys kruyti*, *A. oophorus*, *A. poptae A. roseni*, *Mugilogobius amadi*, *M. sarasinorum*, *Nomorhamphus celebensis*, *Oryzias nebulosus*, *O. nigrimas*, and *O. orthognathus* [6-12].

In recent decades, the abundance of endemic fish populations has declined, due to similar threats to those faced by fish in many other lakes such as eutrophication, intensive fishing,

*Corresponding author: kurniawanbandjolu@gmail.com

invasive non-native species, and pollution [12,13]. One of the endemic fish affected was *A. roseni*; this fish was collected in September 1978, but until now there have been no further reports [7-9,12]. In addition, the status of Sarasin's goby *M. sarasinorum* in the International Union for Conservation of Nature (IUCN) Red List was changed from Vulnerable to Endangered in 2019 [8,9,14]. This goby is known to be traded as ornamental fish [13, 15], although the impacts of this trade have not been quantified [14,15]. The Sarasin's goby is also known as the Black Toraja goby in the ornamental fish trade, and in the area surrounding it's native (endemic) habitat (Lake Poso) it is known and sold on local markets as *bungu masiwu*. This fish exhibits sexual dimorphism, similar to other species in this genus [16]; adult male *M. sarasinorum* in breeding condition appear very dark brown to jet black, while the fuller-bodied females are paler and retain a brownish coloration with some mottling. *M. sarasinorum* is a medium-sized benthic fish typically found in rocky and sandy areas; it tends to be active in shallow areas at night but is rarely seen during the day [9, 16].

In 2013 Lake Poso was assessed as Key Biodiversity Area (KBA) by Burung Indonesia [14] and Participatory Conservation Planning (PCP) has been carried out in 4 villages (Meko, Salukaia, Owini, and Uranosari) since 2017 [17]. Clearly the conservation of endemic fish in Lake Poso is important, including through addressing identified threats such as pollution and the introduction of non-native (alien) invasive species [12]. Conservation actions have been implemented in the area, including raising public awareness and reducing the use of chemical herbicides and pesticides in the surrounding watershed [14]. With regards to fishing intensity and sustainability, the Pamona ethnic community has been using traditional wisdom which can be considered as a conservation effort since the earliest times, using a traditional fishing trap known as *inanco* or *nanco* [17]. The Pamona ethnic group is one of the indigenous peoples in the Poso area, and they are known to have a lot of local wisdom [18]. The use of this fishing trap, specifically used for catching Sarasin's goby, has been considered environmentally friendly [17]. The objective of this study was to describe *inanco* fishing gear and evaluate the catch composition.

2 Materials and methods

2.1 Study site and sample collection

This study was conducted in Bo'e Village River (inlet Lake Poso), Central Sulawesi Province, Indonesia in March-June 2020. The site coordinates were -2°2'45.561''N and 120°42'36.094''E. Sampling was carried out several times over one day. In this study we used eight *inanco* fish traps, placed separately at a distance from one another. Each *inanco* was placed close to the edge of a Lake Poso river inlet at a depth of approximately 33 cm.

Ecological parameters measured at each *inanco* experimental fishing site were temperature, pH, substrate, clarity, and the presence of aquatic plants. All fish caught from each *inanco* were collected. The fish samples obtained from each catch were separated by species and the specimens were then preserved in a 1000 mL sample bottle containing 70% alcohol, labeled with the name of the fish species, station, date of collection, collector's name, and other information required [9]. The samples were then taken to the laboratory for further analysis.

2.3 Data analysis

Fishing gear is described based on how it is used and the dimensions of the fishing gear. Selectivity evaluation is carried out based on the number of target fish caught. All fish caught were identified using standard taxonomic keys [16, 19]. The sex of each Sarasin's goby specimen was noted, as male and female fish can be distinguished based on their morphology

(Figure 1). Specimens were analyzed at the Animal Biosystematics and Evolution Laboratory, Department of Biology, Faculty of Mathematics and Natural Sciences, Tadulako University, Palu, Indonesia.

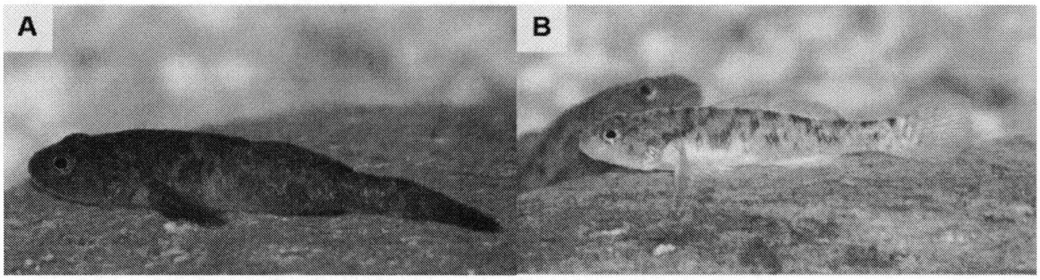

Fig. 1. Live color of *Mugilogobius sarasinorum* male (A) and female (B).

3 Results and discussion

3.1 Background and description of *inanco* fishing method

The Pamona ethnic is an indigenous ethnic group living in the areas around Lake Poso [18]. They have developed and retained a varied body of local wisdom, including the use of local plants in traditional medicines [20]. In addition, this ethnic group is famous for the traditional ceremony known as "Padungku". Padungku is a harvest ceremony as a thanksgiving to God in the area [17,21]. To prepare for the ceremony, the community usually process their rice harvest, and also catch fish using traditional fishing gears such as *inanco*. *Inanco* is traditionally used to catch the gobies *Mugilogobius amadi* and *M. sarasinorum*. These fish are caught as a popular food fish during the wet season, comprising the transition and east monsoon (March to August), but not during the dry season because the rivers dry out then. However, based on Bandjolu et al., research [9], there is currently only *M. sarasinorum* because *M. amadi* has not been caught since 1987 in Lake Poso.

The Pamona ethnic community generally install the *inanco* fish traps in the afternoon and then collect the catch on the next day. Although in general fish are collected every morning and afternoon, the best catches are most often in the morning, considering that *M. sarasinorum* is mostly active at night. The *inanco* are typically placed in unshaded areas where the sunlight can penetrate the riverbed to a depth of 3-5 meters, as previously reported by [17]. Each *inanco* is made from sugar palm (*Arenga pinnata*) leaves which are placed horizontally with the ribs aligned in the direction of the flow of water in the slow-flowing part of the river. One *inanco* unit was made from 20 sugar palm leaves, which were cut into several parts. The middle part of the leaves were bundled and bound using strips of the rotan cane *Calamus minahassae* (Indonesian names rotan tikus or rotan patani). Each *inanco* unit measures less than one meter in width and has a cross-section of approximately 20 cm × 20 cm (Figure 2).

Fig. 2. *Inanco* made with palm leaves (*Arenga pinnata*).

The *inanco* are ballasted with round stones, so that they will not be washed away. The fish are then attracted to the trap, and come to hide in the leaves as a shelter. Fishermen will harvest fish for household consumption by inserting a fishnet at the bottom end of the *inanco*, so that when the ballast is removed, the fishnet can be moved towards the top end as the *inanco* rises up. The fish hiding in the *inanco* will then enter completely into the net. In effect, the *inanco* functions similarly to the bamboo fish traps called *bubu*. It has been proposed that the *inanco* can be used as a model of modern environmentally friendly fishing gear that can be adopted for conservation purposes through the development of local wisdom [17].

3.2 Ecological parameters and catch composition

The water quality parameters measured at the study site were pH 8.3 and temperature 26°C. Visibility could not be measured because it exceeded the water depth due to the high water clarity at the sampling locations. The waters tend to be clear and transparent, most likely due to the generally oligotrophic conditions [12] and because turbidity does not last long in the streams flowing into the lake [9,19]. The substrate at the study site was dominated by coarse sand, pebbles, and rounded stones, and aquatic plants were absent.

The *inanco* catches comprised four species with a total volume of 510 individuals (Table 1). The target species, *M. sarasinorum*, comprised nearly 95% of the catch with 483 individuals, demonstrating the high selectivity of the *inanco* fishing gear. The other fish caught included a small proportion of one other endemic species, the endangered *anasa* (vernacular name) *N. celebensis* (1.37%). A few individuals from two introduced species were also caught: *kepala timah* (vernacular name) *Aplocheilus panchax* (1.76%) and or *nilem* (vernacular name) *Osteochilus vittatus* (2.16%). Both introduced species are of low-value locally. Furthermore, *O. vittatus* is considered invasive [12], such as in Lake Talaga, Lake Rano, and Lake Lindu, Central Sulawesi [8].

Table 1. Fish caught using *inanco* in Bo'e Village River, Lake Poso

Species	Family	n	Common name	IUCN Red List status	Human uses
Aplocheilus panchax	Aplocheilidae	9	Blue Panchax	Least concern	OF
Nomorhamphus celebensis	Zenarchopteridae	7	Poso Halfbeak	Data deficient	OF
*Mugilogobius sarasinorum**	Gobiidae	483	Sarasin's Goby	Endangered	OF & FF
Osteochilus vittatus	Cyprinidae	11	Bonylip barb	Least concern	FF

*: target species, OF: ornamental fish, FF: food fish

The total catch of 483 *M. sarasinorum* in this study comprised 140 males and 343 females. More males than females were caught in each *inanco*. The number of males is 6.83-14.49%, while the number of females is 2.48-5.59% (Figure 3).

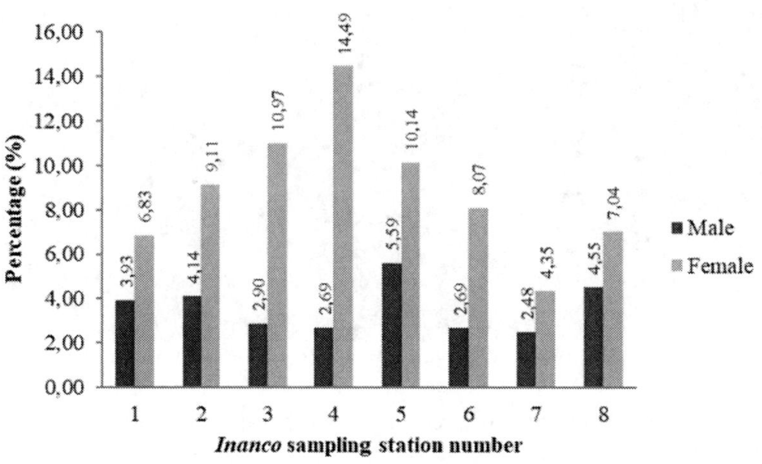

Fig. 3. Percentage of male and female *M. sarasinorum* caught using *inanco* fishing gear.

According to [16], *M. sarasinorum* can mostly be found in open areas on mud and gravel substrates with some aquatic plants, but also on steep gravel slopes and among large limestone rocks. In this study we only found *inanco* being used in Bo'e Village. According to [17], this gear is also used in the rivers of Meko and Salukaia villages, but the coordinates were not provided. Based on a gut content study, the feeding habit of *M. sarasinorum* is predominantly carnivorous with prey dominated by small fish and insect larvae [16].

The *inanco* fishing gear is highly selective, which means that it is unlikely to be a threat to the non-target fish in Lake Poso. This is important for the conservation of the other native fish living in the lake, especially the other endemic species in IUCN Red List at risk categories, several of which tend to be caught in other fisheries [12]. With respect to the introduced species, especially those considered invasive, it is unlikely that *inanco* can contribute to the control of these alien fish populations. With respect to the target species, *M. sarasinorum* is itself considered endangered [14], raising the question as to whether this fish should be targeted for human consumption or indeed for the aquarium trade.

In this respect, the *inanco* can be considered as a potentially sustainable approach to the exploitation of *M. sarasinorum* for several reasons. Firstly, the *inanco* gear used readily available, renewable and non-polluting materials and does not damage the habitat. Secondly, the fish are caught alive and in good condition, thereby potentially allowing further selectivity such as the release of juveniles or fish in breeding condition. Furthermore, in addition to using *inanco* traps as an ephemeral structure to harvest *M. sarasinorum* with minimal adverse side-effects, the *inanco* itself can also be used to help maintain *M. sarasinorum* and other fish populations. As explained by [17], *inanco* structures can be placed to provide protective habitat, resulting in increased *M. sarasinorum* abundance.

4 Conclusion

The *inanco* is a traditional fishing gear for Sarasin's goby *Mugilogobius sarasinorum* (local name *bungu masiwu*), made from sugar palm (*Arenga pinnata*) leaves. One *inanco* unit is made from 20 palm leaves and weighted with stones. The fish taking refuge in the *inanco* can be readily harvested. The gear was highly selective (≈ 95% target species), and the *M. sarasinorum* was male-dominated. The *inanco* shows promise for a sustainable *M. sarasinorum* fishery and, with some modification, for supporting conservation of this endangered species in Poso Lake.

Acknowledgements

Many thanks to Papa and Mama Monge (the *inanco* fishermen) for their information about the *inanco*. The authors also thank Aditya A. P. Aboki, Raynaldi Kayupa and Marno A. Dempaulu for their help and support during research activities, Dr. Abigail Mary Moore for English language proof-reading, and the head of Animal Biosystematics and Evolution Laboratory, Department of Biology, Faculty of Mathematics and Natural Sciences, Tadulako University for in-kind support.

References

1. T.V. Rintelen, K.V. Rintelen, M. Glaubrecht, C.D. Schubart, F. Herder. Biotic Evolution and Environmental Change in Southeast Asia (2012)
2. K. Martens Trends in Ecology and Evolution **12** 177-182 (1997)
3. M. Wilson, S. Moss Palaeogeography, Palaeoclimatology, Palaeoecology **145** 303-337 (1999)
4. M.E. Cristescu, S.J. Adamowicz, J.J. Vaillant, D.G. Haffner Molecular Ecology. **19** 4837-4851 (2010)
5. J.J. Vaillant, G.D. Haffner, M.E. Cristescu Integrative and Comparative Biology **51** 634-643 (2011)
6. M. Kottelat Ichthyological Exploration of Freshwaters **1** 49-67 (1990)
7. L.R. Parenti, B. Soeroto Ichthyological Research **51** 10-19 (2004)
8. M. Herjayanto, A. Gani, Y.S. Adel, N. Suhendra Journal of Aquatropica Asia **4** 1-9 (2019)
9. K.P. Bandjolu, E. Madiyono, M. Herjayanto, A. Gani, M. Nur, S. Laheng, M.T. Gundo IOP Publishing **869** 1-10 (2021)
10. A.M. Meisner Zoological Journal of the Linnean Society **133** 199-283 (2001)
11. L.R. Parenti Zoological Journal of the Linnean Society **154** 494-610 (2008)
12. N. Serdiati, D. Arfiati, M.S. Widodo, T.D. Lelono, S. Ndobe, K. Mansyur, A.M. Moore Revista de Biología Tropical **69** 139-152 (2021)
13. E.K. Sunardi, Sari, R. Partasasmita Journal of Biodiversity and Environmental Sciences **9** 86-95 (2016)
14. Z. Jaafar The IUCN Red List of Threatened Species (2019)
15. M.T. Gundo, Andriani Warta Iktiologi **5** 20-26 (2021)
16. H. Larson Records of the Western Australian Museum **62** 1-233 (2001)
17. Theopillus, S. Maubu, Jihad, F. Alwi IOP Publishing **473** 1-4 (2020)
18. L.V. Aragon Indonesia. **72** 45-79 (2001)
19. M. Kotellat, A.J. Whitten, S.N. Kartikasari, S. Wirjoatmodjo Periplus Editions (1993)
20. N. Rahmawati, Y. Widyastuti, R. Purwanto, S.S. Lestari, I.H.A. Sene, Y. Bakari Advances in Health Sciences Research **22** 460-468 (2020)
21. M.A. Jennah, K. Nawing, R. Kulyawan Jurnal Kreatif Online **9** 61-74 (2021)

Effect of turmeric probiotics on the survival rate and growth of domesticated endemic betta fish (*Betta dennisyongi*)

Fazril Saputra[1*], *Yusran* Ibrahim[1], *Khairul* Samuki[1], *Alfis* Syahril[1], *Muhammad Barru Airil Fizra Hasibuan*[1], and *Omar* Muktaridha[1]

[1]Department of Aquaculture, Faculty of Fisheries and Marine, Teuku Umar University, Aceh Barat 23615, Indonesia.

Abstract. The endemic betta fish (*Betta dennisyongi*) was a species of freshwater ornamental fish native to Indonesia which has been successfully domesticated but it has low survival rate and growth. One way to increase the survival rate and growth of farmed fish is by administering turmeric probiotics. This study aims to analyze the best dosage of turmeric probiotics to increase the survival rate and growth of domesticated betta fish. This research was carried out experimentally. The treatments were control, turmeric probiotic 1.5 mL 100 grams feed, turmeric probiotic 3 mL 100 grams feed, turmeric probiotic 4.5 mL 100 grams feed. The parameters tested during the research were survival rate, daily growth rate, and feed conversion ratio and water quality. The analysis of variance test on the data was analyzed using the SPSS 25.0 program with a 95% confidence interval. Treatments that are significantly different will be further tested using the Duncan test to determine the best treatment. The results of the study showed that giving turmeric probiotics for survival rate gave results that had a significant effect ($P<0.05$) but had no significant effect ($P>0.05$) on the growth of betta fish. The use of turmeric probiotics 3 mL 100 grams feed can increase weight gain, length gain and survival rate of domesticated betta fish.

1 Introduction

Implementing large-scale social restrictions and social distancing in Indonesia has forced workers to work from home [1]. The implementation of work-from-home by the Indonesian government has not only produced positive impacts but also negative impacts. One of the negative impacts of working from home is that many employees experience boredom and stress due to restrictions on activities ranging from work to worship, all of which are carried out at home [2]. Employees have new hobbies, such as keeping ornamental fish to reduce boredom and stress. One type of ornamental fish that is excellent for keeping is the ornamental betta fish [3]. Betta fish is an option to keep because this type of ornamental fish does not require significant capital in maintenance or cultivation [4][5]. One type of ornamental betta fish most sought after is the one endemic to the western and southern regions of Aceh Province, which has the local name jielabu and the Latin name *Betta dennisyongi* [6].

* Corresponding author: fazrilsaputra@utu.ac.id

During the COVID-19 pandemic, the jielabu betta fish (*Betta dennisyongi*) is one of the most popular ornamental betta fish. So far, they are still caught in the wild to meet the demand for jielabu betta fish (*Betta dennisyongi*). The ongoing arrests have been deeply disturbing. This fishing can make this fish increasingly rare in nature and can damage the environment where this fish lives. One way to reduce catching in the heart is to domesticate the jielabu betta fish (*Betta dennisyongi*). Fish domestication is the taming/ adaptation of fish from wildlife to cultivation environments over different periods. Domestication can take years to decades [7]. The obstacles in domesticating fish are that the difficulty of adaptation for fish/ water animals is more serious than that of land/ terrestrial animals and slow growth [8]. One solution to resolve obstacles in domestication is by administering turmeric probiotics.

Turmeric probiotics are probiotics that have been combined with herbal plants. The advantage of this turmeric probiotic is increasing the production performance and health status of farmed fish [9]. This is because turmeric contains curcumin compounds, which help increase fish appetite and maintain fish health. Several previous studies have proven that giving turmeric probiotics to farmed fish can improve production performance, for example, in tilapia [10], catfish [11], freshwater pomfret [12], local snakehead fish [9]. Based on the explanation above, turmeric probiotics have never been tested for the endemic jielabu betta fish (*Betta dennisyongi*). Therefore, research needs to be carried out to evaluate the use of herbal probiotics on the production performance and color quality of the endemic betta fish (*Betta dennisyongi*).

2 Material and Methods

2.1 Time and place

The research was conducted from March 1 to July 31, 2023, at the Integrated Fisheries Laboratory, Faculty of Fisheries and Marine Sciences, Teuku Umar University. Laboratory analysis was carried out at the Integrated Mathematics and Natural Sciences Laboratory, Teuku Umar University, and the Laboratory of the Faculty of Veterinary Medicine, Syiah Kuala University.

2.2 Tools and materials

The tools used in this research include a fiber box for fish keeping, a blender, an aerator, a digital scale, a pH meter, a thermometer, a DO meter, a ruler, and a measuring pipette. The ingredients include Betta fish seeds (*Betta dennisyongi*), turmeric, banana stems, fine bran, sugar, yeast, label paper, commercial feed, and EM4 probiotics.

2.3 Experimental design

This research was carried out using experimental methods. The experimental design was completely randomized, with four treatments and three replications. The research design can be seen in the Table below. The dosage for this study was taken from a combination of research conducted previously [10].

Table 1. Turmeric probiotic treatment design

No	Code	Information
1	P0	Treatment without adding turmeric probiotics to feed (control)
2	P1	Treatment dose is 1.5 ml of turmeric probiotics 100 grams of feed
3	P2	Treatment dose is 3 ml of turmeric probiotics 100 grams of feed
4	P3	Treatment dose is 4.5 ml of turmeric probiotics100 grams of feed

2.4 Research procedure

2.4.1 Preparation of rearing containers and maintenance of test animals

The containers are 12 fiber boxes measuring 50 x 30 x 40 cm3. The fiber box is cleaned with soap and disinfected with potassium permanganate at 15 mg L^{-1}. After that, the fiber box is soaked for three days. After three days, the fiber box is rinsed and left to dry. After the disinfection process, the fiber box is filled with 40 L of water.

The Betta fish (*Betta dennisyongi*) used measures 2-3 cm. Before being stocked in the rearing media, Betta fish (*Betta dennisyongi*) are acclimatized first. Fish rearing is carried out for 60 days. They fed fish using the ad satiation method (according to fish consumption/needs). The feeding frequency is three times daily, namely at 08.00, 12.00 and 16.00 WIB. During maintenance, siphoning and water changes of 10-20% are carried out every ten days.

2.4.2 The process of making turmeric probiotics

According to [10], the process of making turmeric probiotics is by putting 5 liters of clean/sterile water into a container, then adding 500 grams of fine turmeric, 125 grams of chopped banana stems, 25 grams of granulated sugar, 500 grams of fine bran, 200 ml EM4 and 15 grams of yeast. After that, all the ingredients are combined and mixed well. The mixing results are then put into a jerry can/container and closed tightly. Place these materials in a room not exposed to the sun and leave for ± 14 days. Every 2-3 days, the lid of the jerry can/ container is opened for a few minutes to release the fermentation gas. After ± 14 days, this turmeric probiotic is filtered using a filter. Signs of a successful probiotic are that it produces a tape. According to [12], sufficient sugar solution is given during maintenance to keep microorganisms alive.

2.4.3 Feed enrichment process

The commercial feed used is commercial feed, namely Prima Feed 500, with a minimum protein content of 39%, 5% fat, 4% crude fiber, 11% ash, and 10% water. The feed enrichment process is carried out using the spray method. The feed is sprayed with turmeric probiotics according to the treatment. Namely, 1.5 ml, 3 ml, and 4.5 ml for every 100 grams of feed, add egg white as much as 10% of the weight of the feed for all treatment. Then, the feed is air-dried for ± 5 minutes. Then the feed can be given to Betta fish (*Betta dennisyongi*).

2.4.4 Research Test Parameters

a. Weight Gain (WG)
Weight gain was calculated using the following formula [13]:

$$WG = Wt - Wo$$

Remarks: WG = Weight gain (g), Wt = average body weight of fish at the end of the experiment (g), Wo = average body weight of fish at the start of the experiment (g)

b. Length Gain (LG)
Length gain was calculated using the following formula [13]:

$$LG = Lt - Lo$$

Remarks: LG = length gain (cm), Lt = average body length of fish at the end of the experiment (cm), Lo = average body length of fish at the start of the experiment (cm)

c. Daily Growth Rate (DGR)
The daily growth rate was calculated using the following formula [14]:

$$DGR=\frac{(Wt - Wo)}{t}$$

Remarks: DGR= daily growth rate (g day^{-1}), Wt = average body weight of fish at the end of the experiment (g), W0 = average body weight of fish at the start of the experiment (g), t = duration of the feeding experiment (day)

d. Feed conversion ratio (FCR)
The feed conversion ratio was calculated using the following formula [15]:

$$FCR = \frac{KP}{\Delta W}$$

Remarks: FCR=Feed conversion ratio, KP= Total amount of feed consumption (g), ΔW= Weight gain (g)

e. Survival Rate (SR)
The survival rate was calculated using the following formula [16]:

$$SR = \frac{Nt}{No} x\ 100\ \%$$

Remarks: SR = Survival rate (%), Nt = Total of fish at the end of the experiment, No = Total of fish at the start of the experiment

f. Water Quality Parameters
Water quality parameters measured include temperature, pH, and dissolved oxygen. temperature, pH and DO measurements were carried out every 10 days after fish sampling.

3 Result and Discussion

The results of the analysis of variance and Duncan's further tests showed that the addition of turmeric probiotics to feed had a significant effect (P<0.05) on length increase, weight gain, and survival rate but had no significant impact (P>0.05) on the daily growth rate, and ratio conversion feed. The values for length gain, weight gain, survival rate, daily growth rate, and feed conversion ratio can be seen in the table 2 below. Proximate test results showed that feed supplemented with turmeric probiotics had a higher protein content than the control. The comparative test results can be seen in Table 3 below. Water quality measurements show that water quality during maintenance is still within normal limits. The results of the water quality test can be seen in Table 4.

Table 2. Production performance of betta fish (*Betta dennisyongi*) given turmeric probiotics

Perlakuan	Parameter					
	WG (cm)	LG (gram)	SR (%)	DGR (%)	FCR	Glukose
P0	0,26 ± 0,15^{a}	0,10 ± 0,05^{a}	86,67 ± 2,88^{a}	0,39 ± 0,01^{a}	1,22 ± 0,10^{a}	95,3 ± 31,65^{a}

P1	0,49 ± 0,06^{b}	0,22 ± 0,04^{b}	90 ± 5^{ab}	0,60 ± 0,03^{a}	1,08 ± 0,09^{a}	61,3 ± 24,44^{a}
P2	0,22 ± 0,05^{a}	0,14 ± 0,01^{a}	96,67 ± 5,77^{b}	0,53 ± 0,02^{a}	1,18 ± 0,01^{a}	62,0 ± 26,90^{a}
P3	0,28 ± 0,07^{a}	0,15 ± 0,05^{ab}	96,67 ± 2,88^{b}	0,51 ± 0,14^{a}	1,18 ± 0,16^{a}	79,7 ± 12,34^{a}

Note: Different superscript letters in the same column show significantly different (P <0.05).

Table 3. Results of proximate tests on feed added with turmeric probiotics

Treatment	Water content (%)	Ash level (%)	Fat (%)	Protein (%)	Crude fiber (%)	Carbohydrate (%)
P0	9,93	5,71	5,88	32,42	2,99	43,07
P1	8,22	3,54	5,84	34,47	3,13	44,82
P2	6,61	3,40	5,65	36,62	3,29	44,42
P3	6,26	4,69	4,22	37,12	3,59	44,11
Turmeric Probiotic	52,81	2,19	6,86	15,27	5,24	17,63

Table 4. Water quality parameters during maintenance

Water Quality Parameters	Treatment			
	P0	**P1**	**P2**	**P3**
pH	7,5- 8,4	7,4-8,5	7,5-8,4	7,3-8,5
DO (mg/L)	5,20-5,72	5,06-5,68	5,04-5,58	5,25-5,61
Temperature (°C)	30,1-30,4	30,2-30,9	30,6-30,9	30,4-30,8

The results of this study show that giving turmeric probiotics influences the growth in weight and length of betta fish. The best weight and length growth values for betta fish were found in the treatment dose of 1.5 ml of turmeric probiotic 100 grams feed (P1). The weight growth value of P1 is 0.49 ± 0.06 cm, and the length growth value of P2 is 0.22 ± 0.04 grams. The effect of giving turmeric probiotics in this study on weight and length growth is in line with the opinion of [17], which states that the compounds contained in turmeric can increase metabolism and the performance of the digestive system. Turmeric also contains the compound curcumin, which has the function of increasing the absorption of food substances, which is done by stimulating the walls of the gallbladder to release bile into the small intestine, thereby improving digestion [18];[19]. Turmeric also contains vitamin C [20];[21], which functions as a catalyst that can speed up reactions in the body so that it can absorb and utilize it for growth [22]. Apart from that, the probiotic content in the treatment also helps increase food absorption in the fish's intestines. This is thought to be due to the probiotic content of beneficial bacteria to suppress pathogenic bacteria [23]. Proximate results also showed that the administration of turmeric probiotics increased crude protein compared to controls. This is because fermentation is changing complex substances into simpler ones by enzymes produced by microbes. This process will reduce and even eliminate harmful substances such as antitrypsin in feed ingredients by microbes. The fermentation process will also degrade complex materials into amino acids so that microbes can use these amino acids to reproduce [24,25]. Increasing the number of microbial colonies during fermentation can indirectly increase crude protein content because microbes are a source of single-cell protein [26].

The survival rate results also showed that the administration of turmeric probiotics influenced the survival rate of betta fish. The best survival rate value was found in treating 3 ml of turmeric probiotic 100 grams feed (P2). This is because turmeric contains active compounds, namely curcumin and flavonoids. Curcumin and flavonoids play a role in keeping fish healthy. Fish in healthy conditions and with good water quality will have a high survival rate [27];[22]. Apart from that, adding probiotics can increase beneficial microbes, which have a protective function, namely the ability of microbes to inhibit pathogenic microbes in the fish's body [28]. The microbes in probiotics also produce substances that are not harmful to fish but destroy pathogenic microbes that disrupt the digestive system [12].

The results of the parameters of daily growth rate, and feed conversion ratio did not show results that were not significantly different compared to the control. However, adding turmeric probiotics to fish feed made the daily growth rate, and feed conversion ratio higher than the control. This is thought to be because the addition of probiotics to fish feed creates a synergy of probiotics working in the digestive tract and increases the digestibility of the meal, thus stimulating fish growth [9]. The results of this study are similar to research conducted by [29]. Adding probiotic bacteria to *Poecilopsis gracilis* fish resulted in a tendency for daily growth rates to be slightly higher than controls. Research showed that adding probiotic bacteria can affect the fish's digestive tract. Microba function to increase fish digestibility of feed to stimulate fish growth [30][31][32]. Adding probiotics to fish feed can create a synergistic composition of probiotics working in the digestive tract and increase the digestibility of the feed, thus stimulating the growth of local snakehead fish [5], according to [33], where the addition of probiotics can reduce the feed conversion ratio value. However, in the research, the feed conversion ratio value was not significantly different from the control, but the value was higher than the control. According to [34], this occurs because there are still very few probiotic bacteria that reach the fish's digestive tract and have not reproduced very much, so time is needed to affect the fish's digestive process.

The water quality measured during the research was still within the limits betta fish can tolerate. Temperature parameters during the study ranged from 30.1-30.9 °C. According to [35], Betta fish prefer a warm water climate to other tropical fish, namely in the 25 -30°C. The pH parameters during the research were in the range of 7.4-8.5. According to [36], Betta fish can live in water that has a pH range of 7.9 -8. DO parameters are still in the normal range, namely >5 mg/L. According to [37], Betta fish are able to live in water with dissolved oxygen of 3 mg/L because Betta fish are labyrinth fish, which are able to take oxygen directly from the air.

4 Conclusion

Different doses of turmeric probiotics gave different responses to survival rate parameters but gave the same response to growth parameters. A dose of 3 mL turmeric probiotic 100 grams of feed is the best dose to increase the survival rate of betta fish (*Betta dennisyongi*).

References

1. N. Rindam, I. Haq, J. Sos. Budaya. Syar-i. **7**(7), 639-648 (2020)
2. D. Ma'rifah, J. Civ. Serv. **14**(2), 1-10 (2020)
3. E. Setiawan, R.F. Suwarman, A. Firmansyah, M.D. Saputra, J. Abdi. Sang. Buana. **2**(1), 20-28 (2021)
4. F. Saputra, T.R. Efianda, J. Mar. Kreat. **II**(1), 44-49 (2018)
5. F. Saputra, M.A. Thahir, M. Mahendra, Y. Ibrahim, M.A. Nasution, T.R. Efianda, J. Perikan. Trop. **7**(10), 85-97 (2020)
6. T.H. Hui, Raffles. Bull. Zool. **61**(1), 323-330 (2013)
7. F. Teletchea, Anim. Front. **11**(3), 87-91 (2021)

8. Saraiva, L. João, M.F. Castanheira, P. Arechavala-López, J. Volstorf, B.H Studer, IntechOpen. 109-135 (2019)
9. F. Saputra, Y. Ibrahim, (Undana Press, Kupang, 2021)
10. R. Arsyad, A. Muharam, S. Syamsuddin, J. Ilm. Perikan. Kelaut. **3**(2), 51-57 (2015)
11. D. Puspitasar, J. Ilmu. Manaj. **5**(1), 53-9 (2017)
12. J.T. Saselah, J. Mandeno, Budid. Perair. **5**(3), 50-6 (2017)
13. F. Saputra, D. Wahjuningrum, K. Tarman, I. Effendi, J. Ilmu. Teknol. Kelaut. Trop. **8**(2), 747-755 (2017)
14. Z.A. Muchlisin, A.A. Arisa, A.A. Muhammadar, N. Fadli, I.I. Arisa, M.N. Siti-Azizah, Arch. Polish. Fish. **24**(1), 47-52 (2016)
15. N. Zonneveld, E.A. Huisman, J.H Boon, (Gramedia Pustaka Utama, Jakarta, 1991)
16. F. Saputra, M. Mahendra, J. Iktiologi. Indones. **19**(2), 195-203 (2019)
17. M.M.A. Mahmoud, M.M.M. El-Lamie, A.A. Dessouki, M.S. Yusuf, Glob. Res. J. Fish. Sci. Aquac. **1**(12), 26-33 (2014)
18. I.W. Putri, M. Setiawati, D. Jusadi, J. Iktiologi. Indones. 17(1), 11 (2017)
19. D. Kusbiantoro, Y. Purwaningrum, J. Kultiv. **17**(1), 544-549 (2018)
20. M.F. Agustin, A.G.D. Kartika, M. Effendy, I. Maflahah, J. Trunojoy **1**(4), 468-476 (2020)
21. P.A.N.K. Permatananda, A.A.S.A. Aryastuti, P.N. Cahyawati, D.P.C. Udiyani, D. Wijaya, I.G.S.P Pandit, J. Bal. Memba. Bal. **1**(2), 135-142 (2020)
22. L. Linayati, Y.T. Mardiana, F.A. Feni, Z.M. Yahya, J. Sains. Akuakultur. Trop. **6**(1), 75-80 (2022)
23. H. Ahmadi, I. Iskandar, N. Kurniawati, J. Perikan. Kelaut. **3**(4), 99-107 (2012)
24. E.P Rahayu, D. Saefulhadjar, H. Supratman, J. Sumber. Daya. Hewan. **4**(1), 17–20 (2023)
25. V. Muhardina, P.M. Sari, Y. Aisyah, S.Haryani, S.A. Akbar, Rasayan J. Chem **13(1)**, 240-244(2020)
26. E.R.S. Dewi, A.S Nugroho, U. Maria, J. Biol. Trop. **22**(2), 390-397 (2022)
27. Andesra, R. Morina, S. Henni. J. Online. Mhs. UNRI. **1**(1), 1-17 (2019)
28. A. Umasugi, R.A. Tumbol, R.L. Kreckhoff, H. Manoppo, N.P.L. Pangemanan, E.L. Ginting, e-Journal. Budid. Perair. **6**(2), 39-44 (2018)
29. L.H.H. Hernandez, T.C Barrera, J.C. Mejia, G.C Mejia, M. Del Carmen, M. Dosta, Aquac Nutr. **16**(4), 407-411 (2009)
30. F.J. Gatesoupe, Aquaculture. **119,** 147-65 (1999)
31. X. Zhou, Y. Wang, IntechOpen. 215-226 (2012)
32. H. Manoppo, R.A. Tumbol, H.J. Sinjal, J. Sumaraw, AACL Bioflux. **12**(6), 2188–2194 (2019)
33. E.D. Harmilia, H. Helmizuryani, A. Ahlan, J. Fiseries. **8**(1), 9-13 (2019)
34. N. Ardita, A. Budiharjo, S. Lusi, A. Sari, Bioteknologi. **12**(1), 16-21 (2015)
35. S. Biokani, S. Jamili, S. Amini, J. Sarkhosh, **4**(2), 33-37 (2014)
36. A. Mochammad, Suminto, T. Yuniarti T, J. Aquac. Manag. Technol. **6**(3), 179-186 (2017)
37. D. Rachmawati, F, Basuki, T. Yuniarti, J. Aquac. Manag. Technol, **5**(1), 130-136 (2016)

Climbing Perch Fish Sperm DNA (*Anabas testudineus*) Protection Using Cryoprotectant Agent

Siti Maulida[1,2], *Kartini* Eriani[4], *Nur* Fadli[3], *Luvi* S. Handayani[1], and *Zainal* A. Muchlisin[2*]

[1]Graduate School of Mathematics and Applied Sciences, Universitas Syiah Kuala, Banda Aceh 23111, Indonesia
[2]Department of Aquaculture, Faculty of Marine and Fisheries Sciences, Universitas Syiah Kuala, Banda Aceh 23111, Indonesia
[3]Department of Marine Science, Faculty of Marine and Fisheries Sciences, Universitas Syiah Kuala, Banda Aceh 23111, Indonesia
[4]Master Program of Biology, Faculty of Mathematics and Natural Sciences, Universitas Syiah Kuala, Banda Aceh 23111, Indonesia

Abstract. Sperm cryopreservation is a sperm storage method that needs to be developed to ensure the availability of climbing perch fish (*Anabas testudineus*) sperm considering the difficulty of providing quality broodstock. This study aims to determine the effect of several types and concentrations of cryoprotectants on the DNA integrity of climbing perch fish sperm after cryopreservation for 14 days at a temperature of -196 °C. Analysis of sperm DNA integrity used the gel electrophoresis method. This research consists of four types of cryoprotectants; Each type of cryoprotectant consists of two concentration levels. The cryoprotectant concentrations used were DMSO (concentrations 10 and 20%), Methanol (concentrations 10 and 20%), Glycerol (concentrations 5 and 20%), and Ethanol (concentrations 5 and 15%) as well as control treatment and fresh sperm treatment. Based on the results of the DNA integrity analysis of climbing perch fish sperm, it showed that all sperm DNA samples did not form fragments, meaning that there was no damage to climbing perch fish sperm DNA in all samples tested, both fresh sperm samples and post-cryopreserved sperm samples.

1 Introduction

Climbing perch fish *Anabas testudineus* is one of the freshwater fish that has economic value in Indonesia [1]. This fish has been cultivated [2], but the technology for hatching climbing perch fish, especially sperm storage, is still in the development stage [3, 4]. Cryopreservation not only plays a role in fish farming activities, but also for the conservation and improvement of genetic resources [5]. Cryopreservation can make an important contribution to storing genetic material over a long period of time [6]. This technique has been established in several species of freshwater fish using a variety of materials (extenders and cryoprotectants) used

* Corresponding author: muchlisinza@usk.ac.id

 https://doi.org/10.1051/bioconf/20248703010

to maintain post-thawing sperm quality [7]. However, the process often triggers cell injury (cryo-injury) which is caused by mechanical damage due to the formation of ice crystals both inside the cell and outside the cell, and also by osmotic pressure or oxidative stress [8]. This can affect the structure of the plasma membrane, mitochondria and chromatin [9].

One of the parameters in determining sperm quality after cryopreservation is motility [10]. There are several aspects that influence sperm cell motility including mitochondrial function, plasma membrane channel activity, ATP production and flagella structure [11]. For successful egg fertilization, spermatozoa need to maintain plasma membrane integrity, osmotic control, and movement capacity to reach the oocyte [5].

Apart from motility, analysis of sperm DNA integrity is also an important parameter to evaluate the success of established sperm cryopreservation protocols so as to select the best quality samples for freezing. Studies related to fish sperm DNA analysis have been carried out using several different approaches such as the gel electrophoresis method [12], the comet test (single cell gel electrophoresis) [13], TUNEL (terminal deoxynucleotidyl transferase-nick-end-labelling) [14], and analysis of specific DNA sequences using RT-PCR [15, 16]. Sperm DNA damage has been closely linked to various indicators of reproductive health including fertilization, embryo quality, and congenital abnormalities [17, 18]. Thus, sperm DNA integrity must be guaranteed for successful embryo development after fertilization [19]. This research was carried out as a follow-up observation to the previous publication by [20] regarding sperm DNA integrity parameters after cryopreservation with the addition of cryoprotectant using the gel electrophoresis method. Research analyzing the DNA integrity of climbing perch fish sperm after cryopreservation has never been reported before. This is important to know in order to obtain an overview of the influence of several types and concentrations of cryoprotectants on the DNA integrity of climbing perch fish sperm after cryopreservation.

2 Methods

2.1 Time and Site

This research was conducted from September to November 2022. The maintenance and treatment of test animals were carried out at the Fish Breeding and Rearing Laboratory of the Faculty of Marine and Fisheries, Syiah Kuala University. DNA integrity analysis was conducted at the Brackishwater Aquaculture Center (BPBAP) Ujung Batee, Banda Aceh.

2.2 Tools and Materials

Tools used in this research include digital scales, container boxes, cryotubes, micropipettes (Eppendorf), blue tips and yellow tips (Onemed), spuids. Meanwhile, the materials used are climbing perch fish parent, liquid nitrogen, 96% alcohol, glucose-base, DMSO, methanol, ethanol, glycerol (Sigma-Aldrich), genomic DNA purification kit (Promega), agarose gel (Himedia), tris Base BUF, DNA ladder and Loading Dye (GeneAid).

2.3 Research Design

This research uses the gel electrophoresis method, which consists of four types of cryoprotectant with each type having 2 concentration levels, including: DMSO (concentration 10 and 20%), Methanol (concentration 10 and 20%), Glycerol (concentration 5 and 20%), and Ethanol (concentration 15 and 5%), which will be combined with 5% egg yolk and glucose-base and there was a control treatment which only added glucose-base and

5% egg yolk. Apart from that, there is also fresh sperm treatment which is used as a reference in this research.

2.4 Work Procedures

2.4.1 Broodstock Maintenance and Sperm Collection

Climbing perch fish broodstock were collected from the waters of Aceh Besar, consisting of 20 male broodstock with a length of 10-14 cm and a body weight of 39-43 g. The collected broodstock were acclimated for 14 days in a broodstock pond and fed with commercial feed with a protein content of >30% at 5% of body weight. Feeding was provided twice daily at 06:00 and 18:00 WIB. Climbing perch were reared based on the guidelines of the Institutional Animal Care and Use Committee (IACUC, 2018) and have ethical approval B/034SN/XII/2022 from the animal ethic committee of the Marine and Fisheries Faculty, Syiah Kuala University. Mature male broodstock were selected, and each received an intramuscular injection of ovaprim at a dosage of 0.5 ml/kg body weight. Ovaprim was injected below the dorsal fin, with the injection needle at a 45° angle to the fish's body surface or on the dorsal fin of the fish [3]. Subsequently, the fish were aerated for 7 hours. During the conditioning, the broodstock were not given any feed. After 7 hours, sperm collection was performed on the male broodstock. The sperm collection process involved cleaning the fish's genital opening using tissue, followed by gently stripping the fish's abdomen towards the genital area. The collected sperm was then transferred to a 2 ml cryotube and placed in a styrofoam container with a temperature of 4 °C [12].

2.4.2 Extender and Cryoprotectant Preparation

Based on previous research, the most suitable extender for climbing perch fish sperm is Glucose base [4], Therefore, this extender was used in this research, with the following combination of ingredients: NaCl 0.725 g; KCL 0.04 g; $NaHCO3$ 0.080 g; glucose 0.20 g, mixed with aquades to a total volume of 100 ml. The previously obtained sperm was mixed with the extender at a dilution ratio of 1:60 v/v (sperm: glucose base extender).

The cryoprotectants tested were DMSO (10% and 20% concentration), Methanol (10% and 20% concentration), Glycerol (5% and 20% concentration), and Ethanol (15% and 5% concentration). The control group had 0% cryoprotectant. Each treatment was combined with 5% egg yolk. To obtain the test concentrations of 0%, 5%, 10%, 15%, and 20% cryoprotectant, 100 ml of glucose base extender was added to 1.67 ml of sperm, resulting in a sperm and glucose base extender ratio of 1:60 v/v. Then, 5 ml of egg yolk was added to this dilution. Subsequently, the dilution of sperm and egg yolk was distributed in 1.5 ml portions to 9 cryotubes. Each cryotube was then filled with the tested cryoprotectant (DMSO, methanol, glycerol, and ethanol) according to the specified test doses: 0.075 ml of cryoprotectant was added to obtain a 5% concentration, 0.15 ml to obtain a 10% concentration, 0.225 ml to obtain a 15% concentration, and 0.3 ml to obtain a 20% concentration. The 0% concentration (control) was not supplemented with cryoprotectant.

2.4.3 Process of Freezing and Thawing sperm samples

All cryotubes containing sperm samples were placed in an icebox at a temperature of 4 °C for 5 minutes for the initial equilibration (pre-freezing). The samples were then evaporated at a distance of 5 cm above the surface of liquid nitrogen at -79 °C for 5 minutes [21], and then transferred into liquid nitrogen (-196 °C) and stored for 2 weeks. After two weeks, the thawing process was performed. Cryotubes containing frozen sperm were taken from the container and immersed in a water bath at a temperature of 30-32 °C for 5 minutes. Subsequently, DNA integrity analysis was conducted.

2.4.4 DNA laddering Analysis

a. Sperm DNA extraction

The sperm sample was placed in a sterile microcentrifuge tube and centrifuged to obtain a 100 μl pellet. The resulting pellet was transferred into another sterile microcentrifuge tube to which 300 μl of Cell Lysis Solution had been added, the tube containing the mixture was stirred 6 times until homogeneous, then incubated for 10 minutes at room temperature. After incubation, it was centrifuged at a speed of 13,000-16,000 rpm for 20 seconds. The supernatant layer was then discarded, then the tube containing the pellet was vortexed for 10-15 seconds until the pellet was resuspended, then 100 μl of Core Lysis Solution was added to the tube and homogenized again using a vortex to produce a lysate. 35 μl of Protein Precipitation Solution (RNase Solution) was added and homogenized using a vortex for 20 seconds. Next, it was centrifuged at a speed of 13,000-16,000 rpm for 3 minutes. The supernatant was transferred into a microcentrifuge tube containing 100 μl of isopropanol, then the solution was mixed and centrifuged at a speed of 13,000-16,000 rpm at room temperature for 1 minute. The supernatant was discarded and 70% ethanol was added to the DNA sample volume at room temperature, then centrifuged again for 1 minute. The ethanol solution was discarded and the pellet was dried in air for 10-15 minutes. DNA rehydration was carried out by adding 35 μl of DNA Rehydration Solution to the tube containing the pellet, then incubating for 1 hour at 65 °C. Finally, DNA is stored at 2-8 °C [22]. Next, DNA concentration was measured using a Nanodrop 2000c [23].

b. Gel electrophoresis

2 g of agarose was weighed, then 1 x 100 ml of TBE buffer solution was added. Heat the agarose solution and TBE buffer until the solution becomes clear and boils. Once the solution is clear and boiling, add 2 μl of DNA dye. Next, the agarose solution is poured into the prepared gel mold and left for 30 minutes until the agarose gel hardens. Next, 1.5% agarose gel was put into the electrophoresis chamber, 1x TBE buffer solution was added. 10 μl of sample and 2 μl of loading dye were mixed on parafilm paper then injected into the gel well, including 5 μl of DNA marker. Next, close the electrophoresis and turn on the electricity with a voltage of 135 V for 30 minutes. Reading of electrophoresis results was carried out using a UV transilluminator.

2.5 Data Analysis

Data on sperm DNA integrity after cryopreservation were presented descriptively.

3 Result

The analysis of climbing perch fish sperm DNA integrity was conducted using gel electrophoresis with four different cryoprotectants, each at two concentration levels: DMSO (10% and 20%), Methanol (10% and 20%), Glycerol (5% and 20%), and Ethanol (15% and 5%). As a comparison, testing was also performed on the control group (without the addition of cryoprotectant) and fresh sperm. The electrophoresis results showed that DNA integrity was expressed in all treatment lines and fresh sperm. The electropherogram images showed that no DNA fragments were formed in all samples of climbing perch fish sperm, indicating that there was no sperm damage in the tested samples, both in fresh sperm and sperm after thawing.

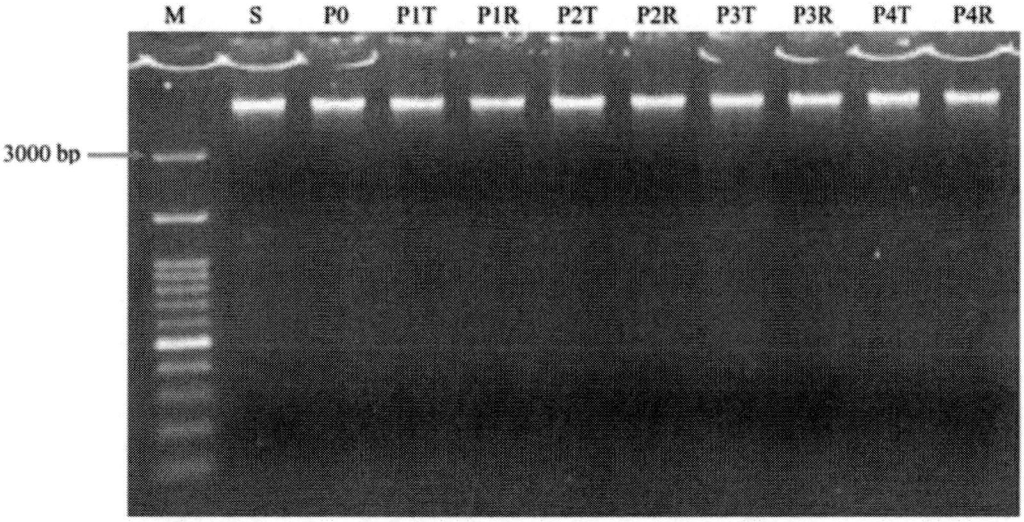

Fig. 1. Results of electrophoresis of climbing perch fish sperm samples. Note: M= Marker; S=Fresh sperm; P0= control; P1T= 10% DMSO; P1R= 20% DMSO; P2T= 10% Methanol; P2R= 20% Methanol; P3T= 5% Glycerol; P3R= 20% Glycerol; P4T=15% Ethanol; P4R=5% Ethanol.

4 Discussion

DNA is one of the cell components susceptible to cryodamage, so it is essential to maintain DNA integrity in the sperm cryopreservation process [24]. The analysis of climbing perch fish sperm DNA integrity using gel electrophoresis showed that the smear pattern in the fresh sperm, control treatment, DMSO (10% and 20% concentration), Methanol (10% and 20% concentration), Glycerol (5% and 20% concentration), and Ethanol (15% and 5% concentration) all exhibited the same results, indicating the absence of DNA fragmentation in the tested sperm samples with fresh sperm samples which were used as a reference in this study. The presence of cryoprotectants in the test samples is believed to have been able to maintain the integrity of sperm DNA during the freezing process. Bhattacharya [25] added that the use of cryoprotectants can slow down the cell freezing process, and sperm cells have a longer adaptation time to extreme temperature changes, preventing cell damage due to freezing. According to [26] cryopreservation can have side effects, such as genetic changes within cells that can damage cell integrity. Therefore, the addition of cryoprotectants plays a crucial role in preserving sperm DNA integrity, reducing the risk of genetic changes and cell damage during the cryopreservation process.

Conclusion

In conclusion, no damage to climbing perch fish sperm DNA was observed in all tested samples, both in fresh sperm and sperm after cryopreservation.

Acknowledgments

The authors are grateful to the Ministry of Education, Culture, Research and Technology of Indonesia for funding and supporting this study through the National Competitive Applied Research scheme (Contract No: 062/E5/PG.02.00.PL/2023).

References

1. A. Asyari, Bawal Widya Riset Perikanan Tangkap. **1**, 5 (2017)
2. Z.A. Muchlisin, J. Iktiologi Indones. **13**, 1 (2013)
3. K. Shahriar, K.M. Nazrul, M.R. Sarder, S. Umma, *Sperm Cryopreservation Protocol of Anabas testudineus (Bloch 1792)* (Bangladesh, Lambert Academic Publishing, 2014)
4. S. Maulida, K. Eriani, N. Fadli, F.M. Nur, Aquac. Res. **53**, 12 (2022)
5. E. Cabrita, C. Sarasquete, S. Martínez-Páramo, V. Robles, J. Appl. Ichthyol. **26**, 5 (2010)
6. D.E. Pegg, *Principles of Cryopreservation*, in *Cryopreservation and Freeze-Drying Protocols*. 2015, Springer: Totowa. p. 3-19.
7. S. Maulida, F.M. Nur, K. Eriani, Z.A. Muchlisin, Depik. **9**, 2 (2020)
8. P. Watson, G. Morris, *Cold shock injury in animal cells*. in *Symposia of the Society for Experimental Biology*, (1987)
9. P. Watson, W. Holt, Cryobanking the genetic resource: wildlife conservation for the future. (2001)
10. Z.A. Muchlisin, M.N.S. Azizah, Cryobiology. **58**, 2 (2009)
11. V. Dzyuba, J. Cosson, Reprod. Biol. **14**, 3 (2014)
12. S. Maulida, K. Eriani, F. Nur, N. Fadli, Braz J Vet Res Anim Sci. **58**, (2021)
13. L. Fraser, J. Strzeżek, Theriogenology. **68**, 2 (2007)
14. C. Martins, M. Dode, S. Báo, R. Rumpf, Genet. Mol. Res. **6**, 1 (2007)
15. R.M. Cawthon, Nucleic Acids Res. **30**, 10 (2002)
16. S. Pérez-Cerezales, S. Martínez-Páramo, E. Cabrita, F. Martínez-Pastor, Theriogenology. **71**, 4 (2009)
17. S.E. Lewis, I.M. Agbaje, Mutagenesis. **23**, 3 (2008)
18. M.B. Shamsi, S.N. Imam, R. Dada, J. Assist. Reprod. Genet. **28**, 11 (2011)
19. E. Figueroa, M. Lee-Estévez, I. Valdebenito, J.G. Farías, Rev. Aquac. **12**, 1 (2020)
20. S. Maulida, K. Eriani, N. Fadli, F.K. Kocabaş, Theriogenology. **201**, (2023)
21. Z.A. Muchlisin, W.N. Nadiah, N. Nadia, N. Fadli, Czech J. Anim. Sci. **60**, 1 (2015)
22. Promega, *Wizard Genomic DNA purification KIT*. 2017, Promega: USA.
23. W.C. Chen, R. Kerr, A. May, B. Ndlovu, Biopreserv. Biobank. **16**, 2 (2018)
24. S. Pérez-Cerezales, S. Martínez-Páramo, J. Beirão, M.P. Herráez, Theriogenology. **74**, 2 (2010)
25. S. Bhattacharya, Cryoprotectants and their usage in cryopreservation process. (London, IntechOpen, 2018)
26. S. Leibo, T.B. Pool, Fertil. Steril. **96**, 2 (2011)

Providing different natural feeds on the growth rate of *tor soro* larvae

Ismarica Ismarica[1], *Mohammad* Hafiansyah[1], *Iko Imelda* Arisa[1*]

[1]Department of Aquaculture, Faculty of Marine and Fisheries, Universitas Syiah Kuala, Banda Aceh, Indonesia

Abstract. *Tor soro* is an Indonesian freshwater fish that is threatened with extinction. Preventing the extinction of the *Tor soro* fish is by carrying out cultivation activities. This research aims to determine the impact of providing different varieties of natural feed on the growth rate of *Tor soro*. Completely Randomised Design (CRD) consisting of four treatments (A; *Artemia salina* 21 days, B; *Artemia salina* 7 days and *Tubifex* sp. 14 days, C; *Daphnia* sp. 21 days, D; *Daphnia* sp. 7 days and *Tubifex* sp.14 days) was the method used in this research. Fish maintenance is for 21 days with a feeding frequency of twice (morning and evening). The results of the ANOVA test showed that giving different natural feeds impacts absolute length growth and specific growth rate, but does not significantly differ the survival rate of *Tor soro* fish larvae. The highest absolute length growth was 1.07 ± 0.01 cm and 0,036±0,025 g of the absolute weight growth. the difference in days of feeding impacts fish growth because the protein contained in *Tor soro* fish feed varies (Treatment B). The conclusion obtained from this research was that the highest value was found in treatment B.

1 Introduction

Tor soro is a freshwater fish commodity that is very prospective for aquaculture [1][2]. *Tor Soro* is one of the fish species whose population has declined in nature [3]. One alternative that can be done to increase the population is to increase the growth rate at the larval stage.

The larval stage is the most vulnerable. That is because the larvae must be given appropriate food to support growth after the yolk runs out. [4] stated that the critical phase occurs is the larval phase because there are many deaths in that phase. Death in the larval phase occurs because the fish larvae have not found the appropriate feed. Fish larvae feed needed for growth is protein ranging from 35-50% [5]. [6] states that feeding with high protein values and following the larvae's mouth opening can increase growth and shorten larval rearing. Natural feeds with a high protein content and size suitable for larval mouth openings include *Artemia salina*, *Tubifex sp.* and *Daphnia* sp.

Artemia salina contains the nutrients needed to help fish larvae grow and survive [7]. The nutrient content in *Artemia salina* is very high, one of which is protein which is 41.21%, fat 29.10%, and ash 11.29% [8]. *Artemia* merupakan pakan alami yang disukai oleh larva *Tor soro* [4]. *Tubifex* sp. is a natural food that is easily digested and has active movement. The nutrient content of *Tubifex* sp. is 3% crude fibre, 11.21% water, 64.47% crude protein, 17.63% crude fat, 7.48% ash, and 10.06% betaine [9]. In addition to *Tubifex* sp., *Daphnia* sp. is also a natural feed that is easily digested with nutrient content suitable for fish larvae. [10] stated that *Daphnia* sp. has a protein content of 39.24%, fat of 98%, carbohydrates of 4.32%,

* Corresponding author: ikoimeldaarisa@usk.ac.id

and ash content of 14.63%. Based on research [4], the average weight gain of *Tor soro* larvae was 0.35 grams, the average length gain of *Tor soro* larvae was 1 cm, and the growth rate of *Tor soro* larvae was 0.01% after 29 days of feeding with *Artemia*. In this way, this research needs to be done to determine the amount and type of natural feed that is effective and suitable for *Tor soro* fish larvae.

2 Material and Method

2.1 Location and Time for Research

This research was carried out at the Cijeruk Fresh Fish Germplasm Research and Development Facility in January - February 2022.

2.2 Research Methodology

The method of Completely Randomized Design (CRD) is the method used in this study. The treatment in CRD method used consisted of four treatments (A; *Artemia salina* 21 days, *B; Artemia salina* 7 days and *Tubifex* sp. 14 days, *C; Daphnia* sp. 21 days, D; *Daphnia* sp. 7 days and *Tubifex* sp. 14 days).

The container used in larvae rearing is a 30x25x20 cm aquarium with a water volume of 15 L. The number of containers is 16 aquariums that have been cleaned beforehand and equipped with an aerator. Larvae were obtained from the Cijeruk Freshwater Fisheries Germplasm Research and Development Institute. *Tor soro* larvae used as test animals are larvae that are 5 days old after hatching with a length of ± 0.1 cm. *Tor soro* were reared based on the guidelines of the Institutional Animal Care and Use Committee (IACUC, 2018) and have ethical approval B/034SN/XII/2022 from the animal ethics committee of the School of Marine and Fisheries Faculty, Universitas Syiah Kuala. *Tor soro* larvae were stocked at 1 fish/L [11] with the larvae were reared for 21 days from day 5 after hatching to day 25 after hatching. The feed given is by the treatment in this study. The frequency of feeding was 2 times a day with the Ad satiation method. The parameters in this study are as follows:

Absolute Length Growth (L), absolute weight growth (W), and survival rate (SR) are measured using the following formula [12]:

$L = L_t - L_0$
$W = W_t - W_0$
$SR = N_0 - N_t/N_0 x\ 100\%$

Pieces of information: L=Absolute length growth (cm), L_t= Fish length at the end of study (cm), L_0= Fish length at the start of the study (cm), W=Absolute length growth (g), W_t= Fish length at the end of study (g), W_0= Fish length at the start of the study (g), SR= Survival rate (%), W_t= Number of fish at the end of study (fish), W_0= Number of fish at the start of study (fish).

The water quality measured is acidity (pH) measured using LAMOTTE pH 5. Temperature and oxygen content (DO) were measured using a Ysi Pro DO Meter. Water quality measurements were taken every 7 days during the study. Sampling at the beginning and end of the study for absolute length growth, absolute weight growth, and survival rate.

2.3 Data analysis

The analysis used is ANOVA with a 95% confidence level. If different natural diets had an effect ($P < 0.05$) on absolute length growth, absolute weight growth and survival rate, the analysis was continued with Tukey's post hoc test. Descriptive analyses were performed on data from water quality measurements.

3 Results and discussion

The results of the ANOVA analysis test showed that different natural diets had an effect ($P<0.05$) on the absolute length growth of *Tor soro* fish, but did not affect the absolute weight growth and survival of *Tor soro* fish. The research data can be seen in Table 1 as follows.

Table 1. Analysed data on all research parameters

Treatment	Parameters		
	Absolute Length Growth (cm)	**Absolute Weight Growth (g)**	**Survival Rate (%)**
A	0.98 ± 0.02^b	0.02±0.02	95.00±3.33
B	1.07 ± 0.01^c	0.04±0.03	96.67±3.85
C	0.82 ± 0.03^a	0.02±0.01	90.00±3.85
D	1.00 ± 0.03^b	0.03±0.02	91.67±3.33

Note. A; *Artemia salina* 21 days, B; *Artemia salina* 7 days dan *Tubifex sp.* 14 days, C; *Daphnia* sp. 21 days, D; *Daphnia* sp. 7 days dan *Tubifex sp.* 14 days)

Data analysis (Table 1) shows that the treatment with the highest value in all parameters is treatment B. In treatment B, the absolute length growth value was 1.07 ± 0.01 cm, absolute weight growth 0.036 ± 0.025 g and survival rate 96.67 ± 3.85%. The lowest value of all parameters was in treatment C. The value of each parameter was 0.82 ± 0.03 cm (absolute length growth), 10.25 ± 0.27 %/day (Absolute weight growth) and 90 ± 3.85 % (survival rate).

The results of the water quality measurements are shown in Table 2. Temperature, pH and dissolved oxygen (DO) were measured as water quality data.

Table 2. Measurement data of water quality parameters

Treatment	Water Quality Parameters		
	Temperature (^{0}C)	**pH**	**DO (mg/L)**
A	23.5-25.4	7.59-7.93	4.56-4.96
B	23.2-25	7.51-7.72	4.49-4.67
C	23-25.1	7.46-7.78	4.48-4.74
D	23.4-25	7.54-7.72	4.53-4.68

Tor soro fish larvae during 21 days of rearing with different natural feeds experienced differences in absolute length growth, Absolute weight growth, and survival between all treatments. The treatment that had the highest absolute length growth value was treatment B with an absolute length growth value of 1.07 ± 0.01 cm and absolute weight growth of 0.036 ± 0.025 g. The highest absolute length growth value was treatment B with an absolute length growth value of 1.07 ± 0.01 cm. Research by [4] resulted in an average length growth of 1 cm and an average weight growth of 0.35 g of *Tor soro* fish during 29 days of maintenance and feed given to *Tor soro* fish in the form of *Artemia*. Based on the above statement, feeding in treatment B (*Artemia salina* 7 days and *Tubifex* sp. 14 days) during 21 days of maintenance showed the best results. This is because during the 21-day rearing showed the same results as the 29-day rearing and the protein content in the feed given was high. In addition, the difference in days of feeding impacts fish growth because the protein contained in *Tor soro* fish feed varies (7 days *Artemia salina*, 14 days *Tubifex* sp.). According to [8], *Artemia* contains protein, namely 41.21%, while the protein content of Tubifex sp. is crude protein 64.47% [9]. Protein is a nutrient that plays a very important role in fish growth and protein is the the largest component of meat and serves as the building blocks of body tissues [13].

Survival rate is the percentage of the number of live fish reared from the start of rearing to the end of rearing. In this study, the best survival rate of *Tor soro* larvae was obtained in treatment B (*Artemia salina* 7 days and *Tubifex* sp. 14 days) during 21 days of rearing, namely 96.67 ± 3.85%. According to [14] survival rate is considered good when the survival rate of fish is greater than 50%. The survival rate of fish in the larval stage is influenced by the availability of food following the required nutrients and the mouth opening of fish larvae. [15] stated that the survival and growth of fish are influenced by the food given to fish larvae. [16] also stated the same, namely the results of their research, a combination of 50% *Artemia* and 50% *Tubifex* feed showed the highest survival results. [17] also stated that the combination of 3 weeks of *Artemia* and 3 weeks of *Tubifex* feed showed a high survival rate of 86%.

The results of quality measurements (Table 2) during 21 days of fish rearing are within the optimal range for fish rearing. [18] stated that a good pH value for fish farming is 6.5-9 and DO ranges from 4.48-4.96 mg/L. [19] stated that if DO value in the rearing tank exceeds 5 mg/L, the fish experience stress and die. [20,21] stated that water quality that is not in the optimal range can cause death, stress and disturbance of metabolic processes.

4 Conclusion

Feeding different natural diets affected absolute length growth ($P<0.05$). However, it did not affect absolute weight growth and fish survival. The best treatment was treatment B (*Artemia salina* 7 days and *Tubifex* sp. 14 days). The value of absolute length growth was 1.07 ± 0.01 cm, absolute weight growth was 0.036 ± 0.025 g and survival rate was 96.67 ± 3.85%.

References

1. E. Nugroho, M.F. Sukadi, G.H. Huwoyon. Media Akuakultur. **7**, 1 (2012)
2. Muchlisin Z. A., A.S. Batubara, M.N.S. Azizah, M. Adlim, A. Hendri, N. Fadli, A. A., Muhammadar, S. Sugianto. Biodiversitas. **16,** 1 (2015)
3. T. F. Haser, E. Supriyono, K. Nirmala, Widanarni, T.H. Prihadi, T. Budiardi, F. Azmi, M. S. Nurdin. IOP Conf. Series: Earth and Environmental Science. EMBRIO-2021. **1033**, 012008 (2022)
4. Risky, Thaib, Nurhayati. Jurnal TILAPIA. **1,** 2 (2020)
5. D. Radona, J. Subagja, I.I Kusmini. Media akuakultur. **12.** 1 (2017)
6. Effendi. Biologi Perikanan Yayasan Pustaka Nusantara Yogtakarta. (2002)
7. D. K. Radhakrishnan, I. Akbar Ali, B. V. Schmidt, E. M. John, S. Sivanpillai, S. T. Vasunambesan. Aquaculture Research. **51** (2020)
8. M. Y. Firmansyah, R. Kusdarwati, Y.Cahyoko. Jurnal Fakultas Perikanan dan Kelautan Universitas Airlangga. **5,** 1 (2013)
9. Alfikri, Rilo, E.I. Raharjo, E. Prasetyo. Jurnal Ruaya FPIK UNMUH. **6**. 1(2018)
10. Bogut, Z. Adámek, Z. Puškadija, D. Galović., D. Bodakoš. Journal Ribarstvo. **68,** 1 (2010)

 https://doi.org/10.1051/bioconf/20248703011

11. T.H Prihadi, T. F. Haser, B. Pantjara, Y. R. Widyastuti, O. Z. Arifin, W. Cahyanti, I. I. Kusmini, D. Radona, K. Kurniawan, J. Subagja, A. Saputra1, V.A. Prakoso, I. Ardi. Journal of Hunan University (Natural Sciences). **49**, 3 (2022)
12. Effendie M. I. Yayasan Pustaka Nusantama. **155** (1997)
13. J.E. Halver, R.W. Hardy. Third edition. London: Academic Press, 823 pp. (2002).
14. A. Nursani. IJAS. **2,** 1 (2012).
15. V.E. Herawati, J. Hutabarat, S. B. Prayitno. J. Coast. Dev. **16,** 1 (2012)
16. T. Risaldo, J.K. Ockstan, Ch. W. Juliaan. Jurnal Budidaya perairan. **4,** 2 (2016)
17. Harisman, Sukendi, A. Nur. Jurnal Akuakultur sebatin. **2,** 2 (2021)
18. F. Cholik, A.G. Jagatraya, R.P. Poernomo, A. Jauzi. Masyarakat perikanan Nusantara dan taman akuarium air-TMII, Jakarta. 415
19. F. Tatangindatu, Kalesaran, R. Rompas. Jurnal Budidaya Perairan. **1,** 2 (2013)
20. E. Emaliana, S. Usman, I. Lesmana. Aquacoastmarine. **13,** 3 (2016)
21. K. Melanie, Z. Zilfira, S.A. Akbar,. Arwana: Jurnal Ilmiah Program Studi Perairan, **5(2)** 162-168 (2023)

Effect of addition of red spinach leaf extract (*Amaranthus tricolor* L.) in feed against the level of color brightness of sword platy fish (*Xiphophorus helleri*)

Adli Waliul Perdana[1*], *Siti* Maulida[1], *Rika* Indriani[1]

[1]Department of Aquaculture, Faculty of Marine and Fisheries, Syiah Kuala University, Banda Aceh, 23111. Indonesia

Abstract. Color is important factor ornamental fish, especially sword platy fish, because can affect selling value. One factor influences color brightness is carotenoids. Carotenoids cannot synthesized by fish's body so must be supplied through feed provided. Red spinach is a natural ingredient that contains carotenoids. Aim of research was to determine effect of adding red spinach leaf extract to artificial feed to increase color brightness, growth and survival of *Xiphophorus helleri*. Research method used a completely randomized design with 4 treatments and 4 repetitions. Concentrations of red spinach leaf extract are 0%, 3%, 6% and 9%. Parameters observed were color brightness, absolute weight, absolute length, survival rate and water quality. Results showed that addition of red spinach leaf extract had a significant effect ($P<0.05$) on brightness of fish color but had no significant effect ($P>0.05$) on growth, weight, length and survival rate. Highest increase in color brightness and absolute weight growth was obtained in 9% treatment with an average color brightness score of 6.66, an absolute weight of 0.36 g. Highest absolute length growth occurred treatments C (6%) and D (9%) at 0.41 cm and highest survival rate was obtained treatments A, B and C with a survival percentage of 88%.

1 Introduction

Freshwater ornamental fish are one of the ornamental fish that have quite good opportunities for the economy of people in Indonesia. The attractiveness of ornamental fish can be measured from their brilliant color, physical shape and completeness, behavior, and health condition or stamina [1]. One ornamental fish that is very popular with the public is the sword platy fish. The sword platy fish (*Xiphophorus helleri*) is a freshwater ornamental fish that is widely cultivated by ornamental fish farmers because it has a unique tail (like a sword) and beautiful body color [2].

One of the problems faced by farmers is that the color of the fish's body can fade. This can be caused by several factors, such as light conditions, environmental stress, disease attacks and nutrients in the feed [3]. Pigmentation substances, such as carotenoids, are needed by the fish's body to increase the brightness of their body color. However, aquatic animals cannot synthesize carotenoids in their bodies, so they need to provide feed containing carotenoids which can be obtained from natural ingredients.

* Corresponding author: adliwaliul@usk.ac.id

Carotenoids are yellow, orange to red pigments that come from plants, animals, fungi, humans and bacteria [4]. One natural ingredient that contains carotenoids is red spinach (*Amaranthus tricolor L.*). Red spinach is very easy to find and also has a relatively cheap price [3]. The carotenoid content found in red spinach is the dye lutein (as the main component), zeaxanthin, violaxanthin, neoxanthin, and β-carotene [5]. One of the functions of the lutein pigment is as a pigment to produce a yellow color, but in the fish's body lutein is converted into astaxanthin as a source of pigment to form a red color [6].

The use of red spinach in ornamental fish feed to increase the brightness of fish color has been tested by several researchers. Stated that adding 3% red spinach flour to the feed was able to increase the brightness of the color of goldfish (*Carassius auratus*) [7]. Research by [3] on clown loach fish (*Chromobotia macracanthus Bleeker*) reported that feeding food containing 6% red spinach flour was able to increase the color pigmentation of clown loach fish. Furthermore, [5] reported that red spinach flour added to feed at a dose of 6% was able to increase the color intensity of albino Sumatran fish (*Punitigrus tetrazona*). However, so far research regarding increasing color brightness in sword platy fish with the addition of red spinach leaf extract has not been reported. Therefore, it is necessary to conduct research to examine the effect of red spinach leaf extract as a source of natural dye in increasing the brightness of the color of sword platy fish (*Xiphophorus helleri*).

2 Materials and Methods

2.1 Time and place

This research was carried out from January to February 2023 in the Fish Hatchery and Breeding Laboratory building, Faculty of Marine Affairs and Fisheries, Syiah Kuala University, Banda Aceh.

2.2 Research methods

The research method used in this study was an experimental method using a completely randomized design (CRD) with 4 treatments and 4 repetitions. The treatment in this experiment was the addition of red spinach leaf extract (*Amaranthus tricolor L.*) to commercial feed and to sword platy fish (*Xiphophorus helleri*). Determination of the dose of red spinach leaf extract in this study was a modification of research by [5]. The treatments used are:

1. Treatment A: Addition of 0% concentration of red spinach leaf extract
2. Treatment B: Addition of red spinach leaf extract with a concentration of 3%
3. Treatment C: Addition of red spinach leaf extract with a concentration of 6%
4. Treatment D: Addition of red spinach leaf extract with a concentration of 9%

2.3 Research procedure

2.3.1 Preparation of Research Containers

The containers used were 16 plastic jars measuring 10 liters. The prepared plastic jars are cleaned first and then dried, this is done to avoid the presence of disease pathogens such as bacteria and fungi. The maintenance containers are then installed with an aeration installation and filled with clean water up to 60% of the height of the container [1], before use the water is aerated for 24 hours [8].

2.3.2 Making Extracts

Making red spinach (*Amaranthus tricolor L.*) extract refers to research [9], namely that red spinach leaves are washed until clean, then dried indoors to avoid sunlight. The drying process is carried out for 8 days until the red spinach leaves are dry. The dried red spinach leaves are then crushed using a blender until you get red spinach leaf powder (simplisia). The red spinach leaf simplicia powder was then taken to the Chemistry Laboratory, Faculty of Teacher Training and Education, Syiah Kuala University for extraction. Extraction is carried out by maceration using a solvent, namely 96% ethanol. Put 263 grams of red spinach leaf simplicia into a plastic jar and add 4.5 liters (until submerged) of 96% ethanol. The red spinach leaf simplicia is macerated for 3 days with regular stirring every morning and evening. The extract obtained was then filtered and the filter results obtained were then evaporated using a rotary evaporator at a temperature of 67°C until a concentrated extract was obtained. The resulting red spinach leaf extract was put into a dark bottle and stored in the refrigerator at 4°C.

2.3.3 Adding Extracts to Feed

The addition of red spinach leaf extract to feed was carried out by referring to research by [10], namely that red spinach leaf extract is measured according to a predetermined concentration and the extract is diluted using distilled water. After dilution, the extract is then sprayed evenly on the feed using a spray bottle. The feed that has been sprayed is then air-dried until dry, then the dried feed is put into labeled plastic bottles. Feed is stored in the refrigerator (refrigerator) until it is ready to be used.

2.3.4 Pisciculture

The test fish used in this research were sword platy fish (*Xiphophorus helleri*) which came from local farmers in Banda Aceh. The test fish used had an average size of 3.33 ± 0.16 cm. Before the fish are put into the test container, the fish are acclimatized for one hour to adjust the temperature to the rearing medium [11], then the fish are fasted for 24 hours with the aim of eliminating the influence of residual food in the fish's body [12]. Each container is filled with 6 fish. Fish are kept for one month (30 days). Measurements of fish length, weight and survival were carried out at the beginning and end of the study [13].

2.3.5 Feeding

The feeding of platy swordfish (*Xiphophorus helleri*) in this study was fed ad satiation (until the fish were full) with a feeding frequency of twice a day, namely at 09.00 and 16.00 WIB [14].

2.3.6 Siphoning

The water quality of the rearing media is a factor that has an important influence on fish survival and can indirectly influence the color of sword platy fish (*Xiphophorus helleri*). One way to maintain good water quality is through siphoning. Syphoning is carried out every 7 days by reducing the water by 20% of the height of the water volume in the maintenance container [15]. Siphoning is carried out with the aim of removing food and feces residues so that accumulation and rot does not occur in the rearing container.

3 Research Parameters

3.1 Identification of compounds using a Fourier Transform Infrared (FTIR) spectrophotometer

Identification of functional groups of compounds in red spinach leaf extract using FTIR. FTIR analysis was carried out at the Environmental Quality Testing Engineering Laboratory located in the Faculty of Engineering, Chemical Engineering Department, Syiah Kuala University. FTIR spectrophotometer is used to determine the functional group of an isolate. FTIR spectrophotometers can be used to analyze solid, liquid and gas samples. A small amount of red spinach leaf extract was taken and analyzed using FTIR, then an FTIR spectrum was created in the wave number range of 4000 – 500 cm-^{1}.

3.2 Color brightness

Assessment of the color brightness of sword platy fish (*Xiphophorus helleri*) was carried out using a questionnaire on 11 respondents to assess the increase in color brightness of sword platy fish [16]. The selected respondents consisted of laboratory assistants and students who did not have visual impairments (color blindness and nearsightedness). Each respondent assessed the brightness level of the fish's color based on the code on the Toca Color Finder (TCF) paper which had been given a score of 1 – 9. Measuring body color started from the smallest value to the largest by comparing the color of the fish with the TCF paper. The color observations carried out only focused on one color that was close to the body color of the test fish, namely orange. The determinant of the color brightness level is that the higher the color assessment given, the brighter the color brightness level of the sword platy fish (Figure 1). Observations of the color quality of sword platy fish were carried out every 10 days for 30 days [8].

Fig. 1. Fish Color Measuring Tool

3.3 Absolute weight growth

Body weight measurements were carried out to determine the increase in fish weight during rearing [17]. The absolute weight growth of fish is calculated based on the [18] formula:

$$B = W_t - W_0 \quad (1)$$

Information:

B = Increase in absolute biomass of test fish (g)
W_t = Biomass of test fish at the end of the experiment (g)
W_0= Biomass of test fish at the start of the experiment (g)

3.4 Absolute length growth

Length growth is the amount of increase in the average length of fish during the rearing period. The increase in fish length can be calculated using the formula [19]:

$$P = P_t - P_0 \qquad (1)$$

Information:

P = Fish length growth (cm)
P_t = Length of test fish at the end of the study (cm)

P_0 = Length of test fish at the start of the study (cm)

3.5 Survival rate

Survival rate is a comparison of the fish population at the end of rearing to the beginning of rearing expressed in percent (%). Measuring the degree of survival of test fish can be calculated using the formula of [20], namely:

$$SR = \frac{N_0 - N_t}{N_0} X\ 100\% \qquad (1)$$

Information :
SR = Survival of test fish (%)
N_t = Number of test fish that died during the study (tails)
N_0 = Number of test fish at the beginning of the study (tails)

3.6 Water quality

Observing water quality is a supporting factor for success in fish farming. The water quality observed is temperature, pH (degree of acidity) of water and dissolved oxygen (DO). Temperature is measured using a thermometer, pH is measured using a pH-meter and DO is measured using a DO meter. The water samples analyzed were taken from fish rearing containers. Water quality parameter measurements are carried out once a week [11].

3.7 Data analysis

Data obtained from the research results were analyzed using analysis of variance (ANOVA) with a confidence level of 95% to determine the effect of treatment on the observed parameters. If there are differences between treatments, further tests will be carried out. The further test used is determined based on the value of the Diversity Coefficient (KK). The Duncan test is used as a follow-up test if the KK value is large, namely at least 10%. The Least Significant Difference (BNT) test is used as a follow-up test if the KK value is moderate, namely in the range of 5-10%. The Honestly Significant Difference (BNJ) test is used as a follow-up test if the KK value is small, namely a maximum of 5% [21].

4 Results and Discussion

4.1 Identification of compounds using a Fourier Transform Infrared (FTIR) spectrophotometer

The results of analysis of red spinach leaf extract compounds using an FTIR spectrophotometer show an infrared spectrum which can be seen in Figure 2, while the analysis results showing the peak of carotenoid absorption from red spinach leaf extract at FTIR wave numbers can be seen in Table 1.

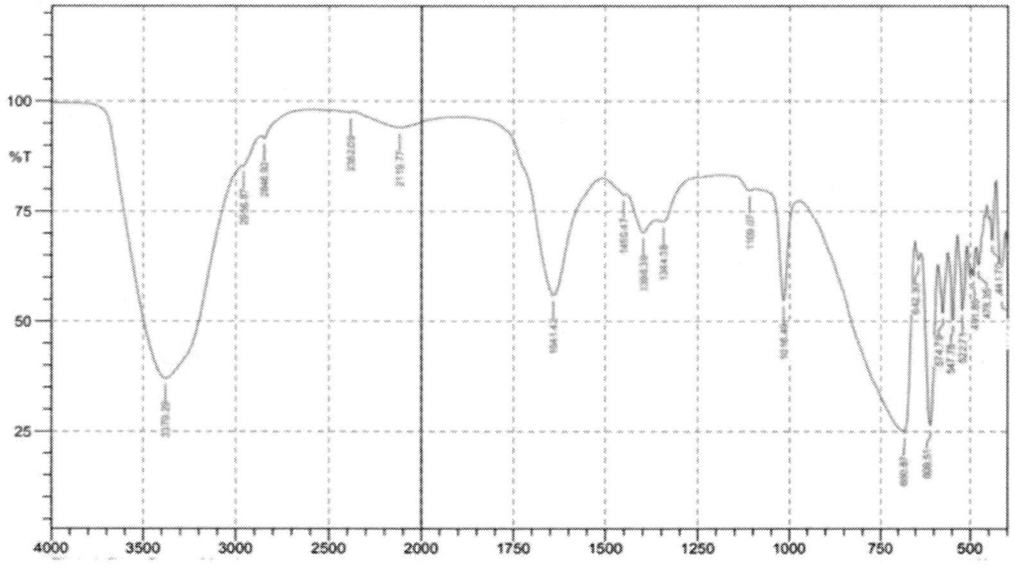

Fig. 2. FTIR Spectrum of Red Spinach Leaf Extract (*A. tricolor* L.)

Table 1. FTIR Spectrum Analysis Results

No	Wave Number (cm^{-1})		Functional groups
	Spectra	Literature	
1.	642.3 680.87	610 - 700	C-H bend alkene
2.	1344.38 1398.39 1450.47	1340 - 1470	C-H alkane
3.	1641.42	1600 – 1680	-C = C alkane
4.	2846.93 2956.87	2750 - 3000	C – H alkyl
5.	3379.29	3200 - 3500	-OH hydroxyl
6.	1016.49 1109.07	1000-1300	C-O

4.2 Color brightness, absolute weight growth, absolute length and survival rate

The results of the Analysis of Variance (ANOVA) test showed that the addition of red spinach leaf extract in the feed had a significant effect ($P<0.05$) on the color brightness level of sword platy fish (*Xiphophorus helleri*), but did not have a significant effect ($P>0.05$) on absolute weight, absolute length and survival rate of sword platy fish. Data on color brightness, absolute weight, absolute length and survival rate of sword platy fish can be seen in Table 2.

Table 2. Data on color brightness (score), absolute weight growth (g), absolute length (cm) and survival rate (%) of sword platy fish

Treatment	Parameter			
	Color Brightness (Skor)	**Heavy Absolute (g)**	**Absolute Length (cm)**	**Survival rate (%)**
A	3,28 ± 0,82[a]	0,17 ± 0,42[a]	0,30 ± 0,12[a]	87 ± 8,5[a]
B	4,44 ± 1,52[ab]	0,20 ± 0,12[a]	0,33 ± 0,23[a]	88 ± 15,84[a]
C	5,49 ± 0,52[bc]	0,25 ± 0,11[a]	0,41 ± 0,20[a]	88 ± 15,84[a]
D	6,66 ± 0,20[c]	0,36 ± 0,22[a]	0,41 ± 0,39[a]	83 ± 13,47[a]

Note: *Superscripts followed by different letters in the same column indicate significant differences.*

Observations of the color of sword platy fish kept for 30 days showed an increase in color brightness from each treatment during the study. Changes in increasing color brightness in sword platy fish from the beginning to the end of the study can be seen in Figure 3 and Figure 4.

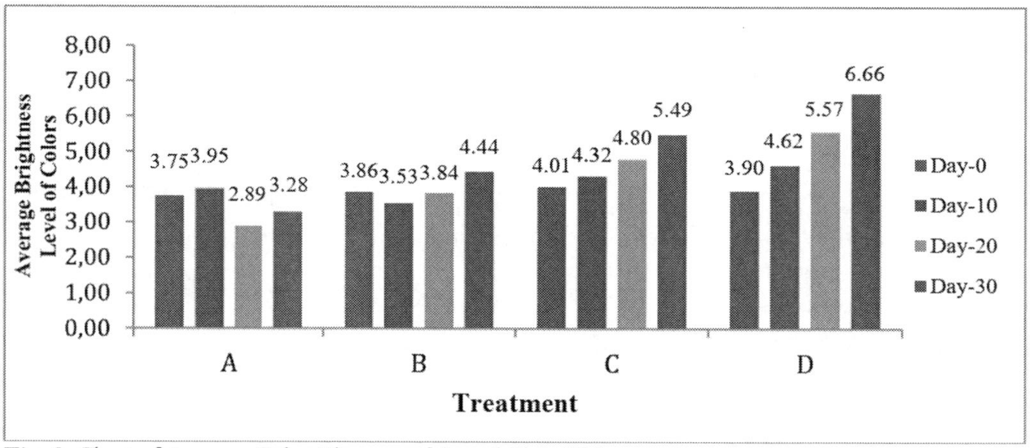

Fig. 3. Chart of Average Color Changes of Sword Platy Fish

Treatment	Before	After
A		
B		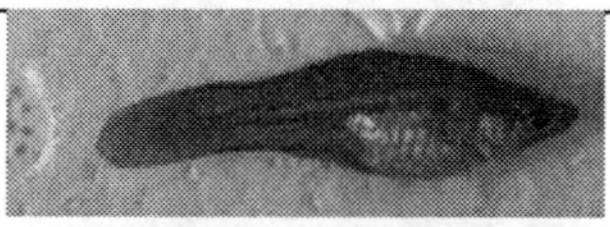
C		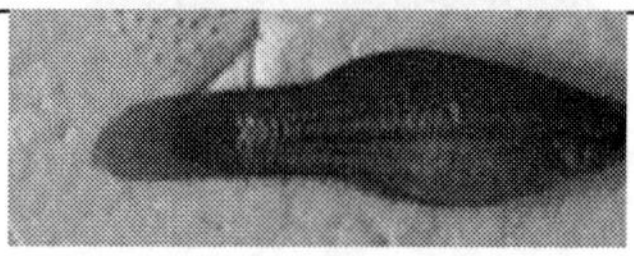

https://doi.org/10.1051/bioconf/20248703012

D	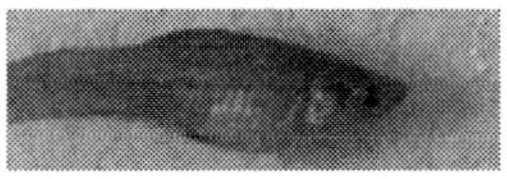	

Fig. 4. Sword Platy Fish Color Improvement

4.3 Water quality

Based on the research that has been carried out, the results of water quality measurements, namely temperature, pH and dissolved oxygen (DO), are presented in Table 3.

Table 3. Data on the range of air quality parameters when rearing platy swordfish (*Xiphophorus helleri)*

No	Parameter	Range	Optimum Value
1.	Temperature (°C)	27,3-28,4	22-29 [2]
2.	pH	7,89-8,46	7-8,5 [22]
3.	*Dissolved oxygen* (DO) (mg/L)	3,30-3,45	>2 [2]

4.4 Discussion

The vibration mode obtained from the results of Fourier Transform Infrared (FTIR) analysis shows the character of the bonds between atoms in compounds from red spinach leaf extract. The broad peak at 3379.29 cm^{-1} is the vibrational character of the –OH hydroxyl group [23]. Then, the absorption peaks are 2846.93 and 2956.87 cm^{-1} is the vibration of the C-H alkyl, then the C=C stretching of the alkene is seen at 1641.42 cm^{-1} [24]. The absorption peaks at 1344.38, 1398.39 and 1450.47 cm^{-1} are the C-H bending of alkanes, then the C-O stretching is seen at 1016.49 and 1109.07 cm^{-1}, then the peaks at 642.3 and 680.87 cm^{-1} are the C-H bending vibrational character of alkenes [25]. Based on the vibration results, it shows that all carotenoid components in red spinach leaf extract can be produced, namely the compounds lutein, zeaxanthin, violaxanthin, neoxanthin, and β-carotene. The chemical structure of the carotenoid compounds in red spinach leaf extract can be seen in Figure 5.

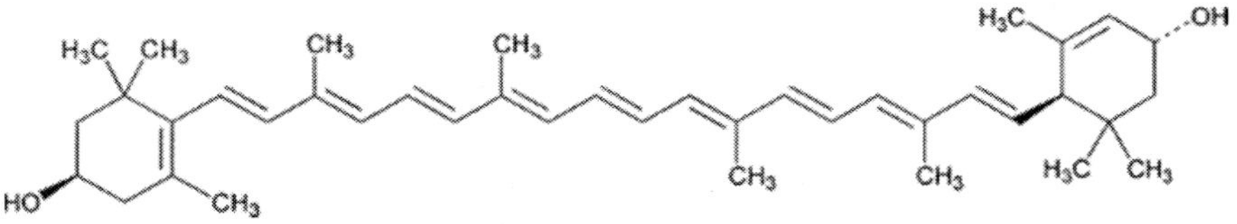

Lutein

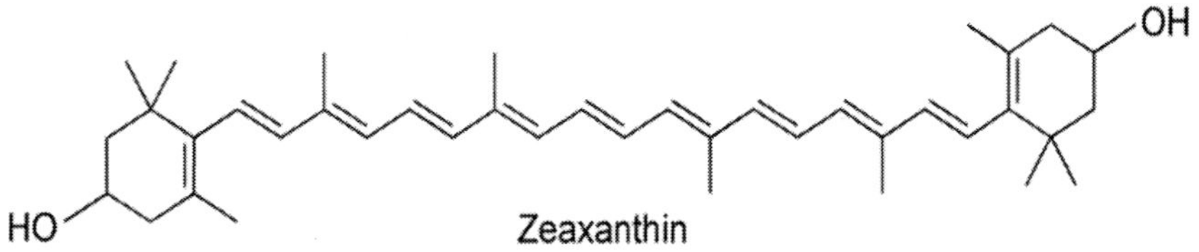

Zeaxanthin

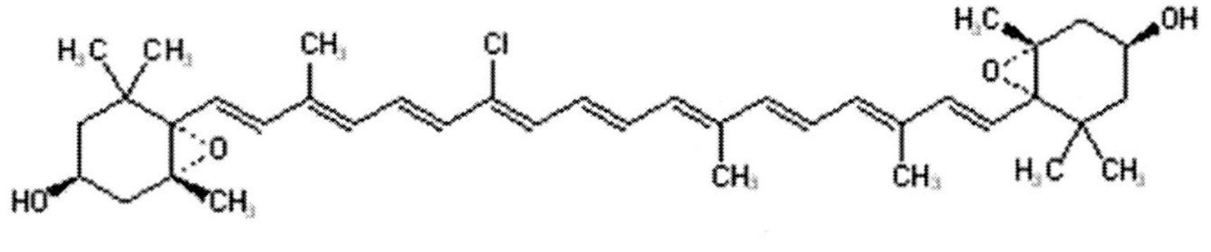

Violaxanthin

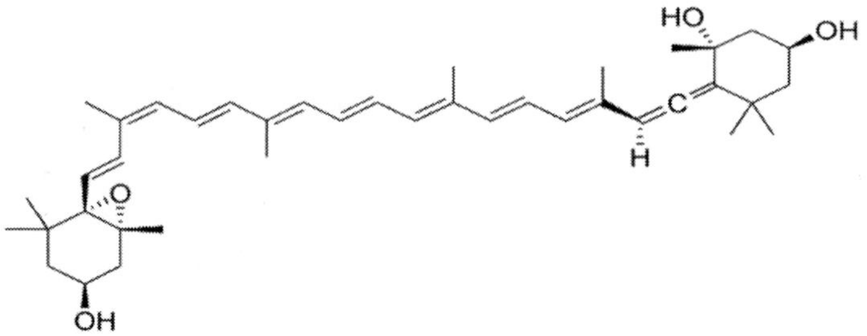

Neoxanthin

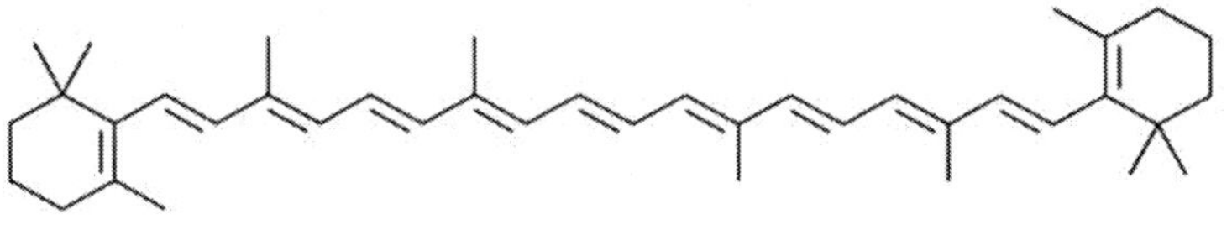

β-Carotene

Fig. 5. Structure of Several Types of Carotenoids in Red Spinach Leaf Extract

Lutein is a yellow carotenoid pigment and belongs to the xanthophyll family [26]. Based on statements from [27], the main carotenoid component in red spinach (*Amaranthus tricolor L.*) is lutein. Lutein is the main source of pigmentation in freshwater fish. In the fish's body, lutein is converted into canthaxanthin and astaxanthin [28] so that lutein is a type of carotenoid that is used to improve the color quality of fish. The results of the 30 day study showed that the addition of red spinach leaf extract (Amaranthus tricolor L.) in the feed was able to improve the color quality of sword platy fish (*Xiphophorus helleri*) in all treatments compared to the control treatment. The color intensity of sword platy fish in treatments C (6%) and D (9%) had increased on the 10th day of observation, while treatment B (3%) had only increased on the 20th day and treatment A (0%) color quality tends to be unstable (Figure 3). The increase in color that occurs in the body of the sword platy fish is caused by the carotenoids provided through feed being able to be absorbed by the body of the sword platy fish.

Treatment D with a concentration of 9% red spinach leaf extract mixed into the feed showed the highest final color change with an average score of 6.66 on a scale of 9. Furthermore, treatment C with an extract concentration of 6% showed an average value of color brightness. of 5.49. Then, treatment B with an extract concentration of 3% showed an average color change value of 4.44. The lowest color change was in treatment A, namely commercial feed without the addition of red spinach leaf extract with an average color brightness score at the end of rearing of 3.28. The maximum increase in fish color brightness was obtained in treatment D, this was due to the carotenoid content at a concentration of 9% greater than the amount of carotenoids in the other treatments so that the increase in color brightness was more significant. This statement is in accordance with what was stated by [29][30][31] that the percentage of carotenoids mixed in the feed is directly proportional to the intensity of the color brightness of the fish. However, the addition of carotenoids to feed has a maximum limit, meaning that if carotenoids are added to feed in excess amounts, at a certain point it will not provide a better color change and may even reduce the color value of the fish [32].

The increase in color which continued to increase in treatments B (3%), C (6%) and D (9%) until the 30th day of rearing is believed to be because sword platy fish still need carotenoid content in feed to improve the quality of their color, because of the concentration used in accordance with the ability of fish to synthesize carotenoids in red spinach leaf extract, so that fish metabolism and absorption work optimally. This is reinforced by the statement made by [33] who stated that, providing carotene levels that are appropriate to the fish's ability to synthesize carotene, can make the fish's absorption and metabolism work optimally. Based on the results of the analysis of variance (ANOVA), the brightness level of the color of the sword platy fish shows a significance value of 0.001, which is a value smaller than 0.05, so it can be concluded that the addition of red spinach leaf extract in the feed has a real influence on improving the color quality of the sword platy fish. Duncan's further test results showed that the addition of red spinach leaf extract at a concentration of 9% gave a higher color brightness score for sword platy fish. It is suspected that treatment D (9%) had a better effect on the color quality of sword platy fish because more red spinach leaf extract was added to the feed. This can be seen from the average color brightness score in treatment D (9%) of 6.66, which is the highest score among the other treatments (Figure 3).

The results of absolute weight analysis during the research showed that the addition of red spinach leaf extract to the feed did not have a significant effect on the absolute weight of sword platy fish ($P > 0.05$). Absolute weight measurements of sword platy fish were carried out at the beginning of rearing and at the end of rearing. During the research, sword platy fish experienced an increase in body weight with each treatment. Based on the data in Table 2, the highest average absolute weight growth of sword platy fish occurred in treatment D (9%) with an average value of 0.36 gr, followed by treatment C (6%) with an average absolute weight value of 0 .25 gr, then followed by treatment B (3%) with an average value of 0.20 gr, and the lowest absolute weight gain occurred in treatment A (control) with an average absolute weight value of 0.17 gr.

Although using the analysis of variance test, the treatment with the addition of red spinach leaf extract in the feed did not have a significant effect on the weight gain of sword platy fish, but biologically it can be seen that there were differences in weight gain at each treatment level. This is thought to be because the carotenoid content of red spinach leaf extract in treatments B, C and D stimulates the fish's appetite so that fish weight gain is better compared to the treatment without the addition of red spinach leaf extract in the feed. According to [6] carotenoids from fish meal contained in commercial feed as test feed can stimulate fish appetite but do not affect color changes.

The absolute length growth of sword platy fish in this study showed results that were not significantly different between treatments. The results of the analysis of variance (ANOVA) showed a significance value greater than 0.05, indicating that the addition of carotenoids to the feed did not have a significant effect on the length growth of the sword platy fish. Based on the data in Table 2, it can be seen that the highest absolute length increase occurred in treatments C (6%) and D (9%) of 0.41 cm, followed by treatment B (3%) 0.33 cm and the lowest absolute length increase in treatment A (control) was 0.3 cm.

The increase in length and weight of sword platy fish (*Xiphophorus helleri*) was not influenced by the carotenoids contained in red spinach leaf extract. These results are in accordance with the statement of [34][35][17][31] who stated that adding carotenoids to feed does not affect growth, because fish use these dyes more to improve the quality of their body color. The lowest increase in weight and length was in treatment A (control), namely 0.17 grams in weight and 0.3 cm in length. This is thought to be because treatment A does not meet the nutritional requirements of sword platy fish. This is supported by [36] who stated that increased fish growth has been obtained from food with added pigments, because carotenoids are known to play a positive role in fish metabolism and can increase nutrient utilization which has a good impact on increasing growth. Based on the results of research conducted by [31] reported that the addition of red spinach flour to the feed resulted in higher absolute weight growth of koi fish compared to treatment without the addition of red spinach flour.

The addition of red spinach leaf extract in feed at various concentrations had no significant effect ($P > 0.05$) on the survival rate of sword platy fish. The highest average survival rate for sword platy fish was in treatments B (3%) and C (6%) with an average survival rate of 88%, followed by treatment A (control) with an average survival rate of 87% and treatment D (9%) had the lowest average survival value, namely 83%. [31], reported that the survival rate of koi fish (*Cyprinus carpio L.*) fed with the addition of red spinach flour at different concentrations (0%, 3%, 6% and 9%) did not have a significant effect on fish survival, where in this study the average survival rate for koi fish was between 90-100%.

The death of the fish at the start of the research was thought to be because during the research the fish were not used to measuring color, weight and length, which caused stress. This can be seen from the fish moving passively at the bottom of the container and not responding to the food given. Fish survival is more determined by optimal water quality as a cultivation medium for platy swordfish. The water quality in this study was considered quite good because it was still within optimal limits for cultivating platy swordfish. The water quality parameters measured in this study were temperature, pH and dissolved oxygen (DO). The temperature during the research ranged between 27.3-28.4°C, this shows that the temperature in the maintenance container during the research was still in optimal conditions. According to [2], the optimum temperature for keeping platy swordfish ranges from 22-29°C.

According to [37], temperature plays an important role in the continued growth of fish, fish metabolic activity is directly proportional to water temperature. The warmer the water, the more active the fish's metabolism will be, and vice versa. Temperature conditions have a significant influence on fish life. Low temperatures make fish lose their appetite and become more susceptible to disease. On the other hand, temperatures that are too high can cause respiratory problems for fish and even permanently damage their gills.

The results of pH measurements during the research showed that the pH in the rearing container was still within the tolerance limits of the fish, where the pH ranged between 7.89-8.46. This is supported by [29] statement which states that the optimal pH for keeping sword platy fish is in the range between 7-8.5. According to [38], low pH will reduce dissolved oxygen levels which will result in reduced oxygen consumption, increased respiratory activity, reduced appetite, stunted fish growth, increased susceptibility to bacterial and parasitic infections, and will even cause fish death.

Dissolved oxygen (DO) is the main factor that must be present so that fish can continue to survive. Dissolved oxygen is very important for fish to be able to breathe and maintain their metabolic cycles. The dissolved oxygen value during cultivation of sword platy fish ranges from 3.30-3.45 mg/L. This value is still within the optimum range for maintaining sword platy fish. According to [2], the optimal DO for rearing sword platy fish is >2 mg/L. This is in line with the statement made by [39] that dissolved oxygen levels <2 ml/L will result in a decrease in appetite and poor development in fish.

4 CONCLUSION

Based on the results and discussion, the following conclusions can be drawn, The addition of red spinach leaf extract (*A. tricolor L.*) to the feed had a significant effect on the color brightness of sword platy fish (*X. helleri*) where the best treatment was obtained by treatment D (9%) with the final color brightness score namely 6.66. The addition of red spinach leaf extract (*A. tricolor L.*) to the feed had no significant effect on the growth in absolute weight, absolute length and survival of sword platy fish (*X. helleri*).

References

1. Z. Phonna., S.P. Febri, Hanisah. Jurnal Mahseer, **4** (1), 17-26 (2022)
2. A. D.Aprilia., W. Lili, S. Astuty, R. Grandiosa. World Sci. News, **160,** 217-231 (2021)
3. A. Yustiati., S.I. Zurwana, A. Rizal, Y. Andriani. Asian J. Fish. Aquat. Res.**9** (3), 21-30 (2020)
4. R. Idris., I. Riniatsih, D. Pringgenies. J. Mar. Res., **3** (3), 244-253 (2014)
5. G. Koncara., N.B.P. Utomo, M. Setiawati, M. Yamin. Jurnal Iktiologi Indonesia, **19** (1), 53-64. (2018)
6. E. Widinata., K. Muslih, A. Kurniawan. Jurnal Sumberdaya Perairan, **10** (2), 62-71 (2016)
7. S. Wulandari. Program Studi Manajemen Sumberdaya Perairan. Fakultas Pertanian. Universitas Sumatera Utara (2021)
8. R. Irawan. Jurusan Perikanan dan Kelautan. Fakultas Pertanian. Universitas Lampung (2017)
9. Alpina. Program Studi Pendidikan Kelautan dan Perikanan. Universitas Pendidikan Indonesia (2022)
10. M.M. Wantah., S.N.J. Longdong, R.L. Kreckhoff, D.M.H. Mantiri, R.A. Tumbol, H. Manoppo. Budidaya Perairan, **6** (2), 32-38 (2018)
11. I. Alfandi., S. Mellisa, I.I. Arisa. Jurnal Ilmiah Mahasiswa Kelautan dan Perikanan Unsyiah, **4** (4), 210-217 (2019)
12. I. Mokoginta., N.P. Utomo, A.D. Akbar, M. Setiawati. Jurnal Akuakultur Indonesia **2** (2), 79-83 (2003)
13. J.D.A. Magno., Y. Jasmanindar, P. Santoso. JVIP, **2** (2), 56-59 (2022)
14. N.P.U. Merlin., I. Samidjan, Pinandoyo. JAMT, **6** (3), 214-223 (2017)
15. F. Amin., S.A.E. Rahimi, S. Mellisa. Jurnal Ilmiah Mahasiswa Kelautan dan Perikanan Unsyiah, **4** (3), 152-160 (2019)
16. S.M. Simbolon., C. Mulyani, S.P. Febri. Jurnal Kelautan dan Perikanan Indonesia, **1** (1), 1-9 (2021)
17. L.A. Nur., M.A. Liliyanti, L.A.T.T.W.S. Kalih. Indonesian Journal of Aquaculture and Fisheries, **2** (1), 40-43 (2020)
18. N. Zenneveld., E.A. Huisman, J.H. Boon. PT Gramedia Pustaka Umum, Jakarta (1991)
19. M.I. Effendie, M.I. Dwi Sri: Bogor (1979)
20. Z. A. Muchlisin., F. Afrido, T. Murda, N. Fadli, A.A. Muhammadar, Z. Jalil, C.Yulvizar. Biosaintifika, **8** (2), 172-177 (2016)
21. K. A. Hanafiah. PT Raja Grafindo Persada: Jakarta (1997)
22. H. Effendie. Kanisius: Yogyakarta (2003)
23. Hasby, Nurhafidhah, G. Pamungkas, S.A. Akbar, Rasayan J. Chem **13(1)** 38-43 (2020)
24. S.A. Akbar, M. Hasan, S. Afriani, C. Nuzlia, Biodiversitas Journal of Biological Diversity **24(10)**, 5283 (2023)
25. S.A. Akbar, A. Rochliadi, V. Suendo, N. Saidi, Lelifajri, A. Mardhiah, Rasayan J. Chem **11(4)** 1525-1531(2018)

26. S. Raut., Thaneshwari. Ecology, Environment and Conservation. 315-320 (2022)
27. D. B.R. Amaya. ILSI Press. Washington, D.C (2001)

28. Sukarman, Chumaidi. Prosiding Forum Inovasi Teknologi Akuakultur: 803-807 (2010)
29. V. Maolana., S.O. Madyowati, N. Hayati. Jurnal TECHNO-FISH, **1** (2), 78-85 (2017)
30. M. Mustaqim., L. Rosdiana, E. Ayuzar, Muliani. Fanik: Jurnal Faperta Uniki, **2** (1), 23-29 (2021)
31. M. A. D. Putra., S.Y. Lumbessy, D.N. Setyowati. Samakia: Jurnal Ilmu Perikanan, **13** (2), 134-146 (2022)
32. M. Hadijah,. Junaidi, D. P. Lestari. Jurnal Perikanan, **10** (1), 41-49 (2020)
33. D. M. Malini., T.D.K. Pribadi, R. Agustin. Jurnal Pro-Life. **5** (2), 579-588 (2018)
34. A. Diansyah., M. Amin, Yulisman. Jurnal Akuakultur Rawa Indonesia, **7** (2), 149-160 (2019)
35. S.I. Apriliani., A. Djunaedi, C.A. Suryono. J. Mar. Res, **10** (4), 551-559 (2021)
36. S. Kaur., V.I. Kaur, S.A. Holeyappa, S.O. Khairnar. IJAN, **33** (4), 448-455. (2016)
37. G. P. Wangni., S. Prayogo, Sumantriyadi. Jurnal Ilmu-ilmu Perikanan dan Budidaya Perairan, **14** (2), 21-28 (2019)
38. R. B. K. Haris, I.A. Yusanti. Jurnal Lahan Suboptimal, **8** (1), 20-30 (2019)
39. S. P. Sari., J.M. Amelia, G.I Setiabudi. Jurnal Perikanan, **12** (3), 346-354 (2022)

Morphometric analysis of spiny lobster *(Panulirus penicillatus)* and bamboo lobster *(Panulirus versicolor)* in Sabang Waters

Muhammad Iqbal[1], *Muhammadar* Abdullah Abbas [1,2*], *Firdus* Firdus[1,3], *Widya* Sari[3]

[1] Master Program in Integrated Coastal Resource Management, Graduate School, Universitas Syiah Kuala, Banda Aceh 23111, Indonesia.
[2] Department of Aquaculture, Faculty of Marine and Fisheries Universitas Syiah Kuala Banda Aceh 23111, Indonesia
[3] Department of Biology, Faculty of Mathematics and Natural Sciences, Universitas Syiah Kuala, Banda Aceh 23111, Indonesia

Abstract. This research aims to analyze morphometrics of *(P. penicillatus*) and (*P. versicolor*) in waters of Sabang. A total of 50 samples from each species of *P. penicillatus* and *P. versicolor* were used for morphometric analysis using Traditional Morphometric and Truss Morphometric approaches. Morphological analysis indicated that both lobster species are categorized into two distinct groups. Traditional morphometric character testing using ANOVA showed that species differences significantly influence the variation in morphological characters ($P<0.05$) in both lobsters. Further Duncan post hoc tests revealed that four characters (HL), (CHW), and (CHL)) differed significantly between the species. *P. penicillatus* shared three characters similar to *P. versicolor,* namely (HL), (CHW), and (CHL). This similarity is based on the shape of HL, CHW, and CHL in both lobsters. In the Truss Morphometric character testing with ANOVA, it was found that species differences affect the characters significantly ($P<0.05$). The Duncan post hoc test revealed that five characters ([A1, A3], [B1, B3], [C5, C6], D4, [D5, D6]) differed significantly between species. DFA resulted in 1 function, with an eigenvalue of 27.595, indicating that it explains 100% of the total variance. The morphological characters that significantly contribute to the differentiation between species are A1, B1, C5, A5, and D5.

1 Introduction

The diversity of lobsters on the island of Sumatra, especially in Aceh, is quite comprehensive. A total of 6 lobster species were found [1]. The most dominant species found in the waters of Aceh include *Panulirus ornatus, Panulirus versicolor, Panulirus penicillatus, Panulirus homarus, Panulirus longipes, and Panulirus polyphagus.* Sabang's waters are located in the coastal area and are surrounded by various small islands. This extensive area includes many beaches, which are the main locations for tourism development and the potential for abundant marine resources, including lobster *(Panulirus spp.)* [2]. There are five regions in Indonesia known for lobster production, including Cilacap, Yogyakarta, Gresik, Aceh, and Lombok [3], and the lobster business is a sector that can help improve and enhance people's livelihoods [4].

* Corresponding author: muhammadar@usk.ac.id

The Stone Lobster *(Panulirus penicillatus)* is characterized by a body color ranging from green to brown. It has long antennae with branched antennules. This lobster species lacks claws, but in females, there are imitation claws at the fifth walking leg's end, which function to tear the sperm pouch during fertilization. The carapace is somewhat round and spiky. The tips of the carapace and the area between the eyes are orange and blue. The frontal horns have a black pattern with white lines encircling them. The antennular plate has two pairs of separated spines, with several small spines among the main spines. The antennular flagella are brown with circular white lines. The differentiation between males and females is based on the gonopore's location on the walking legs and swimmerets [5]

The Bamboo Lobster is characterized by a distinct bright green color with transverse white lines bordered by black. There are no transverse grooves and hair on the surface of the abdominal segments, except at the rear edge and indentation on the sides. The antennae are red, and each leg has horizontal black to yellowish lines [6].

The differences between male and female lobsters can be clearly observed from their body structure. In male lobsters, the gonopore is located at the base of the fifth walking leg. All walking legs of male lobsters have the same shape, including the fifth walking leg which has sharp claws at its tips. The first swimmeret in male lobsters consists of a single plate. In contrast, in female lobsters, the gonopore is located at the base of the third walking leg. The fifth walking leg in female lobsters has an imperfect claw and branches into three, serving the function of carrying fertilized eggs into the brood pouch during the hatching process. The first swimmeret in female lobsters consists of two plates, with the outer plate larger than the inner one. The inner plate has two branches, being rigid and thin [7].

2 Material and Method

2.1 Time and Site

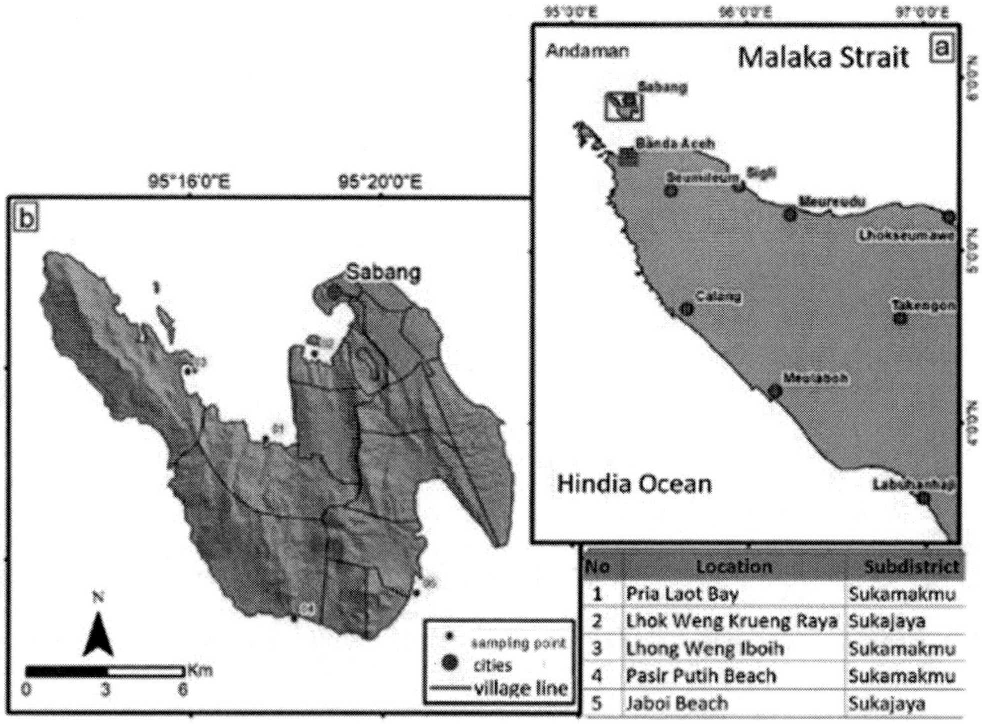

No	Location	Subdistrict
1	Pria Laot Bay	Sukamakmu
2	Lhok Weng Krueng Raya	Sukajaya
3	Lhong Weng Iboih	Sukamakmu
4	Pasir Putih Beach	Sukamakmu
5	Jaboi Beach	Sukajaya

Fig. 1 Map of lobster fishing locations in Sabang waters

This research was conducted in the Coastal area of Sabang City using the Purposive Sampling method, with samples obtained from the catches of local fishermen. The sampling locations were in the areas of Pria Laot Bay, Lhok Krueng Raya, Lhok Weng (Iboih), and Pasir Putih Beach in the waters of Sabang. The research lasted for 6 months from February to July 2023. A total of 50 samples for each species were collected for morphometric measurements. Lobster harvesting in the waters of Sabang employs a very simple method. The most common technique involves diving using basic scuba diving equipment such as an oxygen tank, diving mask, and webbed fins. Activities are conducted at night and utilize illumination to attract lobsters out of coral crevices. The capture process involves using hands to catch the lobsters. This technique requires caution to avoid injuries from the sharp claws of the lobsters. Once obtained, the lobsters are placed in a net before being transported to the shore.

The lobster fishing in several locations is conducted while simultaneously measuring the morphometrics of permitted lobsters in accordance with the government's efforts to address the sustainability of lobster harvesting through the Minister of Marine Affairs and Fisheries Regulation Number: 12/PERMEN-KP/2020 Regarding Lobster Fishing *(Panulirus spp.)*. This regulation governs the capture of lobsters, allowing those with a carapace length greater than 8 cm to be caught. It also mandates the release of lobsters in a gravid condition and requires the documentation of each captured lobster.

The lobster samples were previously placed in Polystyrene boxes, and they were then transported to the Laboratory of Marine Biology, Faculty of Marine and Fisheries, Syiah Kuala University for morphometric measurements and documentation. During the documentation process, it was crucial for the samples to remain intact for a detailed morphometric examination of *P. penicillatus* and *P. versicolor*.

Measurements are taken using calipers with a precision of 0.01 cm. Morphometric measurements follow the method proposed by Brezky and Doyle in 1988. The observed parameters include the head, body, and tail sections of the lobster [8]

Morphometric data were analyzed using univariate analysis (one-way ANOVA) followed by post hoc test using Duncan's and multivariate analysis (Discriminant Function Analysis, DFA) with SPSS. The following are the general steps for morphometric analysis using univariate analysis (one-way ANOVA) and multivariate analysis (Discriminant Function Analysis) in SPSS.

2.2 Collection of Lobster Samples

Additionally, there is a significant role in the policy regarding adult lobster populations to obtain an overview of gender ratios, growth, condition indices, and mortality in specific spatial and temporal contexts. Research on high-value economic lobsters is also essential [5]. Previously, fishermen placed trap nets daily in proximity to suitable crevices along the rocky seabed, secured with weights and floats. In the morning, fishermen visited the fishing locations and brought the nets along with the captured lobsters to the shore. Night trapping was also employed as lobsters are more active during the night [6].

2.3 Lobster Morphometric Measurements

Morphometric measurements (truss morphometric) divided the lobster's body into four parts, denoted as A, B, C, and D (truss cells). From each part, six longitudinal and diagonal lines (truss lengths) were measured using calipers with a precision of 0.01 cm. The morphometric measurements followed the method [7]. The observed parameters included parts of the head, body, and tail.

Truss morphometrics is a technique used to measure the distances at specific points on the body, based on reference points. These Truss morphometric points are connected by horizontal, vertical, and diagonal Truss morphometric distances, allowing for a detailed and specific analysis of the fish's body shape. Therefore, Truss morphometrics is preferred over conventional morphometrics because it provides a more comprehensive picture of body shape due to the limited number of distances in conventional morphometrics [8].

Truss Morphometric measurement analysis, statistical methods used are generally similar to those used in conventional morphometric analysis, including descriptive statistics, univariate analysis, multivariate analysis, or discriminant analysis [9]. Lobsters are suitable subjects for repeated body measurements due to their hard exoskeleton and body segments. Digital techniques have also been used to measure lobsters. A total of 24 body parts were measured in lobster samples [10].

Several of the observed samples focused on the morphological characteristics of the cephalothorax and abdomen carapace. Morphological features examined included antenna type, antennula type, the number of spines on the antennula plates, color patterns on the carapace, abdomen, and pereiopod, eye location, and antenna type. Gender determination through observation of genital opening placement on the base of lobster pereiopod legs, the presence or absence of branching development on the 5th pereiopod leg (second walking leg), and the presence or absence of endopleopod development was conducted. [14]

Morphometric observations on lobsters were also performed by measuring total length, carapace (cephalothorax) length, abdomen segment length, and tail length. Total length measurement started from the anterior end of the carapace near the antennula plate spine to the telson's tip along the midline. Carapace length measurement began from the anterior end of the carapace near the antennula plate spine to the carapace's posterior edge along the midline. [15].

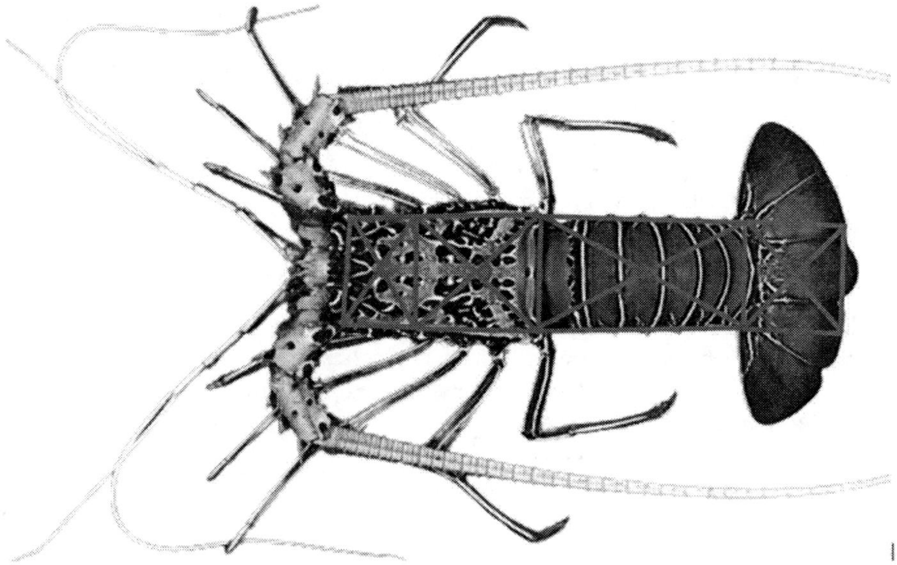

Fig. 2. Truss Morphometric Method for *Panulirus spp*

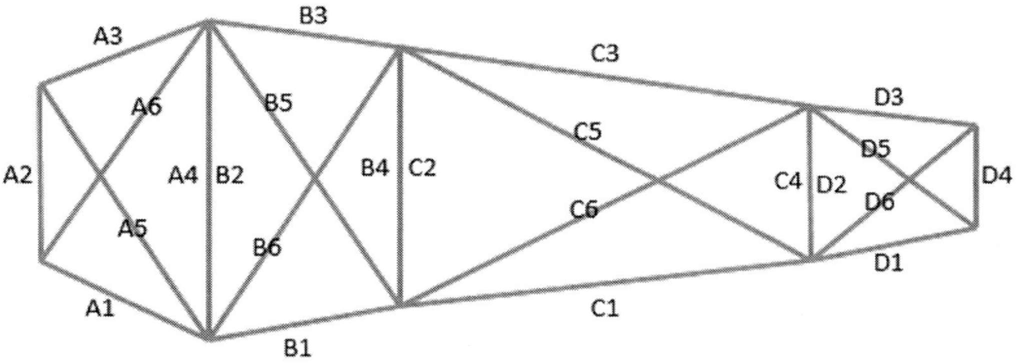

Fig. 3. Lobster parts in morphometric truss measurements

Table 1. Parameters for Truss Morphometric Observation of Lobsters

Body part	Symbol	Information
Front Head	A1	Distance from the widest left carapace point to the base of the left eye stalk.
	A2	Distance between the right and left eye stalks
	A3	Distance from the widest right carapace point to the base of the left eye stalk
	A4	Distance between the widest carapace points (front of the head)
	A5	Diagonal from the left rear carapace point to the right eye
	A6	Diagonal from the right rear carapace point to the left eye
Back of Head	B1	Distance from the rear carapace point to the widest left carapace point
	B2	Distance between the widest carapace points (front of the head)
	B3	Distance from the rear carapace point to the widest right carapace point
	B4	Distance between the two rear carapace points
	B5	Diagonal from the left rear carapace point to the widest right carapace point
	B6	Diagonal from the right rear carapace point to the widest left carapace point
Abdomen	C1	Distance from the left rear abdomen point to the front left abdomen point
	C2	Distance between the two rear abdomen points
	C3	Distance from the right rear abdomen point to the front right abdomen point
	C4	Distance from the rear right abdomen point to the left side
	C5	Diagonal from the left rear abdomen point to the right front abdomen point
	C6	Diagonal from the right rear abdomen point to the left front abdomen point

Tail (Telson)	D1	Distance from the left rear telson point to the front left telson point
	D2	Distance between the two rear abdomen points
	D3	Distance from the right rear telson point to the front right telson point
	D4	Distance from the rear right telson point to the left side
	D5	Diagonal from the left rear telson point to the right front telson point
	D6	Diagonal from the right rear telson point to the left front telson point

(Abinawanto *et, al,* 2018)

The morphometric character measurements were conducted based on the following parameters: Ocular carapace length (OCL), Propodal length (PL), Dactyl length (DL), Chela length (CL), Chela width (CW), Cephalon width (CW), Thorax width (TW), Carapace length (CL), Carapace width (CW), Abdominal length (AL), Telson length (LT), Telson width (TW), Maximum abdomen width (MAW) dan Total weight (W) [11].

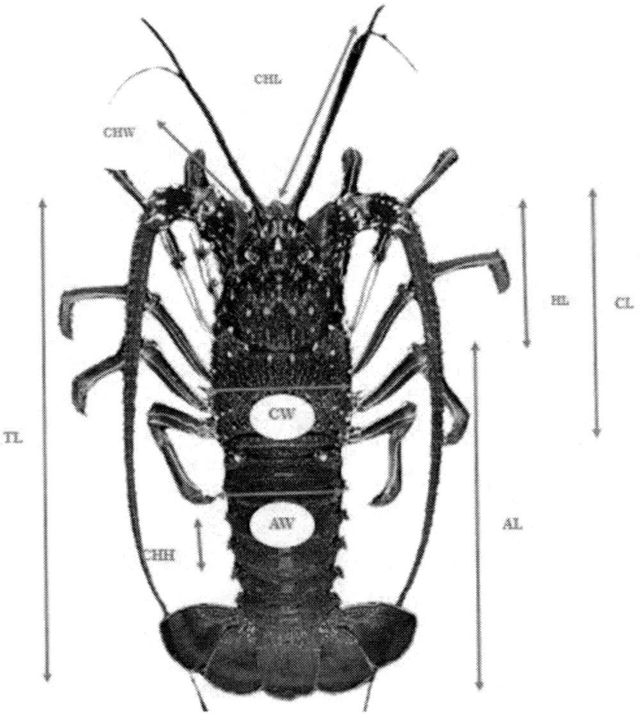

Fig. 4. Traditional Morphometric Method for *Panulirus spp* Lobsters

Table 2. Parameters for Traditional Morphometric Observation of Lobsters

Symbol	Information	Morphometric Character
TL	*Total Length*	Measurement from the tip of the rostrum to the tip
CL	*Carapace Length*	Measurement from the tip of the rostrum to the tip of the carapace
CW	*Carapace Width*	Measurement of the maximum width of the carapace on the dorsal side of the body
HL	*Cheliped Length*	Measurement from the base of the cheliped (claw) to the tip of the claw
CHH	*Abdomen Length*	Measurement of the length from the tip of the carapace to the tip of the telson
CHL	*Antenna Length*	Measurement of the length from the base of the antenna to the tip of the antenna
CHW	*Pereopod Length*	Measurement of the length from the base of the pereopod (abdominal leg) to its tip
AL	*Pleopod Length*	Measurement of the length from the base of the pleopod (abdominal fin) to its tip
AW	*Uropod Length)*	Measurement of the length from the base of the uropod (tail fin) to its tip

3 Result and Discussion

The ANOVA test indicates that the species' differences significantly influence the variation in morphological characteristics ($P < 0.05$) of both lobsters. Duncan's post hoc test revealed that four characteristics (Cheliped Length (HL), Pereopod Length (CHW), and Antenna Length (CHL)) differed significantly among species (Table 2).

In *P. penicillatus,* three characteristics are similar to *P. versicolor,* namely Cheliped Length (HL), Pereopod Length (CHW), and Antenna Length (CHL). This is based on the shape of HL, CHW, and CHL in both lobsters, which have similarities. The form of each segment of the lobster in morphometric truss measurements reveals a trapezium shape divided into four parts: the upper head, lower head, abdomen, and tail. In several segments, there are similarities, especially based on morphometric measurements forming triangles in each section.

In specific areas of each segment, there are similar sizes in lobster morphometrics. In the upper head section, parts A1 and A3 exhibit similarities; in the lower head section, there are similarities in B1 and B3; the abdomen section includes C1 and C3 similarities, and finally, the tail section shows similarities in D1 and D3.

The results of the Discriminant Function Analysis (DFA) on the truss morphometric characters divide the two lobster species in the waters of Sabang. The findings indicate an overlapping degree that signifies morphological similarity. Meanwhile, the morphological differences observed in lobsters are attributed to phenotypic changes influenced by the environment.

Moreover, in each segment, these similarities serve as markers for significant differences in lobster sizes, particularly between *P. penicillatus* and *P. versicolor.* While there are differences in length, the dimensions of each species within the same type have identical values, whereas those of different types exhibit significant differences. Another characteristic morphometric feature is the length of the antennae, with P. versicolor having longer antennae compared to the significantly shorter antennae of *P. penicillatus.* Additionally, *P. penicillatus* has a notably larger head with more defense mechanisms, especially in facing predators in the wild.

Table 3. Mean transformed values of morphometric characters in each lobster species. Values followed by different superscripts indicate significant differences ($p < 0.05$).

No.	Character (code)	Estimate Type Lobster	
		P. penicillatus	*P. versicolor*
1	Total Length *(TL)*	214,002 ± 67,8225[a]	195,692 ± 26,9264[a]
2	Carapace Length *(CL)*	69,954 ± 20,9053[a]	68,684 ± 12,9089[a]
3	Carapace Width *(CW)*	54,328 ± 14,5502[a]	55,914 ± 11,9376[a]
4	Cheliped Length *(HL)*	34,977 ± 10,4526[a]	27,474 ± 5,1618[b]
5	Pleopod Length *(AL)*	107,008 ± 33,9098[a]	117,414 ± 16,1554[b]
6	Uropod Length *(AW)*	44,328 ± 14,5502[a]	46,914 ± 11,9376[a]
7	Abdomen Length *(CHH)*	108,656 ± 29,1004[a]	111,828 ± 23,8752[a]
8	Pereopod Length *(CHW)*	20,496 ± 5,6155[a]	22,540 ± 3,2085[b]
9	Antenna Length *(CHL)*	99,854 ± 54,1513[a]	126,456 ± 26,3484[b]

The DFA analysis resulted in 1 function, which has an eigenvalue of 36.436, indicating that it accounts for 100% of the total variance. The morphological characteristics that contribute significantly among species are Cheliped Length (HL), Carapace Length (CL), Antenna Length (CHL), Pleopod Length (AL), and Total Length (TL) (Table 3).

Table 4. Eigenvalues, % variance, and structure matrix of Traditional Morphometric characters.

Function	**DF1**
Eigenvalue	36,436[a]
% of Variance	100,0
Canonical Correlation	0,987
Carapace Width (CW)[a]	0,082
Abdomen Length (CHH)[a]	0,082
Uropod Length (AW)[a]	0,082
Cheliped Length (HL)	-0,076
Pereopod Length (CHL)	0,052
Antenna Length (CHW)[a]	0,040
Pleopod Length (AL)	0,033
Total Length (TL)	-0,030
Carapace Length (CL)	-0,006

Table 5. of results of Eigenvalue Analysis on Morphometrics in Lobsters

Function	Eigenvalue	% of Variance	Cumulative %	Canonical Correlation
1	27.595[a]	100.0	100.0	.982

Meanwhile, there is a significant difference in carapace length and total length between the two lobster species, especially in the total length of the captured lobsters. It was found that *P. penicillatus* has a larger total length compared to P. versicolor. In terms of weight, *P. penicillatus* is also heavier. The main factors underlying this are, firstly, the higher catch numbers of *P. penicillatus,* partly due to their more extensive habitat, and the capturing process of *P. penicillatus* is much easier. However, *P. penicillatus* can be more aggressive during the capture process.

The population of *P. penicillatus* in the waters of Sabang is significantly higher than that of P. versicolor. The abundant population results in larger sizes observed in *P. penicillatus* compared to P. versicolor. Another contributing factor is the feeding zone of the two lobsters. *P. penicillatus* has a feeding zone much closer to the shore than P. versicolor. Additionally, the dominance in lobster species corresponds to their respective habitats.

3.1 Truss Morphometrics in *P. versicolor* and *P. penicillatus*

The ANOVA results show that species differences affect Truss Morphometric characteristics ($P < 0.05$). Duncan's post hoc test revealed that five characteristics ([A1, A3], [B1, B3], [C5, C6], D4, [D5, D6]) differ significantly among species (Table 3). DFA analysis resulted in 1 function, which has an eigenvalue of 27.595, indicating that it accounts for 100% of the total variance. The morphological characteristics that contribute significantly among species are A1, B1, C5, A5, D5 (Table 5).

Table 6. Mean transformed values of truss morphometric characters in each lobster species. Values followed by different superscripts indicate significant differences ($p < 0.05$).

No	Character (Kode)	Estimate Type Lobster	
		P. penicillatus	*P. versicolor*
1	A1	34.977 ± 10.4526^a	27.474 ± 5.1618^b
2	A2	43.462 ± 11.6402^a	44.731 ± 9.5501^a
3	A3	34.977 ± 10.4526^a	27.474 ± 5.1618^b
4	A4	54.328 ± 14.5502^a	55.914 ± 11.9376^a
5	A5	50.524 ± 13.5348^a	50.323 ± 10.7438^a
6	A6	50.524 ± 13.5348^a	50.323 ± 10.7438^a
7	B1	34.977 ± 10.4526^a	41.210 ± 7.7454^b
8	B2	54.328 ± 14.5502^a	55.914 ± 11.9376^a
9	B3	34.977 ± 10.4526^a	41.210 ± 7.7454^b
10	B4	44.328 ± 14.5502^a	46.914 ± 11.9376^a
11	B5	49.436 ± 13.2354^a	50.323 ± 10.7438^a
12	B6	49.438 ± 13.2407^a	50.323 ± 10.7438^a
13	C1	74.906 ± 23.7368^a	82.196 ± 11.3161^a
14	C2	44.328 ± 14.5502^a	46.914 ± 11.9376^a
15	C3	74.906 ± 23.7368^a	82.196 ± 11.3161^a
16	C4	37.681 ± 12.3685^a	42.223 ± 10.7438^a
17	C5	65.914 ± 20.8940^a	78.080 ± 10.7400^b
18	C6	65.914 ± 20.8940^a	78.080 ± 10.7400^b
19	D1	32.102 ± 10.1729^a	32.106 ± 10.1759^a
20	D2	37.681 ± 12.3685^a	42.223 ± 10.7438^a
21	D3	32.102 ± 10.1729^a	32.106 ± 10.1759^a
22	D4	54.648 ± 17.9398^a	63.332 ± 16.1220^b
23	D5	43.710 ± 14.3462^a	53.832 ± 13.6963^b
24	D6	43.710 ± 14.3462^a	53.832 ± 13.6963^b

Tabel 7. *Eigenvalues*, *% variance*, dan *structure matrix* Character Truss Morphometric

Function	**DF1**
Eigenvalue	27,595[a]
% of Variance	100,0
Canonical Correlation	0,982
A1	**-0,088**
A3[a]	-0,088
C5	**0,070**
C6[a]	0,070
D5	**0,069**
D6[a]	0,069
B1	**0,065**
B3[a]	0,065
C1[a]	0,048
C3[a]	0,048
D4[a]	0,046
C4[a]	0,033
D2[a]	0,033
D1[a]	-0,021
D3[a]	-0,021
B4[a]	0,011
C2[a]	0,011
A4[a]	0,011
B2[a]	0,011
A2[a]	0,011
B5[a]	0,007
B6[a]	0,007
A6[a]	-0,002
A5	**-0,002**

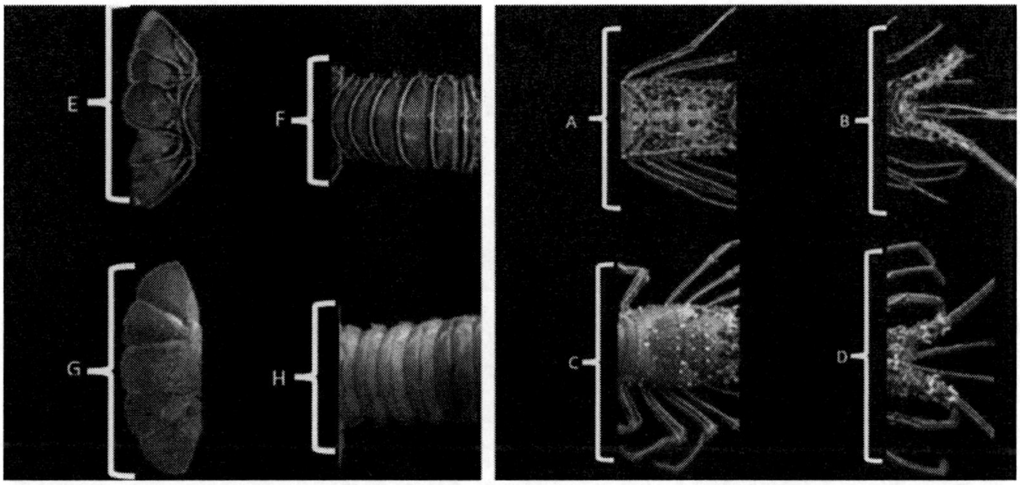

Fig. 5. illustrates the process of identifying the two most dominant lobster species in Sabang, namely *P. penicillatus* and *P. versicolor.* The comparison process focuses on the shapes of the upper head, lower head, abdomen, and tail.

4 Conclusion

In the waters of Sabang, it has been observed that there are morphological differences between the two lobster species, *P. penicillatus* and *P. versicolor.* These differences primarily pertain to the length of the antennae, the size of the claws, and the shape of the uropods. In the case of *P. penicillatus,* the body size is significantly larger when compared to P. versicolor. Physically, *P. penicillatus* holds an advantage in terms of habitat and food competition. As a result, *P. versicolor* tends to inhabit deeper waters.

References

1. Irfanur, R.I. Wahju, M. Riyanto M. Albacore. **1**(2): 211–223 (2017)
2. D. Wijaya, A Nurfiarini. Journal of Indonesian Fisheries Research. (2018)
3. A. Tjoneng, A. Journal of Business and Investment Law, **7**(1), 47 (2017)
4. Y. Akmal, I. Irfannur, M. Muliari, A.S. Batubara, M. Yunus, H. Plumeriastuti, & Y. Dhamayanti. 264, Article 106731 (2023)
5. Y. Wardiatno, A. Hakim, A. Mashar, N. Butet, L. Adrianto, A. Farajallah. Biodiversity Data Journal 4: e8069 (2016)
6. D. Wijaya, and N.N. Wiadnyana. Ministry of Maritime Affairs and Fisheries (2018)
7. A. Setyanto, N.A. Rachman, and E.S. Yulianto. Journal of Fisheries, **20**(2), 49. (2019)
8. Girsang, E., Kristanto, A.H., Hadi, W., and Mardlijah, S. (2018)
9. K. Alka. Amity Institute of Marine Science and Technology (2016)
10. E. Girsang, A.H. Kristanto, W. Hadi, and S. Mardlijah (2018)
11. V.J. Brezky, R.W. Doyle, Department of Fisheris, Bangkok, Thailand & International Center for Living Aquatic Resources Management, Manila. (1988)
12. I.L. Dryden and K.V. Mardia. Statistical shape analysis: with applications in R. John Wiley & Sons. (2016)
13. N. Rycroft, K. Radcliffe, and J. Atema. Canadian Journal of Fisheries and Aquatic Sciences. **70**(11): 1658-1665 (2013)

14. Jumeini, J.D Saadah. Paransa, N.W. Joshian. Schaduw, M.H. Desy. Mantiri, E. Wilmy . Pelle, D. Gaspar, Mamu. Platax Journal. Sam Ratulangi University. (2019)
15. Idami, Z. 2020. State Islamic University of North Sumatra (UINSU), Medan.
16. B. Widigdo, A.A. Hakim, A. Mashar, I.P. Sari, and Wardiatno. Tropical Biology Journal (2020)
17. Pratiwi, R. Biosphere Journal. Research Center for Oceanography, Indonesian Institute of Sciences (LIPI). (2018)

Status of mackerel scad (*Decapterus* sp.) resource based on Ecosystem Approach (EA) at PPS Kutaraja, Banda Aceh

Ratna Mutia Aprilla[1,2*], *Rosnita* Rosnita[1], *Edy* Miswar[1,2], *Imelda* Agustina[1], *Alvi* Rahmah[1,2]

[1]Department of Fisheries Resources Utilization, Faculty of Marine and Fisheries, Universitas Syiah Kuala, Banda Aceh, 23111, Indonesia
[2] Research Center for Marine and Fisheries, Universitas Syiah Kuala, Banda Aceh, 23111 Indonesia

Abstract. The Kutaraja Ocean Fishery Port (PPS) is the largest port serving as the center of fisheries in Aceh Province, with a potential and diverse fishery, including the mackerel scad (*Decapterus* sp.). The fluctuation of the mackerel scad population on an annual basis, which raises concerns about the sustainability of the stock. Therefore, a structured management effort is required, employing an Ecosystem Approach to Fisheries Management (EAFM) to address this problem. The study aimed to assess the management status of mackerel scad resources based on PPS Kutaraja. The research was carried out from February to April 2023. Data collection was carried out through direct observation and interviews with 73 respondents. Data analysis for each indicator was performed using a multi-criteria approach with composite index assessment and flag model visualization. The research results indicate a 20% annual decline in the CPUE trend, a relatively consistent fish size trend, a 4% proportion of juvenile fish caught, a catch composition dominated by target species at 77%. Based on these EAFM indicators, the management status of mackerel scad resources in PPS Kutaraja falls under the 'good' category.

1 Introductionnn

The location of Kutaraja Ocean Fishing Port (PPS) is in WPP 571 (Malacca Strait and Andaman Sea) and WPP 572 (Western Indian Ocean and Sunda Strait) [1]. It is the largest port serving as the center of fisheries in Aceh Province. The fish landed at PPS Kutaraja are potential marine fisheries products, especially for small and large pelagic fish such as tuna, mackerel tuna, skipjack tuna, and mackerel scad. The fishermen at PPS Kutaraja use purse seines and handlines to catch mackerel scad. Mackerel scad is the focus of this research because their catch production fluctuates, while other potential fish did not experience significant fluctuations. In addition, at PPS Kutaraja, research related to mackerel tuna fish

* Corresponding author: ratnamutia@usk.ac.id

resources has been carried out by Salmarika et al. [2], and research related to skipjack tuna fish resources has also been carried out by Sari et al. [3].

Annual variations occur in the production of mackerel scad catches. In 2019, the production of purse seine catches reached 4,952,976 kg/year; in 2020, it decreased to 4,105,500 kg/year; and in 2021, it increased again to 4,711,464 kg/year. Whereas in hand line fishing gear in 2019, the catch production reached 692,076 kg/year, in 2020 it decreased to 554,176 kg/year and increased again in 2021 to reach 689,724 kg/year [4]. The amount of mackerel scad production has fluctuated over the past 3 years. Therefore, it is necessary to assess mackerel scad management at PPS Kutaraja to know the sustainability of mackerel scad stocks.

FAO [5] developed a concept of fisheries management through a holistic approach known as fisheries management with an ecosystem approach (Ecosystem Approach to Fisheries Management/ EAFM). This approach aims to manage fisheries sustainably by ensuring a balance between ecosystem health (ecosystem well-being), actor welfare (human well-being), and harmonious governance [6]. There are six domains for implementing EAFM, one of which is the fish resources domain [7]. Therefore, with the increasing number of fishing fleets and mackerel scad fluctuating every year, it is necessary to research the assessment of mackerel scad fisheries management with EAFM in the fish resource domain in Kutaraja PPS. This study aims to assess the management status and formulate recommendations for the management of mackerel scad resources based on PPS Kutaraja.

2 Materials and Methods

2.1 Location and time research

This research was conducted from February to April 2023. The research location is located in the Kutaraja Fishing Port (PPS), Lampulo Village, Banda Aceh, Aceh.

2.2 Data collection methods

The data used in this research are primary and secondary. The primary data used consisted of the size and length of the fish, the proportion of juvenile mackerel scad, the species composition of the catch, as well as the results of interviews from questionnaires that were directly observed in the field. The interview is face-to-face between interviews with the fishermen as respondent, and the activity is carried out orally [8]. The data utilized for the fish length indicator came from direct measurements of landed mackerel scad fish using a tape measure (cm) and the accidental sampling method, specifically the landing technique for mackerel scad fish. According to Masyhuri and Zainuddin [9], the accidental sampling technique involves selecting a sample of the population based on chance encounters made by researchers in the field. The sampling technique was chosen because of the uncertainty of the number of fish populations and the ease of obtaining samples. The total fish samples measured were 1.211 fish. Meanwhile, the number of respondents to the questionnaire was determined using the Slovin formula [10] as follows:

$$n = \frac{N}{1 + Ne^2}$$

Where:

n : Sample size

N : Population size

e : The percentage of inaccuracy due to sampling errors that can still be tolerated or desired

According to data from PPS Kutaraja, the population of purse seine captains is 269 people [4]. Then a sample of 73 people was taken with a data accuracy rate of 90% and an error margin of 10% because the total population is less than 1000 populations. Then secondary data is needed to determine the CPUE trend indicators, namely mackerel scad production and purse seining efforts from 2017–2021 obtained directly from relevant agencies, such as data from the Regional Technical Implementation Unit (UPTD) and the Department of Marine and Fisheries of Aceh Province (DKP Aceh).

2.3 Data analysis

2.3.1 Identify EAFM Indicators

1. Raw Catch Per Unit Effort (CPUE)

Catch Per Unit Effort (CPUE) is defined as the fish catch rate per year which is obtained using time series data. CPUE trend indicator according to Gulland (1983) in Febriani et al. [11] can be calculated using the following formula:

$$\text{CPUE} = \frac{Catch}{Effort}$$

Where:
CPUE : Number of catches per unit of fishing effort (kg/trip)
Catch : Catch/production (kg)
Effort : Attempt to capture (trip)

2. Fish Size Trends

The size of the mackerel scad in this study was carried out by measuring the fork length (FL) because it was feared that the morphology of the fish's fins was not in good condition and intact which resulted in not being able to measure the total length (TL). The results of mackerel scad measurements are included in the class interval. Sturges' rule (1962) is referred to in Leone et al. [13] explains that determining the number of classes is calculated using the following equation:

$$K = 1 + 3.3 \text{ Log } n$$

Where:
K : Number of classes
N : Number of samples

Then determine the class interval using the following equation:

$$P = R/K$$

Where:
P : Class gap
R : Range (highest fish length – lowest fish length)

3. Proportion of Juveniles Fish Caught

According to the NWG EAFM Module [7], the proportion of juvenile fish is calculated in the following way:

$$Pcy = Cyi/Ctot \times 100$$

Where:
Pcy : Proportion of juveniles caught (%)
Cyi : Yuwana caught (ind)
Ctot: Total catch (ind)

4. Species Composition of the Catch
Analysis of species composition aims to see comparisons between target and non- target fish catches [14]. The species composition of the catch is calculated using the following equation:

$$Pi = ni/N \times 100\%$$

Where:
Pi : Proportion of catch of the i species (%)
ni : Total Catch of fish species I (kg)
N : Total number of catches (Kg)

5. Range Collapse of Fish Resources
The determination of range collapse can be seen based on the condition of the fishing area. The easiest indicator in determining range collapse is to see if there are any indications that it is becoming more difficult to find fishing grounds by interviewing fishermen.

6. ETP (Endangered, Threatened, and Protected) species
Observation of ETP (Endangered, Threatened, and Protected) species was carried out by directly observing the catches of fishermen in the field and collecting data from fishermen who were interviewed as respondents.

2.3.2 EAFM Indicators Assessment

The assessment of indicators was analyzed using a multi-criteria analysis (MCA) approach with the development of a composite index using a Likert scale score (based on ordinal 1, 2, and 3). The following are the stages of the assessment:

1. Determine the criteria for each fish resource domain indicator, its seen in Table 1

Table 1. Criteria and Weight of Fish Resources Domain Indicators in EAFM

Number	Indicato r	Descriptio n	Criteria	Weight
1.	Raw CPUE	the number of fish caught by an amount of effort	1 = CPUE decreased sharply (average decrease >25% per year) 2 = CPUE decreased slightly (average <25% per year) 3 = CPUE stable or increasing	40
2.	Fish size trend	Change in fish size	1 = The size of the fish is getting smaller 2 = Relatively fixed fish size 3 = Fish size increased	20
3	The proportio n of juvenile	Percentage of fish caught before reaching maturity	1 = A lot (> 60%) 2 = Many (30 -60%) 3 = Few (<30%)	15

https://doi.org/10.1051/bioconf/20248703014

Number	Indicator	Description	Criteria	Weight
4.	Species composition of the catch	Comparison of target and non-target fish species	1 = Lower proportion of target fish (<15% of total volume) 2 = The proportion of target fish is the same as non-target 3 = Larger proportion of target fish (>31% of total volume)	10
5.	Range collapse	The fishing location is getting farther away	1 = Fishing ground is getting very far depending on the species 2= Fishing ground further away, according to the target species	10
			3= Fishing grounds are relatively fixed in distance depending on the target species	
6.	ETP species	Species whose populations are decreasing and whose existence is threatened with extinction so it needs to be protected	1 = There are ETP individuals caught but not released 2 = Caught but released 3 = No ETP individuals caught	5

Source: NWG EAFM Module (2014)

2. Provide a score for each indicator (based on ordinal 1,2,3)
3. Determine the weight of each indicator
4. Calculate the index value for each indicator. According to NWG EAFM Module [7], the indicator index value can be calculated using the following formula:

$$Cat\text{-}i = Sat\text{-}i \times Wat\text{-}i$$

Where:
Cat-i : EAFM total value of one indicator in the domain
Sat-i : The score of the I attribute/indicator
Wat-i : Weight of the I attribute/indicator

5. Develop a composite index for each indicator. According to NWG EAFM Module [7], the composite value can be calculated using the following formula:

$$\text{NK-i} = \frac{Cat-i}{Cat-max} \times 100$$

Where:
NK-i : Composite value to i
Cat-i : The total value of all the i indicators
Cat-max: The maximum value of all the i indicators

Furthermore, the composite values obtained were visualized using the flag model which is presented in Table 2. This model describes the condition of mackerel scad resource management in the categories of bad, not good, medium, good, and very good.

Table 2. Composite Index Value Classification and Flag Model Visualization

Composite Score Value	Model Flag	Description
1-20		Bad
21-40		Not good
41-60		Medium
61-80		Good
81-90		Very good

Source: NWG EAFM Module (2014)

3 Result and Discussions

3.1 EAFM fish resource management status

a. Raw Catch Per Unit Effort (CPUE)

Before carrying out CPUE calculations, it is necessary to standardize fishing gear. This is because each fishing gear has a different ability to catch mackerel scad. Mackerel scad landed at Kutaraja PPS are caught using two types of fishing gear, namely purse seines and hand lines. Therefore, standardization of fishing gear needs to be done to determine the number of standard trips so that the CPUE value can be determined. CPUE values for each fishing gear can be seen in Table 3 below.

Table 3. CPUE of scads from purse seines and handlines

Year	Purse seines			Handlines		
	Total Production (kg)	*Effort* (trip)	CPUE (kg/trip)	Total Production (kg)	*Effort* (trip)	CPUE (kg/trip)
2017	3.474.548	3.038	1.143.69	355.460	1.124	316.24
2018	4.696.748	2.883	1.629.11	599.872	1.119	536.07
2019	4.952.976	2.321	2.133.98	692.076	1.131	611.91
2020	4.105.500	2.344	1.751.49	554.176	1.244	445.47
2021	4.711.464	2.312	2.037.83	689.724	1.471	468.88
Amount	21.941.236	12.898	8.696	2.891.308	6.089	2.378
Average	4.388.247	2.579	1.739	578.262	1.218	476

Source: PPS Kutaraja (2021)

Based on the calculation of the CPUE value of the two fishing gears, the purse seine gear has the highest CPUE value. In the standardization of fishing gear, the calculation of the Fishing Power Index (FPI) value is carried out, which begins with determining the standard fishing gear. Based on the mackerel scad data that has the greatest productivity value is the purse seine fishing gear so it becomes a standard fishing gear that has an FPI value equal to one, while the FPI value of other fishing gear is obtained from the CPUE value of other fishing gear divided by the CPUE value of the fishing gear used as the standard.

The next step is to calculate the standard trip with the formula: FPI of the i-year purse seine x effort of the i year of the purse seine, FPI of the i year of hand-line fishing x of the i year of hand-line fishing. After knowing the number of standard efforts, the CPUE value is recalculated using the catch formula divided by the value of the new fishing effort or standard trip. The results obtained can be seen in Table 4 below.

Table 4. Calculation Results for Total Catch, Standard Effort, and CPUEs

Year	Total Production (kg)	Standard Effort (trip)	CPUEs (kg/trip)
2017	3.474.548	3.038	1.143.69
2018	4.696.748	2.883	1.629.11
2019	4.952.976	2.321	2.133.98
2020	4.105.500	2.344	1.751.49
2021	4.711.464	2.312	2.037.83
Amount	21.941.236	12.898	8.696
Average	4.388.247	2.579	1.739

Source: PPS Kutaraja (2021)

Based on the research results, it can be seen that the CPUE trend of mackerel scad based on PPS Kutaraja has decreased by 20% per year. However, this is different from research conducted by Wahyudi [15] in the northern waters of Sumenep Regency, where the CPUE trend for mackerel scad is stable or increases slightly so the score given is 3. Based on data for the last 5 years from PPP Pasongsongan, namely 2017-2021. The lowest CPUE value is in 2018 and the highest is in 2021. The declining trend of CPUE is an indication of a decrease in stock availability from fish resources [2]. According to Nugraha et al. [16] that the decreasing trend of CPUE indicates that the utilization of fish resources is already high.

b. Fish Size Trends

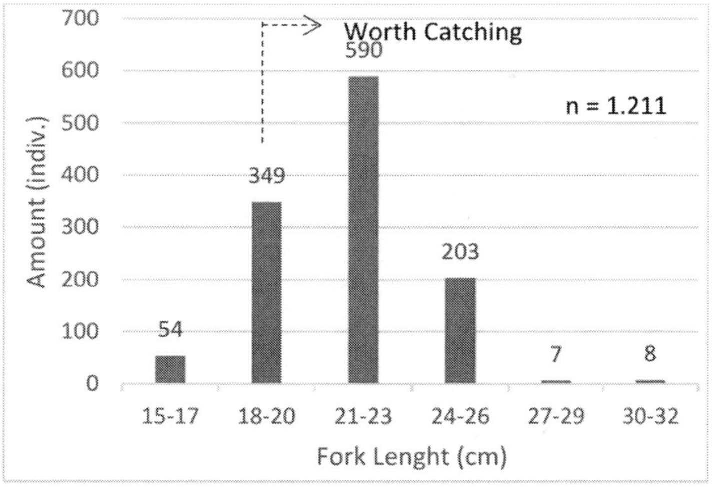

Fig. 1. Frequency distribution of maackerel scad

Based on measurements, the mackerel scad in this study had a size distribution of 15-35 cm FL with a dominant size of 21-25 cm FL. This size is almost the same as the size found in Hariati's research [17] in the waters of Banda Aceh, the fork length of the mackerel scad measures between 16-32 cm with a mode of 28,5 cm. While in the waters west of North Sumatra the size is between 16-26 cm with a mode of 20,5 and 23,5 cm. The length distribution of mackerel scad caught by purse seines shows various lengths. Iksan and Irham [18] stated that mackerel scad caught from the waters of the Maluku Sea ranged in size from 21,1-31,1 cm with a dominant size of 24,1-25 cm. Fadila et al. [19] stated that the mackerel scad that landed at Kendari PPS had a length distribution of 18,2-29,8 cm with a dominant size of 21-22,4 cm.

The difference in the length distribution of mackerel scad is influenced by several factors, namely related to the operation of purse seine fishing where variations in the length of the day of operation will affect the wider fishing area. This can result in the size of the fish caught being more diverse. In addition, it is also influenced by environmental conditions, abundance and availability of food, temperature, and light in each different water [20]. This is in line with Sasmito et al. [21] that the aquatic environment is very limited by the abundance of food. The abundance of food is an important factor in the waters that will affect the growth of fish.

c. Proportion of Juvenile Fish Caught

Based on the results of sample measurements in the study, the size of the mackerel scad landed at PPS Kutaraja was 4% still classified as juvenile fish because the size at first maturity (Lm) for mackerel scad is found in the [22] database, namely at size 17,6 cm. This can be caused by fishing operations carried out around FADs which result in small mackerel scad around FADs being caught too. According to Sudirman and Mallawa [23], FADs are in principle a gathering place for plankton and other small fish, thus inviting large fish to gather in search of food. Wiadnya et al. [24] also stated that FADs can attract juvenile-sized and adult-sized fish in a variety of abundance and diversity. The proportion of juvenile fish caught was also found at several different locations. Research by Sitanggang [25] conducted by PPN Sibolga, North Sumatra with the result that the proportion of juvenile fish caught was 44%, and research by Widodo et al. [26] at PPS Cilacap, Central Java, the proportion of juvenile fish caught was 31,5%.

d. Species composition of the catch

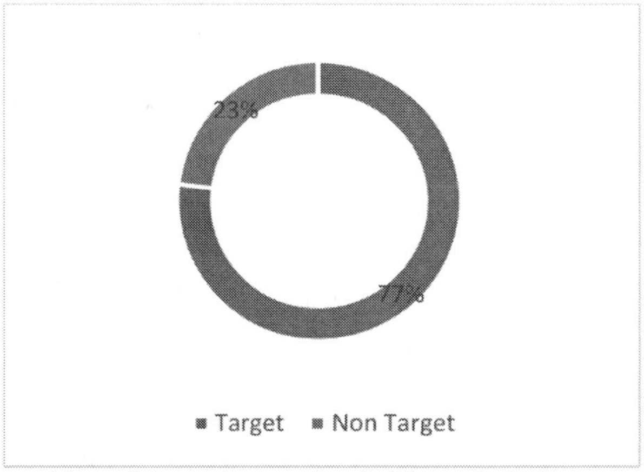

Fig. 2. Species composition of ring seine catches in Kutaraja PPS

The species composition of the catch in this study revealed that the target fish caught by purse seiners dominates compared to the non-target fish. Target fish has a percentage of 77% and non-target fish with a percentage of 23%. Target fish catches include mackerel scad (*Decapterus* sp.), tuna (*Auxis* sp.), skipjack tuna (*Katsuwonus pelamis*), and baby tuna (*Thunnus* sp.). Meanwhile, non-target fish catches include yellowfin tuna (*Thunnus albacares)*, coral leubim (*Abalistes stellaris*), siro (*Amblygaster sirm*), setuhuk (*Makaira* sp.), selar (*Selaroides leptolepis*), sunglir (*Elagastis bipinnulatu)*, kwee/rambe (*Caranx ignobilis*), squid (*Loligo* sp.), swordfish (*Xiphias gladius*), and the rat shark (*Alopias pelagicus*). The results of interviews with respondents obtained that 100% of fishermen stated that the species composition of catches at PPS Kutaraja was dominated by target fish. This condition is similar to a study conducted by Putra et al. [27] in Seraya Timur Village, Bali where the species composition of gill net and troll line fishing gear was dominated by target species with percentages of 98% and 90,7%, respectively.

e. Range Collapse of Fish Resources

The fishing areas for fishermen in the Kutaraja PPS include WPP 571 (the Andaman Sea and Malacca Strait) and WPP 572 (the Western Indian Ocean and Sunda Strait). The results of interviews with respondents obtained as many as 99% of fishermen stated that the fishing area was getting farther away. Currently, fishermen have to do fishing activities as far as 50-300 nautical miles from the port. However, this condition is different from research conducted by Arianto et al. [28] in the waters of Ternate Island where the results of interviews obtained 100% of respondents stated that fishing conditions in the last 5 years were not difficult, and the fishing ground is still the waters south of the island of Ternate.

According to Adrianto et al. [29], the cause of the increasing difficulty in finding fishing areas can be caused by the spatial reduction of the biomass of fish stocks as a catch target due to the impact of increasing fishing pressure. The difficulty of finding fishing areas for fishermen is also caused by the ability of fishermen to carry out fishing activities, and the limited information that fishermen have in determining fishing areas. Himelda et al. [30] changes in the catchment area are influenced by the main factor of the nature of the flying fish whose life is straddling. Roaming is the movement of fish due to the need to carry out the reproductive process, the interest in looking for food, and roaming occurs due to changes in the aquatic environment.

f. ETP (Endangered, Threatened, and Protected) species

Based on observations and interviews with fishermen, there are several types of ETP species caught by fishermen, including dolphins (*Delphinus* sp.), turtles (*Chelonia mydas*), and rat sharks (*Alopias pelagicus*). The results of interviews with respondents showed that 47% of the ETP species caught were released and 53% of the ETP species caught were sold/consumed themselves. Based on PERMEN KP RI No. 26/2013 concerning Capture Fisheries Business in WPP-RI stated that several protected species in the form of sharks, sea turtles, and marine mammals including whales or non-fish which are caught accidentally must be carried out as conservation efforts in the form of releasing the caught species if they are still caught alive and handling if the species is caught dead. However, the rat sharks that were caught during fishing activities were not released by fishermen at PPS Kutaraja.

Based on the results of interviews, PPS Kutaraja fishermen also stated that they had caught turtles and dolphins but were released. This is similar to research Putra et al. [27] which stated that there were accidentally caught endangered (EN) species, namely the green turtle (*Chelonia mydas*) and dolphins (*Delphinus* sp.). The cause of the capture of this ETP species was that the species was accidentally trapped when fishermen used gill nets. If this condition occurs, fishermen will try to free the ETP species in the fishing gear used.

3.2 EAFM Assessment

This assessment illustrates that the status of the mackerel scad fishery resource in Kutaraja PPS is in good condition, where the visualization of the flag model is light green. This condition still has the potential to decrease if efforts to improve and implement policies are not implemented in management. Care needs to be taken in terms of its utilization so that these conditions can be improved and its sustainability maintained, bearing in mind that the results of the assessment of each indicator in the fish resources domain still have indicators with poor and moderate status. An assessment of the status of the fish resource domain for mackerel scad fisheries with an ecosystem approach has also been carried out by Puansalaing et al. [31] in the waters of the Sulawesi Sea, North Sulawesi with a composite value of 56,67. This value is in the range of 41-60 which is classified as a moderate criterion marked by a visualization of a yellow flag model. The results of the assessment of each indicator can be seen in Table 5 below.

Table 5. Analysis of EAFM Indicator Assessment with the Flag Model

Number	Indicator	Score	Weight	Value
1	Raw CPUE	2	40	80
2	Fish size trend	2	20	40
3	The proportion of juvenile	3	15	45
4	Species composition of the catch	3	10	30
5	*Range collapse*	2	10	20
6	ETP species	1	5	5
Composite value				73.33

Based on the current condition of mackerel scad resources in Kutaraja PPS, the recommended management is to control fishing efforts. Controlling fishing efforts can be done by setting fishing gear, location, and time as well as setting limits on fishing quotas [32]. Umamah et al. [33] also suggest that regulation of fishing efforts can be carried out by regulating ships, fishing gear, or the number of fishermen participating in fishing activities. Reducing fishing efforts aims to increase the efficiency of purse seine fishing so that it is expected to increase CPUE per year [34].

The next management recommendation is to closely monitor fishing activities, which are related to the use of selective fishing gear and environmentally friendly fishing methods. This is important so that the target and size of the catch are still by the rules. Then, regulate the distance of illegal FADs by KP Regulation Number 26, 2014, namely not less than 10 nautical miles so as not to hurt fish migration. Nurdin et al. [35] state that the impact of installing FADs that are too close causes damage to the migration patterns of fish that migrate far away, thereby disrupting the balance of the ecosystem. The last recommendation is related to ETP species caught, it is necessary to educate fishing communities regarding ETP species and their handling along with conservation efforts that can be carried out.

4 Conclusions

Based on research that has been carried out at PPS Kutaraja, it can be concluded that The status of mackerel scad resource management based on an ecosystem approach in the domain of fish resources in Kutaraja PPS is included in the 'good' category, meaning that caution is still needed in its utilization. From several indicator values, there are management actions that need to be improved so that this condition can be maintained for the better. Recommendations for the management of mackerel scad resources based on an ecosystem approach in Kutaraja PPS that can be suggested include controlling purse seine fishing efforts by regulating the number of vessels, fishing gear, time, and area of fishing as well as limiting fishing quotas; strict supervision of fishing activities, related to the use of selective fishing gear and environmentally friendly fishing methods; controlling the distance between FADs for purse seine fishermen where the installation distance between FADs is not less than 10 nautical miles; as well as providing education to fishing communities regarding ETP species and their handling and conservation efforts that can be carried out.

References

1. The Marine and Fisheries Ministry, *Estimation of the potential of fish resources, the amount of fish catch that is allowed, and the level of utilization of fish resources in the fisheries management area of the Republic of Indonesia*, in KEPMEN-KP No. 50/ 2017, Jakarta (2022)
2. S. Salmarika, A.A. Taurusman, S. H. Wisudo, JPPI, **24,** 263-272 (2019)
3. R.N. Sari, E. Miswar, C. Makwiyah, JIM Kelautan dan Perikanan Unsyiah, **2**, 423-428 (2017)
4. Regional Technical Implementation Unit (UPTD) Kutaraja, *Capture Fisheries Statistic Banda Aceh,* Banda Aceh (2021)
5. FAO, *Food and Agriculture Organization of the United Nations,* **4** Suppl. 2, 112, Rome (2003)
6. U. Muawanah, G. Yusuf, L. Adrianto, J. Kalther, R. Pomeroy, H. Abdullah, T. Ruchimat, Marine Police, **91,** 150-160 (2018)
7. NWG EAFM, *Fisheries management indicator module using the EAFM,* Directorate of Fish Resources, The Marine and Fisheries Ministry RI, Jakarta (2014)
8. P. J. Subagyo, *Research Methods: In theory and practice*, Rineka Cipta, Jakarta (2011)
9. M. Masyhuri, Z. Zainuddin, *Research methods: practical and applicable approaches,* Aditama Refika, Bandung (2011)
10. A. R. A. Nalendra, Y. Rosalinah, A. Priadi, I. Subroto, R. Rahayuningsih, R. Lestari, S. Kusamandari, R. Yuliasari, D. Astuti, J. Latumahina, M. W. Purnomo, V. A. Zede, *Basic series statistics with SPPS*, Indonesian Science Media, Bandung (2021)
11. P. R. Febriani, A. K. Mudzakir, A. Asriyanto, JFRUMT, **3**, 208-217 (2014)
12. H. Wahyudi, *Level of utilization and pattern of fishing season for lemuru (Sardinella lemuru) in the Bali Strait waters,* in Thesis, IPB, Bogor (2010)
13. A. Leone, R. Romaniello, R. Zagaria, E. Sabella, L. De Bellis, A. Tamborrino, Eur. J. Lipid Sci. Technol, **117**, 1271–1279 (2015)
14. H. Latuconsina, M. N. Nessa, R. A. Rappe, Journal of Tropical Marine Science and Technology, **4**, 35-46 (2012)
15. A. D. Wahyudi, A. F. Syah, Juvenil Journal, **3**, 155-158 (2022)
16. E. Nugraha, B. Koswara, Y. Yuniarti, *Journal of Fisheries and Maritime Affairs* **3**, 91-98 (2012)
17. T. Hariati, News of Indonesian Fisheries Research: Resources and Arrests Edition, **11**, 15-18 (2004)
18. K. H. Ikhsan, I. Irham, Jurnak Iktiologi Indonesia, **9**, 163-174 (2017)
19. M. Fadila, A. Asriyana, M. Tadjuddah, Jurnal Manajemen Sumberdaya Perairan, **1**, 343-353 (2016)
20. A. Zamroni, A. Kuswoyo, U. Chodrijah, *Bawal,* **11**, 137-149 (2019)
21. H. Sasmito, A. I. Nur, A. Abdullah, Jurnal Manajemen Sumberdaya Perairan, **1**, 275 – 284 (2016)
22. https://fishbase.mnhn.fr/summary/SpeciesSummary.php?ID=1938&AT=layang Retrieved 24 May 2023
23. H. Sudirman, A. Mallawa, *Fishing techniques*, Rineka Cipta, Jakarta (2012)
24. D. Wiadnya, A. Damora, M. M. Tamanyira, D. Nugroho, A. Darmawan, IOP Conf. Ser. :Earth Environ. Sci, **139**, 1-6 (2018)

25. O. S. Deysi, *Evaluation of fishing grounds through analysis of catches in Sibolga Waters, North Sumatra*, in Thesis, IPB, Bogor (2015)
26. A. A. Widodo, F. Satria, Ind. Fish. Res.J., 19, 73-79 (2012)
27. I P. Y. P. Putra, I W. Arthana, M. A. Pratiwi, Journal of Tropical Fisheries Management, **4**, 29-37 (2020)
28. M. Arianto, U. Tangke, S. S. Titaheluw, S. Masiyah, AGRIKAN, **15**, 255-264 (2022)
29. L. Adrianto, A. Habibi, A. Fahruddin, A. Azizy, H.A. Susanti, I. Musthofa, *Indicator module for fisheries management with an ecosystem approach* Marine and Fisheries Ministry, Jakarta (2014)
30. H. Himelda, E. S. Wiyono, A. Purbayanto, M. Mustaruddin, Marine Fisheries, **2**, 165-176 (2011)
31. D. M. Puansalaing, J. Budiman, F. B. Boneka, D. M. Makapedua, M. Lasut, E. L. A. Ngangi, D. A. Sumilat, O. P. Darmono, Aquatic Science & Management, **9**, 7-16 (2021)
32. S. Alhuda, Z. Anna, I. Rustikawati, Journal of Marine Fisheries, **7**, 30-40 (2016)
33. M. Umamah, S. H. Wisudo, R. I. Wahyu, *Albacore*, **1**, 245-255 (2017)
34. D. Simbolon, B. Wiryana, P. I. Wahyuningrum, H. Wahyudi, Bulletin PSP, **19**, 293 – 307 (2011)
35. E. Nurdin, A. A. Taurusman. R. Yusfiandayani, J. Lit. Perikan. Ind. **18**, 53-60 (2012)

The effect of frangipani leaf extract (*Plumeria* sp.) on the hatchabiiity and growth of sangkuriang catfish (*Clarias gariepinus*) larvae

Gunawan Gunawan[1], *Ozi* Misbahul Huzy[1], *Zulkarnain* Jalil[2] and *Dedi* Fazriansyah Putra[1*]

[1]Department of Aquaculture, Faculty of Marine and Fisheries, Syiah Kuala University, Banda Aceh 23111
[2]Department of Physics, Faculty of Mathematics and Natural Science, Syiah Kuala University, Banda Aceh 23111

Abstract. Catfish (*Clarias gariepinus*) or in local name *Lele sangkuriang* is a type of freshwater fish that is often cultivated in the worldwide community particularly Indonesia. The market demand for this fish is increasing every year so that fish farmer can get better profit in the future. In order to prevent a decrease in seed production in the future, the seed procurement process must be carried out. Moreover, to ensure the production levels must be continued to increase both in quantity and quality. This research aimed to determine the effect of frangipani leaf extract (*Plumeria* sp.) on hatchability rate and survival of the larvae of catfish (*Clarias gariepinus*). The experimental method in this research was completely randomized design with 5 treatments and 3 replications. The treatments used were P0 without frangipani extract (control), P1 giving frangipani extract 20 ppm, P2 giving frangipani extract 40 ppm, P3 giving frangipani extract 40 ppm and P4 giving frangipani extract 80 ppm. The results showed that different doses of frangipani leaf extract produced different effects on hatchability and the quality of the larvae produced. The highest percentage of hatchability results was obtained in treatment P1 (85.33%) and the best percentage of larvae survival rate was in treatment P2 (72.67%).

1 introduction

Public demand for the consumption of catfish from aquaculture is increasing every year [1]. In order to fulfil Indonesian domestic demand, an increase in the production of catfish should be carried out every year [2]. The total production of cultured catfish in 2020 was 347.511.48 tons, this production capacity tends to decrease compared to previous years, namely in 2018 it produced 1.027.032,54 tons and in 2019 as much as 981.623,40 tons [3]. Catfish (*Clarias gariepinus*) or *lele sangkuriang* in local name is a popular fishery commodity among the public [4, 5, 6]. This fish originated from Africa and was imported to Indonesia in 1984. Catfish is one type of fish that is most easily accepted by the public because of its advantages [7]. The nutritional composition of catfish is protein (17.7%), fat (4.8%), mineral (1.2%), and water (76%) [8]. Catfish have advantages compared to other types of fish, namely having faster growth, having high tolerance for poor water quality, and being more resistant to disease attacks and can be kept in all cultivation containers [9].

[*] Corresponding author: dfputra@usk.ac.id

However, there are many problems that arise such as disease attacks in the egg and larval phases [10]. One of disease that is often and commonly found is Saprolegniasis caused by the *Saprolegnia* fungus. It is often found that the Saprolegnia fungus infects catfish in the age phase to the adult phase, *Saprolegnia* is a disease that is often found attacking fish in fresh waters [11]. In the intensive hatchery stage, problems are often found, namely disease attacks by *Achyia* fungi and *Saprolegnia* sp. which cause a decrease in the hatching rate of the eggs, and during the incubation period many dead fish eggs are found due to fungi [12].

The use of chemical drugs has a negative impact on fish life, namely the killing of non-target organisms, the emergence of immunity to pathogens, affected growth and reproduction and causing pollution. So that a safe alternative is to utilize herbs that are more environmentally friendly and easily decomposed by water, and also herbs have relatively small side effects [13, 14]. The use of herbs as a therapy for fish diseases has been carried out in previous research. Uses of *Lonicera japonica* Thunb. components can increase the resistance of grouper fish in vitro and in vivo [14]. Then extract *Illicium verum* Hook. f. has been proven to be able to resist grouper virus attacks [14]. *Curcuma kwangsiensis* ingredients have also shown significant results against grouper iridoviral infection in vitro and in vivo [15]. Apart from that, several marine based plants and their active ingredients have also been proven to increase the immunity of *Litopenaeus vanname* shrimp against disease attacks such as *Sargassum oligocystum*, carrageenan, *Spirulina* sp. and fucoidan [16, 17, 18, 19]. The use of *Plumeria* sp. as a therapeutic treatment and to help in the hatching and larval capacity of catfish has never been done before. In this study, we used Plumeria species that contain tannins, flavonoids, steroids, saponins, terpenoids that show antibacterial activity [20]. Therefore, the aim of this current study is to examine the effect of different concentrations of frangipani leaf extract *frangipani* leaf extract on the hatching rate and survival rate of catfish, *Clarias gariepinus*.

2 Materials And Methods

2.1 Experimental Design

This research was conducted in December 2022. Located in the laboratory of the Faculty of Marine and Fisheries of Syiah Kuala University, Darussalam, Banda Aceh, Aceh. The tools and materials used in this research were pH meter, thermometer, DO meter, aerator, refractor, evaporator, blender, NaCl, chicken, digital camera, basin, plastic cup, 70% ethanol, shaker, pH meter, scales, ruler, frangipani leaf extract. The experimental design used was a complete randomized design (CRD) using 5 treatments and 3 replications. The treatments are as follows:

P0 = without using frangipani leaf extract (*Plumeria* sp)
P1 = using 20 ppm frangipani leaf extract (*Plumeria* sp.)
P2 = using 40 ppm frangipani leaf extract (*Plumeria* sp.)
P3 = using 60 ppm frangipani leaf extract (*Plumeria* sp.)
P4 = using 80 ppm frangipani leaf extract (*Plumeria* sp.)

2.2 Research Procedure

A 15 units aquarium sized 80 x 40 cm were used during the study. Catfish eggs were stocked as many as 100 fish/container. The fish eggs were first soaked in a geyser for 1 minute with a solution of frangipani leaf extract and placed on pet topes. For the extraction process, frangipani leaves were washed thoroughly so that impurities such as dust that stick to the frangipani leaves are removed. Next, the frangipani leaves were dried without exposure to sunlight. The dried frangipani leaf samples were made into powder by pulverizing. Extraction was carried out using maceration method with 70% ethanol solvent as modified method by previous work [21]. A total of approximately 800 g of frangipani leaves were soaked with 1,600 mL of 70% ethanol for 3 days, covered, stored in a dark room and then shaken with a 120 rpm mixer for 7 days. Afterward, the filtrate was removed and the residue was re-macerated with 50 mL of 70% ethanol. Subsequently, the filtrate was removed every day for 7 days. After that, maceration ended. After the extract filtrate from frangipani leaves is obtained, it was concentrated by using a rotary-evaporator at a temperature of ± 40°C, until a ready-to-use extract is obtained.

3 Research Parameters

3.1 Hatching Rate (HR)

To measure hatching eggs divided by the number of fertilized eggs multiplied by one hundred percent and expressed in (%). According to [22] the degree of hatching of eggs can be calculated using the formula:

$$HR = \frac{number\ of\ hatches}{number\ of\ fertilized\ eggs} x100$$

3.2 Survival Rate

SR calculation was done at the end of the study after 35 days. According to [23], the survival rate is expressed in terms of the percentage of living fish at the beginning and end of the study and is formulated as:

$$SR = \frac{Nt}{N0} 100\%$$

Description:
SR = Survival Rate (%)
Nt = Number of fish at the end of the study (ind)
N0 = Number of fish at the begin of the study (ind)

3.3 Length Gain

Length gain is defined as the total gain in length and final weight minus the length of begin weight. Mutual length gain was calculated based on the formula used by [24] :

$$Lm = Lt - Lo$$

Description :
Lm = Length gain (cm)
Lt = Average length of fish at the end of the study (cm)
Lo = Average length of fish at the beginning of the study (cm)

3.4 Abnormality

Abnormality in the larval phase is in the form of bending of the body and tail [25]. Observations of abnormality in this study include head shape, body shape and tail shape. Calculation of abnormality can be done using the formula [26], which is as follows:

$$Abnormality = \frac{number\ of\ abnormal\ larvae}{number\ of\ normal\ larvae} x100$$

3.5 Data Analysis

All data in this study were analyzed using analysis of variance (ANOVA) in accordance with a complete randomized design (CRD). The average treatment will then be further tested using the Duncan test with a determination of $p < 0.05$ to determine the effect of differences between treatments.

4 Results And Discussion

Result showed that the treatment frangipani leaf extract immersion has a very significant effect on the hatching rate of catfish eggs. This is assumed due to frangipani leaf extract has antifungal properties. According to [27] frangipani leaves showed positively contain flavonoids, alkaloid and saponins. Flavonoid compounds have activity as anti-bacteria with the mechanism of inhibiting the work of nucleic acid synthesis, inhibiting the function of cell membranes, and inhibiting energy metabolism. The lowest value was obtained in the P4 treatment (80 ppm) with value 34.67%. According to [28] because the higher the concentration of the extract, the greater the risk of egg damage. Excessive levels of antioxidants cause toxicity and mortality of fish eggs. In the treatment of P0 (control) has a low value of 15.67%. This is due to the growth of mold that causes further damage to healthy eggs, causing eggs to die and not hatch. According to [29]eggs attacked by mold will show signs of fine threads, such as cotton and white paint, formed around the egg.

Previous results reported by [30] on the effect of soaking doses of *Syzygium aromaticum* leaf extract on the hatchability of catfish, *Clarias gariepinus*, where the use of clove leaf solution at certain concentrations has a significant effect ($P > 0.05$) on hatchability. The similar result were in line with [13] studied the effect of using *Avicennia marina* leaf extract on the hatchability of *Clarias gariepinus*. *Avicennia marina* leaf extract immersion proved to have a significant effect on hatchability of catfish ($P < 0.05$). The use of Moringa leaves (*Moringa olifera* I.) at different concentrations has also significant effect on the infection rate and hatchability of *Clarias gariepinus* [31]. The use of kirinyuh leaf extract (*Euphatorium odoratum* l.) had significant effect on the hatching rate of milkfish *Chanos chanos* [21]. Moreover, egg hatchability and survival larvae of Pomfret Fish (*Colossoma Macropomum*) has also stimulated by turmeric extract (*Curcuma Domestical*) [32]. Therefore, based on previous research, it is suspected that the addition of frangipani leaf extract with a certain concentration has a significant effect on hatching fish eggs as was done in this research.

Table 1. Hatchability, survival, length gain, and abnormality results

Frangipani leaf extract treatment					
Parameter	P0 (0 ppm)	P1 (20 ppm)	P2 (40 ppm)	P3 (60 ppm)	P4 (80 ppm)
HR (%)	15.67±9.59^{a}	85.33±10.06^{de}	80.00±10.06^{d}	53.33±10.01^{c}	34.67±9.89^{b}
SR (%)	13.67±5.61^{a}	71.33±5.88^{d}	72.67±5.88^{de}	47.33±5.85^{c}	29.33±5.78^{b}
Length gain (cm)	0.77±0.04^{c}	0.72±0.04^{a}	0.72±0.04^{a}	0.75±0.04^{bc}	0.73±0.04^{ab}
Abnomality (%)	0	0	0	1±1.50^{d}	3.67±1.50^{e}

Table 1 showed that the highest larva survival was found in the P2 treatment (40 ppm) with a value of 72.67%, this value is not much different from the P1 treatment (20 ppm) with a value of 71.33% and the lowest average in the P4 treatment (80 ppm dose) with a value of 29.33%. In this study, the survival of larvae was influenced by the hatchability of the eggs themselves, and apart from that the mortality of larvae according to [33] is usually due to the age and size of the fish able to survive. According to [34] stated that environmental factors such as handling and water quality affect the survival rate of fish. In this study, water quality is always maintained by drinking regularly to remove feces and uneaten fish feed residues that affect water quality. This research has results that are in line with the research of [35], namely feeding the type of beut fish extract provides a very significant difference in survival (SR) of catfish larvae and research from [36] on the effectiveness test of papaya leaves (*Carica papaya*) to treat *Aeromonas hydrophila* bacterial infection in carp (*Carassius auratus*). Then also in line with [37] on soaking catfish (*Clarias* sp.) in guava fruit juice for the treatment of *Aeromonas hydrophila* infection. Another herb like ginger (*Zingiber officinale*) can enhance the immunity of Tilapia (*Oreochromis niloticus*) [38]. But different result were reported by [39] on the effect of kersen leaf extract (*Muntingia calabura*) with different doses on incubation time, hatchability and larval life of catfish (*Clarias gariepinus*). Based on previous research, it is suspected that the addition of frangipani leaf extract with a several concentration active compound has a significant effect on the survival of fish larvae.

It can be seen that the highest growth (length gain) was obtained in P3 treatment or dosing of 60 ppm extract with a value of 29.33 cm and the lowest average was obtained in P1 and P2 or dosing of 20 ppm and 40 ppm extracts with an average value of 0.72 cm. All larvae in this study had similar growth. Larvae development under captive condition are influenced by several factors like nutrition, environment and rearing technique [40]. According to [41] expressed that the balanced nutritional content (protein, fat and fiber) in fish feed accelerates relatively faster fish growth. The use of supplementation has also significant effect in supporting fish (including shrimp) growth [7, 42, 43]. Previous result has reported the administration of probiotic could boost the significant growth in aquaculture [44, 45, 46, 47]. However some supplementation could also had negative impact on fish or shrimp culture [48].

The results showed that the highest percentage of abnormality rate with a value of 3.67% was obtained from the P4 treatment followed by P3 with a size of 1%, while for treatments P0, P1, and P2 no abnormality larvae were found. The success of a hatchery is not only seen from its egg-laying capacity, but also from the quality of the resulting larvae, such as the level of abnormality of the larvae. This research can be considered to have a better percentage than the research of [49] on the effect of soaking clove leaf extract (*Syzygium aromaticum*) on the hatchability of peres fish (*Osteochillus kappeni*) and the research of [50] the utilization of pineapple leaf extract in determining the survival rate of milkfish larvae. But it is not the same as the research of [12] on the effectiveness of kecombrang flower extract (*Nicolaia speciosa horan*) for the prevention of *Saprolegnia* sp. attack on catfish which did not have any abnormality in its research.

Table 2. Water Quality Parameters

	Water Quality Parameters		
Treatment	**DO (ppm)**	**Temperature °C)**	**pH**
P0	4-7	27-31	6.8-7.2
P1	4-7	27-31	6.8-7.2
P2	4-7	27-31	6.8-7.2
P3	4-7	27-31	7-7.2
P4	4-7	27-31	7-7.2
Reference [9]		25-32	6-8.5

Table 2 showed the water quality parameters during the study. The dissolved oxygen (DO) values obtained ranged from 4-7 ppm. The dissolved oxygen content that can be tolerated by catfish in aquaculture is a minimum of 3 mg/L [6]. The temperature values obtained during the research ranged from 27-31°C. This is in line with the research of [51] on different feeding on the growth and survival of catfish larvae *Clarias gariepinus* can survive at temperature range of 27-30°C. The measurement of the pH value of the water during the study was 6.8-7.2 which is also in line with report from [52]. Thus it can be concluded that the water quality parameters in this research have been well organized and optimal.

5 Conclusion

The administration of frangipani leaf extract has significant effect on the hatchability of catfish *Clarias gariepinus*. Moreover, high dose of frangipani leaves inhibits the hatchability of the eggs themselves due to the concentrated antioxidant content. Frangipani leaf extract affects the survival of the resulting larvae, where at appropriate doses a high survival rate is obtained, but compared to high doses the resulting larvae have abnormalities that affect their development.

References

1. D. F. Putra. Dasar - dasar Budidaya Perairan. Syiah Kuala University Press. (2021)
2. M. H. F. Sitio, D. Jubaedah, M. Syaifudin. JARI. **5**, 83-96. (2017)
3. E. Harianto, T. Budiardi. Jurnal Akuakultur Sungai dan Danau, **6**, 50-57. (2021)
4. D. F. Putra, R. Adriman, N. Abubakar. J . pengabdi. kpd. masy. (Indones. j. community engagem.), **8**, 89. (2022)
5. A. Salsabilla, D. F. Putra, C. Octavina, R. Maulana. IOP Conference Series: Earth and Environmental Science, **869**. (2021)
6. S. Suratno, D. F. Putra. JVIP, **2**, 9–13. (2022)
7. D. F. Putra, M. A. Abbas, N. Solin, N. Othman. Elkawnie. **7**, 395. (2022)
8. D. Zulistyanto, P. H. Riyadi, U. Amalia. JPBHP, **5**, 26-32. (2016)
9. D. P. Hariani. S. W. Kusuma. STIGMA: Jurnal Matematika dan IImu Pengetahuan AIam Unipa. **9**, 1-5. (2016)
10. D. F. Putra, D. Azhar, A. W. Perdana, I. I. Arisa, K. Melanie. Egypt. J. Aquat. Biol. Fish. **26**, 1329–1347. (2022). 2013
11. H. Harun, M. Abdan. JP2V. **2**, 115-119. (2021)
12. M. N. Lingga, I. Rustikawati, I. D. Buwono. Jurnal Perikanan Kelautan, **3**, 75-80. (2012)
13. D. Rahmi, S. Karina, I. Dewiyanti. JIM FKP Unsyiah, **1**, 252-261. (2016)
14. M. Liu, Q. Yu, Y. Yi, H. Xiao, D. F. Putra, K. Ke, Q. Zhang, & P. Li. Aquaculture, **519**. (2020)
15. L M. Liu, H. Xiao, Q. Zhang, S. Wu, D. F. Putra, X. Xiong, M. Xu, L. Dong, S. Li, Q. Yu, P. Li, Aquaculture Research, 1–11. (2019)
16. F. N. Baleta, Y. Lin, Y. Chen, J. Chen, S. Yeh, D. F. Putra, C. Huang. J. Fish. Soc. **40**, 241–256. ()
17. Y. Y. Chen, J. C. Chen, Y. C. Lin, D. F. Putra, S. Kitikiew, C.C. Li, J. F. Hsieh, C. H. Liou, S. T. Yeh. Fish Shellfish Immunol. **36**. (2014)
18. Y. Y. Chen, J. C. Chen, C. M. Tayag, H. F. Li, D. F. Putra, Y. H. Kuo, J. C. Bai, Y. H. Chang. Fish Shellfish Immunol. **55**, 690–698. (2016)
19. S. Kitikiew, J. C. Chen, D. F. Putra, Y. C. Lin, S. T. Yeh, C. H. Liou. Fish Shellfish Immunol. **34**, 280–290. (2013)

20. M. A. Hanif, H. Nawaz, M. Khan, H. Byrne. Medicinal Plants of South Asia. **1**, 1-699. (2020)
21. E. Evendi, S. Karina, D. F. Putra. JIM FKP Unsyiah. **2**, 33–40. (2017)
22. B. A. Murtidjo 2001. Beberapa Metode Pembenihan Ikan Air Tawar. 34-36 (2001)
23. M. Effendie. BioIogi perikanan Diktat pengantar perkuliahan Fakultas Perikanan. In: Bogor: IPB. (1978)
24. Effendi, M. I. Budidaya perikanan. (1997).
25. E. Supriono, I. Lisnawati, D. Djokosetiyanto. JA. **4**, 69-78. (2005)
26. I. Wirawan. Diss. Universitas Airlangga, Surabaya. (2005)
27. S. Erikania, Y. Hariningsih. Edu Masda JournaI, **1**, 74-81. (2017)
28. D. D. Ariyani, E. I. Raharjo. Jurnal Agroqua. **3**,17-24. (2016)
29. Y. Siregar. I, A. Adelina. J. Natur Indones. **12**, 75-81. (2009)
30. B. N. D. Yulihastiana, N. Cokrowati, A. R. Scabra. Jurnal Perikanan Unram, **11**, 89-97. (2021)
31. E. A. A. J. Yohanes, S. Jumiati, Rahmaningsih. J.Miy. **2**, 1-6. (2022)
32. Z. Zulfahmi, D. F. Putra, I. I Arisa. JKPI. **3**, 30–35. (2023)
33. I. E. Berampu, E. Patriono, R. AmaIia. Sri. Bios. **2**, 1-15. (2021)
34. A. Mustofa, S. Hastuti, D. Rachmawati. PA, **17**, 45. (2018)
35. Y. A Nuraini,. Sunaryo. JPPIK. **5**, 28-34. (2011)
36. A. Haryani, R. Grandiosa, I. D. Buwono, A. Santika. JPK. **3**, 213-220. (2012)
37. I. Purnamasari. JARI. **3**, 82-93. (2015)
38. S. Afriani, D. F. Putra, Mursalin. IOP Conf. Series: Earth and Environmental Science, **1221**, 1–6. (2023)
39. H. S. Mulyani, T. I. Johan. Dinamika Pertanian, **36**, 99-110. (2020)
40. D. F. Putra, A. B. Abol-Munafi, Z. A. Muchlisin, J.C. Chen. AACL Bioflux. **5**, 29–35. (2012)
41. R. Syaputra, I. S. Santoso, T. Tarsim. JSTA. 2, 1-11. 2018.
42. D. F. Putra, L. Armaya, S. A. ElRahimi, N. Othman. *BIOTROPIA*, **26**, 136–142(2019).
43. D. F. Putra, M. Rahmawati, M. Z. Abidin, R. Ramlan. IOP Conference Series: Earth and Environmental Science. **348**. (2019)
44. A. L. Dachi, A. A. Muhammadar, I. Sahidhir, D. F. Putra, Z. A. Irwan. IOP Conference Series: Earth and Environmental Science. **348**. 2019)
45. A. A. Muhammadar, M. A. Chaliluddin, D. F. Putra, M. S. Asmawati. IOP Conference Series: Earth and Environmental Science. **216**. (2018)
46. A. A. Muhammadar, Z. A. Muchlisin, F. Firdus, D. Aliza, R. S. Aminah, D. F. Putra, M. S. Asmawati, S. Satria, B. Boyhaqi, K. Razi, R. Ramlan. IOP Conference Series: Earth and Environmental Science. **348**. (2019)
47. D. F. Putra, A. Qadri, S. A. El-Rahimi, N. Othman. E3S Web of Conferences. **151**, 1–4. (2020)
48. I. Safriani, D. F. Putra, S. A. El-Rahimi, N. Othman. IOP Conference Series: Earth and Environmental Science. **348**(1), 0–6. (2019)
49. K. I. Azizi. JISA, **5**, 54-59. (2021)
50. W. Walidin, R. Humairani, T. F. Haser. JISA. **1**, 31-38. (2017)

51. M. Madinawati, N. Serdiati, Y. Yoel. Media Iitbang Sulteng. **4**, 11-19. (2011)
52. S. C. Purwanti, A. Sudaryono. JAMT. **3**, 53-60. (2014)

Effect of herbal leaf extract on *Oreochromis niloticus* immunity infected with *Streptococcus agalactiae*

Iko Imelda Arisa[1,2,3*], *Julianda* Fatma[1], *Nurfadillah* Nurfadillah[1,2], *Cut* Nuzlia[1] and *Ulfa* Rahmi[1]

[1]Departement of Aquaculture, Marine and Fisheries Faculty, Universitas Syiah Kuala, Banda Aceh, Aceh, 23111Indonesia
[2]Research Center for Marine and Fisheries, Universitas Syiah Kuala, Banda Aceh, Aceh, 23111Indonesia
[3]Animal Histology and Fish Nutrition Laboratory, Marine and Fisheries Faculty, Universitas Syiah Kuala, Banda Aceh, Aceh, 23111 Indonesia

Abstract. Streptococcosis disease caused by the pathogen *S. agalactiae* is a serious problem in freshwater fish farming. This research aims to use multiherbal ingredients to stimulate the improvement of the fish's immune system in fighting infection by the pathogenic bacteria *S. agalactiae*. The research was carried out at the FKIP Chemistry Laboratory, the Fish Hatchery and Breeding Laboratory of the FKP, and the Laboratory of the Faculty of Medicine, Syiah Kuala University. The test fish used were tilapia measuring 7-8 cm length. The research method was carried out experimentally using a Completely Randomized Design consisting of 5 treatments and 3 replications, namely treatment A (negative control), B (positive control), C (addition of *C. gigantea*), D (addition of *M. oleivera*), E (addition of *C. alata* L). The test result data is analyzed using Analysis of Variance. The results of research on immersion in multiherbal extracts in fish infected with *S. agalactiae* showed an increase in the immune response as seen in blood parameters (leukocytes, hemoglobin, hematocrit) and a higher survival rate compared to those without immersion in the extract. The highest survival rate was obtained in the treatment of 10 ppm of *C. alata* L leaves extract at 83.33%.

1 Introduction

Tilapia (*Oreochromis niloticus*) has high economic value [1], and is an important commodity in the freshwater fish business in Indonesia to meet animal consumption needs. Commercial cultivation of tilapia can trigger stress, which can increase the fish's susceptibility to disease [2]. According to Nasution *et al.* [3], one of the obstacles in the cultivation business, especially tilapia, is the disease Streptococcosis, which is caused by bacterial infection.

* Corresponding author: ikoimeldaarisa@usk.ac.id

Furthermore, Heckman *et al.*, [4] mentioned that *Streptococcus iniae* is a bacterial pathogen that has reappeared in freshwater and marine aquaculture the worldwide.

Diseases caused by pathogenic bacteria will reduce the value of tilapia production because the survival rate of the fish will be low. Treatment of the disease that has been applied includes giving antibiotics. The existence of resistance to various antibiotics shows that Streptococcosis is no longer effective with various types of antibiotics and also that the use of antibiotics has been banned for treating diseases in farmed fish [5]. Therefore, alternative medicine is needed using natural plants as herbal medicine, including *Moringa oleivera* leaves, *Calotropis gigantea* leaves, and *Cassia alata* L. leaves. This natural plant is known to contain antibacterial compounds to treat fish and shrimp infected with bacterial diseases [6, 7, 8].

Previous research has classified herbal ingredients as rich in substances immune enhancer or immunostimulan, where the immunostimulant is herbs can occur by the mechanism of modulating non-specific immune responses and is currently quite widely used to control diseases in fish. Traditionally, herbs and their extracts have been proven effective as performance enhancers for the immuune system and are also recommended as good alternative immunostimulants in aquaculture. According to Muahiddah & Diamahesa [9], the use of natural ingredients for immunostimulants can increase the non-specific immunity of fish toward off disease attacks and environmentally friendly. This research aims to use multiherbal ingredients to stimulate the improvement of the fish's immune system in fighting infection by the pathogenic bacteria *S. agalactiae.*

2 Material and Methods

This study used a completely randomized design (CRD) with three repetitions and five treatments as its experimental setup. The following can be used to express how extract administration is combined for each treatment:
P1 = Fish free of *S. agalactiae* infection
P2 = Fish with *S. agalactiae* infection
P3 = Fish soaked in 800 ppm of *C. gigantea* leaves extract after being infected with *S. agalactiae*
P4 = Fish soaked in 1000 ppm of *M. oleivera* leaves extract after being infected with *S. agalactiae*
P5 = Fish treated with *C. alata* L leaves extract (10 ppm) after being infected with *S. agalactiae*

2.1 Preparation of Extracts

2.1.1 Cassia alata L leaves extract

Cassia alata leaves were washed with clean water and dried in the sun until the leaves were easily crushed. The dried leaves are ground into flour. A total of ±500 g of *C. alata* leaves powder was soaked in 3 L 96% PA absolute ethanol for 24 hours. The results of the immersion are filtered (Whattman No.41), the results are extracted using a rotary evaporator, then the extract is stored at -5 °C until it is used for testing.

2.1.2 Moringa aloivera leaves extract

Moringa aloivera leaves are dried at room temperature for approximately 4 days with supervision. The dry sample (simplisia) was cut into pieces and then ground using a blender until it became simplicia powder. Simplicia powder was weighed 130 g and put into an Erlenmenyer. Then soak (maceration) with 2 liters of ethanol solution and soak for 24 hours. After 24 hours, the solution was filtered using a filter and evaporated using an evaporator [6].

2.1.3 Calotropis gigantea leaves extract

Calotropis gigantea are dried at room temperature for approximately 4 days under supervision. The dried sample (simplisia) is cut into pieces then ground using a blender until it becomes simplicia powder. 500 g of simplicia powder was weighed and put into a container. Then it was soaked (maceration) with 4 liters of 95% absolute ethanol solution and soaked for 24 hours. After 24 hours, the solution was filtered using a filter and evaporated using an evaporator. Next, the resulting extract is weighed for specific gravity and stored in a freezer until used for testing. According to the level of treatment [10].

2.2 Application of multi-herbal leaves to tilapia infected with the pathogen *S. Agalactiae*

The maintenance container used in the experiment was a 12 L jar filled with 10 L of water. The number of container prepared was 15 units and each was filled with 10 fish/container. The container used is a clean or sterile bucket. The test tilapia fish measuring 7-8 cm. The feed used during maintenance is commercial feed with a protein content of 30%. Tilapia fish are fed three times a day at 07.00, 12.00 and 17.00 WIB as much as 5% of body weight.

Before applying the multiherbal extract, healthy fish were first taken that had been adapted to the rearing container for one week, then the fish were injected with the S. agalactiae pathogen at 105 CFU/ml, which is the LD50 dose [11]. After injection of *S. agalactiae*, the fish were left for 2 hours to observe for clinical symptoms of Streptococcosis. After that, soak it with multiherbal extracts in a container according to the dosage for each treatment for ± 15 minutes. Next, the fish were transferred to a rearing test container and then kept for 14 days and observations were made regarding clinical symptoms, blood picture, survival rate and weight growth of the fish. Control water quality by siphoning 10% of the total water volume of each container every day.

2.3 Observed parameters

Parameter measurements in the study included observing clinical symptoms of sick fish, blood picture, survival rate and growth of tilapia fish. Clinical symptoms of sick fish are observed by observing the fish's behaviour and changes that occur in the fish's body. The fish blood features observed include leukocytes, erythrocytes, hematocrit and haemoglobin. Blood parameter measurements were carried out 4 times, namely on days 0, 1, 3 and 5 after the challenge test/pathogenic bacterial infection

a. Total leukocyte counts were calculated according to Blaxhall and Daisley [12]. The calculation is done by: the blood sample is sucked with a pipette containing a white stirrer to a scale of 0.5 ml, then turk's solution is added to a scale of 11, shaken or swung in a number 8 shape for 5 minutes to mix homogeneously. The first drop was discarded, then dropped into a hemacytometer and covered with a cover glass, observed under a microscope. calculations are carried out on a small box hemacytometer.

$$\sum Leukocyte = Leukocyte \ x \ 50 \ cell/mm^3 \quad (1)$$

b. Hematocrit (He) levels were measured according to Anderson and Siwicki [13]. The He level is determined by: the blood sample is put into a microhematocrit tube until approximately ¾ of the tube, then the end is plugged with crytoseal to a depth of 1 mm. After that, it was centrifuged at a speed of 5000 rpm for 5 minutes. After that, the length of the blood that settles (a) and the total length of the blood volume contained in the tube (b) are measured. He levels are expressed as % volume of blood cell solids and are calculated by

$$He \ (\%) = \left(\frac{a}{b}\right) x100\% \quad (2)$$

c. Hemoglobin (Hb) levels used the Sahli method with a salinometer [14]. First, the blood is sucked with a sahli pipette to a scale of 20 mm^3 or a scale of 0.2 ml, then the tip of the pipette is cleaned with tissue paper. After that, the blood in the pipette is transferred to an Hb-meter tube filled with 0.1 N HCL to a scale of 10 (red), stir and leave for 3-5 minutes. Add distilled water until the color of the blood and HCl is the color of the standard solution in the Hb meter. Then read the scale, namely by looking at the surface of the liquid and checking it with the Sahli tube scale which is seen on the gr% line scale (yellow) which means the amount of hemoglobin in grams per 100 ml of blood.

d. Survival rate

$$SR = \frac{Nt}{No} x100\% \quad (3)$$

Note: SR = Fish survival (%); Nt = Number of live fish at the end of rearing (fish), No = Number of fish at the beginning of rearing (fish).

2.4 Data analysis

Data from testing results of multiherbal extracts for each treatment for tilapia survival rate parameters were analyzed using the ANOVA test, if it had an effect it would be tested further. Furthermore, the results of observations of clinical symptom parameters of sick fish and hematology (leukocytes, hematokrit, hemoglobin) are presented descriptively.

3 Results and Discussion

3.1 Clinical symtoms

The results of observations on tilapia that had not been infected with *S. agalactiae* bacteria showed symptoms of high appetite, active swimming, bright body color and scales that did not peel. According to Arie [15] states that in general the characteristics of healthy tilapia are active movement, high appetite, bright skin color, scales that do not peel off and clear eye color. Meanwhile, after bacterial infection, tilapia showed symptoms such as decreased appetite, dark body color, peeling scales, swimming at the bottom of the container, thin tail fins, hemorrhage on the skin, eyes and pale corneas. A longer time of S. agalactiae bacterial infection in the blood will affect clincal symptoms and changes in the profile and composition of tilapia blood cells [16]. Clinical symptoms of fish infected with *S. agalactiae* bacteria and after soaking with multiherbal extracts (*C. gigantea* leaves extract, *C. alata L* and *M. oleivera*) are presented in Table 1. After soaking with multiherbal leaf extract, it is known that the fish's condition recovers slowly. Herbal planys have properties that enhance the host's immunity and act as antibacterial and antiviral agents [17].

Table 1. Clinical symptoms before, after bacterial infection and after immersion in multi-herbal leaf extract

No	Characteristics of Fish	Picture
1	Characteristics of healthy fish (before bacterial infection and extract soaking): - the movement is active - high appetite - bright skin color - scales do not peel off - clear eye color	
2.	Characteristics of sick fish (after bacterial infection and before extract soaking) : - Scales peel off - Thin caudal fin - The eyes and corneas are pale - Hemorrhage of the skin - Protruding eyes	 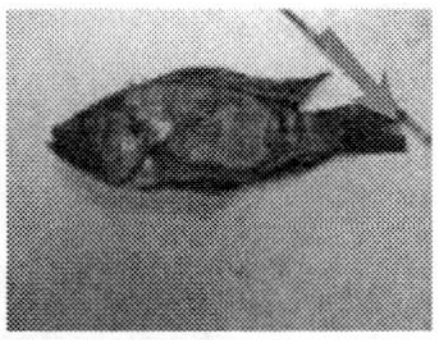
3.	Characteristics of fish (after extract soaking) - The fish's body color returns to normal/bright - Hemarogy on the skin began to improve - The fish's fins begin to grow normally - The eyes have started to returned to normal	

3.2 Tilapia blood parameters

Leukocytes are the most active unit of the body's defense system and circulate in the blood circulation in various types. Leukocytes in tilapia are part of the body's non-specific defense system. A decrease in the number of leukocytes is called leukopenia while an increase in the number of leukocytes is called leukocytosis. The leukocytes produced will be high if there is a viral infection in the tilapia fish's body and the fish's body tries to fight it [18,19,20].

Changes in the number of leukocytes in *O. niloticus* blood observed in the negative control treatment showed that the leukocyte value was 146.2×10^3 cells mm^{-3} and the leukocyte value in the positive control treatment infected with *S. agalactiae* bacteria was 283.8 -273.3×10^3 cells mm^{-3}. In the P1 treatment, the leukocyte value was obtained at (146.2×10^3 cells mm^{-3}) and P2 (283.8×10^3 cells mm^{-3}) showed that the leukocyte value increased, this was due to infection with the *S. agalactiae* bacteria and stress in the fish. The increase in leukocytes is the fish's response to increase the body's defense against bacterial attacks. Leukocytes have various functions related to the removal of foreign objects (including pathogenic microorganisms). The increase in leukocytes in the body is caused by leukocytes acting as the body's defense which quickly reacts to the penetration of antigens into the fish's body. please follow the writing format Description of tilapia leukocyte values during the study is in Table 2.

Table 2. Description of tilapia leukocyte values

Treatment	Leukocyte (x 10^3 cells mm^{-3})			
	D0	D1	D3	D5
P1	146.2	146.2	146.2	146.2
P2	146.2	283.8	279.1	273.3
P3	146.2	269.3	248.7	246.2
P4	146.2	269.3	218.9	189.3
P5	146.2	269.3	160.1	159.5

Hemoglobin (Hb) is a metalloprotein in blood plasma which functions as a transporter of oxygen from the gills which is pumped by the heart to all cells and organs of the body. In the infection treatment and after soaking in herbal extracts, the hemoglobin value at the start of the study was 6 g $100ml^{-1}$ then decreased after being injected with the pathogen *Streptococcus agalactiae*, on day 1 the hemoglobin value was 3.3 g 100 ml^{-1} then gradually increased until the 5th day with a value of 3-4,8 g $100ml^{-1}$.

Table 3. Hemoglobin

Treatment	Hemoglobin Value (g $100ml^{-1}$)			
	D0	D1	D3	D5
P1	6	6	6	6
P2	6	3.3	3	3.3
P3	6	3.3	4.9	3
P4	6	3.3	4.6	4.8
P5	6	3.3	4.6	4.2

The results of hematocrit observations on day 0 did not cause symptoms because the fish were still in normal condition with a value of 21%. Normal hematocrit levels in teleost fish range from 20-30% [21]. Day 1 after bacterial infection in tilapia which had been carried out for 24 hours showed clinical symptoms such as reduced appetite, swimming at the bottom of the container, thin tail fins, hemorrhoids on the skin and eyes and pale corneas as indicated by a decreased hematocrit value with value of 13%, after the fish causes these symptoms, it is then given extracts from multi-herb leaves in different doses, after which it is soaked for 24 hours. The results of observations on days 3 and 5 showed that the fish were still in the healing process because the extract was still reacting.

Table 4. Hematocrit

Treatment	Hematocrit Value (%)			
	D0	D1	D3	D5
P1	21	21	21	21
P2	21	13	15	14
P3	21	13	15	14
P4	21	13	16	17
P5	21	13	6	18

3.3 Survival rate

Based on the results of ANOVA analysis of research data, the effect of giving multiherbal extracts to tilapia infected with the *S. agalactiae* pathogen showed that giving this extract had a significant effect (P<0.05) on survival (Table 5).

Table 5. Survival rate tilapia fish for 14 days of rearing

Treatment	Survival Rate (%)
P1	70.00 ± 10.00^{c}
P2	36.67 ± 5.77^{b}
P3	0^{a}
P4	80 ± 5.77^{c}
P5	83.33 ± 10.00^{c}

In treatment P3 an average value of 0.00% was obtained, in which case all the fish in the rearing container died. This is caused by soaking a little longer, when soaking is too long, the substances contained can turn into poison which causes the digestive system to be disturbed. In treatment P5 was the best treatment with an average value of 83.33%. Meanwhile, the P2 treatment (positive control) showed a decrease in the viability of *O. niloticus* fish by 36.67% due to infection by the pathogen *S. agalactiae* and no treatment was carried out (Table 5). Haq and Prasetio [22] stated that one of the causes of low fish survival could be parasitic, bacterial or viral infections. And the fish immune system plays an important role ini fish survival when fighting pathogenic bacteria during infection [23, 24].

4 Conclusion

The use of herbal ingredients containing active antimicrobial compounds can suppress the development of disease in tilapia caused by the pathogenic bacteria *Streptococcus agalactiae*. Based on research results, the use of *C. alata* L leaf extract at a dose of 10 ppm had the highest survival rate for tilapia with a value of 83.33%. The description of the blood parameters (leukocytes, hemoglobin and hematocrit) of tilapia fish soaked with multiherbal extract shows that the fish's immune response is better than infected fish without soaking in the extract.

Acknowledgment

The author would like to thank the Research and Community Service Institute of Syiah Kuala University for financial support for this research and all parties who have assisted in the process this research.

References

1. Abarike E D, J Cai, Y Li, H Yu, L Chen, J Jian, J Tang, L Jun, F K A Kuebutornye, *J Fish & Shellfish Immunology* **82,** (2022)
2. Doan H V, S H Hoseinifar, S Jaturasitha, M A O Dawood, R Harikrishnan, *Aquaculture 520,* (2020)
3. Nasution A, Basuki F, Hastuti S. Program Studi Perikanan. Jurusan Perikanan. Fakultas Perikanan dan Ilmu Kelautan. UNDIP **3,** (2014)
4. Heckman T I, K Shahin, E E Henderson, M J Griffin, E Soto, *J Fish and Shellfish Immunology* **121** (2022)

5. Hardi E H, Sukenda, E Harris, A M Lusiastuti. *Jurnal Veteriner* **12,** (2011)
6. Arisa I I. S Agustina, C Mutia, N Nurfadillah, S Karina. *Depik Jurnal Ilmu-Ilmu Perairan, Pesisir dan Perikanan***10,** (2021)
7. Arisa I I, F Fitriani, S Agustina, S Karina, C N Devira. *IOP Conference Series: Earth and Enviromental Science* **869,** (2021)
8. Arisa I I. S Agustina, L Handayani. *E3S Web of Conferences*. **339,** 01003 (2022)
9. Muahiddah N, W A Diamahesa. Clarias : Jurnal Perikanan Air Tawar **3,** (2022)
10. Gavamukulya Y, F Abou-Elella, F Wamunyokoli, H Ael-Shemy. *Asian Pacific Journal of Tropical Medicine,* **7**, (2014)
11. Taukhid, U Purwaningsih, D Sugiani, T Sumiati, A M Lusiastuti. *J Riset Akuakultur* **10**, (2015)
12. Baxhall P C, KW Daisley. *J Fish Biology* **5, (1973)**
13. Anderson D P, A K Siwicki. Basic Hematology and Serology For Fish Health Programs. (Paper Presented in Second Symposium on Diseases in Asian Aquaculture "Aquatic Animal Health and the Environment". Phuket, Thailand. 25-29th October 1993. pp17, 1993)
14. Wedemeyer G A, Yasutke. Clinical Methods for the Assesment on the Effect of Environmental Stress on Fish Health. (Technical Paper of the US Departement of the interior fish and the wildlife service, 89 :1-17, 1977)
15. Arie U. Pembenihan dan Pembesaran Ikan Nila Gift. Jakarta : Penebar Swadaya (2007)
16. Suhermanto A, Suhermin, Ridwan, I Astuti, I Nurmawati. *J Riset Akuakultur* **15,** (2020)
17. Hai, N V. 2015. *J Aquaculture*, **446** (1) :88-96
18. Azhar, F. Pengaruh pemberian probiotik dan prebiotic terhadap performan juvenil ikan kerapu bebek (*Cromileptes altivelis*). (Laboratorium Kesehatan Ikan, Institut Pertanian Bogor. Bogor 16680, 2013)
19. V. Muhardina, P.M. Sari, Y. Aisyah, S.Haryani, S.A. Akbar, Rasayan J. Chem **13(1)**, 240-244(2020)
20. K. Melanie, Z. Zilfira, S.A. Akbar, Arwana: Jurnal Ilmiah Program Studi Perairan, 5(2), 162-168 (2023)
21. Bond C E. Biology of Fishes. Philadelphia : Saunders Colege Publishing. Hlm 514 (1979)
22. Haq, I. A, dan E, Prasetio. *J Borneo Akuatika* **2,** (2020)
23. Semple S I, B Dixon, *Biology* (*Basel*) **9,** (2020)
24. Xia Y T, Cheng E H C, Y J Xia, Q Y Wu, L H L Zhang, S Y Lin, T T X Dong, Q W Qin, W X Wang, K W K Tsim. 2021. *JFish and Shellfish Immunology Report* **2,** (2021)

The use of *Chromolaena odorata* leaves to prevent fungal infection on eggs of *Pangasius* sp.

Siska Mellisa[1,2*], *Windy* Mirnawati A.[1], and *Sari* Afriani[1]

[1]Aquaculture Department, Marine and Fisheries Faculty, Universitas Syiah Kuala Banda Aceh 23111, Indonesia
[2]Marine and Fisheries Research Center, Universitas Syiah Kuala Banda Aceh 23111, Indonesia

Abstract. There are several obstacles in *Pangasius* hatchery such as low hatching rate and poor survival. One of the factors that cause low hatching rate on *Pangasius* sp. eggs is the fungal infection of *Saprolegnia* sp. Using plant extracts such as *Chromolaena odorata* leaves can prevent fungal infection because they have antifungal compounds. The objective of the study was to investigate the effect of *C. odorata* leaves extracts on the hatching rate of *Pangasius* sp. eggs. This research was conducted in March 2023 at Fish Seed Center, Southwest Aceh Regency. The method used in this study was Completely Randomized Design (CRD) with 5 levels of treatment and 4 replications. The concentration *C. odorata* leaves extracts used were 0 ppm (control), 100 ppm, 120 ppm, 140 ppm and 160 ppm. Results revealed that the addition of *C. odorata* leaves extracts have a significant effect ($P < 0.05$) on the hatching rate of *Pangasius* sp. eggs and the survival of *Pangasius* sp. larvae. Furthermore, it was found that the best treatment in this study was at concentration of 140 ppm of *C. odorata* leaves extracts, which resulted in a hatching rate of 76.25% and survival of 90.83%.

1 Introduction

Pangasius sp. is a freshwater fish commodity that is increasingly demand in Indonesia. This fish is generally found in large rivers in Sumatera, Kalimantan, Java and is a freshwater fish that is cultivated intensively by rearing primarily in ponds [1, 2]. However, *Pangasius* hatchery are still facing some problems, one of the obstacles to success in fish farming *Pangasius* sp. is a disease that attacks both adult fish, seeds, even in the egg phase. There is a disease attack caused by fungus during egg hatching has an impact on egg damage that causes a decrease in hatchability [3, 4]. Moreover, [5] mentioned that low hatching rate is a problem that often occurs in *Pangasius* hatcheries. Additionally, previous research emphasized that the fungus that attacked fish eggs in cultivation activities was *Saprolegnia* sp. which can reduce hatching rates [6]. *Saprolegnia* is one of the most common aquatic diseases in freshwater fish and affects all stages of the fish life cycle [7, 8]. The characteristics

* Corresponding author: siska_mellisa@usk.ac.id

of fish eggs attacked by the *Saprolegnia* fungus are that the fish eggs grow white threads like cotton on the surface of the fish eggs, so that the eggs rot and do not hatch [9, 10]. Furthermore, eggs that have been infected with the fungus *Saprolegnia* sp. there will be changes in morphology or growth of fungus on the egg, marked by the growth of fine threads resembling cotton on the surface of the egg in the form of white fibers or what are usually called hyphae [11, 12].

Prevention of fungal attacks on eggs has been tried using chemicals. However, the risk of using chemicals or antibiotics can cause water pollution, because these chemicals are difficult to decompose and they can have dangerous effects on the organism that use them and the environment itself [11, 13]. Therefore, natural ingredients are used that have anti-fungal properties. Besides being cheap, natural materials are also safe for the aquatic environment because they decompose easily. One way to control *Saprolegnia* fungus attacks is to use plant extract that have antifungal properties. Moreover, various types of plant extract have been studied to prevent fungal infections on egg stage such as *Cassia alata* L. [4], tea leaf [3], *Allium sativum* [14-16], *Piper betle* L. [11, 17] and other plants that can be used as an anti-fungal agent in fish eggs namely *Chromolaena odorata*. *C. odorata* are a type of plant from the Asteraceae family. According to phytochemical test, *C. odorata* leaves contain several main compounds such as tannins, phenolics, flavonoids, saponins, steroids, alkaloids, flavanone, essential oils and terpenoids [18-22]. Previous works on *C. odorata* has been carried out particularly on other species such as *Chanos chanos* [23], *Clarias gariepinus* [24]. There is no available information currently about the use of *C. odorata* leaves extracts for fungal prevention on *Pangasius* eggs. Therefore, the objective of this study was to investigate the effect of *C. odorata* leaves extracts on the hatching rate and survival of *Pangasius* eggs.

2 Material and Methods

2.1 Experimental design

This research was conducted at Fish Seed Center, Southwest Aceh Regency in March 2023, using experimental methods which designed as completely randomized design (CRD) consisting of 5 treatment levels and 4 repetitions. The treatments tested in this study were different concentration of *C. odorata* leaves extract, namely treatment A (0 ppm), B (100 (ppm), C (120 ppm), D (140 ppm), and E (160 ppm).

2.2 *Chromolaena odorata* leaves extract preparation

Extracts of *C. odorata* leaves was made based on study by [23] which used maseration method. The *C. odorata* leaves washed and dried, then blended until they became powder, and weighed as a dry weight of 300 grams of flour. *C. odorata* leaves which have become flour were soaked in 96% ethanol for 24 hours and evaporated using a rotary evaporator until they became a filtrate. The filtrate was then stored in the refrigerator.

2.3 Eggs Immersion

The eggs used for this research were obtained from the artificial spawning of *Pangasius* sp. *Pangasius* sp. were reared based on the guidelines of the Institutional Animal Care and Use Committee (IACUC, 2018) and have ethical approval B/034SN/XII/2022 from the animal ethic committee of Marine and Fisheries Faculty, Universitas Syiah Kuala.

The number of eggs used in this research was 2000 eggs, each experimental unit consisted of 100 healthy fertilized eggs. Fertilized eggs are characterized by the presence of a developing nucleus and a transparent color. Meanwhile, unfertilized eggs are white color [25, 14]. The eggs were transferred to each soaking container which had been given *C. odorata* leaves extract according to the specified dose. This soaking was carried out for 30 minutes for each treatment [23]. Next, the eggs are removed and transferred to the hatching container. Then the development of the eggs was measured using a microscope. The embryogenesis process of egg development was observed once every hour.

2.4 Parameters

2.4.1 Hatching Rate (HR)

Hatching Rate (HR) were calculated by following formula [26]:

$$\text{HR}\ (\%) = \frac{\text{Total number of eggs hatched}}{\text{Total fertilized eggs}}\text{x}100 \qquad (1)$$

2.4.2 Larval Survival Rate (SR)

Larval survival (SR) can be determined at the end of the study. Observations of larval survival were carried out for 10 days from the start of larval rearing. The survival rate of larvae can be calculated using the [27] formula as follows:

$$\text{SR}\ (\%) =((\text{No-Nt}))/\text{No x}100 \qquad (2)$$

Where Nt is number of larvae at the end of study and No is initial number of larvae.

2.4.3 Water Quality

The water quality observed was temperature, pH, and DO. Water quality was measured in the morning and evening.

2.4.4 Statistical Analysis

Collected data were analysed with statistical software using one-way analysis of variance (ANOVA). Then, the further test, least significant difference (LSD) was applied when statistical differences were observed ($p < 0.05$). Embryonic development of *Pangasius* was explained descriptively.

3 Result

3.1 Embryonic Development

The results of the development of *Pangasius* sp. eggs soaked using *C. odorata* extract can be seen in Table 1.

Table 1. Observation of embryonic development

Embryonic development	Stages	Time	Explanation
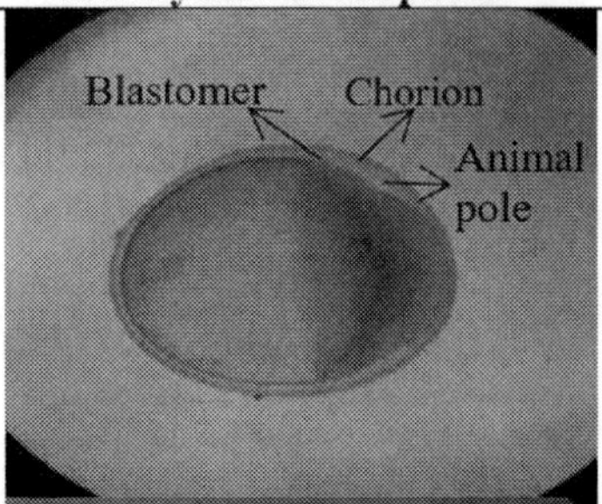	Morula	2nd-4th hour (9 am-11 am)	The formation that occurs after the fertilization phase, then the egg cell undergoes division.
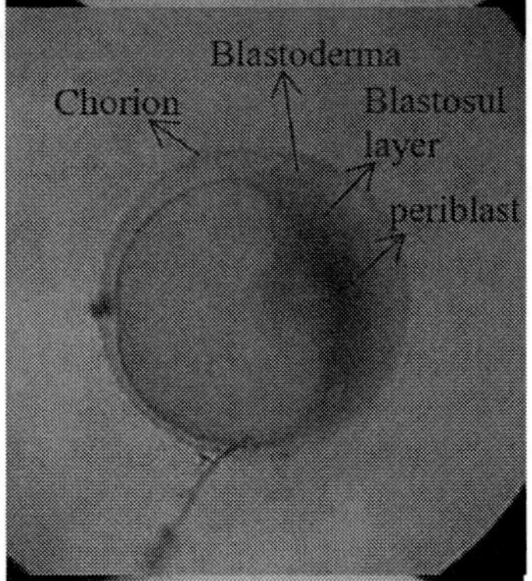	Blastula	5th-6th hour (12 pm-2 pm)	Formation of the blastosul layer
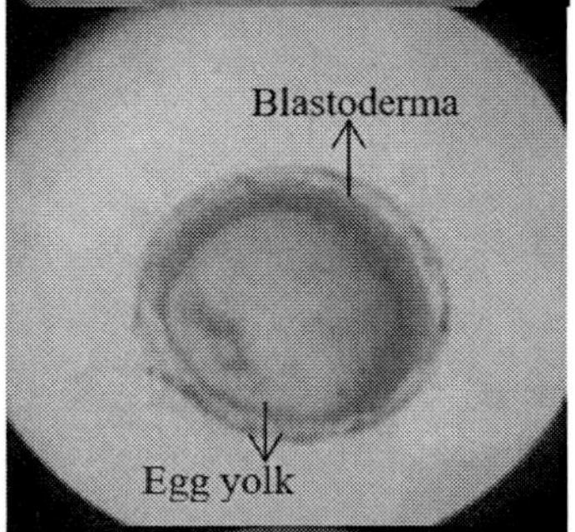	Gastrula	8th-11th hour (3 pm-6 pm)	Closure of the yolk sac by the blastoderma
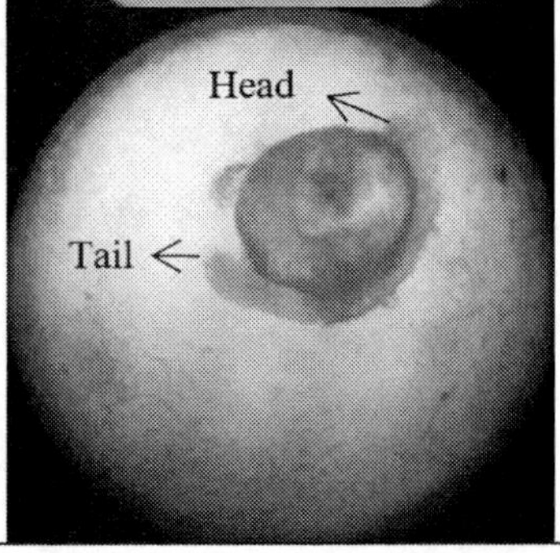	Organogenesis	12th-20th hour (7 pm- 3 am)	The embryo begins to mature, marked by the formation of a head and tail

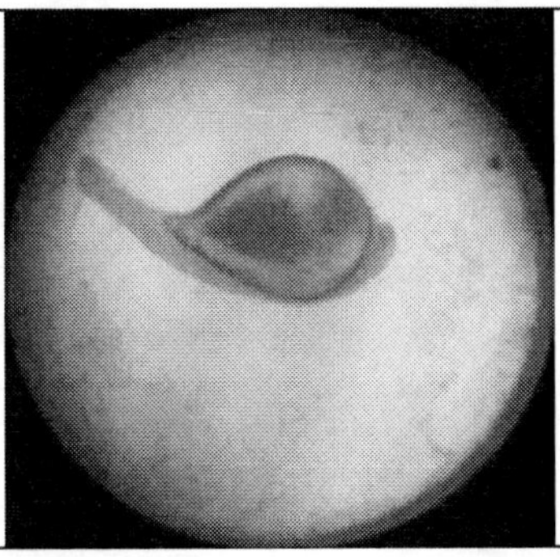	Eggs hatching	21st-24th hour (4 am-7 am)	The process of movement of the embryo out of its shell

3.2 Hatching rate of eggs and Survival of Larvae of *Pangasius* sp.

Use of *C. odorata* leaves extract on this study resulted the hatching rate of eggs ranged from 41.50 ± 3.41% to 76.25 ± 3.86% and survival of larvae ranging from 78.54 ± 6.58% to 90.83 ± 1.68%.

Table 2. Hatching rate and Larval survival

Treatments	**Parameters**	
	Hatching rate (%)	**Larval survival (%)**
A (control)	41.50 ± 3.41^{a}	78.54 ± 6.58^{a}
B (100 ppm)	61.00 ± 3.65^{b}	83.49 ± 5.59^{a}
C (120 ppm)	71.00 ± 6.68^{c}	88.84 ± 4.20^{ab}
D (140 ppm)	76.25 ± 3.86^{c}	90.83 ± 1.68^{b}
E (160 ppm)	57.50 ± 5.00^{b}	86.04 ± 1.76^{a}

Note : The average values in the same column with different superscripts indicate that they are significantly different and the values with the same superscript in the column indicate that they are not significantly different.

3.3 Water Quality

The results of water quality measurements during the research included measurements of temperature, pH and DO. During research, it showed that the water quality was still relatively good.

Table 3. Water quality measurement

Treatment	**Water quality parameters**			
	Suhu (°C)	**pH**	**DO (mg/L)**	
A (Control)	26-28	6,7-7,1	5-6	
B (100 ppm)	26-28	6,8-7,5	5-6	
C (120 ppm)	26-28	6,8-7,5	5-6	
D (140 ppm)	26-28	6,8-7,5	5-6	
E (160 ppm)	26-28	6,8-7,5	5-6	

4 Discussion

4.1 Embryonic development

The development egg stages into a larva starts with the cell division (cleavage) phase of the morula, blastula, gastrula, organogenesis and the egg hatches into a larva. *Pangasius* eggs were observed under a microscope with a magnification of 10x40μ, it showed that the development of embryo had entered the morula phase, because the cell division phase had passed. According to [28], the cleavage phase is characterized by blastocyst formation at the animal pole. Complete blastocyst formation occurs 60 minutes after fertilization. Then this blastocyst will divide to form 2, 4, 8, 16 and 32 cells within 55 minutes, and will enter the morula phase after dividing into 32 cells. Basically, cleavage is a process of cell division during embryonic development, the size of which becomes smaller and smaller over time or becomes small units called blastomeres [29].

The morula is the phase where the blastomeres that are formed will condense at the animal pole to form two cell layers [30]. According to [31] the morula stage is the end of the cleavage phase which will then be continued with the stage of embryonic organ formation. Another opinion stated that the morula phase ends when blastomeres are produced. The blastomeres then condense into small blastodes forming two cell layers and at the end of division two groups will be produced [32]. In this study, the morula phase lasted 3 hours. In the research of [33], this phase lasts 2 hours.

The *Pangasius* embryo continues to undergo cell division until it enters the blastula phase. In this phase the blastomeres divide into smaller ones, so that in the morula stage the blastomeres which were originally solid will form an empty space called the blastosule [30]. [34] stated that the blastomeres moved in both the side of animal pole covering 30-40% area over the yolk sphere. The blastula phase in this study occurred for 3 hours. This is in line with research by [33] which states that this phase lasts 3 hours.

The next phase is the gastrula, according to [30] who said that the gastrula phase is characterized by the process of expanding and closing the egg yolk by the blastoderma towards the blastopore, so that the entire yolk is covered by the blastoderma. [35] stated that the gastrula phase is characterized by the blastoderma covering almost the entire egg yolk, the part of the egg yolk that is not covered is called the blastopore. In this phase, you can see the rotating movement of the cells, the organs of the catfish begin to form in this phase. This phase lasts 4 hours. In the research of [33], this phase lasts 3 hours.

Organogenesis is characterized by the formation of a head and tail. The results of observations in this phase show that there is movement of the embryo. In this study, the organogenesis phase occurred for 9 hours. In the research of [33], this phase occurred lasts 8 hours. The occurrence of the organogenesis phase according to [30] is characterized by the formation of the head and tail, spine segments, eyes, heart, egg yolk pigmentation and hatching into larvae.

The final phase, namely egg hatching, is marked by the embryo actively moving around to change its position, this is because the space for movement is smaller compared to the growing size of the embryo, this phase lasts 4 hours. This is in accordance with what was stated by [36] that when an egg is about to hatch, the embryo often changes its position due to lack of space in its shell. In the research of [33], egg hatching occurred lasts 2 hours. In this study, *Pangasius* eggs hatched for 24 hours, this is in accordance with the statement of [37] which stated that the hatching time for catfish eggs ranged from 20-26 hours. [38] also revealed that eggs hatching of *Pangasionodon hypopthalmus* took 24 hours. However, the length of time for egg hatching in this study is different from the research of [33], namely for 25 hours. According to [39], the newly hatched larvae of Thai *Pangasius* are quite transparent and light yellowish in color with a body length of 2.98–3.10 mm. [40] mentioned that newly hatched larvae of *Pangasius hypophthalmus* ranged from 2.74-3.00 mm total length.

4.2 Hatching rate and larval survival of *Pangasius* sp.

Based on Analysis of Variance (ANOVA) test, use of *C. odorata* leaves extract have a significant effect ($P<0.05$) on hatching rate of eggs and larval survival of *Pangasius* sp. The further test of Least Significant Different (LSD) revealed that treatment D (140 ppm extract of *C. odorata* leaves) had the highest egg hatchability and larval survival of *Pangasius* sp. with values of 76,25% and 90,83% respectively. It was assumed that 140 ppm extract was the optimum concentration for inhibiting growth of fungus on eggs and larvae. Furthermore, this statement is in accordance with [4, 15] that the use of anti-fungal agents at normal concentrations inhibits fungal growth, but at higher concentrations it can damage egg cell tissue. Another statement was revealed by [41] which reported that *C. odorata* leaves contain alkaloids, flavonoids, tannins and saponins as antifungals in fish eggs. Fungistatic antimicrobial substances inhibit the work of certain enzymes which result in disruption of fungal cell metabolism, so that the process of elongation of fungal hyphae (mycelium) is hampered. If the growth of fungal cells, which is characterized by the elongation of hyphae (mycellia), is hampered, the fragmentation of the hyphae becomes disturbed, so it can be said that the fungal cells cannot reproduce. Hyphae or mycelium which cannot experience fragmentation are caused by damage to the hyphal tissue, so that the fungal cells die easily [4]. This showed that *C. odorata* leave extract is able to inhibit the growth of fungus in *Pangasius* eggs.

However, at the highest concentration, treatment E (160 ppm), showed that the egg hatchability value decreased and this was followed by a decrease in larval survival. This is thought to be because the extract concentration is too high, causing the fish eggs to not hatch. This is in line with research by [16], namely the use of garlic extracts on the hatchability of milkfish eggs experienced a decrease in the percentage of egg hatchability (63.66%) at high concentrations (2.500 ppm). Research by [17] explained that the use of *Piper betle* L extract at a high concentration, namely 1.2 g/L, caused a decrease in the hatchability of catfish eggs (50.88%), the higher the concentration of *Piper betle* leaf extract will not only kill the fungus but will also be absorbed by the egg within an intolerable limit so that it is toxic to the egg and causes the egg to not hatch. Moreover, the use of antifungal agents at normal concentrations inhibits the growth of fungi, but at higher concentrations it can damage the chorion layer on the egg, causing the egg to shrink so that it cannot hatch. Apart from that, active compounds such as saponins, which are anti-microbial compounds, can be toxic to organisms at high concentrations. Saponin compounds are also widely used as fish poison [42, 43].

According to the previous study on other species, found that the best concentration of *C. odorata* leave extract was at 120 ppm on *Chanos chanos* with 85% of hatching rate [23], whereas [24] reported that the highest hatching rate of *Clarias gariepinus* was achieved at 200 ppm *C. odorata* leave extract with 80.66% of hatching rate. The difference in optimum concentration obtained on different species.

As shown in Table 2, treatment A as a control without used *C. odorata* leave extract had the lowest egg hatchability, namely 41.50%. It was observed that eggs were attacked by fungus. This is in line with [4, 14] statement, which stated that eggs that were not soaked with antifungal agents accelerated the attack of the *Saprolegnia* fungus and covered the chorion, *Saprolegnia* sp. can weaken the stiffness of the chorion so that it wrinkles and then dies, causing the eggs could not hatch. A previous study of [44] described that other factors that can cause low levels of egg hatching are eggs not developing after being fertilized, changes in the physiological abilities of eggs during embryogenesis. Another statement by

[45] mentioned that not all fertilized eggs will hatch into larvae. Eggs that do not hatch can also be caused by poor condition of the eggs due to the presence of a mixture of water when taking the eggs.

The percentage survival value of catfish larvae in this study was 90.83%. The results in this study were classified as good for all treatments. This statement is in accordance with [46], which stated that the average survival rate for catfish ranged between 70 and 80%. The survival results in this study were not much different from those obtained by [47], namely the percentage of catfish survival obtained was 97.31%. Meanwhile, in research [3], the highest survival was obtained at 89,27%. According to [48] the survival rate of fish is influenced by two factors, namely internal factors and external factors. Internal factors that influence are competitors, parasites, population density, adaptability, and human handling. External factors that influence are the physicochemical properties of waters.

4.3 Water Quality

Water quality parameters are important in the process of hatching eggs and rearing larvae. This is in accordance with the statement of [49] who said that water quality parameters are external factors that influence fish reproduction and egg hatching. The water quality parameters observed during the research were temperature, pH and dissolved oxygen (DO). The average temperature values obtained during the research were 26-28℃, pH 6.8-7.5 and DO 5-6 mg/l (Table 3). The parameter values obtained are in accordance with the statement by [50], who said that in order for catfish eggs to hatch well, the eggs must always be submerged in water and the water temperature must be kept stable in the range of 24-30°C. If the temperature is too low, egg hatching will take longer, while if it is too high, the egg will not hatch. In this study, egg hatching occurred within 24 hours. The pH value obtained in this study is in accordance with [51] statement which states that a good pH for hatching fish eggs ranges from 6.5-8. DO value in this study is in accordance with the statement by [52] which says that dissolved oxygen is good for hatching fish eggs, namely >5 mg/l.

Conclusion

Use of *Chromolaena odorata* leaves extract affects the hatchability and larval survival of *Pangasius* sp. The highest percentage of egg hatchability and larval survival were obtained at 140 ppm extract, namely 76.25% and 90.83%, respectively.

References

1. R. M. Harmain, F.A. Dali. Buku Ajar Ilabulo Ikan Patin (*Pangasius*, sp.). UNG Press Gorontalo. Gorontalo (2017)
2. P.N. Trang, T.T.A. Ngoc. Heliyon 9, 1-8 (2023)
3. Muhlis, M. Ghofur, M. Sugihartono. Jurnal Akuakultur Sungai dan Danau 4 (1), 9-14 (2019)
4. A. T. Bowo, Sunarto, Rachimi. Jurnal Ruaya 1 (1) (2014)
5. P. L. Putra, D. Jubaedah, M. Syaifudin. Jurnal Akuakultur Rawa Indonesia 8 (1), 37-49 (2020)
6. M.N. Lingga, I Rustikawati, I. D. Buwono. Jurnal Perikanan dan Kelautan 3 (4), 75-80 (2012)
7. S. Liu, P. Song, R. Ou, W. Fang, M. Lin, J. Ruan, X. Yang, K. Hu. Aquaculture and Fisheries 2, 227-233 (2017)

8. S. M. Aly, M. A. Elatta, A. A. Nasr, M. Fathi. Aquaculture and Fisheries (In Press)
9. D. R. Widya, D. Suryanto, Desrita. Jurnal Mikrobiologi 2 (1), 7-17 (2014)
10. A. Zubaidah, S. Samsundari, M. A. Q. Jaelani. Jurnal Akuakultur Rawa Indonesia 9 (2), 109-117 (2021)
11. F. Fanitalya, A. A. Damayanti, S. Sudirman. Jurnal Perikanan Unram 1 (1), 22–29 (2018)
12. A. H. Rukka, S. F. Mangitung, A. Fauzan. J. Agrisains 23 (2), 67-76 (2022)
13. M. Muhardisah. Arwana : Jurnal Ilmiah Program Studi Perairan 2 (1), 63-72 (2020)
14. M. Khasanah, R. A. Nugroho, T. Susilowati. Jurnal Sains Akuakultur Tropis 6 (2), 236-245 (2022)
15. F. Rahmayanti, F. Diana, S. Rosa. Jurnal Akuakultura 1 (1), 19-23 (2017)
16. F. Saputra, T. R. Effianda, S. A. E. Rahimi, N. Nurfadillah. Jurnal Akuakultura 2 (1), 9-18 (2017)
17. Rivanto, I. Sidabalok, H. Hasan. Jurnal Ruaya 1 (1), 17-22 (2014)
18. A. Sirinthipaporn, W. Jiraungkoorskul, Pharmacognosy review 11 (21), 35-38 (2017)
19. Fadia, Nurlailah, T. E. Herlina, L. Lutpiatina. Jurnal riset kefarmasian Indonesia 2 (3), 158-168 (2020)
20. D. Panda, S. K. Dash, G. K. Dash. International Journal of Pharmaceutical Sciences Review and Research 1 (2), (2010)
21. S. E. Ukwueze, O. M. Duru, O. A. Shorinwa. Indo American Journal of Pharm Research 3 (4), (2013)
22. Ernawati, N. Jannah, Jurnal Kedokteran dan Kesehatan 17 (2), 137-144 (2021)
23. Evendi, S. Karina, D. F. Putra. Jurnal Ilmiah Mahasiswa Kelautan dan Perikanan Unsyiah 2 (1), 33-40 (2017)
24. H. Harun, M. Abdan. Jurnal Pendidikan dan Pengabdian Vokasi 2 (1), 115-119 (2021)
25. J. Subagja, R. Gustiano. Jurnal Riset Akuakultur 1, 219-225 (2006)
26. M. I. Effendi. Biologi Perikanan. Yayasan Pustaka Nusantara. Bogor (1979)
27. M. I. Effendi. Pengantar akuakultur. Penebar Swadaya. Jakarta (2004)
28. D. Nugraha, M. N. Supardjo, Subiyanto. Journal of Management of Aquatic Resources 1 (1), 1-6 (2012)
29. D. R. Affandi, S. Sjafei, M. F. Rahardjo, Sulistiono. Fisiologi ikan, pencernaan dan penyerapan makanan. Managemen Sumberdaya Perairan. FPIK IPB. Bogor (2015)
30. Ardhardiansyah, U. Subhan, A. Yustiati. Jurnal Perikanan dan Kelautan 7 (2), 17-27 (2017)
31. L. Budianita, M. Junaidi, Nurliah. Jurnal Perikanan 9 (1), 7-16 (2019)
32. A.R. Redha, E. I. Raharjo, H. Hasan. Jurnal Ruaya 4 (1), 1-7 (2014)
33. L. Sitinjak, S. Y. H. Purba, D. Y. Jurnal Penelitian Terapan Perikanan dan Kelautan 1 (2), 76-79 (2019)
34. S. Ferosekhan, S.K. Sahoo, S.S. Giri, A. Saha, M. Paramanik. Journal of Aquaculture Research and Development 6 (6), (2015)
35. Z. Alfath, F. Basuki, R. A. Nugroho. Jurnal Sains Akuakultur Tropis 4 (2), 129-138 (2020)
36. M. I. Effendie. Biologi perikanan. Yayasan Pustaka Nusatama. Yogyakarta. 48-67 (1997)

37. Standar Nasional Indonesia (SNI) 01-6483.4-2000
38. N. K. P. Astuti, A. Tarmizi, M. Ikromin. Jurnal Ganec Swara 17 (3), 785-790 (2023)
39. A. Islam. Development Growth and Differentiation 47 (1), 1-6 (2005)
40. A. P. Baidya, S. Senoo. Suisanzoshoku 50 (4), 423-432 (2002)
41. M. Huzni, B. T. Rahardo, H. Tarno. Jurnal HPT 3 (1), 93-101 (2015)
42. M. Musman. S. Karina, C. N. Defira, N. Fadhillah, A. Kayana, N. Hasballah, A. R. Faunanda, R. Putra. International Journal of sciences: Basic and Applied Research (IJSBAR) 21(1), 78-85 (2015)
43. D. Gunawan, S. Mulyani. Ilmu obat alam (Farmakognosi). Penebar Swadaya. Jakarta. (2004)
44. Z. Jr. Muhammad, R. K. Sari, M. Jurnal Akuakultur Indonesia 4 (2), 103-108 (2005)
45. B. Setyono. Jurnal GAMMA 5 (1), 1-12 (2009)
46. Standar Nasional Indonesia (SNI) 01-6463.3.2002
47. Patricius., Rachimi, E. Prasetio. Borneo Akuatika 1 (1), 18-24 (2019)
48. M. I. Effendi. Biologi Perikanan. Yayasan Pustaka Nustama. Yogyakarta (2002)
49. R. Ulfani, C. N. Defira, Hasanuddin. Jurnal Ilmiah Mahasiswa Kelautan dan Perikanan Unsyiah 3 (1), 135-142 (2018)
50. H. Susanto, K. Amri. Budidaya Ikan Patin. Penebar Swadaya. Jakarta (2002)
51. I. Reny. Jurnal Berkala Ilmiah Perikanan 3 (1), 13 (2008)
52. V. O. Manantung, H. J. Sinjal, R. Monijung. Jurnal Budidaya Perairan **3** (1), 14-23 (2013)

Additional of crocot flour (*Portulaca oleracea l.*) and Mannanoligosaccharides (MOS) in feed for improving growth performance windu shrimp (*Penaeus monodon*)

Cut Dara Dewi[1*], *Cut Widya* Aprilia[1], *Siska* Mellisa[1,2], and *Mahdaliana* Mahdaliana[3]

[1]Department of Aquaculture, Faculty of Marine and Fisheries, Universitas Syiah Kuala, Banda Aceh 23111, Indonesia
[2]Marine and Fisheries Research Center, Universitas Syiah Kuala Banda Aceh 23111, Indonesia
[3]Department of Aquaculture, Faculty of Agriculture, Universitas Malikussaleh

Abstract. This study aims to determine the effect of giving purslane flour and MOS to feed on increasing growth performance in Windu shrimp (*Paneus monodon*). The research was carried out at the Balai Perikanan Budidaya Air Payau Ujung Batee (BPBAP) in Ujung Batee. This study was conducted using a Complete Randomized Design (CRD) method with 4 treatments used, namely A (control), B (30 g/kg purslane flour and 10 g/kg MOS), C (40 g/kg purslane flour and 10 g/kg MOS), and D (50 g/kg purslane flour and 10 g/kg MOS). The results showed that the addition of 40 g/kg purslane flour and 10 g/kg MOS resulted in an absolute weight growth of 0.60±0.05 grams, a specific growth rate 16.00±0.24% per day, an absolute length growth of 4.33±0.06 cm, a feed conversion ratio of 0.52±0.03, feed efficiency of 192.08±12.21%, moulting frequency of 1.50 times per head, and a survival rate of 93.20±9.33%.

1 Introduction

Windu Shrimp *(Penaeus monodon)* is one of the leading commodities in Asia. This is a result of the various benefits that Windu Shrimp offer, such as their higher harvest size, savory and sweet flavor, and high nutritional value [1]. Another favorable trait is its resistance to temperature changes known as eurythemal [2]. This commodity is known to have a high economic value compared to several other commodities, both for local consumption and for export markets. Windu Shrimp farming has been widely practiced in several regions of Indonesia still faces some obstacles. The main problem encountered in shrimp farming activities is the low growth rate [3] Some factors that affect the growth and survival of Windu Shrimp include seed quality, feed type, water quality, disease, and moulting success [4]. Stated that during the moulting process, shrimp absorb calcium and magnesium. The content of these substances is needed in high quantities because this molting is an indicator of growth.

One solution that can be done to support growth and increase Windu Shrimp production is the addition of purslane flour and MOS to feed as an alternative raw material. Purslane leaves and stems contain a high potassium content of 46,000 and 68,600 mg/kg, respectively; magnesium averages 4400 mg/kg; and calcium 25,400 and 60,000 mg/kg [5], the content contained in purslane plants is as functional as the formation of the exoskeleton in shrimp during the moulting process. According to [6], purslane has a protein content of 2-2.5%, which is useful for supporting shrimp growth.

* Corresponding author: cutdd@usk.ac.id

In order to optimize digestibility in shrimp, it is necessary to add supplements to the feed such as the addition of prebiotics. One of the prebiotics that has been studied and applied in aquaculture is mannanoligosaccharide (MOS) [7]. Prebiotic mannanoligosaccharide (MOS) is a prebiotic that is able to spur the growth of beneficial microflora in the gastrointestinal tract so as to increase the growth of organisms [8]. The provision of MOS has so far been applied to several types of shrimp, one of which is vaname shrimp. [9,10] reported that mannanoligosaccharide (MOS) has been tested to improve the growth and survival of vaname shrimp juveniles. [11] reported that giving 1% MOS can effectively improve the growth performance of vaname shrimp. [12] also stated that the provision of prebiotics at a dose of 0.8% can effectively improve the immune response and growth performance of whiteleg shrimp, and the addition of prebiotics is better and safer than antibiotics to be applied in aquaculture because it does not cause residues. Therefore, the addition of purslane flour and MOS to Shrimp shrimp feed is expected to increase growth in these shrimps.

Given above, the research was carried out to determine the effect of giving of purslane flour and MOS to commercial feed will have a significant effect on the growth performance, survival rate, feed conversion, feed efficiency and moulting frequency of shrimp fry.

2 Material and Method

2.1 Location and time of research

The study was conducted from February to March 2023. The research was conducted at the Balai Perikanan Budidaya Air Payau (BPBAP) in Ujung Batee, District Aceh Besar. Experimental Shrimp and Research Containers.

2.2 Experimental Shrimp and Research Containers

The test shrimp used were PL 12 stage shrimp prawn fry obtained from the Balai Perikanan Budidaya Air Payau (BPBAP) Ujung Batee, District Aceh Besar. As many as 900 individuals with each treatment container containing 45 (3 individuals/L) Shrimp prawn fry. The selected fish were allowed to adapt before being treated for approximately 24 hours. The research containers consisted of 20 units with a volume of 15 liters of water and were placed at random.

2.3 Making Purslane Flour

The process of making purslane flour begins with collecting purslane plants. Then the collected purslane is washed thoroughly with running water, separating unwanted parts such as the roots and flowers. The drying process was carried out with the help of an oven at 500°C for 60 minutes. The dried purslane is then ground with a blender and sifted through a sieve for smoother results.

2.4 Preparation of Test Feed

The feed to be used in this study is commercial feed with a protein content of 40%. Furthermore, the feed is added with CMC at a dose of 5%. Then it was added to purslane flour that has been weighed according to the dose of each treatment. Then, MOS was added, which has been weighed according to the treatment dose. Furthermore, all the ingredients are stirred until evenly mixed and then dried by aerating. After the feed is dry, it is put in a container and stored until use.

2.5 Experimental Design and Fish Rearing

The research method used was a Completely Randomized Design (CRD) consisting of four treatments with five replications. The manipulated factors were the addition of purslane flour and MOS. Determining the dose used as treatment refers to the dose modification used by [4],[9]. The treatments given to Windu shrimp fry were, A = 0 ml/kg feed, B = purslane flour 30g/kg+MOS 10g/kg, C = purslane flour 40g/kg+MOS 10g/kg and D purslane flour 50g/kg+ MOS 10g/kg.

Before being put into the rearing container, the Windu shrimp prawn fry were counted, weighed, and recorded as initial research data. Windu shrimp were reared based on the guidelines of the Institutional Animal Care and Use Committee (IACUC, 2018) and have ethical approval B/034SN/XII/2022 from the animal ethic committee of the School of Marine and Fisherie Faculty, Universitas Syiah Kuala. During the rearing period, the Shrimp prawn fry were fed five times a day, starting at 06.00, 10.00, 14.00, 18.00, and 22.00 WIB, with feed as much as 5% of the body weight of the Windu shrimp fry throughout a rearing period of 40 days.

Growth data was collected by calculating the weight of the fish and the length of the fish. During the sampling period, the number of fish was regularly counted to calculate survival and overall weight (biomass) of fish that lived and died in each period in order to measure feed efficiency.

Data were collected once every 10 days, on days 0, 10, 20, 30, and 40, by weighing the biomass from each research container. Parameters observed were absolute length gain, absolute weight gain, specific growth rate, feed conversion ratio, feed efficiency, and survival. Then, the observations of water quality observed in this study includes temperature, pH, and dissolved oxygen. These parameters were measured every day during the study.

2.6 Test parameters

Growth performance parameters include Specific Growth Rate (SGR), absolute length gain, absolute weight growth, and Feed Conversion Ratio (FCR) according to [13], feed use efficiency (EP) according to [14], Moulting frequency according to [15], and survival rate (SR) according to [16]. The equation is as follows:

$SGR = (\ln Wt - \ln Wo)/t \times 100\%$
$PPM = Lt - L0$
$W = Wt - Wo$
$FCR = F/Wt-W0 \times 100$
$EP = Wt-W0/F \times 100\ \%$
$MFq = Xmolt/Ntot$
$SR = Nt/No \times 100\%$

Note: PPM = Absolute length growth (cm); Lt = The average length of the end of the study (cm); Lo : The average length of the start of the study (cm); W = Weight gain (gram); Wt = Weight of biomass at the end of the study (gram); W0 : Weight of biomass at the beginning of the study (gram); t = Time (maintenance period); F = Amount of feed consumed (gram); Xmolt = Overall number of moulting shrimp (time); Ntot = number of Shrimp prawns used in the study; Nt = Number of live fish at the end of rearing; N0 = Number of fish that live at the beginning of maintenance.

2.7 Data analysis

The research data were analyzed by one-way analysis of variance (ANOVA). Significant results of ANOVA were followed up with Duncan's multiple distance test (P<0.05).

3 Result

The 40-day research's showed that MOS and purslane flour considerably the growth performance of Windu shrimp fry. The parameters of absolute weight growth, specific growth rate, absolute length growth, feed conversion ratio, feed utilization efficiency, and moulting frequency were all affected by the addition of purslane flour and MOS, but had no effect on survival rate. This is shown in Table 1 down below:

Table 1. Measurement results of absolute weight growth, absolute length growth, specific growth rate (SGR), feed conversion ratio (FCR), feed efficiency (EP), moulting frequency, and survival rate (SR).

Parameters	**Treatments of purslane flour and MOS**			
	A (0g/kg)	**B (30g+10g/kg)**	**C (40g+10g/kg)**	**D (50g+10g/kg)**
Absolute Weight Growth (G)	0.36 ± 0.06^{a}	0.47 ± 0.00^{b}	0.54 ± 0.00^{c}	0.60 ± 0.05^{d}
Specific Growth Rate (%/Day)	14.71 ± 0.43^{a}	15.43 ± 0.05^{b}	15.75 ± 0.02^{bc}	16.00 ± 0.24^{d}
Absolute Length Growth	3.46 ± 0.17^{a}	3.77 ± 0.01^{b}	3.96 ± 0.05^{c}	4.33 ± 0.06^{d}
Feed (Cm)Conversion Ratio	0.75 ± 0.19^{a}	0.58 ± 0.02^{b}	0.57 ± 0.03^{b}	0.52 ± 0.03^{b}
Feed Efficiency (%)	139.76 ± 36.57^{a}	170.85 ± 7.44^{b}	175.36 ± 10.21^{b}	192.08 ± 12.21^{b}
Moulting Frequency (Time/Tail)	0.69 ± 0.02^{a}	0.91 ± 0.15^{b}	1.23 ± 0.17^{b}	1.50 ± 0.23^{c}
Survival Rate (%)	88.40 ± 9.07^{a}	89.20 ± 7.79^{b}	90.60 ± 9.73^{a}	93.20 ± 9.33^{a}

Note: Significant differences (P<0.05) are indicated by different superscript letters in the same column.

The ANOVA (Analysis of Variant) test results showed that while the addition of MOS and purslane flour in the feed had no significant effect (p>0.05) on survival, it did have a significant effect (P<0.05) on the parameters of absolute weight growth, specific growth rate, absolute length growth, feed conversion ratio, feed efficiency, and moulting frequency

The highest absolute weight growth of Windu Shrimp fry was obtained in the D treatment (purslane flour 50 g/kg + MOS 10 g/kg) with a value of 0.60 ± 0.05 grams and the lowest was obtained in the A treatment which was 0.36 ± 0.06 grams. Based on the results of the BNT further test showed that the D treatment was significantly different from all treatments. The specific growth rate value in the D treatment (purslane flour 50 g/kg + MOS 10 g/kg) had the highest value of 16.00 ± 0.24% / day and the lowest was obtained in the P0 treatment with an average value of 14.71 ± 0.43% / day. Based on the results of the BNJ further test showed that the D treatment (purslane flour 50 g/kg + MOS 10 g/kg) was significantly different from all treatments. The highest absolute length growth of Windu Shrimp fry was obtained in the D treatment feed with a value of 4.33 ± 0.06 cm and the lowest value was obtained in the A (control) treatment feed which was 3.46 ± 0.17 cm. Based on the results of the BNJ further test showed that the D treatment was significantly different from all treatments. The best feed conversion ratio was obtained in the D treatment (purslane flour 50 g/kg + MOS 10 g/kg) with a value of 0.52 ± 0.03, and the highest in the P0 treatment with a value of 0.75 ± 0.19 Based on the results of Duncan's further test, it shows that the D treatment is significantly different from the P0 treatment (control) but not significantly different from the B and C treatments.

The highest feed efficiency was obtained in the D treatment (purslane flour 50 g/kg + MOS 10 g/kg) with a value of 192.08 ± 12.21% and the lowest was obtained in the A treatment (control) with an average value of 139.76 ± 36.57%. Based on Duncan's further results showed that the D treatment was significantly different from the A treatment (control) and but not significantly different from B and C. The highest moulting frequency was obtained in the D treatment (purslane flour 50 g/kg + MOS 10 g/kg) with a value of 1.50 ± 0.23 times/head and the lowest was obtained in the A treatment (control) with an average value of 0.69 ± 0.02 times/head. Based on the results of Duncan's further showed that the treatment D is significantly different from the other treatments. Windu Shrimp survival ranged from 88.40%-93.20%. Although each treatment was not significantly different, but each treatment showed an increase as the dose given increased and the best value was in D (purslane flour 50 g/kg + MOS 10 g/kg).

The results of water quality measurements during the study include physical and chemical parameters and show that water quality is still in the normal range. The results of water quality tests during this research process can be seen in table 2 below:

Tabel 2. Water quality parameters were measured during the study

Chemical physics parameters	Measurement data	Literature data	Source literature
Temperature (^{0}C)	29	28-30	[17]
DO (mg/L)	7,5-7,7	4-8	
pH	7,27-7,28	7,5-8,5	
Salinity	18	15-20	

4 Discussion

Growth can be increased by feeding high-protein foods. Low protein levels in feed can affect shrimp growth [18]. The results showed that the D treatment with a dose of 50 g/kg purslane flour and 10 g/kg MOS had the highest results in absolute weight growth (0.60 grams) and specific growth rate (16.00%/day). It is suspected that the test shrimp in D were able to utilize the nutrients in the feed well compared to the other treatments. The more feed that can be absorbed, the more protein that can be metabolized in order to produce energy. If the energy produced exceeds the need for body maintenance, the more energy will be used for growth [19]. This is in accordance with the statement of [4] that, in his research, showed a significant effect ($p < 0.05$), and the addition of purslane flour 40 g/kg in vannamei shrimp feed also gave the highest absolute weight of 1.08 ± 0.25 grams compared to control and other treatments. The increase in the growth rate of Windu Shrimp fry was also due to the provision of purslane flour and MOS. This indicates that the application of purslane flour and MOS can increase the utilization of nutrients for growth. The addition of purslane flour and MOS in artificial feed as a whole caused an increase in the specific growth rate of Windu Shrimp fry, which, based on the results of the BNJ further test in this study, showed significantly different results

(P <0.05). Feeding with the addition of purslane flour and MOS was able to increase the specific growth rate of Windu Shrimp fry to a value of 16.00% / day. The same thing also happened to the absolute weight growth of Windu Shrimp fry fed with the addition of 50 g/kg purslane flour and 10 g/kg MOS. There was an increase in absolute weight of 0.60 grams when compared to the absolute weight of shrimp that were not given purslane flour and MOS (control). The increase in specific growth rate and absolute weight of Windu Shrimp fry fed with purslane flour and MOS is thought to be because MOS is a food that cannot be digested but is able to increase the growth and activity of beneficial microflora in the shrimp digestive tract.

Adding MOS to feed can increase its digestibility, which in turn affects the growth and survival of shrimp larvae [20]. The same statement was also stated by [21]: the adding prebiotics to feed can enhance the host's gastrointestinal microflora so as to spur growth. The addition of purslane flour and MOS in artificial feed as whole causes an increase in absolute length growth in Windu Shrimp fry, as can be seen in Table 1. that the absolute length growth in each treatment is significantly different (P <0.05), the highest length treatment is in the D treatment with a dose of 50 g/kg purslane flour and 10 g/kg MOS, and the area with the least absolute length growth is the A treatment with a dose of 0 (control). The provision of purslane flour and MOS in different doses had a significant effect on the length growth of Windu Shrimp fry. Under optimal environmental conditions and adequate nutritional needs, the seeds will grow optimally, and vice versa.

The best feed conversion ratio results were found in the D treatment with a dose of 50 g/kg purslane flour and 10 g/kg MOS, with an average value of 0.52 ± 0.03. The provision of purslane flour and MOS with different doses had a significant effect on the feed conversion ratio of Windu Shrimp fry. This is in accordance with the opinion of [22], who states that the value of the feed conversion ratio is small. It can be interpreted that the amount of feed consumed is well met for growth, so the value of the feed efficiency given is good.

The highest feed efficiency results in this study were found in D with a dose of 50 g/kg purslane flour and 10 g/kg MOS with a value of 192.08 ± 12.21%. The high feed efficiency in D indicates that the feed can be utilized well by shrimp. This is in accordance with what is stated by [23]: high feed efficiency indicates the use of feed is also efficient in growth, so that only a little protein will be broken down to meet energy needs and the rest for growth.

The addition of purslane flour in feed can accelerate the moulting of Windu Shrimp fry. The lower the dose of purslane flour given, the fewer shrimp moulted, and it depends on the amount of calcium content in the flour. The calcium and protein content in purslane flour can affect carapace formation in shrimp during moulting [4]. Based on the results of the research that has been done, the highest moulting frequency is obtained in the P3 treatment with a dose of 50 g/kg purslane flour and 10 g/kg MOS with an average value of 1.50 ± 0.23 times/head. The high number of shrimp moulting is due to the amount of protein and calcium content in D of 45.06% and 4.1% respectively. This is clarified by [24]. Overall, giving 4% calcium with high feed protein (45%) can increase calcium stores in body tissues, although directly aquatic animals can absorb calcium from the media environment. Increased calcium retention indicates that feed protein and calcium treatments have an effect mainly on maintaining calcium homeostasis in shrimp.

Based on the results of this study, the value of the survival range of Windu Shrimp fry in a row starting from treatments A, B, C, and D is 88%, 89%, 90%, and 93%. Based on the ANOVA test, the addition of purslane flour and MOS in the feed did not have a significant effect (p>0.05) on the survival of Windu Shrimp fry. This is because there is no significant difference between the treatment of the addition of purslane flour and MOS with the control treatment, but the survival rate of Windu Shrimp fry during this study was classified as good. This is in accordance with the opinion of [25], which states that the specific growth rate is categorized as good if the specific growth rate value is> 70%, for a medium specific growth rate of 50-60%, and a low specific growth rate value < 50%. This is because the appropriate feeding ensures that the availability of shrimp feed is sufficient, and the value of good water quality during the study greatly supports the survival of the Windu Shrimp fry. This is in accordance with the statement in [26], which states that the factors that most influence the survival of shrimp are management in feeding and good water quality management in the maintenance medium. The death of shrimp during this study is thought to be due to stress due to lack of oxygen during the study. This is in accordance with the statement [27], which states that the level of shrimp consumption will decrease if the need for oxygen in the water is not fulfillet resulting in decreased health conditions in shrimp that will cause death.

Based on the results of protein and calcium tests of purslane flour, it shows that purslane flour has a protein content of 40.75% and calcium 2.4%, from the results of protein and calcium tests that have been carried out on each treatment feed, the best protein and calcium values for Windu Shrimp fry growth are shown in the D treatment with a dose of 50 g / kg purslane flour and 10 g / kg MOS with a content value of 45, 06% and 4.1%, followed by treatment with a dose of 40 g / kg purslane flour and 10 g / kg MOS with a value of 44.84% and 3.5%, then A treatment with a dose of 30 g / kg purslane flour and 10 g / kg MOS with a value of 43.27% and 3.1% and the last at P0 treatment (control) with a protein content of 40%.

During the 40-day research period, water quality measurements were taken, including temperature, DO, pH, and salinity. The temperature in this study was 29^{0}C, DO ranged from 7.5-7.7 mg/l, pH ranged from 7.27-7.28 and salinity was 18 ppt. Based on the results of the water quality measurements obtained, it can be seen that the percentage of water quality is still in a safe range for shrimp survival and shrimp growth. As stated by [28], a good temperature for the life of widu shrimp is in the range of 28-30^{0}C, DO 4-8, pH 7.5-8.5 and salinity of 15-25 ppt. The checking of water quality parameters is done every 10 days to keep water quality under control.

5 Conclusions

Based on the research that has been done, the conclusion is that the addition of purslane flour with different doses and the addition of MOS 10 g/kg into the feed has a significant effect on absolute weight growth, absolute length growth, specific growth rate, feed conversion ratio, feed efficiency, and moulting frequency but does not have a significant effect on the survival of Windu Shrimp fry. The results during the study showed that the best treatment in all parameters was obtained in the D treatment (purslane flour 50 g/kg + MOS 10 g/kg) with an absolute weight growth value of 0.60 grams, a specific growth rate of 16.00%/day, an absolute length growth of 4.33 cm, a feed conversion ratio of 0.52, a feed efficiency of 192.08%, a moulting frequency of 1.50 times/head and a survival rate of 93.20%.

References

1. Amri, K. 2003. Budidaya udang windu secara intensif. Agromedia Pustaka, Jakarta. 98 hlm.
2. Suyanto, R. dan Mujiman A. 2004. Budidaya udang windu, Penebar Swadaya. Jakarta. 19: 207-211.
3. Darmono. 1993. Budidaya udang Penaeus. Penerbit Kanisius, Yogyakarta.
4. Prabowo, A., Rusliadi, N.A. Pamukas. 2019. *J. Aquaculture Tecnology Laboratory Marine And Fishery*. Universitas Riau. Pekan Baru. Hal 1-13.

5. Gonnella, M., & Aldo. B. 2010. *The European Journal of Plant Science and Biotechnology*. 4(1): 131-136.
6. Rahardjo, M., 2007. *Warta Penelitian Dan Pengembangan Tanaman Industri*. 13(1): 1-4.
7. Ringo E., Olsen R. E., Gifstad T. Ø., Dalmo A. M. 2010. J.Aquaculture Nutrition 16(2): 117-136.
8. V. Muhardina, P.M. Sari, Y. Aisyah, S.Haryani, S.A. Akbar, Rasayan J. Chem **13(1)**, 240-244(2020)
9. S.A. Akbar, R.R. Fazli, R. R, Meta-Analysis: The effectiveness of artificial diets to increase on survival rate of vannamei shrimp (Litopenaeus vannamei). In *IOP Conference Series: Earth and Environmental Science,* IOP Publishing. **1221(1)** 012066 (2023)
10. Zhang, J., Liu Y., Tian L., Yang H., Liang G., Xu D. 2012. *Fish & Shellfish Immunology,* 33(4) :1027-1032.
11. Hamsah, Darmawati, dan S. Nurhijrah. 2020. *Jurnal Ilmu Perikanan*. 9(2): 81-87.
12. Widanarni, M. Yuhana, Linuwih A.P. 2018. *Jurnal Akuakultur Indonesia*. 17(1): 81-86.
13. Effendie, M. I. 1997. Biologi perikanan. Yayasan Pustaka Nusantara. Yogyakarta. 163 hlm.
14. Tacon, A.G.J. 1987. The nutrition and fedding of farmed fish and shrimp-A training mannual. FAO of the United Nation, Brazil. 2: 106-109
15. Handayani, L., Syahputra, F., 2018. *Depik*, 7(1): 76–83.
16. Effendie, M.I. 1999. Pengantar akuakultur. Penebar Swadaya, Depok. 216 hlm.
17. Soetomo, H.A.1990. Teknik budidaya udang windu. Sinar Baru. Bandung. Hal 179-180.
18. Hanafiah, K. A., 2002. Rancangan percobaan teori dan aplikasi. PT Rajagafindo Persada. Jakarta. 2: 203-238.
19. Darwatin, K.,S. Romziah, M. Gunanti. 2016. *Jurnal Biosains Pascasarjana*, 18(2): 123.
20. Mukti, R., D. Yonarta, A. Pangawikan. 2019. *Depik*, 8(1): 18-25.
21. Hamsah, Widanarni, Alimuddin, Yuhana M, Junior MZ. 2017. *AACL-Bioflux*.10(1): 8-17.
22. Gibson, G.R, Robert HM, Loo JV, Rastall RA, Roberfroid MB. 2004. *Nutrition Research Reviews*. 17: 259-275.
23. [Ihsanudin, I., S. Rejeki, T. Yuniarti. 2013. *Journal of Aquaculture Management and Technology*, 2 (3): 76-85.
24. Masitoh, D., Subandiyono, Pinandoyo. 2015. *Journal of Aquacultur Management and Technology,* 4(4): 95-100.
25. Davis, D.A. and D. Gatlin III. 1991. *Procedings of the Aquaculture Feed Processing and Nutrition Workshop.* American Soybean Association. 23: 49-67.
26. Widigdo, B. 2013. Bertambak udang dengan teknologi biocrete. Kompas Media Nusantara: Jakarta. 22(6): 1-104.
27. Yustianti, M. N. Ibrahim, dan Ruslaini. 2013. *Jurnal Mina Laut Indonesia*, 01(01): 93-103.
28. Budiardi, T., T.Batara dan D. Wahjuningrum. 2005. *Jurnal Akuakultur Indonesia*, 4(1): 89-96.

An update review: several extraction methods for collagen isolation in vertebrate fish

Nadia Isnaini[1,2,3], *Vicky* Prajaputra[2,3,4*], *Siti Khalila* Luthfanna[1], *Irma* Sari[1]

[1] Department of Pharmacy, Faculty of Mathematics and Natural Sciences, Universitas Syiah Kuala, 23111 Banda Aceh, Indonesia
[2] ARC-PUIPT Nilam Aceh, Universitas Syiah Kuala, 23111 Banda Aceh, Indonesia
[3] Research Center for Marine Sciences and Fisheries, Universitas Syiah Kuala, 23111 Banda Aceh, Indonesia
[4] Department of Marine Sciences, Faculty of Marine Sciences and Fisheries Universitas Syiah Kuala, 23111 Banda Aceh, Indonesia

Abstract. Fish are vertebrate animals with a backbone that live in marine environments, presenting a potential source of collagen due to their high protein content. Collagen is a structural protein in the body's connective tissue, contributing 25 to 30% of the total body protein. Collagen sources in fish can be found in the skin, scales, and bones. The isolation of collagen from fish involves several stages, including separation and cleaning, size reduction, removal of non-collagenous components, collagen extraction, and precipitation. The extraction method utilized in collagen isolation impacts the quantity of collagen produced. The aim of this literature review is to identify extraction methods applicable for collagen isolation in vertebrate fish, serving as a reference for selecting extraction methods in future research. The updated review method involved collecting journal data from various literature sources, obtained from national and international journals discussing collagen extraction methods in vertebrate fish. Based on the update review, the isolation of collagen from vertebrate fish is carried out using several extraction methods, including acid soluble collagen extraction, pepsin soluble collagen, water soluble collagen, organic acid ultrasound-assisted, extrusion hydro extraction, deep eutectic solvent, and supercritical fluid extraction.

1 Introduction

Collagen is the primary protein found in animal connective tissues, accounting for approximately 25% to 30% of the total body protein. Collagen plays a significant role in various aspects of our daily lives, including biomedical applications, pharmaceuticals, the food industry, the pharmaceutical industry, and cosmetics. Collagen is naturally produced within the body by fibroblast cells. However, the body's ability to produce collagen decreases with age and due to factors such as poor lifestyle choices. Therefore, there is a need for external sources of collagen. Collagen can be extracted from various animal sources, with the current market primarily relying on collagen sourced from pigs and cows. However, this raises concerns about halal certification for pig-derived collagen and potential disease transmission for collagen sourced from cows. An alternative source of collagen is fish. Fish are cold-blooded vertebrate animals that inhabit aquatic environments. Common sources of fish collagen include bones, skin, and scales, which are byproducts of fish processing. The isolation of collagen from these byproducts involves several steps, including preparation,

* Corresponding author: vicky_prajaputra@usk.ac.id

extraction, and recovery. The extraction methods can vary depending on the marine source [1].

The aim of this review is to provide the latest information on collagen extraction methods applicable to vertebrate fish. The information presented in this review is gathered from both national and international journals that discuss collagen extraction from vertebrate fish, as well as various reference books.

1.1 Vertebrate Fish

Vertebrates are animals characterized by having an internal backbone or spinal column. Vertebrate animals encompass more than 85,000 species [2], and they are categorized into mammals, fish (pisces), reptiles, amphibians, and birds. Among vertebrate animals, fish constitute the largest group, with approximately 25,000 recorded species. Fish are cold-blooded creatures known for their distinctive features, including a backbone, gills, fins, an aquatic habitat, and the use of fins for maintaining balance in water [3]. Various types of fish, such as tilapia, catfish, cod, short-finned fish, and others, offer significant health benefits and can also serve as alternative sources of collagen.

1.2 Collagen

Collagen, derived from the Greek words "cola," meaning glue, and "genno," meaning birth, plays a crucial role in connecting cells to form the structural framework of tissues and organs within the body. Collagen molecules have a diameter of 1.5 nm, a length of 280 nm, and a molecular weight of 290,000 Daltons. The composition of collagen consists of three polypeptide chains, each containing over 1,000 amino acids within [4]. As a protein compound commonly found in vertebrate connective tissues such as skin, scales, and bones, collagen contributes to approximately 25-35% of the total protein content in the body. Collagen, being a long-chain protein, is composed of amino acids such as alanine, arginine, lysine, glycine, proline, and hydroxyproline. Its properties involve easy absorption by the body, non-toxicity, high water affinity, biocompatibility, biodegradability, relative stability, ease of shaping, and water solubility [5]. There are 28 types of collagen varying based on structural domains and suprastructural organization. However, 80-90% of collagen in the body consists of types I, II, and III. Type I collagen, the most abundant, can be found in bones, skin, tendons, and organs. Type II is present in cartilage, while type III is found in reticular fibers, blood, and skin [1].

Collagen type I functions as a membrane for tissue regeneration, antioxidant, anti-inflammatory, anti-aging, and wound healing. Collagen type II acts as a chondroprotective agent, reducing pain in osteoarthritis. Meanwhile, collagen type III plays a role in preventing diastasis recti, wound healing, and participates in bone formation [6-11]. Collagen plays a vital role in building the body's structure, including bones, teeth, joints, muscles, and skin. However, factors such as aging, exposure to ultraviolet radiation, smoking, and diabetes can lead to a decrease in collagen, resulting in skin aging, inflammation, slower wound healing, decreased muscle mass, and weakened cartilage (joint pain or osteoarthritis). Therefore, obtaining collagen externally is crucial for the body. Collagen has several pharmacological activities so it can be used as a raw material in pharmaceutical products such as cosmetics, biomedical products and supplements [12].

1.3 Collagen Isolation

Collagen from vertebrate fish can be isolated through several initial stages. The initial stages involve preparing the fish parts intended for isolation, including separation and cleaning, size reduction, and the removal of non-collagen components. Parts of the fish such as the skin and bones, which serve as collagen sources, are separated from the meat or other impurities and then cleaned with water. Subsequently, the sample is reduced in size or ground to facilitate the purification process. The removal of non-collagen components such as fat, minerals, pigments, and odors is carried out using NaOH. After the initial or pre-treatment stages, the extraction phase is conducted using an extraction method. This is followed by collagen precipitation and drying to obtain collagen yield [13].

Tabel 1. Several extraction methods in vertebrate fish collagen isolation

<table>
<tr><th colspan="2">Extraction Methods</th><th>Fish Species</th><th>Source Collagen</th><th>Yields</th><th>References</th></tr>
<tr><td colspan="2">Acid soluble collagen (ASC)</td><td>Shortfin scad (Decapterus macrosoma)</td><td>Skin
Bone</td><td>3.35 ± 3.43%</td><td rowspan="2">[14]</td></tr>
<tr><td colspan="2">Pepsin soluble collagen (PSC)</td><td>Shortfin scad (Decapterus macrosoma)</td><td>Skin
Bone</td><td>0.10 ± 0.13%</td></tr>
<tr><td colspan="2" rowspan="2">Hydro-extraction</td><td>Catfish (Pangasius sp.)</td><td>Skin</td><td>12,15%</td><td>[15]</td></tr>
<tr><td>Alaska Pollock (Theragra chalcogramma)</td><td>Skin</td><td>3.13 ± 0.04%</td><td>[16]</td></tr>
<tr><td rowspan="2">Ultrasound -assisted (UA) extraction</td><td>Acetic acid</td><td>Sharpnose stingray (Dasyatis zugei)</td><td>Skin</td><td>11.49 ± 0.03%</td><td rowspan="2">[17]</td></tr>
<tr><td>HCl</td><td>Sharpnose stingray (Dasyatis zugei)</td><td>Skin</td><td>3.50 ± 0.01%</td></tr>
<tr><td colspan="2">Deep eutectic solvent (DES)</td><td>Atlantic codfish (Gadus morhua)</td><td>Skin</td><td>5.2%</td><td>[18]</td></tr>
<tr><td colspan="2">Supercritical fluid extraction (SFE)</td><td>Atlantic codfish (Gadus morhua)</td><td>Skin</td><td>13.8 ± 0.013%</td><td>[19]</td></tr>
</table>

2 Collagen extraction method in vertebrate fish

2.1 Acid soluble collagen

The use of acetic acid as a solvent in collagen extraction is promising in the extraction process [5]. This is because acetic acid can disrupt the salt bonds between molecules and Schiff bases, which causes the expansion and dissolution of collagen fibers. In addition, acetic acid has the ability to break the hydrogen bonds that initially stabilize the triple helix structure of collagen. As a result, the hydrolysis process is accelerated in the acid method. The procedure for extracting acid-soluble collagen (ASC) from the samples involved using a 0.5 M acetic acid solution at a ratio of 1 part sample to 10 parts solution (w/v) for a 24-hour period while continuously stirring. Afterward, the resulting extracts were subjected to centrifugation at 10,000 x g for 30 minutes at 4°C, leading to the separation of the supernatant. Subsequently, the remaining residues underwent a second extraction using 0.5 M acetic acid at the same 1:10 (w/v) ratio for a duration of 12 hours, followed by centrifugation at 10,000 x g for 30 minutes at 4°C. The two sets of supernatants were combined, and sodium chloride (NaCl) was added to reach a final supernatant concentration of 0.7 M to induce precipitation. Another round of centrifugation at 2,500 x g was performed to collect the precipitate. Finally, the precipitate was freeze-dried [20]. Based on research results, the collagen yield from shortfin scad skin and bone was 3.35 ± 3.43% [14].

2.2 Pepsin soluble collagen

Extraction using pepsin can increase collagen extraction results. However, this reaction can only act on nonhelical peptide chains, and usually does not affect helical peptide chains peptide chain. Therefore, this method will be retained part of the triple helix in collagen protein, so collagen still biologically active. The reactions that occur in this enzyme method are mild, do not corrode the equipment used and consume low energy. The procedure for extracting pepsin soluble collagen from the samples involved using of 0.5 M acetic acid with a pepsin concentration of 1.5% (w/w) for a duration of 30 hours at 4°C, with continuous stirring. After this stage, the resultant extract underwent centrifugation at 10,000 x g for 30 minutes at 4°C, separating the supernatant. Subsequently, the remaining residue underwent a second extraction using 0.5 M acetic acid with a pepsin concentration of 1.5% (w/w) for 12 hours before being subjected to centrifugation at 10,000 x g for 30 minutes at 4°C. The two batches of supernatants were combined, and sodium chloride (NaCl) was gradually added until the final supernatant concentration reached 0.7 M, facilitating the precipitation process. The supernatant then underwent another round of centrifugation at 2,500 x g to isolate the precipitate. Finally, the precipitate was freeze-dried. According to the research findings, the collagen yield from shortfin scad skin and bone was 0.10 ± 0.13% [14].

2.3 Hydro-extraction

The hydroextraction method is an isolation technique that uses water as a catalyst (acting as heat as a transfer medium) and temperature as a determining factor. This method has several advantages, such as shorter production time, lower production costs, high yields, and ease in controlling the amount of waste, which is in line with the principles of clean technology and sustainable production. The collagen produced in this way is water-soluble and safe for long-term consumption. The water extraction method involved first rinsing the sample with distilled water until the pH approached neutrality. Subsequently, the sample was extracted using water, specifically aquabides, at a 1:1 ratio (by weight/volume) at a temperature of 40°C for a duration of 2 hours. The extraction yielded water-soluble collagen, which was then dried using a freeze dryer to obtain collagen in powdered form. The heating process of the sample in warm water leads to the continuation of the disruption of hydrogen and covalent bonds that had previously occurred during the acetic acid soaking process. The choice of a temperature of 40°C is made with the aim of preventing the degradation of collagen into gelatin during the ongoing extraction. Based on the results of research that has been carried out, the yield of collagen from catfish skin was 12.15% [15] and Alaska pollock skin was 3.13 ± 0.04% [16].

2.4 Ultrasound-assisted extraction

Ultrasonic-assisted extraction (UAE) utilizes high-intensity sound waves surpassing the normal human hearing range (20 kHz) to expedite mass transfer, improve mixing, drying, homogenization, and extraction processes. Moreover, UAE has gained widespread use in collagen extraction due to its capacity to enhance yields and reduce processing time. The mechanism relies on wave propagation, creating zones of high and low pressure that are directly proportionate to the applied energy in the system. Typically, collagen extraction employs organic acid solutions such as acetic acid, citric acid, and lactic acid, as they demonstrate higher extraction efficiency compared to inorganic acids.

The extraction procedure commenced by immersing the pre-treated sample (100g) in a 0.5M acetic acid solution, maintaining a sample-to-acid solution ratio of 1:20 (w/v), for 30 minutes at 4°C. Subsequently, ultrasound treatment was administered using an ultrasound processor operating at 20 kHz and 500W for a duration of 30 minutes. To prevent overheating, the ultrasound operated in a pulse mode, featuring 5-second treatment intervals followed by rest periods. Following this, the resulting extract underwent centrifugation at 6,500×g for 15 minutes at 4°C, effectively separating the supernatant from the residue. The residual material then underwent a second extraction using 0.5 M acetic acid at a sample-to-solution ratio of 1:10 (w/v), employing the same ultrasonic processor at 20 kHz and 500W for 30 minutes. Subsequently, another round of centrifugation was conducted at 6,500×g for 15 minutes at 4°C. The two batches of supernatants were amalgamated, and the solution was salted with NaCl to achieve a final concentration of 0.7 M. This mixture underwent centrifugation once more at 5000×g for 10 minutes at 4°C, resulting in a precipitate that was subsequently frozen overnight at -80°C before undergoing the drying process. These extraction steps were replicated using hydrochloric acid. According to the research findings, the ultrasound-assisted method with acetic acid yielded 11.49 ± 0.03%, while with hydrochloric acid, a yield of 3.50 ± 0.01% was obtained [17].

Higher collagen yields were achieved by extraction with acetic acid compared to hydrochloric acid, due to the better extraction ability of acetic acid. Acetic acid plays an important role in determining the pH of the extraction medium and controlling the collagen loading density. These factors influence the structure, solubility, and extraction of collagen [21]. The increase in collagen yield associated with acetic acid extraction can be further explained by the phenomenon of cavitation, which results in loss of pretreated skin. This loss facilitates more effective penetration of acetic acid through the skin, thereby increasing collagen extractability [19].

2.5 Deep eutectic solvent

Deep Eutectic Solvents (DES) and eutectic mixtures have been effectively employed for the dissolution and extraction of various bioactive compounds. DES typically consists of hydrogen bond acceptors (HBA) and hydrogen bond donors (HBD), which, when combined, form robust hydrogen bond interactions, resulting in the creation of eutectic mixtures that often exhibit liquid properties at temperatures near room temperature. DES is utilized to regulate temperature during extraction, favoring lower temperatures to prevent thermal degradation of collagen. The DES method presents an environmentally friendly, efficient, and straightforward approach to collagen recovery. Collagen extraction was carried out using 0.5 M acetic acid as the control solvent, and the results were compared with those obtained using a Deep Eutectic Solvent (DES) solution in water. This DES comprised betaine (Bet), urea (U), cholinium chloride (CC) as hydrogen bond acceptors (HBA), and formic acid (FA), acetic acid (AA), propanoic acid (PA), and lactic acid (LA) as hydrogen bond donors (HBD). Parameters such as the solid-liquid ratio, pH, and DES concentration were further explored and optimized. The molar ratio of DES components used was typically 1:2, except for instances where U : FA (1 : 4) and U : AA (1 : 5) were employed. The preparation of DES involved several steps. Initially, the water content in the starting material was determined using a Metrohm 831 Karl Fisher coulometer. Subsequently, the two-component mixture (HBD and HBA) was precisely weighed and placed in a round-bottom flask. The mixture was then stirred in an oil bath at 60 ± 2°C with a stirring speed of 500 rpm until it became a homogeneous and transparent liquid. The pH of all aqueous solutions was measured at room temperature. All procedures were conducted at 4°C.

The collagen extraction process commenced with the pre-treatment of the sample, involving cleaning, washing, size reduction, and the removal of non-collagen proteins. This included immersing the skin in a 0.1 M sodium hydroxide (NaOH) solution at a solid-to-solvent ratio of 1:10 (w/v) for 24 hours, with periodic replacement of the NaOH solution every 5-6 hours. Subsequently, the sample was rinsed with cold distilled water until a neutral pH was achieved. Deproteinized samples were then defatted using a 10% (v/v) butyl alcohol solution at a solid-to-solvent ratio of 1:10 (w/v) for 48 hours, with solvent changes every 6-7 hours. Once again, the samples were washed with cold distilled water. Extraction was performed at 4°C for 48 hours using an optimized solid-liquid ratio of 1:10 (w/v) in either 0.5 M acetic acid or an aqueous DES solution, utilizing an orbital shaker. The resulting dense solution was subjected to centrifugation to eliminate insoluble components. The supernatant was salted by adding NaCl to a final concentration of 0.9 M, followed by collagen precipitation by introducing NaCl (final concentration 2.6 M) at a neutral pH (0.05 M TrisHCl, pH 7.5). The resulting precipitate was obtained by centrifugation at 15,000 g for 1 hour and dialyzed against deionized water at 4°C for 72 hours. Finally, the collagen was dried through a process of lyophilization. The DES method is an environmentally friendly method for restoring collagen, efficient and simple. According to research results, the yield of Atlantic codfish skin was 5.2% [18].

2.6 Supercritical fluid extraction

The extraction of collagen using supercritical fluid technology involves employing acidified water under CO_2 pressure for the extraction process. Carbon dioxide is the preferred choice in this technology due to its advantageous properties, such as low toxicity, non-flammability, cost-effectiveness, high availability, stability, and environmental acceptability. Furthermore, carbon dioxide's suitability stems from its operational conditions at moderate temperatures and pressures. Additionally, carbon dioxide serves as a versatile acidifying agent, released from the aqueous medium post-extraction, resulting in a purified compound. The use of CO_2-acidified water entails a one-step extraction with mild operational conditions and eliminates the need for organic solvents.

The extraction process begins with a thorough cleaning of the sample using distilled water to eliminate any remaining impurities, followed by grinding the sample into small pieces. The weighed wet samples are then placed in a high-pressure vessel (30 cm3) containing distilled water (1 g per 20 ml). The vessel is heated to 37°C, and the system is pressurized with carbon dioxide up to 50 bar. The extraction is conducted in batches, utilizing an intermittent mode without external exchanges, for 3 hours. Subsequently, the high-pressure vessel is depressurized efficiently, and the extract is obtained through a double filtration process, initially with filter paper and then with a 0.45 μm syringe filter. The resulting acidic water-treated collagen (AWC) is frozen, subjected to freeze-drying, and stored at room temperature for future use. According to the research findings, the collagen yield from Atlantic codfish skin was 13.8 ± 0.013%. The method operates under mild conditions and employs acidified and pressurized water. The obtained results were higher compared to the conventional acetate-based methodology [22].

3 Collagen yield

Yield is the percentage of collagen obtained from the initial raw material. The yield shows the portion of raw materials that can be utilized and functions as an important parameter for assessing the economic value and efficiency of a material. The percentage yield calculation is obtained by the ratio of the dry weight of the isolated collagen to the weight of the raw material before isolation. The calculation of the % yield value utilizes the AOAC method with the formula [23] :

$$\text{Collagen yield (\%)} = \frac{Dry\ weight\ of\ collagen}{Weight\ of\ raw\ materials} \text{ x } 100\%$$

The percent yield differs between various vertebrate fish species using different extraction methods. This variation is caused by several factors such as differences in protein content in fish, pretreatment conditions [24], temperature, time, and extraction method [15]. Fish with high protein content will yield a higher collagen yield compared to those with low protein content. The initial processing of raw materials using a basic solution has proven to be more efficient than an acidic solution in separating non-collagen proteins, resulting in increased collagen production. Temperature and duration in the collagen isolation process also significantly influence the amount of collagen produced. An increase in temperature above 40°C can lead to the denaturation of collagen into gelatin. Additionally, the duration of soaking the raw materials during the pretreatment process also contributes to the solubility of collagen in the basic solution [15, 24].

4 Conclusions

Based on the journal review above, we can conclude that vertebrate fish can be used as a source of collagen which comes from fish waste in the form of skin, bones and scales. Collagen isolation from vertebrate fish can use different methods depending on the collagen source, in the form of acid soluble collagen extraction, pepsin soluble collagen, hydro-extraction, organic acid ultrasound-assisted, deep eutectic solvent, and supercritical fluid extraction. The extraction method that produces the highest collagen yield is supercritical fluid extraction, where this method uses CO_2 pressure and is more efficient.Collagen yield results produced using different methods have differences because they are influenced by protein content, pre-treatment, temperature and time.

Acknowledgment

This study is a part of the university's internal research, which is supported by the Institute for Research and Community Service (LPPM) and the Faculty of Marine and Fisheries, Universitas Syiah Kuala.

References

1. H. Jafari, A. Lista, M.M. Siekapen, P. Ghaffari-Bohlouli, L. Nie, H. Alimoradi, A. Shavandi, Polymers, **12,** 1–37 (2020).
2. A.I. Burhanuddin, Vertebrata Laut, Yogyakarta, Deepublish (2018).
3. A.I. Burhanuddin, Ikhtiologi : Ikan dan Segala Aspek Kehidupannya, Yogyakarta, Deepublish (2014).
4. H. Setyowati, W. Setyani, Jurnal Farmasi Sains Dan Komunitas, **27**, 58–66 (2019).
5. V. Prajaputra, N. Isnaini, S. Maryam, E. Ernawati, F. Deliana, H. A. Haridhi, N. Fadli, S. Karina, S. Agustina, N. Nurfadillah, I.I. Arisa, L.S. Desiyana, T.K. Bakri, S. Afr. J. Chem. Eng. **47**, 197–211 (2024).
6. M. Borumand, S. Sibilla, Clinical Interventions in Aging, **9**, 1747–1758 (2014).
7. F. Subhan, Z. Hussain, I. Tauseef, A. Shehzad, F. Wahid F, Critical Reviews in Food Science and Nutrition, **61**, 1–11 (2020).
8. Q.A. Dar, E.M. Schott, S.E. Catheline, R.D. Maynard, Z. Liu, F. Kamal, et al, PLoS ONE, **12**, 1– 24 (2017).
9. L. Nakchum, S.M. Kim, Preparative Biochemistry and Biotechnology, **46**, 123–30 (2016).
10. R.M. Blotta, S.D.S. Costa, E.N. Trindade, M.R. Maciel-Trindade, Clinics, **73**, 1–5 (2018).
11. F. Bakilan, O. Armagan, M. Ozgen, F. Tascioglu, O. Bolluk, O. Alatas, Eurasian Journal of Medicine, **48**, 95–101 (2016).
12. V.R. Rahman, M.A. Bratadiredja, N.M. Saptarini, Majalah Farmasetika, **6,** 253-286 (2021).
13. L. Salvatore, N. Gallo, M.L. Natali, L. Campa, P. Lunetti, M. Madaghiele, A. Sannino, Materials Science and Engineering C, **113**, 1–18 (2020).
14. A.W. Sulaiman, N.M. Sarbon, Food Research, **4**, 2272–2280 (2020).
15. P. Suptijah, D. Indriani, S. E. Wardoyo, Jurnal Sains Natural, **8**, 8–23 (2018).
16. Patmawati, A.R. Ergion, L. Sulmartiwi, S. Raseetha, D. Nirmala, Y. Waiprib, S. WijayantI, Scientific Journal of Fisheries and Marine, **15,** 468–477 (2023).

17. M.I. Shaik, A. N. F. Effendi, N. M. Sarbon, Biocatalysis and Agricultural Biotechnology, **35**, 1–17 (2021).
18. M. Bisht, M. Martins, A.C.R.V. Dias, S.P.M. Ventura, J.A.P. Coutinho, Green Chemistry, **23,** 8940–8948 (2021).
19. Y. Zou, P. Xu, P. Li, P. Cai, M. Zhang, Z. Sun, D. Wang, Food Science and Technology, **82**, 72–81 (2017).
20. N.A. Baderi, N.M. Sarbon, International Food Research Journal, **26**, 451–458 (2019).
21. D. Liu, G. Wei, T. Li, J. Hu, N. Lu, J.M. Regenstein, P. Zhou, Food Chemistry, **172**, 836–843 (2015).
22. R.O. Sousa, E. Martins, D.N. Carvalho, A.L. Alves, C. Oliveira, A.R.C. Duarte, R.L, Journal of Polymer Research, **27**, 1–9 (2020)
23. Association of Official Analytical Chemist, Official Methods of Analysis of the Association of Official Analytical Chemist, Virginia (US), The Association of Analytical Chemist, Inc (2005).
24. I. Astiana, N. Nurjanah, T. Nurhayati, Jurnal Pengolahan Hasil Perikanan Indonesia, **19,** 79–93 (2016).
25. C.Y. Huang, J.M. Kuo, S.J. Wu, H.T. Tsai, Food Chemistry, **190,** 997–1006 (2016).

Recent update: collagen extraction from marine resources as a promising natural antiaging

Nadia Isnaini[1,2,3], *Vicky* Prajaputra[2,3,4*], *Trivadya* Syafhira[1], *Siti* Maryam[5], *Iko Imelda* Arisa[3,6], *Sofyatuddin* Karina[3,4], *Sri* Agustina[3,4], and *Haikal Azief* Haridhi[3,4]

[1] Department of Pharmacy, Faculty of Mathematics and Natural Sciences, Universitas Syiah Kuala, Banda Aceh 23111, Indonesia

[2] ARC-PUIPT Nilam Aceh, Universitas Syiah Kuala, Banda Aceh 23111, Indonesia

[3] Research Center for Marine Sciences and Fisheries, Universitas Syiah Kuala, Banda Aceh 23111, Indonesia

[4] Department of Marine Sciences, Faculty of Marine and Fisheries Universitas Syiah Kuala, Banda Aceh 23111, Indonesia

[5] Department of Family Welfare Vocational Education, Faculty of Teacher Training and Education, Universitas Syiah Kuala, Banda Aceh 23111, Indonesia

[6] Department of Aquaculture, Faculty of Marine and Fisheries Universitas Syiah Kuala, Banda Aceh 23111, Indonesia

Abstract. Aging is a natural process that occurs in every human. Aging can be prevented or slowed down through the appropriate and suitable use of collagen as an anti-aging treatment. Collagen is typically derived from raw materials such as pig or cow bones/skin, but recently there have been frequent outbreaks of infectious diseases among livestock, prompting the exploration of raw materials sourced from the sea to address this issue and create promising collagen for anti-aging purposes. The objective of this literature review research is to explore the potential of collagen extracts from marine resources as an anti-aging solution through an examination of relevant literature, which can serve as a reference for future research. A systematic review method was employed by collecting data from various literature sources, including research journals published both nationally and internationally, which discuss various marine resources containing collagen with anti-aging properties. Based on the systematic review, several natural resources containing collagen with anti-aging properties were identified, including sea cucumbers (*Holothuroidea* sp.), milkfish (*Chanos chanos*), tilapia (*Oreochromis niloticus*), jellyfish (*Rhopilema esculentum*), starfish (*Asterias pectinifera*), mackerel (*Decapterus macarellus*), squid (*Todarodes pacificus*), silver carp (*Hypophthalmichthys molitrix*), and blue sharks (*Prionace glauca*).

* Corresponding author: vicky_prajaputra@usk.ac.id

1 Introduction

Premature aging is an inevitable physiological process [1]. Skin aging occurs due to intrinsic and extrinsic factors. Intrinsic aging naturally occurs with age, resulting from the slowing of cell renewal processes and changes in hormone production [2]. Extrinsic aging is caused by chronic exposure to light, pollution, ionizing radiation, chemicals, and toxins [3]. The skin is the body part most frequently exposed to external factors [1]. The effects of skin aging include thinning, dryness, reduced elasticity, rough texture, wrinkles, and dark pigmentation. Collagen is one of the essential components of the extracellular matrix that plays a crucial role in maintaining youthful skin [2]. As the primary structural component of the skin, collagen decreases by 1% each year. Therefore, there is high demand for collagen due to its bioactive effects in preventing skin aging [4].

Collagen is the main structural protein and the most abundant protein found in the animal kingdom, constituting nearly 25-30% of the total protein content in the entire body [5]. One of the numerous benefits of collagen is its anti-aging properties [6]. Over the years, most of the available collagen has been extracted from byproducts of cattle and pigs processing industries. However, in recent decades, the use of collagen from these sources has remained limited due to religious restrictions (for Muslims, Jews, and Hindus) and the risk of transmitting diseases such as Bovine Spongiform Encephalopathy (BSE), Transmissible Spongiform Encephalopathy (TSE), and Foot and Mouth Disease (FMD) that have occurred in cattle in recent years. Therefore, there is an urgent need to discover new alternative sources of collagen [4].

Marine resources have garnered significant attention in the last two decades as a safe and abundant alternative source for collagen extraction. Marine collagen (MC) poses a lower risk of infectious diseases and does not raise religious concerns [7]. Additionally, a substantial number of byproducts generated from the fish processing industry, accounting for approximately 70-85% of the total catch weight, has become a primary driver for research aimed at transforming cost-effective byproducts (such as fish skin and scales) into collagen-based products with high added value and minimal environmental impact [7, 8].

Therefore, this literature review is conducted to explore various marine resources and their collagen extractions with anti-aging properties through a literature search, which can serve as references for further research.

1.1 Aging

Aging is the gradual alteration of the anatomical structure of the skin, resulting in a decline in its physiological functions, which occurs as individuals age and due to environmental influences. Skin aging can be attributed to both intrinsic and extrinsic factors. Intrinsic factors include genetics, age, hormones, and ethnicity. Intrinsic signs of aging encompass fine wrinkles, a thinning and increased transparency of the skin due to a reduction in epidermal cells and melanocytes, as well as skin dryness. Extrinsic factors contributing to skin aging comprise exposure to UV radiation, contact with harmful substances, exposure to electronic device light, and lifestyle factors such as an unhealthy diet, sleep patterns, stress, lack of physical activity, and smoking [9].

The dermis layer is a crucial part of the skin that plays a vital role in maintaining skin elasticity and softness [10]. When collagen production in the dermis layer decreases, as occurs with aging and due to environmental factors, the skin can become dry and lose its elasticity. One of the factors that can lead to reduced collagen production in the skin's dermis layer is excessive exposure to sunlight [7]. UV radiation from the sun can increase the levels of free oxygen radicals (ROS) in the skin, which, in turn, can cause oxidative stress to the skin tissues. This results in damage to the extracellular matrix (ECM) of the skin [11].

Damage to the ECM directly contributes to the skin aging process and is responsible for the increased activity of enzymes such as collagenase, elastase, and tyrosinase. Activation of these three enzymes leads to a decrease in elastin and collagen levels in the skin, causing it to lose its elasticity, strength, and the development of wrinkles [12].

Therefore, efforts need to be made to prevent and slow down skin aging, one of which is through the use of cosmetic products. The use of cosmetic products is one option to avoid skin aging because it can improve skin texture and function [13]. One available solution to address skin aging is the use of anti-aging products. Anti-aging products work to inhibit skin damage and have the ability to reduce signs of aging on the skin [14]. Cosmetic products designed for anti-aging purposes have proven their benefits in reducing wrinkles, blemishes, and skin discolorations caused by pigmentation [15].

1.2 Collagen

Collagen is an abundantly present structural protein in the animal kingdom, and in the human body, it contributes to approximately 30% of the total protein [16]. Collagen plays a crucial role as the main component of the extracellular matrix that forms the skin, bones, ligaments, cartilage, and tendons [17]. There are nearly 28 types of collagens that can be classified based on differences in their molecular chains [18]. More than 85% of the collagen in the human body is of type I, while other types of collagens include type II, III, and IV. Collagen itself is a trimer consisting of three α polypeptide chains. The structure of type I collagen is characterized by a double helix and is rich in glycine, proline, and hydroxyproline residues [17]. Collagen provides a number of health benefits, including [19]:

- Plays a role as an aid in the digestion process.
- Exhibits anti-aging properties.
- Contributes to weight loss.
- Supports joint and bone health.
- Enhances athletic performance by increasing muscle mass.
- Improves sleep quality.
- Strengthens the health of hair, nails, and teeth.

Natural collagen is originally produced in the body by fibroblast cells. However, the body's ability to produce collagen can decrease with age and due to unhealthy lifestyles. Therefore, there is a need for collagen intake from external sources. The raw materials commonly used for this are usually extracted from mammals and poultry. However, concerns about infectious diseases such as Bovine Spongiform Encephalopathy (mad cow disease) and bird flu have raised questions about the safety of collagen from these sources [20]. Therefore, there is a need for new alternatives in the form of collagen sources derived from marine resources. Extracting collagen from marine sources not only addresses concerns of religious significance but also offers unique benefits [18].

1.3 Mechanisms of Antiaging in Collagen

Collagen has numerous applications in the cosmetics industry, especially in its role as an anti-aging ingredient. Collagen can penetrate into the skin layers and continuously form a colloid system on the skin's surface, resulting in a smooth and soft skin sensation. Considering collagen's capabilities, it can be used as an ingredient in cosmeceutical-based cosmetic products with anti-aging purposes [21].

2 Collagen from Marine Sources with Potential Anti-Aging Activity

Table 1. Collagen from marine sources with potential anti-aging activity

Marine Resources	Source of Collagen	Collagen Yield	Anti-Aging Activity Test	Ref.
Sea cucumber (*H. leucospilota*)	Body wall	7.92%	Antityrosinase and antielastase activity	[22]
Milkfish (*C. chanos*)	Scale	19.96 g of 300 g dry sample weight	Measurement of the water content and oil content of the skin	[1]
Tilapia (*O. niloticus*)	Bone	41.5%	Measurement of moisture, pore size, blemishes and number of wrinkles	[23]
Tilapia (*Oreochromis sp.*)	Skin	-	Observation of skin superficial characteristics, histopathological observation, determination of moisture content, Hyp content, and SOD content	[24]
Jellyfish (*R. esculentum*)	Umbrella	50 mg	Antityrosinase	[25]
Starfish (*A. pectinifera*)	All parts	3.8%	Measurement of reduction in MMP expression	[26]
Mackerel (*D. macarellus*)	Skin	6.39±0.97% (pepsin-soluble collagen)	Antiglycation and antityrosinase activity	[27]
Squid (*T. pacificus*)	Outer skin and inner skin	75.4% and 68.9%	Antityrosinase and antielastase	[28]
Silver Carp (*H. molitrix*)	Skin	-	Measurement of inhibition of MMP-1 secretion	[29]
Blue Shark (*P. glaucia*)	Cartilage	-	Measurement of moisture, sebum, texture and blemishes	[30]

2.1 Sea cucumber (*Holothuria leucospilota*)

The study Addillah et al. [22] utilized anti-tyrosinase and anti-elastase activity tests to determine the anti-aging activity of sea cucumber collagen. The results showed that in the anti-tyrosinase activity test, the IC50 value was 1.20 mg/mL, while in the anti-elastase activity test, the IC50 value was 125 µg/mL. Collagen isolated from sea cucumbers exhibited weak anti-tyrosinase activity and moderate anti-elastase activity.

2.2 Milkfish (*Chanos chanos*)

The anti-aging activity of collagen from milkfish scales in the study by Wahid et al. [1] was assessed by measuring the moisture and oil levels in the skin using a skin analyzer. This study formulated the collagen obtained into a cream preparation with concentrations of 5%, 6%, and 7%. The results of the moisture (moist) and oil level activity tests for preparations with concentrations of 5%, 6%, and 7% demonstrated anti-aging effects in improving skin conditions, with significance values for each concentration obtained (P>0.05), indicating no significant differences among the concentrations, although visually, the 7% concentration showed effectiveness in skin moisture levels. Therefore, it can be concluded that the cream containing 5% milkfish scale collagen (*Chanos chanos*) already possesses anti-aging effectiveness in improving the condition of the back of the hand skin of volunteers during 4 weeks of treatment, with an average change in moisture levels from 12.03% in week 0 to 26.98% in week 4, and an average change in oil levels from 18.03% to 24.53%.

2.3 Tilapia (*Oreochromis niloticus*)

The anti-aging activity of collagen cream derived from tilapia fish bones in the study by Ginting et al. [23] was assessed through measurements of moisture (moist), pore size, the presence of blemishes, and the number of wrinkles on human skin. This research analyzed the anti-aging effectiveness using collagen cream from tilapia fish bones with various concentrations, but the concentration that showed the best anti-aging effectiveness was 3.5% compared to other cream formulas and a comparative cream (Ponds cream). The application of this cream provided a moisture level of 50.2%, categorized as "moist," with a recovery percentage of 24.7%, a reduction in pore size of 21.5%, a decrease in the number of blemishes by 22.1%, and a reduction in the number of wrinkles by 22.2%. Furthermore, in a study conducted by Pu & Qu et al. [24], collagen from tilapia fish skin was administered in low, moderate, and high doses to a group of rats. The research results revealed an increase in dermis thickness with a denser structure, along with a significant elevation in water content, Hyp levels, and SOD activity. This effectively mitigated the skin aging process. The anti-aging effect of tilapia fish skin collagen correlated positively with the dosage administered.

2.4 Jellyfish (*Rhopilema esculentum*)

The anti-aging activity of jellyfish umbrella collagen in the study by Zhuang et al. [25] was evaluated through anti-tyrosinase activity. This research conducted anti-tyrosinase activity tests on several different fractions, namely HF-1 (>3000 Da), HF-2 (1000~3000 Da), and HF-3 (<1000 Da). The study results indicated that HF-2 exhibited an inhibitory effect of over 50% at a concentration of 5 mg/ml. Therefore, it can be concluded that HF-2 has the ability to inhibit tyrosinase activity. This suggests that HF-2 derived from jellyfish umbrella collagen hydrolysate may serve as a potential new natural inhibitor in the pharmaceutical and food industries.

2.5 Starfish (*Asterias pectinifera*)

The anti-aging activity of collagen from starfish (*Asterias pectinifera*) in the study by Han et al. [26] was assessed through the measurement of MMP (Matrix Metalloproteinase) expression reduction. The research results showed that in comparison to MMP-1 expression levels under control conditions without collagen peptides (indicating 100%), there was a decrease in MMP-1 expression levels in cells treated with nanoliposomes containing collagen peptides (less than 100%), specifically around 40%. More importantly, nanoliposomes containing collagen peptides extracted from *Asterina pectinifera* demonstrated a more significant reduction in MMP-1 expression compared to nanoliposomes containing collagen peptides from pig and fish meat, which only experienced about a 10% reduction. These results indicate that nanoliposomes containing collagen peptides from *Asterina pectinifera* have excellent potential as anti-aging agents.

2.6 Mackerel (*Decapterus macarellus*)

The anti-aging activity of collagen from mackerel skin in the study by Herawati et al. [27] was evaluated through antiglycation and antityrosinase activities. In this research, *D. macarellus* PSC (Partially Soluble Collagen) demonstrated the ability to inhibit glycation reactions by up to 50% at a concentration of 239.29 ppm. However, after collagen was hydrolyzed into HC (Collagen Peptides), the concentration required to achieve the same level of inhibitory activity significantly decreased to 68.43 ppm. Therefore, in hydrolyzed form, the inhibitory activity against glycation was higher than that of intact collagen. A similar trend was observed in the antityrosinase activities of PSC and HC from *D. macarellus*. The antityrosinase activity of *D. macarellus* PSC could inhibit tyrosinase activity by 50% at a concentration of 234.66±0.185 ppm. In this case, the antityrosinase activity of HC increased threefold compared to the PSC activity of *D. macarellus*, with an inhibition rate reaching 50% at a concentration of 79.35±0.5 ppm. Overall, these findings indicate that mackerel skin collagen has the potential as a material with anti-aging properties that can be further explored for use in the pharmaceutical and cosmetic industries.

2.7 Squid (*Todarodes pacificus*)

The anti-aging activity of collagen from the outer and inner parts of squid skin (*Todarodes pacificus*) in the study by Nam et al. [28] was evaluated through anti-tyrosinase and anti-elastase activity tests. In this research, inhibition of tyrosinase activity was tested on peptide fractions at five different concentrations. The results showed that the F3 fraction in both hydrolysates had the ability to inhibit tyrosinase activity, although its ability was significantly lower compared to L-ascorbic acid (positive control). Additionally, the testing of elastase inhibition activity on peptide fractions was carried out at three different concentrations. The research findings indicated that the F2 and F3 fractions in both hydrolysates had significantly higher elastase inhibition capabilities compared to the F1 fraction, although their elastase inhibition activity remained lower than that of urosolic acid (positive control).

2.8 Silver Carp (*Hypophthalmichthys molitrix*)

The results of measuring the inhibition of MMP-1 secretion by silver carp skin collagen polypeptide (SCSCP) in the study by Haung et al. [29] showed that SCSCP at a concentration of 10 mg/mL exhibited a high level of activity in combating DPPH and eliminating hydroxyl radicals. Meanwhile, at a concentration of 0.1 mg/mL, SCSCP significantly improved the survival of UVB-exposed L929 cells and inhibited the production of MMP-1 in L929 cells after UVB exposure. These findings indicate that SCSCP has the potential as a skin anti-aging agent with the ability to eliminate DPPH and hydroxyl radicals as well as inhibit MMP-1 production.

2.9 Blue Shark (*Prionace glauca*)

The study Lu et al. [30] employed 11 different concentration variations in testing anti-aging activity. Anti-aging activity was assessed through measurements of moisture, sebum, skin texture, and blemishes. Sample F11 resulted in improvements in skin moisture levels, skin texture, and skin color pattern. Additionally, Sample F11 was also effective in regulating skin oil production. Over a period of 10 and 20 minutes of use, Sample F11 demonstrated a reduction in skin wrinkles. Based on these positive results regarding skin quality and function, a cosmetic gel formulation containing collagen hydrolysate from P. glauca could be a promising choice for skincare applications.

3 Conclusion

Based on the above journal review, we can conclude that collagen can be sourced not only from livestock but also from marine resources. Marine resources containing collagen can be found in sea cucumbers, milkfish, tilapia, jellyfish, starfish, mackerel, squid, silver carp, and blue sharks. The type of animal that contains the most collagen from marine sources is Squid (*Todarodes pacificus*) in the outer skin and inner skin. Collagen derived from these marine sources is also considered safe and has been researched for its anti-aging properties over the past few years. Therefore, it can be considered an alternative source of collagen in the cosmetics industry.

Acknowledgment

The current research was strongly supported and facilitated by Universitas Syiah Kuala and received financial support from the Ministry of Education, Culture, Research, and Technology of the Republic of Indonesia under the Matching Fund-Kedaireka, Grant Program Number 28/E1/PPK/KS.03.00/2023. This grant was awarded to the Research Center for Marine Sciences and Fisheries at Universitas Syiah Kuala.

REFERENCES

1. H. Wahid, S.F. Karim, N. Sari, Jurnal Sains Dan Kesehatan. **4**, 428–436 (2022).
2. A.M. Cruz, M.C. Goncalves, M.S. Marques, F. Veiga, A.C. Paiva-Santos, P.C. Pires, Cosmetics. **10**, 1–21 (2023).
3. E. Cevenini, L. Invidia, F. Lescai, S. Salvioli, P. Tieri, G. Castellani, C. Franceschi, Expert Opin. Biol. Ther. **8**, 1393–1405 (2008).
4. D. Coppola, M. Oliviero, G.A. Vitale, C. Lauritano, I.D'Ambra, S. Iannace, D. Pascale, Marine Drugs. **18**, 1–23 (2020).
5. M. Meyer, Biomed Eng Online. **18**, 24 (2019).
6. S. Geahchan, P. Baharlouei, A. Rahman, Marine Drugs. **20**, 1–16 (2022).
7. V. Prajaputra, N. Isnaini, S. Maryam, E. Ernawati, F. Deliana, H. A. Haridhi, N. Fadli, S. Karina, S. Agustina, N. Nurfadillah, I.I. Arisa, L.S. Desiyana, T.K. Bakri, S. Afr. J. Chem. Eng. **47**, 197–211 (2024).
8. G.K. Pal, P.V. Suresh, Innov. Food Sci. Emerg. Technol. **37**, 201–215 (2016).
9. O.R. Adianingsih, O.E. Puspita, D.R. Rububiyah, Kosmetologi. Universitas Brawijaya Press, Malang. (2022).
10. C.P.S. Syaharani, N. Isnaini, E. Harnelly, V. Prajaputra, S. Maryam, F.A. Gani, Journal of Patchouli and Essensial Oil Product. **2**, 9–15 (2023).
11. R. Khare, N. Upmanyu, M. Jha, Current Aging Science. **14**, 46–55 (2019).
12. I.S. Hwang, J.E. Kim, S.I. Choi, H.R. Lee, Y.J. Lee, M.J. Jang, H.J. Son, H.S. Lee, C.H. Oh, B.H. Kim, S.H. Lee, D.Y. Hwang, International Journal of Molecular Medicine. **30**, 392–400 (2012).
13. E. B. Souto, A.R. Fernandes, C. Martins-Gomes, T.E. Coutinho, A. Durazzo, M. Lucarini, S.B. Souto, A.M. Silva, A. Santini, Appl. Sci. **10**, 1594 (2020).
14. S. Fitrianingsih, L.N. Nafi'ah, K. Ismah, Cendekia Journal of Pharmacy. **6**, 318–325 (2022).
15. B. Sharma, A. Sharma, International Journal of Pharmacy and Pharmaceutical Science. **4**, 57–66 (2012).

16. M.E. Nimni, R.D. Harkness, Molecular Structures and Functions of Collagen. CRC Press, US. (1998).
17. N. Xu, X.L. Peng, H.R. Li, J.X. Liu, J.S.Y. Cheng, X.Y. Qi, S.J. Ye, H.L. Gong, X.H. Zhao, J. Yu, G. Xu, D. X. Wei, Front. Nutr. **8**, 702108 (2021).
18. W. Jankangram, S. Chooluck, B. Pomthong, Afr. J. Biotechnol. **15**, 642–648 (2016).
19. D.P. Joy, K.K. Kumar, B.D. Kumar, K.S. Silvipriya, International Journal of Current Pharmaceutical Research. **9**, 24–26 (2017).
20. N. Arfiani, I. Rahmat, Rezki, Jurnal Suara Kesehatan. **9**, 29–35 (2023).
21. G. Secchi, Clin in Dermatol. **26**, 321–325 (2008).
22. S. Abdillah, G. Wijiyanti, M. Setiawan, S. U. Noor, M. Nurilmala, African Journal of Biotechnology. **16**, 771–776 (2017).
23. E. Ginting, N. F. Zebua, Khalisa, Journal of Pharmaceutical and Sciences. **5**, 329–337 (2022).
24. X. Pu, Y. Qu, Journal of Cosmetic Dermatology. **22**, 2436-3444 (2023).
25. Y. Zhuang, X. Zhao, B. Li, Journal of Zhejiang University Science. **10**, 572–579 (2009).
26. S.B. Han, B. Won, S. Yang, D.H. Kim, Journal of Industrial and Engineering Chemistry. **98**, 289–297 (2021).
27. E. Herawati, Y. Akhsanitaqwim, P. Agnesia, S. Listyawati, A. Pangastuti, A. Ratriyanto, Marine Drugs. **20**, 516–529 (2022).
28. K.A. Nam, S.G. You, S. M. Kim, Journal of Food Science. **73**, 249–255 (2008).
29. J. Haung, H. Li, G. Xiong, J. Cai, T. Liao, X. Zu, LWT. **173**, 1–12 (2023).
30. W.C. Lu, C.S. Chiu, Y.J. Chan, T.P. Guo, C.C. Lin, P.C. Wang, P.Y. Lin, A.T. Mulio, P.H. Li, Marine Drugs. **20**, 633 (2022).

Morphometry characters comparison of the painted spiny lobster, *Panulirus versicolor*, from three locations in Simeulue Island

Mulya Rahman Harahap[1], *Sultan* Sufie[2], *Suraiya* Nazlia[1], *Mutia* Ramadhaniaty[2], *Andina Ramadhani Putri* Pane[3], *Adrian* Damora[1*]

[1]Department of Aquaculture, Faculty of Marine and Fisheries, Universitas Syiah Kuala, Banda Aceh, Indonesia
[2]Department of Marine Science, Faculty of Marine and Fisheries, Universitas Syiah Kuala, Banda Aceh, Indonesia
[3]Research Center for Fishery, National Research and Innovation Agency (BRIN), Jakarta, Indonesia

Abstract. The painted spiny lobster, *Panulirus versicolor*, is an economically spiny lobster species on Simeulue Island. Fishing pressure on this species is suspected of experiencing overfishing, with indications that its size is getting smaller during the fishing season. Therefore, it is necessary to manage existing lobster resources. Several aspects that need to be considered in the sustainability of fisheries management are related to biological aspects, including the study of morphometry characters. This study aims to compare the morphometric characteristics of *P. versicolor* from three locations in Simeulue Island. The research was conducted from March to May 2023 with three sampling locations, including East Simeulue, Teluk Dalam, and South Teupah. A total of 90 samples were sampled using the random sampling method. To differentiate or classify the population, 15 parameters of spiny lobster morphometry were studied using discriminant function analysis. The results led to the assertion that *P. versicolor* from three separate places belonged to the same population but had distinct subpopulations. This condition is affected by the lobster samples collected from nearby areas. The investigation of the relationship between length and weight revealed that the growth pattern of *P. versicolor* in three locations were negative allometric, with length increasing faster than weight increasing.

1 Introduction

Indonesia has vast marine resources, including six types of *Panulirus* spp. lobsters with significant economic potential. The habitat conditions of each lobster vary depending on the type [1]. The spiny lobsters found in Indonesian waters are scalloped spiny lobster (*P. homarus*), ornate spiny lobster (*P. ornatus*), pronghorn spiny lobster (*P. penicillatus*), mud spiny lobster (*P. polyphagus*), longlegged spiny lobster (*P. longipes*), and painted spiny

* Corresponding author: adriandamora@usk.ac.id

lobster (*P. versicolor*). This six types of lobster are evenly distributed and are found almost in all coral waters in Indonesia [2]. The six lobster species found in Indonesian waters are also found in Aceh waters, especially the southwest coast and *P. versicolor* is one of the fishing targets of local fishermen, because of its dominance [3].

The painted spiny lobster, scientifically known as *P. versicolor*, is a species of lobster that can be found in coral or rocky reef areas at depths of 5 to 20 meters [4]. Unfortunately, this type of lobster is being overfished, leading to a reduction in size during the fishing season. To protect this valuable resource, it is important to manage it sustainably. In order to do so, various aspects must be taken into consideration, including biological factors such as the study of morphometric characters and condition factors [5].

Morphometric studies play a crucial role in lobster resource management efforts as they aid in determining the diversity of a species by examining general morphological characteristics. The marine and fisheries sector heavily relies on morphometric research. particularly in aquaculture, to ensure that a species genetics are suitable for domestication. This information is also crucial in the initial steps of cultivators. Domestication is possible through the collection of broodstock from various locations as different lobster parents have varying genetic diversity. Genetic diversity can occur between geographically separated populations and even within a single population [6]. This diversity can be reflected in the lobster's phenotype and genotype. Morphometric measurements of the lobster's body structure aid in phenotypic observations. Morphometric studies help in understanding lobster taxonomy and also provide insights into the variations that occur in measurements, which are related to the habits and habitat of different variants of the species [7].

2 Material and Methods

3.1 Data collection

The research was conducted between March to May of 2023 in three locations in Simeulue Island: East Simeulue, Teluk Dalam, and South Teupah. The simple random sampling method was used to collect samples, with lobster sampling stations being determined based on fishing areas and landing sites for *P. versicolor* in the three locations above. Thirty samples were taken at each station.

3.2 Data analysis

2.2.1 Morphometric measurements

The body of *P. versicolor* was measured using a measuring instrument and weighed using a digital scale to determine its body mass. In total, 15 morphometric characters were measured in lobsters, including several commonly used ones as described by [8].

2.2.2 Phenotypic diversity

The diversity value is a percentage of the standard deviation of fish length. To see phenotypic diversity in *P. versicolor*, the following formula is used [9]:

$$kk = \frac{S}{Y} x\ 100\% \quad (1)$$

The equation involves three variables: kk, which is the diversity coefficient, S, which is the standard deviation, and Y, which is the median.

2.2.3 Discriminant Function Analysis (DFA)

The study conducted data analysis using Discriminant Function Analysis (DFA). This involved obtaining eigenvalues, total variance, canonical correlation, matrix structure and morphological analysis. The aim of DFA was to group each population in a canonical correlation graph of function 1 and function 2. All data were analyzed using IBM SPSS Statistics software version 26, according to [10].

2.2.4 Length-weight relationship

To analyze the relationship between the length and weight of *P. versicolor*, [11] used the following formula:

$$W = a\ L\ b \quad (2)$$

Given weight (W), length (L), and the intercept (a) of the length and weight relationship curve with the y-axis, as well as the regression coefficient (b).

To test the value of b = 3 or b ≠ 3, a t-test (partial test) is carried out, with a hypothesis [9].

H_0 : b = 3, the relationship between length and weight is isometric

H_1 : b ≠ 3, the relationship between length and weight is allometric, namely:

1. Positive allometric, if b>3 (weight increase is faster than length increase), and
2. Negative allometric, if b<3 (length increase is faster than weight increase),

$$t_{count} = \frac{b_1 - b_0}{Sb_1} \quad (3)$$

With b_1 representing the b value obtained from the relationship between length and weight, b_0 set at 3, and Sb_1 indicating the deviation coefficient of b.

It is important to compare the t_{count} value with the t_{table} value at the 95% confidence interval. To determine the growth pattern of lobsters, the decision rule taken is:

$t_{count} > t_{table}$: reject the null hypothesis (H_0)

$t_{count} < t_{table}$: fails to reject the null hypothesis

3 Results and Discussion

3.1 Phenotypic diversity

Each character in the phenotypic diversity table displays the highest value for the body weight (BW) variable, which is 0.44. On the other hand, the lowest value of phenotypic variation can be observed in the variables of abdominal width (AL), chest width (ChW), and antennule peduncular length (APL), which have a value of 0.11. Table 1 presents the phenotypic diversity of painted spiny lobsters (*P. versicolor*) found in East Simeulue, Teluk Dalam, and South Teupah.

Table 1. Phenotypic diversity of morphometric characters of painted spiny lobster (*P. versicolor*) in East Simeulue, Teluk Dalam, and South Teupah.

Morphometry Characters	Station			Phenotypic Diversity
	East Simeulue	Teluk Dalam	South Teupah	
Total length	27.58 ± 2.77	21.58 ± 4.14	23.5 ± 2.97	0.17*
Carapace Length	12.54 ± 1.14	10.88 ± 1.51	11.40 ± 0.98	0.12*
Ocular Carapace Length	10.48 ± 0.79	8.71 ± 1.20	9.45 ± 0.95	0.13
Carapace Width	7.20 ± 0.73	6.08 ± 0.93	6.19 ± 0.60	0.14
Abdomen Length	14.72 ± 1.88	12.55 ± 2.35	13.47 ± 1.80	0.16

Abdominal Segment Length	11.05 ± 0.93	8.95 ± 1.94	10.22 ± 0.83	0.16
Abdomen Width	6.48 ± 0.46	5.68 ± 0.69	5.98 ± 0.63	0.11
Telson Length	5.20 ± 0.58	4.89 ± 0.68	4.79 ± 0.44	0.12
Telson Width	6.13 ± 0.93	5.55 ± 0.85	5.66 ± 0.58	0.14
Head Width	6.7 ± 0.60	5.80 ± 0.51	5.97 ± 0.25	0.09
Chest Width	6.5 ± 0.77	6.39 ± 0.80	6.35 ± 0.41	0.11
Periopod Length	13.86 ± 1.56	12.53 ± 2.47	12.54 ± 1.05	0.15
Antennule Length	27.24 ± 2.24	26.23 ± 3.68	27.29 ± 1.74	0.10
Antennule Pedencula Length	8.34 ± 0.45	7.72 ± 1.15	7.79 ± 0.69	0.11
Body Weight	0.62 ± 0.21	0.39 ± 0.23	0.44 ± 0.10	0.44**

* : Shows low phenotypic diversity
** : Shows high phenotypic diversity

Phenotypic characteristics of the painted spiny lobster (*P. versicolor*) can be identified through morphometric measurements of its body structure [7]. Each morphometric character has several phenotypic variations. The extent of diversity of a phenotypic character indicates the genetic diversity of the population. Based on the phenotypic diversity values obtained for *P. versicolor*, it can be concluded that the level of genetic diversity is relatively low, less than 20%.

3.2 Discriminant Function Analysis (DFA)

Discriminant Function Analysis (DFA) was conducted on 15 morphometric characteristics of *P. versicolor*, with total length as the comparison feature, The analysis produced two functions, with the first function having *Eigenvalues* greater than the second. This means that the first function plays a significant role in distinguishing the three stations. The first function accounts for 91.2% of the total variance with *Eigenvalues* of 3.549, while the second function with *Eigenvalues* of 0.344 accounts for the remaining 8.8% of the total variance. The results of the analysis are shown in Table 2.

Table 2. Results of morphometric character analysis of painted spiny lobsters (*P. versicolor*) in East Simeulue, Teluk Dalam, and South Teupah using Discriminant Function Analysis (DFA).

Function	**1**	**2**
Eigenvalues	3.549	0.344
Percentage of variance	91.2	8.8
Carapace Length (CL)[b]	0.085	0.035*
Ocular Carapace Length (OCL)[b]	0.111	0.338*
Carapace Width (CW)[b]	-0.063	0.269*
Abdomen Length (AL)[b]	-0.002	0.276*
Abdominal Segment Length (ASL)	0.106	0.690*
Abdomen Width (AW)[b]	0.117	0.275*
Telson Length (TL)[b]	0.167*	0.086
Telson Width (TW)[b]	0.211*	0.076
Head Width (HW)	0.352*	0.237
Chest Width (ChW)	0.251*	-0.070
Periopod Length (PL)[b]	0.116	0.256*
Antennule Length (AnL)	0.403*	0.231
Antennule Pedencula Length (APL)[b]	0.113*	0.074
Body Weight (BW)	-0.151	0.217*

* The largest real correlation between each variable and the discriminant function

There are certain characteristics of a creature that perform different functions. These include Antenula Length (AnL), Head Width (HW), Chest Width (ChW), Telson Width (TW), Telson Length (TL), and Antenula Peduncular Length (APL). Another set of characteristics that perform different functions include Ocular Carapace Length (OCL), Carapace Length (CL), Abdomen Length (AL), Abdomen Width (AW), Carapace Width (CW), Periopod Length (PL), and Body Weight (BW).

Based on morphometric analysis using Discriminant Function Analysis (DFA) in East Simeulue, Teluk Dalam, and South Teupah, several characters are found to be different. The first function has the character with the highest value, namely the length of the antennula (AnL) with a value of 0.403, while the second function has the character with the highest value, namely the length of the abdominal segment (ASL) with a value of 0.690.

The results of morphological analysis indicate that the three measured stations of *P. versicolor* can be classified into 2 distinct groups, Teluk Dalam, and Teupah Selatan have similar discriminant functions, which are different from East Simeulue. The morphological analysis was conducted using Discriminant Function Analysis (DFA) to analyze morphometric characters and the results can be seen in Figure 1.

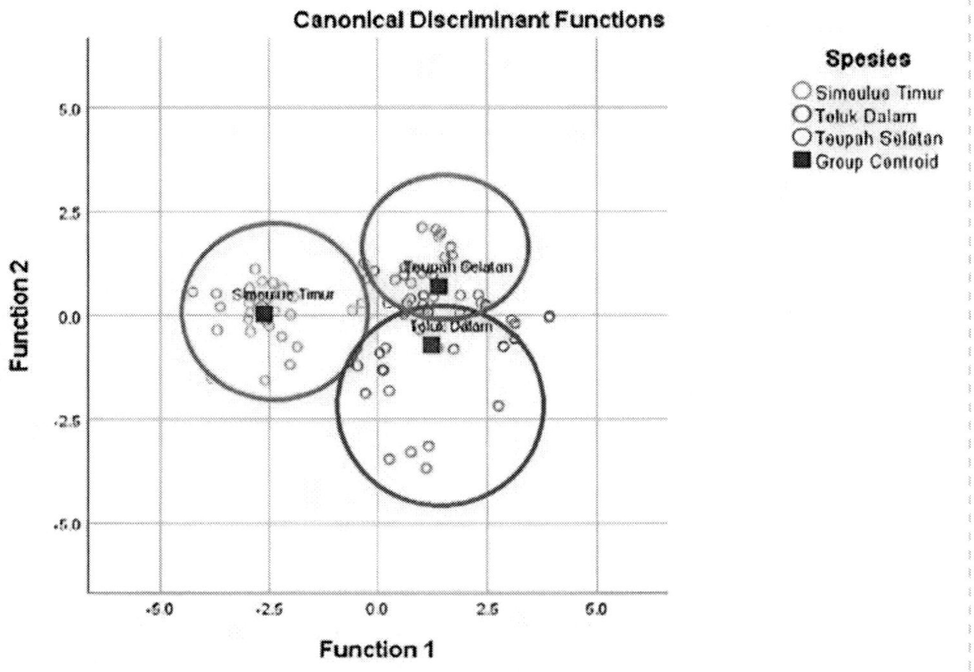

Fig. 1. Morphological analysis of *P. versicolor* morphometric characters based on multi variate analysis.

The Discriminant Function Analysis (DFA) results indicate the distinguishing characteristics of each location. The highest value for character function 1 is the length of the antennula (PAn) with a value of 0.4. This suggests that the antennula length (PAn) is a key characteristic that sets it apart from other locations. On the other hand, the highest value for character function 2 is abdominal segment length (PSA) with a value of 0.69. Therefore, it can be inferred that the character of abdominal segment length (PSA) is another key characteristic that differentiates it from other locations. These differences in characteristics are likely due to physical isolation, ecological barriers, geographic distance or behavioral influences [12].

According to morphological analysis, the three locations can be classified into two groups, Teluk Dalam station, and South Teupah station belong to the same group, while East

Simeulue station belongs to a different group. The analysis of morphometric data suggests that *P. versicolor* found in all three locations belong to the same population but consist of different sub-populations, likely due to the nearby locations. The differences in their morphology and species can be attributed to their overall body shape. Quantitative differences are expressed through statistical values such as average, range, variance, and correlation, obtained by taking various measurements from individual lobsters [13].

3.3 Length-weight relationship

Panulirus versicolor's length and weight are related by the equation y = 2.3914x – 3.0355. with a coefficient of determination (R^2) of 0.82. The b value of the equation is 2.39. indicating a negative allometric growth pattern in length and weight. This means that the increase in length is faster than the increase in weight.

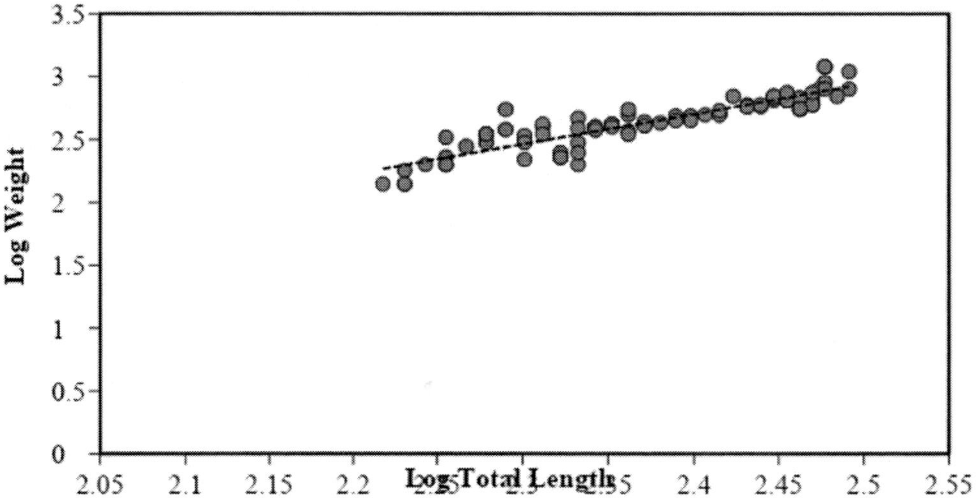

Fig. 2. Graph of the relationship between length and weight of *P. versicolor* at all stations.

In the East Simeulue, the relationship between length and weight was analyzed. resulting in the equation y = 2.9862x – 4.4941 with an R^2 value of 0.80. The b value of 2.98 indicates that the growth pattern in length and weight is negative allometric, meaning that the increase in length is faster than the increase in weight.

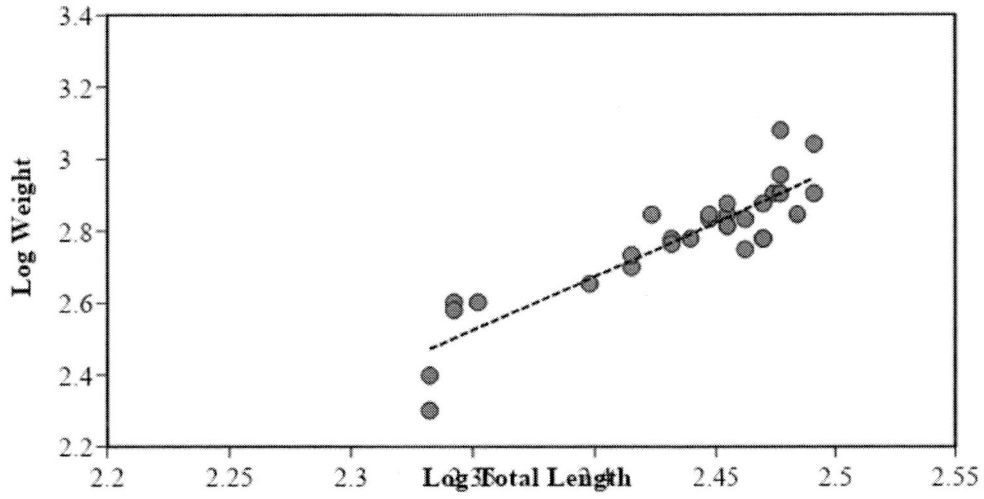

Fig. 3. Graph of the relationship between length and weight of *P. versicolor* in East Simeulue.

In Teluk Dalam, the equation for the relationship between length and weight was determined to be y = 2.671x – 3.6831, with a coefficient of determination (R^2) of 0.82. The b value of the equation for the relationship between length and weight is 2.67. This indicates that the pattern of growth in length and allometric is negative, meaning that the increase in length is faster than the increase in weight.

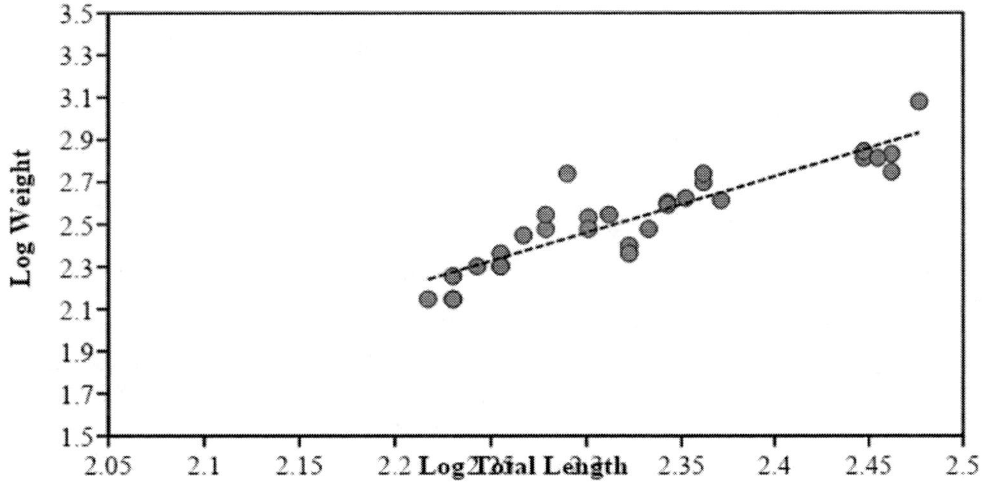

Fig. 4. Graph of the relationship between length and weight of *P. versicolor* in Teluk Dalam.

The study carried out in South Teupah found that the equation representing the relationship between length and weight is y = 1.5956x – 1.1447. The coefficient of determination (R^2) was calculated as 0.64. The b value of the equation is 1.59, indicating a negative allometric growth pattern. This means that the increase in length is faster than the increase in weight.

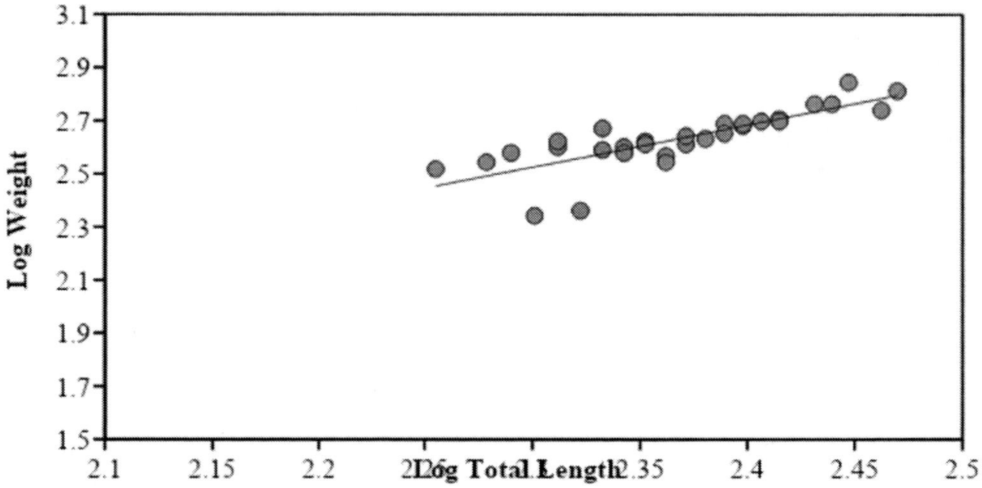

Fig. 4. Graph of the relationship between length and weight of *P. versicolor* in South Teupah.

The relationship between the length and weight of painted spiny lobsters (*P. versicolor*) has been studied at different stations. At the overall station. the equation for the relationship is y = 1.5956x – 1.1447 with a coefficient of determination (R^2) of 0.81, indicating that the model can explain the estimated values by 81%. This suggests a strong correlation between the length and weight of *P. versicolor* at all stations. In East Simeulue, the equation for the relationship between length and weight is y = 2.9862x – 4.4941 with an R^2 of 0.79. This model can explain the estimated values by 79%. suggesting a close relationship between the length and weight of *P. versicolor* in this. Similarly, Teluk Dalam found an equation for the relationship between length and weight, which is y = 2.671x – 3.6831 with an R^2 of 0.81. This model can explain the estimated values by 81%. indicating a strong correlation between the length and weight of *P. versicolor* in this location as well. Finally, Teupah Selatan found the equation for the relationship between length and weight to be y = 1.5956x – 1.1447 with an R^2 of 0.64. This model can explain the estimated values by 63%, suggesting a moderate relationship between the length and weight of *P. versicolor* in this location.

According to [14], painted spiny lobsters (*P. versicolor*) in East Simeulue, Teluk Dalam, and South Teupah exhibit negative allometric growth patterns (b<3). This means that their length increases faster than their body weight. The b value. which represents this growth pattern, is mainly influenced by environmental factors. Similar research findings were also observed for the same species in several other regions including West Aceh, Cilacap, Pangandaran, Ekas Bay, and Thoothukudi-India [15, 16].

4 Conclusions

The results led to the assertion that *P. versicolor* from three separate places belonged to the same population but had distinct subpopulations. This condition is affected by the lobster samples collected from nearby areas. The investigation of the relationship between length and weight revealed that the growth pattern of *P. versicolor* in three locations were negative allometric, with length increasing faster than weight increasing.

References

1. R. Pratiwi. Oseana **38**. 2 (2013)
2. T. Ernawati. D. Kembaren. S. Suprapto. B. Sumiono. Bawal **6**. 3 (2014)
3. R. I. Wahju. M. Riyanto. ALBACORE Jurnal Penelitian Perikanan Laut **1**. 2 (2017)
4. D. Milton. C. Proctor. F. Satria. R. West. Australian National Centre for Ocean Resources and Security. (2012)
5. Y. T. Adiputra. M. Z. Junior. M. A. Suprayudi. W. Manalu. W. Widanarni. Prosiding Simposium Nasional Kelautan dan Perikanan **5**. (2018)
6. H. Baur. C. Leuenberger. Systematic Biology **60**. 6 (2011)
7. S. Langer. N. Tripathi. B. Khajuria. Research Journal of Animal Veterinary and Fishery Sciences **1**. 7 (2013)
8. C. M. Jones. I. M. Ruscoe. F. Fisheries. Freshwater Fisheries and Aquaculture Centre. Department of Primary Industries 1996.
9. R. Steel. J. Torrie. Sumantri Cet ke-2 PT Gramedia. Jakarta. (1991)
10. N. Asiah. J. Junianto. A. Yustiati. S. Sukendi. Jurnal Perikanan dan Kelautan **23**. 1 (2018)
11. M. I. Effendie. Yayasan Pustaka Nusatama Yogyakarta **163**. (1997)
12. K. Sumantadinata. N. Taniguchi. Aquaculture **88**. 3-4 (1990)
13. Imron. (1998)
14. Z. Muchlisin. Diversity of freshwater fishes in Aceh Province with emphasis on several biological aspect of the depik (*Rasbora tawarensis*) an endemic species in Lake Laut Tawar. Thesis. Universiti Sains Malaysia. Penang. 2010
15. M. Junaidi. N. Cokrowati. Z. Abidin. Jurnal Kelautan: Indonesian Journal of Marine Science and Technology **3**. 1 (2010)
16. T. Vaitheeswaran. N. Jayakumar. V. Venkataramani. Tamilnadu Journal of Veterinary & Animal Sciences **8**. 1 (2012)

The impact of fresh and artificial diet on growth and survival rate of spiny lobster, *Panulirus homarus* reared in floating net cage

Ni Wayan Widya Astuti[1*], *Sudewi* Sudewi[1], *Bejo* Slamet[1], *Fahrudin* Fahrudin[1], and *I Nyoman Adiasmara* Giri[1]

[1]Research Center for Fishery, National Research and Innovation Agency, Cibinong, Bogor, Indonesia

Abstract. Some research found that the growth of lobster was higher when fed with fresh diet than artificial diet. Therefore, the growth of lobster that is fed with combination of fresh and artificial diets need to be observed. This study aimed to evaluate different kinds of feed and its combination on growth and survival rate of spiny lobster *Panulirus homarus*. This research was designed with three treatments and three replicates, namely: lobster fed with fresh diet (A), lobster fed with artificial diet (dry pellet) (B), and lobster fed with combination of fresh and artificial diets (C). Lobsters with an average weight 100.30 ±0.79 g were reared in floating net cage with dimension 2×2×2 m^3 at density 40 lobsters/cage. Lobsters were fed twice a day at 08.30 am and 03.30 pm. Fresh diets that were given contained fresh fish, shrimp, crabs and mussels (3:1:1:1). The observation was conducted for 18 weeks. Results of the experiment were lobsters fed with fresh diets (A) and combination of fresh and artificial diets (C) gave higher specific growth rate (0.62±0.03 %/day and 0.61±0.03 %/day) than lobster fed with artificial diets (B) (0.33±0.01 %/day). There were not significantly differences in survival rate of lobster ($p<0.05$).

1 Introduction

Lobsters are known as a valuable crustacean that has high economical value in the market. The demand of lobsters are high not only in domestic, but also in international market [1]. To fulfil the demand of the market, lobster supply still depend on capturing activities. Because of the unselective fishing potentially could reduce lobster population in Indonesia, The Ministry of Marine Affairs and Fisheries published a regulation about the restriction of lobster (*Panulirus* spp.), crabs (*Scylla* spp.) and swimming crab (*Portunus pelagicu*s spp.) in 2015. That regulation leads aquaculturist to concern about lobster farming and increase the production of lobster from aquaculture [2].

Lobster farming in Indonesia develop as grow-out system, which the seed suplies come from capturing activity. The undersized lobster that are caught by the fisherman are reared in floating net cage until meet the market size. Lobster farming in Indonesia are develop in

[*] Corresponding author: nw.widyaastuti@gmail.com

https://doi.org/10.1051/bioconf/20248703022

Lombok, Bali, Java, Sumbawa, Sumatera and Sulawesi [3]. Lobster that are caught in Indonesia are *Panulirus ornatus*, P. *homarus*, P. *polyphagus* and P. *versicolor*. The main lobster that are captured is spiny lobster (*Panulirus homarus*) [4].

Lobster grow-out in floating net cage depends on fresh fish as diets. Crustaceans and mollusks also could be feed source of the lobster [5]. For sustainable reason, artificial feed for lobster had been developed by some researcher. The artificial feed in form of pellets, were studied to find the right formula to support lobster growth. The challenge in developing lobster feed is artificial feed gives lower growth performance [6]. Therefore, feeding management in lobster aquaculture is essential.

A combination of fresh and artificial feed for grow-out of lobster had been experimented in controlled tanks [1]. The result of this experiment showed that the combination of fresh food and formulated diet gave better growth performance, but the survival rate was better when the lobster were fed formulated diet. Although culturing of lobster in indoor tanks provides advantages in terms of controlling the environmental factors most importantly on maintaining water quality, however, farmers prefer to grow lobsters in floating net cages for economic reasons. Hence, a further study which performed in floating net cages is required. A previous study reported that this method has some challenges, such as fluctuation of water quality, high labour of handling, difficult to control cannibalism and also hard to ensure feed is eaten efficiently by lobster [2]. Consequently, it is important to manage grow-out of lobster in floating net cages mostly the feeding regimes for the efficient use of feeds in order to have optimal growth and survival. Therefore, the present study was designed to evaluate the combination of fresh food with artificial diet for growing-out lobster and its impact on survival rate.

2 Materials and methods

This research was conducted in floating net cage of Institute for Mariculture Research and Fisheries Extension (IMRAFE) in Pegametan Bay-Bali, Indonesia. This stage of research including preparation of artificial diets, feeding trial and parameters measurement and data analysis.

2.1 Preparation of artificial diet

The artificial diets were formulated based on the previous research [1]. The ingredients of the diet showed in table 1. The feed was prepared using pelleting machine with 3 mm of diameter. The pellet was dried in oven with temperature 65°C and stored in refrigerator.

Table 1. Composition of artificial diet

Ingredients	Proportion (%)
Fish meal	78.3
Wheat flour	6
Wheat gluten	6
MOS	0.5
Fish oil	2.6

Astaxanthin	1
Cholesterol	0.5
Lecithin	1.7
Mineral premix	0.6
Vitamin premix	1.1
Stay C	0.4
CMC	1.3
Total	100

2.2 Proximate analysis

Nutrient values of artificial diets and fresh feed were analyzed by proximate analysis. The parameters of proximate that were analyzed i.e.: protein, lipid, ash and moisture. Moisture content was analyzed to count dry matter nutrients

2.3 Feeding trial

Lobsters that were reared in this experiment were captured at Southern Bali Coastal Area. Before stocking in the net cage, lobster was acclimated in fiberglass tank for 7 days. A number of 40 lobsters were allocated into cages of 2×2×2 m. Initial carapace length of the lobster was 6.19±0.04 cm, and the initial weight were 100.3±0.79 g. Floating net cage were equipped with squared parallon with dimension 1×1×0.3 m. Net were added on the top of the parallon. Gracillaria were added on the top of the net in amount 5 kg/cage. This parallon and gracillaria function were as shelter for lobsters (Fig 1).

Fig. 1. Square parallon with net (left) and gracillaria (right) as shelter for lobster

This research used completely randomize design (CRD) with 3 treatments and 3 replicates for each treatments. The feed treatments were (A) lobster fed with fresh food, (B) lobster fed with artificial diet, (C) lobster fed with combination of fresh food and artificial diet. Lobster were fed two times daily, at 09.00 am and 15.00 pm. Feed were allocated 30% from total feed daily in the morning and 70% from total feed daily in the evening. Fresh food were given 10% from biomass, meanwhile artificial diets were given 3% from biomass. For the combination diets treatments, fresh food were given 5% and artificial diets were given 1.5% daily. The fresh food that was given to the lobster was a mixture of fish, crab, shrimp and mussel with the ratio 3:1:1:1.

2.4 Parameters observed and data analysis

The parameters observed in this research were lobster growth (carapace length and body weight), specific growth rate, survival rate, BRIX Index and total haemocyte count (THC). BRIX index represents serum protein contained in the lobster's haemolymph. Length and weight were observed every 3 weeks, meanwhile Brix index and THC were measure at the end of treatment. The feeding trial was conducted for 127 days. Water quality that were monitored during treatment i.e. temperature, pH, salinity, ammonia, nitrite and phosphate.

Carapace length increment, weight gain, specific growth rate and survival rate were calculated with this formula:

$$Carapace\ length\ increment\ (cm) = L_t - L_0 \quad (1)$$

Where L_0 : initial length (cm)
L_t : final length (cm)

$$Weight\ gain\ (g) = W_t - W_0 \quad (2)$$

Where W_0 : initial weight (cm)
W_t : final weight (cm)

$$Specific\ Growth\ Rate\ (\%/day) = \frac{lnW_t - lnW_0}{t} \times 100 \quad (3)$$

Where W_0 : initial weight (cm)
W_t : final weight (cm)
t : treatment period (days)

$$Survival\ Rate\ (\%) = \frac{N_t}{N_0} \times 100 \quad (4)$$

Where N_0 : initial number of lobster
N_t : final number of lobster

Data were presented in table and analyzed by Analysis of Variance (ANOVA) followed by Duncan test with 95% confidence interval using SPSS 14.0 for windows.

3 Result and Discussion

Feed is energy source for lobster. Lobster is carnivorous crustacean that consume feed which contain high protein, low lipid, moderate to high carbohydrate [7]. In this study, lobster were fed mixture fresh food or artificial diet or combination of the two feed. Nutrient content of the feed presented in table 2.

Table 2. Nutrient composition of artificial diet and fresh food (% dry matter)

Nutrient	Artificial diet	Fresh Food			
		Fish	Shrimp	Crab	Mussel
Protein (%)	58.98	69.63	72.24	51.18	70.55

Lipid (%)	15.65	10.35	6.97	3.08	5.38
Ash (%)	15.45	17.84	11.54	35.27	15.20

The nutrient analysis showed that protein, lipid and ash value are vary between all kind of feed that were given to lobster. But in general, all the feed had high protein content. The highest protein content was found in shrimp and the lowest was found in crab. The highest lipid content was found in artificial diet, and the lowest was found in crab. Ash analysis showed the opposite than protein, the highest ash content was found in crab, and the lowest was found in shrimp. Crab that were given to lobster was not only the meat, but also some shells. The shells contain mineral, that caused high value of ash. Highest lipid content in artificial diet could be come from fish meal and fish oil.

Growth of lobster in this experiment presented in table 3. Different feed resulted different lobster growth. Carapace length grew from 6.19 cm in average to 7.01±0.07 cm in treatment artificial diet (A), 7.90±0.11 cm in treatment fresh food (B) and 7.92±0.11 cm in treatment combination of fresh food and artificial diet (C). The highest carapace length increment was found in treatment C (1.76±0.14 cm) and the lowest was found in treatment B (0.83±0.05 cm). Carapace length increment in treatment A were statistically different with treatment B but was not statistically different with treatment C.

The same pattern also found in parameter weight gain and specific growth rate. Lobster final weight were higher in treatment A (120.0±8.2 g) and C (118.0±8.4 g) and lower in treatment B (51.3±2.3 g). Specific growth rate was statistically different between treatment A (0.62±0.03 %/day) with treatment B (0.33±0.01 %/day), but was not statististically different with treatment C (0.61±0.03 %/day). This result showed that lobster growth was better while fed fresh food or combination between fresh food and artificial diet.

Table 3. Lobster growth performance.

Parameters	**A**	**B**	**C**
Initial carapace length (cm)	6.24±0.03	6.18±0.05	6.16±0.05
Final carapace length (cm)	7.90±0.11	7.01±0.07	7.92±0.11
Carapace length increment (cm)	1.66±0.14^{a}	0.83±0.05^{b}	1.76±0.14^{a}
Initial weight (g)	101.2±1.0	99.7±2.2	100.0±1.7
Final weight (g)	221.1±7.3^{a}	151.0±4.0^{b}	217.9±7.7^{a}
Weight gain (g)	120.0±8.2^{a}	51.3±2.3^{b}	118.0±8.4^{a}
Specific growth rate (%/day)	0.62±0.03^{a}	0.33±0.01^{b}	0.61±0.03^{a}

Note: different superscript letters in the same column show significant differences ($P<0.05$)

Lobster growth were higher when the feed contains fresh food and grew slowly when fed artificial diets [3]. This result also reported in lobster with the same feed that were reared in fiber tank with controlled environment [1]. Juvenile *P. argus* that fed with artificial diets showed lower growth rate than lobster that fed with fresh squid [8]. High level of chemo-attractant substances are found in fresh ingredients, which could promote feeding responses in crustaceans [9]. Higher growth in treatment combination of feed could be caused by this treatment has better nutrient balance that come from artificial feed and fresh food [1]. For better growth and survival rate, a study of feed for lobster suggested to include fresh fishery products in formulating diets [10].

BRIX index of lobster that were fed with artificial feed (12.95±0.31 %) is lower than the lobster that fed with fresh food (17.60±0.63%) and combination of fresh food and artificial diet (18.02±2.03%) (table 4). BRIX index describes sugar content in aqueous solution, and it is correlated with serum protein consentration in blood. Nowadays, this measurement widely used in laboratory or industry to asses health status of lobster [11]. The high BRIX index in treatment fresh food and combination of fresh food and artificial diet correlated with better growth. Presumably it is because the lobster got better nutrient and could metabolized the nutrient especially protein so the growth could be better than the one that fed with artificial feed. The high BRIX index showed higher protein level in haemolymph and also correlated with molting process. The BRIX index will reach the maximum value before the molting period, and then will decrease after molting [12].

There was no significant different in all the treatment for total haemacyte count. THC for treatment fresh food, artificial diet, and combination of fresh food and artificial diet were 83.67 $\times 10^6$ cell/mL, 96.83 $\times 10^6$ cell/mL, and 97.33 $\times 10^6$ cell/mL respectively (table 4). THC is associated with health status and immunity of lobster [13]. High THC value represented good immune status. In normal condition, THC concentration of *P. cygnus* was in the range 5.6±0.7 x 10^6 cells/mL [13]. In this research, THC value in all treatment are higher. It means the immunity respons of the lobster were active to protect lobster from stress condition. Stress status in lobster can be caused by several condition, i.e.: environmental factors, handling and bacterial infection [14]. Increasement of THC concentration indicated an immune response to defend the lobster from disease [15].

Table 4. BRIX Index and Total Haemocyte Count of lobster at the end of treatment.

Parameters	A	B	C
BRIX Index (%)	17.60 ± 0.63^a	12.95 ± 0.31^b	18.02 ± 2.03^a
Total Haemocyte Count ($\times 10^6$ cell/mL)	83.67 ± 17.97^a	96.83 ± 3.88^a	97.33 ± 6.25^a

Note: different superscript letters in the same column show significant differences ($P<0.05$)

THC value was related with survival rate of lobster. Lobster mortality were very low until 63 days of treatment, and increased especially in treatment fresh food until lobster were treated for 127 days (fig. 2). Survival rate (SR) of lobster was not different significantly among treatments. SR values for treatment fresh food, artificial feed and combination of fresh food and artificial feed were 73.3±29.0%, 82.5±9.0%, and 85.0±0% respectively.

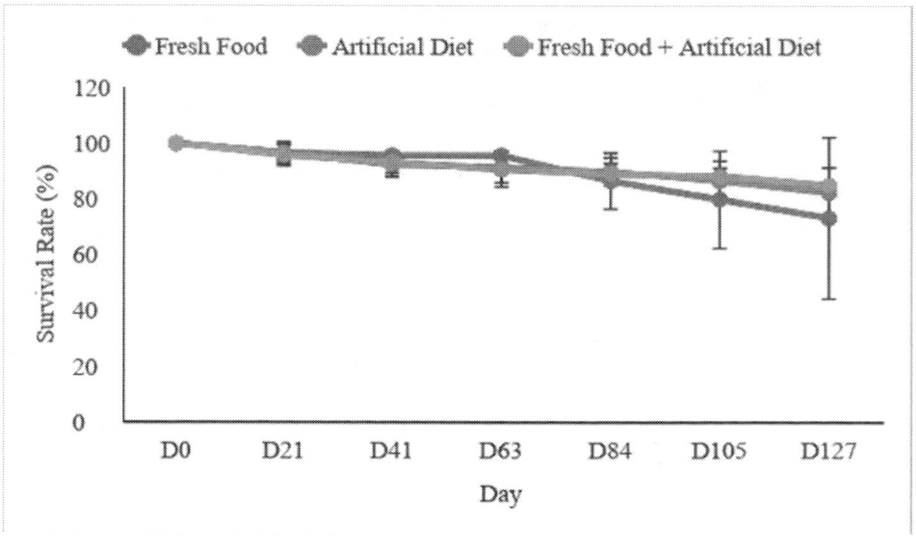

Fig. 2. Survival rate of lobster fed fresh food and artificial diet

Mortalities of lobster in this treatment were caused by cannibalism and disease. The disease that infected lobster after D63 was milky haemolymph disease (MHD). MHD is categorize as fatal disease for lobster farming [16]. MHD is caused by rickettsia like bacteria which spreading in hemolymph and show milky appearance [17]. Fresh food could be disease agent that caused higher mortalities in treatment fresh food. There was addition of vitamin, mineral and other nutrient in artificial feed. This nutrition resulted in better nutritional condition in treatment artificial feed and combination of fresh food and artificial feed.

In floating net cage, the water distribution relative similar from one cage to another. Water quality in floating net cage cannot be controlled as in the tanks. Water quality values during treatment were shown in table 5. Compared with another research, this range of water quality classified as optimum for lobster rearing [1, 3, 17, 18, 19, 20]

Table 5. Water quality in floating net cage during treatment

Parameters	**Values**	
	Result	**Optimum**
Dissolved oxygen (mg/L)	5.05-5.28	2.7-5.4
Temperature (°C)	28.9-29.5	27-32
Salinity (‰)	32-33	30-35
Ammonia (mg/L)	0.0592-0.0621	<1
Nitrite (mg/L)	<0.0087	<5
Phosphate (mg/L)	0.0059-0.00425	<0.015
pH	8.09-8.21	7.0-8.5

4 Conclusion

This study found that fresh diets (A) and the combination of fresh and artificial diets (C) resulted in higher specific growth rate (0.62±0.03 %/day and 0.61±0.03 %/day) than artificial diets (B) (0.33±0.01 %/day). While the survival rates of lobster were not different between the treatments.

Acknowledgement

This study was funded by Ministry of Marine Affairs and Fisheries of The Republic Indonesia in 2020. The author would like to thank to Selamat, Mujiono, Made Sulandra, Made Sutama, Ida Putu Suta, Sumardi, Made Suparya

References

1. Sudewi, N.W.W. Astuti, Fahrudin, I.N.A. Giri, B. Slamet, Growth and survival of spiny lobster, Panulirus homarus fed fresh food and formulated diet, in Proceedings of the 2nd International Conference on Fisheries and Marine, ICFM, 15 July 2021, Ternate, Indonesia (2021)
2. R.Y. Subhan, E. Supriyono, Widanarni, D. Djokosetiyanto, Jurnal Akuakultur Indonesia **17** 1, 53-60 (2018)
3. N.W.W. Astuti, Sudewi, N.A Giri, Fahrudin, Y.N. Asih, B. Slamet, Fish and shrimp hydrolysate as additives in lobster feed formulation: effect on growth, survival, feed consumption of spiny lobster (Panulirus homarus), in Proceedings of the International Seminar Marine Biodiversity. Utilization, Conservation and Management, MarBioUtiCoM, 8-9 August 2022, Bogor, Indonesia (2023)
4. C.M. Jones, T.L. Anh, B. Priyambodo, Lobsters: Biology, Fisheries and Aquaculture, eds E. Radhakrishnan, B. Phillips, G. Achamveetil, Singapore, 541-570 (2019)
5. B. Slamet, I. Rusdi, N.A. Giri, Haryanti, Post puerulus of scalloped spiny lobster, Panulirus homarus (Linnaeus 1758) rearing in floating net cage with different artificial diet, in Proceedings of the Ecological intensification: A new paragon for sustainable aquaculture, 28-30 October 2019, Bogor, Indonesia (2020)
6. E. Perera, C. Simon C, Rev. Aquacult. **7**, 243-261 (2014)
7. K.C. Williams, Aquac. **263**, 1–14 (2007)
8. L. Rodriguez-Viera, E. Perera, Revista de Investigaciones Marinas **32** 1, 9–15, (2012)
9. S.J. Irvin, S Shanks, Tropical Spiny Lobster Feed Development: 2009 – 2013, ACIAR Proceeding 145, Canberra, Australia (2015)
10. G. Wang, I.J. Mc Gaw, J. Shellfish Res. **33** 3, 805-813 (2014)
11. K. Vidya, S. Joseph, Indian. J. Fish. **59**, 113-118 (2012)
12. H.M. Sang, R. Fotedar, Fish Shellfish Immunol. **28** 3, 483-489 (2010)
13. J. Jusilla, J. Jago, E. Tsvetnenko, B. Dunstan, L.H. Evans, Mar. Freshw. Res. **48**, 863-868 (1997)
14. S. Lorenzon, M. Martinis, D. Borme, E.A. Ferrero, Fish. Res. **137**, 9-17 (2013)
15. E. Supriyono, R. W. Prihardianto, K. Nirmala, AACL Bioflux **10** 2, 147-155 (2017)
16. Sudewi, Z. Widiastuti, B. Slamet, K. Mahardika, Indones. Aquac. J. **13**, 31-40 (2018)
17. B. F. Phillips, J. Kittaka, Spiny lobster: fisheries and culture (Osney Mead, Blackwell Science, 2000)
18. M. Junaidi, N. Diniarti, N. Cokrowati, A. Mukhlis, B. H. Astriana, F. D .L. Conway, Ilmu Kelautan: Indonesian Journal of Marine Science 27 4, 315-329 (2022)

19. J. F. Wickins D.O. C. Lee, Crustacean farming ranching and culture 2nd edition (United Kingdom, Blackwell Science Ltd, 2002)
20. Drengstig, A. Bergheim, Aquacult. Eng **53** 14-18 (2013)

The addition of papaya leaf powder (*Carica papaya* L.) in feed to increase digestibility of milkfish fry (*Chanos chanos*)

Sari Afriani[1*] and *Dwi* Rizky Oktari[1]

[1]Department of Aquaculture, Faculty of Marine and Fisheries, Universitas Syiah Kuala, Banda Aceh, 23111, Indonesia

Abstract. The purpose of this study was to see the effect of adding papaya leaf powder with different doses to increase the growth rate and digestibility of milkfish fry (*Chanos chanos*). The method in this study used a completely randomized design (RAL) with 4 treatments each with 5 repetitions. The treatment used in this study was feed without the addition of papaya leaf powder or control treatment (A), the addition of papaya leaf powder 2% (B), 2.5% (C) and 3% (D). The feed that has been made is added with chromium which functions as an indicator of digestibility. Feeding is carried out for 40 days. The results showed that the addition of papaya leaf powder had a significant effect ($P<0.05$) on increasing the specific growth rate, growth in absolute length and absolute weight, FCR and digestibility of milkfish fry, but had no significant effect ($P>0.05$) on survival rate (SR). The best results in this study were shown in treatment D (with the addition of 3%/kg papaya leaf powder to the feed) showed a specific growth rate 0.93%, absolute length and absolute weight growth respectively 1.12 cm and 0.37 g, FCR 1.86%, and digestibility were 77.7%.

1 Introduction

The development of milkfish cultivation in Indonesia has increased, including in Aceh Province. Based on KKP statistical data, the production volume of milkfish cultivation in Aceh has increased over the last ten years, namely from 18,492 tons in 2012 increasing to 20,315.14 tons in 2022 [1]. This condition provides opportunities for farmers to developing milkfish cultivation business. But in cultivation milkfish, farmers often face the problem of the length of time fish farming [2]. Feed is one of the determining factors for fish survival and growth milkfish. High prices of feed ingredients such as fish meal as a source of protein are the main obstacle in fish farming activities because they cause high costs for this cultivation [3]. To decrease feed costs and improve feed quality which can increase the survival value and growth of milkfish can be done by utilizing natural ingredients that are available to be used as substitutes or additions to feed [4]. One of natural ingredient that can be used and contains nutrients is papaya leaf powder which is added to milkfish feed. Papaya leaf powder contains protease enzymes [5]. This enzyme is used to completely break down

* Corresponding author: sariafriani.86@usk.ac.id

or decompose peptide bonds in proteins so that the proteins break down into simpler peptide bonds [6]. The addition of papaya leaf powder to commercial feed is thought to be able to enrich the enzymes in the feed. Papaya leaves contain the enzyme papain which can help break down protein so that it can increase feed digestibility and increase fish growth [7].

Previous research was conducted by [8] regarding the addition of papaya leaf powder which had a significant effect on the growth rate of length and weight of catfish (*Clarias* sp.) with the best treatment being the addition of 3.5% papaya leaf powder which resulted in growth in length and weight 2.69 g and 3.51 cm, specific weight and length 3.89 g and 5.89 cm, and FCR 0.517. Previously, [7] had also conducted research on the potential of papaya leaf powder on tilapia (*Oreochromis niloticus*) and showed that papaya leaf powder addition of as much as 2% can improve the efficiency of feed utilization in tilapia fish farming amounted to 36.65%, can increase the protein efficiency ratio amounted to 0.55%, could increase the growth rate relative to the cultivation of tilapia by 2,725%, can increase the protein content (17.98%) in the flesh of tilapia. Then, the treatment addition of papaya leaf powder 3% can increase the thickness of the flesh of tilapia (38.09%). Research on the addition of papaya leaf flour to milkfish seeds has never been carried out, therefore it is necessary to carry out this research which aims to see the effect of adding papaya leaf powder with different doses to increase the growth rate and digestibility of milkfish fry (*Chanos chanos*).

2 Materials and methods

2.1 Time and places

This research was carried out at the Hatchery and Breeding Laboratory, Faculty of Marine and Fisheries, Universitas Syiah Kuala, Banda Aceh for 40 days, starting from February 25 to March 10 2023.

2.2 Materials and equipment

The materials and equipment used for this research were milkfish fries, buckets, aerators, scales, rulers, blenders, filters, stationery, digital cameras, papaya leaf flour, tapioca flour, commercial fish food and water, Cr_2O_3.

2.3 Research design

This research used a Completely Randomized Design (CRD) method with 4 treatments and each treatment was replicated 5 times. Milkfish fry was reared based on the guidelines of the Institutional Animal Care and Use Committee (IACUC, 2018) and have ethical approval B/KEH/034SN/XII/2022 from the animal ethic committee of Marine and Fiseheries Faculty, Universitas Syiah Kuala. Determining the dose of papaya leaf flour used in this study refers to the treatment used by [8]:
Treatment A: feed without adding papaya leaf powder (control)
Treatment B: adding papaya leaf powder 2% of feed
Treatment C: adding papaya leaf powder 2.5% of feed
Treatment D: adding papaya leaf powder 3% of feed

2.4 Research procedure

2.4.1 Preparation of water and cultivation containers

The containers used in this research were 20 plastic buckets with a volume of 20 liters. Before use, the container to be used is cleaned and dried. Each container is labeled according to the treatment and repetition. The sea water used in this research was sourced from Alue Naga Beach, Banda Aceh. The water is first settled for 2 days, then diluted until the salinity is 14-15 ppt. Then, each container is filled with ±10 liters of water and given aeration.

2.4.2 Preparation and maintenance of milkfish fry

The milkfish fries used in this research were healthy, not sick or deformed, had no wounds, had active movement, and had a proportional body shape. The milkfish fry used measure 3-5 cm with a weight of ± 1-3 grams and are stocked at 1 fish/1 liter. So, the stocking density for each container is 10 fishes/10 liters. Fish acclimatization is carried out by adapting the fish from their place of origin to a new place for 15-20 minutes and then placing them in a container. The fish adaptation process was carried out for 3 days, until the fish consumed normal food by giving pellets (without treatment) 3 times a day, in the absence of death it was considered that the fish had adapted and were ready to be tested.

2.4.3 Preparation of papaya leaf flour and feeding

The way to make papaya leaf powder is to choose young papaya leaves that are not too old. Then the papaya leaves that have been collected are cleaned with a cloth. After dry, papaya leaves should be cut to ± 3 cm. Then air-dry the papaya leaves until dry and grind them using a blender to get papaya leaf flour, sift the flour using a tight sieve to get maximum and smooth results. Next, the addition of tapioca flour at 5% per kg of feed is used to glue the feed and papaya leaf flour. In addition, 0.5% chromium per kg of feed is added as an indicator of feed digestibility in fecal tests.

The feed used in this research is commercial feed with a protein content of 25-30% to which papaya leaf flour is added. Feed is given adlibitum by spreading it into the rearing media and treating it. During rearing, milkfish fry are fed with a feeding frequency of 3 times a day, namely at 08.00, 12.00 and 16.00 WIB, with a dose of 5% of the fish biomass.

2.4.4 Parameter of reserach

The parameters observed in this research: Specific Growth Rate (SGR), absolute length, absolute weight, Feed Conversion Ratio (FCR), feed digestibility, and Survival Rate (SR) and the quality of the water in the cultivation area.

3 Result and discussion

The results of the Analysis of Variance in this study showed that the addition of papaya leaves to milkfish fry feed had a significant effect ($P<0.05$) on specific growth rate, absolute length and weight, feed convertion ratio (FCR) value and digestibility, but had no significant effect ($P>0.05$) on the survival value (SR) of milkfish seeds as shown in Table 1.

Table 1. The effect all of treatment on Specific Growth Rate (SGR), absolute length, absolut weight, Feed Convertion Ratio (FCR), feed digesbility, and Survival Rate (SR)

Treatment	Specific Growth Rate (%/day)	Absolute Length (cm)	Absolute Weight (g)	Feed Convertion Ratio (%)	Digesbility (%)	Survival Rate (%)
A (0%)	0.49 ± 0.07^{a}	0.78 ± 0.16^{a}	0.19 ± 0.03^{a}	3.18 ± 0.48^{a}	65.8 ± 0.33^{a}	90.0 ± 7.07^{a}
B (2%)	0.66 ± 0.11^{b}	0.94 ± 0.16^{ab}	0.26 ± 0.04^{b}	2.46 ± 0.60^{ab}	72.7 ± 0.34^{b}	88.0 ± 4.47^{a}
C (2.5%)	0.77 ± 0.03^{c}	1.01 ± 0.15^{ab}	0.31 ± 0.01^{c}	2.10 ± 0.06^{b}	75.3 ± 1.30^{c}	86.0 ± 11.40^{a}
D (3%)	0.93 ± 0.03^{d}	1.12 ± 0.18^{b}	0.37 ± 0.01^{d}	1.86 ± 0.10^{c}	77.7 ± 0.16^{d}	86.0 ± 7.86^{a}

The values with different superscript letters in a column are significantly different ($p<0.05$)

The results showed that the specific growth rate values of milkfish seeds were significantly different between treatments. The addition of papaya leaf flour to feed at a dose of 3% or the highest dose in this study is thought to further speed up the food digestion process in the bodies of milkfish fry. This is in accordance with the opinion of [6] and [7] stated that the presence of the papain enzyme in papaya leaves can help fish digest food, because the more papain enzyme added to the feed will cause more protein hydrolysis into amino acids, thereby increasing the growth rate and feed digestibility of fish.

Based on the research results, it was found that the absolute growth values for length and weight of milkfish fry were significantly different between those given the highest dose of papaya leaf powder (treatment D) and those without papaya leaf powder (treatment A/Control). This shows that the dose in treatment D is thought to be more effective in increasing the growth in length and weight of seeds compared to feed that does not contain added papaya leaf powder. This is in accordance with the opinion of [8], the use of papaya leaves in feed is good for increasing the efficiency of protein utilization and feed conversion in fish to increase the growth rate of fish. [9] stated that the papain enzyme contained in papaya leaf powder can form and speed up the digestive process of fish so that the nutrients in the feed can be sufficiently available and more easily absorbed so that it can increase the weight and length of the fish.

In this study, it was seen that the higher the percentage of papaya leaf powder given to the feed, the lower the FCR value. This can also show that the presence of papaya leaf flour in research can affect feed quality. [10] stated that the smaller the FCR value means the higher the quality of the feed, this shows that the amount of feed consumed is greater than the amount of feed remaining. The FCR value is still considered efficient if it is less than 3. [11] explained that the addition of papain to fish feed can increase the amino acid content in the feed so that the nutritional content in the feed can be utilized optimally by fish to meet physiological and growth needs.

The research results showed that the addition of papaya leaf powder to treatment D could increase the percentage of the best total digestibility value, namely 77.7%, this value according to [12] is still within the normal protein digestibility range of milkfish seeds, namely 77.2-78.2%. This statement is reinforced in the [13], that protein digestibility by fish is generally 75-95%. The high digestibility value of milkfish seed feed is thought to be because papaya leaf powder contains the enzyme papain which can help the digestibility of the feed given. This is in accordance with the opinion of [14], the papain enzyme content is able to hydrolyze the protein contained in feed, so that the feed given to fish can be optimally beneficial and affect the efficiency value of feed utilization.

The results of the ANOVA test analysis in this study showed that administration of papaya leaf powder at different doses had no significant effect (P>0.05) on the survival percentage (SR) of milkfish fry. The survival rate of milkfish fry at the time of the study ranged from 86-90%. The survival value is in accordance with [15], namely the survival value of milkfish seeds measuring 3-4 cm is 85%. The survival rate of milkfish seeds during the research was still relatively good, ranging between 86-90%. According to [16], there are 3 levels of fish survival, namely survival of ≥ 50% is classified as good, survival of 30-50% is moderate and less than 30% is not good.

4 Conclusions

The addition of papaya leaf powder had a significant effect (P<0.05) on increasing the specific growth rate, growth in absolute length and absolute weight, FCR and digestibility of milkfish fry, but had no significant effect (P>0.05) on survival rate (SR). Treatment D or the addition of 3%/kg papaya leaf powder to the feed is the best treatment in this study. It showed the values of specific growth rate, absolute length growth and absolute weight, FCR and digestibility respectively were 0.93%, 1.12 cm, 0.37 g, 1, 86% and 77.7%.

References

1. Statistik-KKP, *Data Volume Produksi Perikanan Budidaya Pembesaran Komoditas Bandeng per Provinsi (Ton)*. https://statistik.kkp.go.id/home.php?m=prod_ikan_prov&i=2#panel-footer-kpda (2023)
2. V. Cardoso, M. D. R. Oedjoe, N. Dahoklory, Jurnal Aquatik, **3**, 2 (2020)
3. V. T. F. Prajayati, O. D. S. Hasan, M. Mulyono, Jurnal Perikanan Universitas Gajah Mada, **22,** 1 (2020)
4. N. Usman, N. Palinggi, A. Laining, *Aplikasi Pakan Berbasis Bahan Baku Lokal dan Hasil Samping dalam Pakan Pembesaran Ikan Bandeng di Lahan Pembudidaya*, in Proceedings of Forum Inovasi Teknologi Akuakultur (2014)
5. Reski, *Suplementasi Tepung Pepaya dalam Pakan dengan Konsentrasi yang Berbeda terhadap Pertumbuhan dan Sintasan Benih Ikan Patin* (*Pangasius* sp.) (2019)
6. A. P. Nugraha, *Pengaruh Penggunaan Papain terhadap Pemanfaatan Protein Pakan dan Pertumbuhan Ikan Mas* (*Cyprius capio*.) (2016)
7. N. Isnawati, R. Sidik, G. Mahasri, JIPK, **7**, 2 (2015)
8. M. B. Mubaraq, M. Marzuki, F. Azhar, JAS, **6**, 2 (2021)
9. D. Khodijah, D. Rachmawati, JAMTech, **4**, 2 (2015)
10. C. O. Olaniyi, B. A. Salau, Transnatl, **3**, 7 (2013)
11. R. Widaryati, 2018, JIHT, **7**, 2 (2018)
12. E. Listiowati, T.B. Pramono. Berk. Perikan. Terubuk. **42**, 2 (2014)
13. M. Arief, A. Manan, dan C.A. Pradana, JIPK, **8**, 2 (2016)
14. N. Usman, N. Palinggi, E. Haris, D. Jusadi, E. Supriyono, M. Yuhana, JRA, **5**, 3, (2016)
15. *National Research Council [NRC], 1993. Nutrient Requirements of Fish Subcomittee on Fish Nutrition, National Research Council (1993)*
16. K. Melanie, Z. Zilfira, S.A. Akbar, Arwana: Jurnal Ilmiah Program Studi Perairan, 5(2), 162-168 (2023)

The effect of banana peel flour (*Musa paradisiaca*) as additional feed raw materials for common carp (*Cyprinus carpio*)

Kavinta Melanie[1], *Suraiya* Nazlia [1*], *Juwanda* Saputra[1], *Suri* Purnama Febri[2], and *Ika* Rezvani Aprita[3]

[1]Department of Aquaculture, Faculty of Marine and Fisheries, Universitas Syiah Kuala, Banda Aceh 23111, Indonesia
[2]Department of Aquaculture, Faculty of Agriculture, University Samudra, Langsa City 24416, Indonesia
[3]Department of Agro-Industry, Politeknik Indonesia Venezuela, Aceh Besar 23372, Indonesia

Abstract. Banana peel flour has a high nutritional content and can be used as a raw material for fish feed. The study was to find out the effect of banana peel flour as a raw material for feed on enhancing common carp fish growth. The study was conducted from February to April 2023 at the Fish Hatchery and Breeding Laboratory, Faculty of Marine and Fisheries, Universitas Syiah Kuala. The experimental group used a CRD with 5 treatments and 3 replications. The treatments included T0 (0%), T1 (2.5%), T2 (5%), T3 (7.5%), and T4 (10%) doses of additional Kepok banana peel flour. The sample fish used were an average weight of 1.64 ± 0.24 g fish.$^{-1}$, which was cultured for 40 days. The parameters observed included survival, growth rate, and feed conversion ratio. The results revealed that adding Kepok banana peel flour as a raw material for common carp feed had a significant influence on absolute weight growth, specific growth rate, and feed conversion ratio ($P<0.05$), but had no effect on survival ($P>0,5$). The optimal percentage obtained was the addition of 7.5% banana peel flour in the feed formulation (T3) which was able to produce 1.73 ± 0.12% day-1 SGR. Hence, the addition of Kepok banana peel flour as a feed raw material is quite effective in supporting the growth of common carp.

* Corresponding author: suraiya_nazlia@usk.ac.id

1 Introduction

Common carp (*Cyprinus carpio*) is one of the leading commodities in the freshwater fisheries industry. This fish was previously known and consumed by many people, so demand is increasing in Indonesia [1]. However, the current high market demand is occasionally not satisfied because the number of fish produced is insufficient. According to [2], the challenges faced in carp business activities include a low survival rate, a relatively slow growth rate, and comparatively high feed prices.

Feed is extremely important in the development of fish. The availability of sufficient feed in both quality and quantity has a significant impact on fish farming activities [3]. However, feed costs are quite high, covering 60%-80% of total operational costs [4]. In other words, feed contributes to nearly all of the farming costs. As a result, the raw materials used to make feed must not only have nutrient content for the fish being farmed, but also be easily obtained on a sustainable and affordable basis. The fish can then use the energy source from this feed for basal energy, activity, reproduction, and growth.

Scientists' interest in agricultural waste has grown in recent years, and the waste has become more attractive to explore and benefit from rather than being ignored. According to [5], 100 g of banana peel contains 89 Kcal of energy, 2.6 g of fiber, 22.84 g of carbohydrates, 0.33 g of fat, and up to 9.86 g of protein. Banana peels, according to [6], can help excrete feces and nutrients whereas providing aroma to feed. As a result, banana peels can be used as a fish feed additive [7],[8]. Aside from that, the use of agricultural waste is expected to reduce existing feed used, resulting in relatively high farmer profits.

The natural wealth of plants in Indonesia offers opportunities to provide raw materials for feed, particularly from waste products such as banana peels. According to [7], Kepok banana peels contain 4.40% fat, 9.86% protein, and 82.59% carbohydrates; these contents can be used as supplementary components in the production of feed in the form of flour. The carbohydrates are the highest component value. [9] state carbohydrates serve to meet energy and food supply needs while also helping to save protein (protein sparring effect). To determine how much banana peel flour is used, it must be tested in a feed formulation on the growth and feed efficiency values of common carp.

Previous studies on consumption fish, such as tilapia and rohu fish, have been conducted. According to [10], [11], and [12], using banana peels flour in feed formulations can increase absolute growth, daily growth, and fish feed efficiency values, as well as disease resistance. Based on the content of Kepok banana peel flour, it can be used as feed additive for fish growth and survival. The purpose of this research was to determine the effect of adding Kepok banana peel flour on feeding and to find the appropriate dose of Kepok banana peel flour to increase common carp fry growth and survival.

2 Material and methods

2.1 Materials

The materials used in this research are as follows: a) Common carp size 3-7 cm obtained from local fish shop in Aceh Besar; b) Peel of Kepok banana which comes from processed banana waste; c) other feed ingredients include soybean flour, fish meal, fine bran, tapioca flour, water, fish oil, and vitamins.

2.2 Research methods

This study was conducted from February to April 2023 at the Fish Hatchery and Breeding Laboratory, Faculty of Marine and Fisheries, Universitas Syiah Kuala, Banda Aceh.

Proximate analysis was performed at the Food and Agricultural Product Analysis Laboratory, Faculty of Agriculture, Syiah Kuala University, Banda Aceh. A Completely Randomized Design (CRD) with five treatments and three repetitions was used. Each treatment container was filled as many as 10 fish. Common carp was reared based on the guidelines of the Institutional Animal Care and Use Committee (IACUC, 2018) and have ethical approval B/034SN/XII/2022 from the animal ethic committee of Marine and Fisheries Faculty, Universitas Syiah Kuala.

The treatment for this research was given in the form of different doses of Kepok banana peel flour used in fish feed, as follows: T0=Feed without an additional Kepok banana peel flour (control); T1=Feed with an additional of 2,5% Kepok banana peel flour; T3=Feed with an additional of 5% Kepok banana peel flour; T4=Feed with an additional 7,5% Kepok banana peel flour; T5=Feed with an additional 10% Kepok banana peel flour.

2.3 Research Procedure

The research was conducted in several stages, including: a) Preparation of Kepok banana peel flour in feed. Banana peels were washed and dried in an oven at 50°C for 24 hours [12]. The flour was then ground and combined with the raw ingredients until the ingredients are combined and do not crumble when clenched. The feed is then molded using a feed molding machine and dried in an oven at 70°C for 30 minutes [10]. After drying, the feed was stored in a dry and closed container; b) Preparation of test feed; and c) Preparation of fish sample, which included one-day acclimatization as a process of adapting fish to the environment [13], before being cultivated in a container.

The observation phase was carried out for 40 days where the fish sample were given fish feed as much as 5% of the weight of fish biomass. Feeding was carried out three times a day [13]. Siphoning was done to clean the remaining feed and dirt. Furthermore, sampling was carried out every ten days including weight measurements and measurements of water quality such as temperature, pH and dissolved oxygen.

2.4 Research Parameters

The feed formulation results were analyzed descriptively using proximate data, which included changes in nutritional content without and with Kepok banana peel flour. The following parameters were observed in this study:

2.4.1 Absolute weight growth

According to [14], the growth is calculated based on the formula:

$$GR = Wt - Wo$$

Where: GR= Growth rate (g); Wt = Average weight of fish at the end of the experiment (g); Wo = Average fish weight at the beginning of the experiment (g)

2.4.2 Specific growth rate

The specific daily growth is calculated based on the formula [15]:

$$SGR = \frac{(InWt - InWo)}{t} \times 100\%$$

Where: SGR= Specific growth rate (%); Wt = Average weight of fish at the end of the experiment (g); Wo = Average fish weight at the beginning of the experiment (g); t = Rearing period (days)

2.4.3 Feed conversion ratio

The feed conversion ratio is calculated based on the formula [16]:

$$FCR = \frac{F}{(Wt + D) - Wo}$$

Where: FCR = Feed conversion ratio; Wt = Average weight of fish at the end of the experiment (g); Wo = Average fish weight at the beginning of the experiment (g); D = Weight of dead fish (g); F = Amount of feed consumed (g)

2.4.4 Survival rate

According to [17], survival rate is calculated based on the formula:

$$SR = \frac{Nt}{No} \times 100\% \ [16]$$

Where: SR = Survival rate (%); Nt = Number of fish at the end of maintenance (fish); No = Number of fish at the beginning of maintenance (fish)

2.5 Data analysis

The data is analyzed using Analyze of Variance (ANOVA) in SPSS software to determine the effect of the treatment given. If there are significant results, the calculation is continued with the Duncan Multiple Range Test.

3 Result and discussion

Based on the results of the ANOVA test, adding Kepok banana peel flour to common carp feed for 40 days had a significant ($P<0.05$) effect on absolute growth, relative growth, and feed conversion ratio but had no significant ($P>0.05$) effect on survival (Table 1).

Table 1. The results of measuring research parameters of common carp for 40 days

Treatments	Research Parameters			
	Absolute Weight growth (g)	Spesific Growth rate (%/day)	Feed conversion ratio	Survival rate (%)
T0 (0%)	$0{,}92\pm0{,}04^{a}$	$1{,}39\pm0{,}12^{b}$	$1{,}7\pm0{,}12^{bc}$	$63{,}33\pm11{,}54^{a}$
T1 (2,5%)	$0{,}93\pm0{,}08^{a}$	$1{,}06\pm0{,}14^{a}$	$1{,}7\pm0{,}17^{bc}$	$73{,}33\pm5{,}77^{a}$
T2 (5%)	$1{,}19\pm0{,}02^{b}$	$1{,}25\pm0{,}04^{ab}$	$1{,}6\pm0{,}10^{ab}$	$70{,}00\pm10{,}00^{a}$
T3 (7,5%)	$1{,}60\pm0{,}08^{c}$	$1{,}73\pm0{,}12^{c}$	$1{,}4\pm0{,}06^{a}$	$76{,}66\pm5{,}77^{a}$
T4 (10%)	$1{,}03\pm0{,}17^{ab}$	$1{,}15\pm0{,}19^{ab}$	$2{,}0\pm0{,}24^{c}$	$56{,}67\pm5{,}77^{a}$

Note: Different superscript letters in the same column show significant differences ($P<0.05$).

Table 2. Proximate test results of feed with and without Kepok banana peel flour (KBPF)

Sampel	Rate (%)				
	Water	Ash	Fat	Protein	Carbohydrate
T0 (0%)	13,51	11,39	16,09	20,89	38,12
T1 (2,5%)	6,81	12,65	14,62	22,63	43,29
T2 (5%)	7,26	13,82	18,37	26,45	34,10
T3 (7,5%)	5,69	12,61	17,08	26,98	37,64
T4 (10%)	5,51	12,58	19,13	26,11	36,68
KBPF	8,64	12,70	9,39	13,92	55,34

Source: * Proximate Test Results of Food and Agricultural Product Analysis Laboratory, Faculty of Agriculture, USK

After 40 days of observing the addition of Kepok banana peel flour to feed raw materials, the treatment with a dose of 7.5% (T3) had the highest absolute weight growth value, namely 1.60 g. Meanwhile, the control treatment (T0) had the lowest absolute weight growth of 0.92 g. [18] stated that the dietary requirements of common carp are 30–35% protein, 30–40% carbohydrates, and 5–15% fat. According to the proximate test in Table 2, carbohydrates are the most abundant nutritional content in Kepok banana peel flour. Carbohydrates provide energy to the fish's body; they are also required for the continuation of the fat metabolism process and can reduce protein consumption (protein sparring effect) [19]. In other words, common carp get enough energy from non-protein nutrients to allow protein to be utilized to support the fish's growth. In addition, [20] found that common carp-fed pellets provided the highest absolute growth value of 1.55 g. As a result, the addition of Kepok banana peel is thought to be advantageous not only for the dietary requirements of fish but also for reducing the consumption of commercial feed.

The specific growth rate (SGR) of a fish describes its growth rate over time. This refers to the ability of fish to convert feed nutrients into energy. In this study, common carp SGR showed different results depending on the treatment. The treatment with 7.5% Kepok banana peel flour (T3) showed the highest SGR value, 1.73%/day. Meanwhile, the treatment with 2.5% Kepok banana peel flour (T1) resulted in the lowest SGR, 1.06%/day. The greatest SGR results in this study were considered quite good and correlated with the SGR results in [21], who obtained an SGR value of 1.75%/day in common carp-fed pelleted diets. [12] obtained SGR results ranging from 2.50 to 2.80% after feeding a banana peel flour diet to L. rohita, strengthening its immunity. According to [22], the protein and non-protein content (fat, carbohydrates, water, and ash) of the feed affect the fish growth rate.

The feed conversion ratio (FCR) for common carp reared for 40 days resulted in different results for each treatment. The results revealed that the treatment with 7.5% Kepok banana peel flour (P3) had the best FCR value, 1.4. Meanwhile, the treatment with 10% Kepok banana peel flour (T4) produced the lowest FCR value, 2.0. According to [23], the FCR value for rearing common carp on commercial feed was 1.90. This indicates that the common carp feed in this study could be better utilized for growth. In other words, the lower the FCR value, the more efficiently the fish utilize the feed provided, contributing to lower feed production costs. If plant protein sources can be used as a supplement to animal protein sources, not only will production costs be reduced, but growth and production will also be increased [24].

The survival results showed that the feed formulation by adding banana peel flour had no significant effect ($P > 0.05$) on *C. carpio* survival. During 40 days of observation, the survival rate (SR) of common carp ranged from 56.67 to 76.66%. [According to [25], feed with different protein contents but the same protein energy value has an effect on growth but has no effect on survival rates. Water quality is one of the biotic and abiotic factors that influence survival. Survival, according to [26], can be influenced by both biotic and abiotic factors. Abiotic factors include the availability of food and the water quality of the inhabiting media, whereas biotic factors include the age and ability of the fish to adapt to the environment. [12] showed the addition of banana peels flour can be used as a feed additive in aquaculture to increase the SR of rohu fish about >90%.

Table 3. Water quality parameters

Treatments	Research Parameters		
	Temperature (°C)	**pH**	**DO (mg/L)**
T0 (0%)	26-29	6,5-8,1	4,6-6
T1 (2,5%)	26-30	6,7-8	4,5-6
T2 (5%)	26-30	6,9-8	4,8-6,1
T3 (7,5%)	26-29	6,8-8	4,8-6,2
T4 (10%)	26-30	6,9-8,3	4,4-5,7

Water quality is essential in common carp farming because it is the primary inhabiting media. Water quality that complies with the requirements can help fish grow and survive. Table 3 shows the range of water quality values obtained during this study. The optimum temperature range for fish life, according to [27], is between 25 and 30°C. Meanwhile, a pH of 6.5-8.5 is appropriate for common carp farming [28]. Dissolved oxygen levels in common carp-friendly waters are greater than 5 mg/L [29]. In general, the physical parameters of the water during maintenance show the range of living conditions for common carp.

4 Conclusion

Kepok banana peel flour has great potential as a fish feed ingredient because of its nutrients as a source of carbohydrates for herbivorous and omnivorous fish. The T3 treatment with a dose of 7.5% Kepok banana peel flour produced the highest growth, with an absolute weight of 1.60 g, an SGR of 1.73%/day, an FCR of 1.4, and an SR of 76.66%.

References

1. E. S. Kartamihardja, K. Purnomo, C. Umar, J. Kebijakan Perikanan Indonesia **1**, 1 (2017)
2. E. K. Simamora, C. Mulyani, M. F. Isma, J. Ilmiah Samudra Akuatika 5, 1(2021)
3. Handajani, Laporan Penelitian Hibah. Fapetrik Universitas Muhammadiyah Malang 7 hlm, (2004)
4. U. Karimah, I. Samidjan, Pinandoyo, J. of Aquaculture Management and Technology **7**, 1 (2018)
5. USDA Nutrient Database, Wheat flour, white, all-purpose, unenriched, (2018)
6. D.B. Argo, [Thesis] Univ. Brawijaya (2014)
7. T. Aryani, I. A. U. Mu'awanah, A. B. Widyantara, J. Riset Sains dan Teknologi **2**, 2 (2018)
8. V. Heuzè, G. Tran, H. Archimède, Banana peels. Feedipedia, (2016)
9. G. Y. Putra, H. Sudarwati, Mashudi, J. Nutrisi Ternak Tropis **2**, 1 (2019)
10. A. A. Y. Jeharu, C. Lumenta, J. Sampekalo, J. Budidaya Perairan **3**, 3 (2015)
11. R. M. Nanariain, C. Lumenta, H. Pangkey, J. Budidaya Perairan **5**, (1) (2017)
12. S. S. Giri, J. W. Jun, V. Sukumaran, S. C. Park, J. Immunol Res 2016, (2016)
13. S. Serlina, W. Wahidah, D. Dahlia, J. Agrokompleks **22**, 2 (2022)
14. M. I. Effendie, Biologi perikanan (Yayasan Pustaka Nusatama), (1997)
15. S. S. De Silva, T. A. Anderson, Fish Nutrition in Aquaculture, ACUL 1, 320 pp (1994)
16. H. Djajasewaka, Pakan ikan [CV. Yasaguna] (1985)
17. S. Goddard, Feed management in Intensive Aquaculture, 194 pp (1996)

18. T. Takeuchi, S. Satoh., V. Kiron, Common Carp; Cyprinus carpio (2002)
19. A. Khaeriyah, [Thesis] Univ. Hasanuddin (2018)
20. M. T. I. Fajar, Insologi: J. Sains dan Teknologi **1**, 5 (2022)
21. F. Sulawesty, T. Chrismadha, E. Mulyana, J. Limnotek. **21**, 2 (2014)
22. K. Haetami, J. Akuatika. **3**, 2 (2012)
23. Haerudin, Z. Abidin, A. A. Darmayanti. [Thesis] Univ. Mataram (2017)
24. A. Schieber, F. C Stintzing, R. Carle, Trends in Food Science and Technology. **12**, 11 (2001)
25. G. P. Putranti, Subandiyono, Pinandoyo, J. of Aquaculture Management and Technology **4**, 3 (2015)
26. T. Watanebe, Fish Nutrition and Mariculture. JICA Texbook The General Aquaculture Course, 348p (1988)
27. I. B. B. Suryadi, U. Subhan, P.P. Kelana, Aurelia J.: Authentic Research of Global Fisheries Application Journal **4**, 1 (2022)
28. Y. Wihardi, I. A. Yusanti, R. B. K. Haris, J. Ilmu-ilmu Perikanan dan Budidaya Perairan **9**, 1 (2014)
29. K. Melanie, Z. Zilfira, S.A. Akbar, Arwana: Jurnal Ilmiah Program Studi Perairan, 5(2), 162-168 (2023)

Effect of *carica papaya* leaf extract toward hatching and survival rate of nilem fish (*Osteochilus hasselti*) larvae

Cut Nuzlia[1,*], *Ajral* Muhsinin[1], and *Cut* Dara Dewi[1]

[1]Aquaculture Department, Marine and Fisheries Faculty, Universitas Syiah Kuala, Banda Aceh 23111, Indonesia.

Abstract. One of the problems in nilem fish (*Osteochilus hasselti*) harvesting is the unfulfilled availability of seed that is caused by low hatchability of eggs and larvae survival as well as high larvae abnormalities due to attacks by microorganisms such as fungi and bacteria. Papaya leaf was known as one alternative to increase the egg's hatching rate because it has antifungal and antibacterial bioactive compounds. This study aimed to determine the effect of papaya leaf extract on egg hatchability, larvae abnormalities, and larvae survival. This study used an experimental method using Completely Randomized Design (CRD) with 4 treatments and 4 repetitions. The treatment doses of the papaya leaf extract that were tested were control (A), 0.75 mg/L (B), 1 mg/L (C), and 1.25 mg/L (D). The results showed that the addition of papaya leaf extract had a significant effect ($P<0.05$) on egg hatchability, larvae abnormalities, and survival rate of *Osteochilus hasselti* larvae. The hatching rate results for each treatment were 76.50% (A), 79.50% (B), 83.50% (C), and 87.25% (D). The results of larvae abnormalities showed 9.16% (A), 7.24% (B), 5.09% (C), and 4.30% (D). Meanwhile, the results of larvae survival showed 73.87% (A), 76.13% (B), 77.25% (C), and 80.55% (D). Treatment D (1.25 mg/L) was the best treatment in this study, resulting in egg hatchability of 87.25%, larvae abnormality of 4.30%, and larvae survival of 80.55%.

1 Background

Nilem fish (*Osteochilus hasselti)* is one of the freshwater fish commodities that has great potential to be developed into a superior product for aquaculture. One of the problems in cultivating *Osteochilus hasselti* is the inadequate availability of its eggs and larvae. This is because *Osteochilus hasselti* eggs often do not hatch well [1]. Internal and external factors might cause failure to hatch the fish eggs. Internal factors include hormones, egg yolk volume, and mechanical work from the larva's activity and from the egg's enzymatic work. Meanwhile, external factors include inappropriate media, mechanical disturbances such as shocks, low water quality, the presence of unfertilized eggs, and attacks by microorganisms such as fungi and bacteria [2-4]. One effort to increase egg hatchability is to prevent microorganism attacks by adding chemicals, such as Methylene blue, formalin, Malachite green, NaCl, or $KMnO_4$ [4, 5].

The use of synthetic chemicals can cause resistance to pathogenic microbes and cause environmental pollution [6]. Therefore, natural ingredients derived from plants are needed which are antifungal and antibacterial. So far, plant extracts used as antifungal and

* Corresponding author: cutnuzlia@usk.ac.id

antibacterial ingredients in fish eggs include papaya leaves [7], betel leaves [8], pineapple leaves [9], cherry leaves [10], ketapang leaves [11], and many others.

Papaya leaves (*Carica papaya L*) contain natural compounds such as alkaloids, triterpenoids, steroids, flavonoids, saponins, tannins, and quinones which have antifungal and antibacterial activity [12, 13]. Apart from that, papaya leaves also contain the papain enzyme which has a proteolytic activity that can be used as an immunostimulant, can maintain egg health, stimulate faster embryo growth, and increase fish immunity both at the larval and parent stages [12-15].

The use of papaya leaf extract in the field of cultivation has been carried out in several studies, including research on milkfish (*Chanos chanos* forskall) eggs, which was produced the highest hatchability up to 89.94% with the lowest abnormality as much 6.54% with 4 mL papaya leaf extract [15]. Meanwhile, research on gourami (*Osphronemus gouramy*) fish eggs produced the highest hatchability up to 96% with the highest larvae survival up to 97% with 3000 ppm papaya leaf extract [13]. Then a study on punten carp (*Cyprinus carpio*) eggs produced the highest hatchability up to 89.5% with a survival up to 94.68% at a papaya leaf extract dose of 1 mg/L [7]. However, the use of papaya leaf extract in hatching *Osteochilus hasselti* eggs has never been done. Therefore, this research was carried out to see the effect of papaya leaf extract in supporting the successful hatching of *Osteochilus hasselti* eggs.

2 Material and Method

This research was conducted in Fish Seed Centre, Krueng Batee, Kuala Batee District, Southwest Aceh Regency, Aceh Province, Indonesia. The experimental method used a Completely Randomized Design (CRD) with 4 treatments and 4 repetitions. The treatment concentrations of the papaya leaf extract were control (A), 0.75 mg/L (B), 1 mg/L (C), and 1.25 mg/L (D) [7]. The papaya leaf was taken from Banda Aceh City, Aceh Province, Indonesia. The fresh papaya leaf was dried at room temperature for 5 days. The dried papaya leaf was then smashed into powder. As much as 200 g of dried papaya leaf powder was extracted using maceration with 800 mL of 70% ethanol for 3 days at room temperature. The macerate was then filtered and the filtrate was dried using a rotary evaporator [16]. The papaya leaf extract then was screened to identify the phytochemical compounds that consisted of alkaloid, saponin, flavonoid, steroid, polyphenols, glycoside, tannin, quinone, and triterpenoid [17]. Next, the research continued with spawning the Nilem fish (*Osteochilus hasselti*) using a semi-artificial method. The mature female of *Osteochilus hasselti* was injected with 0.5 mL/Kg Salmon Gonadotropin Releasing Hormone (SGnRH-a) and domperidone, and 0.3 mL/Kg for the male, then incubated for 9 hours [3]. Each 100-fertilized of *Osteochilus hasselti* eggs were immersed with 4 treatments of papaya leaf extract for 20 minutes [7], and then transferred to the aquarium with an aerator as a hatching container. During the hatching process, random eggs of *Osteochilus hasselti* were collected to observe the embryogenesis process (cleavage, morulation, blastulation, gastrulation, organogenesis, and hatching) under a microscope every 2 hours [18]. Then calculated the fertilization and hatching rate using the following formula [19]:

$$\text{Fertilization Rate (\%)} = \frac{\text{Number of Fertilized Eggs}}{\text{Total Number of Fertilized Eggs}} \times 100 \quad (1)$$

$$\text{Hatching Rate (\%)} = \frac{\text{Total Number of Hatched Eggs}}{\text{Total Number of Fertilized Eggs}} \times 100 \quad (2)$$

Eggs that have hatched into larvae will be kept for 14 days in the aquarium. During the rearing period, *Osteochilus hasselti* larvae will be fed with artemia. The larvae are ready to be fed with artemia after they are 3-4 days old, 5 times during the day every 2 hours [20]. At the end of rearing *Osteochilus hasselti* larvae, a calculation will be carried out to determine the survival and abnormalities rate of the larvae using the following formula [19, 21]:

$$\text{Survival Rate (\%)} = \frac{\text{Total Number of Larvae} - \text{Total Number of Dead Larvae}}{\text{Total Number of Larvae}} \times 100 \quad (3)$$

$$\text{Abnormalities Rate (\%)} = \frac{\text{Total Number of Abnormal Larvae}}{\text{Total Number of Hatched Larvae}} \times 100 \quad (4)$$

The results of hatching, survival and abnormalities rate then continued to be analyzed using Analysis of Variance (ANOVA).

3 Results and Discussion

3.1 Phytochemical Screening

Based on the phytochemical screening, the papaya leaf extract in this research contained alkaloid, saponin, flavonoid, steroid, polyphenols, and glycoside as shown in Table 1. These results could be different from other papaya leaf extracts from different resources. As phytochemical results of papaya leaf extract that was taken in Malang Regency showed it contained alkaloid, triterpenoid, steroid, flavonoid, saponin, and tannin [12]. Meanwhile in this research, it was negative results for tannin, triterpenoid, and also quinone.

Table 1. Phytochemical results of papaya leaf extract.

Phytochemical Compounds	Results
Alkaloid	Positive
Saponin	Positive
Flavonoid	Positive
Steroid	Positive
Polyphenols	Positive
Glycoside	Positive
Tannin	Negative
Quinone	Negative
Triterpenoid	Negative

3.2 Embryogenesis Observation

The fertilization rate (FR) in this research was obtained 76%. The fertilization rate of nilem fish (*Osteochilus hasselti*) eggs can reach 49.25-88.50%. Several factors that can affect the fertilization rate in fish are egg quality (egg maturity level), male sperm quality, environment or water quality including dissolved oxygen, pH, and temperature, as well as the effect of human handling such as hormone injection doses. Fertilized *Osteochilus hasselti* eggs have clear and transparent characteristics, while unfertilized eggs are cloudy, and milky white [2], as shown in Figure 1.

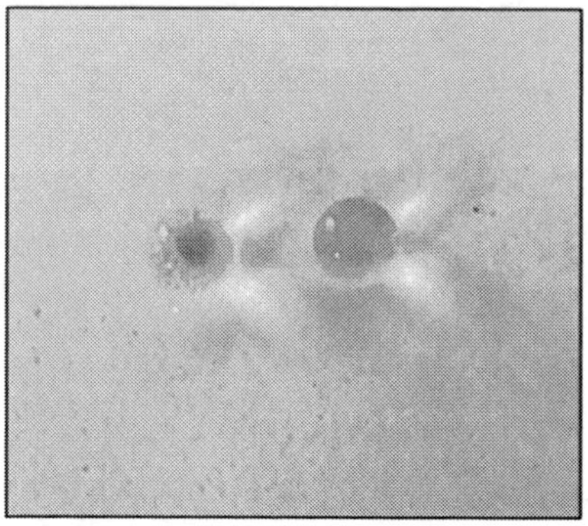

Fig. 1. Fertilized (left) and Unfertilized (right) Egg of Nilem Fish (*Osteochilus hasselti*).

Observation of the embryogenesis phase of *Osteochilus hasselti* eggs lasted 20-24 hours. The embryonic development phase begins right after fertilization occurs. The embryogenesis phase consists of cell division (cleavage), morula, blastula, gastrula, and organogenesis, then hatching into larvae. These observations were made by taking egg samples and then observed them under a microscope with 4x magnification [22]. On examination under a microscope, it was found that the development of the egg had entered the morula phase. This shown that the cleavage phase has passed and cannot be observed under a microscope, because this phase has taken place during the egg transfer process from the hatchery ward to the egg hatching laboratory.

Based on observations, 19 hours after the fertilized, it was found that some of the *Osteochilus hasselti* eggs had hatched into larvae in treatments D, C, and B, while in treatment A (control) there was no eggs had yet hatched. This shows that papaya leaf extract can speed up the hatching process of *Osteochilus hasselti* eggs [15, 20]. Based on the results of phytochemical tests in this study, papaya leaf extract was known contain alkaloid, saponin, flavonoid, steroid, polyphenols, and glycoside. Those components might be giving some affection toward the hatching process. Such as saponin, it can degrade the membrane or wall of egg cells so that the larvae in the eggs will emerge quickly. This saponin can damage the egg cell membrane so that there is a change in the structure of the egg cell wall which causes the fluid in the egg cell to come out [20]. However, on the other hand, high concentration of saponin also can cause the egg to die when the cells in the egg are not yet fully formed and not ready to hatch. Apart from that, the proteolytic enzyme content in papaya leaves can also speed up egg hatching. The proteolytic enzymes in papaya leaves work well, thereby encouraging faster embryo growth [15]. After 23 hours, the eggs were completely hatched in all treatments. Generally, *Osteochilus hasselti* eggs will hatch within 18-24 hours [23].

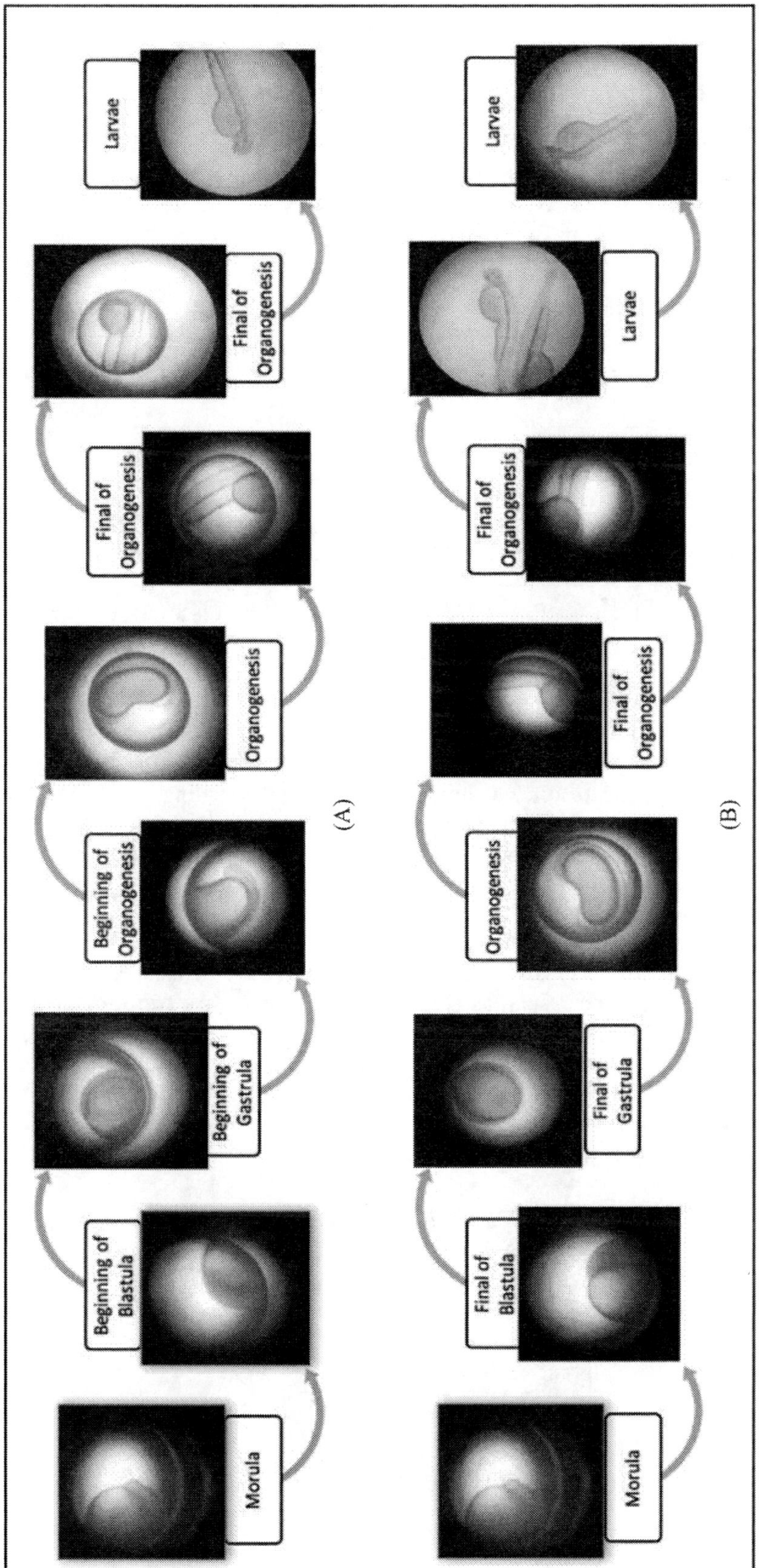

Fig. 2. Observation of Embryogenesis Phase. (A) Egg in Treatment A, (B) Egg in Treatment D.

3.3 Results of Hatching Rate, Larvae Abnormalities and Survival Rate

Hatchability is the rate at which fish eggs hatched. Observation of the hatchability of *Osteochilus hasselti* eggs begins when the eggs first hatch. The egg can be said to have hatched if the chorionic membrane breaks due to the movement of the larva's tail and body [18]. Based on the Honest Significant Difference (HSD) test on egg hatchability, there are significant differences between each treatment ($P<0.05$) which can be seen in Table 2.

Table 2. Results of hatching rate, larvae abnormalities, and survival rate of *Osteochilus hasselti*.

Treatments	Hatching Rate (%)	Larvae Abnormalities Rate (%)	Survival Rate (%)
A (Control)	$76,50 \pm 1,29^{a}$	$9,16 \pm 1,18^{c}$	$73,87 \pm 2,44^{a}$
B (0.75 mg/L)	$79,50 \pm 1,29^{b}$	$7,24 \pm 1,28^{b}$	$76,13 \pm 2,86^{a}$
C (1 mg/L)	$83,50 \pm 1,29^{c}$	$5,09 \pm 1,15^{a}$	$77,25 \pm 1,95^{ab}$
D (1.25 mg/L)	$87,25 \pm 2,21^{d}$	$4,30 \pm 1,14^{a}$	$80,55 \pm 2,62^{b}$

Note: Different superscript letters in the table above indicate a significantly different effect between treatments and the ± sign indicates the standard deviation.

The highest hatching rate of *Osteochilus hasselti* eggs was found in treatment D (1.25 mg/L) with 87.25% as the result. Meanwhile, the lowest hatching rate was obtained in treatment A, without additional of papaya leaf extract, which showed 76.50%. It showed that papaya leaf extract could be able to prevent the eggs from being attacked by microorganisms such as fungi and bacteria which caused the eggs not to hatch. It is because papaya leaf extract contains active compounds that has antifungal and antibacterial activity, such as flavonoid, saponin, tannin, quinone, steroid, triterpenoid, and alkaloid [13].

Other factors that influence the hatchability of eggs are fertilized eggs, inappropriate media, mechanical disturbances such as shaking, mechanical work from the activity of the larva itself and enzymatic work produced by the egg, light intensity, salinity, hormones produced by the pituitary, and thyroid plays a role in metamorphosis, and egg yolk volume [2, 3]. Besides, immotile sperm and poor quality of spermatozoa and eggs also related to the embryo development [24]. From the observation, the eggs that did not hatch showed undeveloped egg cell nuclei as shown in the following Figure 3.

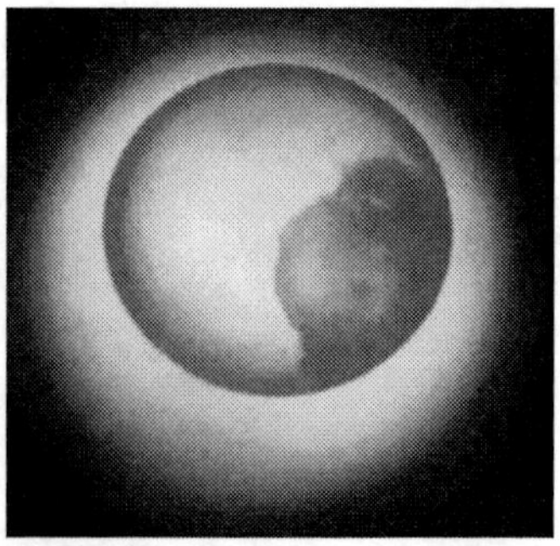

Fig. 3. The Unhatched Egg of *Osteochilus hasselti*.

Based on the results of the Duncan test ($P<0.05$) on larvae abnormality, treatment D did not show significantly different results from treatment C but was significantly different from treatment B and A. The overall results of the Duncan test for larvae abnormalities can be seen in Table 2. Larvae abnormalities in treatments D, C, and B which were given papaya leaf extract was given lower larvae abnormality rate compared to treatment A that was without any papaya leaf extract addition. This is thought to be because treatment without papaya leaf extract caused the eggs to be attacked by disease. Eggs infected with the disease will produce unhealthy and abnormal larvae. Many larvae produced from unhealthy eggs have abnormal body conditions in the form of bent bodies, incomplete body organs, and others [3]. Apart from that, the proteolytic enzyme content in papaya leaf can also stimulate faster embryo growth and help the embryo escape from the chorion layer during the hatching process, thereby reducing abnormalities in the larvae [15]. The results of the observations showed that abnormal larvae were visible in the shape of the spine and tailbone which were bent as in the following Figure 4.

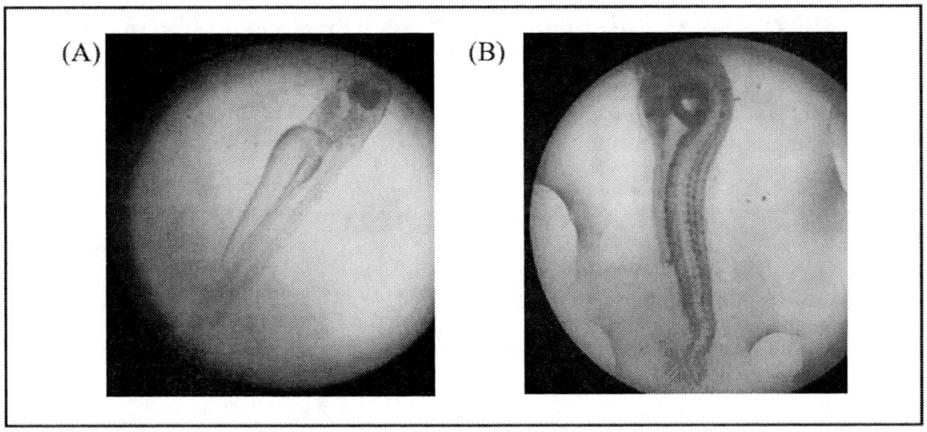

Fig. 4. Normal (A) and Abnormal Larvae (B) of *Ostechilus hasselti.*

Based on the Honestly Significant Difference (BNT) test ($P<0.05$) on larvae survival rate, treatment D showed different results to treatments C, B, and A. The results of the Honestly Significant Difference Test (BNT) on overall larvae survival rate were seen in Table 2. Giving papaya leaf extract to *Ostechilus hasselti* eggs had an effect on larvae survival because in treatments D, B, and C the level of larvae survival was better compared to treatment A (control) which was without addition papaya leaf extract. This is could be because the leaf extract contains flavonoid compounds which can function as immunostimulants. Flavonoid compounds have the potential to work as immunostimulants and work on lymphokines produced by cells so that they will stimulate phagocytic cells to carry out a phagocytic response in forming the body's immune system against fish larvae [8].

In addition, larvae survival was lower in treatment A (control) compared to treatments B, C, and D, because treatment A (control) had a high level of larvae abnormality rate compared to other treatments. Abnormal larvae usually cannot survive long and will die. Abnormal larvae will spend a lot of energy obtained from food to improve body condition and will not be able to grow properly [26]. This condition causes the abnormal larvae to not survive long so the larvae will die [27].

https://doi.org/10.1051/bioconf/20248703025

4 Conclusion

Based on the research and ANOVA test, addition of papaya leaf extract had a significant effect ($P<0.05$) on hatching, abnormalities, and survival rate of nilem fish (*Osteochilus hasselti*) larvae.Treatment D (1.25 mg/L) was the best treatment that produced the highest hatchability, as much 87.25%, the lowest larvae abnormality rate, as much 4.30%, and the highest larvae survival rate, as much 80.55%.

References

1. J. Subagja, *Increasing production technology through gonad maturation, incubation of eggs, and provision of prospective female nilem fish (Osteochilus hasselti) as future freshwater commodities*, in Proceedings of Seminar Nasional Ikan VI, SNI VI, **6**, 315-322 (2011)
2. W. Andriyanto, S. Bejo, I.M.D.J. Ariawan, JITKT, **5**, 192-203 (2013)
3. A. Rosyida, F. Basuki, R.A. Nugroho, T. Yuniarti, S. Hastuti, IJTA, **5**, 97-106 (2021)
4. R. Mahendra, S. Titik, B.P. Slamet, IJTA, **7**, 45-55 (2023)
5. Rivanto, I. Sidabalok, H. Hasan, J. Ruaya, **1**, 17-22 (2014)
6. T.I. Sabrina, Sudarno, H. Suprapto, JIPK, **6**, 7-12 (2014)
7. A. Zubaidah, S. Samsundari, M.A.Q. Jaelani, JARI, **9**, 109-117 (2021)
8. M. Ghofur, M. Sugihartono, R. Thomas, JIUBJ, **14**, 37-44 (2014)
9. M. Walidin, M. Admi, Arwana: JIPSP, **2**, 80-89 (2020)
10. H.S. Mulyani, T.I. Johan, DP, **36**, 99-110 (2020)
11. Saenal, S. Yanto, Amirah, JPTP, **6**, 115-124 (2020)
12. Q. A'yun, A.N. Laily, *The phytochemical analysis of papaya leaf (Carica papaya L.) at the research center of various bean and tuber crops kendalpayak, malang,* in Proceedings of Seminar Nasional Konservasi & Pemanfaatan Sumber Daya, KPSDA, **1**, 134-137 (2015)
13. L.G. Sumahiradewi, N.D. Sulystyaningsih, Y. Pratama, J. Perikanan, **12**, 86-96 (2022)
14. S.A. Akbar, M. Hasan, S. Afriani, C. Nuzlia, Biodiversitas Journal of Biological Diversity **24(10)**, 5283 (2023)
15. V. Muhardina, P.M. Sari, Y. Aisyah, S.Haryani, S.A. Akbar, Rasayan J. Chem **13(1)**, 240-244(2020)
16. R.S. Sugiyanto, R. Khamsita, M. Labertus, R.Y. Utomo, A.A. Susidarti, IJCC, **3**, 384-390 (2012)
17. J.R. Shaiks, M.K. Patil, Int. J. Chem. Stud., **8**, 603-608 (2020)
18. S.Y.T. Ina, S. Tangguda, N.P.D. Kusuma, JURRIH, **2**, 14-24 (2023)
19. Y.B. Esa, A.M. Dadile, F. Syukri, A. Cchristianus, M. Y. Diyaware, Animals, **13**, 1723 (2023)
20. A.F.N. Inaya, Kismiyati, S. Subekti, JIPK, **7**, 159-164 (2015)
21. A. Rolton, J. Vignier, P. Soudant, S.E. Shumway, V.M. Bricelj, A.K. Volety, J. Aquatox, **155**, 199-206 (2014)
22. S. Setiawati, H. Latuconsina, H.D. Prasetyo, Agrikan, **15**, 426-431 (2022)
23. P. Sumedi, Sudiana, *Bul. Tek. Lit.* Akuakultur, **12**, 123-127 (2014)
24. Rachimi, E.I. Raharjo, M. Syaidi, J. Ruaya, **4**, 39-44 (2016)

25. Azwar, Arwana: JIPSP, **2**, 73-79 (2020)
26. D. Nugraha, M.N. Supardjo, Subiyanto, Maquares, **1**, 33-37 (2012)
27. G. Saptiani, E.H. Hardi, C.A. Pebriyanto, Agustina, J. Veteriner, **17**, 285-291 (2016)

BIO Web of Conferences **87**, 03026 (2024) https://doi.org/10.1051/bioconf/20248703026
ICFAES 2023

Efficiency of drift gillnets in nagan raya district, aceh, indonesia

Alvi Rahmah[1,3*], *Ratna* Mutia Aprilla[1,3], *Syahrul* Okta Yufriada[1], *Imelda* Agustina[1,3], *Salmarika* Salmarika[2], and *Yusuf* Fathanah[4]

[1]Fisheries Resources Utilization Department, Faculty of Marine and Fisheries, Universitas Syiah Kuala, Banda Aceh 23111, Indonesia
[2]Department of Marine Science, Faculty of Agriculture, Malikussaleh University, North Aceh 24351, Indonesia
[3]Marine and Fisheries Research Center, Universitas Syiah Kuala, Darussalam, Banda Aceh.
[4]Palabuhanratu National Fishing Port, Directorat of Capture Fisheries, Ministry of Fisheries and Marine Affairs, West Java 43364, Indonesia

Abstract. The achievement of technical and economic efficiency is one indicator in assessing the fishing process's capability. The purpose of this study was to formulate production factors of drift gillnets at The Kuala Tadu and The Kuala Tuha fishing base and to calculate the technical and economic efficiency of the use of production factors for drift gillnets at both locations. Production factors that support the success of drift gillnets operation are net depth, net length, boat size, amount of fuel, trip duration, mesh size, and engine power analyzed using the Cobb-Douglas approach. The results showed that the equation of production function for drift gillnet is $LnY = 3.110 + 0.112\ LnX_2 + 0.191\ LnX_3 + 0.105\ LnX_4 + 0.251\ LnX_6$ with a value of R^2 is 0.833. The technical efficiency study of 20 units of drift gillnets showed that the value of production elasticity (EP) for using production factors for net length (0.112), vessel size (0.191), amount of fuel (0.105), and mesh size (0.251) was < 1. It means that the use of factors of production is no longer efficient. The economic efficiency of using the variable production factors of drift gillnets is no longer efficient because the value of NPMxi/BKMxi <1.

1 Introduction

Nagan Raya Regency is a new district formed as a result of the expansion of West Aceh Regency. This area has a lot of potential, especially those related to marine products, such as fisheries resources [1]. The water area in Nagan Raya Regency is a coastal area that is active as a fish catcher. This region has an excellent diversity of natural resource potential, both biological natural resources and non-biological natural resources. These potential resources can be utilized by residents living in coastal areas who earn their living as fishermen [2].

Fishermen in Nagan Raya Regency have a high dependence on fisheries resources as their main source of livelihood and almost all fisheries activities are based in coastal areas [3]. Fishermen's catches in Nagan Raya Regency will be landed at fish landing sites. The fish landing site (TPI) is one of the fisheries facilities in Nagan Raya Regency. These landing places include The Kuala Tadu fishing base and The Kuala Tuha fishing base.

*Corresponding author: alvi_rahmah@usk.ac.id

One of the fishing tools used at The Kuala Tadu fishing base and The Kuala Tuha fishing base is drift gillnets. The number of fishing gear used at The Kuala Tadu fishing base and The Kuala Tuha fishing base is 12 and 8 units respectively. The total catch production at The Kuala Tadu fishing base decreased in April by 48.05% and at The Kuala Tuha fishing base in March by 38.07%. The decrease in the amount of catch production is due to March and April entering the transition season and western season [4].

Based on previous research, the fishermen's problem always face is the lack of knowledge about how to use production costs to achieve efficient and optimal fishing [5]. According to [6] stated that production factors experienced an increase in prices which caused catches to tend to be uncertain so that fishermen's income tended to be uncertain. Fishing success is the main factor that is influenced by the level of fishing effort carried out by drift gill net fishermen in the use of production factors.

Achieving technical and economic efficiency, especially at The Kuala Tadu fishing base and The Kuala Tuha fishing base, can be achieved if fishermen know the production factors that influence their fishing efforts. If the use of production factors can be carried out appropriately and pursued well, then the fishing operation can be carried out carried out more efficiently, so that optimal fishing success can be achieved. Based on this description and supported by the existence of The Kuala Tadu fishing base and The Kuala Tuha fishing base which have the potential to be developed, research is needed regarding the technical and economic efficiency of production factors in drift gill net fishing units at The Kuala Tadu fishing base and The Kuala Tuha fishing base.

2 Material and Methods

2.1 Location and time

This research was carried out for 1 month from February to March 2022. Located at the Kuala Tadu and Kuala Tuha fishing base, Nagan Raya district, Aceh, Indonesia.

2.2 Sampling procedure

The research method used in this research is a survey method and interviews with 20 fishermens at the both fishing base.

2.3 Data Analysis

2.3.1 Formulation of Production Factors

This research can formulate production factors for drift gillnet fishing units. Before researchers can formulate it, this research must be tested first. The testing stages are carried out in 2 steps, namely:

1. Classic Assumption Test:
    a. Normality test
    b. Heteroscedasticity Test
    c. Multicollinearity Test
    d. Autocorrelation Test

2. Analysis of Production Factors

Production factor analysis is an analysis that explains the relationship between production and the production factors that influence it. According to [7], for observing the influence of certain factors of production on overall output in actual circumstances is impossible. Therefore, the relationship between production factors and output needs to be simplified in the form of a model. The variables used in drift gillnet must have different sizes so that the influence of these variables can be seen. Quantitative relationships between production factors obtained at The Kuala Tadu fishing base and The Kuala Tuha fishing base, namely net depth (X_1), net length (X_2), boat size (X_3), amount of fuel (X_4), trip duration (X_5), mesh size (X_6), engine power (X_7), with production can be calculated based on the Cobb-Douglas production function. For this reason, the Cobb-Douglas Logarithmic Model can be seen as follows:

$$\text{LnY} = \text{Lna}_1 + b_1\text{Ln}X_1 + b_2\text{Ln}X_2 + \ldots + b_n\text{Ln}X_n + \text{Ln e} \qquad (1)$$

Where:

a	= Intercept
b_1, b_2	= Production elasticity
Y	= Catch production (kg)
X_1	= Net depth (m)
X_2	= Net length (m)
X_3	= Boat size (GT)
X_4	= Amount of fuel (liters)
X_5	= Trip duration (hours)
X_6	= Mesh size (inches)
X_7	= Engine power (PK)

To take into account the test, the coefficient of determination is then calculated, the percentage of influence of the independent variable on the dependent variable as a whole is shown by the coefficient of determination (R^2). This test is intended to find out how much influence the independent variable can have on the dependent variable. Then a statistical test is carried out on the production function. The tests carried out in this case are testing the estimator model (f test) and testing the regression parameters (t test).

2.3.2 Technical and Economic Efficiency

1. Technical Efficiency

The regression coefficients b1, b2…bn from the Cobb Douglas production function are the production elasticities of the input variables. The magnitude of production elasticity (Ep) can be used to measure the level of technical efficiency and economic efficiency of using variable inputs. The level of technical efficiency in input use is achieved if Ep = 1, if the value of Ep < 1 then the use of the input is inefficient and if the value of Ep > 1 then the use of the input is inefficient [7].

2. Economic Efficiency

Economic efficiency can be achieved if we can maximize profits. This means equating the marginal product of each production factor with its price [8].

According to [9], economic efficiency is achieved by looking at the comparison between the value of marginal product (NPM) of each input and the input price or marginal factor cost (BKM) = 1.

$$\frac{bx\,Y\,x\,Py}{X} = Px \qquad (2)$$

$$NPM = \frac{b\,x\,Y\,x\,Px}{X} \qquad (3)$$

$$BKMx_i = Px \quad (4)$$

$$\frac{NPMx_i}{BKMx_i} = 1 \quad (5)$$

Where:

NPM	= Marginal product value
BKM	= Marginal factor costs
Px	= Price production factors or marginal factor costs x_i (BKM x_i)
x_i	= Production input
Py	= Production cost
Y	= Production
X	= Number of production factors
b	= production elasticity

In many cases the reality of NPMxi/BKMxi is not always equal to 1, what often happens is as follows [8]:

a. $NPMx_i/BKMx_i > 1$: meaning that the use of input X is not yet efficient, to achieve efficiency input X needs to be increased.
b. $NPMx_i/BKMx_i < 1$: meaning that the use of input X is inefficient, to achieve efficiency, the use of input X needs to be reduced.

3 Results and Discussions

3.1 Formulation of production factors

Production factor data obtained from the drift gillnet was tested using the classic assumption test, which consists of the normality test, heteroscedasticity test, multicollinearity test, and autocorrelation test. The results are as follows:

1. Normality Test

The Normality Test can be said to be normally distributed if the points are spread evenly on a diagonal line. The test results in this research can be seen in Figure 1. Figure 1 shows the normality test for the dependent variable, that the points are spread evenly around the diagonal line and the distribution follows the direction of the diagonal line or in other words has a normal distribution. This means that the model is suitable to be used to predict production results using Cobb Douglass regression analysis.

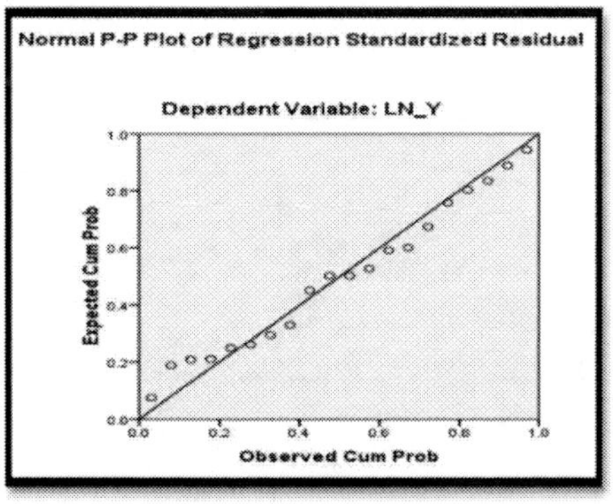

Fig. 1. Normality test

2. Heteroscedasticy Test

The heteroscedasticity test functions to see whether the variance of the observation residuals is related between variables or not. If these variables are related, then the independent variable of drift gillnets has symptoms of heteroscedasticity. The test results show the dots resulting from processing between the independent variable and the dependent variable spread over the Y axis and do not have a clear pattern (wavy, widening, then narrowing) in the Scatterplot image, the dots spread above and below the number 0 on the Y axis. It is concluded that this research meets the requirements of the heteroscedasticity test because the independent variables do not show symptoms of heteroscedasticity (Figure 2).

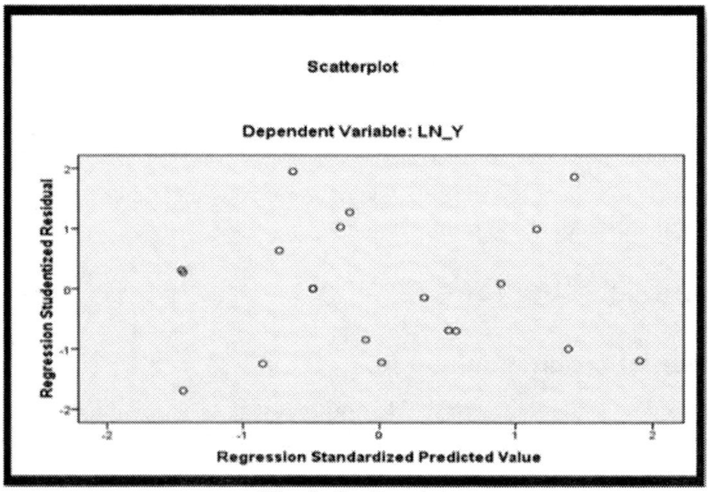

Fig. 2. Scatter plot

3. Multicollinearity test

The multicollinearity test can determine the relationship between independent variables by using a table of statistical coefficients on the VIF value. The multicollinearity test was carried out using SPSS version 16.0. The results of the multicollinearity test can be seen in Table 1.

Tabel 1. Statistical Coefficient

No.	Collinearity Statistics		
	Model (Constant)	**Tolerance**	**VIF**
1.	Ln X_1	0.622	1.608
2.	Ln X_2	0.499	2.003
3.	Ln X_3	0.301	3.324
4.	Ln X_4	0.367	2.722
5.	Ln X_5	0.765	1.308
6.	Ln X_6	0.684	1.462
7.	Ln X_7	0.469	2.134

Based on Table 1 above, the results of the calculations in the multicollinearity test show that almost all of the independent variables do not contain multicollinearity interference so that the regression model is suitable for use. The requirements for the multicollinearity test is value VIF < 10 and tolerance value > 0.100.

4. Autocorrelation test

A good regression equation does not have autocorrelation problems. If autocorrelation occurs then the equation is not good and is not suitable for prediction. The autocorrelation test was carried out using SPSS version 16.0 and can be seen in Table 2 below:

Table 2. Durbin Watson

Number	Durbin Watson	Terms Description
1	1.992	1< DW < 3 Autocorrelation does not occur

From the results of Table 2 above, the Durbin-Watson test value is found to be 1.992 and the condition is that there is no autoscorrelation if the DW value is between values 1 to 3, then with the DW results that have been calculated, namely 1.992, it is indicated that there is no correlation. The calculation of the coefficient of determination (R^2) essentially aims to measure how capable the model is of dependent variation. If the R^2 value is small, the ability of the independent variables to explain variable variations is limited. The magnitude of the influence of the independent variable on the value of the dependent variable can be determined by looking at the determination value from the calculation results using SPSS version 16.0. Based on these calculations, the output can be seen in Table 3 as follows:

Table 3. Coefficient of Determination

Model	No	R	R *Square*	*Adjusted R Square*	*Standard error*
1.	1	0.913a	0.833	0.736	0.079

a. Predictors: (Constant), engine power, mesh size, trip duration, net length, net depth, amount of fuel, boat size
b. Dependent variable: Fishing catch

Based on the dependent variable in Table 3 above, there is a coefficient of determination (R^2) of 0.833 or 83.3% and a correlation coefficient of 0.913. The coefficient of determination value shows that the independent variable used in the model influences 83.3%. Meanwhile, the remaining 16.7% is influenced or explained by other variables not included in this research model. These other factors are thought to include environmental factors or conditions of fishing areas such as weather, fishing season and resource conditions. The standard error of estimation (0.079) measures the level of error of the regression model in predicting the Y value.

The results of the Cobb-Douglas analysis which are tested simultaneously will provide information regarding whether the independent variables have a joint influence on the dependent variable. For more details, see Table 4 as follows:

Table 4. ANOVA table analysis of production function

Source	df	Sum of Squares	Mean squared	F-test	F-table	*P-value*
Regression	7	0.382	0.550	8.540	2.910	0.001
Residual	12	0.77	0.006			
Total	19	0.459				

Based on Table 4 above, it shows that Fcount (8.54) > Ftable (2.91) with a confidence level of 95%. Based on the results of these calculations, the gill net fishing unit (Xi) has a significant influence simultaneously (together) on the catch (Y) of drift gill nets. In partial testing, meaning individually, the t-test is used as a basis for decision-making to see the influence of each production factor variable on catch results.

Table 5. Regression coefficient values and production function T-test

Source	Coefficient	Standard Error	T-test	Sig.
Intercept	3.110	0.863	3.602	0.004
Ln X_1	-0.024	0.139	-0.170	0.868
Ln X_2	0.112	0.790	1.423	0.180
Ln X_3	0.191	0.115	1.658	0.123
Ln X_4	0.105	0.308	0.339	0.740
Ln X_5	-0.479	0.163	-2.943	0.012
Ln X_6	0.251	0.068	3.686	0.003
Ln X_7	-0.339	0.179	-1.892	0.083

T-table with a 95% confidence interval is 2.09

The coefficient value of net depth (X_1) with a coefficient value of -0.024 shows the variable does not affect the catch production of gillnet assuming other variables have a fixed value. However, different from the research of [10], the deeper the net, the more fish can be caught, this makes it more difficult to escape because the net coverage traps the fish quite deeply.

The net length coefficient value (X_2) shows the influence on the production of gillnet catches with a coefficient value of 0.112, which means that every 1 unit of net length (meter) will cause the production of catches to increase by 0.112% assuming the other variables are constant. According to [11], the length of the net determines how much space the fish caught in the net can move, so the longer the net, the greater the chance that the fish will be caught. However, in this study, the effect of net length did not have a significant effect on catch results.

The coefficient value of ship size (X_3) shows that there is an influence on the production of gillnet catches with a value of 0.191, which means that every increase in ship size (GT) affects the catch by 0.191%. During the fishing process, the boat is in a stationary position so that it does not affect fishing operations [12]. In this study, small boats were also able to get more catches compared to large boats.

The coefficient value for the amount of fuel (X_4) shows the influence on the production of gill net catches with a coefficient value of 0.105, which means that every additional 1 liter of fuel will cause the additional production of catches to increase by 0.105%, assuming the other variables are fixed. According to [13], fuel is a very important production factor in the fishing business, because the greater the amount of fuel used, it is hoped that fishermen will have more freedom to go to fishing grounds and get more catches. However, in contrast to this research, the fuel variable itself does not have a real effect on the increase in catch production.

The coefficient for the duration of the trip (X_5) shows the influence on the production of gillnets catches with a coefficient value of -0.479, which means that any time spent at sea will cause the production of the catch to not affect production assuming other variables have a fixed value. This decrease was based on the fact that the trip length variable had been excessive. The increase in trip length showed that it had no real effect on catch results. This is slightly inconsistent with research by [5], which states that fishing time is an important part of a fishing business, therefore the variable trip duration statistically has a real and positive effect on catch results. Other research by [13], stated that the longer the time the net is submerged or the longer the time at sea, the more fish will be caught.

The mesh size coefficient value (X_6) shows the influence on the production of gill net catches with a value of 0.251, which means that each additional mesh size can influence catch results. The addition of every 1-inch mesh will provide an increase of 0.251% in the production of drift gillnet catches. According to [14], stated that the larger the mesh size, the larger the size of the fish caught, and the fewer the number, conversely the smaller the mesh size, the smaller the size of the fish caught but the greater the number. Therefore, the use of small nets must be avoided to minimize the catch of juvenile fish [15]. However, the difference thing with this research is that a large number of mesh sizes are used to provide varying catch results starting from the type and size. The value of the mesh size variable shows a real influence on catch results, which means that the number of types of mesh sizes used will have a real effect on the production of drift gill net catches.

The engine power coefficient value (X_7) also shows an influence on the production of gillnet catches with a value of -0.339, which means that any increase in engine power will not affect the catch. According to [16], the engine power variable does not have much influence on ship speed because the gillnet operates passively which does not require ship movement during fishing operations. Ship movements are usually carried out during the hauling process, where when the ship is moved slowly it will make the process of lifting the net onto the ship easier.

Cobb-Douglass function regression equation on changes in catch production (Y) has the model $LnY = 3.110 - 0.024\ LnX_1 + 0.112\ LnX_2 + 0.191\ LnX_3 + 0.105\ LnX_4 - 0.479\ LnX_5 + 0.251\ LnX_6 - 0.339\ LnX_7$. This equation shows that only 4 variables have positive values or in other words have an influence on increasing catches. These four variables are net length, ship size, amount of fuel, and mesh size. However, of these four variables, only the mesh size variable has a real effect. This is following research by [17], where the mesh size variable has a real influence, so the latest Cobb-Douglass function obtained is $LnY = 3.110 + 0.112\ LnX_2 + 0.191\ LnX_3 + 0.105\ LnX_4 + 0.251\ LnX_6$.

3.2 Technical and economic efficiency

Technical efficiency analysis focuses on independent variables that have significant values, while the variables that have positive values are net length (X_2), boat size (X_3), amount of fuel (X_4), and mesh size (X_6). The independent variable can be said to have achieved technical efficiency if the EP value = 1, if EP > 1 then the use of the input is not efficient, and if the EP value < 1 then the use of the variable input is also inefficient. Table 6 shows that the variables net length, ship size, amount of fuel, and net mesh size have significant positive values, so the level of technical efficiency is seen based on these four variables. The production elasticity values for net length, ship size, amount of fuel, and mesh size show EP values < 1, which means that the production input variables have not yet reached technical efficiency.

Table 6. Technical efficiency of the drift gillnet

No.	Variable	Average	E.P
1.	Net depth	3 m	-0.024
2.	Net length	117 m	0.112
3.	Boat size	3.4 GT	0.191
4.	Amount of fuel	26 Litres	0.105
5.	Trip duration	6 hours/trip	-0.479
6.	Mesh size	2 inches	0.251

7.	Engine power	25 PKs	-0.339

Viewed from an efficiency perspective, based on Table 6, the technical efficiency of the drift gillnet for production factors, variables of net depth, trip duration, and engine power has a negative production elasticity value (Ep<1) which indicates that the use of production factors is no longer efficient. This means that there has been excessive use of production factors by the drift gillnet fishing unit in its fishing operations during the western season. According to [6], it states that increasing the use of these production factors can result in total production decreasing so to achieve efficiency in the use of these production factors, it is necessary to reduce the use of production factors. Efficiency depends on reducing the amount of resources used, so the use of net depth, trip duration, and engine power needs to be considered.

The production factors for net length, ship size, amount of fuel, and drift gill net size are at the rational production stage because they are between 0<Ep<1, which means that with the use of production factors, the length of nets used by fishermen when fishing is appropriate. and obtain maximum catches without the need to reduce or add production factors. The production factors of vessel size and the average amount of fuel of 3.4 GT and IDR 181,650 used by fishermen at TPI have also entered the appropriate and balanced stage, however, these production factor variables do not have a significant effect on the production of drift gillnet catches. Meanwhile, the variable average mesh size used is 2 inches, which has also entered the appropriate and balanced stage and this variable has a real influence on the amount of catch production. This is also following research conducted by [18], that a mesh size of 2 inches has a real influence on gillnet catches.

Using the Cobb-Douglas production function model with regression coefficient values, apart from showing the elasticity of each variable concerned, the total of the regression coefficient values of the 7 production factor variables is an estimate of the state of the business scale of the ongoing production process. The total production elasticity in the model is 0.659, this shows that the business scale level is on the decreasing return to scale, meaning that the proportion of additional production factors exceeds the proportion of additional production. If too many things are added to the production process, the number of products produced will not increase [19]. This shows the resources used in the fishing process using drift gill net fishing gear.

The analysis of economic efficiency in this research also only focuses on the use of statistically significant variables, namely positive elasticity values, namely the variables net length (X_2), ship size (X_3), amount of fuel (X_4), and mesh size (X_6). The four variables analyzed show that the length of the net, the size of the ship, the amount of fuel, and the mesh size are not economically efficient (NPMxi/BKMxi < 1), the use of input X needs to be subtracted. The estimated results of calculating the economic efficiency of using inputs in the drift gill net fishing unit can be seen in Table 7 below. Economically, efficiency will be achieved in conditions where the price is equal to the value of the marginal product [20]. If the price of each production factor i (Pxi) is the marginal factor cost (BKM) and the marginal product multiplied by the price level of the catch (Y) is the marginal product value (NPM), then the condition of economic efficiency is achieved at NPMxi=BKMxi [8].

Based on Table 7, it can be seen that the use of production factors for drift gillnet fishing units is in an economically inefficient condition. Where the variables net length, ship size, amount of fuel, and mesh size have NPM/BKM values smaller than one. So it is necessary to review how production factors are used during the fishing process so that operational costs can be saved when using them. According to [7], the failure to achieve economic efficiency in the use of production factors causes a tendency for drift net fishermen to not carry out the fishing process. According to [21], weather is also a factor that fishermen consider when facing the western season which causes fishermen not to go to sea. Usually, they will only repair damaged fishing equipment. However, based on interviews, the decline in fish prices from drift gill net catches is also the main reason fishermen do not use gillnets. This is what causes catch production to decrease.

Table 7. Economic efficiency of drift gillnet

Description	Variables			
	Net lenght	**Boat size**	**Mesh Size**	**Fuel**
• Average use	117	3	2	181,650
• Elasticity of catch	0.112	0.191	0.251	0.105
• Marginal product	0.009	0.554	1.238	5.705
• Average price of catch (Rp)	30,500	30,500	30,500	30,500
• Value of marginal product (NPM) (Rp)	288.170	16,911.08 3	37,779.89 9	174,002 .5
• Input prices (BKM) (Rp)	45,833	26,384.37 5	235,750	181,650
• Economic efficiency index	0.001	0.001	0.166	0.966
Decision	Not efficient	Not efficient	Not efficient	Not efficient

4 Conclusions

The conclusions that can be obtained from this research are: Production factors for fishing using gill nets are depth net (X_1), net length (X_2), ship size (X_3), amount of fuel (X_4), trip time (X_5), mesh size (X_6), and engine power (X_7) simultaneously or together influence the catch (Y). However, partially or individually only the mesh size (X6) has a real influence on the catch (Y). So the formulation obtained from the Cobb-Douglass production factor is $LnY = 3.110 + 0.112\ LnX_2 + 0.191\ LnX_3 + 0.105\ LnX_4 + 0.251\ LnX_6$.Technical efficiency in the variables net length, ship size, amount of fuel and mesh size is at a production elasticity of $0<Ep<1$ which states that these variables are inefficient. Meanwhile, the economic efficiency of the variables net length, ship size, amount of fuel, and mesh size is at a value of $NPM/BKM < 1$, where the use of input X is not efficient. To achieve efficiency, the use of input X needs to be reduced.

References

1. Muhtarom, A. J. Penelitian Ekonomi dan Akuntansi (JPENSI), **2**, 1-15 (2017)
2. Lasabuda, R. J. Ilmiah Platax, **1**, 92-101 (2013)
3. Diana, F., M. Rizal. J. Perikanan Tropis, **3**, 153 (2016)
4. Dinas Kelautan, Perikanan dan Pangan Kabupaten Nagan Raya (Dinas Kelautan Perikanan dan Pangan, Nagan Raya, 2022)
5. Sukiyono, K., Romdhon, M. M. IJFST, **11**, 99-104 (2016)
6. Aprilla, R.M., Mustaruddin, E.S. Wiyono, N. Zulbarnaini. J. Teknologi Perikanan dan Kelautan, **4**, 9-20 (2013)
7. Soekartawi, Teori ekonomi produksi dengan pokok bahasan analisis fungsi Cobb-

Douglas (Raja Grafindo Persada, Jakarta, 1994)
8. Soekartawi, Prinsip ekonomi pertanian (Rajawali Press, Jakarta, 2003)
9. Nicholson W, Teori Mikro Ekonomi, Prinsip Dasar dan Perluasan (BinaRupa Aksara, Jakarta, 1995)
10. Pratama, M. A. D., T. D. Hapsari, I. Triarso. Jurnal Saintek Perikanan, **11**, 120-128 (2016)
11. Kurnia, D. R., P. Sukardi, A.Iqbal. Marlin, **2**, 131-140 (2021)
12. Aji, I.N., B. A. Wibowo, Asriyanto. J. of Fisheries Resources Utilization Management and Technology, **2**, 50-58 (2013)
13. Widiyanto, A.T., P. Pramonowibowo, I. Setiyanto. J. of Fisheries Resources Utilization Management and Technology, **5**, 19-26 (2016)
14. Irpan, A., D. Djunaidi, R. Hertati. SEMAH Jurnal Pengelolaan Sumberdaya Perairan, **2**, 12-15 (2018)
15. Fachrudin., H. Hudring, Identifikasi jaring insang (gillnet) (Balai Besar Pengembangan Penangkapan Ikan, Jakarta, 2012)
16. Muna, N., B.Ismail, B. Jayanto. J. of Fisheries Resources Utilization Management and Technology, **5**, 38-47 (2016)
17. Putri, V.L, F. Khuromah, A. D. P. Fitri. Indonesian Journal of Fisheries Science and Technology, **13**, 126-132 (2018)
18. Anggrayni, F. D., M. Zainuri. J. Ilmiah Kelautan dan Perikanan, **3**, 85-92 (2022)
19. Damayanti, H. O. Jurnal Litbang, **12**, 83-92 (2016)
20. Lestari, H. S., E. Lisarini, A.S. Alam, R.T.D. Jatmika. J. of Agricultural Sciences and Veteriner, **10**, 260-268 (2022)
21. Tiro, A. R., M. Irianti. Jurnal Abdimasa Pengabdian Masyarakat, **6**, 59-64 (2023)

Sea turtle landing and distribution in aceh jaya district

Fauzan Maulana[1,2], *Maria* Ulfah[1*], *Fauzan* Aulia[3], *Murniadi*[4], *Gunawan* Alza[5], *Tutia* Rahmi[5], *Oni* Kandi[6]

[1]Department of Marine Science, Faculty of Marine and Fisheries, Universitas Syiah Kuala, Banda Aceh, Indonesia.
[2]Ocean Diving Club, Universitas Syiah Kuala
[3]*Wildlife Conservation Society Indonesia Program (WCS-IP),* Banda Aceh, Indonesia.
[4]Aroen Meubanja Sea Turtle Conservation Group, Aceh Jaya, Indonesia
[5]Aceh Natural Resources Conservation Agency
[6]Marine and Fisheries Agency of Aceh Province

Abstract. The research in Aceh Jaya District identified nine turtle nesting beaches: Babah Kuala, Patek, Babah Ngom, Kuala Doe, Rigaih, Alue Piet, Nisero, Aroen Meubanja, and Seunebok Padang. Olive Ridley (*Lepidochelys olivacea*) and Leatherback (*Dermochelys coriacea*) turtles inhabit these coastal areas. Nisero beach, Kuta Tuha Village, Panga Sub-District, recorded the highest nest count (12 nests) with both Leatherback and Olive Ridley turtles. Seunebok Padang beach, Seunebok Padang Village, Teunom Sub-District, had the lowest count (2 Olive Ridley turtle nests). Aroen Meubanja beach is the only one with protected status, while the remaining eight face the threat of egg poaching. Unprotected beaches are at high risk, emphasizing the importance of conservation for sea turtle survival and the local economy, particularly through tourism development. Beyond Aceh Jaya, Indonesia features other significant sea turtle nesting sites, like Pangumbahan beach in West Java and Wairundi Island and Venu Island in West Papua Province. The study evaluates various beach characteristics, including width and slope, influencing sea turtle nesting activities. Sediment analysis reveals predominant medium sand textures in Aceh Jaya District, contributing to suitable nesting conditions and emphasizing the need for targeted conservation efforts.

1 Introduction

Sea turtles belong to a group of vertebrates in the class of reptiles that live in the sea. The existence of sea turtles has long been threatened, both from their natural habitat and from human activities. Internationally, sea turtles are included in the IUCN "red list" and appendix I of CITES (Convention on International Trade in Endangered Species), which means that their existence in nature is threatened with extinction [1].

Indonesia's coastal areas are the nesting habitat for five of the world's seven sea turtles, including the Olive Ridley turtle (*Lepidochelys olivacea*), Green turtle (*Chelonia mydas*),

* Corresponding author: mariaulfah@usk.ac.id

Leatherback turtle (*Dermochelys coriacea*), Loggerhead turtle (*Caretta caretta*), and Hawksbill turtle (*Eretmochelys imbricata*) [2]. Aceh Province is a sea turtle nesting habitat such as Lhoknga beach, Lampuuk beach, Lange beach, Gugop beach and Alue Raya beach in Aceh Besar District, Bengkaru beach in Aceh Singkil District [3] and Rantau Sialang beach in South Aceh District [4]. Some beaches in Indonesia are turtle nesting habitats such as the beach of Paloh District, West Kalimantan Province [5–8]. and Jamursba Medi beach in West Papua Province [9,10].

The population of sea turtles in Aceh is currently experiencing an increasingly troubling decline, due to several causes, one of which is the illegal hunting of eggs and capture of mothers by coastal residents who do not understand the importance of preserving this animal [11]. Activities carried out by residents around coastal areas, such as settlements and pond businesses, accelerate the process of damage and change in coastal ecosystems. Changes in the ecosystem environment in coastal areas, both direct and indirect, have resulted in damage to the habitats of biota living in these ecosystems [12].

This study aims to analyse the distribution of sea turtles and their habitat types in the coastal areas of Aceh Jaya District. The information obtained can be a reference for students for further research and the central and local governments in making decisions related to the management of these beaches, especially the Turtle Conservation Area.

2 Research Methods

2.1 Location and Time

This research was conducted from 23 December 2022 to 19 March 2023 along the coast of Aceh Jaya District. This research consists of two stages, namely field data collection and laboratory analysis. Field data collection was carried out along the coast of Aceh Jaya District, Aceh Province with information on beaches that are turtle nesting habitat obtained from residents around the coast, using random sampling techniques. Laboratory analysis was carried out at the Soil and Plant Research Laboratory, Faculty of Agriculture, Syiah Kuala University.

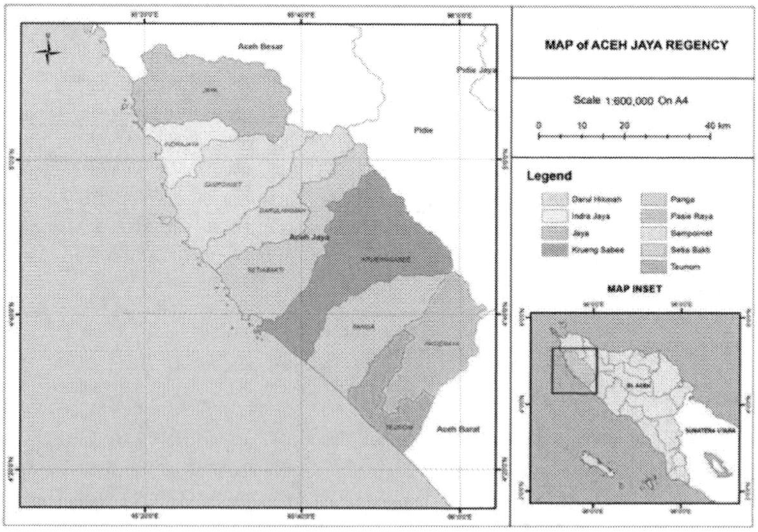

Fig.1. Map of Aceh Jaya District

2.2 Tools and Materials

The equipment used in this study were Global Positioning System (GPS), roll metre, 2 metre scale board, plastic, and stationery. The materials used in this study were data on the location of nests, beach slope, and beach sediment samples.

2.3 How It Works

2.3.1 Nesting Beach Designation

The determination of nesting beaches was carried out by looking for information circulating in the community around the coastal area, after which a sweep of the coastal area is suspected to be a sea turtle nesting habitat in the afternoon until morning. The determination of nests is not easy to do because it is very rare to find sea turtle hatchlings making nests, the determination of nests is done by looking at the flipper traces of sea turtle hatchlings left on the sand. Nests that were found were marked with GPS coordinates and entered into the Google Earth Pro application so that the data could be processed into a map using the ArcMap 10.3 application.

2.3.2 Measurement of Beach Width and Slope

Beach slope was measured using a Roll metre and a 2 metre high scale board. The slope was measured from the highest tide to the outer vegetation boundary. Measurements were made by pulling the roll metre from the outermost vegetation and placing it on the sand until it reached the highest tide limit. After that, a scale board is placed at the highest tide limit to get the height of the beach, so that the slope of the beach can be known by calculating the angle formed between the horizontal line and the vertical line [7].

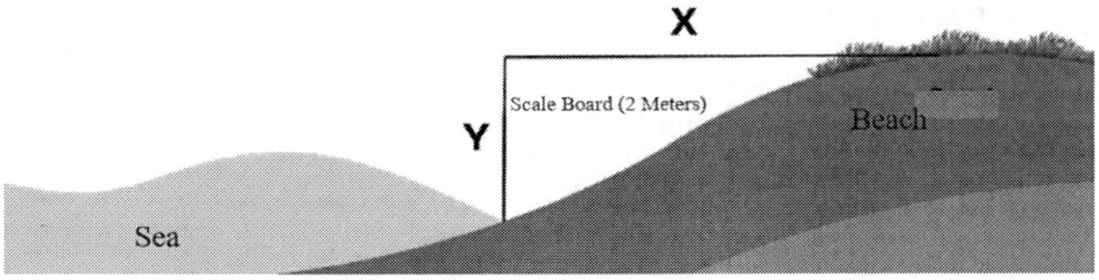

Fig.2. Illustration of beach slope measurement

2.3.3 Beach Sediment Sampling

The first step was to collect 200 grams of sediment from the beach that is the habitat of sea turtle nesting. Sediment has an impact on the ability of sea turtles to dig nests. Sediment measurement was carried out through a sieve process. The method of filtering the sediment substrate was carried out using a sieve shaker [13]. The volume measurement process utilised the soil texture classification [14].

2.4 Data Analysis

2.4.1 Beach Slope

The beach slope angle value is determined by the formula of Kalay et al. [15] as follows:

$$Beach\ Slope(°) = arc\ tan\ tan\ \left(\frac{y}{x}\right) \qquad (1)$$

Description:
x = Horizontal plane distance measured using a roll metre
y = Distance of the vertical plane of the beach placed at the highest tide limit using a scale board

2.4.2 Identification of Sediment Grain Size

According to Agus et al. [14] the calculation of sediment grain size uses the following calculation:

- $Sand = (cup\ weight + sandy\ soil) - weight\ of\ empty\ cup = X$ (1)
- $A = X * 10$
(2)

- $Clay\ dust = (cup\ weight + clay\ dust) - weight\ of\ empty\ cup = Y$ (3)
- $Clay = (cup\ weight + clay) - weight\ of\ empty\ cup = Z$ (4)
- $C = Z * 500$ (5)
- $Dust = clay\ dust - clay$ (6)
- $= (Y - Z) * 500 = B$ (7)
- $A + B + C = W$ (8)
- $\%Sand = \frac{A}{W}100\%$ (9)
- $\%Sand = \frac{B}{W}100\%$ (10)
- $\%Sand = \frac{C}{W}100\%$ (11)

After obtaining the sediment weight percentage for each particle size, the next step was to use the data to graph the soil texture diagram to classify the samples into the appropriate soil type.

3 Results And Discussion

3.1 Distribution of sea turtle nesting beaches in Aceh Jaya District

The results of the research survey showed 9 beaches that are turtle nesting habitat in Aceh Jaya District, including Babah Kuala beach, Patek beach, Babah Ngom beach, Kuala Doe beach, Rigaih beach, Alue Piet beach, Nisero beach, Aroen Meubanja beach, and Seunebok Padang beach. Sea turtle species that can be found in the coastal area of Aceh Jaya District are Olive Ridley turtle (*Lepidochelys olivacea*) and Leatherback turtle (*Dermochelys coriacea*). The beach with the largest number of nests was found in the Nisero beach area, Kuta Tuha Village, Panga Sub-District, the results showed 12 nests found with different types of turtles, 2 leatherback turtle nests and 10 olive ridley turtle nests, and the beach with the smallest number of nests found in the Seunebok Padang beach area, Seunebok Padang Village, Teunom Sub-District, 2 olive ridley turtle nests were found. From the research results, almost all beaches have 2 different species, there are only 2 beaches that have 1 species, namely Rigaih beach and Seunebok Padang beach. The following table shows the location of sea turtle nesting beaches in Aceh Jaya District.

Table 1. Location of sea turtle nesting beaches in Aceh Jaya District

No.	Sub-District	Village	Beach Name	Protection Status	Coordinate Points	Number of Nests Found	
						Olive Ridley turtle (*Lepidochelys olivacea*)	Leatherback turtle (*Dermochelys coriacea*)
1	Indrajaya	Kuala	Babah Kuala	Not protected	5°2'38.43"U 95°21'13.39"T	7	2
2	Darul Hikmah	Patek	Patek	Not protected	4°45'52.02"U 95°28'46.03"T	3	2
3	Setia Bakti	Sawang	Babah Ngom	Not protected	4°44'54.41"U 95°29'40.84"T	5	1
4	Setia Bakti	Lhok Geulumpang	Kuala Doe	Not protected	4°41'45.31"U 95°31'23.89"T	4	1
5	Setia Bakti	Lhok Buya	Rigaih	Not protected	4°39'57.54"U 95°34'3.22"T	3	0
6	Panga	Alue Piet	Alue Piet	Not protected	4°34'10.67"U 95°40'3.26"T	4	2
7	Panga	Kuta Tuha	Nisero	Not protected	4°32'8.38"U 95°42'29.42"T	10	2
8	Panga	Keude Panga	Aroen Meubanja	RMCA*	4°31'27.02"U 95°43'17.22"T	5	2
9	Teunom	Seunebok Padang	Seunebok Padang	Not protected	4°29'0.70"U 95°45'49.63"T	2	0

*Information: Regional Marine Conservation Area

There are a total of 9 sea turtle nesting beaches in Aceh Jaya District, only 1 of which has a protected status, namely Aroen Meubanja beach in Keude Panga Village, Panga Sub-District, the protection status of the area is written in the Decree of the Minister of Marine Affairs and Fisheries Number 76 of 2020, while the other 8 beaches have an unprotected status (Figure 3). Unprotected beaches are at great risk of high egg poaching [3]. While protected beaches can save sea turtles from extinction and benefit the economy of communities around the area, one of the benefits of conservation areas is tourism development [16].

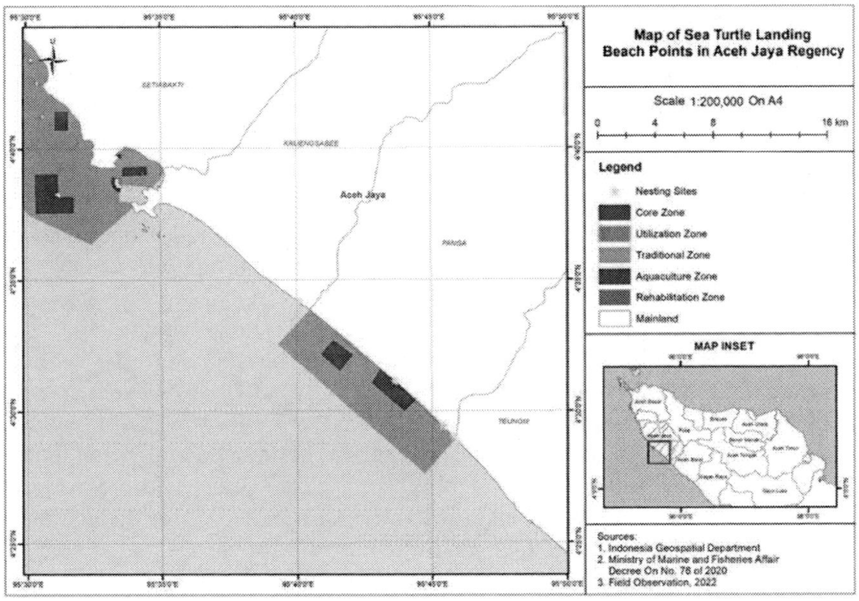

Fig.3. Map of turtle landing sites in Aceh Jaya District

Some beaches in Indonesia are sea turtle nesting beaches, such as Pangumbahan beach in Sukabumi District, West Java, which is a nesting habitat for green turtles (*Chelonia mydas*). In 2017, 26 green turtles were found landing, 25 nesting and 1 that did not nest [17]. Some islands in West Papua Province are also sea turtle nesting habitats, such as Wairundi Island and Venu Island. In 2016, 214 sea turtle nests were recorded on Venu Island, mostly from green and hawksbill sea turtles [18].

3.2 Beach Width and Slope

The results of measuring the width of the beach in Aceh Jaya District at each observation point from the outer vegetation boundary to the highest tide limit show that the width of the beach ranges from 16.35 metres to 53.77 metres. The shortest beach width is at the observation point on Rigaih beach which is 16.35 metres, while the longest width is at the observation point on Aroen Meubanja beach which is 53.77 metres. The results of coastal slope measurements at each observation point in Aceh Jaya District ranged from 1.19° to 6.15°. The lowest beach slope is on Aroen Meubanja beach, which is 1.19°, while the highest beach slope is on Patek beach with a value of 6.15°. The beach area of sea turtle nesting habitat in Aceh Jaya District mostly has a gentle slope category. The results of beach width measurements are shown in Table 2.

Table 2. Slope and width of the coast in Aceh Jaya District

No	Beach Name	Coordinate Point	Beach Width (M)	Beach Slope (°)	Category*
1	Babah Kuala	5°2'38.43"U 95°21'13.39"T	19,35	3,87	Sloping slope
No.	Beach Name	Coordinate Point	Beach Width (M)	Beach Slope (°)	Category*
2	Patek	4°45'52.02"U 95°28'46.03"T	17,82	6,15	Sloping slope
3	Babah Ngom	4°44'54.41"U 95°29'40.84"T	30,1	3,1	Gentle slope
4	Kuala Doe	4°41'45.31"U 95°31'23.89"T	18,5	3,34	Gentle slope
5	Rigaih	4°39'57.54"U 95°34'3.22"T	16,35	2,45	Gentle slope
6	Alue Piet	4°34'10.67"U 95°40'3.26"T	18,15	4,22	Sloping slope
7	Nisero	4°32'8.38"U 95°42'29.42"T	29,34	1,58	Gentle slope
8	Aroen Meubanja	4°31'27.02"U 95°43'17.22"T	53,77	1,19	Gentle slope
9	Seunebok Padang	4°29'0.70"U 95°45'49.63"T	39,35	2,33	Gentle slope

*Source: Kalay et al. [15]

Sea turtles have unique characteristics in choosing their nesting sites [6]. The slope of the beach and the width of the beach have an influence on turtle nests, the slope that is suitable for turtles to make nests is 10° to 30° and the width of the beach that is suitable for turtles to make nests is 20 metres to above 30 metres [19,4]. The slope of the beach has a great influence on sea turtles when landing on the beach to carry out nesting activities. The steeper the beach, the more difficult it will be for the turtle hatchlings to see the object in front of them, so that the greater the energy needed by the turtle hatchlings to climb to the beach [6,20].

Naturally, sea turtles choose beaches that meet criteria such as being wide and sloping, with an average slope below 30°, the beach is also located about 30 metres to 80 metres above the high tide mark [21]. Although some beaches are less than 30 metres wide, in reality, ocean waves at high tide do not reach the area where turtle nests are located, thus keeping the eggs safe. This is consistent with research by Fitriani et al. [4] conducted on Rantau Sialang beach, Pasie Lembang Village, South Aceh, which has an average beach width of 18 metres. The width of Rantau Sialang beach is not included in the ideal category for the characteristics of sea turtle nesting habitat, even so, in fact, sea turtle nests in Rantau Sialang are not submerged during high tide. This is also supported by the opinion of Nuitja [22] which states that a sufficient distance between the nest and sea water will prevent sea water from submerging the turtle nest. If seawater seeps into the nest, it can result in the failure of turtle egg hatching.

3.3 Nesting Beach Sand Texture

Textural composition of sediments according to grain size scale Wentworth [23] showed differences in sediment texture, namely from 9 beaches on the coast of Aceh Jaya District, there are 6 beaches dominated by medium sand texture, 1 beach is dominated by coarse sand texture, and 2 beaches are dominated by fine sand. According to Balaira et al. [24] there is a correlation between the medium sand component found in sea turtle egg nesting sites, among others, by reducing the possibility of sand collapse when sea turtle mothers dig holes and lay eggs, oxygen circulation and water absorption, ease for hatchlings to move out of the hatching hole. Research results Dima et al. [19] conducted on Sosadale beach, Rote-Ndao, that no sea turtle nests were found on beaches with rocky and coarse sand conditions, while on beaches with fine sand textures a number of sea turtle nests were found on the beach, this is because sand that has a fine texture and lighter colour creates an ideal nesting habitat for sea turtles to carry out the process of laying and hatching their eggs. If the nest is dominated by fine sand, it can make it difficult for turtles to dig the nest hole and make it difficult for hatchlings to get out because of the dense texture, whereas if coarse or gravelly sand dominates, the hole can easily collapse [24]. Pradana et al. [5] also stated that the composition of sediments with medium-sized sand tends to dominate around the holes where turtles lay eggs.

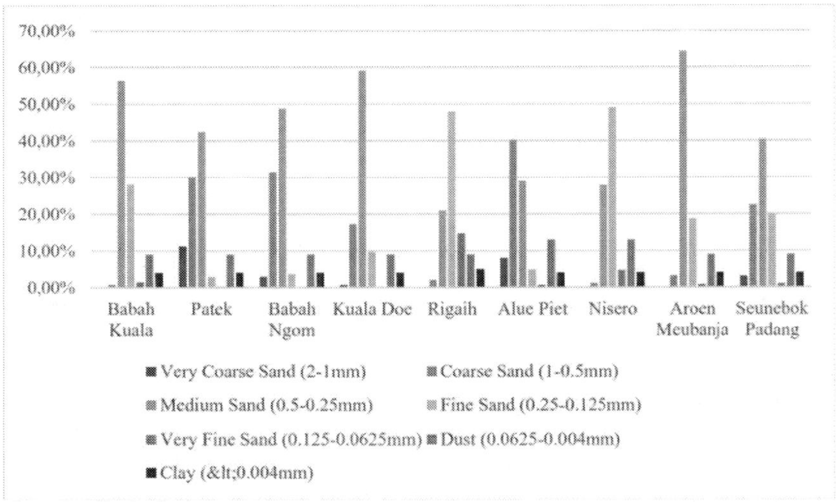

Fig.4. Sediment texture composition at 9 beach points

The composition of sediment texture in the coastal area of Aceh Jaya District is dominated by medium sand with a value of 43% followed by fine sand 21%, coarse sand 16%, dust 10%, clay 4%, and the lowest texture of very coarse sand and very fine sand which is 3%. These results show similarities as in Paloh beach, West Kalimantan, where medium sand texture dominates the beach with a value of 53.11% followed by fine sand 26.91%, coarse sand 12.18%, very fine sand 7.7% and the lowest texture is very coarse sand 0.08% [25].

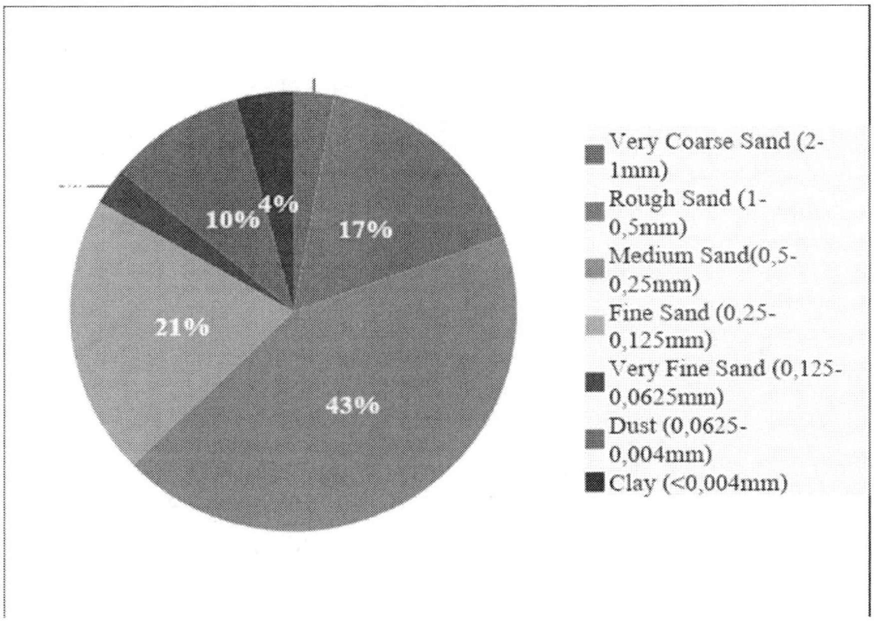

Fig.5. Percentage of beach sediment texture in Aceh Jaya District

4 Conclusion

Aceh Jaya district has 9 turtle nesting beaches, namely Babah Kuala beach, Patek beach, Babah Ngom beach, Kuala Doe beach, Rigaih beach, Alue Piet beach, Nisero beach, Aroen

Meubanja beach, and Seunebok Padang beach. However, of the 9 nesting beaches, only 1 beach is officially protected. The value of physical characteristics of the beach such as beach width ranging from 16.35 metres to 53.77 metres, beach slope ranging from 1.19° to 6.15° and sand sediment texture dominated by medium sand texture (0.5-0.25 mm) with a value of 43%.

References

1. A. Isdianto, O.M. Luthfi, M.A. Asadi, D. Aliviyanti, B. Semedi, Guntur, G. Arfiani. B.M. Putri, M.F. Hayka. Universitas Brawijaya, Malang: UB Press (2022)
2. SWOT. *The State Of The World's Sea Turtles Swot Report Volume IV: Discovering The Flatback Australia's Own Sea Turtle.* State Of The World's Sea Turtles. (2009)
3. Maslim, A. Farajallah. *Zoo Indonesia Jurnal Fauna Tropika.* **25.** 160–164 (2016)
4. Fitriani D, Zurba N, Edwarsyah, Marlian N, Munandar R A and Febrina C D. *Journal of Aceh Aquatic Science.* **5.** 35–45 (2021)
5. Pradana F A, Said S and Siahaan S. *Jurnal Hutan Lestari.* **1.** 156–63 (2013)
6. Anshary M, Setyawati T R and Yanti A H. *Jurnal Protobiont.* **3.** 232–239 (2014)
7. Mursalin, Budhi S and Manurung T F. *Jurnal Hutan Lestari.* **5.** 338–47 (2017)
8. Yusra A A, Fisesa A A, Fachrizal A and Susanto H. 2022. Yayasan WWF Indonesia.
9. Tapilatu R F and Tiwari M. *Chelonian Conservation and Biology.* **6.** 154–8 (2007)
10. Dutton P H, Gjertsen H and Squires D. *Handbook of Marine Fisheries Conservation and Management.* New York, USA: Oxford University Press. pp 195–204 (2010)
11. Juliono and Ridhwan M. *Jurnal Serambi Saintia.* **5.** 45–54 (2017)
12. Mustaqim A. *Jurnal Analisa Sosiologi.* **7.** 224–42 (2018)
13. Benni B, Adi W and Kurniawan K. *Jurnal Sumberdaya Perairan.* **11.** 1–6 (2018)
14. Agus F, Yusrial and Sutono. Balai Besar Litbang Sumberdaya Lahan Pertanian. pp 43–62 (2006*)*
15. Kalay D E, Lopulissa V F and Noya Y A. *Jurnal TRITON.* **14.** 10–8 (2018)
16. Ismane M A, Kusmana C, Gunawan A, Affandi R and Suwardi S. *Journal of Natural Resources and Environmental Management.* **8.** 36–43 (2018)
17. Elfidasari D, Sabil M Q T, Sasaerilla Y and Sugoro I. *AACL Bioflux.* **15.** 2329–38 (2022)
18. Tapilatu R F, Wona H And Batubara P P. *Biodiversitas Journal of Biological Diversity.* **18.** 129–36 (2017)
19. Dima A O M, C K Z, D M E, Fransiskus K D, M A V And N M A. *Biofaal Journal.* **1.** 55–65 (2020)
20. Wicaksono M A, Elfidasari D and Kurniawan A. Seminar Nasional FMIPA-UT 2013 vol **4.** (2013)
21. Dharmadi and Wiadnyana N N. *Jurnal Penelitian Perikanan Indonesia.* **14.** 195–204 (2008)
22. Nuitja I N S. *Biologi dan ekologi pelestarian penyu laut.* IPB Press (1992)
23. Wentworth C K. A Scale of Grade and Class Terms for Clastic Sediments. *The Journal of Geology.* **30.** 377–92 (1922)
24. Balaira E M, Boneka F B and Wagey B T. *Jurnal Pesisir Dan Laut Tropis.* **1.** 20–5 (2017)
25. Nugroho A D, Redjeki S and Taufiq N. *UNDIP.* 422–33 (2017)

Supply chain analysis of coral trout (*Plectropomus leopardus*) in simeulue regency: case study of production value and contribution of PiSiSi MPA

Maria Ulfah[1*], *Amelia* Rizki Ananda[1], *Insyafrizal*[2], *Erdiansyah*[3], Indra[3]

[1]Department of Marine Science, Faculty of Marine and Fisheries, Universitas Syiah Kuala, Banda Aceh, Indonesia.
[2]Marine and Fisheries Agency of Aceh Province
[3]Fauna & Flora Indonesia Programme – Aceh Site Office, Banda Aceh, Indonesia

Abstract. Coral Trout (*Plectropomus leopardus*) is a species of grouper that is highly valued and the main commodity in Simeulue Regency. This research aims to identify the supply chain, calculate the value of fisheries production and the percentage of production of Coral Trout (*Plectropomus leopardus*) in the PiSiSi Simeulue MPA. It is hoped that the benefit of this research will be market literacy so that it can be used as consideration for management policies and optimization of fisheries marketing in Simeulue Regency. The research was carried out in March-April 2023 through interviews, field surveys and collecting data on purchases/sales of Coral Trout from collectors. The results of the research show that there is a production value of Coral Trout that is different for each buyer in Simeulue Regency. The production value generated by buyers (collectors) was IDR 1,293,195,632 with a total weight of 10.23 tons. The fisheries sector in the PiSiSi MPA contributes to Coral Trout fish production of 41.47% or IDR. 536.3 million in 2022. Based on this research, Coral Trout fish is sold to local communities, processing coMPAnies and exporters in Belawan, North Sumatra and Jakarta.

1 Introduction

Coral Trout (*Plectropomus leopardus*) is a species of grouper commonly found in tropical marine waters in the Indo-Pacific region. Coral Trout are often found in coral waters, or in muddy coral areas, living at depths of 40 meters to 60 meters. In their life cycle, juvenile of Coral Trout live in coral waters with a depth of 0.5-3 meters, then they reach adulthood and go deeper, and usually this movement takes place during the day and at dusk [1]. Coral Trout has a red body, therefore it is also called red grouper. There are dark blue spots on the edges of its body. According to Heemstra [2], the scientific classification of Coral Trout (*Plectropomus leopardus*) consists of Kingdom: Animalia, Phylum: Chordata, Class: Actinopterygi, Family: Serranidae, Genus: *Plectropomus*, Species: *Plectropomus leopardus*. Coral Trout is a hermaphrodite protagonist fish which has the female to male sex of a certain size [3]. Coral Trout is a type of demersal fish which is a fish that lives on the seabed, both looking for food and spawning.

* Corresponding author: mariaulfah@usk.ac.id

A commodity is any real economic product that directly or indirectly contributes to fulfilling human satisfaction and needs. So it can be said that a product, commodity is an item or object that is real and has a certain physical substance so that it can be traded or bought and sold. Fish is one of the main commodities traded in the Indonesian domestic market and international markets [4]. One of the fish that has become a commodity is the Coral Trout, with a high selling value and market demand, so this fish has become one of the targets for fishermen's catches. Some areas refer to Coral Trout as Janang grouper and red grouper [5]. Economic analysis of the Coral Trout commodity includes factors such as selling price, market demand, production, distribution and government regulations [6]. The selling price of Coral Trout tends to be high due to large demand from Asian and European markets, as well as limited supply from cultivation and natural catch. The supply chain process for fish consists of all production activities up to the delivery point in safe conditions so as to provide consumers with good service and quality [7]. The supply chain is also a marketing journey that a product must go through from the time it is produced to the final consumer. Supply chains require management as the design, operation, and enhancement of the systems used to create and trade the products and services of a company. Supply chain management relates to the entire system that produces a product [8].

Simeulue Regency has a marine conservation area called Pulau Pinang Pulau Siumat Pulau Simanaha Marine Protected Area (PiSiSi MPA) which was formed in 2020 through the Governor of Aceh Decree Number: 523/1297/2018 with an area of 44,404.10 hectares. The PiSiSi MPA was established by the Decree of the Minister of Maritime Affairs and Fisheries of the Republic of Indonesia NUMBER 78/KEPMEN-KP/2020 concerning the Aceh Besar Marine Conservation Area, the Pinang Island, Siumat Island and Simanaha Island Marine Conservation Area, the Southwest Aceh Water Conservation Area, and the Aquatic Conservation Area South Aceh in Aceh Province. The ecosystems found in the PiSiSi MPA area are mangroves, seagrass and coral reefs [9].

By conducting research on the supply chain of Coral Trout on Simeulue Island, it is hoped that we can increase understanding of Coral Trout fish resources and assist in the development of sustainable fisheries management strategies and can improve the welfare of local communities who depend on fishery resources. This study aims to analyze how the supply chain process and production value and the contribution of PiSiSi MPA in the supply of Coral Trout in Simeulue district. This research is limited to the scope of fishermen and fish collectors. Apart from that, the scope of the discussion was carried out by comparing the performance efficiency of the Coral Trout supply chain between fishermen and collectors in the PiSiSi MPA area.

2 Research Methods

2.1 Location and Time

This research was conducted in March-April 2023 in six fish storage locations in Sinabang and Kuala Makmur, East Simeulue District, Simeulue Regency. All the research site are within PiSiSi MPAand can be seen through Figure 1 below.

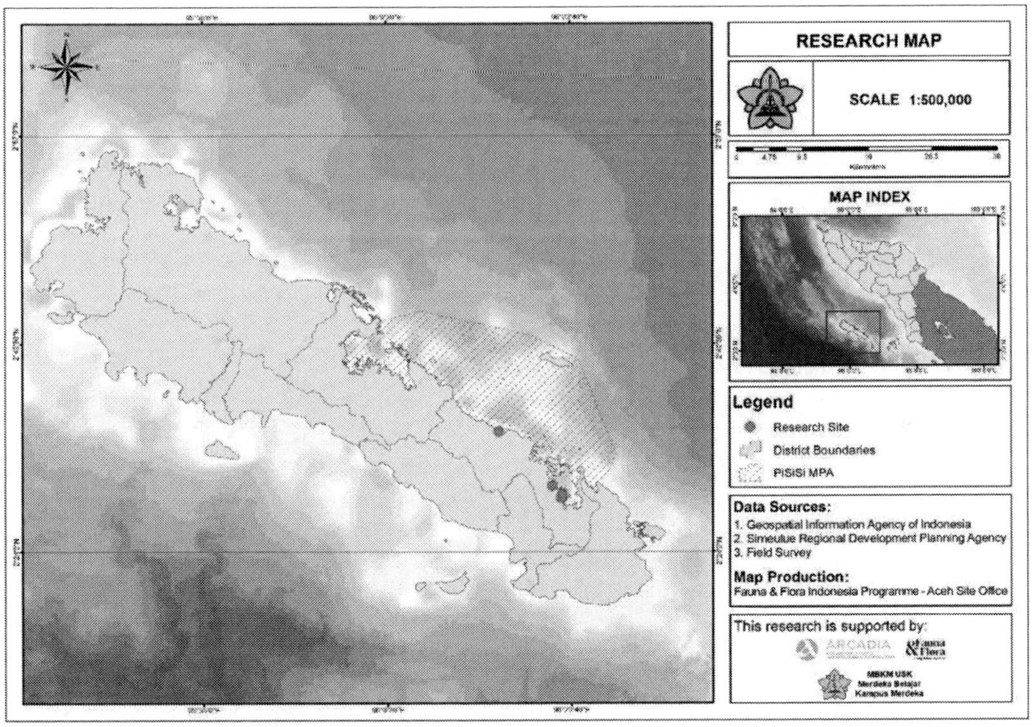

Fig.1. Map of Simeulue Regency

2.2 Tools and Materials

The tools and materials used in the research are writing instruments for recording observations made, cameras for documentation, laptops for data processing, analysis and reporting, and questionnaires for guiding questions in gathering information.

2.3 Data Retrieval

Data collection began by conducting interviews with collectors who were determined using purposive sampling techniques. Interviews were conducted using direct interview techniques with questions that had been prepared in a questionnaire. The questions in the questionnaire are arranged based on the parameters to be measured so that they can produce the required information. The respondents in this study were 6 people who were owners of fish shelters. After the interview was conducted, it was continued with a field survey where the data taken was the size of the Coral Trout, as well as the freezing treatment of the fish.

2.4 Data Analysis

The identification of the supply chain was carried out by direct interviews with the collectors who became respondents. The identification of the supply chain was carried out by direct interviews with the respondent collectors. The identification of the supply chain in this study focuses on how Coral Trout fish travels from producers (fishermen) to consumers (collectors and buyers) [10]. Analysis of supply chain performance to measure production value using measurements that take into account the calculation of fish production issued to market the product and the percentage of each actor (Permen kp No. 1 of 2023).

- Production Value (PV)

To calculate the production value formula, the following formula is used:

$$Fish\ PV = Weight\ of\ caught\ fish\ (kg) \times Reference\ price\ of\ fish\ (IDR) \quad (1)$$

- Percentage of production value

To calculate the percentage of Production Value, the following formula is used:

$$\%\ Production\ Value\ = \frac{Production\ Value\ MPAs\ PiSiSi}{Production\ Value\ Simeulue}\ X\ 100 \quad (2)$$

3 Results And Discussion

3.1 Coral Trout Production Data

During 2022, production of Coral Trout (*Plectropomus leopardus*) will fluctuate every month, with the lowest production values occurring in January and February, each at 0.16 tons. Meanwhile, the highest production value was in December with a total of 1.56 tons. Overall, peak production during the year increased twice, namely in April and December (fig 2).

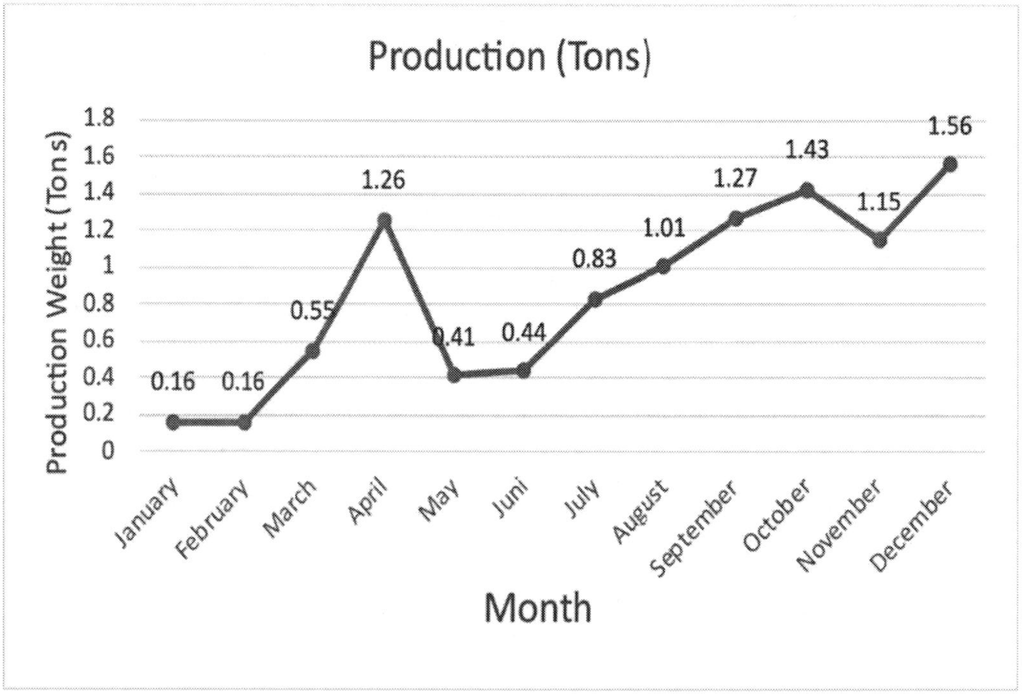

Fig 2. Graph of the amount of Coral Trout production in Simeulue Regency in 2022 in tons

The increase in production value in December occurs because that month enters the fishing season, where the population of Coral Trout increases, which is a reciprocal relationship between water conditions and fish. This increase can also be caused by increased fish activity in searching for food by utilizing low sea currents, so that it is easier for fish to move and some fish can also feed in strong current conditions [11]. One of the reasons for the increase in Coral Trout production in December is supportive water conditions, such as relatively calm waves and tolerable storm conditions, making it easier for fishermen to carry out fishing operations [12]. Meanwhile, the lowest production occurred in January and February, namely 0.16 tonnes. This decrease in production levels was caused by the Coral Trout fish spawning so that their activity decreased. This is in line with Sadovy [13] that when spawning, Coral Trout fish are in groups so that the fish are difficult to catch if the speed of the sea current makes it difficult for the fish to move.

3.2 Percentage of Production Value of Coral Trout (*Plectropomus leopardus*)

Based on the results of observations, it shows that the percentage value of production of Coral Trout (*Plectropomus leopardus*) in Simeulue Regency in 2022. This includes the production value in Simeulue Regency and in the PiSiSi Conservation Area. Production Value Percentage Value Data in table 1 below.

Table 1. Percentage (%) of Production Value in 2022

Month	Production Value Simeulue (IDR)	Production Value PiSiSi MPA (IDR)	% Production Value PiSiSi MPA
January	20,064,200	5,955,200	29.68
February	20,155,950	2,869,450	14.24
March	66,521,000	17,309,600	26.02
April	151,763,057	76,828,030	50.62
May	48,578,750	35,807,150	73.71
June	52,549,250	41,632,600	79.23
July	99,652,055	73,934,555	74.19
August	123,179,400	57,097,150	46.35
September	157,899,350	69,210,800	43.83
October	189,582,650	42,067,250	22.19
November	156,504,000	56,079,550	35.83
December	206,745,970	57,523,270	27.82
Total	1,293,195,632	536,314,605	41.47

Source: Production data (collectors) 2022.

Based on the results of research that has been carried out, it is known that the percentage value of Coral Trout production in 2022 at the PiSiSi Simeulue MPA is 41.47%. Based on Ministerial Decree No. 1 of 2023, the production value is obtained from multiplying the weight of Coral Trout times the price per kg of fish. Meanwhile, the percentage (%) of the PiSiSi MPA Production Value is obtained from the division between the PiSiSi MPA Production Value and the Simeulue Production Value and then multiplied by 100. The difference in fish size greatly influences the price of the fish itself, the higher the type of fish size, the higher the price [14]. Apart from that, variations in fish prices also often occur due to fish condition, collector purchasing prices and production costs which of course will influence the price of fish each month [15]. The production value of Coral Trout (*Plectropomus leopardus*) in Simeulue Regency in 2022 will reach IDR. 1,293,195,632 and the Production Value of the PiSiSi MPA is IDR. 536,314,605.

The highest production value of the PiSiSi MPA was obtained in April with a production value of IDR. 76,828,030. The high production value occurred because in April the PiSiSi

MPA area entered the west season so that fishermen's catches increased. This is in line with Kusnadi [16] During the western season (October to April) fishing activities by fishermen will increase. Meanwhile, the lowest production value of the PiSiSi MPA occurred in February, namely IDR 2,869,450. According to some collectors, this happens because there is a transition between the west and east seasons. The seasonal winds (eastern and western seasons) greatly determine the seeds of Coral Trout in nature, these two seasons influence current conditions, salinity, temperature, and nutrients contained. The spawning season for Coral Trout usually occurs in April and June and between January and September [1].

3.3 Size Grouping Based on Market Sorting

Observation results show that buyers in Simeulue group the Coral Trout (*Plectropomus leopardus*) in Simeulue Regency based on the weight size in 2022, with different weight sizes per fish. Based on the weight sorting of Coral Trout by buyers, it is grouped into four categories, namely Up, A, B and C in grams. The Up size category has a weight range of over 1000 grams/head, the A size category has a weight range of 500 - 1000 grams/individu, the B size category weighs 300 - 500 grams/head, the C size category weighs under 300 grams/head. The weight of Coral Trout in Simeulue Regency in 2022 can be seen in fig 3.

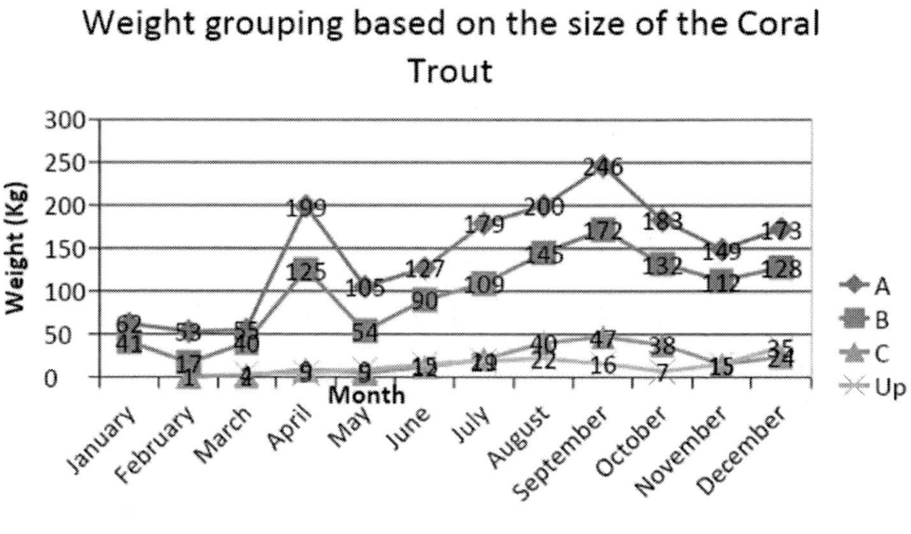

Fig 3. Size Grouping of Coral Trout (*Plectropomus leopardus*) in 2022.

The highest size grouping for Coral Trout in Simeulue Regency is in size group A with a total size weight reaching 1,731 kg/year. The highest catch of A-sized Coral Trout was in September, namely 246 kg. This is in line with Suwirya and Giri [17] in Jakarta. A weight of 500 grams for Coral Trout is a size that meets the requirements for local consumption or export. Apart from that, in Riau the distribution process for Coral Trout fish is carried out every day, especially for Coral Trout fish with a consumption size of 500-800 grams/fish, as well as for fish exported to Singapore or Hong Kong [18].

Meanwhile, the lowest fish catch was in size C (<300 grams/head), namely in February, it was 1 kg. The low catch of C size Coral Trout is caused by the C size fish being very small fish, ranging between 25-29 cm. This is also because many Simeulue district fishermen still use fishing gear, so that if they catch a C size fish they will be released again. This will also minimize the number of small Coral Trout fish caught, thus avoiding indications of growth overfishing [19].

3.4 Fish Market Chain

It can be seen from the market chain for Coral Trout in Simeulue Regency that fishermen are starting to sell their catch of Coral Trout to buyers in the village and also to fish collectors in Sinabang, then grouper that are small in size and do not meet export quality requirements will be sold. sold to the people of Simeulue and Coral Trout fish that meet export standards will be distributed outside areas such as North Sumatra (Belawan), Jakarta and also exported to other countries. However, there are obstacles in this fish market chain, namely difficulties in processing interbank transactions, a lack of ice production during fish exports which causes the fish to no longer be fresh when it arrives at its destination, and also erratic ship schedules so that the delivery process cannot be carried out according to with the specified time.

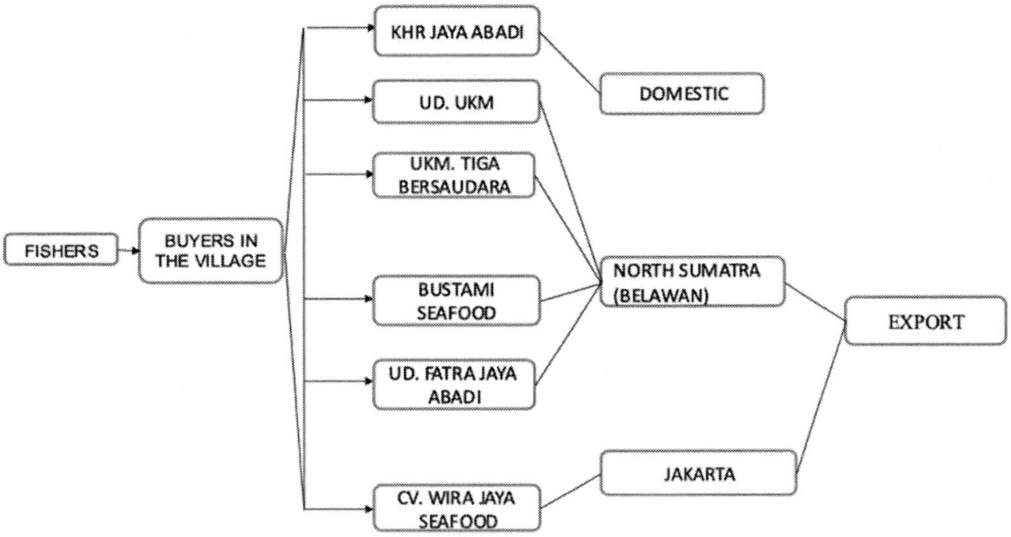

Fig 4. Market Chains for Coral Trout (*Plectropomus leopardus*) in Simeulue

The distribution and marketing process requires special treatment so that the quality and freshness of the fish can be maintained. This special treatment is one of the marketing functions which aims to increase the economic value of fish products, which can be done through increasing the efficiency of the marketing system in the context of coordination mechanisms for fish production, distribution, and consumption activities [20]. Based on the results of the questionnaire, in the marketing process there are several factors that must be taken into account, such as the grouping time according to the type and weight of fish, the storage time of the fish, marketing costs, and the ratio of ice to the number of Coral Trout in one storage fiber tub. For the length of fish storage, collectors usually store Coral Trout before shipping for ± 1 week with the ice ratio for each storage fiber tub ranging from 10-14 sticks of ice for 100-120 kg of Coral Trout. The costs incurred in the marketing process range from IDR. 600,000 - 60 million depending on the number of fish transported. The large marketing costs received are a result of the large costs incurred in marketing the production results, such as transportation and loading and unloading costs, cooling costs and other costs [21].

3.5 Quarantine Production Value in 2022

Based on observations on traffic data on the Coral Trout fishery at the Fisheries Quarantine Work area, it shows that in 2022 Simeulue Regency will have two types of commodities, namely Frozen Coral Trout and Live Coral Trout. The number of Coral Trout fish ranges from 3 - 2,808 kg with a commodity value of around IDR. 180,000 - IDR. 207,000,000. Data on Quarantine Production Values in Simeulue Regency in 2022 can be seen in table 2 below.

Table 2. Quarantine Production Value in 2022

No	Month	Commodity Name	Amount (kg)	Mark Commodities (IDR)	Average Price (IDR)
1	January	Frozen Coral Trout Fish	35	2,100,000	60,000
2	April	Frozen Coral Trout Fish	2,500	150,000,000	60,000
3	May	Frozen Coral Trout Fish	2,808	207,000,000	73,718
4	May	Live Coral Trout Fish	350	33,250,000	95,000
5	June	Frozen Coral Trout Fish	300	18,000,000	60,000
6	July	Frozen Coral Trout Fish	5	300,000	60,000
7	September	Frozen Coral Trout Fish	3	180,000	60,000
8	November	Live Coral Trout Fish	300	28,500,000	95,000

Source: Simeulue Fisheries Quarantine Work area Traffic Commodity Data 2022

Simeulue Fishery Quarantine Work Area Traffic acts as a sign that fishery products supplied outside the area are the result of Simeulue fishery production. This is also an effort to prevent the spread of fish diseases and control the quality and safety of fishery products in traffic [22].

The table presented above illustrates that the highest production output within the quarantine work area data occurred in May 2022, with a total of 2,808 kg, whereas the lowest production output was recorded in September 2022, only 3 kg.

4 Conclusion

It can be concluded that the Coral Trout supply chain in Simeulue Regency can be described as follows: fishermen start selling their catch of Coral Trout to buyers in the village and also fish collectors in Sinabang, then the Coral Trout will be sold to processing companies and exporters in Belawan, North Sumatra, Jakarta, and then exported abroad. the results showed that there were differences in the value of Coral Trout production for each buyer in Simeulue Regency. The total production value generated by buyers (collectors) was IDR 1,293,195,632 with a total weight of 10.23 tons. The contribution of PiSiSi MPA to the production value of Coral Trout in Simeulue Regency was 41.47%. This indicates that the existence of conservation areas is very important to maintain the sustainability of sun grouper stocks that will provide fisheries and economic benefits in Simeulue Regency.

Acknowledgement

Thank you for financial support from Arcadia – a charitable fund of Lisbet Rausing and Peter Baldwin, Fauna & Flora Indonesia Programme - Aceh Site Office, Aceh Fish

https://doi.org/10.1051/bioconf/20248703028

Quarantine and Quality Control Center for Simeulue and the Simeulue Fish Collectors including KHR Jaya Abadi, UD. UKM, UKM Tiga Bersaudara, Bustami Seafood, UD. Fatra Jaya Abadi, CV. Wira Jaya Seafood, Panglima Laot for their full support and the author also expressesour gratitude to the MBKM USK Unggul Program so that this research was carried out well.

Bibliography

1. Mulyani Hitijahubessy. 2021. Pusaka Almaida. **87** pp
2. Heemstra P.C and J.E Randall. 1993. FAO Spesies Catalogue **16**
3. Andamari R. 2017. Jurnal Penelitian Perikanan Indonesia, **11** (7):7
4. Utami T. N. Indrayani E. 2018. *Komoditas Perikanan*, Universitas Brawijaya Press,
5. Ismi S. Asih Y.N. Nasukha A. 2020. Journal of Fisheries and Marine Research. **4**
6. Suryanto M. H. 2018. Gramedia Pustaka Utama. **43**
7. Rosa. F.S.R Farisa. 2020. IAIN Ponorogo.
8. Jacob F.R and Chase R. 2018. McGraw-Hill.
9. DKP Aceh. 2020. Dinas Kelautan dan Perikanan Aceh: Banda Aceh
10. Minarro S. Forero G.N. Reuter H. Van Putten I.E. 2016. *Marine policy*, **69**
11. Lennox R.J., J. Alos, R. Arlinghaus, A. Horodysky, T. Klefoth, C.T. Monk, S.J. 2017. Fish and Fisheries. **18**
12. Nurhayati A. Aisyah I. Supriatna A.K. 2019. *WSEAS Transactions on Environment and Development*. **15**
13. Sadovy Y. Colin P.L. 2005. JFB. **46(6)**: 961-976
14. Firmansyah M.H. 2021. STIE Perbanas Surabaya 56p
15. Bronnmann J. and Asche F. 2016. Journal of Agricultural Economics. **67**
16. Kusnadi M. A. 2002. LKIS Pelangi Aksara. 190p
17. Suwirya K. and Giri N.A. 2010. Prosiding Forum Inovasi Teknologi Akuakultur, **1**
18. Manadiyanto M., Nasution Z., Tajerin, T.. Pranowo S.A. 2017. Jurnal Penelitian Perikanan Indonesia, **9** (6), 65-73
19. Nuraini S. Hartati S.T. 2006. Proceeding of the 2006 National Conference of Fish IV
20. Takalamingan M., Longdong F.V., Jusuf A. 2017. Jurnal Ilmiah Agribisnis Perikanan, **5**(9):705-712
21. Kharie W.R.,. Manoppo V.E, Wasak M. 2017. Jurnal Ilmiah Agribisnis Perikanan. 5(9): 572-580
22. Achmad, H. Rahman H. Rina. 2018. Proseding Penyuluh Perikanan. Bogor. pp. 187-201

Valorization of weed *Portulaca oleracea* L. as an alternative to fish feed ingredient

R Lia Rahadian Amalia[1], *Lusi H* Suryaningrum[1,*], *Sumitro* Sumitro[2], *Budiyanti* Budiyanti[2], *Sulasy* Rohmy[3], *Bastiar* Nur[3] and *Mulyasari* Mulyasari[4]

[1]Research Center for Applied Zoology, Research Organization for Life Sciences and Environment, National Research and Innovation Agency (BRIN), Cibinong, West Java 16911, Indonesia
[2]Department of Aquaculture, Faculty of Fisheries and Marine Sciences, Universitas Dayanu Ikhsanuddin, Baubau, Sulawesi Tenggara 93724, Indonesia
[3]Research Center for Fishery, Research Organization for Earth Sciences and Maritime, National Research and Innovation Agency (BRIN), Cibinong, West Java 16911, Indonesia
[4]Research Center for Marine and Land Bioindustry, Research Organization for Earth Sciences and Maritime, National Research and Innovation Agency (BRIN), North Lombok, West Nusa Tenggara 83352, Indonesia

Abstract. Feed ingredients such as fishmeal and soybean meal are common used in the aquaculture industry. However, they have drawbacks both environmentally and economically and should switch over to more sustainable materials. Using a weed in fish feed opens up potential and environmentally conscious possibilities. The objective of this study is to explore the potential valorization of a common weed plant called *Portulaca oleracea* L., using enzymatic hydrolysis for incorporation into fish feeds. This common weed can thrive by itself without human intervention. The nutrient content of *P. oleracea* L. was 16.33% protein, 0.88% lipid, 10.19 % ash, 26.20% crude fiber, and 46.40% carbohydrates (by differences) (% dry weight). However, the high crude fiber restricts its utilization in feed formulation. Hydrolysis has been employed to reduce crude fiber and enhance its quality. The treatments in doses are as follows: 10%, 20%, 30%. and 40% (v/w). Each treatment was conducted in three triplicates for 60 hours, 55°C, and pH 5.0. The result revealed that the nutrient quality of *P. oleracea* L. was improved. In conclusion, the valorized *P. oleracea* L. is feasible to be an ingredient in fish feed.

1 Introduction

The aquaculture industry faces challenges with conventional feed ingredients such as fish meal and soybean meal, which have ecological and economical drawbacks [1]. One promising and eco-friendly approach is using weed, *Portulaca oleracea* L., as an ingredient

* Corresponding author: lusi006@brin.go.id

in fish feed [2]. This innovation aligns with the circular economy principles, transforming unwanted materials into valuable feed resources.

P. oleracea L., known as common purslane, demonstrates high production efficiency with prolific growth worldwide in tropical and subtropical regions [3] [4] [5]. Its wide global distribution and ability to withstand abiotic stress [6]. Purslane is a versatile plant that can adapt and grow in various arid and moist habitats [7]. Purslane can be found in orchards, gardens, crop fields, roadsides, and even saline-alkali soils [8]. The rapid proliferation and formation of dense coverings by purslane can push aside indigenous plant species, and reducing biodiversity [7]. In agriculture, purslane will compete with cultivated crops to grow, reducing crop yield [9].

Numerous research studies have examined the nutritional profile of purslane and highlighted its potential benefits. [10] and [11] state that purslane contains various phytochemicals and bioactive substances like phenols, antioxidants, ω-3 fatty acids, minerals, flavonoids, alkaloids, organic acids, and polysaccharides. The nutrient composition of purslane can be outlined as follows 10.9% moisture, 2.45% ash, 21.63% crude fiber, 18.2% protein, 5.2% lipid, and 44.3% carbohydrates [8]. Purslane has a relatively high crude fiber content [12] [13]. This is limited its utilization in feed formulation. However, enzymatic hydrolysis could reduce the crude fiber and improve the quality of unconventional fish feed ingredients. Enzymatic hydrolysis breaks down complex fiber structures into simpler components, reducing crude fiber and improving digestibility [14]. [15] demonstrated that enzymatic hydrolysis could reduce the crude fiber content in coconut cake by 67.8%. [16] discovered that hydrolyzed corncob meal had reduced crude fiber by 57.53%, NDF fiber fraction by 38.15%, ADF fiber fraction by 6.43%, and hemicellulose by 61.96%. [17] reported that enzymatic hydrolysis reduces the fiber content of sweet sorghum bagasse. [18] stated that enzymatic hydrolysis can significantly enhance soybean hulls' nutritional value. [19] demonstrated that the bacterial enzyme extract enhances the nutrient value of sugarcane bagasse through a hydrolytic reaction.

This research aims to investigate the valorization of the weed purslane through enzymatic hydrolysis for its incorporation into fish feed. This research is expected to augment the purslane's nutritional content and improve its digestibility for fish. The research effort aspires to contribute to advancing eco-friendly and sustainable fish feed formulations, thus diminishing the dependence on conventional feed ingredients and fostering the growth of a more sustainable aquaculture industry.

2 Materials and methods

2.1 Preparation of unrefined enzyme extract

An aliquot of bacterial culture was introduced into a 10 mL Tryptic Soy Broth (TSB) medium. The cultures were subsequently incubated at 30C in an incubator for 24 hours. After this incubation, aseptically, a portion of the cultured solution, equivalent to 1 mL, was transferred to 9 mL of TSB medium, done threefold. These secondary cultures were incubated again for an additional 24-hour period. Then, the bacterial cultures underwent centrifugation at 10,000 rpm for 45 minutes at a temperature of 40°C. The resulting supernatant was meticulously poured off and subsequently preserved in a refrigerated setting for further analysis.

2.2 Enzymatic hydrolysis on the purslane meal

The purslane meal was obtained from a commercial producer. Enzymatic hydrolysis was carried out by mixing 50 grams of purslane meal, buffer solution, and unrefined enzyme extract with a dose according to the treatment, namely 10%, 20%, 30%, and 40% (v/w). The mixture was diligently stirred to achieve homogenously, securely sealed, and subsequently set for a designated duration of 60 hours, pH 5.0, and temperature of 55°C. The research was executed following a completely randomized design, with each treatment replicated three times. The mixture was heated using a conventional oven at a temperature of 105°C for 5 minutes to stop the reaction. Following that, it undergoes the drying process at 60°C for 24 hours, or until a consistent weight is achieved. The nutrient composition of purslane before and after hydrolysis were all measured.

2.3 The determination of nutrient composition in purslane

The determination of nutrient composition involved both pre- and post-enzymatic hydrolysis measurement using the proximate method, following the AOAC guidelines [20]. The moisture content was determined using the gravimetric analysis technique that entailed drying the sample material in a conventional oven at 105°C up to a constant weight. The quantification of protein content was carried out using the Kjeldahl procedure, which includes sample digestion, distillation, and titration, respectively. Lipid content was determined through organic solvent extraction. The determination of ash content involved subjecting the sample to a 600°C furnace for 3 hours, while the assessment of crude fiber content was conducted through a method with both strong acids and bases. The solubility of proteins underwent evaluation using the Bradford method [21], while the quantification of reducing sugars was conducted through the DNS method [22].

2.4 Data analysis

Data analysis was performed by tabular the obtained data using Microsoft Excel Software, followed by a study using the One-Way Analysis of Variance (ANOVA). In cases where differences were observed, Duncan's test was applied. Statistical data analysis was conducted utilizing Statistical Program for Social Science (SPSS) version 26 for Windows, while maintaining a significance level set at $p<0.05$ to ascertain statistical significance.

3 Result and Discussion

In this research, an examination and evaluation were carried out on the nutrient composition of the purslane plant, which included the levels of moisture, ash content, lipid, crude fiber, protein, and carbohydrates (by differences). Table 1 demonstrates that purslane exhibits high quantities of carbohydrates, crude fiber, and protein, accounting for 46.40%, 26.20%, and 16.33%, respectively. These findings align with [12] [13], who reported that purslane displayed elevated protein and crude fiber levels alongside comparatively lower lipid content.

Following the hydrolysis process, the unrefined enzyme extracts significantly improved purslane nutrient value. It recorded an improvement in protein, crude fiber, and ash content ($p<0.05$). There was a considerable rise in protein content (Figure 1). It began from 16.33% to 16.98%, 17.47%, 18.53%, and 19.28% respectively for the doses of 10%, 20%, 30%, and 40% (v/w). These results emphasize that a considerable enhancement in protein content was conducted by the hydrolyzation process. Lipid content increases from 0.88% to 2.75% (Figure 2). There was a significant reduction in the crude fiber content (Figure 3). Crude fiber decreased from its initial content of 26.20% and fell to a mere 11.43% for the 40% (v/w). A reduction in ash content was witnessed (Figure 4). Particularly, it declined from the initial 10.19% to 8.04% (40% v/w). Reducing sugar and soluble protein also confirm the findings presented in Figure 5 and 6, respectively. There was an increase in the reducing sugar which went from 115.23 mg/L (initial) to 528.94 mg/L (40% v/w). The soluble protein of hydrolyzed purslane markedly increased from 101.28 mg/L (initial) to 427.45 mg/L (40% v/w). The doses of unrefined enzyme extract which results in an improved purslane nutrient value, follow a straight line. The results suggest that the hydrolysis process has improved the nutritional quality of purslane.

Table 1. The results of proximate analysis of purslane

Parameters	**%**
Moisture	8.27 ± 0.14
Protein	16.33 ± 0.13
Lipid	0.88 ± 0.09
Ash	10.19 ± 0.28
Crude fiber	26.20 ± 0.36
Carbohydrates (by differences)	46.40 ± 0.51

The data is presented as the mean ± standard deviation (n = 3)

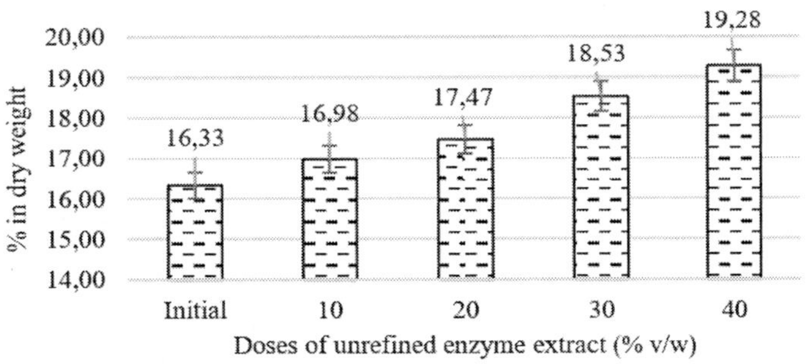

Fig. 1. Protein content of purslane (%)

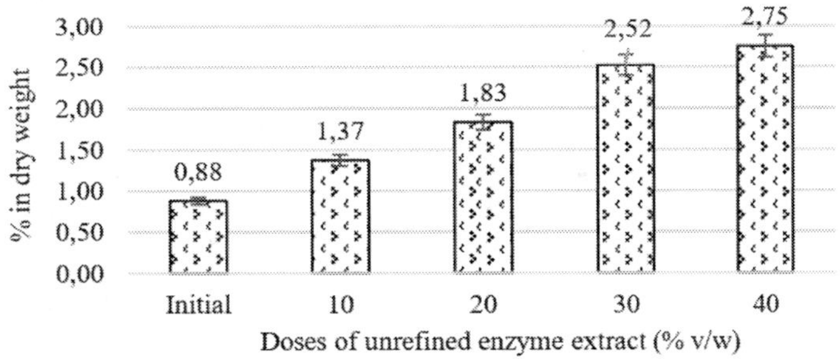

Fig. 2. Lipid content of purslane (%)

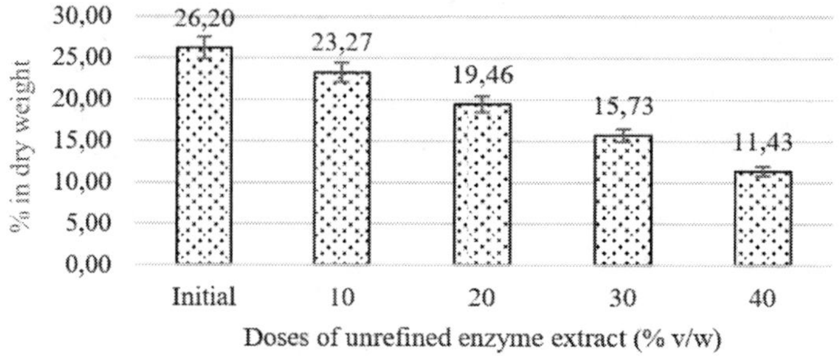

Fig. 3. Crude fiber content of purslane (%)

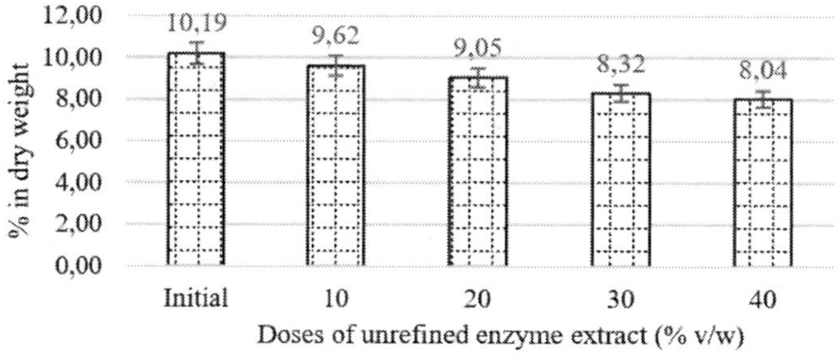

Fig. 4. Ash content of purslane (%)

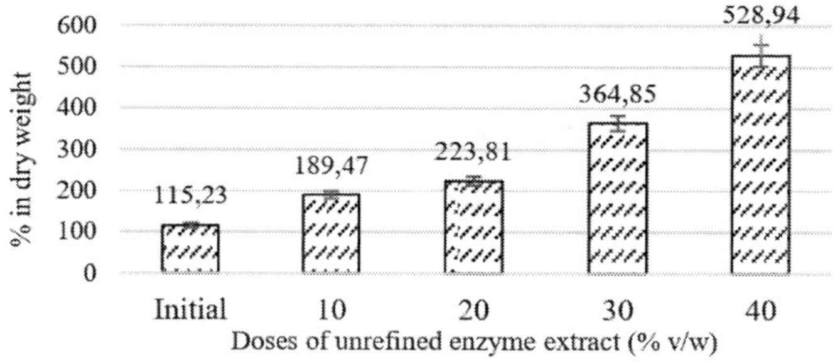

Fig. 5. Reducing sugar in purslane (mg/L)

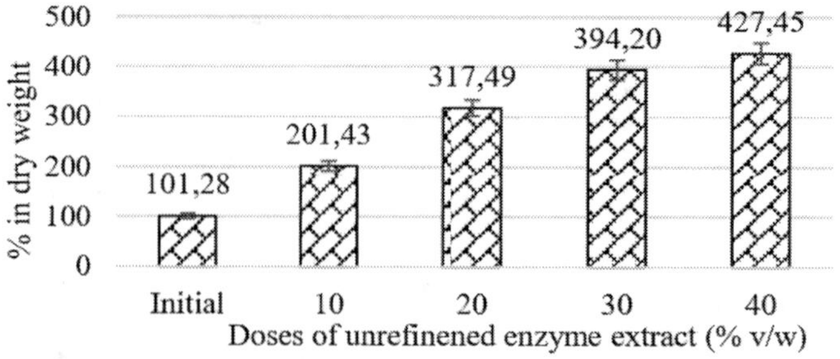

Fig. 6. Soluble protein in purslane (mg/L)

Purslane, with relatively high fiber content, necessitates preprocessing prior to its inclusion as an ingredient in fish feed. Employing enzymatic hydrolysis with an unpurified enzyme extract aims to break down the crude fiber in purslane into simpler components that are more readily digestible by fish. The enhancement in purslane quality exhibited a positive relationship with the quantity of added unpurified enzyme extract, 40% (v/w) yielding the most favorable results. This specific outcome will be elaborated upon in the subsequent discussion.

The protein content in hydrolyzed purslane using the unrefined enzyme extract 40% (v/w) was notably elevated. The protein component in fish feed is pivotal in influencing fish growth, as it is the primary determinant. A higher protein content in the feed corresponds to an increased potential for fish growth. Protein provides amino acids that are crucial for fish growth and development. The National Research Council [23] highlights the importance of protein in fish nutrition and states that protein is the most expensive and limiting nutrient in fish feed. The protein requirements of fish depend on various factors, such as species, life stage, size, and environmental conditions. The feed formulation should consider the amino acids requirement and their bioavailability. It also must examine the balancing of both cost-effectiveness and nutrient adequacy. Using plant-based protein would increase the reliance on fish meal and fish oil, which is categorized as a non-renewable source [24]. In comparison to locally prevalent plant ingredients typically utilized in feed, such as wheat bran (11.05 %) [25], rice bran (15.23%), and fermented rice bran (17.62%) [26], sugarcane bagasse (1.49%) [27], the protein content (19.28%) in hydrolyzed purslane is relatively higher.

The hydrolyzed purslane notably increased the quantity of soluble protein, as depicted in Figure 6. Soluble proteins are either oligopeptides or amino acids that can be readily taken up by the digestive system [28]. Throughout the hydrolysis process, the interplay between the enzyme and substrate streamlined the peptide bonds, ultimately enhancing the solubility of proteins. A greater concentration of soluble proteins indicates that fish will more efficiently utilize protein for their metabolic processes and growth [29].

Purslane has a relatively significant amount of crude fiber, 26.20%. The excessive crude fiber in fish feed can pose challenges for fish as they are monogastric animals with limited ability to digest high crude fiber [30]. Fish have a relative lack of specialized microbial fermentation in their digestive tract compared to terrestrial animals. High levels of crude fiber in fish feed influence the digestion and absorption of other nutrients, leading to non-optimal nutrient utilization and fish growth [31]. Hydrolyzation process with enzyme extract from plant or microbial sources, which contains specific enzymes, such as cellulases or hemicellulases, can break down complex carbohydrates in crude fiber into more straightforward and digestible by fish [19]. In this research, the unrefined enzyme extract reduces the crude fiber of purslane from 26.20% to 11.43%. Lower than the plant-based ingredients commonly used in fish feed, such as fermented palm kernel cake (12.35%) [32] or wheat bran (13.02 %) [25].

This approximation is supported by the formation of reducing sugars in hydrolyzed purslane, as depicted in Figure 5. The elevated levels of reducing sugars in the hydrolyzed purslane indicate the creation of simpler constituents resulting from the degradation of crude fiber by the unrefined enzyme extract [33]. Glucose, the simplest form of carbohydrate, is readily digestible by fish [34].

An excessively high ash content in fish feed can adversely affect the utilization of nutrients by fish. A high ash content can disrupt the fish's digestive system. High ash content can bind essential minerals and other nutrients in the feed, reducing their bioavailability [35]. This means that nutrients may not be efficiently absorbed by the fish, even if they are present in the feed [36].

Consequently, fish may experience nutrient deficiencies despite these nutrients in the feed. High ash content can also reduce the availability of energy in the feed [37]. If ash replaces these energy-rich components in the feed, fish may not receive sufficient energy for optimal growth and metabolism. This can lead to decreased productivity in fish farming [38]. Hydrolyzed purslane through enzymatic reaction resulted in a reduction of ash content from 10.19% to 8.04%. The ash content within the hydrolyzed purslane exhibited a notable decrease in comparison to that of Moringa leaves meal (9.70%) and fermented Moringa leaves meal (14.05%) [39].

Utilizing hydrolyzed materials as ingredients in fish feed has garnered widespread recognition and application. [25] reported that incorporating hydrolyzed wheat bran at levels of up to 20% into Nile tilapia (*Oreochromis niloticus*) diets resulted in elevated protein content, reduced fiber content, and improved growth, feed utilization, and blood parameters as well as intestinal and liver development. [40] reported that enzyme pretreatment successfully enhances the nutritive value of high-fiber plant byproducts, leading to increased growth and feed efficiency in *Mugil cephalus* [41] presented research findings regarding the influences of a carbohydrase-rich extract obtained from hydrolyzed Brewer's spent grain (BSG). This extract was introduced into the European seabass (*Dicentrarchus labrax*) diet to improve the digestibility of dry matter, starch, cellulose, glucans, and energy. This was validated through both in vitro and in vivo assessments of digestibility.

In a separate study, [27] revealed that hydrolyzed sugarcane bagasse had a digestibility of 66.08% for dry matter, 84.35% for protein, 95.26% for lipid, and 70.49% for energy when fed to *Barbonymus schwanenfeldii*. These results suggest that hydrolyzed sugarcane bagasse is a reasonably digestible nutritional resource for *B. schwanenfeldii* and could be incorporated into its diet. Moreover, [42] reported that replacing fish meal with a hydrolyzed plant-based protein concentrate, at levels of up to 27.9% of the fish meal in juvenile olive flounder (*Paralichthys olivaceus*) diet, had no significant impact on survival rates, overall nutritional composition, hematological parameters (such as aspartate aminotransferase, alanine aminotransferase, glucose, total serum protein), and responses of non-specific immune (such as lysozyme and activity of superoxide dismutase).

4 Conclusions

Applying an unrefined enzyme extract of 40% (v/w) led to a substantial increase in protein, soluble protein, and reducing sugars while concurrently decreasing the crude fiber and ash content in purslane. The findings of this investigation indicate that the use of enzymatic hydrolysis with the unrefined enzyme extract has improved the quality of purslane to the extent that its inclusion as an ingredient in fish feed is feasible.

References

1. B. Samuel-Fitwi, S. Meyer, K. Reckmann, J. Schroeder & C. Schulz, J. Clean. Prod. **52**, 225-233 (2013). https://doi.org/10.1016/J.JCLEPRO.2013.02.031
2. Z. Sourani, S. Shirian, S. Shafiei, N. Mosayebi, A. Nematollahi, Mar. Biotechnol. **25**, 214-221 (2023). http://dx.doi.org/10.1007/s10126-022-10195-z
3. R. Srivastava, V. Srivastava, A. Singh, Environ. Manage. **72**, 309–320 (2023). https://doi.org/10.1007/s00267-021-01456-z
4. J. He, S.Y. Leng, L. Qin, Plants, **12**, 3214 (2023). http://dx.doi.org/10.3390/plants12183214
5. A. Carrascosa, J.A. Pascual, M. Ros, S.A. Petropoulos, M. del M. Alguacil, Plants, **12**, 1246 (2023). http://dx.doi.org/10.3390/plants12061246
6. K. Li, T. Xia, Y. Jiang, N. Wang, L. Lai, S. Xu, X. Yue, H. Xin, J. Ethnopharmacol. **319**, Part 2, 117211 (2024), https://doi.org/10.1016/j.jep.2023.117211
7. M. Fernandez-Poyatos, E. Llorent-Martínez, A. Ruiz-Medina, Foods, **10**, 94 (2021). https://doi.org/10.3390/foods10010094
8. S. Karakas, I. Bolat, M. Dikilitas, Hortic. **7**, 63 (2021). https://doi.org/10.3390/horticulturae7040063

9. M. Camalle, D. Standing, M. Jitan, R. Muhaisen, N. Bader, M. Bsoul, Y. Ventura, A. Soltabayeva, M. Sagi, Agron. **10**, 656 (2020). https://doi.org/10.3390/agronomy10050656
10. M. Uddin, A.S. Juraimi, M.S. Hossain, M. Nahar, A. Un, M. Ali, M. Rahman, Sci. World. J 951019 (2014). https://doi.org/10.1155/2014/951019
11. C.O. Montoya-García, R. García-Mateos, E. Becerra-Martínez, R. Toledo-Aguilar, V.H. Volke-Haller, J.J. Magdaleno-Villar, Sci. Hortic. **308**, 111584 (2023). https://doi.org/10.1016/j.scienta.2022.111584
12. D.A. Al-Quwaie, A. Allohibi, M. Aljadani, A.M. Alghamdi, A.A. Alharbi, R.S. Baty, S.H Qahl, O. Saleh, A.O. Shakak, F.S. Alqahtani, O.S.F. Khalil, M.T. El-Saadony, A.M. Saad, Molecules, **28**, 5859 2023. https://doi.org/10.3390/ molecules28155859
13. W. Obied, E. Mohamoud, O. Mohamed, Small. Rumin. Res. **48**, 31–36 (2003)
14. S. Borges, J. Odila, G. Voss, R. Martins, A. Rosa, J.A. Couto, A. Almeida, M. Pintado, Molecules, **28**, 1155 (2023). https://doi.org/10.3390/molecules28031155
15. F.M. Pratiwy, Y. Maulida, Asian J. Fish. Aquat. Res. **20**, 95-101 (2022). https://doi.org/10.9734/ajfar/2022/v20i6521
16. M.A. Fauzan, A. Suprayudi, A. Fauzil, J. Ekasari. JAI, **21**, 1 (2022), 32-40. https://doi.org/10.19027/jai.21.1.32-40
17. R. Haiwei, S. Ruifeng, W. Huiyuan, W. Li, G. Xiaopeng, L. Dong, L. Ruiyuan, L. Jinping, (Transactions of the CSAE), **39**, 224-236 (2023). https://doi.org/10.11975/j.issn.1002-6819.202211017
18. G. Lili, Z. Weiyun, H.H. Suqin, Chinese J. Anim. Nutr. **35**, 3361-3371 (2023). https://doi.org/10.12418/CJAN2023.312
19. L.H. Suryaningrum, R. Samsudin, 2021. *Improvement quality of sugar cane bagasse as fish feed ingredient.* in Proceedings of the International Conference on Biotechnology and Food Sciences, INCOBiFS, 11 September 2020, Surabaya, Indonesia (2020)
20. AOAC International. Guidelines for Standard Method Performance Requirements. AOAC Official Methods of Analysis. Appendix F, pp. 1-18 (2016). http://www.eoma.aoac.org/app_f.pdf
21. M.P. Spinola, M.M. Costa, J.A.M. Prates, Agricult. **13**, 221 (2023). https://doi.org/10.3390/agriculture13010221
22. S. Wonorahardjo, D.P. Sari, A. Salsabila, E. Estiyawati, D. Yuliani, A.R. Wijaya, S. Suharti, I.K. Kusumaningrum, C.A. Maharani, T.A. Noviyanti, Makara J. Sci. **27,** 148-159 (2023)
23. National Research Council. Nutrient requirements of fish and shrimp. Washington, DC: The National Academies Press (2011)
24. A.G. Tacon, M. Metian, Aquacult. **285**, 146-158 (2008)
25. E.Y. Mohammady, A.M. Aboseif, M.R. Soaudy, E.A. Ramadan, M.S. Hassaan, Aquacult. **575**, 739755 (2023). https://doi.org/10.1016/j.aquaculture.2023.739755
26. J.J.D. Manlapig, T. Ban-Tokuda, H. Matsui, Anim. Sci. J. **94**, e13860 (2023). https://doi.org/10.1111/asj.13860
27. L.H. Suryaningrum, R. Samsudin, Mulyasari, A. Saputra, *The digestibility of improved sugar cane bagasse on Barbonymus schwanenfeldii,* in Proceedings of the International and National Symposium on Aquatic Environment and Fisheries, ISAEF, 24 September 2020, Banda Aceh, Indonesia (2020)
28. E.T. Moran, E.T. Anim. Feed Sci. Technol. **221**, 284-303 (2016). https://doi.org/10.1016/J.ANIFEEDSCI.2016.05.015
29. Y. Wei, M. Liang, H. Xu, Aquacult. Nutr. 2019; 00: 1–11. https://doi.org/10.1111/anu.12976

30. Y. Sun, X. Zhao, H. Liu, Z. Yang, J. Aquac. Res. Dev. **10**, 1-7 (2019). https://doi.org/10.35248/2155-9546.19.10.577
31. F. Gai, P. Peiretti, A. Brugiapaglia, L. Gasco., J. World. Aquac. Soc. **47**, 496-507 (2016). https://doi.org/10.1111/JWAS.12294
32. M. Firdaus, M.H. Rohaya, M.S. Miskandar, A.A. Astimar,, J. Oil. Palm. Res. **34**, 92-103 (2022). https://doi.org/10.21894/jopr.2021.0022
33. Lusi H. Suryaningrum, D. Jusadi, M. Setiawati, M.T.D Sunarno, AACL Bioflux, **10**, 77-86 (2017)
34. L.H. Suryaningrum, R. Samsudin, EJFA. **32**, 7, 488–494. https://doi.org/10.9755/ejfa.2020.v32.i7.2131
35. S. Sugiura, V. Raboy, K. Young, F. Dong, R. Hardy, Aquacult, **170**, 285-296 (1999). https://doi.org/10.1016/S0044-8486(98)00414-1
36. V. Kaiza, M. Yildiz, V. Eldem, S. Golzaradabi, S. Ofori-Mensah, J. Anim. Physiol. Anim. Nutr. **107**, 1517–1529. https://doi.org/10.1111/jpn.13858
37. R. Ruiji. W. Wang, L. Liu, F. Ma, Q. Li, P. Zhao, W. Ma, J. Cen, X. Liu, Int. J. Biol. Macromol. **250**, 126155. https://doi.org/10.1016/j.ijbiomac.2023.126155
38. R. Defaix, J. Lokesh, M. Ghislain, M. Le Bechec, M. Marchand, V. Veron, A. Surget, S. Biasutti, F. Terrier, T. Pigot, S. Panserat, K. Ricaud, Aquacult, **578**, 740049, https://doi.org/10.1016/j.aquaculture.2023.740049
39. S. Helmiati, A. Rustadi, Isnansetyo, Zuprizal, AACL Bioflux **13**, 1622-1632 (2020)
40. F.P. Martínez-Antequera, I. Barranco-Ávila, J.A. Martos-Sitcha, F.J. Moyano, Fishes, **7**, 4 (2022). https://doi.org/10.3390/fishes7010004
41. H. Fernandes, F. Moyano, C. Castro, J. Salgado, F. Martínez, M. Aznar, N. Fernandes, P. Ferreira, M.M. Goncalves, I. Belo, A. Oliva-Teles, H. Peres, Sci. Rep. **11**, 22946 (2021). https://doi.org/10.1038/s41598-021-02393-x
42. M. Seong, S. Lee, Y. Song, J. Bae, K. Chang, S.C. Bai, Aquacult. **483**, 196–202 (2018). http://dx.doi.org/10.1016/j.aquaculture.2017.10.023

Growth rate and survival of green shells (*perna viridis* l.) Inponds, alue naga village, banda aceh

Ichsan Rusydi[1*], *Imam Safir Alwan* Nurza[3,4,5], *Irma* Dewiyanti[2], *Siska* Mellisa[1], *Siti* Maulida[5], *Cut Maya* Ulfita Shari[5]

[1] Department of Aquaculture, Faculty of Marine and Fisheries, Syiah Kuala University, Indonesia;
[2] Department of Marine Science, Faculty of Marine and Fisheries, Syiah Kuala University, Indonesia,
[3] Research Center for Elephant Conservation and Forest Biodiversity, Syiah Kuala University, Indonesia
[4] Researcher Group of KPM, Jakarta State University, Indonesia
[5] Department of Plant Biology, Faculty of Mathematics and Natural Sciences, Bogor Agricultural University, Indonesia

Abstract. Alue Naga is one of the coastal and marine areas in the city of Banda Aceh which has the potential for aquaculture and high economic value, namely green mussels (*Perna viridis* L). This research aims to determine the growth and survival rate of green mussels with different stocking densities in Alue Naga ponds which are still influenced by the tides of seawater. The layout of the green mussel cultivation container is at the pond water gate. The method used is an experimental test method using a Completely Randomized Design (CRD) model with 3 treatments, each treatment 3 times repeated. Research shows that the best average growth and survival rate for green mussels are found in treatment A with a stocking density of 20 seeds/basket, absolute length growth with an average value of 45.52 ± 0.42 mm, weekly length growth rate with an average value -an average of 3.79 ± 0.03 mm/week and an average weight of 25.92 ± 0.19 grams, while the survival rate reached 91.66%. Differences in green mussel stocking density had a significant effect on absolute length growth (PM), absolute weight (BM), and weekly length growth rate (LPPM), while survival (SR) had no significant effect.

1 introduction

Alue Naga is one of the coastal and marine areas in the city of Banda Aceh which has the potential for aquaculture and has high economic value, namely green mussels (*Perna viridis* L). Green mussels (*Perna viridis* L) are a type of shellfish that can live in bay waters, estuaries, waters around mangrove areas, and river estuaries, with environmental conditions where the bottom of the waters are muddy mixed with sand, with sufficient light, and salt levels that are not too high [1]. Green mussels are widely used locally for aquaculture because of their economic value and delicious taste. The increasingly rapid development of industry causes high levels of pollution in waters caused by the large number of industries dumping waste in waters, thus affecting the biological processes of aquatic biota [2].

The rapid development of green mussel cultivation is due to the ease of cultivation techniques for this species, compared to other biota cultivation technologies. Factors promoting the development of green mussel aquaculture include the relatively fast growth rate of green mussels, which requires a short culture period to reach a consumable size [3,4,5].

Growth rate is an overall measurement of performance which is easy to interpret

*Corresponding author: ichsanrusydi@usk.ac.id

(profitable when increasing or detrimental when decreasing) and growth is related to birth and death rates. This can have important influences on population and community structure [6]. The growth of green mussels depends on energy supply and the body's metabolic processes. Salinity, dissolved oxygen content and media temperature are related to the growth of green mussels [7]. Similar results in the waters of Semarang Bay, the relative growth rate obtained in Tambak Lorok which is on the coast of Semarang Bay for 60 days shows that the average relative growth rate at the first station was 220.69%, the second station was 190.16% and the second station was 190.16%. third amounted to 133.33%. Appropriate environmental parameters will influence the growth of green mussels such as depth, temperature, water, salinity and oxygen content [8].

The potential of ponds in Banda Aceh is very profitable if used well, and there are still many ponds around coastal areas that are neglected, because people have not utilized ponds properly. According to observations in the field, green mussels are one of the biota that can be cultivated in ponds, the same as research studies in Tambak Lorok, Semarang. The cultivation of green mussels in ponds is very controlled compared to in reservoirs, this is because the pond environment is not easily contaminated by other waste, and the optimal salinity does not decrease which could result in green mussels dying as a result of the decrease in salinity, temperature, and DO [9].

2 Material And Method

2.1 Time and Place

Data collection was carried out from January to March 2022. This research was conducted in Alue Naga Waters, Syiah Kuala District, Banda Aceh.

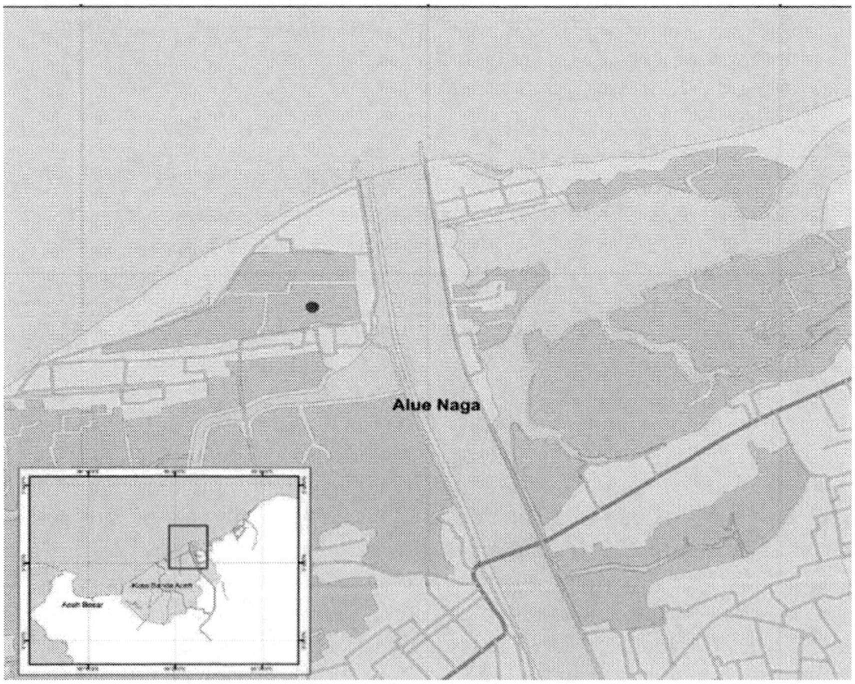

Fig. 1. Map of research locations in Alue Naga, Banda Aceh

2.2 Material and Tools

The main tools used in this research were baskets, hanging ropes for baskets, bamboo. Meanwhile, the material used in this research was green mussel seeds.

2.3 Method of Data Collection

The research method used was an experimental method using a Completely Randomized Design (CRD) model with 3 treatments and 3 replications. Sampling was carried out once a week for 12 weeks. The total number of seeds was 225 seeds, with different densities, namely 20, 25 and 30/basket. Baskets A1, A2, A3 contain 20 seeds, baskets B1, B2, B3 contain 25 seeds, while baskets C1, C2, C3 contain 30 green mussel seeds. Weekly measurements of the length and weight of the clams were measured randomly using dice.

2.4 Container Preparation and Sample Testing

The container used in this research used a rectangular basket measuring (30 x 50 cm), with a plastic basket type, PP Block Copolymer material. The basket is placed on top of the pipe as a support for the basket. Light pipes are installed on the left and right to prevent the basket from shifting and tilting. Then the green mussels are tied using a 60 cm long fiber rope for attachment. The basket is made floating in the middle of the pond, tied to bamboo as a support. The green mussels used in this research were green mussel seeds measuring 27-32 mm obtained from the Alue Naga reservoir. Containers are differentiated based on the density of 20 seeds, 25 seeds, and 30 seeds/basket using the floating method.

2.5 Growth Parameters

Green mussel growth and weight parameters including absolute length, absolute width, LPPM, and survival of green mussels were analyzed using the ANOVA test of variance prog, to see if there was a significant effect ($P<0.05$) and continued with the Duncan further test. to see more clearly the differences.

2.6 Data Analysis

2.6.1 Absolute Length

The absolute length growth of green mussel shells was measured using the following formula [3].

$$L = L_2 - L_1$$

Where,
L = Absolute length (cm)
$L2$ = Final length (cm)
$L1$ = Initial length (cm)

2.6.2 Weekly Length Growth Rate

The weekly length growth rate can be calculated using the formula [10]:

$$WLGR = \frac{Lt - Lo}{t}$$

Where,
WLGR = Weekly length growth rate (cm)

Lt	= Average length at the final of research (cm)
L0	= Average initial length of research (cm)
t	= Sampling time (12 minggu)

2.6.3 Absolute Weight

The weight growth of green mussels was measured using the following formula [11]:

$$\mathbf{W = W_t - W_o}$$

Where,

W	= Absolute weighted average (g)
Wt	= Average weight of individuals at time – t
Wo	= Average individual weight at the start of research

2.6.4 Survival Rate

Calculating the survival of green mussels uses the following formula [12]:

$$\mathbf{SR = \frac{N_t - N_o}{No} \times 100\%}$$

Where,

SR	= Survival rate (%)
Nt	= Number of shellfish at the final of research (sample)
No	= number of shellfish at the start of research (sample)

3 Results And Discussion

3.1 Results

The results of research on the growth of rate increase every week with various treatments. Treatment A (stocking density of 20 green mussels) had the fastest growth rate compared to treatments B (stocking density of 25 green mussels) and C (stocking density of 30 green mussels) can be seen in (Figure 1). Based on the absolute length and weight growth charts, green mussels recorded an increase in growth every week, with the highest growth achieved by treatment A at a stocking density of 20 seeds/basket, with an average absolute length of 45.52 mm and weight 25.92 g.

The test results obtained using ANOVA also showed that stocking density had a significant effect ($P<0.05$), namely on absolute length growth, weekly length growth rate, and weight growth. The highest treatment was obtained in treatment test A, with the lowest stocking density of 20 seeds/basket with an average absolute length value of 45.52 mm, an average absolute weight of 25.92 g, and a long weekly with an average of 3.79 mm. Meanwhile, for survival, the results of the ANOVA test showed that there was no significant effect ($P>0.05$), where the differences in changes were not too much different between treatments, with the highest treatment having a stocking density of 20 seeds/basket reaching 91% (Table 1).

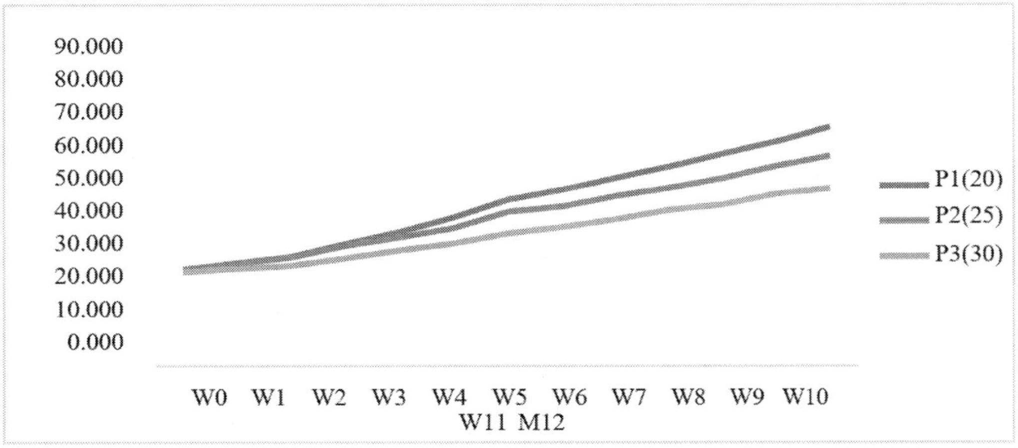

Fig. 1. Graphic growth rate of green mussels with block P1 (treatment A), P2 (treatment B), and P3 (treatment C)

Table 1. Treatment results on the growth of Absolute Length (AL), Absolute Weight (AW), Weekly Length Growth Rate (WLGR), and Survival Rate (SR) of green mussels

Stocking Density Treatment	AL (mm)	AW (g)	WLGR (mm/week)	SR (%)
A (Treatment 20)	45,52±0,42^{c}	25,92 ± 0,19^{c}	3,79 ± 0,03^{c}	91,66 ± 2,88^{a}
B (Treatment 25)	36,34±0,10^{b}	22,69 ± 0,61^{b}	3,02 ± 0,00^{b}	90,66 ± 2,30^{a}
C (Treatment 30)	26,69±0,627^{a}	19,39 ± 0,29^{a}	2,22 ± 0,05^{a}	88,89 ± 1,92^{a}

Treatments A, B, and C differ significantly in absolute length (AL), absolute weight (AW), and weekly length growth rate (WLGR). However, there was no significant difference in the survival rate (SR) of green mussels

Table 2. Water quality parameters in pond Alue Naga

Water Quality	Measurement results	Satuan
Flow	0,16	m/s
Salinity	30	ppt
pH	8,7	-
Temperature	28	°C
Dissolved Oxygen	4,8	mgl

3.2 Discussion

Results from a study conducted over a 12-week period showed that differences in stocking density had a significant impact on absolute linear growth, absolute weight, weekly linear growth rate, and survival rate. was shown. There is no significant effect on survival rate. The best average growth and survival of green mussels was in treatment A with a stocking density of 20 seeds/basket, with an average absolute linear growth rate of 45.52 mm, a weekly linear growth rate of 45.52 mm, and a weekly linear growth rate of 45.52. showed that it was mm. The average weight was 3.79 mm/week, the average weight was 25.92 g, and the survival rate reached 91.66%.

Measurements of green mussel growth in this study included length and weight. The water environmental factors in Alue Naga ponds are very supportive for cultivating green mussels, because the availability of the food needed is sufficient and there is a water flow entrance into the pond so it is possible for food not only to be in the pond. Environmental factors that greatly influence the growth rate of green mussels are aquatic ecological factors. Environmental factors are important parameters because they influence feed requirements and intake, which can influence the growth rate of shellfish. The process of green mussels looking for food is by filtering particles in the water and then filtering the food in the mantle cavity or called a filler feeder [13]. Filtration speed is related to the length of the clam shell, the longer the clam shell, the faster the filtration [14]. Filtration is influenced by environmental factors such as pH, dissolved oxygen, temperature and salinity [15].

High density will result in reduced feed efficiency, increased energy consumption, reduced feed rate, and digestibility of *Labeo victorianuson* [16]. This is inbalance with the results of research cultivation of green mussels (*Perna viridis* L) with different methods and densities in the coastal waters of Kuala Langsa, Aceh which were observed for 3 months or 12 weeks with a treatment density of 20 individuals/ 5.30 L bags had the highest survival, compared to treatment densities of 30 and 40 individuals/5.30 L bags, so it can be concluded that the higher the density, the lower the survival and growth of green mussels [17].

Survival rate is a comparison between the number of individuals alive at the end and the beginning from number of individuals [18]. The survival rate of green mussels in each treatment for 12 weeks showed results that were not too different. The highest survival rate in this study was in treatment A, namely 91%, B 90%, and C 88%. However, based on the results of ANOVA analysis, the differences between treatments were not significantly different ($P>0.05$). The treatment that experienced the most deaths was treatment C with a percentage of 88%. This is thought to be because high density will affect the death rate of green mussels. High density will cause predation or competition, where green mussels compete with each other for space, food, and other factors [19]. Competition in space will lead to aggression between green mussels which leads to death [20].

The survival of green mussels is also greatly influenced by temperature, temperatures that are too high can also have a negative effect on the growth and survival of green mussels. The temperature in the results of this research was 28°C, which is still in good condition for the growth of green mussels. Green mussels are best cultivated at natural temperatures ranging from 23-31°C. pH also greatly influences the growth rate and survival of green mussels, changes in pH can result in the activity and growth of green mussels, because low pH will also have low DO. The results of pH measurements in this study were still relatively good, namely 8.7 [21]. The appropriate and good pH used for cultivating green mussels is around 6.5-9. Another important factor that can influence the success of green mussel cultivation is salinity [22]. The salinity measurement results obtained were 30 ppt. Good salinity for green mussel cultivation is between 27-32 ppt [23].

4 Conclusion

The results showed that average growth and survival of green mussels were best in treatment A with a stocking density of 20 seeds/basket. The average absolute linear growth was 45.52 mm, the weekly linear growth rate was 3.79 mm/week, the average weight was 25.92 g, and the survival rate reached 91.66%. Differences in stocking density of green mussels had a significant effect on absolute body length, absolute weight, and weekly body length growth rate, but did not have a significant effect on survival rate.

References

1. Yonvitner, Sutrisno, Struktur Komunitas Makrozoobentor dan Pertumbuhan Kerang Hijau *(Perna viridis* L) di Perairan Muara Kamal dan Bojonegoro (Institut Pertanian Bogor, Bogor 2009)
2. H. Suseno, Biokumulasi kadmium Melalui Jalur Air Laut pada Kerang Hijau (Pusat

Teknologi Limbah Radioaktif, Batam, 2006)
3. T.K. Soon, J. Ransangan, Advances in Bioscience and Biotechnology, **5**, (2014)
4. N. A. Loekman, A. Manan, M. Arief, P. Prayogo, Journal of Aquculture and Fish Health **7**, 2 (2018)
5. S. H. Liyana, L. A. Sari, N. N. Dewi, E. D. Masithah, A. M. Sahidu, K. T. Pursetyo, In IOP Conference Series: Earth and Environmental Science **236**, 1 (2019)
6. M. P. J. Scarlet, H.P. Halldórsson, Granmo, Åke. Regional Studies in Marine Science, **1**, (2015)
7. T. Yutaka, S. Isamu, I. Gyo, Journal of Experimental Marine Biology and Ecology, **470**, (2015)
8. B. Kingzet, J. of Scientific Research **30**, 4 (2002)
9. Temmy, A. Sutrisno, W. Niniek Widyorini, Tingkat Kerja Osmotik dan Pertumbuhan Kerang Hijau (*Perna viridis*) yang dikultivasi di Perairan Tambak Lorok Semarang (Undip, Semarang, 2016)
10. M. I. Effendie, Metode Biologi Perikanan (Yayasan Pustaka Dwi Sri, Bogor, 1997)
11. G. W. Dewantoro, Jurnal Iktiologi Indonesia **1**, 2 (2001)
12. Z. A. Muchlisin, A. A Arisa, A. A. Muhammadar, N. Fadli, I. I. Arisa, M. N. Siti Azizah, Archives of Polish Fisheries, **23** (2016)
13. J. A. Pattikawa, E. Ferdinandus, Marine Research Indonesia, **34**, (2009)
14. K. V. Rajesh, K. S. Mohamed., V. Kripa, Indian Journal Marine Science **30**, 2 (2001).
15. H.U Riisgard, Marine Ecology Progress Series, **22**, (2001)
16. E.W. Magondu, H. Charo-Karisa, M. C. J. Verdegem, Aquaculture, **410** (2013)
17. Sagita, R. Kurnia, Sulistiono, Jurnal Riset Akuakultur **12**, 1 (2017)
18. S. Djunaidah, M. R. Toelihere, M. I. Effiendie, S. Sukiman, E. Riani, Jurnal Akuakultur Indonesia **9**, 1 (2004)
19. H. E. P. Leimena, T.S. Subahar, Adianto, J Mat Sains, **10** (2005)
20. C. B. Manley, C. F. Rakocinski, P. G. Lee, R. B. Blaylock, Aquaculture, **420**, (2014)
21. Effendie, I. N. J. Bugri, Widanarni, Jurnal Akuakultur Indonesia **5**, 2 (2006)
22. S. H. J. Sari, L. I. Harlyan. Research Journal of Life Science 2, 1 (2015)
23. I. Rusydi, S.A.E. Rahimi, D. Rianjuanda, K. Melanie, S.A. Akbar, The effect of different spat collectors on attachment and growth of oyster spat (Crassostrea gigas) in Alue Naga. In *IOP Conference Series: Earth and Environmental Science,* IOP Publishing. **(1221(1)** 012069). (2023)

Phylogeography pattern of *Lutjanus bengalensis* (bloch, 1790) in indonesia waters and south china sea

Sri Riska Rahayu[1], *Zainal A*. Muchlisin[2,3*], *Nur* Fadli[2,4], *Syamsul* Rizal[2], *Ilham* Raza[2], *Muhammad Isran* Lazuardi[2], *Nanda Muhammad* Razi[1,3,4], *Luvi Syafrida* Handayani[1,3,4], *Firman* M. Nur[5]

[1] Graduate School of Mathematics and Applied Sciences, Universitas Syiah Kuala, Banda Aceh 23111, Indonesia.
[2] Faculty of Marine and Fisheries, Universitas Syiah Kuala, Banda Aceh 23111, Indonesia.
[3] Laboratory of Fish Hatchery and Breeding, Faculty of Marine and Fisheries, Universitas Syiah Kuala, Banda Aceh 23111, Indonesia.
[4] Laboratory of Genetics and Aquatic Biodiversity, Faculty of Marine and Fisheries, Universitas Syiah Kuala, Banda Aceh 23111, Indonesia.
[5]Research center for Biosystematics and Evolution - BRIN, Cibinong Science Center, JI. Raya Jakarta Bogor Km.46 Cibinong 16911, Indonesia.

Abstract. Lutjanus or snappers usually known as economical fish in Indonesia and worldwide. This research aims to observe the different phylogeography correlation of *Lutjanus bengalensis* in Indonesia Waters and South China Sea used COI gene marker. The primary data collected from fish port Pulo Aceh, Indonesia (5 sequences), while the population of others region such as Bali (3 sequences) and Maluku (3 sequences) was retrieved from NCBI GenBank. South China Sea represented by sampled from Taiwan (3 sequences) and Hongkong (3 sequences) that also retrieved from NCBI Genbank. In total 17 sequences were analysed the diversity of haplotypes 0-1 and the diversity of nucleotide 0-0.030. Based on this researched data, it will become basic data for spesies management for Lutjanus bengalensis in Indonesia.

* Corresponding author: muchlisinza@usk.ac.id

1 Introduction

Snapper is a Lutjanidae family that inhabit tropical and subtropical waters [1]. Based on the data cited in [2], there are over 113 species of Lutjanidae family over the world with 58 of them found in Indonesia waters to the South China Sea. One of the most frequently encountered snappers is the *Lutjanus bengalensis* [3, 4]. This species dominates the waters of the Indo-Pacific region [5]. The term Indo-Pacific refers to a geographic area encompassing a significant portion of the waters around the Asian continent, including regions around Indonesia, Malaysia, the Philippines, Australia, Japan, and China [6, 7].

According to Leliaert, Payo, Gurgel, Schils, Draisma, Saunders, Kamiya, Sherwood, Lin and Huisman [8] the Indo-Pacific is a highly productive marine region, harboring diverse fish species and unique marine ecosystems. The marine biodiversity in the Indo-Pacific waters is crucial for global ecological balance and supports the livelihoods and resources for millions of people living in the vicinity. The marine waters of Indonesia and the South China Sea are part of the Central Indo-Pacific region. *L. bengalensis* is often found in the waters of the Malacca Strait, the Java Sea, and the Banda Sea [9-11]. The South China Sea is directly connected to Indonesian waters and borders several Asian countries, including Hong Kong and Taiwan [12, 13]. The South China Sea is considered snappers fishing ground, with Lutjanus bengalensis being one of the widely distributed species in these waters[14].

Due to the similarities and distinctiveness observed in these two marine regions, investigations into the genetic relationships among species exhibiting regional variations have become a compelling subject for exploration. Numerous genetic studies on population diversity have been carried out, encompassing from several species and population such as *Katsuwonus pelamis* and *Thunnus albacares* from North Maluku Sea [15], snapper species *L. fulvus* and *L. kasmira* from Indo-Pacific waters [16], and *L. kasmira* from China waters [17]. Nevertheless, there is an absence of reported studies regarding the correlation between geographical factors and genetic diversity within the *L.bengalensis* population in Indonesian waters and the South China Sea. Phylogeography analysis has been applied to provided valuable insights into the historical biogeography of species and how they have responded to environmental changes over time [18]. This field holds particular significance in comprehending how climate change and the destruction of habitats impact the distribution and biodiversity of life on Earth. Previous studies have delved into and recorded comparisons of phylogeographic patterns, such as those found in reef fish from the western Indian Ocean and Eden Bay, Arab[19,20], distributed populations of snapper *L. jocu* and *L. analis* from Atlantic Ocean [21], mackerel and tuna from Indonesia archipelago [22], reef fish *Cephalopolis argus* from Indo-Pacific sea [23], and *L. kasmira* from the Bengal Strait and Arabian Sea [24].The phylogeography investigation holds significance in uncovering the connection biogeography and the life histories of species, providing a foundation for the development of more effective management strategies. Species with close genetic ties can be incorporated into a unified management and conservation plan, while those with greater genetic divergence should be managed with distinct approaches. Consequently, the primary goal of this research was to assess the phylogeographic patterns of *L. bengalensis* between Indonesia waters and South China Sea populations, serving as fundamental data for devising an improved species management plan in Indonesia.

2 Material and Methods

2.1 Collecting samples of tissues

This research employed primary and secondary data. Specifically, the research utilized primary data from Indonesian populations (Aceh) and secondary data sourced by NCBI Genbank for *L. bengalensis* samples in Indonesian population (Maluku and Bali). In addition, samples from South China Sea were sourced from populations in Hongkong and Taiwan (as depicted in Figure 1), and this data was also obtained by the NCBI Genbank.

A total of five *L. bengalensis* samples from Aceh population were collected at fish landing (Pulo Aceh) in Aceh Province, Indonesia in January 2023. These fish were documented and about 1 cm of their fin (pectoral) was aseptically excised before preserving the samples in a 96% ethanol solution following the Fish-BOL collaborator's protocol [25]. The specimens were conveyed to the laboratory for additional examination. Comparative Phylogeography analysis utilized *L. bengalensis* sequences sourced Hong Kong and Taiwan, obtained from The GenBank as the secondary data. Figure 1 and Table 1 illustrate the origins of the sample sources.

2.2 The extraction of DNA

The process of DNA esxtraction followed established protocols using Cetyltrimethyl Ammonium Bromide (CTAB). Initially, finely chopped pectoral fin was placed in a 1.5 mL sterile tube also added 3 µl of Proteinase K and 700 µl of CTAB. After vortex in 15 seconds, it was cultivated at 60°C for 180 minutes. Following cultivating, mixed 700 µl of Chloroform Isoamyl Alcohol (CIA). Subsequently, the solution underwent centrifugation at 11,000 rpm for 900 seconds. The resulting supernatant was cautiously moved to a new 1.5 mL tube. Then, 100% ethanol was applied, followed by vortexing for 30 seconds and centrifugation at 12,000 rpm for 15 minutes. The supernatant was discarded, and the tube underwent rinsing with 70% ethanol. After inverting the tube and tapping it on a dry tissue to remove excess ethanol, it was left to air-dry at 20-22°C for 600 seconds. Following this, 60 µl of deionized water was added, and the DNA sample was ultimately inventory at -20°C for future utilization.

2.3 The Amplification using PCR method

The PCR protocol was executed using a mixture consisting of 25 µl of DNA template, 8.5 µl of ddH2O, 1 µl of F1 and R1 (forward and reverse) and 12.5 µl of Red Master Mix. The PCR process occurred in a PCR machine (Senso Quest Lab cycler) and comprised 30 cycles. Each cycle initiated with an initial pre-denaturation step at 95°C for 2 minutes, followed by denaturation at 94°C for 45 seconds, annealing at 52°C for 45 seconds, extension at 72°C for 60 seconds, and a final extension at 72°C for 10 minutes, concluding with cooling to 4°C [26]. The PCR product was subjected to electrophoresis in a 2% agarose gel in the following manner: Two microliters of the PCR product were loaded into wells created in the agarose gel, and electrophoresis was carried out at 100 volts for a duration of 30 minutes. Subsequently, DNA visualization was achieved with UVITEC FIRERIDER V10 system. The DNA bands that appeared clean and well-defined were chosen for sequencing to First Base Laboratories, Malaysia.

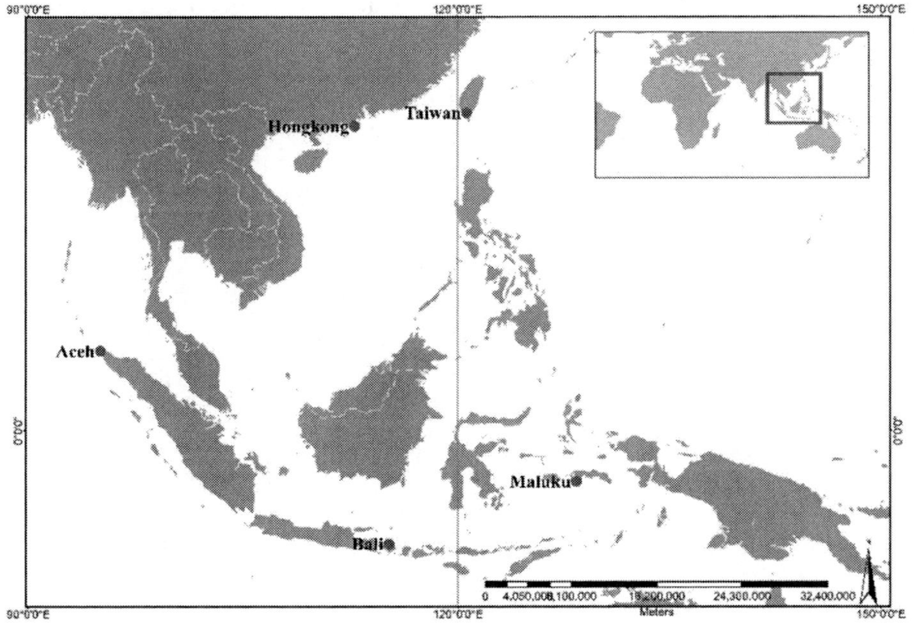

Fig.1. The five population of *L.bengalensis*

Table 1. *L. bengalensis* Sample ID from Indonesian waters and South China Sea

No.	Population	Total of sample	Sample ID
1.	Aceh	5	LPL-K11-01, KKP-01-01, LPL-K11-05 LPL-K11-02, LPL-K11-03
2.	Maluku	3	MH085862, MN870259, MN870315
3.	Bali	3	FOAH914-08, FOAJ199-09, FOAJ200-09
4.	Hongkong	3	EU600137, EU600136, EU600135
5.	Taiwan	3	KU943912, KU943905, KU943892

2.4 Genetic analysis

The sequence data underwent editing and analysis through the utilization of MEGA 6.06. BlAST genebank for validated the seqeuences. Kimura-2-parameter (K2P) model within the MEGA X software [27] for analysis genetic distance. The Maximum Likelihood (ML) approach with replicating 1000 bootstraps and the Hasegawa-Kishino-Yano model for the phylogenetic tree. Additionally, for analys the haplotype distribution used DNASP 5.10.

3 Result and Discussion

There was 17 mitochondrial COI gene sequences of L. bengalensis were obtained from five locations, namely Aceh (5 sequences), Maluku (3 sequences), Bali (3 sequences), Hongkong (3 sequences), and Taiwan (3 sequences). Haplotype diversity within these five populations ranged from 0 to 1, while nucleotide diversity ranged from 0 to 0.030. The genetic diversity of the grouper populations in Bali, Hongkong, and Taiwan exhibited high haplotype diversity

(Hd) values, reaching 1, whereas the Aceh population had a moderate Hd value of 0.6 (Table 2).

The genetic distance that show highest within-population was observed in the Taiwan with 0.032, while the Maluku population shows the lowest with a value of 0 (Table 3). *L. bengalensis* from all five populations fall within the same clade, indicating a common ancestry (Figure 2). Bali and Maluku populations showed connectivity. Ten haplotypes were derived from the 17 sequences, with haplotype 6 distributed in the Hongkong and Taiwan populations, while haplotype 3 was found in the Bali and Maluku populations (Figure 3).

Table 2. Comparing the genetic diversity of snapper *L. bengalensis* among five populations in Indonesian waters and the South China Sea. Evaluating total haplotypes (Hn), nucleotide diversity (π) and haplotype diversity (Hd)

No.	Population	Total sample	Genetic diversity		
			Hn	Hd	π
1	Aceh	5	2	0.6	0.001
2	Maluku	3	1	0	0.000
3	Bali	3	3	1	0.008
4	Hongkong	3	2	1	0.002
5	Taiwan	3	2	1	0.030

Table 3. The genetic distance within population (in bold) and between population of *L. bengalensis* from populations using COI gene

No.	Population	1	2	3	4	5
1	Aceh	**0.001**				
2	Maluku	0.001	**0.000**			
3	Bali	0.005	0.004	**0.008**		
4	Hongkong	0.003	0.003	0.006	**0.002**	
5	Taiwan	0.018	0.017	0.020	0.016	**0.032**

The research demonstrated that the greatest haplotype variety was identified in the Bali, Taiwan and Hongkong population with a value of 1 with 3,2,2 haplotypes, whereas the minimum haplotype variety was identified in the Maluku population with only 1 haplotype, while Aceh with 0.6. Based on [29] Haplotype variety within the range of 0.1 to 0.4 is considered low, while values between 0.5 and 0.7 are classify as moderate, and haplotype diversity ranging from 0.8 to 1.00 was classified to high class. Hence, the diversity of haplotype within the snapper inhabitants in Aceh falls into the moderate classify, whereas Bali, Hong Kong, and Taiwan exhibit high variety. The elevated genetic diversity observed in Hong Kong and Taiwan populations is likely attributed to the currents originating from the South China Sea, flowing directly toward Taiwan and Hong Kong. This facilitates the mixing of *L. bengalensis* larvae from these regions, resulting in increased a close genetic distance and genetic variety. Juvenile of the fish exhibit limited movement capabilities, making them susceptible to being transported by ocean currents [30]. The elevated genetic diversity is attributed to factors like geographic proximity and the influence of ocean currents, whereas diminished genetic diversity within a population results from a combination of factors, including habitat conditions and excessive exploitation [31,32].

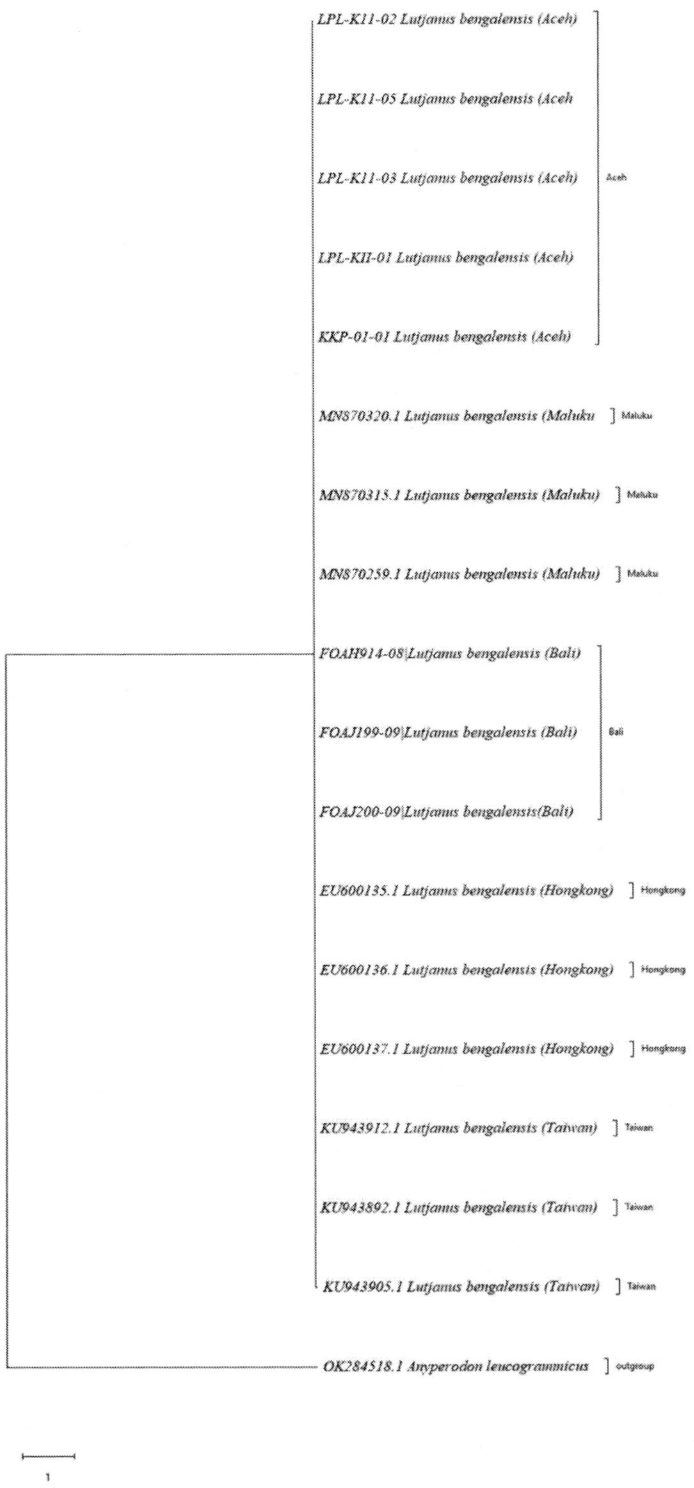

Fig. 2. The recontruction of phylogenetic tree from 17 sequences of *L. bengalensis*

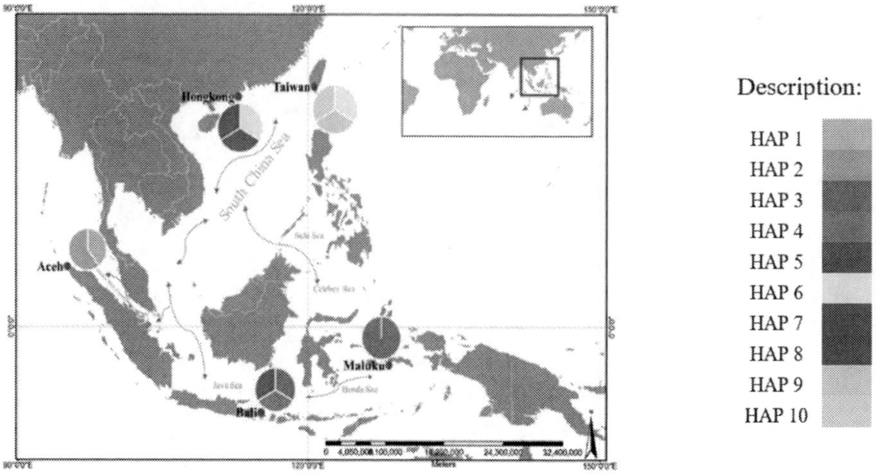

Fig. 3. The map of the haplotypes distribution of *L. bengalensis* from 5 population

The 17 sequences from five populations of Aceh, Maluku, Bali, Hongkong and Taiwan it produces one main clade. This suggests the strong familial relationship among these populations that can be traced back to a common ancestor. The genetic distance between clades provided support for this phylogenetic tree [33]. The Taiwan population exhibited the highest genetic variation among populations (interspecific) at a value of 0.032, while the lowest divergence was observed in Maluku populations with a value of 0. Populations with lower genetic distance values signify a more intimate familial connection, while higher genetic distance values indicate a more distant kinship relationship [34].

The analysis of population connectivity revealed that populations in the South China Sea exhibit interlinked gene flow, as evidenced by the presence the haplotypes were shared among them. For example, haplotype No. 6 was identified in both the Taiwan and Hongkong inhabitants. According to Saleky and Dailami [35], the connecting of the populations results from shared geographic locations and consistent current patterns. Nevertheless, there is no common haplotype shared between the populations in Indonesian waters and the South China Sea.

Conclusion

The genetic variation within the snapper fish population in Indonesian waters surpasses that in the South China Sea, with the Bali population exhibiting the highest genetic diversity. Nevertheless, the genetic diversity within the Indonesian snapper fish population remains at a moderate level. Although there is gene flow connectivity within the snapper fish population in the South China Sea, it is not linked to the population between Indonesian waters and the South China Sea.

Acknowledgment

This research was supported by the PMDSU research scheme year 2022 with the contract number 062/E5/PG.02.00.PL/2023. Therefore the authors thank the Minister of Education, Culture, Research, and Technology of the Republic of Indonesia for funding this research.

References

1. C.B. Wakefield, P.G. Coulson, L. Loudon, S. J. Newman. Fish Res **230**, (2020)
2. Fishbase. (2023)
3. C.H. Downey, M.K. Streich, R.A. Brewton, *et al.* Trans Am Fish Soc **147**, 6 (2018)
4. S.P. Powers, J.M. Drymon, C.L. Hightower, *et al.* Trans Am Fish Soc **147**, 5 (2018)
5. S.L. Coles. Atoll Res Bull, (2003)
6. I. A. Kimirei, I. Nagelkerken, Y.D. Mgaya, C.M. Huijbers. PloS one **8**, 6 (2013)
7. P. H. Wepfer, Y. Nakajima, M. Sutthacheep, *et al.* Mol Phylogen Evol **151**, (2020)
8. F. Leliaert, D.A. Payo, C.F.D. Gurgel, *et al.* J Biogeogr **45**, 10 (2018)
9. N. Fadli, S.A.M. Nor, A.S. Othman, H. Sofyan, Z.A. Muchlisin. PeerJ **8**, (2020)
10. S.R. Rahayu, Z.A. Muchlisin, N. Fadli, *et al.* Zoologis Anz **303**, (2023)
11. G.R. Allen, W.T. White, M. V. Erdmann. J Oc Sci Foundt **6**, (2013)
12. T. Shoji. NIDS Journal of Defense and Security **13**, (2012)
13. N. Kipgen. Strategic Analysis **45**, 2 (2021)
14. T. Arai, R. Amalina, Z. Bachok. SpringerPlus **4**, 1 (2015)
15. N. Akbar, N.P. Zamani, H.H. Madduppa. Depik **3**, 1 (2014)
16. M.R. Gaither, R.J. Toonen, D.R. Robertson, S. Planes, B. W. Bowen. J Biogeogr **37**, 1 (2010)
17. F. Zhao, L. Guo, N. Zhang, *et al.* Res Squ 2022.
18. G. Ocampo, J.T. Columbus. Am J Bot **97**, 11 (2010)
19. P. Borsa, J.D. Durand, W.J. Chen, *et al.* Acta Oecol **72**, (2016)
20. J.D. DiBattista, M.R. Gaither, J.P.A. Hobbs, *et al.* Cor Reef **36**, 2 (2017)
21. A.S.d. Souza, E.A. Dias Júnior, M.F. Perez, *et al.* Front Mar Scie **6**, (2019)
22. A.M. Jackson, M.V. Erdmann, A.H.A. Toha, L.A. Stevens, P.H. Barber. Bull Mar Sci **90**, 1 (2014)
23. M.R. Gaither, B.W. Bowen, T.R. Bordenave, *et al.* BMC Evol Biol **11**, 1 (2011)
24. S. Rahayu, Z. Muchlisin, N. Fadli, *et al.* IOP Conf Ser: Earth Environ Sci **1221**, 1 (2023)
25. D. Steinke, R. Hanner. Mitochondrial DNA **22 Suppl 1**, (2011)
26. Z.A. Muchlisin, Z. Thomy, N. Fadli, M.A. Sarong, M.N. Siti-Azizah. AIeP **43**, 1 (2013)
27. K. Tamura, G. Stecher, D. Peterson, A. Filipski, S. Kumar. Mol Biol Evol **30**, 12 (2013)
28. P. Librado, J. Rozas. Bioinformatics **25**, 11 (2009)
29. M. Nei, *Molecular evolutionary genetics* (Columbia university press, 1987)
30. D.M. Parsons, M.A. Morrison, J.R. McKenzie, *et al.* Can J Fish Aquat Sci **68**, 4 (2011)
31. I.G. A. D. Hendiari, A. Sartimbul, I. W. Arthana, G.R.A. Kartika. Acta Aqua: Aqua Sci J **7**, 1 (2020)
32. Y.W. Chiu, H.Bor, M.S. Tan, H.D. Lin, C.T. Jean. Intern J Mol Sci **14**, 5 (2013)
33. S.K. Thomas, X. Liu, Z.Y. Du, *et al.* Am J Bot **108**, 7 (2021)
34. M. Pilot, M. E. Dahlheim, A.R. Hoelzel. J Evol Biol **23**, 1 (2010)
35. D. Saleky, M. Dailami. J Kelaut Trop **24**, 2 (2021)

Phylogeography of anabas testudineus bloch, 1792 (pisces: anabantidae) in asia

Nanda Muhammad Razi[1], *Muhammad* Nasir[2], *Mutia* Ramadhaniaty[3,5], *Sri Riska* Rahayu [1,5], *Luvi Syafrida* Handayani[1,4], *Ahmad Sazali A* Sitrous[3,5], *Cut Tria* Aswina[3,5], *Nur* Fadli[3,5], *Zainal A.* Muchlisin[3,4*]

[1] Graduate School of Mathematics and Applied Science, Universitas Syiah Kuala, Banda Aceh 23111, Indonesia.
[2] Departement of Biology, Faculty of Mathematics and Natural Sciences, Universitas Syiah Kuala, Banda Aceh, 23111, Indonesia.
[3] Faculty of Marine and Fisheries, Universitas Syiah Kuala, Banda Aceh 23111, Indonesia.
[4] Laboratory of Fish Hatchery and Breeding, Faculty of Marine and Fisheries, Universitas Syiah Kuala, Banda Aceh 23111, Indonesia.
[5] Laboratory of Genetics and Aquatic Biodiversity, Faculty of Marine and Fisheries, Universitas Syiah Kuala, Banda Aceh 23111, Indonesia.

Abstract. *Anabas testudineus*, the climbing perch, is a valuable and well-liked freshwater fish with significant commercial worth. This fish is found in Indonesia and other parts of Asia. Nonetheless, there hasn't been much research done on the genetic relationships across Asian populations of climbing perch. Thus, the aim of this work was to use the mitochondrial COI gene to examine the phylogeography of *A. testudineus* in freshwater Asia. The population sample was taken from South Aceh, Indonesia. The sample was extracted using a modified C-TAB protocol. In all, 48 sequences from the Genbank and three sequences from South Aceh were included in this investigation. Seven samples came from Indonesia, three from Malaysia, nine from Philippines, seven from Vietnam, eight from Thailand, ten from India, and four from Bangladesh. According to the findings, the sequences formed 17 different haplotypes. For every population, the haplotype diversity (Hd) value is 0.897. The genetic distance measuring the closest populations is 0.0007 between Indonesia and Philippines, whereas Indonesia and Vietnam populations are the furthest apart, at 0.0904. Between 0 and 0.0157 is the genetic distance within the population. Thus, the populations of *A. testudineus* in Aceh and Indonesia and the Philippines are genetically more similar.

1 Introduction

The climbing perch, *Anabas testudineus*, are economically significant in Indonesia [1]. *A. testudineus* are typically found in rice fields, swamp, and ditches and ponds that have overflowing water or are connected to open water channels [2, 3]. There is a tendency for *A.*

* Corresponding author: muchlisinza@usk.ac.id

testudineus populations to decline in their natural habitats, and this is assumed to be a result of a number of stresses, including intensive fishing and alterations in the environment [4-6].

This work analyzed the phylogeographic pattern of A. testudineus from Asia using the genomic technique and the COI gene marker. Combining phylogenetics and geography, phylogeography analysis studies how lineages of a species are distributed geographically [7]. Prior research on freshwater fish phylogeographic pattern comparisons in significant basins in the trans-Andean area and Lower Central America has been published [8], Lake Kolleru, Andhra Pradesh, India [9], genetic diversity and phylogenetic relationship among Anabantoidei fish in South Kalimantan [10], the Indonesian phylogeography of Aplocheilus panchax [7], and the Arabian Sea's Bengal Strait's phylogeographic pattern snapper [11].

For the purpose of creating a better management strategy, it is crucial to comprehend how biogeography and species' life histories relate to one another through the study of phylogeography. Fish that are genetically distant from one another need to be maintained differently from species that are closely related to one another. As a result, the current study's objective was to examine the phylogeography of A. testudineus across Asia in order to provide crucial data for creating an improved management plan for this species in Indonesia.

2 Material and Methods

2.1 Collection of tissue sample

In this work, two different kinds of sequencing data were used: primary and secondary data. The samples of *A. testudineus* populations from Aceh (as primary data), As secondary data from the NCBI Genbank, populations from Indonesia, Malaysia, the Philippines, Vietnam, Thailand, India, and Bangladesh were used to represent samples from Asia (Figure 1).

Three samples of the Aceh population's *A. testudineus* were taken from the Singkil peat swamp in South Aceh, Aceh Province nn July 2023, Indonesia. After the sample was taken for documentation and around 1 centimeter of the pectoral fins were removed with sterile scissors, it was preserved in a 96% ethanol solution. Fish tissues are documented and collected using the protocol created by the Fish-BOL cooperation [12]. The samples were brought to the lab so that they could be examined further. For Comparative Phylogeography study, secondary data on A. testudineus sequences from Bangladesh, Vietnam, Thailand, India, Malaysia, Indonesia, and the Philippines were obtained from The GenBank. The sources of the sample sources are displayed in Figure 1 and Table 1.

2.2 DNA isolation

Modified Cetyltrimethyl Ammonium Bromide (CTAB) was used to extract DNA [13, 14]. In a 1.5 mL sterile tube, chopped fin tissue was added, which 700 microliters of Cetyltrimethyl Ammonium Bromide (CTAB) and 3 microliters of Proteinase K were added, after which it was vortexed for 15 seconds and incubated for 3 hours at 60°C. Following incubation, 700 µl of Chloroform Isoamyl Alcohol (CIA) were added, and after 30 seconds of revortexes, the mixture was centrifuged for 15 minutes at 11000 rpm. After transferring the supernatant to a new 1.5 mL tube and adding 100% ethanol, the tube was vortexed for 30 seconds again and centrifuged for 15 minutes at 12000 rpm. After discarding the produced supernatant and rinsing the tube with 70% ethanol, the tube was tapped on a dry tissue to remove any remaining ethanol, and it was allowed to dry for ten minutes at room temperature. Subsequently, the DNA sample was added 60 µl of DDH2O was added. It was then stored at -20°C for future use.

2.3 Amplification of DNA

Amply up samples of the extraction product, Polymerase Chain Reaction (PCR) was utilized in vitro. using both forward and reverse primers: Fish F1 - 5' TCA ACC AAC CAC AAA GAC ATT GGC AC-3', and Fish R1 - 5' TAG ACT TCT GGG TGG CCA AAG AAT CA-3' [15]. 25 μl of DNA template, 8.5 μl of ddH2O, 1 μl of forward primer, 1 μl of reverse primer, and 12.5 μl of red master mix were used in the PCR. A PCR equipment (Senso Quest Lab cycler) with 30 cycles was used to perform the PCR procedure. The steps in each cycle are as follows: two minutes at 95 °C for pre-denaturation, 45 seconds at 94 °C for denaturation, 45 seconds at 49.7 °C for annealing, 60 seconds at 72 °C for extension, and 10 minutes at 72 °C for final extension before cooling at 4 °C [16]. Two microliters of the PCR product were taken, placed in agarose wells, and electrophoresed in 2% agarose for thirty minutes at 100 volts. Using UVITEC FIRERIDER V10, DNA was seen. Selected for sequencing at First Base Laboratories in Malaysia are the clear and bright DNA bands.

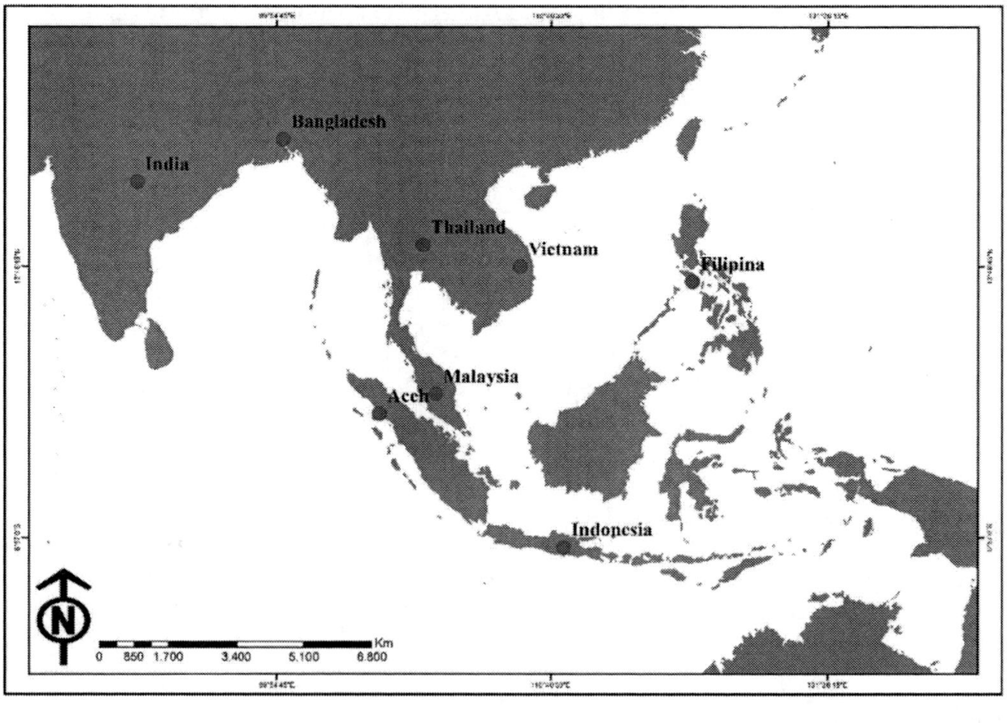

Fig. 1. The research map (the red dot signifying the *A. testudineus* population in Asia)

2.4 Data analysis of mitochondrial DNA

MEGA 11 software was used to analyze and modify the sequencing data in order to verify the taxonomic status. The data from the Basic Local Alignment Search Tool (BLAST) was then compared with the sequences [17], Using MEGA 11, the genetic distance was determined using the Kimura-2 parameter (K2P) model. The phylogenetic tree was created using the Neighbour Joining (NJ) method with a 10.000x bootstrap and the Kimura-2 parameter (K2P) model. Additionally, the distribution of haplotypes was examined using DNASP 5.10 [18]. While DNA polymorphism, haplotype diversity (Hd), and nucleotide diversity (π) were examined using ARLEQUIN software.

Table 1. *Anabas testudineus* sample IDs for main and secondary data from eight Asian populations.

No.	Country/ population	Total Sample	Total Haplotype	Sample ID
1	Aceh	3	2	SM 12, SM 13, SM 14
2	Indonesia	7	4	KU692244, KU692243, KM213039, KM213038, KU692242, KU692240, KU692241
3	Malaysia	3	2	MW590991, MW590990, JF781185
4	Philippines	9	1	MG407353, MG407352, MG407351, MG407350, MG407349, JN021211, HQ682666, HQ682665, HQ682664
5	Vietnam	7	4	MK368521, MK368520, MK368519, MK351908, KF752456, KF752454, KF752455
6	Thailand	8	4	MK628410, MK049492, MK049491, JQ661371, MK448172, MK448171, JQ661370, JQ661369
7	India	10	3	JX260823, JX260824, MK213551, MK213550, MK213552, JX983213, JX983214, KY356781, KY356780, KY356761
8	Bangladesh	4	2	MK572026, KX455903, MN083164, MN083163

3 Results and Discussion

Seventeen haplotypes were obtained from a total of 51 sequences derived from 8 populations, with an average fragment length of 608 bp. Seven population samples from Indonesia yielded four haplotypes; three samples from Malaysia yielded two haplotypes; one haplotype was from the Philippines population; Vietnam yielded four haplotypes; Thailand yielded four haplotypes; India generated three haplotypes; and Bangladesh supplied two haplotypes, all in all. In the meanwhile, 2 haplotypes were generated from 3 samples in the Aceh population. All populations had nucleotide diversity ranging from 0 to 0.04 and haplotype diversity ranging from 0 to 0.3. The populations in Aceh and Malaysia had the most diversity (value > 0.3), while the population in the Philippines had the lowest diversity (value = 0). Table 2 displays the results for nucleotide diversity (π), total haplotype (Hn), and haplotype diversity (Hd). The Malaysian population had the greatest genetic distance (interspecific) within populations (0.016), while the populations in the Philippines had the lowest (value of 0).

Table 3 shows that the Indonesian population with Vietnam had the greatest intraspecific genetic distance (0.090). *A. testudineus* is descended from a single ancestor, according to phylogenetic connections analysis, with all samples belonging to one major clade and three sister clades (Figure 3). Filipina, Aceh, dan Indonesia termasuk dalam sister clade 1 yang sama. India dan Bangladesh termasuk dalam sister clade lainnya, sedangkan Vietnam, Malaysia, dan Thailand termasuk dalam sister clade 2. Figures 2 and 4 depict the connectivity population, with populations in Malaysia and Thailand sharing haplotype No. 1. Populations in Indonesia and the Philippines share Haplotype No. 3. While the other haplotypes are independent, the populations of Bangladesh and India share haplotype No. 16 and Indonesia and Aceh share haplotype No. 4.

Table 2. Analysis of *A. testudineus* genetic diversity among eight Asian populations. Three measures of haplotype diversity (Hd), total haplotype (Hn), and nucleotide diversity (π)

Species	Population	N	Genetic Diversity		
			Hn	*Hd*	π
Anabas testudineus	Aceh	3	2	0.314	0.002
	Indonesia	7	4	0.181	0.001
	Malaysia	3	2	0.314	0.039
	Philippines	9	1	0	0
	Vietnam	7	4	0.181	0.002
	Thailand	8	4	0.139	0.002
	India	10	3	0.104	0.001
	Bangladesh	4	2	0.204	0.001

The population of Aceh and Malaysia had the most haplotype diversity (value = 0.314), while the population of the Philippines had the lowest haplotype diversity (value = 0), according to the research. Based on [19] the haplotype diversity frequency is divided into three categories: low (0.1–0.4), moderate (0.5–0.7), and high (0.8–1.00). Consequently, the total haplotype diversity of the Asian *A. testudineus* population falls into the low range. Furthermore, Hendiari, Sartimbul, Arthana and Kartika [20] claimed that currents and geographic closeness are the main causes of high genetic diversity, whereas habitat conditions and overexploitation are the main causes of low genetic variety in a population [21].

Table 3. The genetic distance between eight populations of *A. testudineus* using the COI gene, as well as the genetic distance within each group (bold).

No	Location	1	2	3	4	5	6	7	8
1	Aceh	**0.002**							
2	Bangladesh	0.084	**0.001**						
3	India	0.084	0.002	**0.001**					
4	Indonesia	0.002	0.083	0.083	**0.001**				
5	Malaysia	0.087	0.081	0.083	0.087	**0.016**			
6	Philippines	0.002	0.082	0.082	0.001	0.086	**0**		
7	Thailand	0.086	0.078	0.080	0.085	0.009	0.084	**0.002**	
8	Vietnam	0.088	0.085	0.087	0.090	0.018	0.090	0.025	**0.002**

https://doi.org/10.1051/bioconf/20248703032

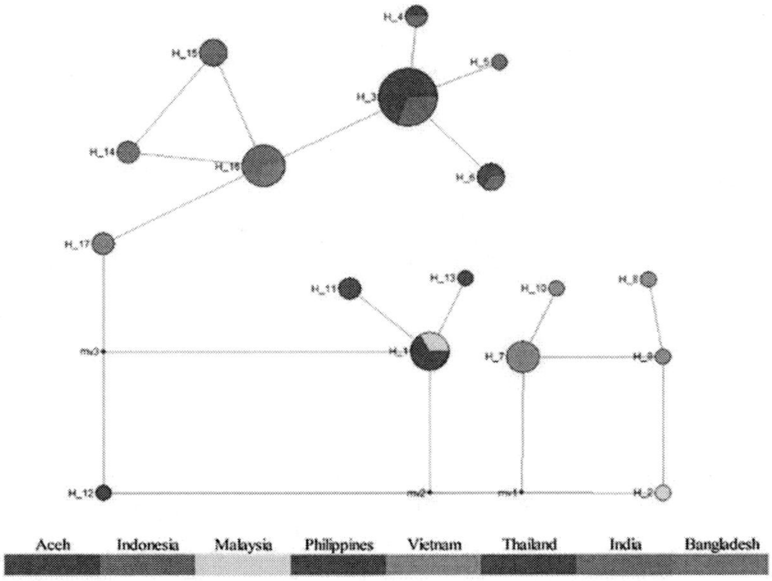

Fig. 2. The haplotype networks of *A. testudineus* from eight Asian populations

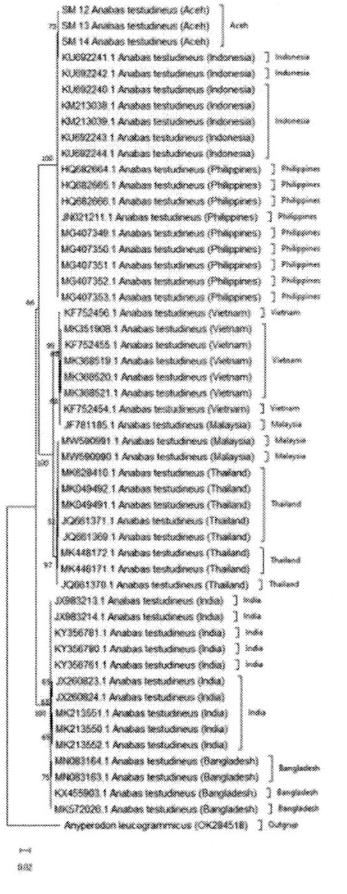

Fig. 3. The phylogenetic tree of 51 *A. testudineus* individuals from eight Asian populations

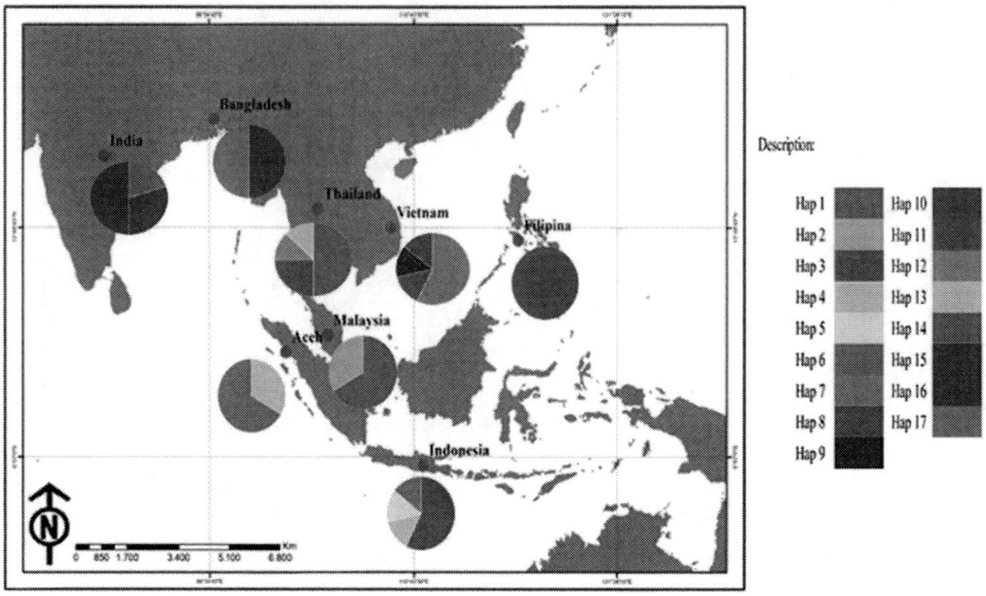

Fig. 4. The distribution map of A. testudineus haplotypes from eight Asian populations

One primary clade that was split into three sister clades is produced by reconstructing the phylogenetic tree using 51 sequences from eight populations: Aceh, Indonesia, Malaysia, Philippines, Vietnam, Thailand, India, and Bangladesh. This suggests that the progenitor was the source of the tight relationship between these groups. The genetic separation between clades provided support for this phylogenetic tree. [19]. The Malaysian population had the maximum genetic distance (interspecific) within populations (0.016), whereas the populations in the Philippines had the lowest (0). According to Pilot, Dahlheim and Hoelzel [20], populations with lower genetic distance values have tighter family ties, whereas populations with greater genetic distance values have farther-off kinship ties.

Populations in Asia are linked by shared haplotypes, as evidenced by the connectivity analysis between populations. For example, haplotype No. 1 was found in populations in Malaysia and Thailand, haplotype No. 3 in populations in Indonesia and the Philippines, haplotype No. 4 in populations in Indonesia and Aceh, and haplotype No. 16 in two populations, India and Bangladesh. The same geographic location is the cause of population connection, according to Saleky and Dailami [21]. Compared to other Asian nations, Aceh and Malaysia have greater levels of genetic variation in their *A. testudineus* populations. The gene flow between the Acehnese *A. testudineus* population and the Indonesian fish population is related.

Acknowledgments

This study was funded by Universitas Syiah Kuala through the Program Riset Unggulan Universitas Syiah Kuala Percepatan Doktor (PRUU-PD) scheme for the corresponding author (contract number: 536/UN11.2.1/PT.01.03/PNBP/2023). The authors are also grateful to the Rector of Universitas Kuala Syiah University for supporting this study.

References

1. S. Maulida, K. Eriani, N. Fadli, *et al.* Theriogenology **201**, (2023)
2. M. Mustakim, M.T.D. Sunarno, R. Affandi, M.M. Kamal. JPPI **15**, 2 (2017)
3. E. Nuralam, M.J.f. Luthfi. Proc Internat Conf Sci Engin **2**, (2019)
4. H. Hasnidar, A. Tamsil, H. Hasrun, A.M. Akram. JII **22**, 1 (2022)
5. M.A. Hossain, Z. Sultana, A.S.M. Kibria, K.M. Azimuddin. Turkish J Fish Aquat Sci **12**, 2 (2012)
6. S. Ndobe, A. Masyahoro, N. Serdiati, A.M. Moore. AACL Bioflux **13**, 1 (2020)
7. D. Mustikasari, A. Nuryanto, S. Suryaningsih. Biodiversitas **23**, 4 (2022)
8. M. Rincon-Sandoval, R. Betancur-R, J.A. Maldonado-Ocampo. Mol Ecol **28**, 5 (2019)
9. G. Srinu, P. Padmavathi, D. Chatla. J Coast Res **86**, SI (2019)
10. [R. Sasmita, M. Mabrur, U.S.A. Rahmy, B. Badruzsaufari. Biodiversitas **20**, 9 (2019)
11. [S.R. Rahayu, Z.A. Muchlisin, N. Fadli, *et al.* IOP Conf Ser: Earth Environ Sci **1221**, 1 (2023)
12. D. Steinke, R. Hanner. Mitochondrial DNA **22**, S1 (2011)
13. A.A. Bakar, E.A.S. Adamson, L.H. Juliana, *et al.* PLoS One **13**, 9 (2018)
14. P.M. Grewe, C.C. Krueger, C.F. Aquadro, *et al.* Can J Fish Aquat Sci **50**, 11 (1993)
15. R.D. Ward, T.S. Zemlak, B.H. Innes, P.R. Last, P.D. Hebert. Philos Trans R Soc B,: Biol Sci **360**, 1462 (2005)
16. Z.A. Muchlisin, Z. Thomy, N. Fadli, M.A. Sarong, M.N. Siti-Azizah. Acta Ichthyol Pisc **43**, 1 (2013)
17. K. Tamura, G. Stecher, D. Peterson, A. Filipski, S. Kumar. Mol Biol Evol **30**, 12 (2013)
18. P. Librado, J. Rozas. Bioinformatics **25**, 11 (2009)
19. S.K. Thomas, X. Liu, Z.Y. Du, *et al.* Am J Bot **108**, 7 (2021)
20. M. Pilot, M.E. Dahlheim, A.R. Hoelzel. J Evol Biol **23**, 1 (2010)
21. D. Saleky, M. Dailami. JKT **24**, 2 (2021)

https://doi.org/10.1051/bioconf/20248703033

Shark and ray monitoring on the marine critical habitat in Aceh Province

Ilham Fajri[1,3], *Benaya* M Simeon[4], *Shoimatul* Ula[5], *Kusuma* Banda Naira[6], *Efin* Muttaqin[5], *Adrian* Damora[2,3*], *Rianjuanda* Djamani[1,3], *Bustamam* Bustamam[3]

[1]Department of Fisheries Resources Utilization, Faculty of Marine and Fisheries, Universitas Syiah Kuala, Banda Aceh, 23111, Indonesia
[2]Department of Aquaculture, Faculty of Marine and Fisheries, Universitas Syiah Kuala, 23111, Indonesia
[3]Scientific Forum, The Agency of Marine Affairs and Fisheries of Aceh, Banda Aceh, 23127, Indonesia
[4]Research Institute of Environment and Livelihoods, Charles Darwin University, Australia
[5]Fisheries Resource Center of Indonesia, Rekam Foundation, Bogor, 16129, Indonesia
[6]The Agency of Marine Affairs and Fisheries of Aceh Jaya, Aceh Jaya, 23655, Indonesia

Abstract. Aceh is the leading province in Indonesia, where the marine protection area for juvenile sharks and rays was first established, precisely in Aceh Jaya waters. Rigaih is the fishing port located closest to Aceh Jaya waters, where the angler fish closely with the marine protected area and even intersect with the juvenile's habitat. This study aimed to examine shark and ray catches in Aceh Jaya before and after establishing the marine protection area. We conducted landing monitoring from 2017 to 2019 (before establishing a conservation area) and monitored in 2020 when the conservation area was established. We recorded that 73% of sharks and ray caught on set gill net. The composition of sharks and rays landed at PPI Rigaih consisted of *Sphyrna lewini* (75%, n=1783), *Chiloscyllium punctatum* (8%), *Neotrygon caeruleopunctata* (7%, n =175), and other species (10%). This study aimed to provide scientific recommendations to support shark conservation action in Aceh Jaya Waters. We found that Aceh Jaya waters are an important area for the life stages of several threatened species, including Scalloped Hammerhead Sharks (S.lewini) and Bottlenose Wedgefish (R.australiae) that are listed in Critically Endangered and listed in Appendix II CITES. Management to support the conservation action needs to be strengthened not only limited to Marine Protected Areas but also fisheries management to reduce the incidental catch of threatened sharks and ray species. Further studies need to be conducted to delineate the habitat range of threatened shark and ray species and identify fishing behaviours that pose a risk for incidental catches of sharks and rays.

* Corresponding author : adamora@usk.ac.id

1 Introduction

Sharks and rays are listed into cartilagous fishes from the Chondrichthyes class, *Elasmobranchii* subclass. This subclass consists of two large groups, namely groups, shark and ray. Shark and ray (Elasmobranchii) fisheries are one of the most essential fisheries commodities in the world. However, this group of fishes has unique biological characteristics such as long longevity, late maturity, and number of pups. Naturally, these groups have high susceptibility due to it given biological factors and are also threatened by habitat loss and fishing pressure [1].

The total catch of Elasmobranch fish in the world is increasing from year to year 1950 - 2010, and 2003 was the highest catch of sharks and rays in the world reached 800,000 tons/year [1]. On other side, research showed that in the last 50 years, the population of pelagic sharks declined reached to 70%, and more than 30% of total species already listed into IUCN redlist threatened species [2, 3].

Shark and ray fisheries in Indonesia existed for a long time, recordlike even before 1940's [4, 5]. The fisheries growth until Indonesia was recorded as the leading country for shark fisheries. Shark fisheries itself started in the 1980's due to the increase of shark fin demands in the international market [5]. The average elasmobranch production annually is not less than 100,000 tons [6, 7]. On a local scale, Aceh Province significantly contributed to the shark fisheries in Indonesia. Within two decades, there has been a decreasing trend in shark and fisheries production rays in Aceh Province. This is shown by [8], in years 1989 - 2000, the *average* production of sharks and rays reached ~3200 tons/year then in From 2001 to 2010, shark and ray fisheries production was only around ~2900 tons/year. There was a 10% decline in production over a decade. This situation indirectly reveals that the waters of Aceh serve as a primary fishing ground and a habitat for sharks and rays.

Sharks and rays that are caught and landed need conservation action with the robust science baseline to support conservation and fisheries management to ensure the sustainability of the populations, considering at least 74 shark species in Aceh waters threatened by extinction referred to IUCN Red-List of Threatened Species [2]. For example, scalloped hammerheads that regularly landed that listed as Critically Endangered species or a step behind extinct in the wild [2]. Aceh Jaya waters were the first marine protected area that protected juvenile scalloped hammerheads in Indonesia. Nevertheless, the water is not only the habitat of juvenile scalloped hammerheads but also other elasmobranch species.

This research provided information about time series data of sharks and rays that were caught and landed at Rigaih landing site. Time series data would be important to show the baseline information to support the impact of marine protected area establishment, considering Rigaih waters had been established as a Marine Protected Area in 2020 [9]. The aim of this research was to provide information about species composition and catch abundance based on fishing power index standardisation.

2 Material and Methods

2.1 Time and Place

We conducted landing monitoring at the Rigaih fishing port 15 days in a month. During the data collection, we also followed and adjusted to the local activity and rules, including respecting Friday as a working day. We collected data from 2017 to 2019 by recording sharks and rays, which landed through census monitoring from 6 am till mid-day, adjusted with fishers' landing time.

2.2 Data and Monitoring Methods

This study used primary and secondary data. Primary data consisted of catch information and fishing behaviour that the enumerator recorded. We recorded sharks and rays that landed on the fishing port, measured the length, and identified species and sexes. We also recorded information about fishing gear, the fishing vessels, and fishing grounds. This information was taken from the boat crew, boat owners, and captains; we interviewed several fishers at the same time to verify the information. To complete the time series information, we also collected statistics data from the government as secondary data.

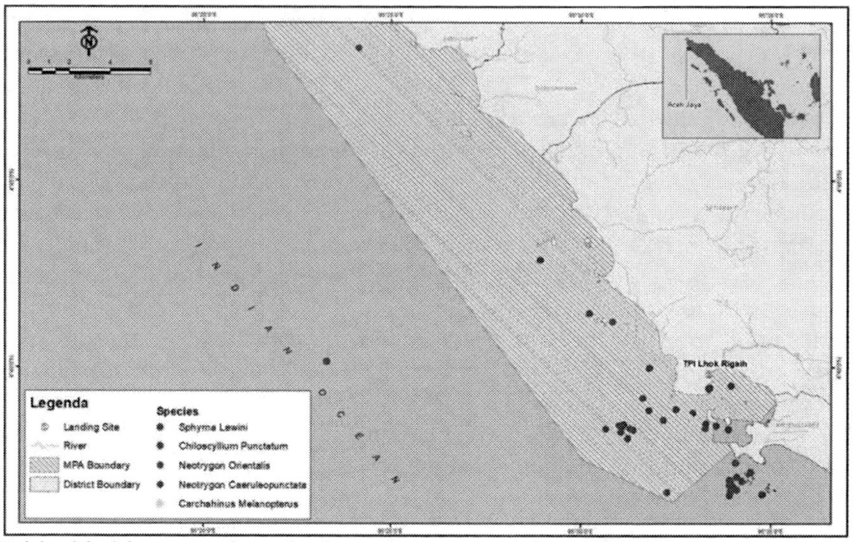

Fig. 1. Critical habitat mapping: sharks and rays catch site at PPI Rigaih - Aceh Jaya.

2.3 Data Analysis

2.3.1 CPUE Standard with Method Fishing Power Index (FPI)

Catch per Unit Effort (CPUE) is known as an approach to assess fish population by the catch abundance considering its uncertainty. CPUE is known as one of population estimation which proposed a relationship between harvest effort and the probability of capture [10]. Population trend often assumedly changes along the trend changes of CPUE [11]. CPUE is calculated with the use formula:

$$CPUE = \frac{Catch}{Effort}$$

Where:
CPUE = Catch per fishing effort (fish/trip)
Catch (C) = Total catch (head)
Effort (E) = Total fishing effort (trips)

2.3.2 Identify The Type of Shark and Ray

We considered analysis until the species level to make the study more applicable for management; however, sharks and rays identification until species level need to be done precisely. Identification is conducted by a protocol which developed by [7], whereas the enumerators need to take pictures of sharks and rays. The images that have been taken are identified using the shark and ray identification book:

1. Sharks of The World [12]
2. Rays of The World [13]
3. Sharks and Rays which have Important Economic Value in Indonesia [14]
4. Descriptions of New Sharks and Rays from Borneo [15]

2.3.3 Composition of Shark and Ray Species

The species composition of sharks and rays caught was analyzed based on the number of individuals, fishing areas and size distribution. There are more than ~200 species of sharks and rays in Indonesia, which are spread across various regions, including Aceh. Considering the high diversity of sharks and rays, several priority species of sharks and rays will be selected. The sharks and rays that are included in the priority analysis are the sharks and rays that are predominantly caught and landed, the types that have been included in Appendix II of the Convention International Trade in Endangered Species (CITES), and species that have a high threat conservation status based on the red criteria list International Union for Conservation of Nature (IUCN). This type of composition analysis aims to determine catch patterns of sharks and rays, which is expected to help explain the type composition, resource conditions and habitat distribution based on size.

3 Result and Discussion

3.1 Site Overview

Rigaih Fishing Port is located on Western Coast of Sumatera Island in Aceh Province. The precise geographical positions are 4° 39,739 N and 95° 33,355 E. The fishing port was built in 2006, named Rigaih Fishing Port, under the district authority. Administratively, PPI Rigaih is located in Gampong Lhok Timon, Setia Bakti District, and is precisely located on the edge of Rigaih Bay.

The main fisheries characteristics are pelagic and demersal that sharks were not the main targeted fish in Aceh Jaya Regency. Conversely sharks and rays regularly landed as additional and valuable catches. We recorded 26 fleets from three landing locations with coastal and small pelagic fishing fleets with various fishing gear such as handlines, gill nets, bottom longlines, and left nets. Sharks and rays that landed were dominated by small size individuals, including juveniles and immatures.

The main fishing grounds were relatively close from the fishing base, which were in small and scattered islands; there are: Raya Island, Keuh Island, and Khek Island. Those islands and the waters surrounding it have distance approximately 30 kilometers parallel from the coastline.

3.2 Sharks and Rays Catch at PPI Rigaih - Aceh Jaya

Based on monitoring data since July 2017, the highest catches were recorded on August 2019. Averagely the number of catches relatively high in the middle of the year counted from July till September. Those months are the west wind season, which made the fishers shift their fishing gear into entangle nets, which also caught more sharks and rays compared with other fishing gear. (Fig. 2).

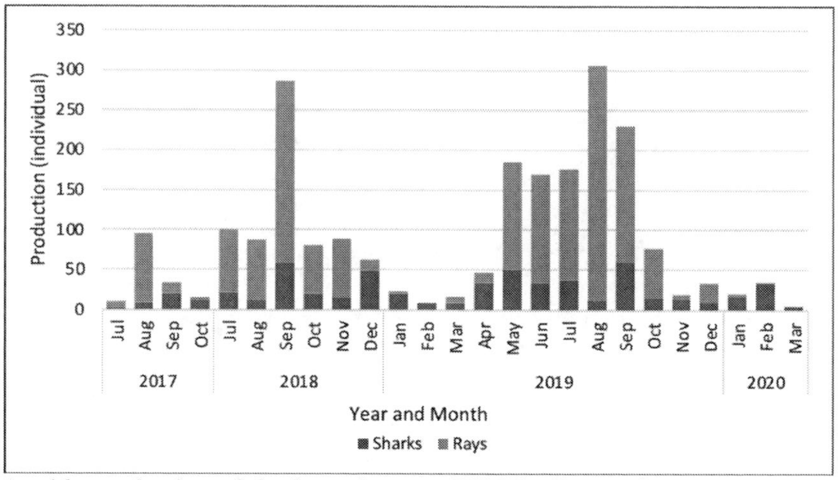

Fig. 2. Monthly production of sharks and rays at PPI Rigaih - Aceh Jaya.

The analysis results showed that the catch abundance based on the fishing power index showed some decreasing and increasing trends yet relatively stable from 2017 to 2019. The number of vessels and efforts that operated in Rigaih fluctuated, however, we used standardisation through fishing power index to reduce the bias due to these fluctuations. The highest CPUE was recorded in 2019, when every boat landed a shark every trip. (Fig. 3).

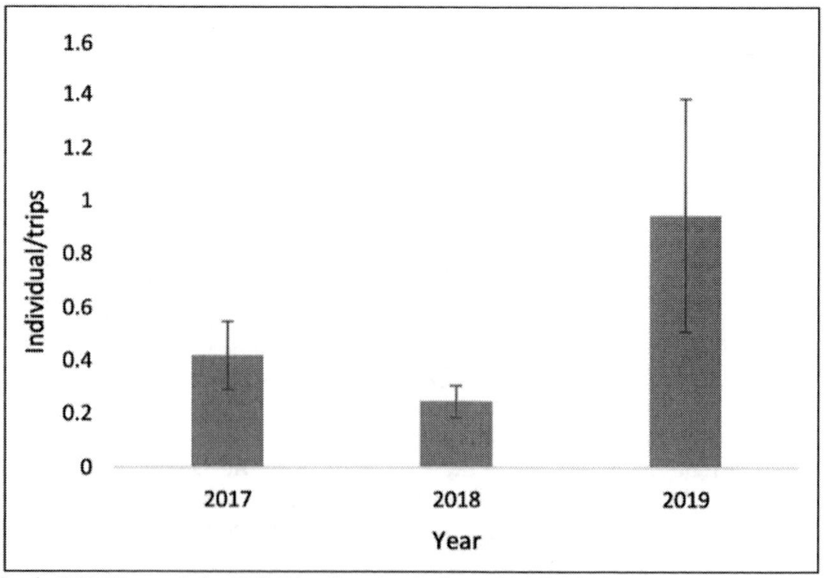

Fig. 3. Shark CPUE graph in PPI Rigaih - Aceh Jaya 2017-2019.

3.3 Composition of shark and ray catches

Sharks and rays that landed were dominantly caught by Gillnet. However, sharks and rays are also caught by other fishing gear such as drift gillnets, longlines, handlines, and trammel nets (Table 1). The species composition that landed consisted by *Sphyrna lewini* (75%),

Chiloscyllium punctatum (8%), and *Neotrygon caeruleopunctata* as much as (7%). Meanwhile, the species composition of sharks and rays caught during the landing survey included *Sphyrna lewini* (80 %), *Loxodon macrorhinus* (4%), *Carcharhinus sorrah* and *R. australiae* as much as 3% (Fig. 4).

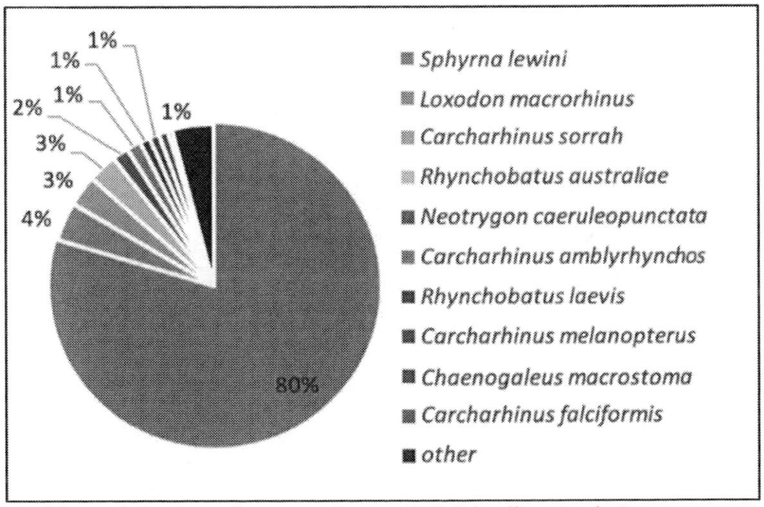

Fig. 4. Composition of shark and ray catches at PPI Rigaih - Aceh Jaya.

73% of the sharks and rays landed at the Rigaih Fishing Port, Aceh Jaya, were caught by gill net fishing gear. Other fishing gear used for catching sharks and rays at PPI Rigaih are drift gill nets, longlines, hand lines and trammel nets (Table 1).

Table 1. Catch composition based on fishing gear at PPI Rigaih.

Fishing Gears	Number of Individuals
Gill Net	1836
Drift Gill Net	137
Trammel Net	35
Tuna drift longline	107
Hand lines and troll lines	39

The Scalloped Hammerhead Shark that caught and landed were classified by catch from gill net fishing gear; in addition the catches were dominated by juveniles and immature individuals. The average total length recorded was 66 cm, and the catches recorded between 66 cm and 240 cm. We also recorded embryos that were possibly accidentally delivered from pregnant females with size 29 cm (Table 2). The hammerhead catches were caught in the coastal area within 12 miles of the coast.

Table 2. Length of sharks and rays caught at PPI Rigaih - Aceh Jaya.

Species	Number of Sharks and Rays	Average total *length*	Total length Maximum	Minimum Total Length
Sphyrna Lewini	1783	66	240	29
Neotrygon caeruleopunctata	175	36.8	81	7
Loxodon macrorhinus	65	63.7	120	43
Rhynchobatus australiae	54	92.8	236	47
Carcharhinus sorry	52	80.6	123	53
Chiloscyllium punctatum	31	55.8	82	40
Carcharhinus amblyrhynchos	24	76.8	120	40
Gymnura poecilura	18	35	45	30
Rhynchobatus laevis	18	105.3	221	64
Carcharhinus melanopterus	17	110.9	156	66
Chaenogaleus macrostoma	15	84	120	46

Based on three years of landing monitoring at Rigaih fishing port, we recorded 1783 individuals of Scalloped Hammerhead (*S. lewini*) that are dominated by juveniles and immature individuals. Then followed by the *Neotrygon caeruleopunctata, which* counted 175 individuals. There are several rays of this type that were caught in juvenile condition, namely 7 cm in size. Other species that were recorded were *Rhynchobatus australiae, Rhynchobatus laevis*, *Chaenogaleus macrostoma*, and *Carcharhinus falciformis*. The composition of shark and ray catches based on IUCN red list status is 47%. The composition of the IUCN red list includes the composition Vulnerable 31%, Endangered 6%, and Critically Endangered 10% (Fig. 5).

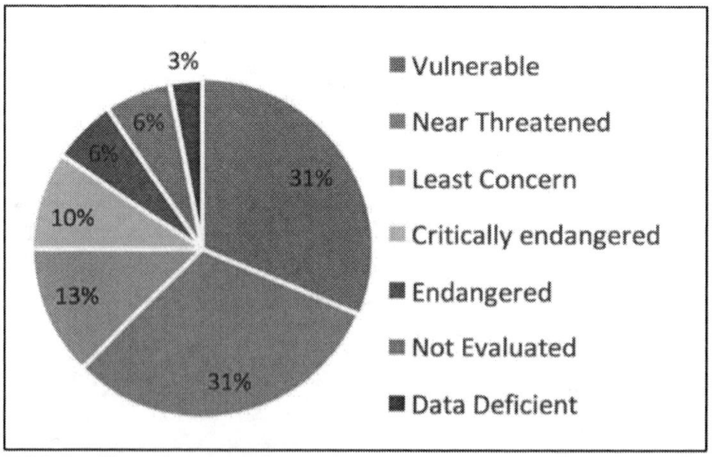

Fig. 5. The IUCN Redlist Threatened Species composition of sharks and rays that landed at Rigaih Fishing Port, Aceh Jaya.

We recorded 14% of shark and ray catches had been listed into CITES Appendix II. The species composition of CITES appendix II list showed that 95% dominated by Scalloped Hammerhead (*Sphyrna* lewini), followed by Bottlenose Wedgefish (*Rhynchobatus australiae*) 3% and other mix species 1% including Silky Shark (*Carcharhinus falciformis*) and Pelagic Thresher Shark (*Alopias pelagicus*). All species recorded dominated by immature rather than mature individuals.

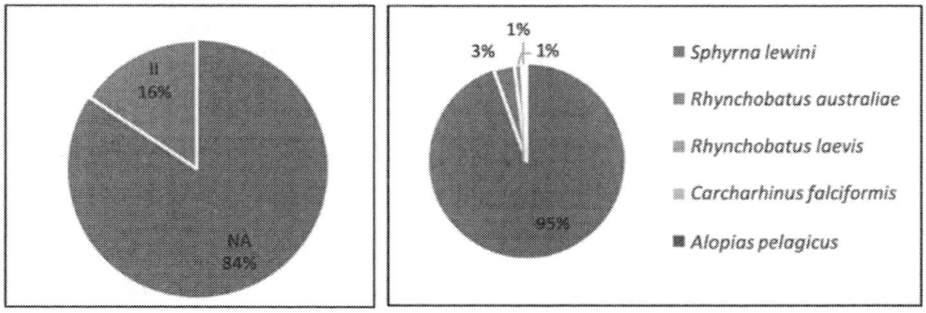

(a) (b)

Fig.6. (a) Catch results sharks and rays based on the status of Appendix II (b) types sharks and rays Appendix II category which was landed at PPI Rigaih - Aceh.

A Sharks and rays catches that landed in Rigaih Fishing Port were dominated by threatened species referred to IUCN Redlist Threatened Species and CITES Appendix II list. The sharks and rays that landed relatively had similar range sizes, there were small species of sharks and rays or juveniles. These conditions commonly found due to sharks and rays tend to aggregate on the same size with the mix species [16].

The catches dominated by Scalloped Hammerhead as Critically Endangered species that also listed into CITES Appendix II. Management and conservation action is needed to ensure its healthy population in the ocean. Some of the species listed in Appendix II CITES, including Scalloped Hammerhead, had been regulated by catch quota and minimum size limit. However, the catches in the Rigaih fishing port were not complied with the size limit that was 200 cm [17].

Scalloped hammerhead known as migratory species however it has specialist behaviors and habitat preferences, the juveniles inhabit in inner bay and coastal environment [18], while adult distributed globally and migrated in schoolings between waters [19-22]. The Scalloped Hammerhead that landed in Rigaih fishing port varied from mature females, juveniles, and embryos. This condition showed that fishing grounds of Aceh Jaya fishers have overlay with area that important to Scalloped Hammerhead for carrying out their vital functions highly probability as reproduction and nursery area. Coastal waters close with Rigaih Fishing Port are shallow continental shelf located with the edge of Indian Ocean that known as suitable habitat for life-stages for Scalloped Hammerhead. This was supported by another research that individuals of Scalloped Hammerhead occupied shelf waters and the juveniles could be found in nearshore bay and estuarine habitats [23-26].

Another dominant threatened species that recorded was the juveniles of Bottlenose Wedgefish (*R. australiae*) that also listed as Critically Endangered by IUCN Redlist Threatened species and listed into Appendix II CITES. The mature individuals could be found encounter across broad habitats; however, juveniles stages tend to be a specialist feeder with preys consisted by small bony fishes and crustaceans [27, 28]. This condition relevant with fisheries that exist in Aceh Jaya, pertaining trammel net targeting crustacea that the main prey of juveniles Bottlenose Wedgefishes.

The leading fishing gears that incidentally caught sharks and rays were gillnet, drift gillnet, drift longline. Although those fishing gears had different operational methods, they were all operated in relatively the same fishing grounds. In fact, the fishing grounds located outside of Marine Protected Area. This showed that MPA that has been established had low impact to reduce the number of sharks and rays that caught and landed. We assumed that based on vertical and horizontal movements of the sharks and rays, yet there is still a high risk of incidentally catching sharks and rays. To determine the swimming layer and overlay between fishing ground and habitat range, further research needs to be done.

The average catch abundance from 2017 to 2019 showed insignificant differences, meanwhile the number of catches increased almost three times higher in 2019 compared with 2017. These conditions demonstrate that fishing effort also influences the number of catches. However, further studies need to be conducted to determine the specific fishing behaviors that impact the catch numbers and identify the necessary fisheries management measures.

4 Conclusion

Aceh Jaya waters are important area for the life stages of several threatened species including Scalloped Hammerhead Sharks (*S. lewini*) and Bottlenose Wedgefish (*R. australiae*) that listed into Critically Endagered and listed in Appendix II CITES. Management to support the conservation action need to be strengthen not only limited on Marine Protected Area but also fisheries management to reduce the incidental catch of threatened shark and ray species. Further studies need to be conducted to delineate the habitat range of threatened shark and ray species and identify fishing behaviours that pose a risk for incidental catches of shark and ray.

References

[1] N. K. Dulvy, S. L. Fowler, J. A. Musick, *et al.* elife **3**, (2014)
[2] IUCN. 2023.
[3] N. Pacoureau, C. L. Rigby, P. M. Kyne, *et al.* Nature **589**, 7843 (2021)
[4] R. H. Barnes, *Sea hunters of Indonesia: Fishers and weavers of Lamalera* (Oxford University Press, 1996)
[5] J. Christensen, M. Tull, *Historical perspectives of fisheries exploitation in the Indo-Pacific* (Springer Science & Business Media, 2014)
[6] F. Dent, S. Clarke. FAO Fisheries and Aquaculture technical paper, 590 (2015)
[7] E. Mutaqqin, B. Simeon, M. Ichsan, *et al.* Rome: FAO 66pp, (2018)
[8] DKP Aceh. 2015.
[9] Ministerial decree Minister of Marine Affairs and Fisheries. 2020.
[10] J. R. Skalski, K. E. Ryding, J. Millspaugh, *Wildlife demography: analysis of sex, age, and count data* (Elsevier, 2010)
[11] S. M. McCluskey, R. L. Lewison. Fish and fisheries **9**, 2 (2008)
[12] D. A. Ebert, S. Fowler, L. Compagno. Wild Nature Press, (2013)
[13] P. Last, G. Naylor, B. Séret, *et al.*, *Rays of the World* (CSIRO publishing, 2016)
[14] W. T. White, P. R. Last, J. D. Stevens, G. Yearsly. 2006.
[15] P. R. Last, W. T. White, J. J. Pogonoski. (2010)
[16] T. Vierus, S. Gehrig, J. M. Brunnschweiler, *et al.* Ecology and Evolution **8**, 14 (2018)
[17] I. Sualia, S. Mardiah, E. Muttaqin, *et al.* Jakarta **43 pp**, (2020)
[18] A. Harry, W. Macbeth, A. Gutteridge, C. Simpfendorfer. Journal of Fish Biology **78**, 7 (2011)
[19] G. H. Burgess, J. Carvalho, J. L. Imhoff. Florida Museum of Natural History Report, (2009)
[20] A. J. Gallagher, A. P. Klimley. Reviews in Fish Biology and Fisheries **28**, 4 (2018)
[21] A. P. KLIMLEY, *Social organization of schools of the scalloped hammerhead shark, Sphyrna lewini (Griffith and Smith), in the Gulf of California* (University of California, San Diego, 1983)

[22] A. P. Klimley. Zeitschrift für Tierpsychologie **70**, 4 (1985)
[23] K. N. Holland, B. M. Wetherbee, J. D. Peterson, C. G. Lowe. Copeia, (1993)
[24] E. M. Hoyos-Padilla, J. T. Ketchum, A. P. Klimley, F. Galván-Magaña. Animal Biotelemetry **2**, (2014)
[25] Y. E. T. Rojas, F. P. Osuna, A. H. Herrera, *et al.* Hydrobiologia **726**, (2014)
[26] I. Zanella, A. López-Garro, K. Cure. Environmental biology of fishes **102**, 10 (2019)
[27] M. D. P. Blanco-Parra, F. Galván-Magaña, J. F. Márquez-Farías, C. A. Niño-Torres. Environmental Biology of Fishes **95**, (2012)
[28] P. M. Kyne, C. L. Rigby, R. W. Dharmadi Jabado. The IUCN Red List of Threatened Species, e T41853A68643043, (2019)

The influence of the number of fishing fleets on the growth of fishery production in an effort to determine shipyard clusters using linear regression

Thaib Rizwan[1,2,*], *Makwiyah. A* Chaliluddin[1], *Ayana* Rizki[2], *Razali* Thaib[3], *Akhyar* Akhyar[3]

[1]Department of Capture Fisheries, Marine and Fisheries Faculty, Universitas Syiah Kuala, Darussalam, Banda Aceh. Indonesia.
[2] Fishing and Navigation Vessel Laboratory, Marine and Fisheries Faculty, Universitas Syiah Kuala, Darussalam, Banda Aceh. Indonesia.
[3] Department of Mechanical Engineering, Engineering Faculty, Universitas Syiah Kuala, Darussalam, Banda Aceh. Indonesia.

Abstract. Shipyards are not only an essential part of the shipping industry but also function as a crucial element for carrying out various tasks in the process of building and repairing ships. The shipyard's involvement in this process makes it the most important element supporting the progress of the maritime industry as a whole. Therefore, this research was designed with the main aim of investigating and analyzing the relationship between the number of ships or fleets and production growth in the fisheries sector. The data used in this research consists of statistical data on capture fisheries covering the period from 2020 to 2022. This data was then analyzed using a simple linear regression analysis method to identify and understand the dynamics of the relationship between the number of vessels and the growth of fisheries production. Through the use of this approach, the research aims to explore the extent of the influence of the number of vessels on the growth of fisheries production in the context of the years studied. However, the results of the analysis show that there is no significant influence between the number of vessels and the growth of fisheries production. Nevertheless, these findings provide an important contribution to a further understanding of factors that may influence fisheries production dynamics. The implications of this research can provide valuable insights for shipping and fishing industry stakeholders in their efforts to optimize their strategies and policies to support sustainable growth in this sector.

1 Introduction

Modern shipyards are entities involved in various activities such as shipbuilding, ship repair, ship modification, and other similar maritime work [1]. Taking into account these activities, shipyards can be categorized into several types, including new building yards, special repair yards, and combined new building and repair yards [2], [3]. Despite their crucial role in the maritime industry, shipyards in Aceh still encounter significant challenges related to human resources, land, facilities, and production capacity.

Several issues have arisen, encompassing a shortage of skilled labor, imperfect organizational structure, and incomplete facilities [4]. Moreover, there is a lack of updated technology [5], irregular working time patterns, and the absence of set annual production

* Corresponding Author : rizwanthaib@usk.ac.id

targets [6]. Additionally, some shipyards are situated in less strategic land locations [7], [8], [9]. The modern shipbuilding industry demands adequate facilities due to its high production complexity, particularly as it generates a substantial number of large ships annually [10].

Meanwhile, shipyards in Aceh still adhere to traditional practices. These traditional shipyards are considered community shipyards that employ traditional construction methods, utilizing traditional tools, and relying on wood as the primary material [11], [12]. Shipbuilding is regarded as a part of the traditional heritage in Indonesia, where handmade wooden boats are constructed using techniques passed down through generations [13], [14]. Given the substantial disparities between modern and traditional shipyards, the challenges in Aceh necessitate adaptation and renewal to ensure the shipbuilding industry can effectively compete in an era marked by increasingly complex and modern ship production Efforts to cluster shipyards are a crucial aspect that must be undertaken to optimize various facets of the industry. Clustering not only shortens processing time in building new ships and repairing old ones but also has a positive impact on reducing ship queues and enhancing efficiency in ship construction and repair processes [15]. Concurrently, establishing a comfortable and organized work environment remains a top priority. An organized workplace facilitates the implementation of comprehensive maintenance activities on the facilities and infrastructure systems within the shipyard. This not only enhances the comfort of the workforce but also minimizes operational costs and improves the efficiency of managing the diverse facilities available at the shipyard.

However, to achieve a more efficient level of clustering, comprehensive data concerning the number of vessel fleets and fishery production results is essential. This information serves as the foundation for clustering management and forms the basis for mapping the relationship between fleet size and fisheries production. Therefore, regression analysis was conducted to gain a more profound understanding of the dynamics of the relationship between these variables. Through regression analysis, it is hoped that significant patterns and trends can be identified between the number of vessel fleets and fisheries production results. The outcomes of this analysis can provide a deeper understanding, offer strategic insights, and serve as a basis for more effective decision-making in managing shipyard clustering. This way, shipyard clustering efforts can become more efficient, positively impacting the shipping industry as a whole, and contributing positively to economic growth in the region.

The purpose of simple regression analysis is to elucidate how one variable can quantitatively influence other variables. In the realm of regression analysis, variables that serve as instigators or catalysts for change are referred to as independent variables. Conversely, variables that undergo changes in response to the independent variable are labeled related variables or dependent variables. When the regression equation encompasses one independent variable and one related variable, it is termed a simple regression equation. In such instances, the analysis centers on the direct relationship between these two variables. For instance, to ascertain the degree to which changes in temperature (independent variable) can impact plant growth (related variable), simple regression analysis can be employed.

However, in scenarios where there are multiple independent variables that could influence related variables, a multiple regression equation becomes necessary. Multiple regression equations enable the consideration of the combined effects of two or more independent variables on related variables. For example, if the aim is to comprehend how not only temperature but also air humidity and sunlight intensity collectively influence plant growth, multiple regression analysis is the appropriate tool.

Simple regression analysis provides an estimate of the extent to which the related variable can alter due to a change in one independent variable. This information proves invaluable in making predictions and comprehending the intricacies of relationships between variables. Through the application of this technique, a more profound exploration of the impact of independent variables on related variables is facilitated, thereby establishing a robust foundation for decision-making that is well-informed [16]

2 Research methods

Simple linear regression analysis involves establishing a linear relationship or model between the independent variable (x) and the dependent variable (y). In this study, the independent variables, referred to as influencing variables, include the growth of fisheries production and the number of fishing fleets. The dependent variable is the district/city name in Aceh. This analysis aims to ascertain the direction of the relationship between the independent variable and the dependent variable in the event of an increase or decrease in the independent variable. The formula utilized in simple linear regression analysis is as follows:

$$Y = \alpha + \beta x$$

Where :
Y = Dependent variable
X = Independent variable
α = Constant
β = Regression coefficient

The coefficient of determination (R^2) in simple linear regression analysis is used to assess how well the regression line fits the data. The range of the coefficient of determination (R^2) is 0 to 1. The ability to explain the dependent variable is severely constrained the smaller the coefficient value. Conversely, a number around 1 indicates that the fluctuations in the dependent variable can be explained by the independent variable.

3. Results and Discussion

Based on capture fisheries statistical data from 2020 to 2022, it is evident that the number of fleets owned by each district/city in Aceh Province exhibits significant fluctuations, as illustrated in Figure 1. These fluctuations reflect variations occurring over time, signifying intricate dynamics within the fishing industry across these regions. It is crucial to emphasize that these fluctuations extend beyond fleet size and are also evident in the results of fishery production. Figure 2 outlines the trends of increases and decreases in the value of fisheries production in each district/city throughout this period. These changes indicate factors that could impact the performance of the fisheries sector, including climate change, resource management policies, or economic variability.

A comprehensive understanding of these fluctuations allows for a more profound analysis to identify the underlying factors causing these changes. For instance, whether an increase in production correlates with successful management interventions or specific external factors, and conversely, whether a decline in production is linked to a particular issue requiring special attention.

A profound comprehension of fluctuations in both the number of fleets and fisheries production in each district/city forms a robust foundation for making more precise and strategic decisions in fisheries resource management. Thus, a more detailed and holistic analysis can aid in formulating policies and strategic measures to foster sustainable growth within the fisheries sector in Aceh Province.

 https://doi.org/10.1051/bioconf/20248703034

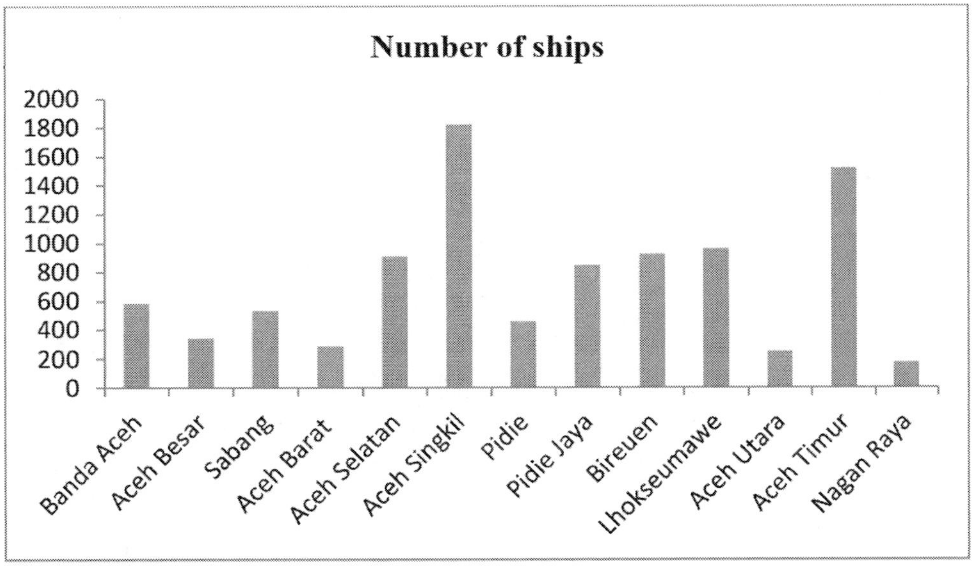

Fig. 1. Number of ships

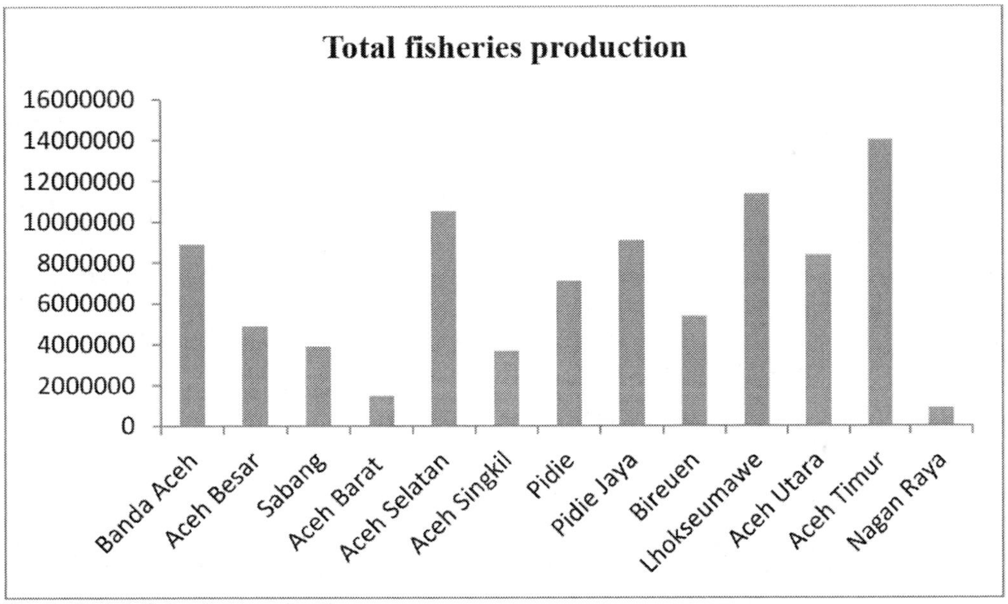

Fig. 2. Total fisheries production

Based on this data, a linear regression analysis was conducted to assess the relationship between variable x (number of vessels) and variable y (amount of fisheries production). The analysis yielded a multiple R value of 0.416, indicating a moderate correlation between the two variables. The coefficient of determination, with a value of 0.098 or 10%, suggests that variable x (number of vessels) can account for approximately 10% of the variation in variable y (amount of fisheries production), while the majority of other variations are influenced by factors not encompassed in the model.

However, for a more comprehensive understanding, it is imperative to proceed with analysis of variance (ANOVA) calculations as the subsequent step. The outcomes of this analysis are presented in Table 1, where ANOVA provides insights into the significance of the relationship between the number of vessels and total fishery production. Thus, delving into regression analysis and ANOVA allows for a thorough comprehension of the impact and significance of variable x on variable y, creating opportunities for enhanced interpretation and a deeper understanding of the factors influencing fisheries production

Table 1. ANOVA Calculation

	df	*SS*	*MS*	*F*	*Significance F*
Regression	1	3,22545E+13	3,22545E+13	2,314282	0,156403
Residual	11	1,53309E+14	1,39372E+13		
Total	12	1,85563E+14			

Based on the table above, it is evident that the significance value of F is greater than 0.05, indicating the absence of influence between variable x (number of vessels) and variable y (amount of fisheries production). This finding affirms that, in the context of this case, the number of vessels does not exert a significant impact on fisheries production. Despite the initial regression analysis suggesting a moderate correlation between the two variables, a significance F value surpassing the significance threshold indicates that this relationship cannot be considered statistically significant.

$$Y = 4448881 + 3292{,}5x$$

Consequently, the resulting regression equation assumes significance in providing a detailed depiction. While this equation holds true under the condition x=0, possibly reflecting a scenario with no ships involved, a noteworthy increase in the y value of 3292.5 occurs when x=1. This seemingly simple equation offers insight into how fishery production may fluctuate with an increasing number of vessels.

In this context, the regression analysis and F significance results offer profound insights into the relationship between variables x and y, while the regression equation further elucidates the changes in fisheries production that may accompany variations in the number of vessels. This comprehensive interpretation establishes a robust foundation for meticulous and strategic decision-making regarding fisheries resource management.

4 Conclusion

The study aimed to examine the correlation between one or more independent variables and the dependent variable. The association among these variables was assessed using simple linear regression analysis. The findings indicate that there is no substantial impact observed between the quantity of fleets and the increase in fisheries production. The implications of this research can provide valuable insights for shipping and fishing industry stakeholders in their efforts to optimize their strategies and policies to support sustainable growth in this sector.

Acknowledgments

Authors greatly thank LP3M USK for funding in this research in the 2023 fiscal year under number 146/UN11.2.1/PT.01.03/PNPB/2023, Shipyard and Navigation Lab, PSP USK for collecting data. Marine and fisheries service officer of Aceh Province.

References

1. R. A. Sari, R. M. Aprilia, R. Rizwan, M. Muhammad, And O. Kandi, **1(1)** (2021).
2. B. Saputra, I. P. Mulyanto., W. Amiruddin. 2017. Jurnal Teknik Perkapalan, **5(2)** 353-366 (2017).
3. Warsilan. Shipyard Industrial Development Studies East Kalimantan. Mimbar, **34(2)** 463-464 (2018).
4. E. Lai, F. Yun, I. Arokiam, And J. Joo, *Oper. Supply Chain Manag. Int. J.*, **13(2)** 166–175 (2020)
5. D. Ford And T. Housel, Increasing Value And Savings In Shipbuilding With Innovative Technologies, *Def. Acquis. Res. J.*, **27** 240–263 (2020)
6. R. D. Putra And J. Koto, Production Process Of Traditional Ship In Bintan-Indonesia, (2015).
7. T. Rizwan, D. Hikmah, M. Muhammad, M. A. Chaliluddin, A. Akhyar, And R. Thaib, *Depik*, **11(2)** (2022).
8. T. Rizwan, H. Husaini, H. Husin, And A. Akhyar, **11(3)** (2022)
9. T. Rizwan, Z. Jalil, A. Akhyar, And H. Husaini, **22(9)** 237–245 (2021)
10. M. Porath, R. Simoni, A. Giovanoni, And J. Amaral, Field Evaluation Of A Hull Block Assembly Process Assisted By Advanced Optical 3d Measurement Systems, Nov. (2016).
11. F. Firmansyah, A. Nurkhalida, And R. O. S. Gurning, A Study On Indonesia Shipbuilding Competitiveness Challenge And Opportunity, Presented At The Proceeding Of Marine Safety And Maritime Installation, Clausius Scientific Press, Sep. (2018).
12. I. S. Wekke, S. Ladiqi, And R. Bustami, *Ulul Albab J. Studi Islam*, **20(1)** (2019)
13. R. Romadhoni And N. Nurhasanah, *Costing J. Econ. Bussines Account* **4(1)** 63–69 (2020)
14. Y. Praharsi, M. A. Jami'in, G. Suhardjito, And H.-M. Wee, *Ocean Eng* **189** 106234, (2019),
15. T. Rizwan, H. Husaini, H. Husin, A. Akhyar, And Z. Jalil, *Pol. J. Environ. Stud.*, **32(2)** 1755–1766 (2023).
16. D. Ispriyanti And D. Safitri, Analisis Regresi Linier Piecewise Dua Segmen, **1(1)** (2012).

AUTHOR INDEX